04 21 93 BO

# MANUAL OF THE GRASSES OF THE UNITED STATES


## A. S. Hitchcock

Second Edition
Revised by AGNES CHASE


IN TWO VOLUMES

VOLUME ONE


DOVER PUBLICATIONS, INC.
NEW YORK

Published in Canada by General Publishing Company, Ltd., 30 Lesmill Road, Don Mills, Toronto, Ontario.

Published in the United Kingdom by Constable and Company, Ltd.

This Dover edition, first published in 1971, is an unabridged republication of the second revised edition, as published by the United States Government Printing Office in 1950 as U. S. Department of Agriculture Miscellaneous Publication No. 200. The first edition of the work was published in 1935.

For convenience in handling, the text is published in two volumes in this paperback edition.

*International Standard Book Number: 0-486-22717-0*
*Library of Congress Catalog Card Number: 70-142876*

Manufactured in the United States of America
Dover Publications, Inc.
180 Varick Street
New York, N. Y. 10014

# MANUAL OF THE GRASSES OF THE UNITED STATES

By the late A. S. HITCHCOCK,[1] *principal botanist, Division of Plant Exploration and Introduction;* second edition revised by AGNES CHASE, formerly *senior botanist* and later *collaborator, Division of Plant Exploration and Introduction, Bureau of Plant Industry, Soils, and Agricultural Engineering, Agricultural Research Administration,* and *research associate, United States National Museum, Smithsonian Institution*

VOLUME I



VOLUME II



## INTRODUCTION

Of all the plants of the earth the grasses are of the greatest use to the human race. To the grasses belong the cereals, sugarcane, sorghum, and the bamboos; and, since they furnish the bulk of the forage for domestic animals, the grasses are also the basis of animal industry.

### USES OF GRASSES

The grasses furnish the principal breadstuffs of the world and a large part of the food of domestic animals; they are also used in the industrial arts and extensively as greensward and ornamentals in parks and gardens.

#### FOOD GRASSES

The most important food plants for the human race are the cereals, including wheat, corn (maize), rice, barley, rye, oats, and many kinds of grain sorghums. For primitive peoples the seed of certain other grasses, such as pearl millet, common millet, broomcorn millet, Japanese millet, and African millet (ragi), have played an important role. The seeds of the cereals are also extensively used as feed for domestic animals.

#### FORAGE GRASSES

Forage grasses are used for hay, pasturage, soiling, and silage.

##### HAY GRASSES

The grasses together with clovers and alfalfa are the basis of permanent

[1] Died December 16, 1935.

meadows. The most important perennial grasses used for tame hay are: Timothy (*Phleum pratense*), redtop (*Agrostis alba*), orchard grass (*Dactylis glomerata*), meadow fescue (*Festuca elatior*), smooth brome (*Bromus inermis*), and Johnson grass (*Sorghum halepense*). A few other species are used occasionally or rarely: Rhodes grass (*Chloris gayana*), Dallis grass (*Paspalum dilatatum*), desert wheatgrass (*Agropyron desertorum*) and crested wheatgrass (*A. cristatum*), velvet grass (*Holcus lanatus*), Natal grass (*Rhynchelytrum repens*), tall oatgrass (*Arrhenatherum elatius*), and slender wheatgrass (*Agropyron trachycaulum*). Some of the grasses used primarily for pasture are also occasionally used for hay.

Market hays from grasses usually consist of timothy, prairie grasses, Johnson grass, or grain (wheat, oats, and wild oats). The prairie hays are divided into upland prairie and midland prairie. The species of most importance in the upland prairie are *Agropyron smithii* and *Stipa comata* (northern Great Plains), *Andropogon gerardi* and *A. scoparius* (eastern Great Plains), *A. saccharoides* (Texas), and *Panicum virgatum* (Kansas to Texas). Midland prairie is invariably composed of *Spartina pectinata*. Tussock sedge (*Carex stricta*) is harvested in large quantities on the marshes of Wisconsin for use as packing hay.

For temporary meadows the grasses most used are the cereals, which, with wild oats, furnish the grain hay of the Pacific coast, the sorghums, including Sudan grass, and millet (*Setaria italica*).

#### PASTURE GRASSES

The more common grasses used for permanent pasture are: Kentucky bluegrass (*Poa pratensis*), Bermuda grass (*Cynodon dactylon*), redtop (*Agrostis alba*), colonial bent (*A. tenuis*), orchard grass (*Dactylis glomerata*), smooth brome (*Bromus inermis*), Italian ryegrass (*Lolium multiflorum*), perennial ryegrass (*L. perenne*), meadow fescue (*Festuca elatior*), Dallis grass (*Paspalum dilatatum*), carpet grass (*Axonopus compressus* and *A. affinis*), Canada bluegrass (*Poa compressa*), and sheep fescue (*Festuca ovina*). Many of the meadow grasses mentioned above are also used for pasture.

Temporary pasture is furnished by the cereals and by rescue grass (*Bromus catharticus*), Italian ryegrass, and Sudan grass.

Two grasses, important in the tropics but in the United States grown only in southern Florida and southern Texas, are Guinea grass (*Panicum maximum*) and Para grass (*P. purpurascens*).

#### SOILING GRASSES

Grasses used for soiling are for the most part the cereals, millet, and other annual grasses used for temporary meadows, and in addition but only locally, pearl millet (*Pennisetum glaucum*), teosinte (*Euchlaena mexicana*), and Napier or elephant grass (*P. purpureum*).

#### SILAGE GRASSES

Any grass may be cut and stored in silos, but corn (maize) and sorghum are the ones most used.

#### RANGE GRASSES

A large number of grasses make up much of the wild pasture, known in the West as the range, only the more abundant and valuable of which are recognized by stockmen as important. Probably the best known range grass is buffalo grass (*Buchloë dactyloides*), a sod-forming "short grass" dominant over much of the Great Plains. Throughout the same region two tufted short grasses, blue grama (*Bouteloua gracilis*) and hairy grama (*B. hirsuta*), are abundant. In Texas the dominant grass over much of the range is curly mesquite (*Hilaria belangeri*), a sod-former similar to buffalo grass.

In the prairie region of the Mississippi Valley and in the eastern part of the Great Plains certain "tall grasses" in earlier days furnished excellent hay and pasture, but in recent times these fertile grasslands have been broken up for cultivated fields. The more important tall grasses are big bluestem (*Andropogon gerardi*), little bluestem (*A. scoparius*), switch grass (*Panicum virgatum*), side-oats grama (*Bouteloua curtipendula*), and Indian grass (*Sorghastrum nutans*).

The marsh hay of the northern Mississippi Valley consists of bluejoint (*Calamagrostis canadensis*), reed canary grass (*Phalaris arundinacea*), and a few other wet-land species.

The forage grasses of the Great Basin include species of *Poa*, *Festuca*, *Bromus*, *Aristida*, and *Stipa*. In the Southwest, the gramas, species of *Bouteloua*, dominate the range. A large bunchgrass, sacaton (*Sporobolus wrightii*), and alkali sacaton (*S. airoides*) furnish much forage.

A few of the many nutritious species found in the Northwestern States are greenleaf fescue (*Festuca viridula*), bluebunch fescue (*F. idahoensis*), pinegrass (*Calamagrostis rubescens*), slender wheatgrass (*Agropyron trachycaulum*), California bromegrass (*Bromus carinatus*), and in the semiarid regions bluebunch wheatgrass (*Agropyron spicatum*).

#### GRASSES IN THE INDUSTRIAL ARTS

The most important species of the industrial arts group is sugarcane (p. 740). This might be included among grasses that furnish food, but sugar is a manufactured product.

The chief fiber-producing grasses are esparto (*Lygeum spartum* and *Stipa tenacissima*), also known as alfa, natives of Spain and north Africa. The leaves and stems are utilized in paper making. The pith of the cornstalk and the oil of the corn grain find many uses in the arts.

Certain aromatic grasses furnish essential oils used in perfumery. The best known are lemon grass (*Cymbopogon citratus*), citronella grass (*C. nardus*), and vetiver (*Vetiveria zizanioides*).

The bamboos, the largest of the grasses, are of vast importance in the Indo-Malay region and are receiving increasing attention in the United States. The larger kinds reach a height of 30 meters and are 15 to 25 or 30 centimeters thick below, tapering to the summit. The culms or stems are very strong and are used in building houses and bridges. When the stems are split, flattened out, and the partitions at the joints removed they make very durable boards, a foot or more wide, for floors and walls. Rafts and floats are made of the hollow stems closed at the joints by natural airtight partitions. With the partitions removed bamboo stems furnish water pipes or conduits. Sections of the stem closed at one end by the partition form convenient vessels for holding water. Much of the furniture and many of the utensils and implements used by the Malays are made wholly or in part of bamboo. Slender bamboo stems are familiar to us in the form of fishing rods and walking canes. Shoots of *Phyllostachys edulis*, *Sinocalamus beecheyanus*, and other species of bamboo are a choice vegetable in the Orient and an expensive dainty in the United States. Paper and rayon are made from the culms of some species.

Brooms are made from the seed heads of broomcorn, a variety of sorghum. Leghorn hats are made of a kind of wheat straw cut young and bleached. Straw of rice and oats is used for matting and for hats.

Starch and alcohol are made from the grain of maize, wheat, and other cereals. The stalks, grain, and cobs of maize furnish a great variety of products, such as wallboard, glucose, oil, red rubber, and corncob pipes.

#### SOIL-HOLDING GRASSES

Grasses used to hold soil in place and prevent erosion possess strong creeping rhizomes.

Sand-binding grasses in addition are able to grow up through the deepening sand. The most effective sand binders for seacoast drifting sand are the European beachgrass (*Ammophila arenaria*) and its American relative (*A. breviligulata*). The dunes of the Netherlands, southwestern France, northern and western Denmark, and other parts of Europe and areas on Cape Cod are planted with beachgrass. These fixed dunes act as barriers, protecting the land behind them. The land now occupied by Golden Gate Park, once an area of drifting sand, was first held in place with beachgrass and later planted to shrubs and trees. *Calamovilfa longifolia* and *Redfieldia flexuosa* are effective native sand binders on sand dunes of the interior.

Grasses with strong rhizomes are used to hold the sides of cuts and banks and to protect them against erosion. Bermuda grass in the South and quackgrass (*Agropyron repens*) in the North have been used successfully for this purpose. Rhizome-bearing species of *Elymus* and *Agropyron* have been used in the Northwest to hold railroad embankments along the Columbia River.

Shallow-water marshes and lagoons are in many places being converted into dry land by native plants growing therein that accumulate soil and gradually raise the level of the bottom. Grasses, especially species of *Spartina* and *Phragmites*, play an important part in the process. Artificial plantings of *Spartina townsendii* have been used with great success in the south of England, northern France, and in parts of the Netherlands to convert marshes and mud flats along the coast into dry land.

### GRASSES FOR LAWNS AND GOLF COURSES

The lawn is a most important part of a well-planned landscape, park, or garden. For the humid regions of the Northern States, Kentucky bluegrass, also used for pasture, is the best-known lawngrass. Rough bluegrass (*Poa trivialis*) is often used as a lawngrass in shady places. In the Southern States Bermuda grass takes the place of bluegrass. Two other species are prominent as grasses for lawns and putting greens, creeping bent (*Agrostis palustris*) and colonial bent (*A. tenuis*). Along southern coasts St. Augustine grass (*Stenotaphrum secundatum*) and centipede grass (*Eremochloa ophiuroides*) are planted, being propagated by cuttings. Some of the fescue grasses are used in mixtures for lawns. These are red fescue (*Festuca rubra*), sheep fescue (*F. ovina*), hard fescue (*F. ovina* var. *duriuscula*), and shade fescue (*F. rubra* var. *heterophylla*).

### ORNAMENTAL GRASSES

Among typical ornamentals the plumegrasses, giant reed (*Arundo donax*), Ravenna grass (*Erianthus ravennae*), eulalia (*Miscanthus sinensis*), and pampasgrass (*Cortaderia selloana*) are the most popular for parks and large areas. Dwarf bamboo (*Bambusa multiplex*) is used for hedges in the South, and the smaller species of *Phyllostachys* for masses of evergreen. foliage. *Pseudosasa japonica*. an aggressively spreading hardy bamboo, is rather common in parks. Fountain grass (*Pennisetum setaceum*) and blue fescue (*Festuca ovina* var. *glauca*) are used for borders. Ribbon grass (*Phalaris arundinacea* var. *picta*) is a familiar grass in old gardens. Basket grass (a variegated form of *Oplismenus hirtellus*) will fall in long festoons from hanging baskets.

## DISTRIBUTION OF GRASSES

One of the most widely distributed of the families of flowering plants, the grasses are found over the land surface of the globe, in marshes and in deserts, on prairies and in woodland, on sand, rocks, and fertile soil, from the Tropics to the polar regions and from sea level to perpetual snow on the mountains.

The different grasses, like other kinds of plants, thrive best under certain conditions of soil, moisture, temperature, exposure, and altitude.

The conditions under which a plant normally grows is its habitat. Some species are narrowly restricted in their habitat—being found only in sand or on rocks, in salt marshes or on alpine summits, for example, whereas others are tolerant of wide variations of habitat. Red fescue (*Festuca rubra*) is an example of wide distribution of a species tolerant of a variety of habitats. It is found from the Arctic regions south at low altitudes to Georgia and central California and in the mountains farther south, and from the seacoast marshes to mountain tops.

Each species is found growing over a rather definite geographic area, but within this area it is confined to its particular habitat.

In mountain regions altitude is an important factor in modifying range, each species thriving within certain limits of altitude. Species found at high altitudes in one range of mountains may reappear at about the same altitude on other ranges. Certain grasses growing at low levels in the north are found in the mountains and at increasingly higher elevations southward.

The geographic range is of importance and is given in some detail for each species in the manual. The range as given is based upon the study of a vast amount of material, both in the herbarium and in the field. For convenience in keeping the records of distribution a series of outline maps, one for each species or variety, has been prepared in the grass herbarium of the United States National Herbarium. The known range of each species is indicated upon these maps by a dot on each State from which specimens are in the herbarium or have been examined by the author. (A few extensions of range have been found since the maps were engraved. These are included in the text.) Local floras, lists, and records of distribution have been checked, and efforts have been made to verify the records that seemed to indicate an extension of range. Other herbaria have been visited or have lent specimens, and many correspondents have submitted specimens for verification. No additions have been made without a study of the specimens. But it must be borne in mind that dots (representing specimens) necessarily indicate where the different specimens have been collected, therefore where botanists have been. Absence of a dot in a state does not necessarily mean the species in question does not grow in that state.

The ranges of native species are usually fairly well defined and continuous. A species of the Coastal Plain extends, for example, from New Jersey to North Carolina or from Virginia to Florida and Texas, without a conspicuous break. Mountain plants extend along mountain ranges where similar conditions prevail. Some species have in the main a continuous range but are found also in isolated and distant localities. *Bouteloua hirsuta* extends over the Great Plains east to Wisconsin and Louisiana, and again occurs abundantly and apparently native on Sanibel Island, Fla. Some Coastal Plain species appear again around the head of Lake Michigan. In these cases it is probable that the species do not occur in the intermediate areas.

Certain arctic or northern species also show interrupted range, being found within the limits of the United States only on isolated mountain tops. The arctic grass, *Phippsia algida*, for example, is known within the United States only from alpine summits in Colorado. What appear to be interrupted ranges along the northern or southern borders are mostly due to extensions into this country from the main ranges in Canada or Mexico.

The distribution of recently introduced species is often very erratic. A single introduction may maintain itself or even spread considerably for several years before coming to the notice of botanists. Introduced species often travel rapidly along railroads by means of cattle cars, or they spread as impurities in the seed of crop plants. That seeds may travel great distances through the air has been shown by experiments in which airplanes have collected seeds, insects, and other objects at varying heights in the atmosphere. For example, spikelets of *Paspalum dilatatum* and *P. urvillei* were taken at altitudes up to 5,000 feet in Louisiana.

Grasses introduced into cultivation may spread or "escape" from cultivation and become established over wide areas. Kentucky bluegrass (*Poa pratensis*) and the ryegrasses (*Lolium perenne* and *L. multiflorum*) are familiar examples. Johnson grass is an excellent forage grass, but if it escapes into cultivated fields may become a troublesome weed.

Other cultivated grasses, such as the grains, frequently spread from fields but are unable to maintain themselves for long. Eulalia (*Miscanthus sinensis*) has been cultivated for ornament in the eastern part of the United States for many years. Recently it has shown a tendency to spread by seed. It is now becoming a nuisance in some localities because of its aggressiveness in old fields.

## MORPHOLOGY OF GRASSES

The organs of grasses undergo many modifications or departures from the usual or typical structure. A knowledge of the structure and modifications of the organs, especially of the parts of the spikelet, is essential for the interpretation of relationships.

### VEGETATIVE ORGANS

In size grasses vary from minute species only 2 or 3 cm. high to the giant bamboos 30 m. tall. The vegetative organs, however, consist, in all cases, of root, stem, and leaves. A single unbranched stem with the attached leaves is a shoot.

#### ROOT

The roots of grasses are fibrous with little modification. The primary root persists only a short time after germination, its place being taken by secondary roots produced from the nodes of the young culm. Besides the original root system at the base of the plant, secondary roots are often formed from nodes above the ground as in maize (prop roots), or from the nodes of creeping culms (rhizomes or stolons). Roots are never produced from the internodes of the culms.

#### STEM

The jointed stem of a grass, called a culm, is made up of a series of nodes and internodes. The internode is hollow (wheat), or solid (maize); the node or joint is always solid. The culm may branch at the base as in wheat (stools) or above the base as in *Muhlenbergia*. Creeping culms, modified for propagation, may be below ground (rhizomes) or above ground (stolons). The lower internodes may thicken into corms (timothy, species of *Melica*, *Arrhenatherum elatius* var. *bulbosum*), sometimes referred to as bulbs. Perennial grasses may form a sod or mass of individuals by means of rhizomes or stolons, or they may form a crown or tuft by the continual formation of upright branches within the lower sheaths.

#### LEAF

The leaves are borne on the culm in two ranks, one at each node. The leaf consists of sheath and blade. The sheath envelops the culm above the node, the margins overlapping (open) or infrequently united into a cylinder for a part or a whole of the distance to the summit (closed).

The blades are typically flat, narrow, and sessile. In dry regions they are usually involute or convolute; in tropical shade they are often comparatively short and wide (lanceolate, ovate, or elliptic); in most of the bamboos they are narrowed into a short petiole articulate with the sheath.

Some grasses (especially the Hordeae) bear, one on either side at the base of the blade, appendages known as auricles. At the junction of the blade and sheath on the inside is a membranaceous or ciliate appendage called the ligule. The region on the back of the leaf at the junction of the sheath and blade is called the collar.

#### PROPHYLLUM

At the point where a branch shoot originates from a main shoot (in the axil of a sheath), there is produced on the side next to the parent shoot a 2-keeled organ (the first leaf of the shoot) called the prophyllum. At first the prophyllum completely covers the bud but later opens as the shoot develops. The organ is usually concave between the keels toward the parent shoot but clasps the new shoot by its margins.

#### FLORAL ORGANS

The floral organs of all flowering plants are modified shoots. The flowers of grasses consist of stamens and pistils with no floral envelopes or perianth, except as they are represented by the lodicules.

#### THE INFLORESCENCE

The unit of the grass inflorescence is the spikelet. The spikelets are nearly always aggregated in groups or clusters which constitute the inflorescence. The tassel of maize, the spike or head of wheat or timothy, and the panicle of the oat or bluegrass are examples of inflorescences.

The simplest inflorescence is the raceme, in which the spikelets are pediceled along an axis. The typical raceme, as in *Pleuropogon*, is rare in grasses. Modified spikelike racemes are characteristic of *Paspalum, Digitaria*, and allied genera, in which the spikelets are paired and short-pedicellate, and of most Andropogoneae, in which the spikelets are paired, one sessile, the other pedicellate. The inflorescences of the groups mentioned may best be considered as specialized panicles.

The spike differs from the raceme in having sessile spikelets. In the Hordeae the spikes are symmetrical, in the Chlorideae they are one-sided.

The panicle is the commonest kind of grass cluster. In this the spikelets are pediceled in a branched inflorescence. The panicle may be open or diffuse, as in *Panicum capillare*, or contracted, as in millet. Compact panicles, especially if cylindric like timothy, are called spikelike panicles.

Numerous small inflorescences may be aggregated into a large or compound inflorescence. Many Andropogoneae have compound inflorescences, for example, the broomsedge (*Andropogon virginicus*).

Panicles often expand at the time of flowering (anthesis). Such expansion or spreading of the branches and branchlets is brought about by the swelling of motor organs (pulvini) in the axils of the inflorescence.

Sometimes the ultimate branches of an inflorescence are sterile instead of bearing spikelets. The sterile branchlets of *Setaria, Pennisetum*, and *Cenchrus* are modified into bristles around the spikelets.

#### THE SPIKELET

A typical spikelet consists of a short axis (rachilla) on which the flowers are borne in the axils of 2-ranked imbricate bracts. The spikelet is, therefore, a reduced modified shoot in which the rachilla is a stem bearing at each node a reduced leaf (bract). The flowers are secondary reduced shoots borne in the axils of the bracts, the first bract (palea) on the secondary shoot being a modified prophyllum and the stamens and pistil being modified leaves or bracts. The bracts of the lowest pair on the rachilla, being always empty, are distinguished as glumes. The succeeding bracts are called lemmas (flowering glumes of some authors). The glumes and lemmas represent the sheath of the leaves, the blades not developing (in proliferous spikelets the parts are partially developed into typical leaves). The lemma, palea, and included flower are called the floret. The branchlet bearing the spikelet is the pedicel.

The spikelet may be reduced to a single floret (Agrostideae), sometimes with a prolongation of the rachilla behind, as in *Calamagrostis*. In *Andropogon* a fertile spikelet is paired with a sterile one in which the pistil or both pistil and stamens are wanting. The upper florets of the spikelet are often reduced in Festuceae, and the lower lemmas may be empty in some genera (*Uniola*, *Blepharidachne*). In *Melica* and *Chloris* the upper florets may be reduced and form a club-shaped body. In *Phalaris* there is one fertile floret with a pair of sterile florets below, each reduced to a small appressed scale. In *Lamarckia* and *Cynosurus* there are prominent sterile spikelets mixed with the fertile ones.

In Paniceae the spikelet has a perfect terminal floret and below this a sterile floret, consisting of a sterile lemma similar to the glumes, either empty or with a hyaline palea or sometimes with a staminate flower.

In a few grasses (*Amphicarpum*, *Chloris chloridea*) there are, in addition to the usual inflorescence above ground, cleistogamous spikelets borne on underground culms.

#### RACHILLA

The axis bearing the florets, the rachilla, usually disarticulates between the florets when the spikelet is more than 1-flowered. In many species of *Eragrostis* it is continuous, usually bearing the persistent paleas, after the remainder of the florets have fallen. When the rachilla disarticulates the break is usually just below the florets so that the rachilla joint remains attached as a little stipe back of the palea. The disarticulation is near the middle of the internode in *Trichoneura* and *Festuca subuliflora*. The rachilla disarticulates just above the floret in *Phragmites*, the rachilla remaining as a plumose stipe below it. The rachilla is short-villous or pilose in many genera of Aveneae (the callus of the floret often pilose also).

In some genera with 1-flowered spikelets (*Calamagrostis*, *Cinna*, *Cynodon*) the rachilla is prolonged behind the floret as a slender, often villous, stipe or bristle, and in several genera with several-flowered spikelets (*Koeleria*, *Poa*) it is prolonged beyond the uppermost floret.

#### GLUMES

The glumes are usually similar in shape and texture, the first often smaller and with fewer nerves. Rarely the first glume is longer than the second (species of *Aristida*). The first may be much reduced or wanting (*Axonopus*, *Paspalum*, *Digitaria*). Rarely both glumes are wanting (*Leersia*, *Reimarochloa*). In *Eriochloa* the first glume is reduced or wanting, the first rachilla joint being a hard ring below the spikelet. In Andropogoneae the first glume is usually indurate, sometimes strongly so. In some Hordeae the glumes are bristle-like.

#### LEMMAS

The lemmas in the more primitive grasses are typically similar to the glumes but may be variously modified. In *Panicum* the fertile lemma is much harder than the glumes; in Andropogoneae they are much thinner than the glumes, often hyaline. The indurate cylindric lemma of *Stipa* and *Aristida* bears a sharp callus at base, formed by the oblique articulation with the rachilla.

#### PALEA

The palea is mostly 2-keeled and often concave between the keels. It is homologous with the prophyllum. Sometimes the 2 nerves of the palea are so close together as to appear like a single nerve (*Cinna*); sometimes the 2 nerves are marginal and widely separated as in rice. The keels may be ciliate (*Eragrostis*), bearded (*Triplasis*), or winged (*Pleuropogon*). The palea is much reduced or wanting in species of *Agrostis*. Usually the palea falls with its lemma, but in many species of *Eragrostis* it persists upon the rachilla after the fall of the lemma.

#### FLOWER

The flower proper consists of the stamens and pistil. The stamens are usually 3 but may be 1 to 6, rarely more. The slender filaments bear 2-celled anthers which are basifixed but so deeply sagittate as to appear versatile. The pistil is 1-celled, with 1 ovule; the styles are usually 2 but may be 1 or 3; the stigmas may arise from a single style or directly from the ovary. The style of *Zea* is greatly elongated and stigmatic over much of the exserted surface.

The lodicules are small organs found at the base of the flower outside the stamens. There are usually 2, rarely 3, the function of which is to open the floret at anthesis by their turgidity. They probably represent much reduced divisions of a perianth.

Typically the grasses are adapted to cross-pollination, but many species are cleistogamous in part. The axillary inflorescences of some species (*Panicum clandestinum* and allies, *Leersia oryzoides*) are enclosed in the sheaths and are self-pollinated. The florets of wheat expand for only a short time, when cross-pollination may take place, but for the most part are self-pollinated.

The fruit of the grasses is usually a caryopsis, in which the single seed is grown fast to the pericarp, forming a seedlike grain. In a few genera (*Sporobolus, Eleusine*), the seed is free from the pericarp. The caryopsis may be free from the lemma and palea, as in wheat, or it may be permanently enclosed, as in the oat and in the Paniceae. The grain (caryopsis) may enlarge during ripening and greatly exceed the glumes, lemma, and palea, as in maize and *Pennisetum glaucum*.

The embryo lies on the side of the caryopsis next to the lemma, and can be easily seen as an oval depression (the "germ" of maize and wheat). The hilum is the dot or line opposite the embryo which marks the point of attachment of the seed to the pericarp. The part of the caryopsis not occupied by the embryo is the endosperm, or nourishment for the germinating seed.

### CLASSIFICATION OF GRASSES

A natural classification of plants is one in which the different kinds or species are arranged in groups according to their resemblances as shown by their structure, especially (in the grasses and other flowering plants) by the structure of their flowers. The plants of today represent a cross section of the lines of descent from countless generations that have preceded them. It is generally accepted that there has been much variation during the evolutionary process, and that all living plants are genetically connected through their lines of descent. Some of the gaps in present-day knowledge of relationship are filled by fossil remains, but relatively few of the ancestors of living plants are represented by fossils. Knowledge of the ancestry of the kinds of plants now on the globe is necessarily very incomplete. Hence, ideas of the relations of groups to each other are largely inferences based upon morphological resemblances. Those individuals which are so much alike as to appear to be of one kind, with, presumably, a common ancestor in recent geological times, are regarded as belonging to the same species. The species is the unit of classification. For convenience, species are grouped into genera and genera into families. For example, the white oak, red oak, black oak, and other kinds or species of oak belong to the oak genus (*Quercus*), all the species of which have one character in common —the fruit is an acorn. The oak genus, the beech genus, the chestnut genus, and a few allied genera are grouped together as a family.

The grass family (Gramineae or Poaceae) is one of the largest in number of genera and species, and, among flowering plants, is probably the largest in the number of individuals and is one of the most widely distributed. Some genera, such as the bluegrasses (*Poa*), the bromegrasses (*Bromus*), and the immense genus *Panicum*, contain numerous often closely allied species. Some genera contain but a few species or only one.

When an attempt is made to classify a group of related variable species the question always arises whether there are several closely related but distinct species or a few distinct species, each of which shows great variation. It is but natural that botanists should differ in their conclusions. This explains in part the different classifications of the same group given by botanists of different periods or even of the same period. A satisfactory classification depends upon the study of abundant material both in the field and in the herbarium. By observation in the field one learns the range of variability of a species, while in the herbarium one can compare plants from different localities, interpreting the dried specimens in the light of field experience.

In the classification of variable species it is found convenient sometimes to separate variants as varieties. A variety comprises those individuals of a species that show a definite tendency to vary in a certain direction, but which are connected with the species by rather numerous intergrades. Sometimes a variety is founded on a single variation which is distinct but trivial, for example, pubescent specimens of a glabrous species. A variation supported by a distinct geographical range or even by a distinct habitat is given greater weight than is a variation found in a few individuals growing among plants of the typical form.

The study of a vast amount of material in field and herbarium during some 40 years has resulted in the recognition of relatively few varieties, the intergrades proving to be more numerous than fairly clear-cut variants. Well-marked varieties are given a separate paragraph in the text, but are not usually given in the keys. Less well-marked varieties are given in the paragraph with the species. Many additional forms are indicated in a descriptive statement without being formally recognized as species or varieties. For example, under *Digitaria gracillima* appears, "A tall plant with * * * has been called *D. bakeri* (Nash) Fernald"; and under *Eriochloa michauxii*, "a form with * * * has been described as *E. mollis* var. *longifolia* Vasey."

The arrangement of the genera in this manual is, in general, from the simple to the complex. It is, of course, impossible to arrange all the genera in linear sequence and at the same time represent a gradual increase in complexity because plants have not developed in a single line, but have diverged in all directions, their relationships being a complex network. The highest genus of one tribe may be much more complex than the lowest genus of the next tribe above. On the average the Bambuseae seem to be the most primitive and the Tripsaceae the most complex. A grass with a spikelet consisting of glumes and several florets, the lemmas and glumes being similar and resembling bracts, is a primitive form. Grasses with spikelets in which the parts are reduced, enlarged, or much differentiated, are derived or complex forms. Derived forms may be simple from the reduction of parts and yet not be primitive. In the main the genera of grasses fall readily into a few large groups or tribes, but several genera of uncertain affinities are, for convenience, placed in the recognized tribes on artificial characters, with the hope that further study and exploration will bring to light their true relationships.

The grasses of the world (some 600 genera) have been grouped into 14 tribes, all of which are represented in the United States.

The sequence of tribes and genera in the manual with a few minor changes, is that found in The Genera of Grasses of the United States.[2]

## NOMENCLATURE

The cooperative study of botany depends for progress and success on definiteness in the application of the names of plants. Research workers in all branches of botany must use the names of plants in the same sense, or serious misunderstandings will result. One of the functions of systematic botany is to determine the correct names of plants. The study of the application of plant names is nomenclature. By common consent of the botanists of the world Latin has been accepted as the language for technical plant names.

Modern nomenclature commences with the publication in 1753 of Linnaeus' Species Plantarum in which the binomial system of naming plants was first proposed. During the nearly 200 years following that date many thousands of plants have been described. During this time there has been a lack of uniformity in the use of names, causing much confusion and resulting in frequent changes. The same species has been described under different names at different times, and the same name has been given to different plants. This confusion has been especially embarrassing to the agriculturist, ranger, seedsman, pathologist, entomologist, and to all others interested in plants but not familiar with nomenclature and the history of the names used.

The difference in the Latin names applied in different books to the same kind of grass is due to several causes.

(1) A species is described as new by one author without knowing that the same species had been previously described by another author. The second name is known as a synonym.

(2) An author applies a new name to a variant of a species already described. The author recognizes the variant as a distinct species. Other botanists may consider it to be only a variety of the older species or may consider it as a variant not sufficiently distinct to be worthy of varietal rank.

(3) Authors have different concepts of the limits of genera. The genus *Triticum* was described by Linnaeus. A later botanist thought that many of the species of this genus were different enough to constitute a distinct genus, *Agropyron*, and transferred quackgrass, first described as *Triticum repens* to *Agropyron*, as *A. repens*.

(4) Authors sometimes misidentify species. Linnaeus described one of the cordgrasses as *Spartina cynosuroides*. Later, Michaux used the specific name for a different species (*Trachynotia cynosuroides*, based on *S. cynosuroides* L.) This error was corrected and the species described by Michaux was given a new name, *S. michauxiana*. Later the loan of the type of *Spartina pectinata* Link, poorly described many years earlier, shows that that name is the valid one for the species.

It will be seen that the differences in names are due in part to differences of opinion as to the generic, specific, or varietal distinctness of forms; in part to lack of knowledge as to what plants have been described previously; and in part to errors of identification.

All the preceding shows the need of rules of nomenclature. To enable users of this manual to coordinate the names published to date a synonymy has been appended in which all the names published for grasses in the United States have been arranged under the names here adopted, that is, under the oldest valid name for each species. In determining the valid names of the species the International Rules of Botanical Nomenclature have been followed. Under these rules certain generic names are conserved though they are not the earliest. The names of genera of grasses on the conserved list are as follows: *Chrysopogon, Tragus, Zoysia, Setaria, Leersia, Ehrharta, Hierochloë, Crypsis, Coleanthus, Corynephorus, Cynodon, Ctenium, Buchloë, Diarrhena, Lamarckia, Glyceria, Scolochloa.*

Certain other names of genera are used for different reasons. *Digitaria* antedates *Syntherisma* with which it is synonymous. It was proposed at the Cambridge International Botan-

[2] HITCHCOCK, A. S. THE GENERA OF GRASSES OF THE UNITED STATES, WITH SPECIAL REFERENCE TO THE ECONOMIC SPECIES. U.S. Dept. Agr. Bul. 772, 307 pp., illus. 1920, revised 1936.

ical Congress (and referred to a committee) that the standard species of *Holcus* be *H. lanatus* and of *Aira* be *A. praecox,* thus leaving *Sorghum* and *Deschampsia* the valid names for their respective genera.

The synonymy attempts to record all the effectively published names given to species and varieties described from the United States or known to grow in the United States. In addition many names are given that have been published as synonyms or without sufficient description (nomina nuda). Whether such names are included depends upon whether they have appeared in such works as the Index Kewensis or have some connection with effectively published names. When a species is transferred from one genus to another, a new name results. The basis of the transfer is given in each case. If the name was published as new the original published locality is given. Statements enclosed in brackets following the original locality are based upon unpublished evidence.

Forms (formae) are included in the synonymy so far as they have been indexed in the grass herbarium. The index includes all forms recently published in this country. Misapplied names have not been included among the synonyms but are mentioned in a paragraph at the end of the synonymy of the valid species, and then only names that have appeared in recent manuals are given. For convenience the names of the genera are arranged alphabetically and under each genus the valid names of the species are given in alphabetic order in boldface type, the synonyms of each species (in italics) being arranged chronologically under the valid name.

So far as possible the names have been confirmed or identified by examination of the types. The type of a species or variety is the specimen which an author had chiefly in mind when he wrote the original description. The type specimen determines the application of the name. The type specimens of the early American botanists are mostly in European herbaria. The types of species described by Vasey and other botanists connected with the Department of Agriculture are mostly in the United States National Herbarium. Types not in Washington have been studied in other herbaria and photographs and drawings made of them by the agrostologists of the Department of Agriculture, or have been lent by the curators of the herbaria in which they are deposited. Through the courtesy of these curators many fragments of types have been deposited in the United States National Herbarium. A few type specimens have not been located, and doubtless in some of these cases there are no types in existence to confirm original descriptions. A relatively small number of published names still remain unidentifiable. These names are listed following the synonymy. Certain exotic species, occasionally cultivated for ornament or for trial, have been included in notes appended to the genera to which they belong. It has not been practicable in all cases to verify the application of the names on a type basis, and the species are admitted under the names they bear in cultivation.

## COMMON NAMES

The common or English names of plants are often uncertain in their application, different plants bearing the same name or the same plant bearing different names in different localities. A recent work, Standardized Plant Names,[3] recently reissued, has coordinated and standardized the common names. One of the authors of this work, Frederick V. Coville, standardized the common names of the grasses for the first edition of this Manual.

[3] AMERICAN JOINT COMMITTEE ON HORTICULTURAL NOMENCLATURE. STANDARDIZED PLANT NAMES. Prepared by Olmsted, F. L., Coville, F. V., and Kelsey, H. P. 546 pp. Salem, Mass. 1923. (Revised by Kelsey, H. P., and Dayton, W. A. 675 pp. Harrisburg, Pa. 1942.)

## SCOPE OF THE MANUAL

The manual includes descriptions of all grasses known to grow in the continental United States, excluding Alaska. There are 169 numbered genera and 1,398 numbered species. Of these, 46 genera and 156 species are introduced, mostly from the Eastern Hemisphere.

In addition to the numbered species, which may be considered permanent constituents of the flora of the United States, there are 16 genera and 120 species that are known only as ballast plants, or as waifs, or are only rarely cultivated. These appear not to be established and are mentioned, without numbers, in paragraphs appended to their nearest allies. They are not included in the keys.

The manual is based mainly on the material in the United States National Herbarium, the grass collection of which is the largest in the world, numbering more than 320,000 sheets. In addition, all the larger collections of grasses in the United States have been consulted and the curators have lent specimens for study and have aided in other ways. Many smaller collections have contributed information, especially on the ranges of species. The cooperation of the Forest Service, United States Department of Agriculture, has been invaluable. The Forest Service maintains in its Washington office a range-plant herbarium consisting of the collections made by forest officers, especially those located in western national forests and forest experiment stations. The grasses of this range-plant herbarium have been examined and have furnished important data on distribution.

Many botanists throughout the country have rendered valuable assistance in recent years by contributing specimens that have added species previously unknown from the United States, have extended ranges, and have helped to solve the position of puzzling species and varieties.[4]

Nearly all the numbered species are illustrated.[5] About half are accompanied by a map, giving the distribution of that species in the United States.

To aid the users of this work in pronouncing the Latin names the accented syllable is indicated. The accent mark is used to show the accented syllable without reference to the length of the vowel.

## GRAMINEAE (POACEAE), THE GRASS FAMILY

Flowers perfect (rarely unisexual), small, with no distinct perianth, arranged in spikelets consisting of a shortened axis (rachilla) and 2 to many 2-ranked bracts, the lowest 2 being empty (the glumes, rarely one or both obsolete), the 1 or more succeeding ones (lemmas) bearing in their axils a single flower, and, between the flower and the rachilla, a second 2-nerved bract (the palea), the lemma, palea, and flower together constituting the floret; stamens 1 to 6, usually 3, with very delicate filaments and 2-celled anthers; pistil 1, with a 1-celled 1-ovuled ovary, 2

---

[4] The more important are: A. A. Beetle, from California; E. E. Berkeley, from West Virginia; H. L. Blomquist, from North Carolina; W. E. Booth, from Montana; Clair Brown, from Louisiana; V. H. Chase, from Illinois, Arkansas, and Idaho; Earl Core, from West Virginia; R. A. Darrow, from Arizona; R. J. Davis, from Idaho; Charles C. Deam and J. E. Potzger, from Indiana; H. I. Featherly, from Oklahoma; M. L. Fernald, from Northeastern States and Virginia; A. O. Garrett, from Utah; L. N. Goodding, from the Southwest; F. W. Gould, from Arizona and California; C. R. Hanes, from Michigan; H. D. Harrington, from Colorado; Bertrand Harrison, from Utah; R. F. Hoover and John Thomas Howell, from California; T. H. Kearney, from Arizona; John and Charlotte Reeder, California to Michigan; and W. A. Silveus, from Texas and other Southern States.

Jason R. Swallen, Curator, Division of Grasses, U.S. National Museum, has given valuable assistance. The bibliography is based on the catalog of grass names maintained in the Division of Grasses, this catalog being the work, over some 35 years, of Cornelia D. Niles, bibliographer. F. A. McClure, bamboo specialist, U.S. Department of Agriculture, contributed the economic notes on bamboos and has aided in the elucidation of the native species of bamboos.

[5] The drawings illustrating the genera (previously published in the U. S. Department of Agriculture Bulletin 772, the Genera of Grasses of the United States . . . ) and nearly half of the others were made by Mary Wright Gill; the rest were drawn by Edna May Whitehorn, Frances C. Weintraub, Leta Hughey, and Agnes Chase. The last-named made most of the spikelet drawings. In each case the specimen from which the drawing was made is cited, for example (Nash 2198, Fla.).

(rarely 1 or 3) styles, and usually plumose stigmas; fruit a caryopsis with starchy endosperm and a small embryo at the base on the side opposite the hilum.

Herbs, or rarely woody plants, with hollow or solid stems (culms) closed at the nodes, and 2-ranked usually parallel-veined leaves, these consisting of 2 parts, the sheath, enveloping the culm, its margins overlapping or sometimes grown together, and the blade, usually flat; between the 2 on the inside, a membranaceous hyaline or hairy appendage (the ligule).

The spikelets are almost always aggregated in spikes or panicles at the ends of the main culms or branches. The perianth is usually represented by 2 (rarely 3) small hyaline scales (the lodicules) at the base of the flower inside the lemma and palea. The grain or caryopsis (the single seed and the adherent pericarp) may be free, as in wheat, or permanently enclosed in the lemma and palea, as in the oat. Rarely the seed is free from the pericarp, as in species of *Sporobolus* and *Eleusine*. The culms of bamboos are woody, as are also those of a few genera, such as *Olyra* and *Lasiacis*, belonging to other tribes. The culms are solid in our species of the tribes Tripsaceae and Andropogoneae and in several other groups. The margins of the sheaths are grown together in some species of *Bromus*, *Danthonia*, *Festuca*, *Melica*, *Glyceria*, and other genera.

The parts of the spikelet may be modified in various ways. The first glume, and more rarely also the second, may be wanting. The lemmas may contain no flower, or even no palea, or may be reduced or rudimentary. Rarely, as in species of *Agrostis* and *Andropogon*, the palea is obsolete.

The division of the family into two subfamilies is somewhat artificial. The tribes Zoysieae, Oryzeae, Zizanieae, and especially Phalarideae, do not fall definitely into either of the recognized subfamilies. They are placed as indicated largely for convenience.

## DESCRIPTIONS OF THE SUBFAMILIES AND KEYS TO THE TRIBES

### SUBFAMILY 1. FESTUCOIDEAE

Spikelets 1- to many-flowered, the reduced florets, if any, above the perfect florets (except in Phalarideae; sterile lemmas below as well as above in *Ctenium*, *Uniola*, and *Blepharidachne*); articulation usually above the glumes; spikelets usually more or less laterally compressed.

*Key to the tribes of Festucoideae*

Plants woody, the culms perennial. Spikelets several-flowered........ 1. BAMBUSEAE (p. 27)
Plants herbaceous, the culms annual (somewhat woody and persistent in *Arundo*).
  Spikelets with 2 (rarely 1) staminate, neuter, or rudimentary lemmas unlike and below the fertile lemma; no sterile or rudimentary floret above......8. PHALARIDEAE (p. 547)
  Spikelets without sterile lemmas below the perfect floret (or these rarely present and like the fertile ones, a dissimilar pair below and a rudimentary floret above in *Blepharidachne*).
    Spikelets unisexual, falling entire, 1-flowered, terete or nearly so. 10. ZIZANIEAE (p. 561)
    Spikelets perfect (rarely unisexual but then not as above), usually articulate above the glumes.
      Spikelets articulate below the glumes, 1-flowered, very flat, the lemma and palea about equal, both keeled. Glumes small or wanting........ 9. ORYZEAE (p. 556)
      Spikelets articulate above the glumes (rarely below, but the glumes, at least one, well developed).
        Spikelets 1-flowered (or the staminate 2-flowered) in groups (short spikes) of 2 to 5 (single in *Zoysia*), the groups racemose along a main axis, falling entire; lemma and palea thinner than the glumes.................................. 6. ZOYSIEAE (p. 482)
        Spikelets not as above.
          Spikelets sessile on a usually continuous rachis (short-pedicellate in *Leptochloa*

and *Trichoneura;* the rachis disarticulating in *Monerma*, *Parapholis*, *Hordeum*, *Sitanion*, and in a few species of allied genera. See also *Brachypodium* in *Festuceae.*)

Spikelets on opposite sides of the rachis; spike terminal, solitary. 3. HORDEAE (p. 230)

Spikelets on one side of the rachis; spikes usually more than 1, digitate or racemose ........ 7. CHLORIDEAE (p. 491)

Spikelets pedicellate in open or contracted, sometimes spikelike, panicles, rarely racemes.

Spikelets 1-flowered (occasionally some of the spikelets 2-flowered in a few species of *Muhlenbergia*) ........ 5. AGROSTIDEAE (p. 313)

Spikelets 2- to many-flowered.

Glumes as long as the lowest floret, usually as long as the spikelet (sometimes shorter in *Sphenopholis*); lemmas awned from the back (spikelets awnless in species of *Trisetum*, *Koeleria*, *Sphenopholis*, and *Schismus*). 4. AVENEAE (p. 280)

Glumes shorter than the first floret (except in *Dissanthelium* with long rachilla joints, and in *Tridens strictus*); lemmas awnless or awned from the tip or from a bifid apex ........ 2. FESTUCEAE (p. 31)

### SUBFAMILY 2. PANICOIDEAE

Spikelets with 1 perfect terminal floret (disregarding those of the few monoecious genera and the staminate and neuter spikelets) and a sterile or staminate floret below, usually represented by a sterile lemma only, 1 glume sometimes (rarely both glumes) wanting; articulation below the spikelets, either in the pedicel, in the rachis, or at the base of a cluster of spikelets, the spikelets falling entire, singly, in groups, or together with joints of the rachis; spikelets, or at least the fruits, more or less dorsally compressed.

*Key to the tribes of Panicoideae*

Glumes membranaceous, the sterile lemma like the glumes in texture.

Fertile lemma and palea thinner than the glumes. Sterile lemma awned from the notched summit ........ 11. MELINIDEAE (p. 569)

Fertile lemma and palea indurate or at least firmer than the glumes. 12. PANICEAE (p. 569)

Glumes indurate; fertile lemma and palea hyaline or membranaceous, the sterile lemmal ike the fertil e one in texture.

Spikelets unisexual, the pistillate below, the staminate above, in the same inflorescence or in separate inflorescences ........ 14. TRIPSACEAE (p. 789)

Spikelets in pairs, one sessile and perfect, the other pedicellate and usually staminate or neuter (the pedicellate one sometimes obsolete, rarely both pedicellate). Lemmas hyaline ........ 13. ANDROPOGONEAE (p. 737)

## DESCRIPTIONS OF THE TRIBES AND KEYS TO THE GENERA

### TRIBE 1. BAMBUSEAE

Culms woody, perennial, usually hollow; spikelets 2- to several-flowered, in panicles or racemes, or in close heads or fascicles; often 1 or more sterile lemmas at base of spikelet; lemmas usually awnless; blades usually articulated with the sheath, flat, rather broad. Only one genus, *Arundinaria*, is native within our limits. Several species of this and other genera are cultivated in the Southern States.

### TRIBE 2. FESTUCEAE

Spikelets more than 1-flowered, usually several-flowered, in open, narrow, or sometimes spikelike panicles (rarely in racemes); lemmas awnless or awned from the tip, rarely from between the teeth of a bifid apex; rachilla usually disarticulating above the glumes and between the florets.

A large and important tribe, mainly inhabitants of the cooler regions. The lemma is divided into several awns in *Pappophorum* and its allies, is deeply 2-lobed in *Triplasis* and in a few species of *Tridens*, 3-lobed in *Blepharidachne*, several-toothed in *Orcuttia*, and slightly 2-toothed in *Bromus* and in a few other genera, the awn, when single, arising from between the teeth. The paleas are persistent upon the continuous rachilla in many species of *Eragrostis*. *Scleropogon*, *Monanthochloë*, *Distichlis*, *Hesperochloa* and a few species of *Poa* and *Eragrostis* are dioecious. *Gynerium*, *Cortaderia*, *Arundo*, *Phragmites*, and *Neyraudia* are tall reeds. In *Blepharidachne* there is a pair of sterile florets at the base of the single fertile floret, and a rudiment above. In some species of *Melica* there is, above the fertile florets, a club-shaped rudiment consisting of 1 or more sterile lemmas. In *Uniola* there are 1 to 4 sterile lemmas below the fertile ones. In *Melica imperfecta* and *M. torreyana* there may be only 1 perfect floret.

*Key to the genera of Festuceae*

1a. Plants dioecious, (sometimes monoecious), the sexes very dissimilar, the pistillate lemmas with 3 long twisted divergent awns, the staminate lemma awnless or mucronate. 41. SCLEROPOGON.
1b. Plants with perfect flowers, or, if dioecious, the sexes not dissimilar in appearance.
  2a. Lemmas divided at the summit into 5 to several awns or awnlike lobes.
    Awnlike lobes 5. Inflorescence an erect raceme or simple panicle........ 36. ORCUTTIA.
    Awns 9 or more.
      Awns unmixed with awned teeth; all the florets falling attached, their awns forming a pappuslike crown, the lower 1 to 3 fertile; panicles narrow.
        Spikelets 3-flowered, the first floret fertile; awns 9, plumose, equal. 40. ENNEAPOGON.
        Spikelets 4- to 6-flowered, the lower 1 to 3 fertile; awns numerous, not plumose, unequal........ 39. PAPPOPHORUM.
      Awns mixed with awned teeth; florets not falling attached, the rachilla disarticulating between them; panicles somewhat open........ 38. COTTEA.
  2b. Lemmas awnless, with a single awn, or, if with 3, the lateral awns minute.
    3a. Tall stout reeds with large plumelike panicles. Lemmas or rachilla with long silky hairs as long as the lemmas.
      Leaves crowded at the base of the culms........ 27. CORTADERIA.
      Leaves distributed along the culms.
        Lemmas naked. Rachilla hairy........ 28. PHRAGMITES.
        Lemmas hairy.
          Rachilla naked........ 26. ARUNDO.
          Rachilla hairy........ 29. NEYRAUDIA.
    3b. Low or rather tall grasses, rarely more than 1.5 m. tall.
      4a. Plants dioecious, perennial.
        Plants densely tufted, rather coarse, erect from short rhizomes; lemmas scabrous; grasses of dry mountain slopes........ 11. HESPEROCHLOA.
        Plants not densely tufted, spreading by stolons or extensively creeping rhizomes; lemmas glabrous; grasses of salt or alkaline soil.
          Plants low, stoloniferous; spikelets obscure, scarcely differentiated from the short crowded rigid leaves........ 20. MONANTHOCHLOË.
          Plants erect from creeping rhizomes; spikelets in narrow simple exserted panicles. 21. DISTICHLIS.
      4b. Plants not dioecious (except in a few species of *Poa* with villous lemmas and in an annual species of *Eragrostis*).
        5a. Spikelets of two forms, sterile and fertile intermixed. Panicle dense, somewhat one-sided.
          Fertile spikelets 2- or 3-flowered; sterile spikelets with numerous rigid awn-tipped lemmas; panicle dense, spikelike........ 24. CYNOSURUS.
          Fertile spikelets with 1 perfect floret, long-awned; sterile spikelets with many obtuse sterile lemmas; panicle branchlets short, nodding.... 25. LAMARCKIA.
        5b. Spikelets all alike in the same inflorescence.
          6a. Lemmas 3-nerved, the nerves prominent, often hairy.
            7a. Inflorescence a few-flowered head or capitate panicle overtopped by the leaves or partly concealed in them. Lemmas toothed or cleft; low plants of the arid regions.
              Inflorescence hidden among the sharp-pointed leaves, not woolly; plants annual (Chlorideae)........ 114. MUNROA.
              Inflorescence a capitate woolly panicle, not concealed; plants perennial.
                Lemmas cleft either side of the midnerve to near the base, the lower two

sterile, the third floret fertile, the fourth reduced to a 3-awned rudiment ........ 37. BLEPHARIDACHNE.
Lemmas 2-lobed but not deeply cleft, all fertile but the uppermost. 33. TRIDENS.
7b. Inflorescence an exserted open or spikelike panicle.
8a. Lemmas pubescent on the nerves or callus (except in *Tridens albescens*), the midnerve usually exserted as an awn or mucro.
Nerves glabrous. Callus densely hairy; lemmas firm; panicle large, diffuse. 19. REDFIELDIA.
Nerves hairy at least below, the lateral ones often conspicuously so.
Palea densely long-ciliate on the upper half ........ 34. TRIPLASIS.
Palea sometimes villous but not long-ciliate on the upper half. Perennials ........ 33. TRIDENS.
8b. Lemmas not pubescent on the nerves nor callus (the internerves sometimes pubescent), awnless.
Glumes longer than the lemmas; lateral nerves of lemma marginal, the internerves pubescent ........ 18. DISSANTHELIUM.
Glumes shorter than the lemmas; lateral nerves of lemma not marginal, the internerves glabrous.
Lemmas chartaceous; grain large, beaked, at maturity forcing the lemma and palea open ........ 17. DIARRHENA.
Lemmas membranaceous; if firm, the grain neither large nor beaked.
Spikelets subterete; palea longer than the lemma, bowed out below. 16. MOLINIA.
Spikelets compressed; palea not longer than the lemma, not bowed out below (except in *Eragrostis oxylepis* and *E. sessilispica*).
Lemmas truncate; spikelets 2-flowered ........ 15. CATABROSA.
Lemmas acute or acuminate; spikelets 3- to many-flowered. Rachilla continuous, the paleas persistent after the fall of the lemmas (rachilla disarticulating in Sect. Cataclastos). 14. ERAGROSTIS.
6b. Lemmas 5- to many-nerved, the nerves sometimes obscure.
Spikelets with 1 to 4 empty lemmas below the fertile florets; nerves obscure; lemmas firm ........ 22. UNIOLA.
Spikelets with no empty lemmas below the fertile florets; nerves usually prominent; lemmas membranaceous (firm in a few species of *Bromus* and *Festuca*).
Lemmas flabellate; glumes wanting; inflorescence dense, cylindric. Low annual ........ 35. NEOSTAPFIA.
Lemmas not flabellate; glumes present; inflorescence not cylindric.
Lemmas as broad as long, the margins outspread; florets closely imbricate, horizontally spreading ........ 13. BRIZA.
Lemmas longer than broad, the margins clasping the palea; florets not horizontally spreading.
Callus of florets bearded.
Lemmas erose at summit, awnless ........ 9. SCOLOCHLOA.
Lemmas bifid at summit, awned ........ 31. SCHIZACHNE.
Callus not bearded (lemmas cobwebby at base in *Poa*). Lemmas not erose (slightly in *Puccinellia*).
9a. Lemmas keeled on the back (somewhat rounded in *Poa scabrella* and its allies).
Spikelets strongly compressed, crowded in 1-sided clusters at the ends of the stiff, naked panicle branches ........ 23. DACTYLIS.
Spikelets not strongly compressed, not crowded in 1-sided clusters.
Lemmas awned from a minutely bifid apex (awnless or nearly so in *Bromus catharticus* and *B. brizaeformis*); spikelets large ........ 2. BROMUS.
Lemmas awnless; spikelets small ........ 12. POA.
9b. Lemmas rounded on the back (slightly keeled toward the summit in *Festuca* and *Bromus*).
Glumes papery; lemmas firm, strongly nerved, scarious-margined; upper florets sterile, often reduced to a club-shaped rudiment infolded by the broad upper lemmas. Spikelets tawny or purplish, usually not green ........ 30. MELICA.
Glumes not papery; upper florets not unlike the others.
Nerves of lemma parallel, not converging at summit or but slightly so.

Spikelets in racemes.
Racemes short, dense, overtopped by the leaves; spikelets awnless........ 8. SCLEROCHLOA.
Racemes elongate, loose, exserted; spikelets awned or mucronate........ 10. PLEUROPOGON.
Spikelets in open or contracted panicles.
Nerves prominent; plants usually rather tall, growing in woods or fresh-water marshes........ 7. GLYCERIA.
Nerves faint; plants low, growing in saline soil.
6. PUCCINELLIA.
Nerves of lemma converging toward the summit, the lemmas narrowed at apex.
Lemmas awned or awn-tipped from a minutely bifid apex (awnless in *Bromus brizaeformis*); palea adhering to the caryopsis.
Spikelets in open to contracted panicles; stigmas borne at the sides of the summit of ovary........ 2. BROMUS.
Spikelets nearly sessile in a strict raceme; stigmas terminal on the ovary........ 3. BRACHYPODIUM.
Lemmas entire, pointed, awnless or awned from the tip (minutely toothed in *Festuca elmeri* and *F. gigantea*).
Spikelets awned (awnless in a few perennial species); lemmas pointed........ 4. FESTUCA.
Spikelets awnless.
Second glume 5- to 11-nerved; spikelets mostly 1 cm. or more long; lemmas broad.
Florets persistent on the continuous rachilla, the caryopsis falling free........ 32A. ECTOSPERMA.
Florets falling together with the joints of the articulate rachilla........ 32. VASEYOCHLOA.
Second glume 1- to 3-nerved; spikelets smaller; lemmas 5-nerved, membranaceous, not pointed.
Spikelets on slender pedicels in compound panicles; perennials........ 12. POA.
Spikelets on thick short pedicels in simple panicles; annual. Rachilla disarticulating at the base, forming a stipe to the floret above.... 5. SCLEROPOA.

## TRIBE 3. HORDEAE

Spikelets 1- to several-flowered, sessile on opposite sides of a jointed or continuous axis forming symmetrical spikes (not 1-sided, but spikelets sometimes turned to one side in some species).

This small but important tribe, found in the temperate regions of both hemispheres, includes our most important cereals, wheat, barley, and rye. The rachis is flattened or concave next to the spikelets, or in some genera is thickened and hollowed out, the spikelets being more or less enclosed in the hollows. In *Triticum* and its allies there is 1 spikelet at each node of the rachis; in *Hordeum* and its allies there are 2 or 3 at each node. In *Lolium* and its allies the spikelets are placed edgewise to the rachis, and the first or inner glume is suppressed except in the terminal spikelet. The rachis of the spikes disarticulates at maturity in several genera. In some species of *Elymus* and especially in *Sitanion* the glumes are very slender, extending into long awns, in the latter genus sometimes divided into several slender bristles. The spikes are rarely branched or compound, especially in *Elymus condensatus*. In this tribe the blades of the leaves usually bear on each side at the base a small appendage or auricle.

*Key to the genera of Hordeae*

1a. Spikelets solitary at each node of the rachis (rarely 2 in species of *Agropyron*, but never throughout).
2a. Spikelets 1-flowered, sunken in hollows in the rachis. Spikes slender, cylindric; low annuals.

Lemmas awned; florets lateral to the rachis........ 53. SCRIBNERIA.
Lemmas awnless; florets dorsiventral to the rachis.
First glume wanting........ 51. MONERMA.
First glume present, the pair standing in front of the spikelet...... 52. PARAPHOLIS.
2b. Spikelets 2- to several-flowered, not sunken in the rachis.
Spikelets placed edgewise to the rachis. First glume wanting except in the terminal spikelet........ 50. LOLIUM.
Spikelets placed flatwise to the rachis.
Plants perennial........ 42. AGROPYRON.
Plants annual.
Spikelets turgid or cylindric........ 44. AEGILOPS.
Spikelets compressed.
Glumes ovate, 3-nerved........ 43. TRITICUM.
Glumes subulate, 1-nerved........ 45. SECALE.
1b. Spikelets more than 1 at each node of the rachis (solitary in part of the spike in some species of *Elymus*).
Spikelets 3 at each node of the rachis, 1-flowered, the lateral pair pediceled, usually reduced to awns........ 49. HORDEUM.
Spikelets 2 or more (sometimes solitary in *Elymus*) at each node of the rachis, alike, 2- to 6-flowered.
Glumes wanting or reduced to 2 short bristles; spikelets horizontally spreading or ascending at maturity. Spikes very loose........ 48. HYSTRIX.
Glumes usually equaling the florets (reduced in *Elymus interruptus*); spikelets appressed or ascending.
Rachis continuous (rarely tardily disarticulating); glumes broad or narrow, entire. 46. ELYMUS.
Rachis disarticulating at maturity; glumes subulate, extending into long awns, these and the awns of the lemmas making the spike very bristly........ 47. SITANION.

### TRIBE 4. AVENEAE

Spikelets 2- to several-flowered in open or contracted panicles, or rarely in racemes (solitary in *Danthonia unispicata*), glumes usually as long as or longer than the first lemma, commonly longer than all the florets; lemmas usually awned from the back or from between the teeth of a bifid apex, the awn usually bent, often twisted, the callus and rachilla joints usually villous.

A rather small tribe widely distributed in both warm and cool regions. In our genera the rachilla is prolonged beyond the upper floret as a slender stipe (except in *Aira* and *Holcus*). The lemma is awnless or nearly so in *Schismus*, two species of *Trisetum*, one species of *Koeleria*, and in most of the species of *Sphenopholis*. *Koeleria* and *Sphenopholis* are placed in this tribe because they appear to be closely allied to *Trisetum* with which they agree in having oblanceolate glumes about as long as the first floret.

*Key to the genera of Aveneae*

Florets 2, one perfect, the other staminate.
Lower floret staminate, the awn twisted, geniculate, exserted.... 63. ARRHENATHERUM.
Lower floret perfect, awnless; upper floret awned........ 64. HOLCUS.
Florets 2 or more, all alike except the reduced upper ones.
Articulation below the glumes, the spikelets falling entire.
Lemmas, at least the upper, with a conspicuous bent awn; glumes nearly alike. 57. TRISETUM.
Lemmas awnless or (in *S. pallens*) the upper with a short awn; second glume much wider than the first........ 56. SPHENOPHOLIS.
Articulation above the glumes, the glumes similar in shape.
Lemmas bifid at apex, awned or mucronate between the lobes. Spikelets several-flowered.
Awns conspicuous, flat, bent. Spikelets 1 cm. or more long........ 66. DANTHONIA.
Awns minute or nearly obsolete.
Spikelets 8 to 12 mm. long........ 65. SIEGLINGIA.
Spikelets not more than 5 mm. long; awns, when present, slender, rounded. 54. SCHISMUS.
Lemmas toothed, but not bifid and awned or mucronate between the lobes.
Glumes 2 to 3.5 cm. long, 7- to 9-nerved; spikelets 2-flowered, or with a rudimentary third floret, pendulous. Plants annual........ 61. AVENA.

Glumes not more than 1 cm. long, 1- to 5-nerved; spikelets not pendulous.
Spikelets 3- to several-flowered, 1 to 1.5 cm. long .......... 62. HELICTOTRICHON.
Spikelets 2-flowered (or 3-flowered in *Trisetum cernuum*), mostly less than 1 cm. long.
Lemmas keeled, the awn when present from above the middle.
Rachilla joints very short, glabrous or minutely pubescent; lemmas awnless or with a straight awn from a toothed apex .......... 55. KOELERIA.
Rachilla joints slender, villous; lemmas with a dorsal bent awn (awnless or nearly so in 2 species) .......... 57. TRISETUM.
Lemmas convex, awned from below the middle.
Rachilla prolonged behind the upper floret; lemmas truncate and erose-dentate at summit.
Awn slender, not jointed .......... 58. DESCHAMPSIA.
Awn clavate, jointed near the middle .......... 60. CORYNEPHORUS.
Rachilla not prolonged; lemmas tapering into 2 slender teeth .......... 59. AIRA.

## TRIBE 5. AGROSTIDEAE

Spikelets 1-flowered, usually perfect, in open, contracted, or spikelike panicles, but not in true spikes nor in 1-sided racemes.

A large and important tribe, inhabiting more especially the temperate and cool regions. The articulation of the rachilla is usually above the glumes, the mature floret falling from the persistent glumes, but in a few genera the articulation is below the glumes, the mature spikelet falling entire (*Alopecurus, Cinna, Polypogon, Lycurus,* and *Limnodea*). The palea is small or wanting in *Alopecurus* and in some species of *Agrostis*. In a few genera the rachilla is prolonged behind the palea as a minute bristle, or sometimes as a more pronounced stipe (*Brachyelytrum, Limnodea, Cinna, Gastridium, Calamagrostis, Ammophila, Lagurus, Apera,* and a few species of *Agrostis*). In some genera the rachilla joint between the glumes and the lemma is slightly elongated, forming a hard stipe which remains attached to the mature fruit as a pointed callus. The callus is well marked in *Stipa* (especially in *S. spartea* and its allies) and in *Aristida*, the mature lemma being terete, indurate, and convolute, the palea wholly enclosed. In many genera the lemma is awned either from the tip or from the back, the awn being trifid in *Aristida*.

### *Key to the genera of Agrostideae*

Glumes wanting. Low annual .......... 73. COLEANTHUS.
Glumes present (the first obsolete in *Muhlenbergia schreberi* and sometimes in *Brachyelytrum* and *Phippsia*).
1a. Articulation below the glumes, the spikelets falling entire.
Spikelets in pairs in a spikelike panicle, one perfect, the other staminate or neuter, the pair falling together .......... 78. LYCURUS.
Spikelets all alike.
Glumes long-awned .......... 77. POLYPOGON.
Glumes awnless.
Rachilla not prolonged behind the palea; panicle dense.
Glumes united toward the base, ciliate on the keel; inflorescence not capitate and bracteate .......... 76. ALOPECURUS.
Glumes not united, glabrous; inflorescence capitate in the axils of broad bracts. 85. CRYPSIS.
Rachilla prolonged behind the palea; panicle narrow or open, not dense; glumes not united, not ciliate on the keel.
Panicle narrow; lemma with a slender bent twisted awn from the bifid apex. 75. LIMNODEA.
Panicle open, drooping; lemma with a minute straight awn just below the entire apex (rarely awnless) .......... 74. CINNA.
1b. Articulation above the glumes.
Fruit dorsally compressed, indurate, smooth, and shining, awnless .......... 88. MILIUM.
Fruit laterally compressed or terete, awned or awnless.
2a. Fruit indurate, terete, awned, the nerves obscure; callus well developed, oblique, bearded.
Awn trifid, the lateral divisions sometimes short, rarely obsolete (when obsolete no

line of demarcation between awn and lemma as in the next).... 92. ARISTIDA.
Awn simple, a line of demarcation between the awn and the lemma.
Awn persistent, twisted, and bent, several to many times longer than the fruit.
Edges of lemma overlapping (rarely only meeting), enclosing the palea; callus sharp-pointed, usually narrow and acuminate.......................... 91. STIPA.
Edges of lemma not meeting, exposing the indurate sulcus of the palea, this projecting from the summit as a minute point; callus short, acutish.
90. PIPTOCHAETIUM.
Awn deciduous, not twisted, sometimes bent, rarely more than 3 or 4 times as long as the plump fruit; callus short, usually obtuse ........ 89. ORYZOPSIS.
2b. Fruit thin or firm, but not indurate; callus not well developed.
Lemma firm, subindurate at maturity, bearing a long delicate straight awn just below the tip; palea about as long as the lemma, the naked rachilla produced back of the palea.......................................................................... 70. APERA.
Lemma thin or membranaceous.
3a. Glumes longer than the lemma (nearly equal in *Agrostis thurberiana* and *A. aequivalis*).
Panicle feathery, capitate, nearly as broad as long; spikelets woolly.
81. LAGURUS.
Panicle not feathery; spikelets not woolly.
Glumes compressed-carinate, stiff-ciliate on the keel; panicle dense, cylindric or ellipsoid.................................................................. 79. PHLEUM.
Glumes not compressed-carinate, not ciliate.
Glumes saccate at base; lemma long-awned; panicle contracted, shining.
80. GASTRIDIUM.
Glumes not saccate at base; lemma awned or awnless; panicle open or contracted.
Floret bearing a tuft of hairs at the base from the short callus; palea well developed, the rachilla prolonged behind the palea (except in *Calamagrostis epigeios*) as a hairy bristle.... 67. CALAMAGROSTIS.
Floret without hairs at the base or with short hairs (nearly half as long as the lemma in *A. hallii*); palea usually small or obsolete (developed and with a minute rachilla back of it in Nos. 1 to 3).
71. AGROSTIS.
3b. Glumes not longer than the lemma, usually shorter (the awn tips longer in *Muhlenbergia racemosa* and *M. glomerata*).
Lemma awned from the tip or mucronate, 3- to 5-nerved (lateral nerves obscure in a few species of *Muhlenbergia*).
Rachilla prolonged behind the palea; floret stipitate; glumes minute or obsolete.............................................................. 87. BRACHYELYTRUM.
Rachilla not prolonged; floret not stipitate................ 82. MUHLENBERGIA.
Lemma awnless or awned from the back.
Floret bearing a tuft of hairs at the base from the short callus; lemma and palea chartaceous, awnless.
Panicle spikelike; rachilla prolonged................................ 68. AMMOPHILA.
Panicle open; rachilla not prolonged............................ 69. CALAMOVILFA.
Floret without hairs at base.
Nerves of lemma silky............................................... 84. BLEPHARONEURON.
Nerves of lemma not silky.
Caryopsis at maturity falling from the lemma and palea; seed loose in the pericarp, this usually opening when ripe; lemma 1-nerved.
83. SPOROBOLUS.
Caryopsis not falling from the lemma and palea, remaining permanently enclosed in them; seed adnate to the pericarp.
Panicle few-flowered, slender, rather loose; glumes minute, unequal, the first often wanting. Low arctic-alpine perennial.
72. PHIPPSIA.
Panicle many-flowered, spikelike; glumes well developed, about equal.
Panicle short, partly enclosed in the sheath; low annual.
86. HELEOCHLOA.
Panicle elongate; perennial.......................... 82. MUHLENBERGIA.

### TRIBE 6. ZOYSIEAE

Spikelets subsessile in short spikes of 2 to 5 (single in *Zoysia*), each spike falling entire from the continuous axis, usually 1-flowered, all perfect, or perfect and staminate together in the same spike; glumes usually firmer than the lemma and palea, sometimes awned, the lemma awnless.

This small and unimportant tribe is known also as Nazieae. In *Zoysia* the spikelets are single and have only 1 glume, this coriaceous, much firmer than the lemma and palea, the palea sometimes obsolete.

*Key to the genera of Zoysieae*

Spikelets single; first glume wanting .......... 94. ZOYSIA.
Spikelets in clusters of 2 to 5; first glume present.
  Spikelets bearing hooked spines on the second glume, the group forming a little bur. .......... 93. TRAGUS.
  Spikelets not bearing hooked spines, the second glume mostly cleft and awned.
    Groups of spikelets erect, the inflorescence not 1-sided .......... 95. HILARIA.
    Groups of spikelets nodding along one side of the delicate axis .......... 96. AEGOPOGON.

## TRIBE 7. CHLORIDEAE

Spikelets 1- to several-flowered, in 2 rows on one side of a continuous rachis, forming 1-sided spikes or spikelike racemes, these solitary, digitate, or racemose along the main axis.

A large and rather important tribe, confined mostly to warm regions. The group is heterogeneous, the only common character of the genera (aside from the characters that place them in Festucoideae) being the arrangement of the spikelets in 1-sided spikes. *Chloris* and the allied genera form a coherent group, in which the spikelet consists of 1 perfect floret and, above this, 1 or more modified or rudimentary florets. *Leptochloa*, *Eleusine*, and their allies, with several-flowered spikelets, are more nearly related to certain genera of Festuceae. The spike is reduced to 2 or 3 spikelets or even to 1 spikelet and is sometimes deciduous from the main axis (*Cathestecum* and Sect. Atheropogon of *Bouteloua*). In *Ctenium* there are 2 sterile florets below the perfect one.

*Key to the genera of Chlorideae*

Plants monoecious or dioecious. Low stoloniferous perennial .......... 115. BUCHLOË.
Plants with perfect flowers.
  1a. Spikelets with more than 1 perfect floret.
    Inflorescence a few-flowered head or capitate panicle hidden among the sharp-pointed leaves. Low spreading annual .......... 114. MUNROA.
    Inflorescence exserted.
      Spikes solitary, the spikelets distant, appressed, several-flowered .......... 99. TRIPOGON.
      Spikes more than 1 (sometimes 1 in depauperate *Eleusine*).
        Spikes numerous, slender, racemose on an elongate axis.
          Rachilla and callus of floret glabrous or nearly so; glumes acute, less than 5 mm. long .......... 97. LEPTOCHLOA.
          Rachilla and callus of floret strongly pilose; glumes long-acuminate, about 1 cm. long .......... 98. TRICHONEURA.
        Spikes few, digitate or nearly so.
          Rachis of spike extending beyond the spikelets .......... 101. DACTYLOCTENIUM.
          Rachis not prolonged .......... 100. ELEUSINE.
  1b. Spikelets with only 1 perfect floret, often with additional imperfect florets above or below.
    2a. Spikelets without additional modified florets, the rachilla sometimes prolonged.
      Rachilla articulate below the glumes, the spikelets falling entire.
        Glumes unequal, narrow .......... 107. SPARTINA.
        Glumes equal, broad, boat-shaped .......... 106. BECKMANNIA.
      Rachilla articulate above the glumes.
        Spike solitary, slender, arcuate .......... 102. MICROCHLOA.
        Spikes 2 to many.
          Spikes digitate; rachilla prolonged .......... 103. CYNODON.
          Spikes racemose along the main axis; rachilla not prolonged.
            Spikes slender, divaricate, the main axis elongating and becoming loosely spiral in fruit .......... 105. SCHEDONNARDUS.
            Spikes short and rather stout, appressed, the axis unchanged in fruit. .......... 104. WILLKOMMIA.

2b. Spikelets with 1 or more modified florets above the perfect one.
Spikelets with 2 sterile florets below the perfect one; second glume bearing a squarrose spine on the back; spike single, arcuate........ 108. CTENIUM.
Spikelets with no sterile florets below the perfect one; second glume without a squarrose spine.
Spikes digitate or nearly so.
Fertile lemma 1-awned or awnless........ 110. CHLORIS.
Fertile lemma 3-awned........ 111. TRICHLORIS.
Spikes racemose along the main axis.
Spikelets distant, appressed; spikes slender, elongate........ 109. GYMNOPOGON.
Spikelets approximate or crowded, not appressed; spikes usually short and rather stout.
Spikelets 3 in each spike, the 2 lateral staminate or rudimentary; spikes falling entire........ 113. CATHESTECUM.
Spikelets 2 to many (rarely 1) in each spike, all alike; spikes falling entire or persistent, the florets falling........ 112. BOUTELOUA.

## TRIBE 8. PHALARIDEAE

Spikelets with 1 perfect terminal floret and, below this, a pair of staminate or neuter florets (1 sometimes obsolete in *Phalaris*).

A small tribe of about 6 genera, 4 of which are found in the United States. In *Phalaris* the lower florets are reduced to minute scalelike lemmas closely appressed to the edges of the fertile floret. In *Hierochloë* the lateral florets are staminate and as large as the fertile floret.

*Key to the genera of Phalarideae*

Lower florets staminate; spikelets brown, shining........ 116. HIEROCHLOË.
Lower florets neuter; spikelets green or yellowish.
Lower florets consisting of awned hairy sterile lemmas exceeding the fertile floret. 117. ANTHOXANTHUM.
Lower florets reduced to small awnless scalelike lemmas, much smaller than the fertile florets........ 118. PHALARIS.

## TRIBE 9. ORYZEAE

Spikelets 1-flowered, perfect, strongly laterally compressed, paniculate; glumes reduced or wanting; palea apparently 1-nerved; stamens 6.

A small tribe whose affinities are not evident. It includes rice, the important food plant.

*Key to the genera of Oryzeae*

Glumes minute; lemma often awned........ 119. ORYZA.
Glumes wanting; lemma awnless........ 120. LEERSIA.

## TRIBE 10. ZIZANIEAE

Spikelets unisexual, the pistillate terete or nearly so; glumes shorter than the lemma, usually 1 or both obsolete, the pedicel disarticulating below the spikelet. Glumes well developed in *Pharus*, a tropical genus placed in this tribe provisionally.

A small tribe of uncertain affinities, aquatic or subaquatic grasses (except *Pharus*) of no economic importance except the Indian rice (*Zizania*).

*Key to the genera of Zizanieae*

Blades elliptic, 2 to 4 cm. wide........ 125. PHARUS.
Blades much longer than wide.
Culms slender; plants low; staminate and pistillate spikelets borne in separate inflorescences.
Inflorescence a few-flowered raceme; floating aquatic........ 124. HYDROCHLOA.
Inflorescence a panicle; plants stoloniferous........ 123. LUZIOLA.
Culms robust; plants tall; staminate and pistillate spikelets borne in the same panicle.
Pistillate spikelets on the ascending upper branches, the staminate on the spreading

lower branches of the panicle; plants annual or perennial.............. 121. ZIZANIA.
Pistillate spikelets at the ends, the staminate below on the same branches of the panicle; plants perennial.............................................................................. 122. ZIZANIOPSIS.

### TRIBE 11. MELINIDEAE

Spikelets disarticulating below the glumes, these very unequal, the first minute, the second and the sterile lemma equal, membranaceous, strongly nerved, the latter bearing a slender awn from the notched summit; fertile lemma and palea thinner in texture, awnless.

A tribe of about a dozen genera represented in the United States by an introduced species, *Melinis minutiflora.*

### TRIBE 12. PANICEAE

Spikelets with 1 perfect terminal floret and below this a sterile floret and 2 glumes; fertile lemma and palea indurate or at least firmer than the glumes and sterile lemma, a lunate line of thinner texture at the back just above the base, the rootlet protruding through this at germination; articulation below the spikelet.

A large tribe, confined mostly to warm regions, and containing relatively few economic species. The first glume is wanting in some genera, such as *Paspalum,* and rarely the second glume also (*Reimarochloa*). The spikelets are usually awnless, but the glumes and sterile lemma are awned in *Echinochloa* and *Oplismenus,* and the second glume and sterile lemma in *Rhynchelytrum.* In *Eriochloa* and in some species of *Brachiaria* the fertile lemma is awn-tipped. In *Setaria* there are, beneath the spikelet, 1 or more bristles, these representing sterile branchlets. In *Pennisetum* similar bristles form an involucre, falling with the spikelet. In *Cenchrus* the bristles are united, forming a bur. The spikelets are of 2 kinds in *Amphicarpum,* aerial and subterranean. The culms are woody and perennial in *Lasiacis* and *Olyra.*

*Key to the genera of Paniceae*

Spikelets of two kinds.
  Spikelets all perfect, but those of the aerial panicle rarely perfecting grains, the fruitful spikelets borne on subterranean branches.................................... 146. AMPHICARPUM.
  Spikelets unisexual, the pistillate above, the staminate below on the branches of the same panicle. Blades broad, elliptic.............................................................. 147. OLYRA.
Spikelets all of one kind.
  Spikelets sunken in the cavities of the flattened corky rachis........ 131. STENOTAPHRUM.
  Spikelets not sunken in the rachis.
    1a. Spikelets subtended or surrounded by 1 to many distinct or more or less connate bristles, forming an involucre.
      Bristles persistent, the spikelets deciduous................................................ 143. SETARIA.
      Bristles falling with the spikelets at maturity.
        Bristles not united at base, slender, often plumose.................. 144. PENNISETUM.
        Bristles united into a burlike involucre, the bristles retrorsely barbed. 145. CENCHRUS.
    1b. Spikelets not subtended by bristles.
      Glumes or sterile lemma awned (awn short and concealed in the silky hairs of the spikelet in *Rhynchelytrum;* awn reduced to a point in *Echinochloa colonum*).
        Inflorescence paniculate; spikelets silky............................... 142. RHYNCHELYTRUM.
        Inflorescence of unilateral simple or somewhat compound racemes along a common axis; spikelets smooth or hispid, not silky.
          Blades lanceolate, broad, thin; culms creeping........................ 140. OPLISMENUS.
          Blades long, narrow; culms not creeping................................ 141. ECHINOCHLOA.
      Glumes and sterile lemma awnless.
        2a. Fruit cartilaginous-indurate, flexible, usually dark-colored, the lemma with more or less prominent white hyaline margins, these not inrolled.
          Spikelets covered with long silky hairs, arranged in racemes, these panicled. 128. TRICHACHNE.
          Spikelets glabrous or variously pubescent but not long-silky (somewhat silky in *Digitaria villosa*).

Spikelets in slender racemes more or less digitate at the summit of the culms. 129. DIGITARIA.
Spikelets in panicles.
Fruiting lemma boat-shaped; panicles narrow........ 127. ANTHAENANTIA.
Fruiting lemma convex; panicles diffuse........................... 130. LEPTOLOMA.
2b. Fruit chartaceous-indurate, rigid.
Spikelets placed with the back of the fruit turned away from the rachis of the racemes, usually solitary (not in pairs).
First glume and the rachilla joint forming a swollen ringlike callus below the spikelet.................................................................. 132. ERIOCHLOA.
First glume present or wanting, not forming a ringlike callus below the spikelet.
First glume present (next to the axis); racemes racemose along the main axis.................................................................. 133. BRACHIARIA.
First glume wanting; racemes digitate or subdigitate...... 134. AXONOPUS.
Spikelets placed with the back of the fruit turned toward the rachis (first glume, when present, away from the rachis) of the spikelike racemes or pedicellate in panicles.
Fruit long-acuminate; both glumes wanting................ 135. REIMAROCHLOA.
Fruit not long-acuminate; at least one glume present.
First glume typically wanting; spikelets plano-convex, subsessile in spikelike racemes................................................................. 136. PASPALUM.
First glume present; spikelets usually in panicles.
Second glume inflated-saccate, this and the sterile lemma much exceeding the stipitate fruit..................................................... 139. SACCIOLEPIS.
Second glume not inflated-saccate.
Culms woody, bamboolike; fruit with a tuft of down at the apex. 138. LASIACIS.
Culms herbaceous; no tuft of down at the apex of the fruit. 137. PANICUM.

## TRIBE 13. ANDROPOGONEAE

Spikelets in pairs along a rachis, the usual arrangement being one of the pair sessile and fertile, the other pedicellate and staminate or neuter, rarely wanting, only the pedicel present; fertile spikelet consisting of 1 perfect terminal floret and, below this, a staminate or neuter floret, the lemmas thin or hyaline, and 2 awnless glumes, 1 or usually both firm or indurate.

A large tribe, confined mostly to warm regions. The rachis is usually jointed, disarticulating at maturity, with the spikelets attached to the joints. In a few genera it is thickened. Sometimes the racemes are shortened to 1 or 2 joints and borne on branches, the whole forming a panicle (as in *Sorghum* and *Sorghastrum*) instead of a series of racemes. In a few genera the spikelets of the pair are alike. In *Trachypogon* the fertile spikelet is pedicellate and the sterile one nearly sessile. The most important economic plants in this tribe are sugarcane and sorghum.

*Key to the genera of Andropogoneae*

1a. Spikelets alike, all perfect. (See also *Arthraxon* and *Sorghastrum* in which pedicellate spikelets are not developed.)
Spikelets surrounded by a copious tuft of soft hairs.
Rachis continuous, the spikelets falling; the spikelets of the pair unequally pedicellate.
Racemes in a narrow spikelike panicle; spikelets awnless.................. 148. IMPERATA.
Racemes in a broad fan-shaped panicle; spikelets awned............ 149. MISCANTHUS.
Rachis breaking up into joints at maturity with the spikelets attached; one spikelet sessile, the other pedicellate.
Spikelets awnless................................................................ 150. SACCHARUM.
Spikelets awned.................................................................. 151. ERIANTHUS.
Spikelets not surrounded by turfs of hairs; racemes few................ 152. MICROSTEGIUM.
1b. Spikelets unlike, the sessile perfect, the pedicellate sterile (sessile spikelet staminate, pedicellate spikelet perfect in *Trachypogon*.
2a. Pedicel thickened, appressed to the thickened rachis joint (at least parallel to it) or adnate to it; spikelets awnless, appressed to the joint.

Rachis joint and pedicel adnate. Annuals.
Perfect spikelet globose; sterile spikelet conspicuous.............. 164. HACKELOCHLOA.
Perfect spikelet oblong; sterile spikelet minute.......................... 162. ROTTBOELLIA.
Rachis joint and pedicel distinct, the sessile spikelet appressed to them, its first glume lanceolate.
Racemes subcylindric; rachis joints and pedicels glabrous, much thicker at the summit, the spikelets sunken in the hollow below; sterile spikelet rudimentary. 163. MANISURIS.
Racemes flat; rachis joints and pedicels woolly, not much thicker at the summit; sterile spikelet staminate or neuter.......................................... 161. ELYONURUS.
2b. Pedicel not thickened (if slightly so the spikelets awned), neither appressed nor adnate to the rachis joint, this usually slender; spikelets usually awned.
3a. Fertile spikelet with a hairy-pointed callus, formed of the attached supporting rachis joint or pedicel; awns strong.
Racemes reduced to a single joint, long-peduncled in a simple open panicle. 158. CHRYSOPOGON.
Racemes of several to many joints, single.
Primary spikelet subsessile, sterile, persistent on the continuous axis after the fall of the fertile pedicellate spikelet.......................................... 160. TRACHYPOGON.
Primary spikelet sessile, fertile; pedicellate spikelet sterile. Lower few to several pairs of spikelets all staminate or neuter.......................... 159. HETEROPOGON.
3b. Fertile spikelet without a callus (a short callus in *Hyparrhenia*), the rachis disarticulating immediately below the spikelet; awns slender.
Blades ovate. Annual.................................................................. 153. ARTHRAXON.
Blades narrow, elongate.
Racemes of several to many joints, solitary, digitate, or aggregate in panicles.
Lower pair of spikelets like the others of the raceme.......... 154. ANDROPOGON.
Lower pair of spikelets sterile, awnless. Racemes in pairs on slender flexuous peduncles...................................................................... 155. HYPARRHENIA.
Racemes reduced to one or few joints, these mostly peduncled in a subsimple or compound panicle.
Pedicellate spikelets staminate.................................................. 156. SORGHUM.
Pedicellate spikelets wanting, the pedicel only present...... 157. SORGHASTRUM.

## TRIBE 14. TRIPSACEAE

Spikelets unisexual, the staminate in pairs, or sometimes in threes, 2-flowered, the pistillate usually single, 2-flowered, the lower floret sterile, embedded in hollows of the thickened articulate rachis and falling attached to the joints, or enclosed in a thickened involucre or sheath or, in *Zea*, crowded in rows on a thickened axis (cob); glumes membranaceous or thick and rigid, awnless; lemmas and palea hyaline, awnless. Plants monoecious.

This small tribe of seven genera is scarcely more than a subtribe of Andropogoneae, differing chiefly in the total suppression of the sterile spikelet of a pair, the fertile spikelet being pistillate only and solitary; staminate spikelets paired. It is also known as Maydeae.

*Key to the genera of Tripsaceae*

Staminate and pistillate spikelets in separate inflorescences, the first in a terminal tassel, the second in the axils of the leaves.
Pistillate spikes distinct, the spikelets embedded in the hardened rachis, this disarticulating at maturity.................................................................. 167. EUCHLAENA.
Pistillate spikes grown together forming an ear, the grains at maturity much exceeding the glumes.................................................................................. 168. ZEA.
Staminate and pistillate spikelets in separate portions of the same inflorescence, the pistillate below.
Spikes short, the 1- or 2-flowered pistillate portion enclosed in a beadlike sheathing bract. 165. COIX.
Spikes many-flowered, the pistillate portion breaking up into several 1-seeded joints; no beadlike sheathing bract.................................................................. 166. TRIPSACUM.

# DESCRIPTIONS OF GENERA AND SPECIES

## TRIBE 1. BAMBUSEAE

### 1. ARUNDINÁRIA Michx. CANE

Spikelets 8- to 12-flowered, large, compressed, the rachilla disarticulating above the glumes and between the florets; glumes unequal, shorter than the lemmas, the first sometimes wanting; lemmas papery, rather fragile, about 11-nerved, acute, acuminate, mucronate or awn-tipped; palea about as long as the lemma or a little shorter, prominently 2-keeled, deeply sulcate between the keels; rachilla joints rather thick, appressed-hirsute; stamens 3; caryopsis narrowly elliptic, terete, 1 to 1.2 cm. long. Shrubs or tall reeds with extensively creeping horizontal rhizomes 5 to 10 mm. thick, the woody perennial branching culms erect, 2 to 5 m., sometimes to 8 m., tall and 2 cm. thick, freely branching, the flowering branchlets borne in fascicles on the main culm or on primary branches, their sheaths bladeless or nearly so, flowering shoots also arising from the rhizomes, their sheaths bladeless; flowering at infrequent intervals, usually each species over a wide area simultaneously, the flowering period apparently continuing for about a year; the flowering culms apparently dying after setting seed; sterile branches numerous and repeatedly branching, the basal shoots and primary branches with 6 to 10 loose, papery culm-sheaths with narrow rudimentary blades 2 to 20 mm. long, not petiolate at base, and 4 to 10 large petiolate tessellate blades toward the ends, their sheaths overlapping, the upper blades crowded, the lower papery sheaths finally falling, the leaf-sheaths bearing several flat scabrous bristles at the summit, these readily falling in age. Type species, *Arundinaria macrosperma* Michx. (*A. gigantea*). Name from Latin *Arundo,* a reed.

Primary branches erect or nearly so, the individual culm with its branches oblong-linear in outline; spikelets usually rather loose; lemmas appressed-hirsute or canescent, at least toward the base, greenish tawny to bronze-russet .......... 1. A. GIGANTEA.
Primary branches ascending at an angle of about 45°, the individual culm with its branches broadly lanceolate in outline; spikelets rather compact; lemmas glabrous or obscurely pubescent at base only, usually livid-purple .......... 2. A. TECTA.

**1. Arundinaria gigantéa** (Walt.) Muhl. GIANT CANE. (Fig. 1.) Culms as much as 2 cm. thick and 2 to 8 m. tall, smooth; lower sheaths about half as long as the internodes, finally falling, the upper 6 to 10 sheaths striate, tessellate, usually hirsute, becoming glabrous or nearly so, densely ciliate, canescent at base, the 10 to 12 bristles at the summit 5 to 9 mm. long, these often borne from the margin of a rather firm auricle, this sometimes prominent but often obscure or wanting, a dense band of stiff hairs across the collar; ligule firm, scarcely 1 mm. long; blades of main culm and primary branches 15 to 27 cm. long, 2.5 to 4 cm. wide, rounded at base (petiole 1 to 2 mm. long), strongly finely tessellate, acuminate, pubescent to glabrous on the lower surface, puberulent to glabrous on the upper, the margin finely serrulate; blades of ultimate branchlets much smaller, often crowded in flabellate clusters, commonly glabrous or nearly so; flowering branchlets finally crowded toward the ends of the branches, the racemes or simple panicles with few to several spikelets on slender angled pedicels 2 to 30 mm. long, hirsute to nearly glabrous; spikelets 4 to 7 cm. long, about 8 mm. wide, mostly 8- to 12-flowered, rather loose; glumes distant, acuminate, pubescent, the lower minute, sometimes wanting; lemmas broadly lanceolate, keeled, mostly 1.5

FIGURE 1.—*Arundinaria gigantea*. Flowering shoot, × ½; summit of culm sheath, outer and inner face, showing auricles and ligule, and two views of floret, × 2. (Swallen 6717, Miss.)

to 2 cm. long, sometimes tapering into an awn 4 mm. long, ciliate, appressed-hirsute to canescent, rarely glabrous except toward the base and margins, faintly to clearly tessellate; rachilla segments densely hirsute; palea scabrous on the keels. ♃ —Forming extensive colonies in low woods, river banks, moist ground, southern Ohio, Indiana, Illinois, Missouri, and Oklahoma to North Carolina, Florida, and Texas, mostly above the Coastal Plain. Livestock eagerly eat the young plants, leaves, and seeds, and canebrakes furnish much forage. The young shoots are sometimes used as a potherb. The culms are used for fishing rods, pipestems, baskets, mats, and a variety of other purposes. Early travelers speak of the abundance of this species and state that the culms may be as much as 2 or even 3 inches in diameter. It is said that the plants are easily destroyed by the continuous grazing of cattle and by the rooting of swine.

**2. Arundinaria técta** (Walt.) Muhl. SWITCH CANE. (Fig. 2.) Similar to *A. gigantea*, the culms usually not more than 2 m. tall, the sheaths more commonly as long as the internodes; auricle at summit of sheaths only rarely developed, the bristles 2 to 6 mm. long, a very short firm erose to ciliate membrane across the collar; blades on the average a little longer and narrower; inflorescence similar, the spikelets 3 to 5 cm. long, 6- to 12-flowered, relatively compact and less compressed than in the preceding; glumes obtuse to acuminate, often glabrous or nearly so; lemmas scarcely keeled, 12 to 15 mm. long, glabrous or minutely canescent at the base, rarely very faintly tessellate toward the summit; the rachilla strigose. ♃ —Forming colonies in swampy woods, moist pine barrens and live oak woods, and sandy margins of streams, Coastal Plain, southern Maryland to Alabama and Mississippi. Two collections from northwest Florida appear to be intermediate between the two species.

A great many exotic species of bamboo have been introduced into cultivation in the United States, particularly from China, Japan, India, and Java. *Arundinaria, Bambusa, Cephalostachyum, Chimonobambusa, Dendrocalamus, Gigantochloa, Guadua, Indocalamus, Lingnania, Oxytenanthera, Phyllostachys, Pleioblastus, Pseudosasa, Sasa, Schizostachyum, Semiarundinaria, Shibataea, Sinarundinaria, Sinobambusa, Sinocalamus*, and *Thamnocalamus* are the principal genera represented. In southern Florida the commonest introduced species are *Bambusa multiplex* (Lour.) Raeusch., *B. bambos* (L.) Voss,[6] *B. vulgaris* Schrad. ex Wendl., and *Sinocalamus oldhami* (Munro) McClure ("*Dendrocalamus latiflorus*" of California and Florida gardens). Farther north, where the minimum winter temperatures are lower, *Arundinaria simoni* (Carr.) A. and C. Riv., *Phyllostachys aurea* A. and C. Riv., and *P. bambusoides* Sieb. and Zucc. are the commonest, and in regions where the winters are still more severe *Pseudosasa japonica* (Sieb. and Zucc.) Makino is the species most commonly found in cultivation in the open air; escaped in Philadelphia. In California, *Sinocalamus oldhami, Bambusa multiplex*, and several species of *Phyllostachys* are about equally popular. The most recent systematic treatment of the species of bamboo cultivated in the United States is that of Rehder.[7]

[6] Contributed by F. A. McCLURE; see also McCLURE, F. A. THE GENUS BAMBUSA AND SOME OF ITS FIRST-KNOWN SPECIES. Blumea Sup. 3. (Henrard Jubilee vol.): 90–112, pl. 1–7, 1946; and YOUNG, R. A. BAMBOOS IN AMERICAN HORTICULTURE. Nat. Hort. Mag. 1945: 171–196; 274–291; 1946: 40–64; 257–283; 352–365, illus.

[7] REHDER, ALFRED. MANUAL OF CULTIVATED TREES AND SHRUBS. Ed. 2, 996 pp. New York. 1940.

FIGURE 2.—*Arundinaria tecta*. Flowering and leafy shoot, × ½; spikelet and floret, × 2. (Chase 5881, Va.); summit of culm sheath, outer and inner face, × 2. (Amer. Gr. Natl. Herb. 498, Va.)

## TRIBE 2. FESTUCEAE

### 2. BRÓMUS L. Bromegrass

Spikelets several- to many-flowered, the rachilla disarticulating above the glumes and between the florets; glumes unequal, acute, the first 1- to 3-nerved, the second usually 3- to 5-nerved; lemmas convex on the back or keeled, 5- to 9-nerved, 2-toothed, awned from between the teeth or awnless; palea usually shorter than the lemma, ciliate on the keels. Low or rather tall annuals or perennials with closed sheaths, usually flat blades, and open or contracted panicles of large spikelets. Standard species, *Bromus sterilis* (type species, *B. secalinus*). Name from *bromos,* an ancient Greek name for the oat, from broma, food.

The native perennial species of bromegrass form a considerable portion of the forage in open woods of the mountain regions of the western United States. *Bromus carinatus,* California brome, and its more eastern ally, *B. marginatus,* are abundant from the Rocky Mountains to the Pacific coast. Before maturity, they are relished by all classes of stock. Horses and sheep are particularly fond of the seed heads. *Bromus anomalus, B. pumpellianus,* and *B. ciliatus,* of the Rocky Mountain region, are abundant up to 10,000-11,000 feet altitude, and are of first rank for all classes of stock. Several other species are nutritious but are usually not abundant enough to be of importance in the grazing regions. The most important species agronomically is smooth brome, *B. inermis,* a native of Eurasia, which is cultivated for hay and pasture in the northern part of the Great Plains. It is more drought-resistant than timothy and can be grown farther west on the Plains, but does not thrive south of central Kansas. It is recommended for holding canal banks. Also called smooth, awnless, and Hungarian brome. Rescue grass, *B. catharticus,* is cultivated for winter forage in the Southern States from North Carolina to Texas and in the coast district of southern California.

The annuals are weedy species introduced mostly from Europe. The best known of these is chess, *Bromus secalinus,* a weed of waste places sometimes infesting grainfields. Formerly it was believed by the credulous that under certain conditions wheat changed into chess or "cheat." Chess in a wheatfield is due to chess seed in the soil or in the wheat sown. This species is utilized for hay in places in Washington, Oregon, and Georgia. On the Pacific coast the annual bromegrasses cover vast areas of open ground at lower altitudes where they form a large part of the forage on the winter range. They mature in spring or early summer and become unpalatable. Those of the section Eubromus are, at maturity, a serious pest. The narrow, sharp-pointed minutely barbed florets (or fruits) with their long rough awns work into the eyes, nostrils, and mouths of stock, causing inflammation and often serious injury. Sometimes the intestines are pierced, and death results. On the Pacific coast, *B. rigidus,* the chief offender, is called ripgut grass by stockmen, and the name is sometimes applied to other species of the section.

Spikelets strongly flattened, the lemmas compressed-keeled...... Section 1. Ceratochloa.
Spikelets terete before anthesis or somewhat flattened, but the lemmas not compressed-keeled.
  Plants perennial........ Section 2. Bromopsis.
  Plants annual. Introduced, mostly from Europe.
    Awn straight or divaricate, sometimes minute or obsolete, not twisted and geniculate; teeth of the lemma sometimes slender but not aristate.
      Lemmas broad, rounded above, not acuminate, the teeth mostly less than 1 mm. long........ Section 3. Bromium.
      Lemmas narrow, with a sharp callus, gradually acuminate, bifid, the teeth 2 to 5 mm. long. Awns usually more than 1.5 cm. long.... Section 4. Eubromus.
    Awn geniculate, twisted below; teeth of the lemma aristate. Section 5. Neobromus.

*Section 1. Ceratochloa*

Lemmas awnless or nearly so........ 1. B. CATHARTICUS.
Lemmas awned, the awn more than 3 mm. long.
Panicle branches elongate, slender, drooping, bearing 1 or 2 large spikelets at the end, the lowermost naked for as much as 10 to 15 cm. Sheaths smooth; Washington and Oregon........ 2. B. SITCHENSIS.
Panicle branches not greatly elongate.
Panicle branches ascending, rather stiff, naked below, bearing 1 or 2 large spikelets. Washington........ 3. B. ALEUTENSIS.
Panicle branches short and ascending or longer and drooping, with some short branches at the base.
Blades canescent, densely short-pilose, 2 to 5 mm. wide, often involute; panicle narrow........ 4. B. BREVIARISTATUS.
Blades not canescent, glabrous to puberulent or sparsely pilose, mostly 5 to 12 mm. wide.
Sheaths strongly to sparsely retrorsely pilose; blades 4 to 12 mm. wide; lemmas usually pubescent, the awns mostly less than 7 mm. long; plants perennial. 7. B. MARGINATUS.
Sheaths scaberulous to pilose.
Plants annual or biennial; culms mostly 30 to 100 cm. tall; spikelets rather open at anthesis, the rachilla joints relatively long; awns 7 to 15 mm. long.
Spikelets 6- to 10-flowered; second glume shorter than the lowest lemma. 5. B. CARINATUS.
Spikelets mostly 5- to 7-flowered; second glume nearly or quite equaling the length of the lowest floret........ 6. B. ARIZONICUS.
Plants perennial; awns mostly less than 15 mm. long.
Culms erect, mostly 80 to 120 cm. tall; panicle mostly open; spikelets rather glossy, loose, the rachilla joints relatively long........ 9. B. POLYANTHUS.
Culms subgeniculate and leafy at base, mostly 25 to 70 cm. tall; panicle rather dense; spikelets closely flowered........ 8. B. MARITIMUS.

*Section 2. Bromopsis*

1a. Creeping rhizomes present; lemmas awnless or short-awned; panicle erect, somewhat open, the branches ascending.
Lemmas glabrous........ 10. B. INERMIS.
Lemmas pubescent near the margins........ 11. B. PUMPELLIANUS.
1b. Creeping rhizomes wanting (base of culm decumbent in *B. laevipes*).
2a. Panicle narrow, the branches erect.
Lemmas glabrous or evenly scabrous........ 12. B. ERECTUS.
Lemmas appressed-pubescent on the margins and lower part........ 13. B. SUKSDORFII.
2b. Panicle open, the branches spreading or drooping.
3a. Lemmas glabrous.
Blades broad and lax, more than 5 mm., at least some of them 10 mm., wide (var. *laeviglumis*)........ 20. B. PURGANS.
Blades narrow, not more than 6 mm. wide........ 23. B. TEXENSIS.
3b. Lemmas pubescent.
4a. Lemmas pubescent along the margin and on lower part of the back, the upper part glabrous.
First glume 3-nerved; plant mostly pale or glaucous. Culms decumbent at base. 17. B. LAEVIPES.
First glume 1-nerved, or only faintly 3-nerved near the base; plants dark green.
Ligule prominent, 3 to 5 mm. long; lemmas narrow; awn usually more than 5 mm. long........ 18. B. VULGARIS.
Ligule inconspicuous, about 1 mm. long; lemmas broad; awn 3 to 5 mm. long. 19. B. CILIATUS.
4b. Lemmas pubescent rather evenly over the back, usually more densely so along the lower part of the margin (glabrous in *B. purgans* var. *laeviglumis*).
Panicle branches short, stiffly spreading; blades short, mostly on lower part of culm........ 14. B. ORCUTTIANUS.
Panicle branches lax or drooping; blades along the culm, mostly elongate.
Panicle small, drooping, usually not more than 10 cm. long. Spikelets densely and conspicuously pubescent.
Sheaths and blades sparsely pilose to subglabrous; blades mostly 2 to 4 mm. wide (rarely 5 to 6 mm.)........ 24. B. ANOMALUS.
Sheaths and blades (except uppermost in some) conspicuously pubescent; blades 5 to 10 mm. wide........ 25. B. KALMII.

Panicle larger, usually erect, the branches more or less drooping. Blades mostly wide and lax.
Ligule 3 to 4 mm. long; blades pilose above, scabrous or smooth beneath; panicle large, open, the slender branches long, drooping. 16. B. PACIFICUS.
Ligule short; blades pubescent or pilose on both surfaces, or glabrous or scabrous.
Blades densely short-pubescent on both surfaces .......... 15. B. GRANDIS.
Blades more or less pilose or glabrous.
Sheaths, at least the lower, retorsely pilose (rarely glabrous in *B. purgans*) blades mostly more than 5 mm. wide.
Sheaths shorter than the internodes. Nodes 4 to 6 .... 20. B. PURGANS.
Sheaths as long as or longer than the internodes.
Second glume 5-nerved; nodes 6 to 8; sheaths without flanges at the mouth .......... 22. B. NOTTOWAYANUS.
Second glume 3-nerved; nodes 10 to 20; sheaths with prominent flanges at the mouth .......... 21. B. LATIGLUMIS.
Sheaths glabrous; blades mostly less than 5 mm. wide .... 26. B. FRONDOSUS.

*Section 3. Bromium*

Panicle contracted, rather dense, the branches erect or ascending.
Lemmas glabrous .......... 37. B. RACEMOSUS.
Lemmas pubescent.
Spikelets compressed; lemmas rather thin and narrow .......... 31. B. MOLLIFORMIS.
Spikelets turgid; lemmas rather thick, broader .......... 30. B. MOLLIS.
Panicle open, the branches spreading.
Awn short or wanting; lemmas obtuse, inflated (see also short-awned forms of *B. secalinus*). 27. B. BRIZAEFORMIS.
Awn well developed.
Foliage glabrous .......... 28. B. SECALINUS.
Foliage pubescent.
Branches of the panicle rather stiffly spreading or drooping, not flexuous; awn straight. 29. B. COMMUTATUS.
Branches lax or flexuous, usually slender, but rather stout in *B. squarrosus*.
Spikelets inflated, 5 to 8 mm. or even 10 mm. wide; awns flattened, strongly divergent, about 1 cm. long; panicle branches stout but flexuous, bearing 1 or 2 spikelets .......... 33. B. SQUARROSUS.
Spikelets not inflated, usually less than 5 mm. wide, if more the spikelets pubescent; awn not strongly flattened, straight or somewhat spreading.
Panicle 8 to 11 cm. (rarely to 15 cm.) long; branches and pedicels conspicuously flexuous or curled; lemmas pubescent .......... 36. B. ARENARIUS.
Panicle 15 to 25 cm. long (smaller in depauperate specimens), the long branches spreading or drooping, somewhat flexuous but usually not curled; lemmas glabrous or scaberulous.
Palea distinctly shorter than its lemma; awn flexuous, usually somewhat divergent in drying; spikelets rather turgid .......... 34. B. JAPONICUS.
Palea about as long as its lemma; awn straight or nearly so in drying; spikelets thinner and flatter, scarcely turgid .......... 35. B. ARVENSIS.

*Section 4. Eubromus*

Panicle contracted, erect; awn 12 to 20 mm. long.
Culms pubescent below the dense panicle .......... 39. B. RUBENS.
Culms glabrous below the scarcely dense panicle .......... 40. B. MADRITENSIS.
Panicle open, the branches spreading.
Second glume usually less than 1 cm. long; pedicels capillary, flexuous. 41. B. TECTORUM.
Second glume more than 1 cm. long; pedicels sometimes flexuous but not capillary.
Awn about 2 cm. long; first glume 8 mm. long .......... 38. B. STERILIS.
Awn 3 to 5 cm. long; first glume about 15 mm. long .......... 37. B. RIGIDUS.

*Section 5. Neobromus*

A single species .......... 42. B. TRINII.

SECTION 1. CERATÓCHLOA (Beauv.) Griseb.

Annuals, biennials, or perennials; spikelets large, distinctly compressed; glumes and lemmas keeled, rather firm.

**1. Bromus cathárticus** Vahl. Rescue grass. (Fig. 3.) Annual or biennial; culms erect to spreading, as much as 100 cm. tall; sheaths glabrous or pubescent; blades narrow, glabrous or sparsely pilose; panicle

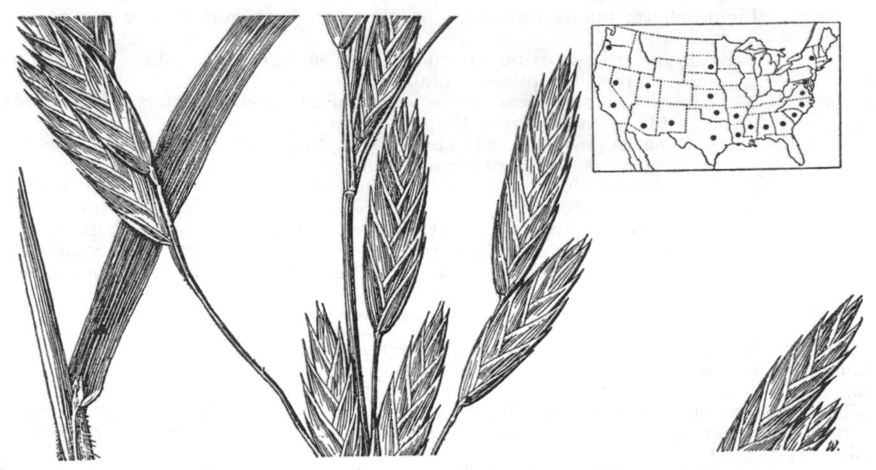

Figure 3.—*Bromus catharticus*, × 1. (Peebles, Harrison, and Kearney 1271, Ariz.)

Figure 4.—*Bromus sitchensis*, × 1. (Piper 3013, Alaska.)

open, as much as 20 cm. long, the branches as much as 15 cm. long, naked at base, in small plants the panicles reduced to a raceme of a few appressed short-pediceled spikelets; spikelets 2 to 3 cm. long, 6- to 12-flowered; glumes acuminate, about 1 cm. long; lemmas glabrous, scabrous, or sometimes pubescent, acuminate, 1.5 cm. long, closely overlapping, concealing the short rachilla joints, awnless or with an awn 1 to 3 mm. long; palea two-thirds as long as the lemma. ⊙ (*B. unioloides* H. B. K.)—Cultivated in the Southern States as a winter forage grass. Escaped from cultivation or sparingly introduced in waste places throughout Southern States and rarely northward. Known also as Schrader's bromegrass. Introduced from South America.

**2. Bromus sitchénsis** Trin. (Fig. 4.) Stout smooth perennial; culms 120 to 180 cm. tall; sheaths glabrous; blades elongate, 7 to 12 mm. wide, sparsely pilose on the upper surface; panicles large, lax, drooping, 25 to 35 cm. long, the lower branches (2 to 4) as much as 20 cm. long, naked below for as much as 10 or 15 cm., few-flowered; spikelets 2.5 to 3.5 cm. long, 6- to 12-flowered, the rachilla joints longer than in *B. catharticus*, exposed at anthesis; lemmas scabrous, sometimes hirtellous toward base; awn 5 to 10 mm. long. ♃ —Woods and banks near the coast, Alaska to Oregon.

**3. Bromus aleuténsis** Trin. ex Griseb. (Fig. 5.) Culms rather stout, erect from a usually decumbent base, 50 to 100 cm. tall; sheaths sparsely retrorse-pilose or glabrous; blades sparsely pilose, 5 to 10 mm. wide; panicle erect, loose, 10 to 20 cm. long, the branches rather stiffly ascending, bearing 1 or 2 (rarely 3) spikelets, the lower as much as 10 cm. long; spikelets 2.5 to 3.5 cm. long, 3- to 6-flowered; glumes subequal, the first 3-nerved, the second 5- or indistinctly 7-nerved; lemmas broadly lanceolate, 7-nerved, scarious-margined, smooth to scabrous-

FIGURE 5.—*Bromus aleutensis*, × 1. (Evans 550, Alaska.)

pubescent, about 15 mm. long; awn mostly about 1 cm. long. ♃ —Open ground, Aleutian Islands to the Olympic Mountain region.

**4. Bromus breviaristátus** Buckl. (Fig. 6.) Erect tufted perennial; culms 25 to 50 cm. tall; sheaths canescent to densely retrorse-pilose; blades narrow, becoming involute, canescent and also pilose with spreading hairs, mostly erect or ascending, often only 1 to 2 mm. wide; panicle narrow, erect, 5 to 15 cm. long, the branches short, appressed, often bearing only 1 spikelet; spikelets 2 to 3 cm. long; lemmas appressed-puberulent; awn 3 to 10 mm. long. ♃ (*B. subvelutinus* Shear.)—Dry wooded hills and meadows, Wyoming to British Columbia, eastern Washington, Nevada, and California.

**5. Bromus carinátus** Hook. and Arn. CALIFORNIA BROME. (Fig. 7.) Erect annual or mostly biennial; culms mostly 50 to 100 cm. (occasionally to 120 cm.) tall; sheaths scabrous to rather sparsely pilose; blades flat, mostly 20 to 30 cm. long, the lower shorter (those of the innovations numerous), scabrous or sparsely pilose, mostly 3 to 10 mm. wide; panicle mostly 15 to 30 cm. long, with spreading or drooping branches, in small plants much reduced; spikelets (excluding awns) 2 to 3 cm. long, mostly 6- to 10-flowered, the florets in anthesis not or scarcely overlapping, exposing the relatively long rachilla joints; glumes acuminate, the first 6 to 9 mm., the second 10 to 15 mm., long; lemmas minutely appressed-pubescent to glabrous, about 2 to 2.5 mm. wide as folded, 10 to 20 mm. long; awn 7 to 15 mm. long; palea acuminate, nearly as long as the lemma, the teeth short-awned. ⊙ —Open ground, open woods, and waste places, at low and middle altitudes, common on the Pacific coast, British Columbia to Idaho and California; New Mexico and Baja California. The species is extremely variable in size, in shape of panicle, and in pubescence, and intergrades freely with the following.[8]

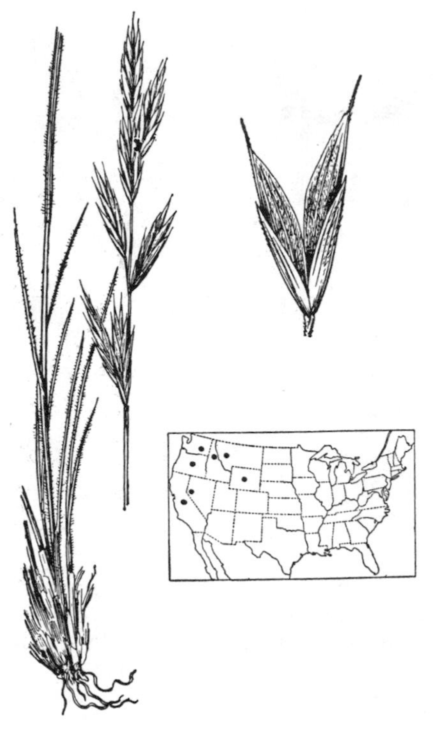

FIGURE 6.—*Bromus breviaristatus*. Plant, × ½; spikelet, × 5. (Nuttall, Rocky Mts.)

**6. Bromus arizónicus** (Shear) Stebbins. Annual, similar to the preceding, commonly shorter; panicle mostly

[8] For variability in *Bromus carinatus* see HARLAN. J. R., Amer. Jour. Bot. 32: 142. 1945. For proposed varieties see SHEAR, C. L., U. S. Dept. Agr., Div. Agrost. Bul. 23. 1900. See also STEBBINS, G. L., TOBGY, H. A., and HARLAN, J. R., Calif. Acad. Sci. Proc. 25: 307–321. 1944.

FIGURE 7.—*Bromus carinatus*, × 1. (Hitchcock 2704, Calif.)

stiff, erect and relatively narrow; spikelets mostly 5- to 7-flowered; glumes less unequal, the second often equaling the length of the lowest lemma; lemmas hirsute toward the margin, occasionally sparsely so across the back, the teeth of the apex 0.7 to 2 mm. long. ⊙ —Open, mostly arid slopes and valleys, western Texas; Arizona to middle California and Baja California. Plants short-lived, flowering in the early spring rains and dying after seeding.

**7. Bromus marginátus** Nees. (Fig. 8.) Perennial, sheaths mostly conspicuously retrorsely pilose; blades commonly pubescent, 6 to 12 mm. wide; panicles usually less open than in *B. carinatus;* spikelets mostly

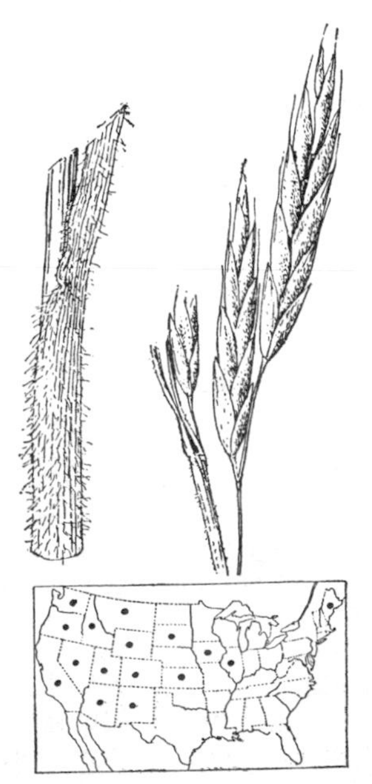

Figure 8.—*Bromus marginatus*, × 1. (Hunter 555, Oreg.)

closely flowered, lemmas more strongly pubescent, awns usually less than 7 mm. long. ♃ —Open woods, open or wooded slopes, meadows, and waste places, British Columbia and Alberta to South Dakota, New Mexico, and California, mostly on the eastern slope; adventive in Maine (in wool waste); introduced in Illinois, Iowa, and Kansas. Variable, intergrading with *B. carinatus* and scarcely distinct, though extremes are very different in appearance.

**8. Bromus marítimus** (Piper) Hitchc. Perennial; culms stout, 25 to 70 cm. tall, geniculate at base with numerous basal shoots; sheaths smooth or scaberulous; blades mostly 6 to 8 mm. wide, scabrous; panicle mostly 10 to 20 cm. long, the branches short, erect; spikelets 3 to 4 cm. long. ♃ (*B. marginatus maritimus* Piper.) —Near the coast, Lane County, Oreg., to Monterey County, Calif.

**9. Bromus polyánthus** Scribn. (Fig. 9.) Perennial; culms robust, mostly 90

Figure 9.—*Bromus polyanthus*, × 1. (Chase 5349, Colo.)

to 125 cm. tall; sheaths glabrous; blades 6 to 15 mm. wide, scabrous; panicles commonly 15 to 25 cm. long, the branches ascending; spikelets glabrous or scaberulous, somewhat glossy, rather loose at anthesis; awns 4 to 6 mm. long. ♃ —Open or sparsely wooded slopes, foothills, moist ground, Montana to Washington, south to Texas and California (Yosemite National Park); Kansas (experiment station).

Bromus laciniatus Beal. Tall slender perennial; blades flat; panicles 20 to 30 cm. long, open, drooping; spikelets flattened, about 3 cm. long, mostly purplish; lemmas keeled, awned. ♃ (*B. pendulinus* Sessé, not

Schrad.)—Occasionally cultivated for ornament; Mexico.

SECTION 2. BROMÓPSIS Dum.

Perennials; panicles mostly open; spikelets rather elongate, subterete or slightly compressed before flowering; florets closely overlapping.

**10. Bromus inérmis** Leyss. SMOOTH BROME. (Fig. 10.) Culms erect, 50 to 100 cm. tall, from creeping rhizomes; ligule 1.5 to 2 mm. long; blades smooth or nearly so, 5 to 10 mm. wide; panicle 10 to 20 cm. long, erect, the branches whorled, spreading in flower, contracted at maturity; spikelets 2 to 2.5 cm. long, subterete before flowering; first glume 4 to 5 mm. long, the second 6 to 8 mm. long; lemmas 9 to 12 mm. long, glabrous or somewhat scabrous, rarely villous, obtuse, emarginate, mucronate, or with an awn 1 to 2 mm. long. ♃ —Cultivated as hay and pasture grass, especially from Minnesota and Kansas to Washington and California, occasionally eastward to Michigan and Ohio and south to New Mexico and Arizona, now running wild in these regions; introduced along roads and in waste places in the northern half of the United States; occasionally southward. Also used for reseeding western mountain ranges. Introduced from Europe.

FIGURE 10.—*Bromus inermis*. Plant, × ½; spikelet, × 2½. (Deam 11633, Ind.)

**11. Bromus pumpelliánus** Scribn. (Fig. 11.) Resembling *B. inermis;* culms 50 to 120 cm. tall, from creeping rhizomes; sheaths glabrous or pubescent; blades rather short, mostly glabrous beneath, scabrous or somewhat pubescent on upper surface; panicle 10 to 20 cm. long, rather narrow, erect, the branches short, erect, or ascending; spikelets 7- to 11-flowered, 2 to 3 cm. long; first glume 1-nerved, the second 3-nerved; lemmas 10 to 12 mm. long, 5- to 7-nerved, pubescent along the margin and across the back at base, slightly emarginate; awn mostly 2 to 3 mm. long. ♃ —Meadows and grassy slopes, Colorado to the Black Hills of South Dakota, Idaho, and Alaska; introduced in Michigan. BROMUS PUMPELLIANUS var. TWEÉDYI Scribn. Differing in having lemmas more densely pubescent. ♃ —Alberta to Colorado.

**12. Bromus eréctus** Huds. Culms tufted, erect, 60 to 90 cm. tall, slender; sheaths sparsely pilose or glabrous; ligule 1.5 mm. long; blades narrow, sparsely pubescent; panicle 10 to 20 cm. long, narrow, erect, the branches ascending or erect; spikelets 5- to 10-flowered; glumes acuminate, the first 6 to 8 mm., the second 8 to

10 mm. long; lemmas 10 to 12 mm. long, glabrous or evenly scabrous-pubescent over the back; awn 5 to 6 mm. long. ♃ —Established in a few localities from Maine to New York; also in Washington, California, Wisconsin, West Virginia, Kentucky, and Alabama; introduced from Europe.

Bromus ramósus Huds. Tall slender perennial; blades flat; panicles 15 to 25 cm. long, open, drooping; spikelets 2 to 3 cm. long, lemmas 12 to 15 mm. long, awned. ♃ —Introduced in Washington; Europe.

**13. Bromus suksdórfii** Vasey. (Fig. 12.) Culms 60 to 100 cm. tall; panicle 7 to 12 cm. long, the branches erect or ascending; spikelets about 2.5 cm. long, longer than the pedicels; first glume mostly 1-nerved, 8 to 10 mm. long, the second 3-nerved, 8 to 12 mm. long; lemmas 12 to 14 mm. long, appressed-pubescent near the margin and on the lower part of midnerve; awn 2 to 4 mm. long. ♃ —Rocky woods and slopes, Washington to the

Figure 12.—*Bromus suksdorfii*, × 1. (Type.)

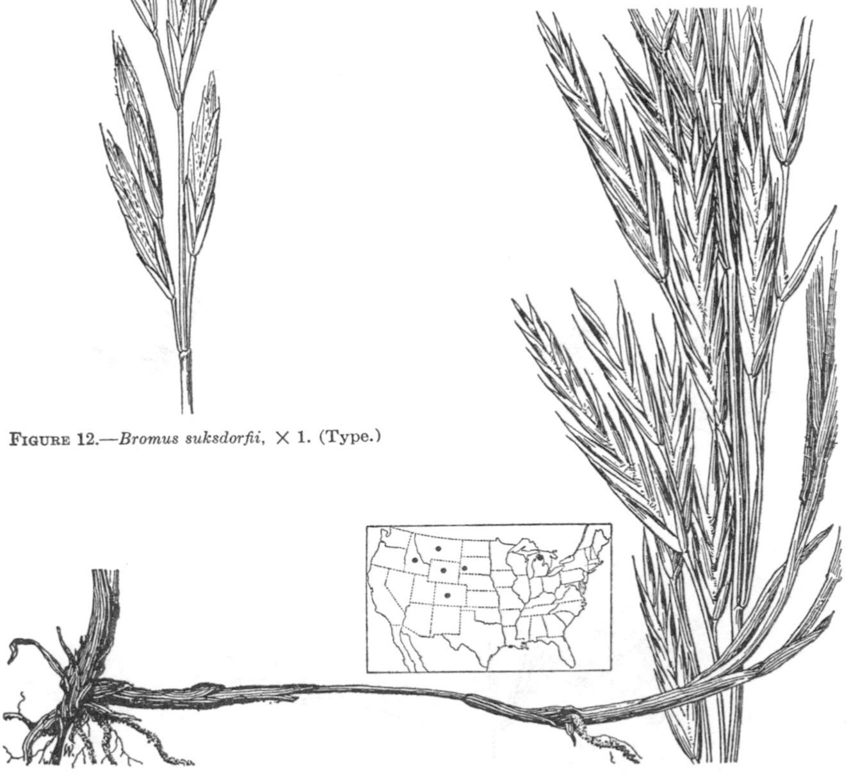

Figure 11.—*Bromus pumpellianus*, × 1. (Umbach 453, Mont.)

southern Sierra Nevada of California; Nevada (Lake Tahoe).

**14. Brómus orcuttiánus** Vasey. (Fig. 13.) Culms 80 to 120 cm. tall, erect, leafy below, nearly naked above, pubescent at and below the nodes; sheaths pilose or more or less velvety or sometimes glabrous; blades rather short and erect; panicle 10 to 15 cm. long, narrow-pyramidal, the few rather rigid short branches finally divaricate; spikelets about 2 cm. long, not much flattened, on short pedicels; glumes narrow, smooth, or scabrous, the first 6 to 8 mm. long, acute, 1-nerved, or sometimes with faint lateral nerves, the second 8 to 10 mm. long, broader, obtuse, 3-nerved; lemmas 10 to 12 mm. long, narrow, inrolled at margin, obscurely nerved, scabrous or scabrous-pubescent over the back; awn 5 to 7 mm. long. ♃ —Open woods, Washington to California; Arizona.

Bromus orcuttianus var. hállii Hitchc. Blades soft-pubescent on both surfaces; glumes and lemmas pubescent. ♃ —Dry, mostly wooded

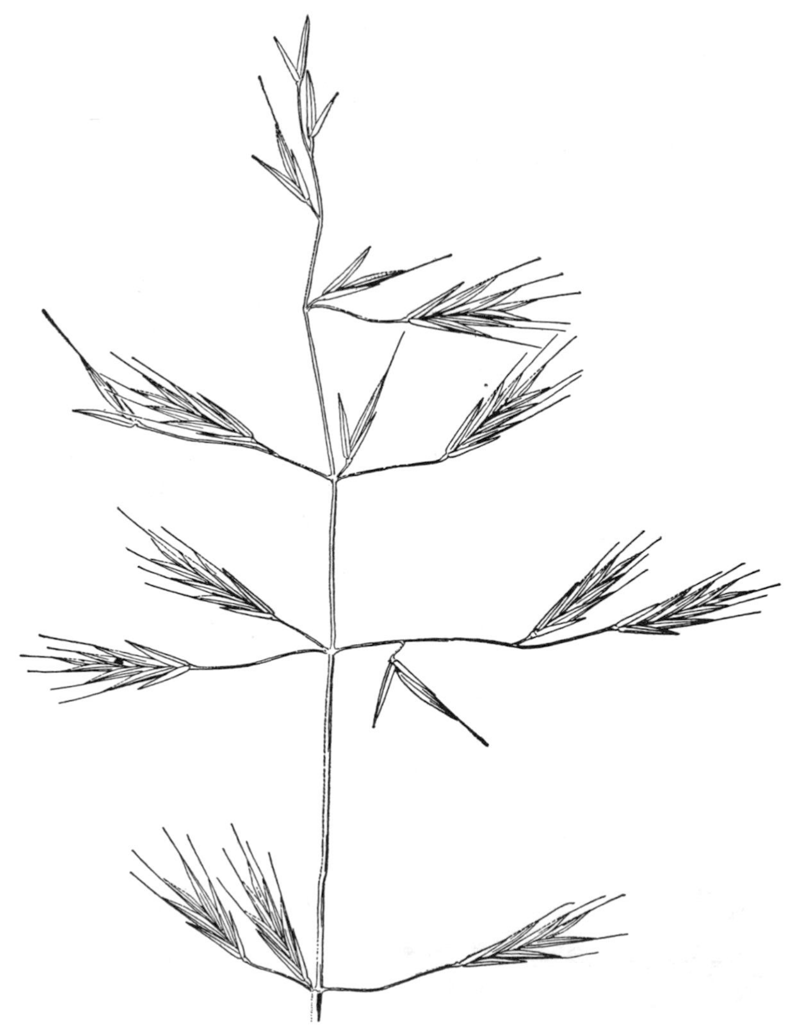

Figure 13.—*Bromus orcuttianus*, × 1. (Type.)

ridges and slopes, 1,500 to 3,000 m. elevation, California.

**15. Bromus grándis** (Shear) Hitchc. (Fig. 14.) Culms 1 to 1.5 m. tall; sheaths softly retrorsely pubescent; blades elongate, rather lax, spreading, densely short-pubescent on both surfaces; panicle 15 to 20 cm. long, broad, open, the branches slender, drooping, naked below, the lower usually in pairs, as much as 15 cm. long; spikelets 2 to 2.5 cm. long, on subflexuous pedicels; first glume usually distinctly 3-nerved, the second 3-nerved; lemmas 12 to 15 mm. long, densely pubescent all over the back; awn 5 to 7 mm. long. ♃ —Dry hills at moderate altitudes, Monterey and Madera Counties, Calif., south to San Diego.

FIGURE 14.—*Bromus grandis*, × 1. (Johnston 1407, Calif.)

FIGURE 15.—*Bromus pacificus*, × 1. (Elmer 1957, Wash.)

**16. Bromus pacíficus** Shear. (Fig. 15.) Culms 1 to 1.5 m. tall, stout, erect, pubescent at the nodes; sheaths sparsely pilose; ligule 3 to 4 mm. long; blades sparsely pilose on upper surface, scabrous or smooth beneath, 8 to 10 mm. wide; panicle very open, 10 to 20 cm. long, the branches slender, drooping; spikelets 2 to 2.5 cm. long, coarsely pubescent throughout; lemmas 11 to 12 mm. long, the pubescence somewhat dense on the margin; awn 4 to 6 mm. long. ♃ —Moist thickets near the coast, southern Alaska to western Oregon.

**17. Bromus laévipes** Shear. (Fig. 16.) Light green or glaucous; culms 50 to 100 cm. tall, from a decumbent base, often rooting at the lower nodes; sheaths and blades glabrous; ligule 2 to 3 mm. long; blades 4 to 8 mm. wide; panicles broad, 15 to 20 cm. long, the branches slender, drooping; first glume 3-nerved, 6 to 8 mm. long, the second 5-nerved, 10 to 12 mm. long; lemmas obtuse, 7-nerved, 12 to 14 mm. long, densely pubescent on the margin nearly to the apex and on the back at base; awn 3 to 5 mm. long. ♃ —Moist woods and shady banks, southern Washington to California.

FIGURE 16.—*Bromus laevipes*, × 1. (Amer. Gr. Natl. Herb. 866, Calif.)

**18. Bromus vulgáris** (Hook.) Shear (Fig. 17.) Culms slender, 80 to 120 cm. tall, the nodes pubescent; sheaths pilose; ligule 3 to 5 mm. long; blades more or less pilose, to 12 mm. wide; panicle 10 to 15 cm. long, the branches slender, drooping; spikelets narrow, about 2.5 cm. long; glumes narrow,

FIGURE 17.—*Bromus vulgaris*, × 1. (Chase 4945, Wash.)

the first acute, 1-nerved, 5 to 8 mm. long, the second broader, longer, obtuse to acutish, 3-nerved; lemmas 8 to 10 mm. long, sparsely pubescent over the back, more densely near the margin, or nearly glabrous; awn 6 to 8 mm. long. ♃ —Rocky woods and shady ravines, western Montana and Wyoming to British Columbia and California. Two scarcely distinct robust varieties have been described: *B. vulgaris* var. *eximius* Shear, a form with glabrous sheaths and nearly glabrous lemmas, Washington to Mendocino County, Calif.; and *B. vulgaris* var. *robustus* Shear, with pilose sheaths and large panicle, British Columbia to Oregon.

**19. Bromus ciliátus** L. FRINGED BROME. (Fig. 18.) Culms slender, 70 to 120 cm. tall, glabrous or pubescent at the nodes; sheaths glabrous or the lower short-pilose, mostly shorter than the internodes; blades rather lax, as much as 1 cm. wide, sparsely pilose on both surfaces to glabrous; panicle 15 to 25 cm. long, open, the branches slender, drooping, as much as 15 cm. long; first glume 1-nerved, the second 3-nerved; lemmas 10 to 12 mm. long, pubescent near the margin on the lower half to three-fourths, glabrous or nearly so on the back; awn 3 to 5 mm. long. ♃ —Moist woods and rocky slopes, Newfoundland to Washington, south to New Jersey, Tennessee, Iowa, western Texas, and southern California (San Bernardino Mountains); Mexico. *B. richardsoni* Link is a form that has been distinguished by its larger spikelets and lemmas and more robust habit, but it grades freely into *B.*

FIGURE 18.—*Bromus ciliatus*. Plant, × ½; spikelet and floret, × 5. (Hitchcock, Vt.)

*ciliatus* and can scarcely be ranked even as a variety. This is the common form in the Rocky Mountains.

**20. Bromus púrgans** L. Canada brome. (Fig. 19.) Resembling *B. ciliatus;* nodes mostly 4 to 6; sheaths, except the lower 1 or 2, shorter than the internodes, more or less retrorsely pilose, or sometimes all glabrous; blades elongate, 5 to 17 mm. wide, narrowed at base, and without flanges or auricles; pubescence of lemma nearly uniform, sometimes more dense on the margins, sometimes sparse and short on the back or scabrous only. ♃ —Moist woods and rocky slopes, Massachusetts to North Dakota, south to northern Florida and Texas.

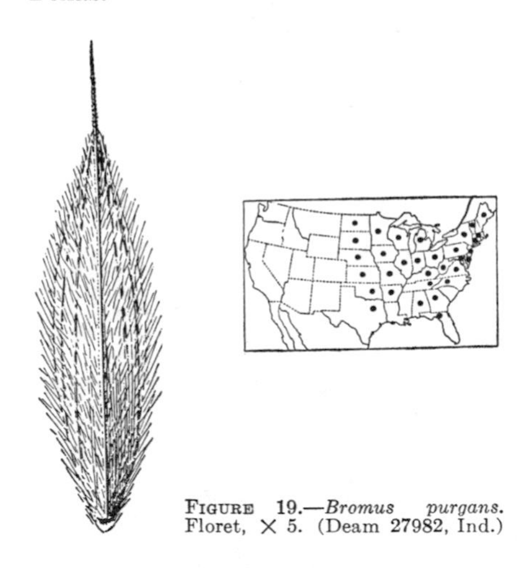

Figure 19.—*Bromus purgans.* Floret, × 5. (Deam 27982, Ind.)

Bromus purgans var. laeviglúmis (Scribn.) Swallen. Culms stout, leafy, mostly more than 1 m. tall; sheaths shorter or longer than the internodes, glabrous to pubescent, not strongly pilose; blades elongate, as much as 1 cm. wide or even wider; panicle large, open; lemmas glabrous or nearly so. ♃ —Woods and river banks, rare. Known from Quebec, Ontario, Maine, Vermont, Connecticut, New York, Michigan, Wisconsin, Maryland, West Virginia, and North Carolina.

**21. Bromus latiglúmis** (Shear) Hitchc. (Fig. 20.) Differing from *B. purgans* in having usually 10 to 20 nodes; sheaths overlapping, more or less pilose, especially about the throat and collar; base of blades with prominent flanges on each side, these usually prolonged into auricles. Where the ranges of *B. purgans* and *B. latiglumis* overlap, the latter flowers several weeks later than the former. ♃ —Alluvial banks of streams, Quebec and Maine to North Dakota, south to North Carolina and Kansas.

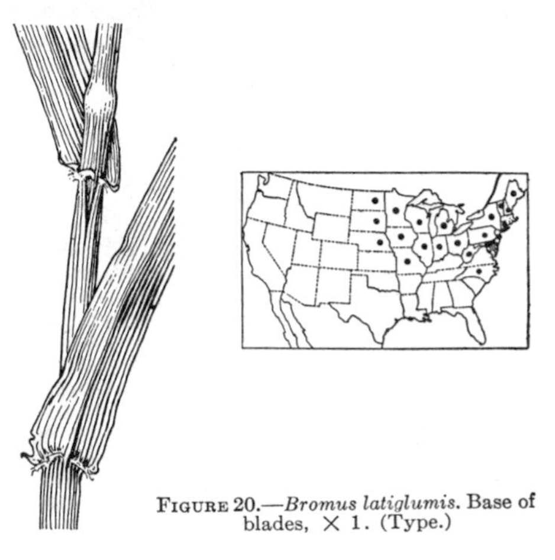

Figure 20.—*Bromus latiglumis.* Base of blades, × 1. (Type.)

Bromus latiglumis f. incánus (Shear) Fernald. Culms 1 to 2 m. tall, decumbent below, mostly somewhat weak and sprawling; sheaths densely canescent; panicles rather heavy. ♃ —Low woods, Indiana, Illinois, Michigan, and Maryland.

**22. Bromus nottowayánus** Fernald. (Fig. 21.) Resembling *B. latiglumis*, but with fewer nodes; sheaths mostly longer than the internodes, usually retrorsely pilose, without flanges at the mouth; ligule very short; blades elongate, 6 to 13 mm. wide, pilose above, some sparsely so beneath; panicles 12 to 22 cm. long, the slender branches drooping, the pulvini inconspicuous; first glume 1- to 3-nerved, the second 5-nerved; lemma 8 to 13 mm. long, densely appressed-pilose, the awn 5 to 8 mm. long. ♃ —Rich woods, Indiana and Illinois; Maryland to North Carolina; Tennessee; Arkansas.

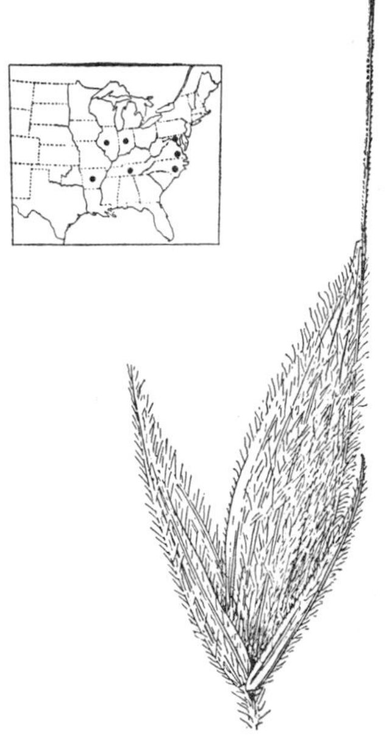

FIGURE 21.—*Bromus nottowayanus.* Glumes and lower floret, × 5. (Type number.)

**23. Bromus texénsis** (Shear) Hitchc. (Fig. 22.) Culms slender, mostly solitary, 40 to 70 cm. tall; sheaths much shorter than the internodes, softly retrorsely pilose; blades pubescent on both surfaces, rarely gla-

FIGURE 22.—*Bromus texensis,* × 1. (Tracy 8881, Tex.)

brous, mostly 3 to 6 mm. wide; panicle mostly not more than 12 cm. long, few-flowered, drooping; lemmas scabrous to nearly smooth; awn 5 to 7 mm. long. ♃ —Among brush, Texas (Bexar County and Corpus Christi) and Cochise County, Ariz.; apparently rare; northern Mexico.

FIGURE 23.—*Bromus anomalus,* × 1. (Pammel, Colo.)

**24. Bromus anómalus** Rupr. Nodding brome. (Fig. 23.) Culms slender, 30 to 60 cm. tall, the nodes pubescent; sheaths sparsely pilose to glabrous; ligule about 1 mm. long; blades scabrous, mostly 2 to 4 mm. wide; panicle about 10 cm. long, often less, few-flowered, drooping; first glume 3-nerved, the second 5-nerved, lemmas about 12 mm. long, evenly and densely pubescent over the back; awn 2 to 4 mm. long. ♃ (*B. porteri* Nash.)—Open woods, Saskatchewan to North Dakota and south to western Texas, southern California, and Mexico.

Bromus anomalus var. lanátipes (Shear) Hitchc. More robust, with woolly sheaths and usually broader blades. ♃ (*B. porteri lanatipes* Shear.)—Colorado to western Texas and Arizona.

Figure 24.—*Bromus kalmii*, × 1. (Chase 1866½, Ind.)

**25. Bromus kálmii** A. Gray. (Fig. 24.) Culms slender, 50 to 100 cm. tall, usually pubescent at and a little below the nodes; sheaths usually shorter than the internodes, pilose or the upper glabrous; blades usually sparsely pilose on both surfaces, 5 to 10 mm. wide; panicle rather few-flowered, drooping, mostly 5 to 10 cm. long, the branches slender, flexuous, bearing usually 1 to 3 spikelets; first glume 3-nerved, the second 5-nerved; lemmas 7 to 10 mm. long, villous over the back, more densely so near the margins; awn 2 to 3 mm. long. ♃ —Dry or sandy ground and open woods, Maine to Minnesota and South Dakota, south to western Maryland and Iowa. Called wild chess.

**26. Bromus frondósus** (Shear) Woot. and Standl. (Fig. 25.) Culms erect to weakly reclining, 80 to 100 cm. tall; sheaths glabrous or the lower pilose; blades pale green, scabrous, mostly less than 5 mm. wide, occasionally to 10 mm., rarely wider; panicle open, drooping, the slender lower branches naked below; first glume 2- to 3-nerved; lemmas pubescent all over, rarely nearly glabrous. ♃ (*B. porteri* var. *frondosus* Shear.) —Open woods and rocky slopes, Colorado, Utah, New Mexico, and Arizona.

Section 3. Brómium Dum.

Annuals; spikelets subcompressed; glumes and lemmas comparatively broad, elliptic or oblong-elliptic. Introduced, mostly from Europe.

**27. Bromus brizaefórmis** Fisch. and Mey. Rattlesnake chess. (Fig.

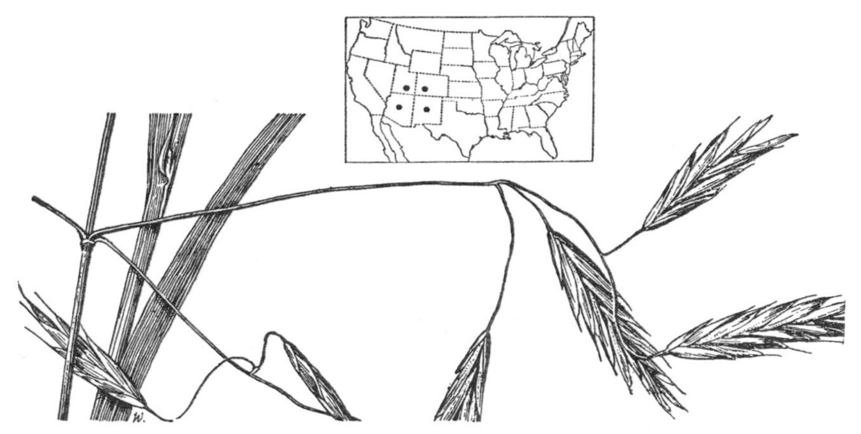

FIGURE 25.—*Bromus frondosus*, × 1. (Hitchcock 13282, N. Mex.)

26.) Culms 30 to 60 cm. tall; sheaths and blades pilose-pubescent; panicle 5 to 15 cm. long, lax, secund, drooping; spikelets rather few, oblong-ovate, 1.5 to 2.5 cm. long, about 1 cm. wide; glumes broad, obtuse, the first 3- to 5-nerved, the second 5- to 9-nerved, about twice as long as the first; lemmas 10 mm. long, very broad, inflated, obtuse, smooth, with a broad scarious margin, nearly or quite awnless. ☉ —Sandy fields and waste ground, Canada and Alaska; occasional from Washington, Montana, and Wyoming to California, rare eastward to Massachusetts and Delaware; introduced from Europe. Sometimes cultivated for ornament.

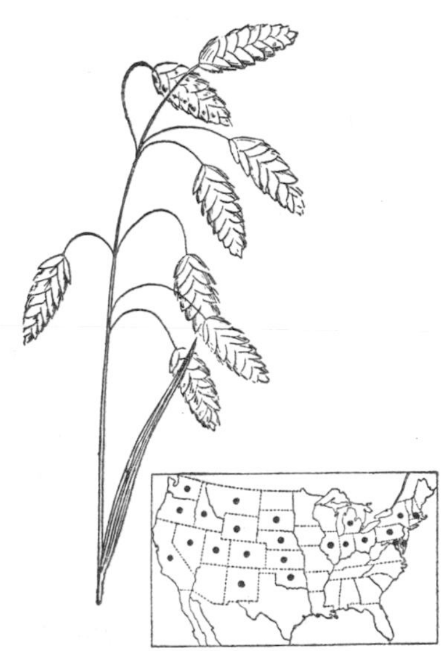

FIGURE 26.—*Bromus brizaeformis*, × ½. (Leckenby 40, Wash.)

**28. Bromus secalínus** L. CHESS. (Fig. 27.) Culms erect, 30 to 60 cm. tall; foliage glabrous or the lower sheaths sometimes puberulent; panicle pyramidal, nodding, 7 to 12 cm. long, the lower branches 3 to 5, unequal, slightly drooping; spikelets ovoid-lanceolate, becoming somewhat turgid at maturity, 1 to 2 cm. long, 6 to 8 mm. wide; glumes obtuse, the first 3- to 5-nerved, 4 to 6 mm. long, the second 7-nerved, 6 to 7 mm. long; lemmas 7-nerved, 6 to 8 mm. long, elliptic, obtuse, smooth or scaberulous, the margin strongly involute at maturity, shortly bidentate at apex, the undulate awns usually 3 to 5 mm. long, sometimes very short or obsolete; palea about as long as lemma. ☉ —Introduced from Europe, a weed in grainfields and waste places, more or less throughout the United States. Also called cheat. Occasionally utilized for hay in Washington and Oregon. In fruit the turgid florets are somewhat distant so that, viewing the spikelet sidewise, the light passes through the small openings at base of each floret. BROMUS SECALINUS var. VELUTÍNUS

FIGURE 27.—*Bromus secalinus*. Plant, × ½; spikelet and floret, × 5. Chase, Ill.)

(Schrad.) Koch. Spikelets pubescent. ⊙ —Oregon (Corvallis, The Dalles). Europe.

The species of the group containing *Bromus secalinus, B. commutatus, B. mollis,* and *B. racemosus* are closely allied, differentiated only by arbitrary characters. The forms are recognized as species in most recent European floras and this disposition is here followed.

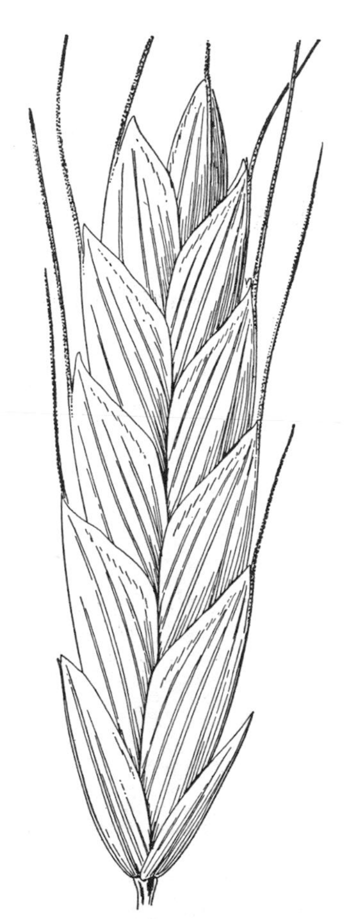

FIGURE 28.—*Bromus commutatus,* × 5. (Amer. Gr. Natl. Herb. 890, Va.)

**29. Bromus commutátus** Schrad. HAIRY CHESS. (Fig. 28.) Resembling *B. secalinus,* but the sheaths retrorsely pilose; the blades more or less pubescent; lemmas at maturity less plump and more overlapping; awn commonly somewhat longer. ⊙ —Introduced from Europe, a weed in fields and waste places, Washington to California, Montana, and Wyoming, eastward through the Northern States, thence less commonly southward. BROMUS COMMUTATUS var. APRICORUM Simonkai. Lemmas pubescent. ⊙ —Washington, Nevada, and California; rare. Introduced from Europe.

**30. Bromus móllis** L. SOFT CHESS. (Fig. 29.) Softly pubescent throughout; culms erect, 20 to 80 cm. tall; panicle erect, contracted, 5 to 10 cm. long, or, in depauperate plants, reduced to a few spikelets; glumes broad, obtuse, coarsely pilose or scabrous-pubescent, the first 3- to 5-nerved, 4 to 6 mm. long, the second 5- to 7-nerved, 7 to 8 mm. long; lemmas broad, soft, obtuse, 7-nerved,

FIGURE 29.—*Bromus mollis,* × 1. (Hall 258, Calif.)

coarsely pilose or scabrous-pubescent, rather deeply bidentate, 8 to 9 mm. long, the margin and apex hyaline; awn rather stout, 6 to 9 mm. long; palea about three-fourths as long as lemma. ⊙ —Weed in waste places and cultivated soil, introduced from Europe, Canada, and Alaska, abundant on the Pacific coast, occasional eastward to Nova Scotia and south to North Carolina. This has been referred to *B. hordeaceus* L., a distinct European species.

FIGURE 30.—*Bromus molliformis*, × 1. (Chase 5564, Calif.)

**31. Bromus mollifórmis** Lloyd. (Fig. 30.) Culms erect, mostly 10 to 20 cm. tall, sometimes taller; lower sheaths felty-pubescent, the upper glabrous; blades narrow, the upper surface with scattered rather stiff hairs; panicle 2 to 4 cm. long, ovoid, dense, few-flowered; spikelets oblong, compressed, 12 to 18 mm. long; glumes about 6 mm. long, the second broader, loosely pilose, the hairs spreading; lemmas thinner and narrower than in *B. mollis*, closely imbricate, about 8 mm. long, appressed-pilose, the margin whitish; awn from below the entire apex, 5 to 7 mm. long; palea a little shorter than the lemma; anthers 0.4 mm. long, about as broad. ☉ —Open ground, southern California; Texas (College Station); introduced from Europe.

**32. Bromus racemósus** L. (Fig. 31.) Differing from *B. mollis* in the somewhat more open panicle and glabrous or scabrous lemmas. ☉ (Including what in this country has been called *B. hordeaceus glabrescens* Shear, *B. hordeaceus* var. *leptostachys* Beck, and *B. mollis* f. *leiostachys* Fernald.)—Weed in waste places, chiefly on the Pacific coast and east to Montana, Colorado, and Arizona; a few points from Wisconsin and Illinois to Maine and North Carolina; introduced from Europe.

**Bromus scopárius** L. Resembling *B. molliformis;* culms 20 to 30 cm. tall; sheaths soft-pubescent; blades glabrous, scabrous or sparingly pilose; panicle contracted, erect, 3 to 7 cm. long; spikelets about 1.5 cm. long, 3 to 4 mm. wide; lemmas about 7 mm. long, narrow, glabrous; awn 5 to 8 mm. long, finally divaricate. ☉ —Introduced from Europe in California (Mariposa), Virginia (Newport News, on ballast), and Michigan (Schoolcraft).

**Bromus macróstachys** Desf. Annual; culms erect, 30 to 60 cm. tall; panicle narrow, compact, consisting of a few large coarsely pilose, awned spikelets about 3 cm. long. ☉ —Wool waste, Yonkers, N. Y., and College Station, Tex. Sometimes cultivated for ornament. Mediterranean region.

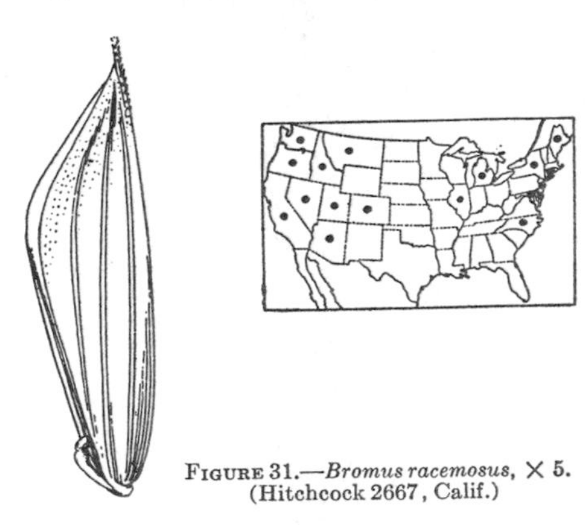

FIGURE 31.—*Bromus racemosus*, × 5. (Hitchcock 2667, Calif.)

**33. Bromus squarrósus** L. (Fig. 32.) Culms mostly 20 to 30 cm. tall, erect; sheaths and blades softly and densely pubescent; blades 5 to 15 cm. long, 2 to 4 mm. wide, usually erect; panicles nodding, the relatively coarse, short branches subverticillate, flexuous, bearing 1 or 2 large spikelets; spikelet about 2 cm. long, 5 to 8 mm. wide, somewhat inflated; awns flat, spreading or recurved, about 1 cm. long. ☉ —Waste places, Michigan and North Dakota. Introduced from Europe.

**34. Bromus japónicus** Thunb. JAPANESE CHESS. (Fig. 33.) Culms erect or geniculate at base, 40 to 70 cm. tall; sheaths and blades pilose; panicle 12 to 20 cm. long, broadly pyramidal, diffuse, somewhat drooping, the slender lower branches 3 to 5, all the branches more or less flexuous; glumes rather broad, the first acute, 3-nerved, 4 to 6 mm. long, the second obtuse, 5-nerved, 6 to 8 mm. long; lemmas broad, obtuse, smooth, 7 to 9 mm. long, 9-nerved, the marginal pair of nerves faint, the hyaline margin obtusely angled above the middle, the apex blunt, emarginate; awn 8 to 10 mm. long, usually somewhat twisted and flexuous at maturity, those of the lower florets shorter than

FIGURE 32.—*Bromus squarrosus*, × 1. (Hanes 688, Mich.)

FIGURE 33.—*Bromus japonicus*, × 1. (Deam 6833, Ind.)

the upper; palea 1.5 to 2 mm. shorter than the lemma. ☉ (*B. patulus* Mert. and Koch)—Weed in waste places, Vermont to Washington, south to North Carolina and California; Alberta; widely distributed in the Old World, whence introduced.

BROMUS JAPONICUS var. PORRÉCTUS Hack. Differing only in straight awns. ☉ —New York to Utah and New Mexico infrequent; more common from Maryland to Alabama. In some mature panicles both straight and flexuous-divergent awns occur. In *B. japonicus* before maturity the awns are straight and identity is often uncertain. Specimens of this have been distributed as *B. japonicus* var. *subsquarrosus*.

**35. Bromus arvénsis** L. (Fig. 34.) Resembling *B. japonicus*, foliage downy to subglabrous; spikelets thinner, flatter (less turgid), often tinged with purple; lemmas acute, bifid; awn

FIGURE 34.—*Bromus arvensis*, × 1. (Gray, Md.)

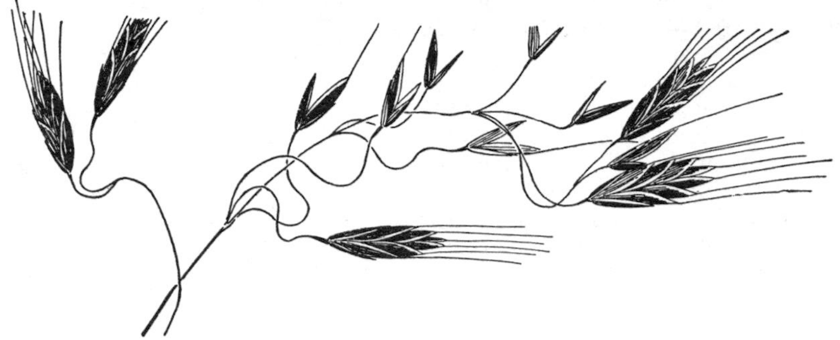

FIGURE 35.—*Bromus arenarius*, × 1. (Pendleton 1459, Calif.)

straight or nearly so in drying; palea as long as the lemma or only slightly shorter. ⊙ —Open ground, cultivated soil, New York, Maryland; North Dakota, Nevada, Arizona, and California.

**36. Bromus arenárius** Labill. AUSTRALIAN CHESS. (Fig. 35.) Culms slender, 15 to 40 cm. tall, sheaths and blades pilose; panicle open, pyramidal, nodding, 8 to 11 (rarely 15) cm. long, the spreading branches and pedicels sinuously curved; glumes densely pilose, acute, scarious-margined, the first narrower, 3-nerved, 8 mm. long, the second 7-nerved, 10 mm. long; lemmas densely pilose, 7-nerved, 10 mm. long; awn straight, 10 to 16 mm. long. ⊙ —Sandy roadsides, gravelly or sterile hills, Oregon, California, Nevada, and Arizona; adventive at Philadelphia, Pa.; introduced from Australia.

**Bromus alopecúros** Poir. Weedy annual 20 to 40 cm. tall; foliage softly pubescent; panicle narrow, dense, 5 to 10 cm. long; spikelets short-pedicellate, about 2 cm. long, the glumes and lemmas softly pubescent, the awn of the lemma, flat, twisted at the base, spreading, 1.5 to 2 cm. long. ⊙ —Adventive in waste ground, Ann Arbor, Mich. Mediterranean region.

SECTION 4. EUBRÓMUS Godr.

Tufted annuals; spikelets compressed; glumes and lemmas narrow, long-awned; first glume 1-nerved, the second 3-nerved; lemma 5- to 7-nerved, cleft at the apex, the hyaline teeth 2 to 5 mm. long; floret at maturity with a sharp hard point or callus. Introduced from Europe.

**37. Bromus rígidus** Roth. RIPGUT GRASS. (Fig. 36.) Culms 40 to 70 cm. tall; sheaths and blades pilose; panicle open, nodding, rather few-flowered, 7 to 15 cm. long, the lower branches 1 to 2 cm. long; spikelets usually 5- to 7-flowered, 3 to 4 cm. long, excluding awns; glumes smooth, the first 1.5 to 2 cm. long, the second 2.5 to 3 cm. long; lemmas 2.5 to 3 cm. long, scabrous or puberulent, the teeth 3 to 4

mm. long; awn stout, 3.5 to 5 cm. long. ⊙ (*B. villosus* Forsk. not Scop.; *B. maximus* Desf., not Gilib.) —Common weed in open ground and waste places in the southern half of California, forming dense stands over great areas in the lowlands, occasional north to British Columbia and east to Idaho, Utah, and Arizona; rare in the Eastern States, Maryland, Virginia, Mississippi, Texas, introduced from Mediterranean region. Distinguished from the other species of the section by the long awns. Bromus rigidus var. gussónei (Parl.) Coss. and Dur. Differing in having more open panicles, the stiffer, more spreading lower branches as much as 10 to 12 cm. long. ⊙ —Weed like *B. rigidus*, growing in similar places, Washington to California, and Arizona; more common than the species in middle and northern California.

**38. Bromus stérilis** L. (Fig. 37.) Resembling *B. rigidus*, less robust; culms 50 to 100 cm. tall; sheaths pubescent; panicle 10 to 20 cm. long, the branches drooping; spikelets 2.5 to 3.5 cm. long, 6- to 10-flowered; glumes lanceolate-subulate, the first about 8 mm. long; lemmas 17 to 20 mm. long, scabrous or scabrous-pubescent, the teeth 2 mm. long; awn 2 to 3 cm. long. ⊙ —Fields and waste places, introduced in a few localities from British Columbia to California and

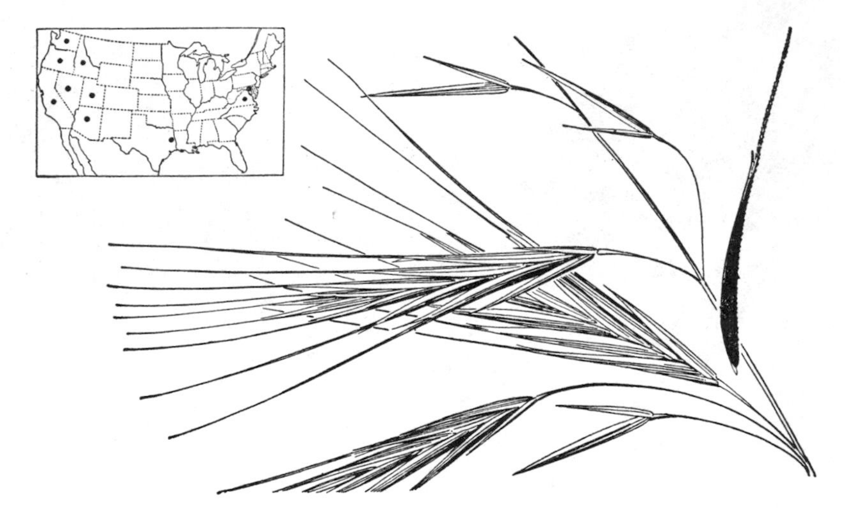

Figure 36.—*Bromus rigidus*, × 1. (Tracy 4702, Calif.)

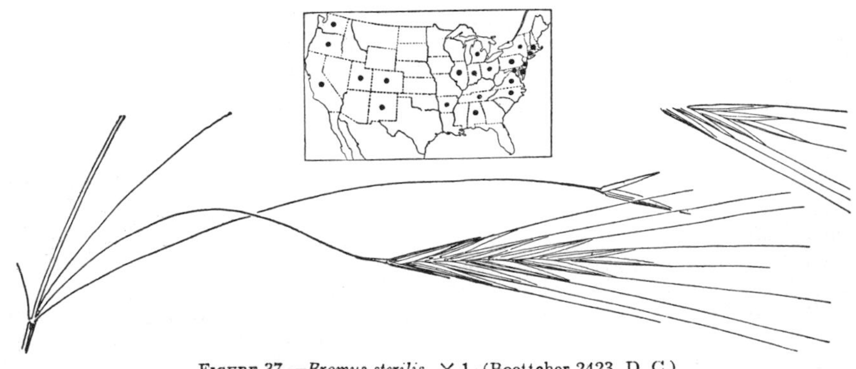

Figure 37.—*Bromus sterilis*, × 1. (Boettcher 2423, D. C.)

Colorado, and New Mexico; in the Eastern States from New England and Illinois to Virginia and Arkansas.

**39. Bromus rúbens** L. FOXTAIL CHESS. (Fig. 38.) Culms 15 to 40 cm. tall, puberulent below the panicle; sheaths and blades pubescent; panicle erect, compact, ovoid, usually 4 to 8 cm. long, usually purplish; spikelets 4- to 11-flowered, about 2.5 cm. long; first glume 7 to 9 mm. long, the second 10 to 12 mm. long; lemmas scabrous to coarsely pubescent, 12 to 16 mm. long, the teeth 4 to 5 mm. long; awn 18 to 22 mm. long, somewhat spreading at maturity. ☉ —Dry hills and in waste or cultivated ground, Washington to southern California, very abundant over extensive areas, and east to Idaho, Utah, and Arizona; Texas; Massachusetts.

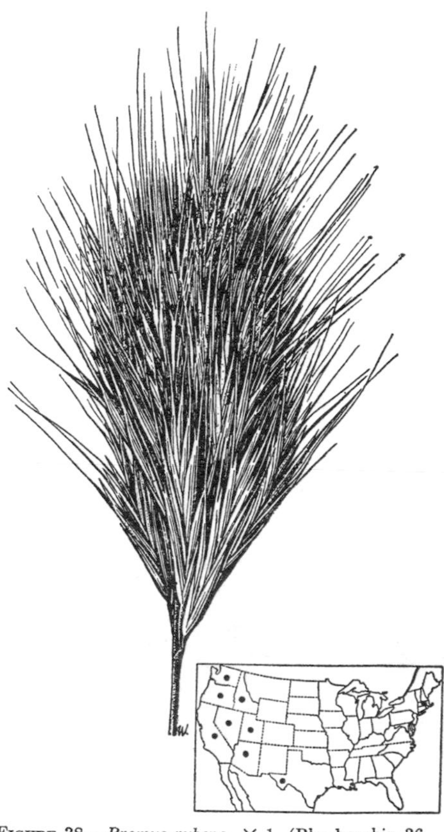

FIGURE 38.—*Bromus rubens*, × 1. (Blankenship 36, Calif.)

**40. Bromus madriténsis** L. (Fig. 39.) Resembling *B. rubens*, but the culms smooth below the less dense panicles; sheaths mostly smooth; blades puberulent to glabrous; panicle 5 to 10 cm. long, oblong-ovoid (in dried specimens more or less fan-shaped); lemmas a little longer than in *B. rubens*, the teeth 2 to 3 mm. long; awn rather stout, 16 to 22 mm. long. ☉ —Open ground and waste places, Oregon and California; less common than *B. rubens*. Occasionally cultivated for ornament.

FIGURE 39.—*Bromus madritensis*, × 1. (Eastwood, Calif.)

**41. Bromus tectórum** L. DOWNY CHESS. (Fig. 40.) Culms erect or spreading, slender, 30 to 60 cm. tall;

FIGURE 40.—*Bromus tectorum*. Plant, × ½; spikelet and floret, × 5. (Chase 2051, Ind.)

sheaths and blades pubescent; panicle 5 to 15 cm. long, rather dense, soft, drooping, often purple; spikelets nodding, 12 to 20 mm. long; glumes villous, the first 4 to 6 mm. long, the second 8 to 10 mm. long; lemmas lanceolate, villous or pilose, 10 to 12 mm. long, the teeth 2 to 3 mm. long; awn 12 to 14 mm. long. ⊙ —Along roadsides, banks, and waste places, common on the Pacific coast, especially in Washington and Oregon; Alberta, and here and there throughout the United States as far south as South Carolina and Texas. BROMUS TECTORUM var. GLABRÁTUS Spenner. Differing in having glabrous spikelets. ⊙ (*B. tectorum* var. *nudus* Klett and Richter.)—About the same range as the species, less common.

SECTION 5. NEOBRÓMUS Shear, as subgenus.

Annual; lemmas lanceolate, deeply bifid, the teeth aristate; awn twisted, geniculate. Approaches *Trisetum*.

FIGURE 41.—*Bromus trinii*, × 1. (Eastwood, Calif.)

**42. Bromus trínii** Desv. CHILEAN CHESS. (Fig. 41.) Culms 30 to 60 cm. or even 100 cm. tall, erect or branched and spreading below, often pubescent at the nodes; sheaths and blades pilose-pubescent to nearly smooth; panicle 8 to 20 cm. long, narrow, rather dense, erect, the branches erect or the lower more or less spreading or flexuous; spikelets narrow, 1.5 to 2 cm. long, 5- to 7-flowered; glumes lanceolate, acuminate, the first mostly 1-nerved, 8 to 10 mm. long, the second mostly 3-nerved, 12 to 16 mm. long; lemmas 5-nerved, 12 to 14 mm. long, pubescent, acuminate, with narrow teeth 2 to 3 mm. long, the teeth aristate; awn 1.5 to 2 cm. long, twisted below, bent below the middle and strongly divaricate when old. ⊙ (Including *B. trinii* var. *pallidiflorus* Desv.) —Dry plains and rocky or wooded slopes, Oregon, California, and Baja California, rarely eastward to Colorado; introduced from Chile.

BROMUS TRINII var. EXCÉLSUS Shear. Differing in having larger spikelets, 7-nerved lemmas, and divaricate but not twisted or bent awns; teeth of the lemma acuminate, but not aristate. ⊙ —A little-known form from the Panamint Mountains, Calif., and Emory Canyon, Lake Mead, Ariz.

## 3. BRACHYPÓDIUM Beauv.

Spikelets nearly sessile, several- to many-flowered, the rachilla disarticulating above the glumes and between the florets; glumes unequal, sharp-pointed, 5- and 7-nerved; lemmas firm, rounded or somewhat flattened on the back, 7-nerved, acuminate, awned or mucronate; palea as long as the body of the lemma, concave, the keels pectinate-ciliate. Annuals or perennials with erect racemes of subsessile spikelets. Type species, *Bromus pinnatus* L. (*Brachypodium pinnatum* (L.) Beauv.) Name from *brachys*, short, and *podion*, foot, alluding to the very short pedicels.

Eurasian species introduced in the United States; two American species only; Mexico to South America.

**1. Brachypodium distáchyon** (L.) Beauv. (Fig. 42.) Annual, branching and geniculate at base, 15 to 30 cm. tall; nodes pubescent; sheaths and blades sparsely pilose to subglabrous; ligule, 1.5 to 2 mm. long, pubescent; blades flat, 2 to 6 cm. long, 3 to 4 mm. wide; raceme strict, the segments of the axis alternately concave; spikelets 1 to 5, imbricate, 2 to 3.5 cm. long, excluding the awns, 5 to 6 mm. wide; florets closely imbricate; lemmas scabrous, the slender scabrous erect awn 1 to 2 cm. long. The spikelets resemble those of some species of *Agropyron*. ☉ —Open ground, Arapahoe County, Colo., Humboldt, Sonoma, and Marin Counties, Calif.; on ballast, Camden, N. J., and Portland, Oreg. Sparingly introduced from Europe, but spreading in Marin County, Calif.

Brachypodium sylváticum (Huds.) Beauv. Perennial, 60 to 100 cm. tall; blades to 25 cm. long and 1 cm. wide; raceme 12 to 20 cm. long, the spikelets 4 to 5 cm. long, subterete, the lower distant, the upper closely imbricate. ♃ Occasionally cultivated for ornament and in grass gardens. Europe.

Brachypodium caespitosum (Host) Roem. and Schult., a tall, leafy perennial, with racemes 8 to 12 cm. long of overlapping spikelets 2.5 to 3 cm. long, the lemmas imbricate, strongly nerved, glabrous, the awns about 5 mm. long. Introduced from Turkey; has been grown at the experiment station, Tucson, Ariz.

Brachypodium pinnatum (L.) Beauv., similar to the preceding, but with pubescent nodes, scabrous laxer foliage, and narrower spikelets with hirsute lemmas. Introduced from Rumania; has been grown in the Grass Garden, Beltsville, Md. The results of both trials are as yet inconclusive.

Figure 42.—*Brachypodium distachyon*. Plant, × ½; glumes and floret, × 5. (J. T. Howell 23186, Calif.)

## 4. FESTÚCA L. Fescue

Spikelets few- to several-flowered (rarely 1-flowered in some of the spikelets of a panicle), the rachilla disarticulating above the glumes and between the florets, the uppermost floret reduced; glumes narrow, acute, unequal, the first sometimes very small; lemmas rounded on the back, membranaceous or somewhat indurate, 5-nerved, the nerves often obscure, acute or rarely obtuse, awned from the tip, or rarely from a minutely bifid apex, sometimes awnless. Low or rather tall annuals or perennials, the spikelets in narrow or open panicles. The blades are sometimes somewhat auriculate as in the Hordeae. Standard species, *Festuca ovina*. Name from *Festuca*, an old Latin name for a weedy grass.

Many of the perennial species of fescue are important forage grasses in the grazing regions of the West. *Festuca arizonica*, Arizona fescue, of northern Arizona and *F. idahoensis*, Idaho fescue, of the region from Colorado to central California and northward, are important, though they become rather tough with age. *F. viridula*, greenleaf fescue, locally called mountain bunchgrass, is an outstanding grass in subalpine regions of the Northwestern States, and *F. thurberi*, Thurber fescue, is important in similar regions from Colorado to Montana. *F. ovina*, sheep fescue, is a good grazing grass though not abundant, but its variety *brachyphylla*, alpine fescue, furnishes much of the forage above timber line from the Rocky Mountains westward. *F. occidentalis*, western fescue, in open woods up to 10,000 feet in the Northwest, and *F. rubra*, red fescue, widely distributed at various altitudes in the West, are valuable in proportion to their abundance.

The most important cultivated species is *F. elatior*, meadow fescue, a native of Europe, used for hay and pasture in the humid region, especially in Tennessee, Missouri, and Kansas. *F. ovina*, and its allies, and *F. rubra*, are cultivated to a limited extent in the Eastern States as lawn or pasture grasses, usually in mixtures.

Plants annual ........ SECTION 1. VULPIA.
Plants perennial ........ SECTION 2. EUFESTUCA.

*Section 1. Vulpia*

1a. Spikelets mostly more than 5-flowered. Lowest lemma 4 to 5 mm. long, the margin inrolled, not scarious ........ 1. F. OCTOFLORA.
1b. Spikelets mostly less than 5-flowered (sometimes 6-flowered in *F. dertonensis* and *F. sciurea*). Lemmas usually scarious-margined.
  2a. Panicle narrow, the branches appressed.
    Lemmas appressed-pubescent over the back, about 3 mm. long ........ 2. F. SCIUREA.
    Lemmas glabrous, scabrous or ciliate, not pubescent over the back.
      Lemmas ciliate toward the apex ........ 3. F. MEGALURA.
      Lemmas not ciliate.
        First glume two-thirds to three-fourths as long as the second. 4. F. DERTONENSIS.
        First glume much shorter than the second, 1 to 2 mm. long ........ 5. F. MYUROS.
  2b. Panicle rather short, the branches and often the spikelets spreading (scarcely spreading in *F. arida*).
    3a. Spikelets glabrous.
      Pedicels appressed; lower branches of the panicle usually finally reflexed; spikelets usually 3- to 5-flowered ........ 6. F. PACIFICA.
      Pedicels or nearly all of them finally reflexed, notably those of the upper part of the main axis; branches of the panicle reflexed; spikelets mostly 1- or 2-flowered. 10. F. REFLEXA.
    3b. Spikelets pubescent, the pubescence on glumes or lemmas or on both.
      4a. Pedicels appressed or slightly spreading; lower branches of panicle usually spreading or reflexed.
        Lemmas glabrous; glumes pubescent ........ 7. F. CONFUSA.
        Lemmas pubescent.
          Lemmas hirsute; glumes glabrous or pubescent; lower branches of panicle spreading or reflexed ........ 8. F. GRAYI.
          Lemmas woolly-pubescent; glumes glabrous; panicle nearly simple, the branches scarcely spreading ........ 9. F. ARIDA.
      4b. Pedicels and panicle branches all finally spreading or reflexed.
        Glumes glabrous; lemmas pubescent ........ 11. F. MICROSTACHYS.
        Glumes pubescent; lemmas pubescent ........ 12. F. EASTWOODAE.
        Glumes pubescent; lemmas glabrous ........ 13. F. TRACYI.

*Section 2. Eufestuca*

1a. Blades flat, rather soft and lax, mostly more than 3 mm. wide.
  Lemmas awned, the awn usually more than 2 mm. long.
    Floret long-stipitate, the rachilla appearing to be jointed a short distance below the floret ........ 14. F. SUBULIFLORA.
    Floret not stipitate.
      Lemmas indistinctly nerved; awn terminal; blades 3 to 10 mm. wide. 15. F. SUBULATA.
      Lemmas distinctly 5-nerved; awn from between 2 short teeth; blades 2 to 4 mm. wide. 16. F. ELMERI.

Lemmas awnless or with an awn rarely as much as 2 mm. long.
Spikelets oblong to linear, mostly 8- to 10-flowered and more than 10 mm. long. 17. F. ELATIOR.
Spikelets ovate or oval, mostly not more than 5-flowered, less than 10 mm. long.
Lemmas acuminate, sometimes with an awn as much as 2 mm. long, membranaceous, distinctly nerved, 6 to 9 mm. long ..... 18. F. SORORIA.
Lemmas awnless, obtuse to acutish, rather firm, indistinctly nerved.
Lemmas 5 to 7 mm. long, acutish ..... 19. F. VERSUTA.
Lemmas about 4 mm. long, relatively blunt, rather turgid.
Spikelets loosely scattered in a very open panicle with long slender branches. 20. F. OBTUSA.
Spikelets somewhat aggregate toward the ends of rather short branches of a less open nodding panicle ..... 21. F. PARADOXA.

1b. Blades involute or if flat less than 3 mm. wide (sometimes flat in *F. californica*, but firm and soon involute).
Ligule 2 to 4 mm. long or longer. Lemmas awnless or cuspidate.
Lemmas 7 mm. long ..... 22. F. THURBERI.
Lemmas 4 mm. long ..... 23. F. LIGULATA.
Ligule short.
Collar and mouth of sheath villous. Culms tall and stout (rather short in var. *parishii*). 25. F. CALIFORNICA.
Collar and mouth of sheath not villous.
Panicle branches densely ciliate on the angles. Blades about 1 mm. wide, flat or folded ..... 26. F. DASYCLADA.
Panicle branches not ciliate on the angles.
Culms decumbent at the usually red, fibrillose base, in loose tufts. Awn of lemma shorter than the body; blades smooth ..... 28. F. RUBRA.
Culms erect.
Lemmas 7 to 10 mm. long, scabrous. Culms densely tufted, rather stout, usually scabrous below the panicle; lemmas acute, rarely short-awned. 24. F. SCABRELLA.
Lemmas mostly not more than 7 mm. long.
Lemmas awnless (see also *F. arizonica*).
Lemmas 6 to 7 mm. long; culms slender, loosely tufted. 27. F. VIRIDULA.
Lemmas about 3 mm. long ..... 31. F. CAPILLATA.
Lemmas awned.
Awn as long as or longer than body of the lemma; blades soft, glabrous, sulcate ..... 29. F. OCCIDENTALIS.
Awn shorter than body of the lemma; blades slender, numerous, usually scabrous.
Blades mostly not more than half as long as the culms; panicle narrow, often almost spikelike, few-flowered, mostly less than 10 cm. long; culms mostly less than 30 cm. tall ..... 30. F. OVINA.
Blades elongate; panicles 10 to 20 cm. long, somewhat open; culms 30 to 100 cm. tall.
Awn 2 to 4 mm. long ..... 32. F. IDAHOENSIS.
Awn short or obsolete ..... 33. F. ARIZONICA.

SECTION 1. VÚLPIA (Gmel.) Reichenb.

Slender annuals; lemmas awned; stamens usually 1, sometimes 3; flowers usually self-pollinated, but young panicles are found with anthers and stigmas exserted. Some of the species, especially numbers 7 to 13, resemble each other closely. The differences, though small, appear to be constant, hence the recognizable forms are maintained as species, rather than reduced to varieties under leading species.

**1. Festuca octoflóra** Walt. SIX-WEEKS FESCUE. (Fig. 43.) Culms erect, usually 15 to 30 cm. tall, sometimes as much as 60 cm.; blades narrow, involute, 2 to 10 cm. long; panicle narrow, the branches short, appressed or spreading; spikelets 6 to 8 mm. long, densely 5- to 13-flowered; glumes subulate-lanceolate, the first 1-nerved, the second 3-nerved, 3 to 4.5 mm. long; lemmas firm, convex, lanceolate, glabrous or scabrous, 4 to 5 mm. long, the margins not scarious; awn commonly 3 to 5, sometimes to 7 mm. long. ☉ —Open sterile ground, New York to Florida, Illinois, Kansas, and Texas; Idaho, Washington. The species and its varieties are found throughout the United States.

Festuca octoflora var. tenélla (Willd.) Fernald. Mostly smaller; panicle usually nearly simple; spikelets smaller; first glume 2.3 to 4 mm. long, awns 1 to 5 mm. Distinctions not constant, many intermediates occur. ☉ —Canada and Connecticut to Washington, south to Virginia, Tennessee, and Oklahoma; Georgia, Alabama, Texas; Colorado, Nevada, and New Mexico.

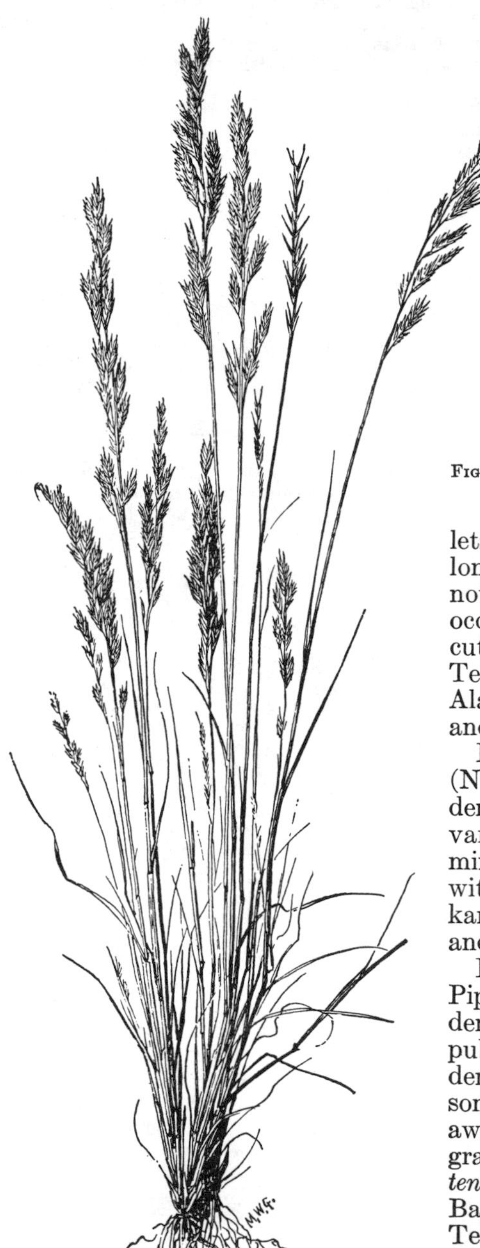

Figure 43.—*Festuca octoflora.* Plant, × ½; spikelet, × 5. (Chase 1776, Ind.)

Festuca octoflora var. glaúca (Nutt.) Fernald. Panicle shorter and denser than in most specimens of var. *tenella;* awn of lemma from minute to 2 mm. long. Intergrades with var. *tenella.* ☉ —Indiana, Arkansas, Nebraska, Kansas, Oklahoma, and Texas.

Festuca octoflora var. hirtélla Piper. Commonly rather low and densely tufted; foliage sometimes pubescent; panicle usually rather dense; lemma hirtellous or pubescent, sometimes strongly scabrous only; awns mostly 2 to 4 mm. long. Intergrades with *F. octoflora* and with var. *tenella.* ☉ —British Columbia to Baja California, east to Kansas and Texas; Florida.

**2. Festuca sciúrea** Nutt. (Fig. 44.) Culms erect, 15 to 50 cm. tall; blades less than 1 mm. wide, often capillary, soft, mostly involute, 1 to 10 cm. long; panicle narrow, 5 to 20 cm. long; spikelets 4- to 6-flowered, 4 to 5 mm. long; first glume 2 mm. long, the second 3.5 mm. long; lemmas 3 to 3.5 mm. long, sparsely appressed-pubescent; awn 6 to 11 mm. long. ⊙ —Open ground, New Jersey and Maryland to Florida, west to Oklahoma and Texas.

FIGURE 44.—*Festuca sciurea.* Panicle, × ½; spikelet, × 5. (Reverchon, Tex.)

**3. Festuca megalúra** Nutt. FOXTAIL FESCUE. (Fig. 45.) Culms 20 to 60 cm. tall; sheaths and narrow blades glabrous; panicle narrow, 7 to 20 cm. long, the branches appressed; spikelets 4- or 5-flowered; first glume 1.5 to 2 mm. long, the second 4 to 5 mm. long; lemmas linear-lanceolate, scabrous on the back especially toward the apex, ciliate on the upper half; awn 8 to 10 mm. long. ⊙ —Open sterile ground, British Columbia to Baja California, common in the Coast Ranges of California, east to Montana and Arizona; introduced in a few localities eastward; Guatemala; Pacific slope of South America. In mature lemmas the cilia may be obscured by the inrolling of the edges; moistening the floret will bring the cilia to view.

FIGURE 45.—*Festuca megalura.* Panicle, × 1; spikelet, × 5. (Leiberg 150, Oreg.)

4. **Festuca dertonénsis** (All.) Aschers. and Graebn. (Fig. 46.) Resembling *F. megalura*, the panicles on the average shorter, usually less dense; glumes longer, the first about 4 mm. long, the second 6 to 7 mm. long; lemma lanceolate, scabrous on the back toward the apex, 7 to 8 mm. long; awn 10 to 13 mm. long. ⊙ —Dry hills and meadows, British Co-

FIGURE 46.—*Festuca dertonensis*. Plant, × ½; spikelet, × 5. (Palmer 2041, Calif.)

lumbia to southern California, Arizona, and Texas; rare as a waif in the Eastern States; introduced from Europe. This species has been referred to *F. bromoides* L. by American authors.

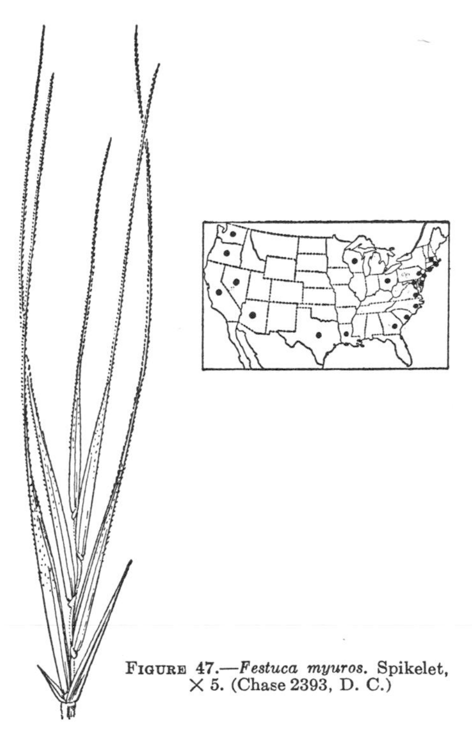

FIGURE 47.—*Festuca myuros*. Spikelet, × 5. (Chase 2393, D. C.)

**5. Festuca myúros** L. (Fig. 47.) Differing from *F. megalura* chiefly in the absence of cilia on the lemma; panicle usually smaller, first glume 1 to 1.5 mm., the second 4 to 4.5 mm. long. ⊙ —Open ground, Coastal Plain, Massachusetts to Texas; Ohio; Wisconsin; Pacific coast, Washington to southern California; Arizona; Mexico and South America; introduced from Europe.

**6. Festuca pacífica** Piper. (Fig. 48.) Culms erect or geniculate at base, 30 to 60 cm. tall; blades soft, loosely involute, glabrous, 3 to 5 cm. long; panicle 5 to 12 cm. long, the lower branches solitary, somewhat distant, subsecund, spreading, 1 to 3 cm. long; spikelets 3- to 6-flowered; first glume subulate-lanceolate, about 4 mm. long, the second lanceolate-acuminate, about 5 mm. long; lemmas lanceolate, glabrous or scaberulous, 6 to 7 mm. long; awn 10 to 15 mm. long. ⊙ —Open ground, mountain slopes, and open woods, British Columbia to Baja California, east to western Montana and New Mexico.

FIGURE 48.—*Festuca pacifica*. Panicle, × 1; floret, × 10. (Type.)

FESTUCA PACIFICA var. SÍMULANS Hoover. All spikelets reflexed or divergent at maturity. ⊙ —Kern and Kings Counties, Calif.

**7. Festuca confúsa** Piper. (Fig. 49.) Resembling *F. pacifica;* sheaths retrorsely pilose; foliage pubescent; spikelets usually 2- or 3-flowered; glumes hirsute with long spreading hairs; lemmas glabrous. ⊙ —Dry hillsides, Washington to southern California.

**8. Festuca gráyi** (Abrams) Piper. (Fig. 50.) Resembling *F. pacifica*, often somewhat stouter; sheaths and sometimes blades pubescent; glumes glabrous to sparsely villous; lemmas pubescent, puberulent or sometimes villous. ⊙ (*F. microstachys* var. *grayi* Abrams.)—Open ground and rocky slopes, Washington to southern California and Arizona.

**9. Festuca árida** Elmer. (Fig. 51.) Culms erect or spreading, mostly less than 15 cm. tall; sheaths and blades

FIGURE 49.—*Festuca confusa*. Plant, × 1; spikelet, × 5. (Type.)

FIGURE 50.—*Festuca grayi*. Plant, × ½; spikelet, × 5. (Pringle, Ariz.)

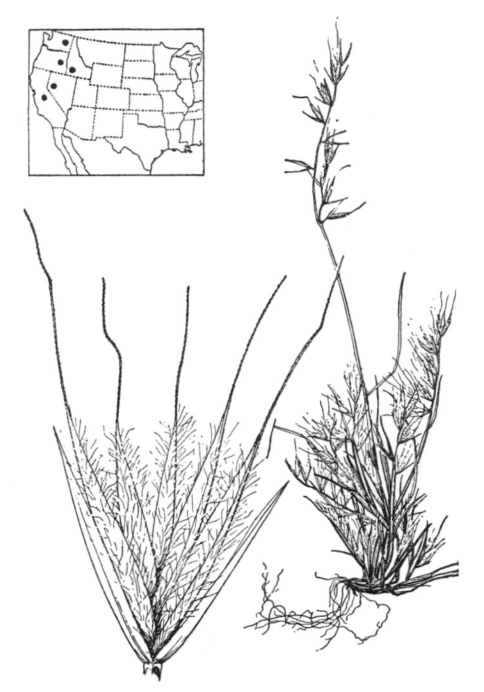

FIGURE 51.—*Festuca arida.* Plant, × ½; spikelet, × 5. (Type.)

glabrous, the blades loosely involute, mostly less than 4 cm. long; panicle narrow, 2 to 5 cm. long, the branches appressed or the lowermost somewhat spreading; glumes about equal, glabrous, 5 to 6 mm. long; lemmas densely woolly, about 5 mm. long; awn 5 to 10 mm. long. ⊙ —Sandy or dry ground, rare, eastern Washington and Oregon, southwestern Idaho, northeastern California, and western Nevada.

**10. Festuca refléxa** Buckl. (Fig. 52.) Culms 20 to 40 cm. tall; sheaths glabrous or pubescent; blades narrow, flat to subinvolute, 2 to 10 cm. long; panicle 5 to 12 cm. long, the solitary branches and the spikelets all at length divaricate; spikelets mostly 1- to 3-flowered, 5 to 7 mm. long; first glume 2 to 4 mm. long, the second 4 to 5 mm. long; lemmas glabrous or scaberulous, 5 to 6 mm. long; awn usually 5 to 8 mm. long. ⊙ —Mesas, rocky slopes, and wooded hills, Washington to southern California, east to Arizona and Utah.

**11. Festuca micrôstachys** Nutt. (Fig. 53.) Resembling *F. reflexa;* glumes glabrous; lemmas pubescent. ⊙ —Open ground, Washington to California; rare.

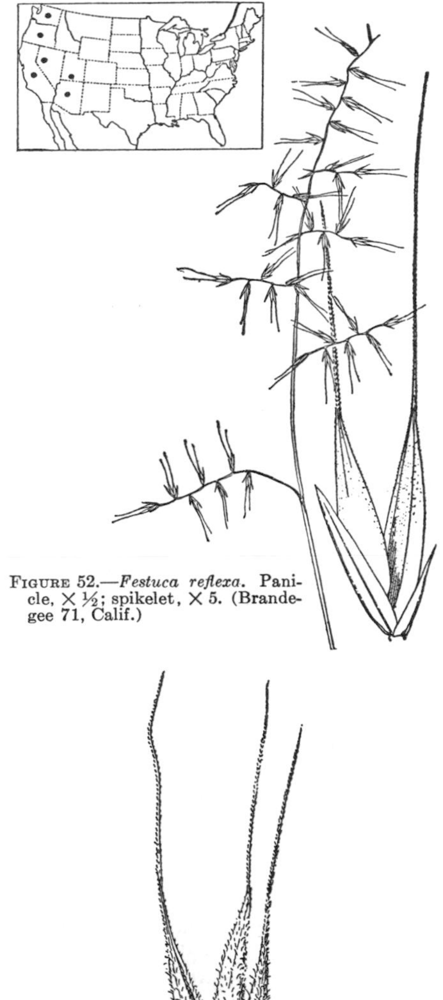

FIGURE 52.—*Festuca reflexa.* Panicle, × ½; spikelet, × 5. (Brandegee 71, Calif.)

FIGURE 53.—*Festuca microstachys.* Spikelet, × 5. (Allen, Calif.)

**12. Festuca eastwoodae** Piper. (Fig. 54.) Resembling *F. reflexa;* glumes hirsute; lemmas hirsute, the awn 4 to 5 mm. long. ⊙ —Open ground, Oregon, Arizona, and California; rare.

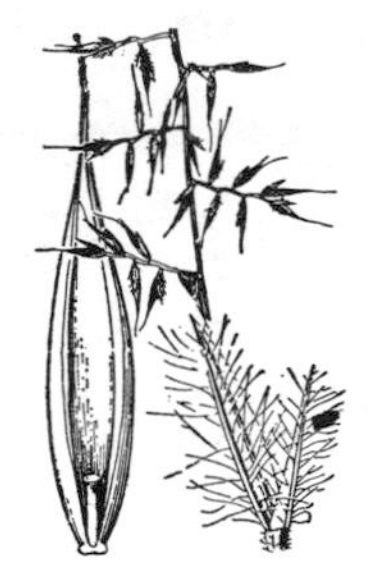

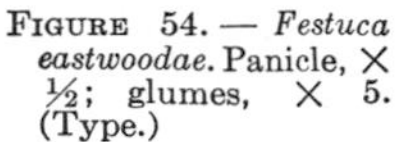

FIGURE 54. — *Festuca eastwoodae*. Panicle, × ½; glumes, × 5. (Type.)

FIGURE 55. — *Festuca tracyi*. Panicle, × ½; glumes, × 5; floret, × 5. (Type.)

**13. Festuca trácyi** Hitchc. (Fig. 55.) Resembling *F. reflexa;* glumes rather sparsely hispid-villous, the first 1.5 to 2 mm. long, acute, the second 3 to 4 mm. long, obtusish or abruptly acute; lemmas glabrous, about 4 mm. long; awn 4 to 7 mm. long. ⊙ —Open rocky ground, Washington (Bingen) and California (Kings and Napa Counties).

SECTION 2. EUFESTUCA Griseb.

Perennials, culms simple, stamens 3.

**14. Festuca subuliflóra** Scribn. (Fig 56.) Culms erect, slender, 60 to 100 cm. tall; blades flat (or loosely involute in drying), lax, pubescent on the upper surface, those of the culm mostly 2 to 4 mm. wide, those of the innovations narrower; panicle loose, lax, 10 to 20 cm. long, nodding, the branches drooping, the lower naked at base; spikelets loosely 3- to 5-flowered, the rachilla pubescent or hispid, the internodes of the rachilla as much as 2 mm. long; floret long-stipitate, the rachilla appearing to be jointed a short distance below the floret; glumes very narrow, acuminate, the first 3 to 4 mm., the second 4 to 5 mm., long; lemmas scaberulous toward the apex, 6 to 8 mm. long; awn somewhat flexuous, 10 to 15 mm. long. ♃ —Moist shady places from sea level to 1,000 m., British Columbia to northern California, mostly near the coast. Peculiar in the stipitate base of the lemma. Aspect of *F. subulata*.

**15. Festuca subuláta** Trin. BEARDED FESCUE. (Fig. 57.) Culms erect, mostly 50 to 100 cm. tall; blades flat, thin, lax, 3 to 10 mm. wide; panicle loose, open, drooping, 15 to 40 cm. long, the branches mostly in twos or threes, naked below, finally spreading or reflexed, the lower as much as 15 cm. long; spikelets loosely 3- to 5-flowered; glumes narrow, acuminate, the first about 3 mm., the second about 5 mm., long; lemmas somewhat keeled, scaberulous

FIGURE 56.—*Festuca subuliflora*. Panicle, × ½; spikelet, × 5. (Howell 19, Oreg.)

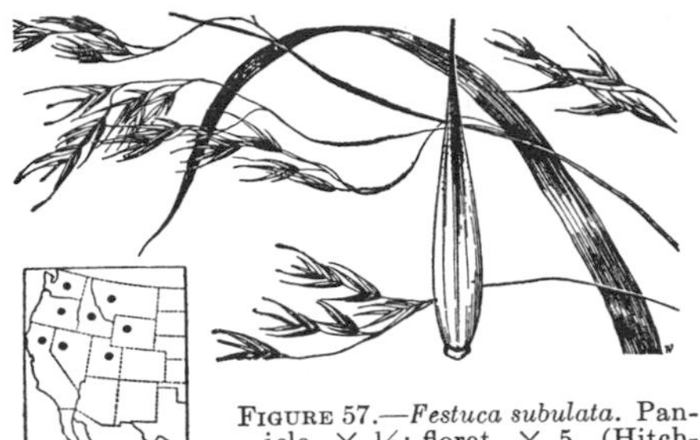

FIGURE 57.—*Festuca subulata*. Panicle, × ½; floret, × 5. (Hitchcock 23511, Oreg.)

toward the apex, the intermediate nerves obscure, the tip attenuate into an awn 5 to 20 mm. long. ♃ —Shady banks and moist thickets, up to 2,000 m., southeastern Alaska to Wyoming, Utah, and northern California.

**16. Festuca élmeri** Scribn. and Merr. (Fig. 58.) Culms loosely tufted, slender, 40 to 100 cm. tall, or even taller; blades flat, scabrous or pubescent on upper surface, 2 to 4 mm. wide, those of the innovations narrower, more or less involute; panicle loose, open, 10 to 20 cm. long, the branches slender, somewhat drooping, naked below, the lower as much as 10 cm. long; spikelets 3- or 4-flowered; glumes lanceolate-acuminate, the first 2 to 2.5 mm., the second 3 to 4 mm. long; lemmas membranaceous, hispidulous, about 6 mm. long, the nerves rather prominent, the apex minutely 2-toothed; awn 2 to 8 mm. long. ♃ —Wooded hillsides, up to 500 m., mostly in the Coast Ranges, Oregon to central California. FESTUCA ELMERI var. CONFÉRTA (Hack.) Hitchc. More luxuriant; spikelets often 5- or 6-flowered and somewhat congested on the panicle branches. ♃ (*F. jonesii* var. *conferta* Hack.)—Coast Ranges of California.

FIGURE 58.—*Festuca elmeri*. Panicle, × ½; spikelet, × 5. (Type.)

**17. Festuca elátior** L. MEADOW FESCUE. (Fig. 59.) Culms 50 to 120 cm. tall; blades flat, 4 to 8 mm. wide, scabrous above; panicle erect, or nodding at summit, 10 to 20 cm. long, contracted after flowering, much-branched or nearly simple, the branches spikelet-bearing nearly to base; spikelets usually 6- to 8-flowered, 8 to 12 mm. long; glumes 3 and 4 mm. long, lanceolate; lemmas oblong-lanceolate, coriaceous, 5 to 7 mm. long, the scarious apex acutish, rarely short-awned. ♃ (*Festuca pratensis* Huds.)—Meadows, roadsides, and waste places; introduced throughout the cooler parts of North America; native of Eurasia. Cultivated for meadow and pasture. Sometimes called English bluegrass.

**Festuca gigantéa** (L.) Vill. Blades broad, flat, thin; panicles open; lemmas long-awned, the awn flexuous and 2 or 3 times as long as the lemma. ♃ —Dobbs Ferry, N. Y.; adventive from Europe.

**Festuca arundinácea** Schreb. REED FESCUE, ALTA FESCUE. Culms somewhat taller and more robust than in *F. elatior*, and without rhizomes; blades longer; panicles 15 to 32 cm. long with more numerous branches and spikelets, the spikelets broader, mostly looser, the lemmas 7 to 10 mm. long. ♃ (*F. elatior* var. *arundinacea* (Schreb.) Wimm.)—Roadsides and meadows; introduced from Europe, sparingly spontaneous, Maine, Massachusetts, New York, Ohio, Michigan, Utah, Washington to California. Recently rather widely cultivated in the Northern States, and also in Kentucky.

FIGURE 59.—*Festuca elatior*. Plant, × ½; spikelet and floret, × 5. (Amer. Gr. Natl. Herb. 488, D. C.)

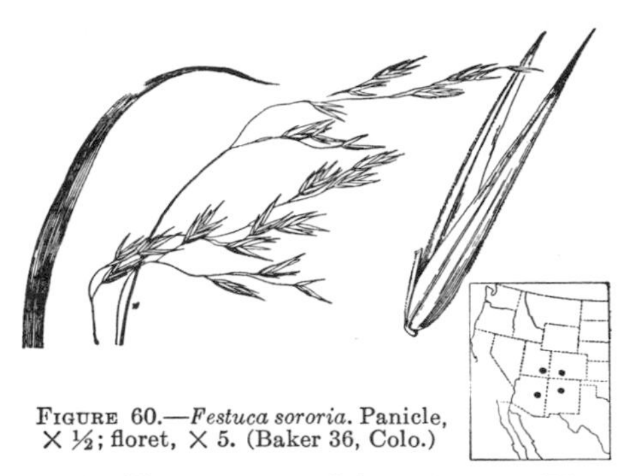

FIGURE 60.—*Festuca sororia*. Panicle, × ½; floret, × 5. (Baker 36, Colo.)

**18. Festuca sorória** Piper. (Fig. 60.) Culms erect, loosely tufted, 60 to 90 cm. tall; blades flat, thin, smooth except the scabrous margins, 3 to 6 mm. wide; panicle loose, open, nodding, or sometimes somewhat condensed, 10 to 15 cm. long, the branches solitary or in twos, naked below; spikelets rather loosely 3- to 5-flowered; glumes lanceolate, the first about 3 mm., the second about 5 mm. long; lemmas membranaceous, somewhat keeled, scaberulous or nearly smooth, the nerves evident but not prominent, the apex tapering into a fine point or an awn as much as 2 mm. long. ♃ —Open woods, 2,000 to 3,000 m., southern Colorado and Utah to New Mexico and Arizona.

FIGURE 61.—*Festuca versuta*. Panicle, × ½; spikelet, × 5. (Johnson, Tex.)

**19. Festuca versúta** Beal. (Fig. 61.) Culms slender, 50 to 100 cm. tall; blades flat, mostly 2 to 5 mm. wide; panicle open, 10 to 15 cm. long, the spreading lower branches bearing a few spikelets above the middle; spikelets 2- to 5-flowered; glumes narrow, acuminate, nearly equal, 5 to 6 mm. long; lemmas firm, obscurely nerved at maturity, 5 to 7 mm. long, acute, awnless, rarely awn-tipped. ♃ (*F. texana* Vasey; *F. johnsoni* Piper.)—Shady banks, Arkansas, Texas, and Oklahoma.

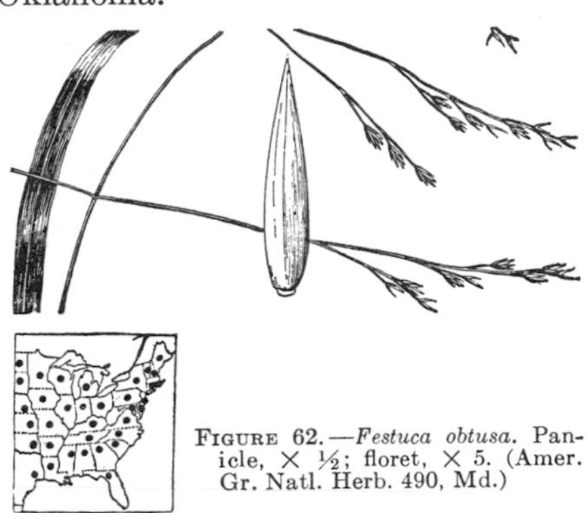

FIGURE 62.—*Festuca obtusa*. Panicle, × ½; floret, × 5. (Amer. Gr. Natl. Herb. 490, Md.)

**20. Festuca obtúsa** Bieler. NODDING FESCUE. (Fig. 62.) Culms solitary or few in a tuft, mostly 50 to 100 cm. tall; blades flat, lax, somewhat glossy, 4 to 7 mm. wide; panicle nodding, very loose and open, the branches spreading, spikelet-bearing toward the ends, the lower usually reflexed at maturity; spikelets 3- to 5-flowered; glumes about 3 and 4 mm. long; lemmas coriaceous, rather turgid, about 4 mm. long, obtuse or acutish, the nerves very obscure. ♃ —Low or rocky woods, Quebec to Manitoba, south to northern Florida and eastern Texas.

**21. Festuca paradóxa** Desv. (Fig. 63.) Culms few to several in a tuft, 50 to 110 cm. tall, widely leaning; blades flat or subinvolute in drying, lax, 4 to 8 mm. wide; panicle 12 to 20 cm. long, heavily drooping, the slender scabrous branches not so long as in *F. obtusa*, the brownish spikelets somewhat aggregate toward the ends; spikelets 3- to 6-flowered, the lemmas more blunt. ♃ (*F. shortii* Kunth) —Prairies, low open ground, and thickets, Pennsylvania and Delaware to South Carolina, Wisconsin, and eastern Texas.

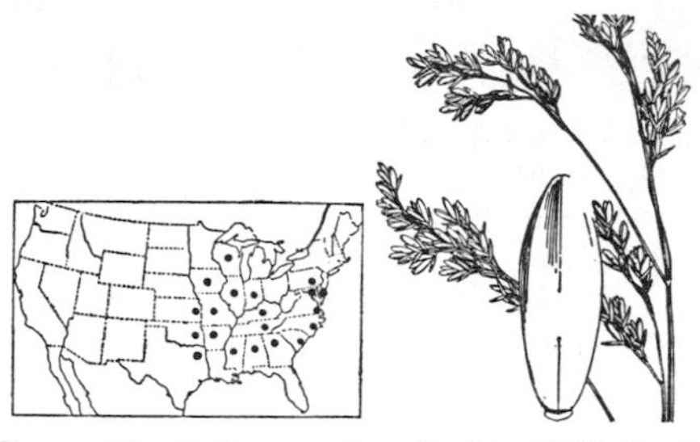

FIGURE 63.—*Festuca paradoxa.* Panicle, × ½; floret, × 5. (Palmer 34672, Mo.)

FIGURE 64.—*Festuca thurberi.* Panicle, × ½; spikelet, × 5. (Pammel, Colo.)

**22. Festuca thúrberi** Vasey. THURBER FESCUE. (Fig. 64.) Culms densely tufted, rather stout, erect, 60 to 90 cm. tall; ligule 2 to 4 mm. long; blades involute, scabrous, firm, erect; panicle 10 to 15 cm. long, the branches usually solitary, somewhat remote, ascending or spreading, naked below; spikelets 3- to 6-flowered; glumes rather broad, about 4 and 5 mm. long; lemmas rather firm, faintly nerved, glabrous or nearly so, acute or cuspidate, 7 to 8 mm. long. ♃ —Dry slopes and rocky hills, 2,500 to 3,500 m., Wyoming to New Mexico and Utah.

**23. Festuca liguláta** Swallen. (Fig. 65.) Culms slender, loosely tufted, erect from a decumbent often rhizomatous base, scabrous below the panicle; sheaths glabrous; blades 6 to 20 cm. long, those of the innovations as much as 30 cm. long, flat and 1 to 2 mm. wide or mostly involute, scabrous, rather firm; ligule 3 to 3.5 mm. long; panicle 6 to 10 cm. long, the 1 or 2 scabrous branches stiffly ascending or spreading, few-flowered, naked below; spikelets 6 mm. long, 2- to 3-flowered, the pedicels (mostly shorter than the spikelets) appressed; glumes acute or acutish, scabrous, the first 3 mm. long, 1-nerved, the second 4 mm. long, 3-nerved; lemmas 4 to 5 mm. long, acutish, scabrous, obscurely nerved, awnless, the paleas slightly longer. ♃ —Moist shady slopes, Guadalupe and Chisos Mountains, Tex.

FIGURE 65.—*Festuca ligulata.* Plant, × ½; floret, × 5. (Type.)

**24. Festuca scabrélla** Torr. ROUGH FESCUE. (Fig. 66.) Culms densely tufted (rarely producing a slender rhizome), erect, 30 to 90 cm. tall; ligule very short; blades firm, erect, scabrous, involute, or those of the culm sometimes flat but narrow; panicle narrow, 5 to 15 cm. long, the branches solitary or in pairs, the lowermost sometimes in threes, appressed or ascending, naked below;

spikelets 4- to 6-flowered; glumes somewhat unequal, lanceolate, 7 to 9 mm. long; lemmas firm, rather strongly nerved, scaberulous, acute to cuspidate or short-awned, 7 to 10 mm. long. ♃ (*F. hallii* Piper; *F. kingii* var. *rabiosa* (Piper) Hitchc.; *Hesperochloa kingii* var. *rabiosa* (Piper) Swallen.)—Prairies, hillsides, and open woods, up to about 2,000 m. (probably alpine in Colorado), Newfoundland to British Columbia, south to Oregon, North Dakota, and Colorado. FESTUCA SCABRELLA var. MÁJOR Vasey. Culms on the average taller; panicle larger and more spreading; lemmas more strongly nerved. ♃ (*F. campestris* Rydb.)—Hills and dry woods, Michigan (Roscommon), Montana to Washington.

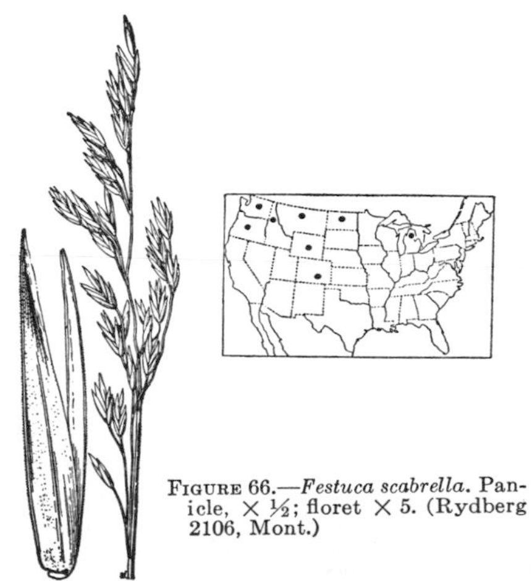

FIGURE 66.—*Festuca scabrella*. Panicle, × ½; floret × 5. (Rydberg 2106, Mont.)

**25. Festuca califórnica** Vasey. CALIFORNIA FESCUE. (Fig. 67.) Culms tufted, rather stout, 60 to 120 cm. tall; sheaths somewhat scabrous, the collar pubescent or pilose; blades firm, usually involute, sometimes flat, scabrous; panicle open, 10 to 30 cm. long, the rather remote branches usually in pairs, spreading or drooping, naked below; spikelets mostly 4- or 5-flowered; glumes somewhat unequal, 5 to 8 mm. long; lemmas firm, faintly nerved, scaberulous, acuminate or short-awned. ♃ (*F. aristulata* Shear.)—Open dry ground, thickets and open woods, up to about 1,500 m., Oregon and California, west of the Sierra Nevada. A smaller form with pubescent lower sheaths, and shorter, mostly glabrous blades, has been segregated as *F. californica* var. *paríshii* (Piper) Hitchc.—Oregon and California (San Bernardino Mountains).

**26. Festuca dasýclada** Hack. ex Beal. (Fig. 68.) Culms 20 to 40 cm. tall; blades folded, about 2 mm. wide when spread, those of the culm 4 to 6 cm. long, those of the innovations 10 to 15 cm. long; panicle open, 7 to 12 cm. long, the branches rather stiffly and divaricately spreading, softly pubescent; angles ciliate; spikelets pale, long-pediceled, 2-flowered; glumes lanceolate, acuminate, the first about 4 mm., the second about 6 mm. long; lemmas rather thin,

FIGURE 67.—*Festuca californica*. Panicle, × ½; floret, × 5. (Elmer 4431, Calif.)

somewhat keeled, rather strongly nerved, scaberulous, about 6 mm. long; awn about 2 mm. long, from between 2 minute teeth. ♃ —Rocky slopes, rare, Utah.

Festuca rigéscens (Presl) Kunth. Densely tufted, about 30 cm. tall; blades firm, involute, sharp-pointed; panicle narrow, few-flowered, 5 to 10 cm. long; spikelets about 3-flowered, 6 to 7 mm. long; lemmas ovate, thick, convex, awnless or mucronate, 4 to 4.5 mm. long. ♃ —There is a single specimen of this species in the United States National Herbarium, labeled "Arizona, Tracy?" On the sheet is a note made by Professor Piper (Feb. 12, 1904) quoting Tracy, "In open pine woods 4 miles southeast of Flagstaff, about June 20, 1887." This agrees exactly with specimens of this species from Peru, whence originally described. Since the species is not known north of Peru, except from this specimen, it seems probable that the label has been misplaced.

Figure 68.—*Festuca dasyclada*. Panicle, × ½; glumes and floret, × 5. (Dupl. type.)

**27. Festuca virídula** Vasey. Greenleaf fescue. (Fig. 69.) Culms rather loosely tufted, erect, 50 to 100 cm. tall; blades soft, erect, those of the culm flat or loosely involute, those of the innovations slender, involute; panicle open, 10 to 15 cm. long, the branches mostly in pairs, ascending or spreading, slender, somewhat remote, naked below; spikelets 3- to 6-flowered; glumes lanceolate, somewhat unequal, 5 to 7 mm. long; lemmas membranaceous, acute or cuspidate, glabrous, 6 to 8 mm. long. ♃ —Mountain meadows and open slopes, 1,000 to 2,000 m., British Columbia to Alberta, south to central California and Idaho; Colorado (Willow Pass). An important forage grass in the mountains of the Northwestern States. *Festuca howellii* Hack. ex Beal, differing from *F. viridula* in having more scabrous lemmas and awns 2 mm. long, does not seem sufficiently distinct to be recognized as a species. ♃ —Known from a single collection (Josephine County, Oreg.).

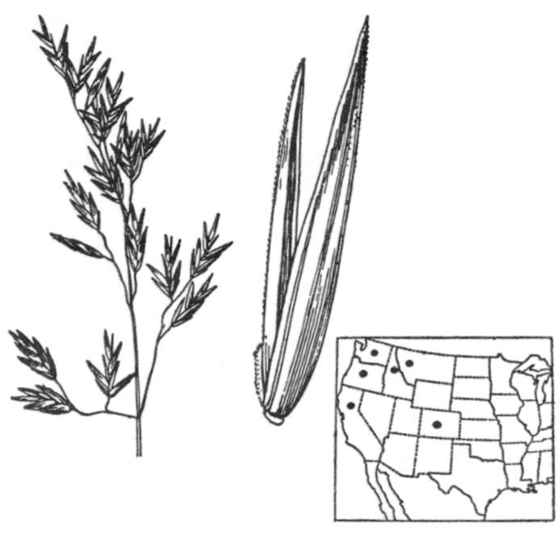

Figure 69.—*Festuca viridula*. Panicle, × ½; floret, × 5. (Cusick 2431, Oreg.)

**28. Festuca rúbra** L. Red fescue. (Fig. 70.) Culms usually loosely tufted, bent or decumbent at the reddish or purplish base, occasionally closely tufted, erect to ascending, 40 to 100 cm. tall; lower sheaths brown, thin, and fibrillose; blades smooth, soft, usually folded or involute; panicle 3 to 20 cm. long, usually contracted and narrow, the branches mostly erect or ascending; spikelets 4- to 6-flowered, pale green or glaucous, often purple-tinged; lemmas 5 to 7 mm. long, smooth, or scabrous toward apex, bearing an awn about half as long. ♃ —Meadows, hills, bogs, and marshes, in the cooler parts of the northern hemisphere, extending south in the Coast Ranges to Monterey, in the Sierra Nevada to the San Bernardino Mountains, in the Rocky Mountains to Colorado and New Mexico, San Francisco Mountains of Arizona; in the Alleghany Mountains and in the Atlantic coastal marshes to Georgia; Mexico,

FIGURE 70.—*Festuca rubra*. Plant, × ½; spikelet and floret, × 5. (Hitchcock 4201, Alaska.)

Eurasia, North Africa. Occasionally used in grass mixtures for pastures in the Northern States. Festuca rubra var. lanuginósa Mert. and Koch. Lemmas pubescent. ♃ —Oregon to Wyoming and northward; Michigan, Vermont to Connecticut; Europe. A proliferous form (*F. rubra* var. *prolifera* Piper, *F. prolifera* Fernald) is found in the White Mountains of New Hampshire, in Maine and northward. Festuca rubra var. commutáta Gaud. (*F. fallax* Thuill.). Chewings fescue. A form with more erect culms, producing a firmer sod, commonly cultivated in New Zealand and occasionally in the United States. ♃ —Festuca rubra var. heterophýlla (Lam.) Mut. Shade fescue. Densely tufted; basal blades filiform; culm blade flat. ♃ —Used for lawns in shady places. Europe.

**29. Festuca occidentális** Hook. Western fescue. (Fig. 71.) Culms tufted, erect, slender, 40 to 100 cm. tall; blades mostly basal, slender, involute, sulcate, soft, smooth or nearly so; panicle loose, 7 to 20 cm. long, often drooping above, the branches solitary or in pairs; spikelets loosely 3- to 5-flowered, 6 to 10 mm. long, mostly on slender pedicels; lemmas rather thin, 5 to 6 mm. long, scaberulous toward the apex, attenuate into a slender awn about as long or longer. ♃ —Dry rocky wooded slopes and banks, British Columbia to central California, east to Wyoming, northern Michigan, and western Ontario.

Figure 71.—*Festuca occidentalis*. Panicle, × ½; spikelet, × 5. (Piper 4908, Wash.)

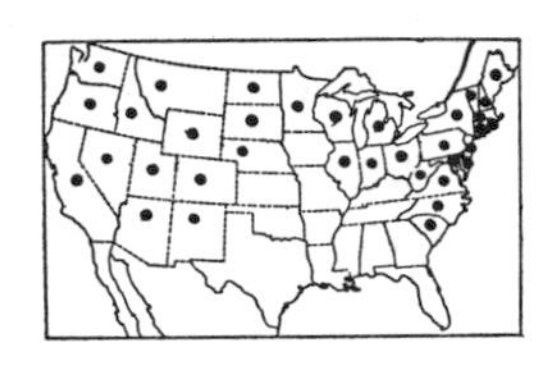

Figure 72.—*Festuca ovina*. Panicle, × ½; floret, × 5. (Robbins 8692, Colo.)

**30. Festuca ovína** L. Sheep fescue. (Fig. 72.) Culms densely tufted, usually 20 to 40 cm. tall; blades slender, involute, from very scabrous to glabrous, the innovations numerous in a basal cluster, 5 to 10 cm. long or sometimes longer; panicle narrow, sometimes almost spikelike, 5 to 8 cm. long, sometimes longer; spikelets mostly 4- or 5-flowered; lemmas about 4 to 5 mm. long, short-awned. ♃ (*F. saximontana* Rydb.; *F. calligera* Rydb.; *F. minutiflora* Rydb., a rare form with small florets; *F. ovina* var. *pseudovina* Hack. of Piper's revision of *Festuca*.)—Open woods and stony slopes, North Dakota to Washington and Alaska, south to Arizona and New Mexico; introduced eastward through Michigan, Maine, Illinois, and South Carolina; Eurasia. *Festuca ovina*, *F. ovina* var. *duriuscula*, and *F. capillata* are occasionally cultivated in lawn mixtures.

FESTUCA OVINA var. DURIÚSCULA (L.) Koch. HARD FESCUE. Blades smooth, wider and firmer than in *F. ovina.* ♃ —Maine to Iowa and Virginia; introduced from Europe.

FESTUCA OVINA var. BRACHYPHÝLLA (Schult.) Piper. ALPINE FESCUE. An alpine and high northern form differing in the lower culms, mostly 5 to 20 cm. tall, and the smooth short rather lax blades. ♃ (*F. brachyphylla* Schult.; *F. ovina* var. *supina* Hack. of Piper's revision of *Festuca.*)—Rocky slopes, at high altitudes, mostly above timber line in the United States, arctic regions south to San Bernardino Mountains, San Francisco Mountains, California, and, in the Rocky Mountains, to northern New Mexico; also in the high mountains of Vermont, New Hampshire, and New York.

FESTUCA OVINA var. GLAÚCA (Lam.) Koch. BLUE FESCUE. Blades elongate, glaucous. ♃ (*F. glauca* Lam.)—Cultivated as a border plant.

**31. Festuca capilláta** Lam. HAIR FESCUE. (Fig. 73.) Densely tufted, more slender and lower than *F. ovina;* blades capillary, flexuous, usually more than half as long as the culm; spikelets smaller; lemmas about 3 mm. long, awnless. ♃ —Lawns and waste places, Newfoundland and Maine to North Carolina and Illinois; Minnesota; Oregon; introduced from Europe.

FIGURE 73.—*Festuca capillata.* Plant, × ¼; floret, × 5. (Hitchcock 23624, Newf.)

**32. Festuca idahoénsis** Elmer. IDAHO FESCUE. (Fig. 74.) Culms usually densely tufted in large bunches, 30 to 100 cm. tall; blades numerous, usually elongate, very scabrous, rarely smooth, filiform, involute; panicle narrow, 10 to 20 cm. long, the branches ascending or appressed, somewhat spreading in anthesis; spikelets mostly 5- to 7-flowered; lemmas nearly terete, about 7 mm. long; awn usually 2 to 4 mm. long. ♃ (*F. ovina* var. *ingrata* Beal.)—Open woods and rocky slopes, British Columbia to Alberta, south to central California and Colorado.

FIGURE 74.—*Festuca idahoensis.* Plant, × ½; floret, × 5. (Heller 3318, Idaho.)

**33. Festuca arizónica** Vasey. ARIZONA FESCUE. (Fig. 75.) Resembling *F. idahoensis;* differing in the stiffer glaucous foliage, somewhat smaller awnless or nearly awnless lemmas. ♃ —Open pine woods, Nevada and Colorado to Texas and Arizona. Often called pinegrass.

FIGURE 75.—*Festuca arizonica.* Panicle, × ½; floret, × 5. (Leiberg 5685, Ariz.)

FESTUCA AMETHÝSTINA L. Slender tufted perennial; blades filiform, 15 to 25 cm. long; panicle 5 to 10 cm. long, rather narrow; spikelets about as in *F. ovina*, often purplish. ♃ —Sometimes cultivated for ornament. Europe.

FESTUCA GENICULÁTA (L.) Cav. Annual; culms slender, geniculate below, 20 to 50 cm. tall; panicle 3 to 6 cm. long, rather compact; spikelets awned. ⊙ —Sometimes cultivated for ornament. Portugal.

FESTUCA VALESIACA Schleich. ex Gaud. Slender densely tufted perennial, 15 to 30 cm. tall; blades very slender, sulcate, scabrous, those of the innovations numerous, 10 to 18 cm. long; panicle 4 to 8 cm. long, narrow, the short branches ascending; spikelets similar to those of *F. ovina*, to which this species is closely related. ♃ —Sometimes cultivated in grass gardens. Europe.

## 5. SCLERÓPOA Griseb.

Spikelets several-flowered, linear, somewhat compressed, the thick rachilla disarticulating above the glumes and between the florets, remaining as a minute stipe to the floret above; glumes unequal, short, acutish, strongly nerved, the first 1-nerved, the second 3-nerved; lemmas nearly terete, obscurely 5-nerved, obtuse, slightly scarious at the tip. Annuals with slightly branched 1-sided panicles. Type species, *Scleropoa rigida*. Name from Greek *skleros*, hard, and *poa*, grass, alluding to the stiff panicle.

**1. Scleropoa rígida** (L.) Griseb. (Fig. 76.) Culms erect or spreading, 10 to 20 cm. tall; blades flat, 1 to 2 mm. wide; panicles narrow, stiff, condensed, 5 to 10 cm. long, the branches short, floriferous to base, these and the thick pedicels somewhat divaricately spreading in anthesis; spikelets 4- to 10-flowered, 5 to 8 mm. long; glumes about 2 mm. long; lemmas about 2.5 mm. long. ⊙ —Waste places and fields, sparingly introduced from Europe, Massachusetts; Florida to Mississippi; Texas; South Dakota; Washington to California.

FIGURE 76.—*Scleropoa rigida*. Plant, × 1; two views of floret, × 10. (Cocks, Miss.)

## 6. PUCCINÉLLIA Parl. ALKALI-GRASS

Spikelets several-flowered, usually terete or subterete, the rachilla disarticulating above the glumes and between the florets; glumes unequal, shorter than the first lemma, obtuse or acute, rather firm, often scarious at tip, the first 1-nerved or sometimes 3-nerved, the second 3-nerved; lemmas usually firm, rounded on the back, obtuse or acute, rarely acuminate, usually scarious and often erose at the tip, glabrous or puberulent toward base, rarely pubescent on the nerves, 5-nerved, the nerves parallel, indistinct, rarely rather prominent; palea about as long as the lemma or somewhat shorter. Low pale smooth tufted annuals or perennials with narrow to open panicles. Type species, *Puccinellia distans*. Named for Prof. Benedetto Puccinelli.

The species of the interior are grazed by stock. One, *P. airoides*, furnishes considerable forage in the regions where it is common. A form of this, called Zawadke alkali-grass, is cultivated in Montana.

Lemmas obtuse, pubescent on the nerves for half or three-fourths their length. Dwarf annual .......... 1. P. PARISHII.
Lemmas glabrous or, if pubescent, the hairs not confined to the nerves.
  Panicles narrow, strict, the branches appressed, mostly with one spikelet; annual, mostly less than 20 cm. tall; lemmas acute, more or less pubescent .......... 2. P. SIMPLEX.
  Panicles narrow or open, not strict; annual or perennial; lemmas glabrous or pubescent only at base.
    Panicles ellipsoid, rather compact, less than 10 cm. long, the branches floriferous nearly to base. Lemmas rather coriaceous; culms rather stout.
      Spikelets 5 to 8 mm. long; lemmas 3 to 3.5 mm. long .......... 3. P. RUPESTRIS.
      Spikelets 3 to 4 mm. long; lemmas 2 to 2.5 mm. long .......... 4. P. FASCICULATA.
    Panicles pyramidal or elongate, some of the branches naked below, or reduced, narrow, and few-flowered.
      Leaves mostly in a short basal tuft, the blades involute, 5 to 10 cm. long. Panicle 5 to 10 cm. long, open and spreading; lemmas 3.5 mm. long, glabrous, acute. 5. P. LEMMONI.
      Leaves distributed, not in a basal tuft.
        Anthers about 2 mm. long; lemmas 4 to 5 mm. long, pubescent at base. 6. P. MARITIMA.
        Anthers 1 mm. long or less.
          Lemmas about 2 mm. long (2 to 3 mm. in *P. airoides*); panicle open; the slender branches spreading or reflexed.
            Lemmas broad, obtuse or truncate, not narrowed above; lower panicle branches usually reflexed .......... 7. P. DISTANS.
            Lemmas narrow, narrowed into an obtuse apex; panicle branches spreading, usually not reflexed .......... 8. P. AIROIDES.
          Lemmas 3 to 4 mm. long; panicle narrow, the branches ascending or finally spreading.
            Plants lax, usually 10 to 30 cm. tall; panicle 5 to 10 cm. long, the branches finally spreading, glabrous .......... 9. P. PUMILA.
            Plants usually 50 to 90 cm. tall; panicle 10 to 20 cm. long, the branches ascending or appressed, scabrous .......... 10. P. GRANDIS.

**1. Puccinellia paríshii** Hitchc. (Fig. 77.) Annual; culms 3 to 10 cm. tall; blades flat to subinvolute, less than 1 mm. wide; panicle narrow, few-flowered, 1 to 4 cm. long; spikelets 3- to 6-flowered, 3 to 5 mm. long; lemmas about 2 mm. long, obtuse to truncate, scarious and somewhat erose at the tip, pubescent on the mid and lateral nerves nearly to the apex, and on the intermediate nerves about half way. ⊙ —Marshes, California

FIGURE 77.—*Puccinellia parishii*. Panicle, × 1; floret, × 10. (Type.)

(Rabbit Springs, San Bernardino County) and Arizona (Tuba City).

**2. Puccinellia símplex** Scribn. (Fig. 78.) Annual; culms 7 to 20 cm. tall; blades narrow, soft, flat; panicle narrow, about half the length of the entire plant, the branches few, short, appressed, mostly with 1 spikelet; spikelets 6 to 8 mm. long, appressed; glumes strongly 3-nerved, 1 and 2 mm. long; lemmas 2.5 mm. long, tapering from below the middle to the acute apex, more or less pubescent over the back. ⊙ —Alkaline soil, California; common in alkaline areas of the San Joaquin Valley.

FIGURE 78.—*Puccinellia simplex.* Plant, × 1; floret × 10. (Type.)

**3. Puccinellia rupéstris** (With.) Fern. and Weath. (Fig. 79.) Annual; culms rather stout, mostly 10 to 20 cm. tall; blades flat, 2 to 6 mm. wide; panicle ellipsoid, glaucous, rather dense, mostly 3 to 6 cm. long, the branches mostly not more than 1.5 cm. long, stiffly ascending, floriferous nearly to base; spikelets 3- to 5-flowered, 5 to 8 mm. long, sessile or nearly so; glumes 3- to 5-nerved, 1.5 and 2.5 mm. long; lemmas 3 to 3.5 mm. long, firm, obscurely nerved, glabrous, obtuse, the apex entire or nearly so. ⊙ —Ballast near New York and Philadelphia. Europe.

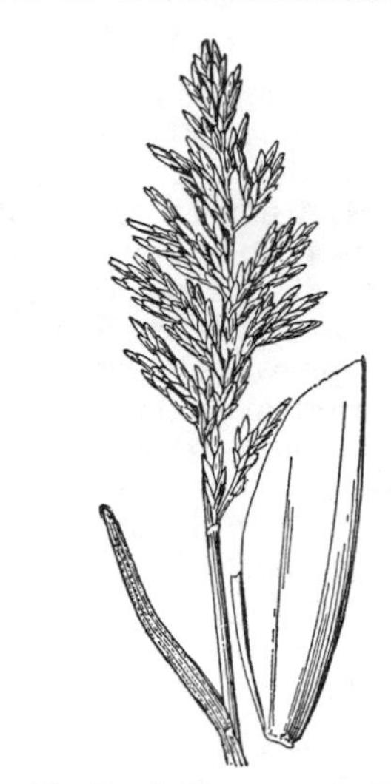

FIGURE 79.—*Puccinellia rupestris.* Panicle, × 1; floret, × 10. (Martindale, N. J.)

FIGURE 80.—*Puccinellia fasciculata.* Panicle, × 1; floret, × 10. (Stebbins, Maine.)

**4. Puccinellia fasciculáta** (Torr.) Bicknell. (Fig. 80.) Apparently perennial; culms rather stout, 20 to 50 cm. tall, sometimes taller; blades flat, folded, or subinvolute, 2 to 4 mm. wide; panicle ellipsoid, 5 to 15 cm. long, the branches fascicled, rather stiffly ascending, some naked at base but with short basal branchlets, all rather densely flowered; spikelets 2- to 5-flowered, 3 to 4 mm. long; glumes ovate, 1 and 1.5 mm. long; lemmas 2 to 2.5 mm. long, firm, obtuse. ♃ (*P. borreri* Hitchc.)—Salt marshes along the coast, Nova Scotia to Virginia; Utah, Nevada and Arizona; Europe.

FIGURE 82.—*Puccinellia maritima*. Plant, × 1; floret, × 10. (Fernald and Long 20051, Nova Scotia.)

**5. Puccinellia lemmóni** (Vasey) Scribn. (Fig. 81.) Perennial; culms erect, slender, 15 to 30 cm. tall; leaves mostly in an erect basal tuft, the slender blades involute, 5 to 10 cm. long; panicle pyramidal, open, 5 to 10 cm. long, the slender flexuous branches fascicled, the lower spreading, the longer ones naked on the lower half; spikelets narrow, 3- to 5-flowered, the rachilla often exposed; glumes about 1 and 2 mm. long; lemmas narrow, acute, glabrous, about 3.5 mm. long; anthers 1.5 mm. long. ♃ —Moist alkaline soil, southern Idaho and Washington to Nevada and California.

FIGURE 81.—*Puccinellia lemmoni*. Panicle, × 1; floret, × 10. (Jones 4115, Nev.)

**6. Puccinellia marítima** (Huds.) Parl. (Fig. 82.) Perennial; culms erect, rather coarse, 20 to 40 cm. tall, sometimes taller; blades 1 to 2 mm. wide, usually becoming involute; panicle mostly 10 to 20 cm. long, the branches ascending or appressed, or spreading in anthesis; spikelets 4- to 10-flowered; glumes 3-nerved, 2 to 3 and 3 to 4 mm. long; lemmas 4 to 5 mm. long, pubescent on the base of the lateral nerves and sometimes sparingly between the nerves; anthers 1.5 to 2 mm. long. ♃ —Salt marshes and brackish shores, Nova Scotia to Rhode Island; Washington; on ballast, Philadelphia and Camden; Europe.

**7. Puccinellia dístans** (L.) Parl. (Fig. 83.) Perennial; culms erect or decumbent at base, 20 to 40 cm. tall, sometimes taller; blades flat or more or less involute, mostly 2 to 4 mm. wide; panicle pyramidal, loose, 5 to 15 cm. long, the branches fascicled, rather distant, the lower spreading or finally reflexed, the longer ones naked half their length or more; spikelets 4- to 6-flowered, 4 to 5 mm. long; glumes 1 and 2 mm. long; lemmas rather thin, obtuse or truncate, 1.5 or usually about 2 mm. long, with a few short hairs at base; anthers about 0.8 mm. long. ♃ —Moist, more or less alkaline soil, Quebec to British Columbia, south to Maryland, Michigan, Wisconsin, and North Dakota; Washington, south to New Mexico and California; introduced from Eurasia. The more slender specimens are the form described as *P. distans* var. *tenuis* (Uechtritz) Fern. and Weath.

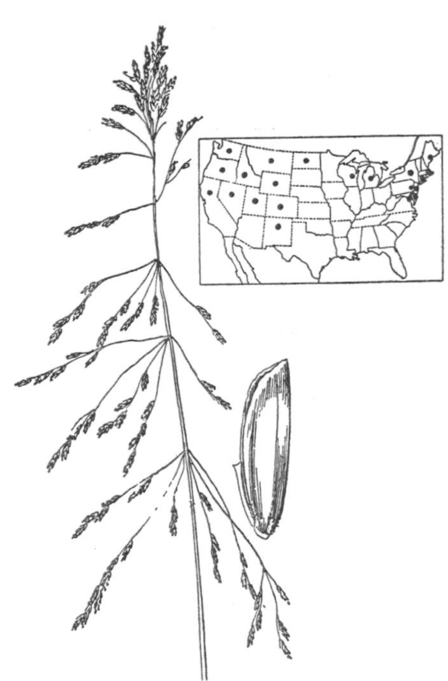

FIGURE 83.—*Puccinellia distans*. Panicle, × ½; floret, × 10. (Schuette, Wis.)

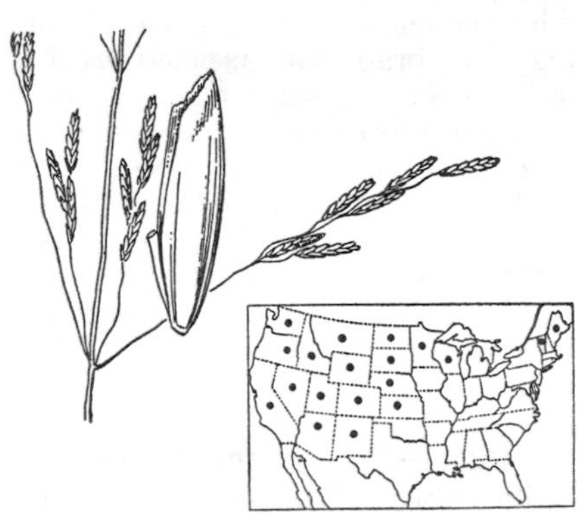

FIGURE 84.—*Puccinellia airoides*. Panicle, × 1; floret, × 10. (Rydberg 2135, Mont.)

**8. Puccinellia airoídes** (Nutt.) Wats. and Coult. NUTTALL ALKALI-GRASS. (Fig. 84.) Perennial; culms usually erect, slender, rather stiff and firm at base, mostly 30 to 60 cm. rarely to 1 m. tall; blades 1 to 3 mm. wide, flat, or becoming involute; panicle pyramidal, open, mostly 10 to 20 cm. long, the distant scabrous branches fascicled, spreading, naked below, as much as 10 cm. long; spikelets 3- to 6-flowered, 4 to 7 mm. long, the florets rather distant, the rachilla often exposed; pedicels scabrous; glumes 1.5 to 2 mm. long; lemmas 2 to 3 mm. long, rather narrow, somewhat narrowed into an obtuse apex; anthers about 0.7 mm. long. ♃ (*P. nuttalliana* Hitchc.) —Moist, usually alkaline soil, Wisconsin to British Columbia, south to Kansas, New Mexico, and California; introduced in Maine and Vermont. The form with lemmas 2.5 to 3 mm. long has been called *P. cusickii* Weatherby. Alberta to Wyoming and Oregon.

**9. Puccinellia púmila** (Vasey) Hitchc. (Fig. 85.) Perennial; culms lax, erect or ascending from a de-

FIGURE 85.—*Puccinellia pumila*. Plant, × 1; floret, × 10. (Type.)

cumbent base, 10 to 30 cm. tall; blades rather soft, mostly flat, 1 to 2 mm. wide; panicle pyramidal, open, mostly 5 to 10 cm. long, the lower branches naked below, usually finally spreading or even reflexed; spikelets 4- to 6-flowered; glumes 1.5 and 2.5 mm. long; lemmas 3 to 4 mm. long, rather broad, narrowed toward the obtuse nearly entire apex, obscurely pubescent near base or glabrous; anthers 0.8 to 1 mm. long. ♃ —Salt marshes and shores, Labrador to Connecticut; Alaska to Oregon.

**10. Puccinellia grándis** Swallen. (Fig. 86.) Culms densely tufted, 50 to 90 cm. tall; sheaths glabrous; ligule 2 to 3 mm. long; blades firm, drying involute, 2 to 3.5 mm. wide, panicles 10 to 20 cm. long, pyramidal, the scabrous branches finally spreading; spikelets 8 to 15 mm. long, 5- to 12-flowered, appressed; lemmas 3 to 4 mm. long, obtuse or subacute, sparsely pilose at the base; anthers 1.3 to 1.5 mm. long. ♃ —Sea beaches, Alaska to central California. This species has been referred to *P. nutkaensis* (Presl) Fern. and Weath., a northern species, not known from the United States.

FIGURE 86.—*Puccinellia grandis*. Panicle, × 1; floret, × 10. (Macoun 66, Br. Col.)

## 7. GLYCÉRIA R. Br. MANNAGRASS

(*Panicularia* Heist.)

Spikelets few- to many-flowered, subterete or slightly compressed, the rachilla disarticulating above the glumes and between the florets; glumes unequal, short, obtuse or acute, usually scarious, mostly 1-nerved (the second 3-nerved in a few species); lemmas broad, convex on the back, firm, usually obtuse, scarious at the apex, 5- to 9-nerved, the nerves parallel, usually prominent. Usually tall aquatic or marsh perennials, with creeping and rooting bases or with creeping rhizomes, simple culms, mostly closed or partly closed sheaths, flat blades, and open or contracted panicles. Type species, *Glyceria fluitans.* Name from the Greek *glukeros,* sweet, the seed of the type species being sweet.

The species are all palatable grasses but are usually of limited distribution, and most of them are confined to marshes or wet land. *Glyceria elata,* tall mannagrass, is a valuable component of the forage in moist woods of the Northwestern States. *G. striata,* fowl mannagrass, widely distributed, *G. grandis,* American mannagrass, in the Northern States, and *G. pauciflora* of the Northwest are marsh species, but are often grazed.

Spikelets linear, nearly terete, usually 1 cm. long or more, appressed on short pedicels; panicles narrow, erect ........ SECTION 1. EUGLYCERIA.
Spikelets ovate or oblong, more or less compressed, usually not more than 5 mm. long; panicles usually nodding ........ SECTION 2. HYDROPOA.

*Section 1. Euglyceria*

Lemmas acute, much exceeded by the palea ........ 1. G. ACUTIFLORA.
Lemmas obtuse; palea about as long as the lemma (or slightly longer in *G. septentrionalis* and *G. fluitans*).
  Lemmas glabrous between the slightly scabrous nerves ........ 2. G. BOREALIS.
  Lemmas scaberulous or hirtellous between the usually distinctly scabrous nerves.
    Lemmas about 3 mm. long, broadly rounded at the summit.
      First glume 1.5 mm. long; lemmas scaberulous ........ 3. G. LEPTOSTACHYA.
      First glume 2 to 2.5 mm. long; lemmas hirtellous ........ 4. G. ARKANSANA.
    Lemmas 4 to 7 mm. long.
      Culms more than 60 cm., commonly more than 1 m. tall, flaccid; sheaths closed from below the summit, blades elongate, mostly more than 5 mm. wide.
        Lemmas pale or green, not tinged with purple, about 4 mm. long; palea usually exceeding the lemma; Eastern States ........ 5. G. SEPTENTRIONALIS.
        Lemmas slightly tinged with purple near the tip, 5 to 6 mm. long; palea about as long as the lemma, sometimes slightly exceeding it; Northeastern States. 6. G. FLUITANS.
        Lemmas usually tinged with purple near the tip, 4 to 6 mm. long; palea rarely exceeding the lemma; Western States ........ 7. G. OCCIDENTALIS.
      Culms 15 to 30 cm. tall, slender but rather firm; sheaths open, the margins overlapping; blades with boat-shaped tip, 3 to 5 cm. long, 2 to 3 mm. wide. 8. G. DECLINATA.

*Section 2. Hydropoa*

Lemmas with 7 usually prominent nerves; second glume 1-nerved; sheaths, at least the upper, closed from below the summit.
  Panicle contracted, narrow.
    Lemmas 3 to 4 mm. long; panicle oblong, dense, usually not more than 10 cm. long. 11. G. OBTUSA.
    Lemmas 2 to 2.5 mm. long; panicle rather loose, nodding, 15 to 25 cm. long. 12. G. MELICARIA.
  Panicle open, lax.
    Nerves of lemma evident but not prominent ........ 13. G. CANADENSIS.
    Nerves of lemma prominent.
      First glume not more than 1 mm. long.
        Blades 2 to 4 mm. wide, sometimes to 8 mm., rather firm, often folded; first glume 0.5 mm. long ........ 14. G. STRIATA.
        Blades 6 to 12 mm. wide, flat, thin, lax; first glume about 1 mm. long. 15. G. ELATA.
      First glume more than 1 mm. long, usually about 1.5 mm. long.
        Glumes subequal, blunt, pale, in striking contrast to the purple florets. 9. G. GRANDIS.

Glumes narrow, acute, the second longer than the first; florets olive green. 10. G. NUBIGENA.
Lemmas with 5 prominent nerves; second glume 3-nerved; sheaths open.
Panicle narrow, the branches ascending ........ 16. G. ERECTA.
Panicle open, lax.
Culms relatively thick, commonly 1 m. tall; blades mostly 8 to 12 mm. wide.
Panicle branches numerous, many-flowered ........ 17. G. PAUCIFLORA.
Panicle branches few, distant, few-flowered ........ 18. G. OTISII.
Culms slender, decumbent, weak.
Blades 4 to 8 mm. wide; anthers 1 mm. long ........ 19. G. PALLIDA.
Blades 1 to 3 mm. wide; anthers 0.2 to 0.5 mm. long ........ 20. G. FERNALDII.

SECTION 1. EUGLYCÉRIA Griseb.

Spikelets linear, nearly terete, usually more than 1 cm. long, appressed on short pedicels; panicles narrow, erect, the branches appressed or ascending after anthesis. The species of Euglyceria, with the exception of *Glyceria acutiflora*, are very closely allied and appear to intergrade.

**1. Glyceria acutiflóra** Torr. (Fig. 87.) Culms compressed, lax, creeping and rooting below, 50 to 100 cm. long; blades flat, lax, 10 to 15 cm. long, 3 to 6 mm. wide, scabrous on the upper surface; panicle 15 to 35 cm. long, often partly included, the branches rather stiff, bearing 1 or 2 spikelets, or the lower 3 or more; spikelets 5- to 12-flowered, 2 to 4 cm. long, 1 to 2 mm. wide, the lateral pedicels 1 to 3 mm. long; glumes about 2 and 5 mm. long; lemmas 7-nerved, acute, scabrous, 6 to 8 mm. long, exceeded by the acuminate, 2-toothed paleas. ♃ —Wet soil and shallow water, New Hampshire to Virginia and West Virginia, west to Michigan, Missouri, and Tennessee; also northeastern Asia.

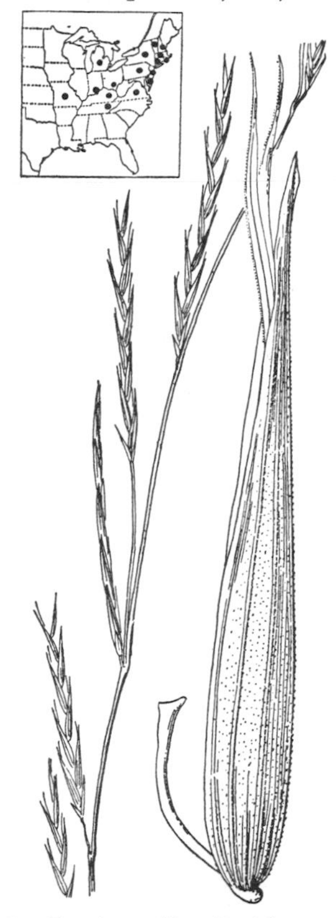

FIGURE 87.—*Glyceria acutiflora*. Panicle, × 1; floret, × 10. (Knowlton 866, Mass.)

**2. Glyceria boreális** (Nash) Batchelder. NORTHERN MANNAGRASS. (Fig. 88.) Culms erect or decumbent at base, slender, 60 to 100 cm. tall, blades flat or folded, usually 2 to 4 mm. wide, sometimes wider; panicle mostly 20 to 40 cm. long, the branches as much as 10 cm. long, bearing several appressed spikelets; spikelets mostly 6- to 12-flowered, 1 to 1.5 cm. long; glumes about 1.5 and 3 mm. long; lemmas rather thin, obtuse, 3 to 4 mm. long, strongly 7-nerved, scarious at the tip, glabrous between the hispidulous nerves. ♃ —Wet places and shallow water, Newfoundland to southeastern Alaska, Pennsylvania to Illinois, Minnesota, and Washington, and in the mountains to New Mexico, Arizona, and central California.

**3. Glyceria leptostáchya** Buckl. (Fig. 89.) Culms 1 to 1.5 m. tall, rather stout or succulent; sheaths slightly rough; blades flat, scaberulous

FIGURE 88.—*Glyceria borealis.* Panicle, × 1; floret, × 10. (Fernald 193, Maine.)

FIGURE 89.—*Glyceria leptostachya.* Panicle, × 1; floret, × 10. (Heller 5606, Calif.)

on the upper surface, 4 to 7 mm. rarely to 1 cm. wide; panicle 20 to 40 cm. long, the branches ascending, mostly in twos or threes, several-flowered, often bearing secondary branchlets; spikelets 1 to 2 cm. long, 8- to 14-flowered, often purplish; glumes 1.5 and 3 mm. long; lemmas firm, broadly rounded toward apex, about 3 mm. long, 7-nerved, scaberulous on the nerves and between them. ♃ (*Panicularia davyi* Merr.)—Shallow water, up to 1,200 m., rare, Washington to central California.

**4. Glyceria arkansána** Fernald. (Fig. 90.) Resembling *G. septentrionalis;* first glume 2 to 2.5 mm. long; lemmas 3 to 3.5 mm. long, hirtellous rather than scaberulous. ♃ —Wet ground, Louisiana and Arkansas.[9]

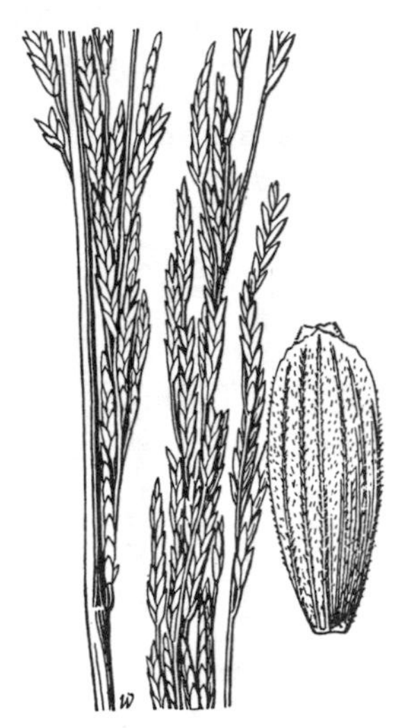

FIGURE 90.—*Glyceria arkansana.* Panicle, × 1; floret, × 10. (Ball 362, La.)

**5. Glyceria septentrionális** Hitchc. EASTERN MANNAGRASS. (Fig. 91.) Culms 1 to 1.5 m. tall, somewhat succulent; sheaths smooth; blades flat, mostly 10 to 20 cm. long, 4 to 8 mm. wide, usually smooth beneath, slightly scaberulous on the upper surface and margin; panicle 20 to 40 cm. long, somewhat open, the branches as much as 10 cm. long, several-flowered, often spreading at anthesis; spikelets 1 to 2 cm. long, 6- to 12-

[9] A specimen labeled "Western part of New-York," 1840, may have a misplaced label.

flowered, the florets rather loosely imbricate; glumes 2 to 3 and 3 to 4 mm. long; lemmas green or pale, about 4 mm. long, narrowed only slightly at the summit, scaberulous, the paleas usually exceeding them. ♃ —Shallow water and wet places, Quebec to Minnesota, south to Georgia and eastern Texas.

FIGURE 91.—*Glyceria septentrionalis*. Panicle, × 1; floret, × 10. (Deam 3184, Ind.)

**6. Glyceria flúitans** (L.) R. Br. MANNAGRASS. (Fig. 92.) Resembling *G. septentrionalis* in habit; first glume usually only one-third as long as the first lemma; lemmas scaberulous, the nerves distinct but not raised prominently above the tissue of the internerves; tip of palea usually exceeding its lemma. ♃ (*Panicularia brachyphylla* Nash.)—Shallow water, Newfoundland to Quebec and New York; South Dakota; Eurasia.

**7. Glyceria occidentális** (Piper) J. C. Nels. (Fig. 93.) Culms flaccid, 60 to 100 cm. tall; blades 3 to 12 mm. wide, smooth beneath, somewhat scabrous on the upper surface; panicle loose, spreading at anthesis, 30 to 50 cm. long; spikelets, 1.5 to 2 cm. long; first glume mostly about 2 mm. long; lemmas usually tinged with purple near the tip, 4 to 6 mm. long, rather strongly scabrous, 7- to 9-nerved, the nerves prominent, raised above the tissue of the internerves; palea about as long as its lemma, sometimes slightly exceeding it. ♃ —Marshes, shallow water, and wet places, Idaho to British Columbia,

FIGURE 92.—*Glyceria fluitans*. Panicle, × 1; floret, × 10. (McIntosh 1076, S. Dak.)

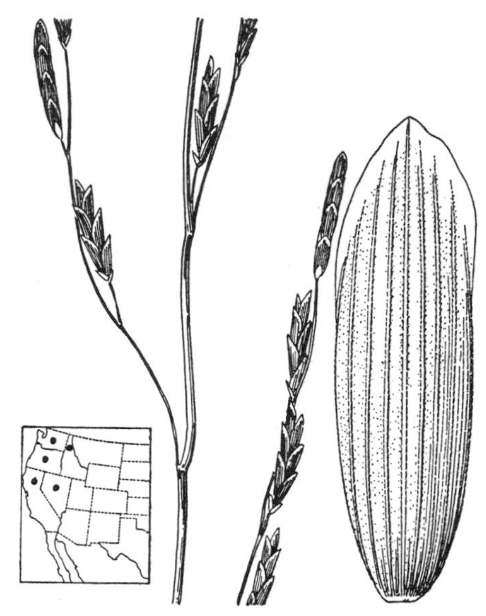

FIGURE 93.—*Glyceria occidentalis.* Panicle, × 1; floret, × 10. (Type.)

northern California and Nevada. The seeds are used for food by the Indians.

8. **Glyceria declináta** Brébiss. (Fig. 94.) Culms 15 to 70 cm. tall, erect from a decumbent branching base; sheaths open, keeled, scaberulous, the margins thin and hyaline; ligule 5 to 7 mm. long; blades 3 to 12 cm. long, 2 to 6 mm. wide, the tip boat-shaped; panicle simple, 6 to 25 cm. long; spikelets 15 to 20 mm. long, appressed; glumes obtuse, the first 1.8 to 2.2 mm. long, the second 3 to 3.5 mm. long; lemma 4 to 5 mm. long, scabrous, 7-nerved, obtuse, irregularly dentate; palea about as long as the lemma, the keels narrowly winged. ♃ —Moist canyons and meadows, Nevada and California; New York (Long Island); Europe, whence probably introduced.

SECTION 2. HYDROPÓA Dum.

Spikelets more or less laterally compressed, ovate to oblong, usually not more than 5 mm. long; panicles open or condensed, but not long and narrow (except in *G. melicaria*).

9. **Glyceria grándis** S. Wats. AMERICAN MANNAGRASS. (Fig. 95.) Culms tufted, stout, 1 to 1.5 m. tall; blades flat, 6 to 12 mm. wide; panicle large, very compound, 20 to 40 cm. long, open, nodding at summit; spikelets 4- to 7-flowered, 5 to 6 mm. long, glumes whitish, about 1.5 and 2 mm. long; lemmas purplish, about 2.5 mm. long; palea rather thin, about as long as the lemma. ♃ (*Panicularia americana* MacM.)—Banks of streams, marshes, and wet places, Prince Edward Island to Alaska, south to Virginia, Tennessee, Iowa, Nebraska, New Mexico, Arizona, and eastern Oregon.

FIGURE 94.—*Glyceria declinata.* Plant, × 1; floret, × 10. (Cooke 15312, Calif.)

FIGURE 95.—*Glyceria grandis*. Panicle, × 1; floret, × 10. (Pearce, N. Y.)

**10. Glyceria nubígena** W. A. Anderson. (Fig. 96.) Culms 1 to 2 m. tall, slender to rather stout, smooth, shining; sheaths glabrous or scaberulous, the lower much longer than the internodes; ligule truncate, 1 mm. long; blades as much as 45 cm. long, 6 to 10 mm. wide, smooth below, scabrous above; panicles 20 to 30 cm. long, the branches stiffly spreading or reflexed; spikelets 3- to 4-flowered, the florets early deciduous; lemmas about 2.5 mm. long, obtuse or subacute. ♃ —Moist ground, balds and high ridges, Great Smoky Mountains, Tennessee and North Carolina.

**11. Glyceria obtúsa** (Muhl.) Trin. (Fig. 97.) Culms erect, often decumbent at base, 50 to 100 cm. tall, rather firm; blades elongate, erect, mostly smooth, flat or folded, 2 to 6 mm. wide; panicle erect, oblong or narrowly elliptic, dense, 5 to 15 cm. long, the branches ascending or appressed; spikelets mostly 4- to 7-flowered, 4 to 6 mm. long, green or tawny, the rachilla joints very short; glumes broad, scarious, 1.5 and 2 mm. long; lemmas

FIGURE 96.—*Glyceria nubigena*. Panicle, × 1; floret, × 10. (Barksdale and Jennison 1970, Tenn.)

firm, faintly nerved, smooth, 3 to 4 mm. long, obtuse, the scarious tip narrow, often revolute. ♃ —Bogs and marshy places, Nova Scotia to North Carolina, mostly near the coast.

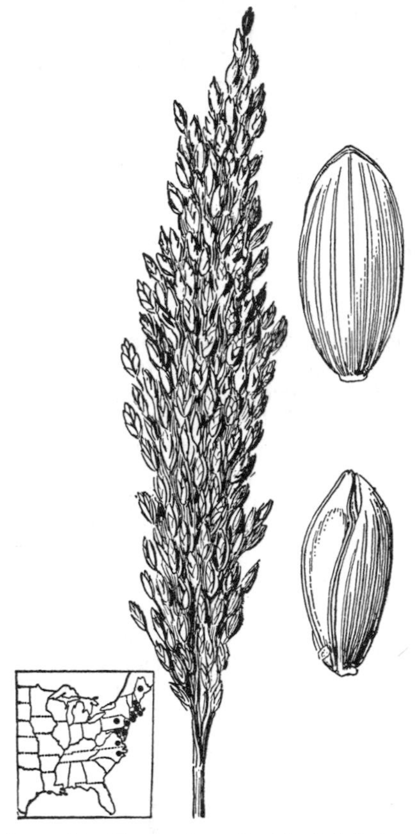

FIGURE 97.—*Glyceria obtusa*. Panicle, × 1; 2 views of floret, × 10. (Miller, N. Y.)

**12. Glyceria melicária** (Michx.) Hubb. (Fig. 98.) Culms slender, solitary or few, 60 to 100 cm. tall; blades elongate, scaberulous, 2 to 5 mm. wide; panicle narrow but rather loose, nodding, 15 to 25 cm. long, the branches erect, rather distant; spikelets 3- or 4-flowered, about 4 mm. long, green; glumes about 1.5 and 2 mm. long, acutish; lemmas firm, 2 to 2.5 mm. long, acutish, smooth, the nerves rather faint. ♃ (*G. torreyana* Hitchc.; *Panicularia torreyana* Merr.)—Swamps and wet woods, New Brunswick to Ohio, south to the mountains of North Carolina.

**13. Glyceria canadénsis** (Michx.) Trin. RATTLESNAKE MANNAGRASS. (Fig. 99.) Culms erect, solitary or few in a tuft, 60 to 150 cm. tall; blades scabrous, 3 to 7 mm. wide; panicle open, 15 to 20 cm. long, nearly as wide, the branches rather distant, drooping, naked below; spikelets ovate or oblong, 5- to 10-flowered, 5 to 6 mm. long, the florets crowded, spreading; glumes about 2 and 3 mm. long; lemmas 3 to 4 mm. long, the 7 nerves obscured in the firm tissue of the lemma; palea bowed out on the keels, the floret somewhat tumid. ♃ —Bogs and wet places, Newfoundland to Minnesota, south to Virginia and Illinois.

GLYCERIA CANADENSIS var. LÁXA (Scribn.) Hitchc. On the average taller, with looser panicles of somewhat smaller 3- to 5-flowered spikelets. ♃ (*Panicularia laxa* Scribn.) —Wet places, Nova Scotia to New York, Michigan, Wisconsin, Maryland, West Virginia, North Carolina, and Tennessee.

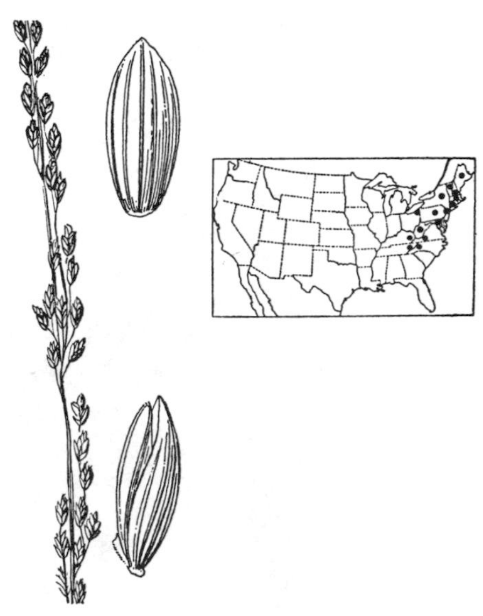

FIGURE 98.—*Glyceria melicaria*. Panicle, × 1; 2 views of floret, × 10. (Harvey 1322, Maine.)

FIGURE 99.—*Glyceria canadensis.* Panicle, × 1; floret, × 10. (Kneucker, Gram. 464, Conn.)

**14. Glyceria striáta** (Lam.) Hitchc. FOWL MANNAGRASS. (Fig. 100.) Plants in large tussocks, pale green; culms erect, slender, rather firm, 30 to 100 cm. tall, sometimes taller; blades erect or ascending, flat or folded, moderately firm, usually 2 to 6 mm. wide, sometimes to 9 mm.; panicle ovoid, open, 10 to 20 cm. long, nodding, the branches ascending at base, drooping, naked below; spikelets ovate or oblong, 3- to 7-flowered, 3 to 4 mm. long, often purplish, somewhat crowded toward the ends of the branchlets; glumes about 0.5 and 1 mm. long, ovate, obtuse; lemmas oblong, prominently 7-nerved, about 2 mm. long, the scarious tip inconspicuous; palea rather firm, about as long as the lemma, the smooth keels prominent, bowed out. ♃ (*G. nervata* Trin.)—Moist meadows and wet places, Newfoundland to British Columbia, south to northern Florida, Texas, Arizona, and northern California; Mexico. A low strict northern form has been called *G. striata* var. *stricta* Fernald (*G. nervata* var. *stricta* Scribn.)

**15. Glyceria eláta** (Nash) Hitchc. TALL MANNAGRASS. (Fig. 101.) Resembling *G. striata;* plants dark green; culms 1 to 2 m. tall, rather succulent; blades flat, thin, lax, 6 to 12 mm. wide; panicle oblong, 15 to 30 cm. long, the branches spreading, the lower often reflexed; spikelets 6- to 8-flowered, 4 to 6 mm. long; glumes and lemmas a little longer than in *G. striata.* ♃ (*Panicularia nervata elata* Piper.)—Wet meadows, springs, and shady moist woods, Montana to British Columbia, south in the mountains to New Mexico and California.

**16. Glyceria eréeta** Hitchc. (Fig. 102.) Culms 10 to 40 cm. tall, sometimes in dense tufts, from slender fragile rhizomes; blades flat, mostly 5 to 12 cm. long, 4 to 9 mm. wide, often equaling the panicle or exceeding it; panicle 3 to 8 cm. long, with ascending or appressed few-flowered

FIGURE 100.—*Glyceria striata*. Plant, × ½; spikelet, × 5; floret, × 10. (V. H. Chase 60, Ill.)

FIGURE 101.—*Glyceria elata*. Plant, × 1; floret, × 10. (Hitchcock 2731, Calif.)

branches; spikelets 3 to 4.5 mm. long; second glume 3-nerved; lemmas 2.5 to 3 mm. long, scaberulous, the tip somewhat erose. ♃ —Springy or boggy places, mostly near or above timber line, Crater Lake, Oreg., to Mount Whitney, Calif., and Glenbrook, Nev.

**17. Glyceria pauciflóra** Presl. (Fig. 103.) Culms 50 to 120 cm. tall; sheaths open, smooth or scaberulous, sometimes inflated in floating plants; blades thin, flat, lax, scaberulous, mostly 10 to 15 cm. long, 5 to 15 mm. wide; panicle open or rather dense, nodding, 10 to 20 cm. long, the branches ascending or spreading, rather flexuous, the spikelets crowded on the upper half, the lowermost usually 2 to 4; spikelets mostly 5- or 6-flowered, 4 to 5 mm. long, often purplish; glumes broadly ovate or oval, about 1 and 1.5 mm. long, the margins erose-scarious, the second 3-nerved; lemmas oblong, 2 to 2.5 mm. long, with 5 prominent nerves and an outer short faint pair near the margins, scaberulous on the nerves and somewhat so between them, the tip rounded, scarious, somewhat erose. ♃ —Shallow water, marshes and wet meadows, Alaska to South Dakota, south to California and New

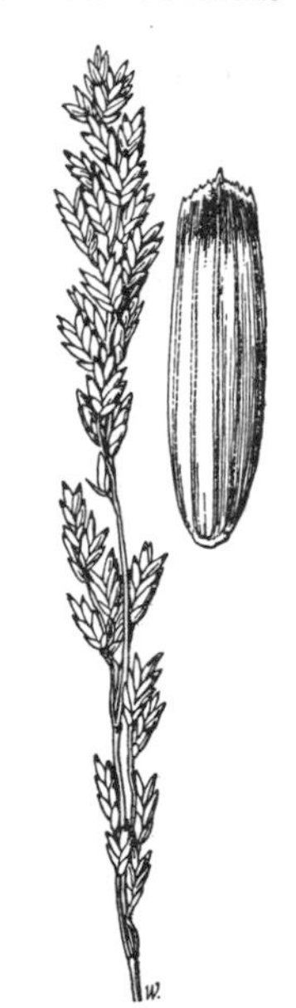

FIGURE 102.—*Glyceria erecta*. Panicle, × 1; floret, × 10. (Hitchcock 3059, Oreg.)

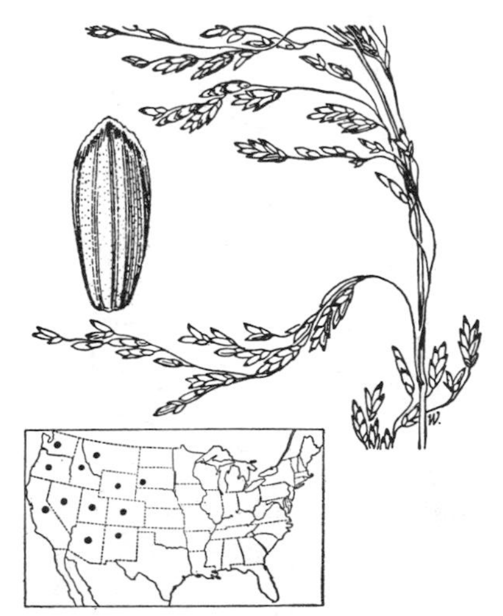

FIGURE 103.—*Glyceria pauciflora*. Panicle, × 1; floret, × 10. (Sandberg, Heller, and McDougal 636, Idaho.)

Mexico, rising in the mountains to timber line.

**18. Glyceria otísii** Hitchc. (Fig. 104.) Culms about 1.25 m. tall; blades flat, lax, 7 to 16 cm. long, 8 to 12 mm. wide; panicle loosely pyramidal, to 18 cm. long, the branches few, drooping; spikelets scarcely compressed, 5- to 6-flowered; glumes 1 and 1.5 mm. long; lemmas broad, especially at the summit, very scabrous, the prominent hyaline tip contrasting with the purple zone just below, the lower part green. ♃ —Timber, Jefferson County, Wash. Known only from the type collection.

**19. Glyceria pállida** (Torr.) Trin. (Fig. 105.) Culms slender, lax, ascending from a decumbent rooting base, 30 to 100 cm. long; sheaths open, blades mostly 4 to 8 mm. wide; panicle pale green, open, 5 to 15 cm. long, the

FIGURE 104.—*Glyceria otisii*. Panicle, × 1; floret, × 10. (Type.)

FIGURE 105.—*Glyceria pallida*. Plant, × 1; floret, × 10. (Pearce, N. Y.)

branches ascending, flexuous, finally more or less spreading; spikelets somewhat elliptic, 4- to 7-flowered, 6 to 7 mm. long; glumes 1.5 to 2 and 2 to 2.5 mm. long, the second 3-nerved; lemmas 2.5 to 3 mm. long, scaberulous, obtuse, the scarious tip erose; anthers linear, about 1 mm. long; caryopsis with a crown of erect white hairs 0.2 to 0.25 mm. long. ♃ —Shallow cold water, Maine to Wisconsin, south to North Carolina and Missouri. Resembles species of *Poa*.

**20. Glyceria fernáldii** (Hitchc.) St. John. (Fig. 106.) Resembling *G. pallida* and appearing to grade into it; culms more slender, 20 to 40 cm. long; blades 1 to 3 mm. wide; panicle on the average smaller, the branches finally spreading or reflexed; spikelets mostly 3- to 5-flowered, 4 to 5 mm. long; glumes and lemmas a little shorter than in *G. pallida;* anthers globose, 0.2 to 0.5 mm. long; crown of hairs of caryopsis 0.1 mm. long. ♃ —Shallow water, Newfoundland to Minnesota, south to Pennsylvania.

FIGURE 106.—*Glyceria fernaldii*. Plant, × 1; floret, × 10. (Collins, Fernald, and Pease, Quebec.)

## 8. SCLERÓCHLOA Beauv.

Spikelets 3-flowered, the upper floret sterile; rachilla continuous, broad, thick, the spikelet falling entire; glumes broad, obtuse, rather firm, with hyaline margins, the first 3-nerved, the second 7-nerved; lemmas rounded on the back, obtuse with 5 prominent parallel nerves and hyaline margins; palea hyaline, sharply keeled. Low tufted annual, with broad upper sheaths, folded blades with boat-shaped tips, and dense spikelike racemes, the spikelets subsessile, imbricate in 2 rows on 1 side of the broad thick rachis. Type species, *Sclerochloa dura*. Name from Greek *skleros*, hard, and *chloa*, grass, alluding to the firm glumes.

**1. Sclerochloa dúra** (L.) Beauv. (Fig. 107.) Culms erect to spreading, 2 to 7 cm. long; foliage glabrous, the lower leaves very small, the upper increasingly larger, with broad overlapping sheaths; blades 7 to 18 mm. long, 1 to 3 mm. wide, the upper exceeding the raceme, the junction with the sheath obscure; raceme 1 to 2 cm. long, nearly half as wide; spikelets 6 to 7 mm. long on very short thick pedicels; first glume about one-third, the second half as long as the spikelets; lower lemma 5 mm. long. ⊙ —Dry sandy or gravelly soil, Washington, Oregon, Idaho, Colorado, Utah, and Texas; New York; introduced from southern Europe.

## 9. SCOLÓCHLOA Link

(*Fluminea* Fries)

Spikelets 3- or 4-flowered, the rachilla disarticulating above the glumes and between the florets; glumes nearly equal, somewhat scarious and lacerate at summit, the first 3-nerved, the second 5-nerved, about as long as the first lemma; lemmas firm, rounded on the back, villous on the callus, 7-nerved, the

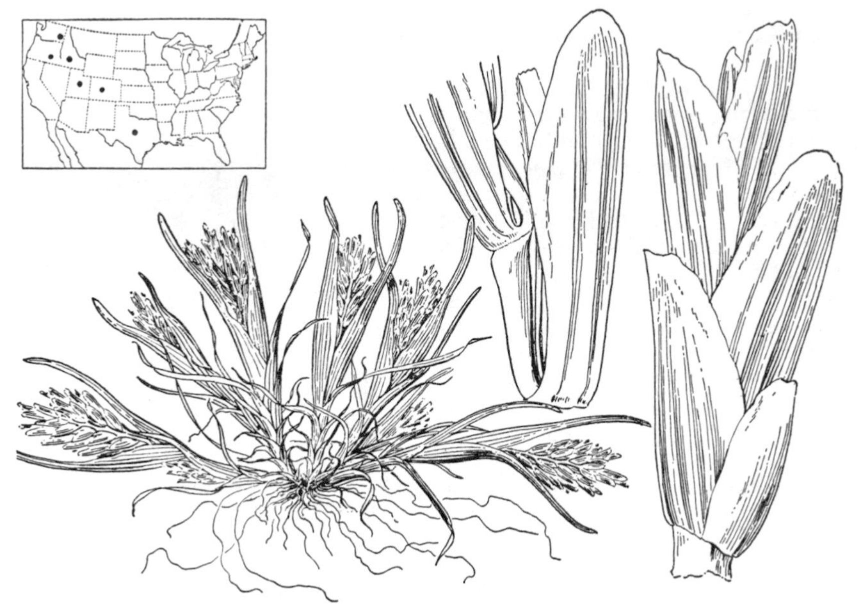

FIGURE 107.—*Sclerochloa dura.* Plant, × 1; spikelet and floret, × 10. (Fallas, Utah.)

nerves rather faint, unequal, extending into a scarious lacerate apex; palea narrow, flat, about as long as the lemma. Tall perennials, with succulent rhizomes, flat blades, and spreading panicles. Type species, *Scolochloa festucacea.* Name from Greek *scolos,* prickle, and *chloa,* grass, alluding to the excurrent nerves of the lemma.

The single species has some value for forage and is often a constituent of marsh hay.

**1. Scolochloa festucácea** (Willd.) Link. (Fig. 108.) Culms erect, stout, 1 to 1.5 m. tall, from extensively creeping, succulent rhizomes; blades elongate, scabrous on the upper surface, mostly 5 to 10 mm. wide, extending into a fine point; panicle 15 to 20 cm. long, loose, the distant branches fascicled, ascending, naked below, the lowermost nearly as long as the panicle; spikelets about 8 mm. long, the florets approximate; lemmas about 6 mm. long. ♃ —Shallow water and marshes, Manitoba to British Columbia, south to northern Iowa, Nebraska, and eastern Oregon; northern Eurasia.

## 10. PLEUROPÓGON R. Br. SEMAPHORE-GRASS

Spikelets several- to many-flowered, linear, the rachilla disarticulating above the glumes and between the florets; glumes unequal, membranaceous or subhyaline, scarious at the somewhat lacerate tip, the first 1-nerved, the second obscurely 3-nerved; lemmas membranaceous, 7-nerved, with a round indurate callus, the apex entire or 2-toothed, the midnerve extending into a short mucro or into an awn; keels of the palea winged on the lower half. Soft annuals or perennials, with simple culms, flat blades, and loose racemes of rather large spikelets on a slender flexuous axis. Type species, *Pleuropogon sabinii* R. Br. Name from Greek *pleura,* side, and *pogon,* beard, the palea of the type species having a bristle on each side at the base.

Palatable grasses, but usually too infrequent to be of economic value.

FIGURE 108.—*Scolochloa festucacea*. Plant, × ½; spikelet and floret, × 5. (Griffiths 870, S. Dak.)

FIGURE 109.—*Pleuropogon californicus*. Plant, × ½; spikelet, × 3; floret, × 5. (Bolander 6075, Calif.)

Keels of palea awned about one-third from the base, the awns 2 to 7 mm. long.
5. P. OREGONUS.
Keels of palea awnless.
Lemmas awnless or mucronate, thick, firm, strongly nerved........................ 4. P. DAVYI.
Lemmas awned, the awns 1 to 12 mm. long.
Lemmas 4 to 6 mm. long, firm, strongly nerved; wings of palea split about half way to the base forming 2 prominent teeth; culms mostly 30 to 60 cm. tall.
1. P. CALIFORNICUS.
Lemmas 8 to 9 mm. long, relatively thin, the nerves evident but not prominent; culms mostly more than 1 m. tall.
Spikelets reflexed or spreading; awn of the lemma 5 to 12 mm. long.
2. P. REFRACTUS.
Spikelets erect or ascending; awn of the lemma 1 to 2.5 mm. long.
3. P. HOOVERIANUS.

**1. Pleuropogon califórnicus** (Nees) Benth. ex Vasey. (Fig. 109.) Annual; culms tufted, erect or decumbent at base, 30 to 60 cm. tall; blades flat or folded, seldom more than 10 cm. long, 2 to 5 mm. wide; raceme 10 to 20 cm. long, with 5 to 10 rather distant short-pediceled spikelets; spikelets 6- to 12-flowered, mostly about 2.5 cm. long, erect, or somewhat spreading; glumes obtuse, erose, 4 to 6 mm. long; lemmas scabrous, 5 to 6 mm. long, the nerves prominent, the tip obtuse, scarious, erose, the awn usually 6 to 12 mm. long; wings of palea prominent, cleft, forming a tooth about the middle. ⊙ —Wet meadows and marshy ground, Mendocino County to the San Francisco Bay region, California.

**2. Pleuropogon refráctus** (A. Gray) Benth. ex Vasey. NODDING SEMAPHORE-GRASS. (Fig. 110.) Perennial; culms 1 to 1.5 m. tall; blades elongate, the uppermost nearly obsolete, 3 to 7 mm. wide; raceme mostly 15 to 20 cm. long, the spikelets as many as 12, about 3 cm. long, 8- to 12-flowered, finally reflexed or drooping; lemmas about 8 mm. long, subacute, less scabrous and the nerves less prominent than in *P. californicus;* awn 5 to 12 mm. long; palea narrow, keeled to about the middle, scarcely or minutely toothed. ♃ —Bogs, wet meadows, and mountain streams, Washington to Mendocino County, Calif., west of the Cascades.

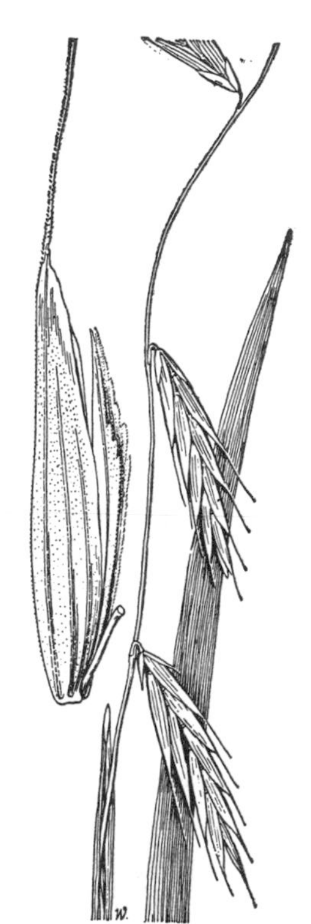

FIGURE 110.—*Pleuropogon refractus.* Plant, × 1; floret, × 5. (Sandberg and Leiberg 734, Wash.)

**3. Pleuropogon hooveriánus** (Benson) J. T. Howell. (Fig. 111.) Similar to *P. refractus,* but the spikelets erect or ascending; lemmas toothed at the broader hyaline summit, the awn 1 to 2.5 mm. long; wings of palea with a single pointed tooth 1 to 1.5 mm. long; rachilla joints swollen and spongy toward the base. ♃ —Grassy wooded flats, Mendocino and Marin Counties, Calif.

**4. Pleuropogon dávyi** Benson. (Fig. 112.) Culms erect from short slender rhizomes, 60 to 100 cm. tall; sheaths

FIGURE 111.—*Pleuropogon hooverianus.* Floret and rachilla joint, × 5. (Dupl. type.)

soft, somewhat inflated, transversely veined; blades 10 to 30 cm. long, 6 to 9 mm. wide, glabrous; raceme 20 to 33 cm. long; spikelets 2 to 5.5 cm. long, 8- to 20-flowered, erect or ascending; lemmas 5.5 to 7.5 mm. long, strongly nerved, obtuse, awnless or mucronate; palea oblong, prominently winged, two-thirds to nearly as long as the lemma. ♃ —Wet ground around marshes and creek beds, Sherwood and Walkers Valley (Mendocino County) to Big Valley (Lake County), Calif.

FIGURE 112.—*Pleuropogon davyi.* Floret and rachilla joint, × 5. (Type.)

FIGURE 113.—*Pleuropogon oregonus.* Plant, × 1; floret, × 10. (Type.)

**5. Pleuropogon oregónus** Chase. (Fig. 113.) Culms 55 to 90 cm. tall, erect from slender rhizomes, soft, spongy, with long internodes; sheaths overlapping, the lower rather loose; ligule 4 to 5 mm. long, lacerate; blades 8 to 18 cm. long, 4 to 7 mm. wide, mucronate, scaberulous; raceme 6 to 16 cm. long; spikelets 1.5 to 4 cm. long, 7- to 13-flowered, ascending; glumes 2 to 4 mm. long, nerveless; lemmas 5.5 to 7 mm. long, obtuse, erose, awn 6 to 10 mm. long; keels of palea with an awn 2 to 7 mm. long, about one-third from the base. ♃ —Wet meadows, Union (Union County) and Adel (Lake County), Oreg.

## 11. HESPERÓCHLOA (Piper) Rydb.

(Included in *Festuca* L. in Manual, ed. 1)

Spikelets 3- to 5-flowered, the rachilla disarticulating above the glumes and between the florets; glumes subequal or the second longer than the first, shorter than the first floret, lanceolate, acute, the first 1-nerved, the second 3-nerved; lemmas rounded on the back, acute or acuminate, awnless, 5-nerved; palea as long as the lemma, scabrous-ciliate on the keels; stigmas sessile, long and slender; grain beaked, bidentate at the apex. Densely tufted, dioecious, rhizomatous perennial with firm, narrow, flat or loosely involute blades, and narrow erect panicles. Type species, *Hesperochloa kingii*. Name from Greek *esperis*, western, and *chloa*, grass.

**1. Hesperochloa kíngii** (S. Wats.) Rydb. (Fig. 114.) Culms in large dense clumps, erect, the rhizomes usually wanting in herbarium specimens; sheaths smooth, striate, the lower reddish brown in age; blades firm, flat, or becoming loosely involute, scabrous on the margins, 3 to 6 mm. wide; panicles 7 to 20 cm. long, the branches short, appressed, floriferous nearly to the base, the staminate inflorescences denser with somewhat larger spikelets than the pistillate; spikelets 7 to 12 mm. long; glumes thin, shining, acute or subobtuse, the first 3 to 4 mm. long, the second 4 to 6 mm. long; lemmas 5 to 8 mm. long, acute or acuminate, scabrous. ♃ (*Festuca confinis* Vasey; *F. kingii* Cassidy.)—Dry mountains and hills, 2,000 to 3,500 m., Oregon to southern California, east to Montana, Nebraska, and Colorado.

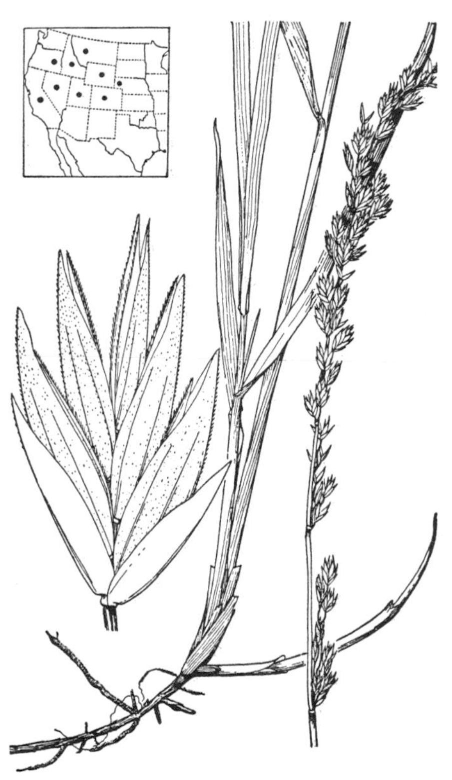

FIGURE 114.—*Hesperochloa kingii*. Plant, × ½; spikelet, × 5. (Osterhout 1897, Colo.)

## 12. POA L. BLUEGRASS

Spikelets 2- to several-flowered, the rachilla disarticulating above the glumes and between the florets, the uppermost floret reduced or rudimentary; glumes acute, keeled, somewhat unequal, the first usually 1-nerved, the second usually 3-nerved; lemmas somewhat keeled, acute or acutish, rarely obtuse, awnless, membranaceous, often somewhat scarious at the summit, 5-nerved (intermediate nerves, that is, the pair between the keel and the marginal nerves, rarely obsolete), the nerves sometimes pubescent, the callus or base of the lemma in many species with scant to copious cottony hairs, termed "web." Low or rather tall slender annuals or usually perennials with spikelets in open or contracted panicles, the relatively narrow blades flat, folded, or involute, ending in a boat-shaped tip. Standard species, *Poa pratensis*. Name from Greek, *poa*, grass.

There are several groups of *Poa* that present many taxonomic difficulties. In the groups containing, for example, *P. nervosa*, *P. arctica*, *P. scabrella*, and *P. nevadensis* many species have been proposed which are not here recognized as valid, because they were based upon trivial or variable characters. The keys are based upon average specimens, but the student may find occasional intermediates between the valid species.

The bluegrasses are of great importance because of their forage value, some species being cultivated for pasture and others forming a large part of the forage on the mountain meadows of the West. The most important is *Poa pratensis*, commonly known as bluegrass or Kentucky bluegrass. In the cooler parts of the United States it is cultivated for lawns and is the standard pasture grass in the humid regions where the soil contains plenty of lime. It has been extensively used in the improvement of badly depleted western mountain ranges. *P. compressa*, Canada bluegrass, is cultivated for pasture in the Northeastern States and Canada, especially on poor soils. *P. trivialis* and *P. palustris* are occasionally grown in meadow mixtures, but are of little agricultural importance. *P. arachnifera*, Texas bluegrass, has been used in some parts of the South for winter pasture and as a lawn grass. *P. annua* is a common weed in lawns and gardens. *P. bulbosa* is cultivated about Medford, Oreg., and elsewhere.

With very few exceptions the bluegrasses are palatable and nutritious and are often the most important grasses in many parts of the West. At high altitudes, *P. alpina*, *P. arctica*, *P. epilis*, and *P. rupicola* are important. In the mountains mostly below timber line are found *P. fendleriana* (mutton grass), *P. longiligula*, *P. nervosa*, *P. secunda* (Sandberg bluegrass), *P. canbyi*, and *P. juncifolia*, all of wide distribution. *P. interior* is most abundant in the Rocky Mountains; *P. scabrella* is probably the most important forage grass of the lower elevations in California; *P. gracillima* and *P. ampla* are mostly in the Northwestern States; *P. arida* is the most valuable bluegrass of the Plains. *P. bigelovii*, an annual, is important in the Southwestern States. *P. macrantha* and *P. confinis* are native sandbinders of the sand dunes on the coast of Washington and Oregon, but are not cultivated.

Spikelets little compressed, narrow, much longer than wide, the lemmas convex on the back, the keels obscure, the marginal and intermediate nerves usually faint. All bunchgrasses.
- Lemmas crisp-puberulent on the back toward the base (the pubescence sometimes obscure or only at the very base ........ 7. SCABRELLAE.
- Lemmas glabrous or minutely scabrous, but not crisp-puberulent ........ 8. NEVADENSES.

Spikelets distinctly compressed, the glumes and lemmas keeled.
- Plants annual ........ 1. ANNUAE.
- Plants perennial.
  - Creeping rhizomes present ........ 2. PRATENSES.
  - Creeping rhizomes wanting.
    - Lemmas webbed at base (web sometimes scant or obscure in *P. interior*). 3. PALUSTRES.
    - Lemmas not webbed at base (sometimes sparsely webbed in *P. fernaldiana* and *P. pattersoni*).
      - Lemmas pubescent on the keel or marginal nerves or both, sometimes pubescent also on the internerves ........ 5. ALPINAE.
      - Lemmas glabrous (minutely pubescent at base in *P. unilateralis* and sometimes in *P. curta*).
        - Blades narrow, usually involute ........ 6. EPILES.
        - Blades flat, 4 to 8 mm. wide, bright green, often splitting at the apex. Panicles about 15 cm. long with slender spreading branches ........ 4. HOMALOPOAE.

*1. Annuae*

Lemmas glabrous, except the scabrous keel, webbed at base. Sheaths glabrous. 1. P. BOLANDERI.
Lemmas pubescent.
  Lemmas pubescent on the back especially toward the base, but not distinctly villous on the keel and nerves, slightly webbed at base. Sheaths usually scabrous; panicle open. 2. P. HOWELLII.
  Lemmas pubescent on the nerves, sometimes also on the internerves.
    Panicle narrow, contracted, usually interrupted; sheaths scabrous. Lemmas webbed, pubescent on the internerves below ........ 3. P. BIGELOVII.
    Panicle oblong or pyramidal, the branches spreading; sheaths glabrous.
      Lemmas with webby hairs at base, distinctly 3-nerved, the intermediate nerves obscure; anthers 0.1 to 0.2 mm. long ........ 4. P. CHAPMANIANA.
      Lemmas not webbed at base, distinctly 5-nerved; anthers 0.5 to 1 mm. long. 5. P. ANNUA.

*2. Pratenses*

1a. Culms strongly flattened, 2-edged ........ 6 P. COMPRESSA.
1b. Culms terete or slightly flattened, not 2-edged.
  2a. Plants dioecious.
    Panicle oblong, the two sexes unlike in appearance, the pistillate spikelets woolly, the staminate glabrous or nearly so. Plains of Texas ........ 7. P. ARACHNIFERA.
    Panicle oblong or ovoid, the two sexes similar. Seacoast, California and northward.
      Glumes and lemmas about 8 mm. long ........ 8. P. MACRANTHA.
      Glumes and lemmas not more than 6 mm. long.
        Panicle densely ovoid; lemmas 6 mm. long, slightly villous below. 9. P. DOUGLASII.
        Panicle somewhat open; lemmas 3 mm. long, scaberulous ........ 10. P. CONFINIS.
  2b. Plants not dioecious, the florets perfect.
    3a. Blades involute. Glumes and lemmas 4 to 5 mm. long ........ 11. P. RHIZOMATA.
    3b. Blades flat or folded.
      4a. Lemmas not pubescent nor webbed.
        Panicle almost spikelike, erect.
          Panicle pale, narrow, linear; lemmas scabrous; leaves crowded toward the base, the blades very firm, conduplicate, pungent, curved. Lower sheaths fibrous. 12. P. FIBRATA.
          Panicle tinged with purple, oblong; lemmas glabrous; leaves not crowded toward the base, the blades flat or sometimes folded, straight, erect. 13. P. ATROPURPUREA.
        Panicle open, nodding; glumes 3 to 4 mm. long.
          Blades broad and short; lower panicle branches reflexed ........ 14. P. CURTA.
          Blades elongate; panicle branches ascending ........ 15. P. NERVOSA.
      4b. Lemmas pubescent.
        5a. Lemmas glabrous except for the web at base ........ 16. P. KELLOGGII.
        5b. Lemmas pubescent on the nerves or back, sometimes also webbed at base.
          6a. Internerves glabrous, the keel and marginal nerves pubescent.
            Lower sheaths retrorsely pubescent, purplish; lemmas pubescent on keel and marginal nerves, not webbed ........ 15. P. NERVOSA.
            Lower sheaths glabrous (scaberulous in *P. laxiflora*); lemmas webbed at base.
              Culms retrorsely scabrous ........ 17. P. LAXIFLORA.
              Culms glabrous.
                Lower panicle branches in a whorl of usually 5; blades mostly shorter than the culm ........ 18. P. PRATENSIS.
                Lower panicle branches usually in twos, spreading, spikelet-bearing near the ends; blades about as long as the culm ........ 19. P. CUSPIDATA.
          6b. Internerves pubescent near base, the keel and marginal nerves pubescent.
            Panicle contracted, the branches ascending or appressed (sometimes open in *P. glaucifolia*).
              First glume 2.5 to 3 mm. long, 1-nerved; first floret about 3 mm. long; anthers 1.5 mm. long. Plains and alkali meadows at medium altitudes. 20. P. ARIDA.
              First glume 4 to 5 mm. long, 3-nerved; first floret 5 mm. long; anthers 2.5 mm. long; spikelets mostly shining ........ 21. P. GLAUCIFOLIA.
            Panicle open, the branches spreading.
              Blades broad and short; lower panicle branches reflexed ........ 14. P. CURTA.
              Blades 2 to 3 mm. wide; panicle pyramidal, the lower branches horizontal 22. P. ARCTICA.

*3. Palustres*

1a. Lemmas glabrous, or the keel sometimes pubescent.
Sheaths retrorsely scabrous. Culms decumbent and often rooting at base; keel of lemma glabrous or slightly pubescent........ 23. P. TRIVIALIS.
Sheaths glabrous.
Panicle narrow, drooping, the branches appressed or ascending........ 24. P. MARCIDA.
Panicle very open, the few branches slender, naked below, spreading or drooping.
Lemmas villous on the keel; panicle branches mostly in fours or fives. 25. P. ALSODES.
Lemmas glabrous on the keel; panicle branches mostly in twos or threes.
Lemmas obtuse........ 26. P. LANGUIDA.
Lemmas acute........ 27. P. SALTUENSIS.
1b. Lemmas pubescent on keel and marginal nerves.
2a. Sheaths distinctly retrorse-scabrous (sometimes faintly so). Culms usually stout, 40 to 120 cm. tall; panicle usually large and open, mostly more than 15 cm. long. 28. P. OCCIDENTALIS.
2b. Sheaths glabrous or faintly scaberulous.
3a. Lower panicle branches distinctly reflexed at maturity.
Panicle oblong, erect, mostly more than 15 cm. long, the branches several (usually more than 3) in a whorl........ 30. P. SYLVESTRIS.
Panicle nodding, mostly less than 15 cm. long, the branches 1 to 3 together. 31. P. REFLEXA.
3b. Lower panicle branches not reflexed.
4a. Panicle narrowly pyramidal, erect, 15 to 20 cm. long. Lemmas 4 mm. long, pubescent on nerves and internerves; webbed at base; New Mexico. 29. P. TRACYI.
4b. Panicle broadly pyramidal, usually nodding.
5a. Intermediate nerves of lemma distinct........ 32. P. WOLFII.
5b. Intermediate nerves of lemma obscure (distinct in *P. leptocoma*).
6a. Lower panicle branches in pairs, elongate, capillary, bearing a few spikelets near the ends.
Spikelets rather broad, the rachilla joints short, hidden by the florets; sheaths smooth; culms in dense tufts; alpine rocky slopes. 33. P. PAUCISPICULA.
Spikelets narrow, the rachilla joints slender, somewhat elongate, usually not hidden by the florets; sheaths minutely roughened; culms solitary or in small tufts; shady bogs.
Intermediate nerves of lemma distinct; uppermost ligule acute, 3 to 4 mm. long; western mountains below timber line........ 34. P. LEPTOCOMA.
Intermediate nerves of lemma obscure; uppermost ligule truncate, 0.3 to 1.5 mm. long; Great Lakes region at low altitudes. 35. P. PALUDIGENA.
6b. Lower panicle branches often more than 2, if only 2 not capillary and elongate.
Florets usually converted into bulblets with dark purple base; culms swollen and bulblike at base........ 36. P. BULBOSA.
Florets normal; culms not bulblike at base.
Glumes narrow, acuminate, about as long as the first lemma; ligule very short........ 37. P. NEMORALIS.
Glumes lanceolate, acute, shorter than the first lemma; ligules rather prominent, those of the culm leaves 1 to 3 mm. or more long.
Spikelets about 6 mm. long; lemmas 4 mm. long........ 38. P. MACROCLADA.
Spikelets about 4 mm. long; lemmas 2.5 to 3 mm. long.
Culms decumbent at the purplish base; panicle 10 to 30 cm. long, large and open........ 39. P. PALUSTRIS.
Culms erect from a green or tawny base; panicle mostly less than 10 cm. long, comparatively small and few-flowered........ 40. P. INTERIOR.

*4. Homalopoae*

One species........ 41. P. CHAIXII.

*5. Alpinae*

Blades folded or involute, firm, rather stiff.
Ligule very short, not noticeable when viewed from the side of sheath. 42. P. FENDLERIANA.

Ligule prominent, easily seen in side view, 5 to 7 mm. long .......... 43. P. LONGILIGULA.
Blades flat or, if involute, rather lax or soft.
Panicle branches slender, spreading or drooping, the lower naked and simple for 3 to 4 cm. or more .......... 44. P. AUTUMNALIS.
Panicle branches not long and spreading.
Panicle broadly pyramidal, condensed, about as broad as long, the lower branches spreading. Spikelets broad, subcordate .......... 45. P. ALPINA.
Panicle longer than broad.
Panicle nodding, the lower branches slender, arcuate-drooping. 46. P. STENANTHA.
Panicle erect, the lower branches short (see also *P. gracillima*).
Panicle rather loose, lower branches naked below, ascending (see also *P. macrolada*).
Plants glaucous, culms flattened; panicle rather narrow.
Spikelets 2- or 3-flowered; panicle 3 to 7 cm. long .......... 47. P. GLAUCA.
Spikelets 3- to 6-flowered; panicle 6 to 16 cm. long .......... 48. P. GLAUCANTHA.
Plants not glaucous; culms terete, rather lax .......... 49. P. FERNALDIANA.
Panicle narrow, condensed, the branches short (see also *P. unilateralis*).
Culms rather lax; ligule minute; glumes about 4 mm. long. 50. P. PATTERSONI.
Culms stiff, ligule about 1.5 mm. long, glumes about 3 mm. long. 51. P. RUPICOLA.

*6. Epiles*

Panicle open, 10 to 15 cm. long. Blades involute, slender .......... 52. P. INVOLUTA.
Panicle contracted or, if open, less than 10 cm. long.
Blades scabrous, filiform, mostly basal.
Spikelets 7 to 9 mm. long; lemmas 4.5 to 6 mm. long, mostly smooth. 53. P. CUSICKII.
Spikelets 6 to 7 mm. long; lemmas about 4 mm. long, scabrous .......... 54. P. NAPENSIS.
Blades glabrous.
Lemmas minutely pubescent at base .......... 55. P. UNILATERALIS.
Lemmas glabrous.
Blades of the culm 2 to 3 mm. wide, flat, those of the innovations slender and filiform. 56. P. EPILIS.
Blades of the culm and innovations similar. Panicle few-flowered.
Panicle short, open, the capillary branches bearing 1 or 2 spikelets. Culms 10 to 20 cm. tall .......... 57. P. VASEYOCHLOA.
Panicle narrow.
Lemmas 5 to 6 mm. long; panicle usually pale or silvery .......... 58. P. PRINGLEI.
Lemmas less than 4 mm. long; panicle usually purple.
Glumes about as long as the first and second florets; panicle mostly not exceeding the short soft blades.
Glumes and lemmas smooth, the lemmas erose at summit. 59. P. LETTERMANI.
Glumes and lemmas scabrous, the lemmas acute, scarcely erose. 60. P. MONTEVANSI.
Glumes shorter than the first floret; panicle usually much longer than the usually stiff blades .......... 61. P. LEIBERGII.

*7. Scabrellae*

Sheaths somewhat scabrous .......... 62. P. SCABRELLA.
Sheaths glabrous.
Panicle rather open, the lower branches naked at base, ascending or somewhat spreading; culms usually decumbent at base .......... 63. P. GRACILLIMA.
Panicle contracted, the branches appressed or at anthesis somewhat spreading.
Culms slender, on the average less than 30 cm. tall; numerous short innovations at base. Blades usually folded .......... 64. P. SECUNDA.
Culms stouter, on the average more than 50 cm. tall; innovations usually not numerous. 65. P. CANBYI.

*8. Nevadenses*

Sheaths scaberulous. Ligule long, decurrent .......... 66. P. NEVADENSIS.
Sheaths glabrous.
Ligule prominent; blades broad and short .......... 67. P. CURTIFOLIA.
Ligule short; blades elongate.
Blades involute .......... 68. P. JUNCIFOLIA.
Blades flat .......... 69. P. AMPLA.

**1. Ánnuae.**—Annuals; culms seldom more than 50 cm. tall; panicles open (contracted in *P. bigelovii*).

**1. Poa bolandéri** Vasey. (Fig. 115.) Culms erect, 15 to 60 cm. tall; sheaths glabrous; blades relatively short, 3 to 5 mm. wide, abruptly narrowed at tip; panicle about half the length of the entire plant, at first contracted, finally open, the branches few, distant, glabrous, stiffly spreading, naked below; spikelets usually 2- or 3-flowered, the internodes of the rachilla long; glumes broad, 2 and 3 mm. long; lemma scantily webbed at base, acute, the marginal nerves rather indistinct, the intermediate nerves obsolete. ⊙ —Open ground or open woods, 1,500 to 3,000 m., Washington and Idaho to western Nevada and the southern Sierras in California.

FIGURE 115.—*Poa bolanderi*. Panicle, × 1; floret, × 10. (Swallen 799, Calif.)

**2. Poa howéllii** Vasey and Scribn. HOWELL BLUEGRASS. (Fig. 116.) Culms 30 to 85 cm. tall; sheaths retrorsely scabrous to glabrous; blades wider than in *P. bolanderi*, gradually acuminate; panicle one-third to half the entire height of the plant, open, the branches in rather distant fascicles, spreading, scabrous, naked below, some short branches intermixed; spikelets 3 to 5 mm. long, usually 3- or 4-flowered; glumes narrow, acuminate, 1.5 and 2 mm. long; lemmas webbed at base, 2 to 3 mm. long, ovate-lanceolate, pubescent on the lower part, the nerves all rather distinct. ⊙ —Rocky banks and shaded slopes, mostly less than 1,000 m., Vancouver Island to southern California, especially in the Coast Ranges.

**3. Poa bigelóvii** Vasey and Scribn. BIGELOW BLUEGRASS. (Fig. 117.) Culms erect, 15 to 35 cm. tall; blades 1 to 5 mm. wide; panicle narrow, interrupted, 7 to 15 cm. long, the branches short, appressed; spikelets about 6 mm. long; glumes acuminate, 4 mm. long, 3-nerved; lemmas about 3 mm. long, sometimes 4 mm., webbed at base, conspicuously pubescent on the lower part of keel and lateral nerves, sometimes sparsely pubescent on lower part of internerves. ⊙ —Open ground, at medium altitudes, Oklahoma and western Texas to Colorado, Nevada, and southern California; northern Mexico.

**4. Poa chapmaniána** Scribn. (Fig. 118.) Plant drying pale or tawny; culms densely tufted, slender, 10 to 30 cm. tall; blades 1 to 1.5 mm. wide; panicle oblong-pyramidal, 3 to 8 cm. long, open, the lower branches spreading; spikelets 3 to 4 mm. long, mostly 3- to 5-flowered; glumes 2 and 2.5 mm. long; lemmas about 2 mm. long, webbed at base, strongly pubescent on the keel and lateral nerves, the intermediate nerves obscure; anthers 0.1 to 0.2 mm. long. ⊙ —Open ground and cultivated fields, Massachusetts and New York; Delaware to Nebraska, Florida and Texas.

FIGURE 116.—*Poa howellii*. Panicle, × 1; floret, × 10. (Suksdorf 10464, Wash.)

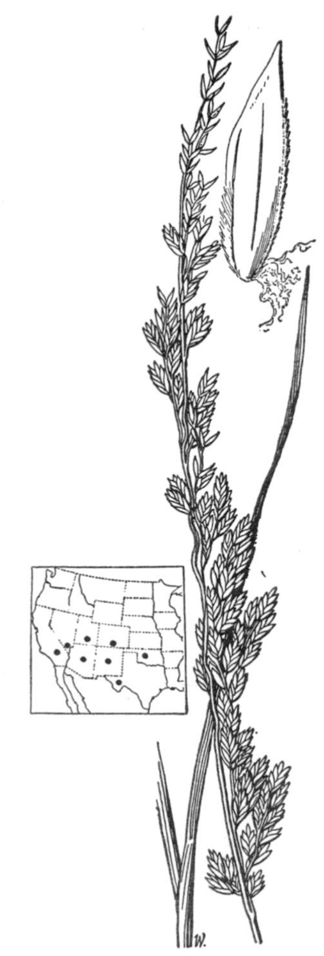

FIGURE 117.—*Poa bigelovii*. Panicle, × 1; floret, × 10. (Fendler 931, N. Mex.)

FIGURE 118.—*Poa chapmaniana*. Panicle, × 1; floret, × 10. (V. H. Chase 3557, Ill.)

**5. Poa ánnua** L. ANNUAL BLUEGRASS. (Fig. 119.) Tufted, bright green, erect to spreading, sometimes rooting at the lower nodes, usually 5 to 20 cm. tall, sometimes taller, forming mats; culms flattened; blades soft, lax, mostly 1 to 3 mm. wide; panicle pyramidal, open, 3 to 7 cm. long; spikelets crowded, 3- to 6-flowered, about 4 mm. long; first glume 1.5 to 2, the second 2 to 2.5 mm. long; lemmas not webbed at base, distinctly 5-nerved, more or less pubescent on the lower half of all the nerves, the long hairs on the lower part of the keel sometimes simulating a web; anthers 0.5 to 1 mm. long. ⊙ —Open ground, lawns, pastures, waste places, and openings in woods, Newfoundland and Labrador to Alaska, south to Florida and California; tropical America at high altitudes; introduced from Europe. In warmer parts of the United States the species thrives in the winter; in intermediate latitudes it is a troublesome weed in lawns, growing luxuriantly in spring, dying in early summer and leaving unsightly patches. Occasionally found in flooded places and stream banks, the culms spreading.

**2. Praténses.**—Perennials with slender creeping rhizomes. Several species dioecious.

FIGURE 119.—*Poa annua*. Panicle, × 1; floret, × 10. (Hitchcock, D. C.)

FIGURE 120.—*Poa compressa*. Panicle, × 1; floret, × 10. (Gayle 750, Maine.)

**6. Poa compréssa** L. CANADA BLUEGRASS. (Fig. 120.) Culms solitary or few together, often gregarious, strongly flattened, wiry, decumbent at base, bluish green, 15 to 70 cm. tall; blades mostly rather short, 1 to 4 mm. wide; panicle narrow, 3 to 10 cm. long, the usually short branches in pairs, spikelet-bearing to the base; spikelets crowded, subsessile, 3- to 6-flowered, 4 to 6 mm. long; glumes 2 to 3 mm. long; lemmas firm, 2 to 3 mm. long, the web at base scant or wanting, the keel and marginal nerves slightly pubescent toward base, the intermediate nerves obscure. ♃ —Open ground, open woods, meadows, and waste places, Newfoundland to Alaska, south to Georgia, Tennessee, Alabama, Oklahoma, New Mexico, and California; introduced from Europe. Cultivated for pastures in poor soil.

**7. Poa arachnífera** Torr. TEXAS BLUEGRASS. (Fig. 121.) Plants dioecious; culms tufted, 30 to 75 cm. tall; blades mostly 2 to 4 mm. wide, scabrous above; panicle narrow, compact, more or less lobed or interrupted, 5 to 12 cm. long; spikelets mostly 5- to 10-flowered, the pistillate conspicuously cobwebby, the lemmas 5 to 6 mm. long, acuminate, copiously long webby at base, the strongly compressed keel and lateral nerves ciliate-fringed along the lower half; staminate lemmas glabrous or with a scant web at base. ♃ —Prairies and plains, southern Kansas to Texas and Arkansas; introduced eastward to North Carolina and Florida; Idaho. Sometimes cultivated for winter pasture.

FIGURE 121.—*Poa arachnifera.* Plant and pistillate (♀) and staminate (♂) panicles, × 1; pistillate (♀) and staminate (♂) florets, × 10. (Blackman, Tex.)

8. **Poa macrántha** Vasey. (Fig. 122.) Plants dioecious; culms erect from a decumbent base, with extensively creeping rhizomes, and also long runners creeping over the sand, 15 to 40 cm. tall; sheaths tawny, papery; blades involute, subflexuous; panicle contracted, sometimes dense and spikelike, 5 to 12 cm. long, pale or tawny; spikelets about 12 mm. long, about 5-flowered; glumes 3-nerved, or the second indistinctly 5-nerved, about 8 mm. long; lemmas about 8 mm. long, short-webbed at base, pubescent on the keel and marginal nerves below, slightly scabrous on the keel above; pistillate florets with abortive stamens. ♃ —Sand dunes along the coast, Washington to northern California.

FIGURE 122.—*Poa macrantha*. Plant, × 1; floret, × 10. (Hitchcock 2822, Oreg.)

**9. Poa douglásii** Nees. (Fig. 123.) Plants dioecious, the two kinds similar; culms ascending from a decumbent base, usually less than 30 cm. tall; rhizomes slender; sheaths glabrous, tawny and papery; blades involute, some of them usually exceeding the culm; panicle ovoid, dense, spikelike, 2 to 5 cm. long, 1 to 2 cm. wide, pale or purplish; spikelets 6 to 10 mm. long, about 5-flowered; glumes broad, 3-nerved, 4 to 6 mm. long; lemmas 6 to 7 mm. long, slightly webbed at base, pubescent on the lower part of the keel and marginal nerves, scabrous on the upper part of the keel, usually with 1 to 3 pairs of intermediate nerves. ♃ —Sand dunes near the coast, California, Point Arena to Monterey.

**10. Poa confínis** Vasey. (Fig. 124.) Plants dioecious, the two kinds similar; culms often geniculate at base, usually less than 15 cm. tall, sometimes as much as 30 cm.; blades involute, those of the innovations numerous; panicle narrow, 1 to 3 cm. long, tawny, the short branches ascending or appressed; spikelets 4 to 5 mm. long, mostly 3- or 4-flowered; glumes unequal, the second 3 mm. long; lemmas 3 mm. long, scaberulous, sparsely webbed at base, the nerves faint; pistillate florets with minute abortive anthers, the staminate often with rudimentary pistil. ♃ —Sand dunes and sandy meadows near the coast, British Columbia to Mendocino County, Calif.

FIGURE 123.—*Poa douglasii*. Plant, × 1; floret, × 10. (Bolander 6074, Calif.)

**11. Poa rhizómata** Hitchc. (Fig. 125.) Culms tufted with numerous innovations, 40 to 60 cm. tall; lower sheaths usually scaberulous with a puberulent collar; ligule rather prominent on the culm leaves, inconspicuous on the leaves of the innovations; blades involute or sometimes flat, firm, less than 1 mm. thick, flexuous, mostly basal, 2 on the culm, usually puberulent on the upper surface; panicle open, 5 to 8 cm. long, the lower branches mostly in pairs, 2 to 3 cm. long; spikelets, 3- to 5-flowered, 6 to 10 mm. long; glumes 3 to 5 mm. long; lemmas 4 to 5 mm. long, with a rather short web at the base, scaberulous at least on the rather distinct nerves, pubescent on the lower part of keel. ♃ —Dry slopes, southwestern Oregon and northwestern California; apparently rare.

FIGURE 124.—*Poa confinis*. Plant, × 1; floret, × 10. (Piper 4910, Wash.)

FIGURE 125.—*Poa rhizomata*. Plant, × 1; floret, × 10. (Type.)

**12. Poa fibráta** Swallen. (Fig. 126.) Culms 15 to 35 cm. tall, erect from an ascending base; lower sheaths thin, smooth and shining; ligule 1 to 1.5 mm. long; blades 4 to 8 cm. long, firm, conduplicate, curved, pungent, scabrous; panicles 4 to 10 cm. long, dense, the short appressed branches floriferous to the base; spikelets 3- to 4-flowered, 5 to 6 mm. long; lemmas 2.5 to 3 mm. long, acute or subobtuse, glabrous or obscurely pubescent toward the base. ♃ —Saline flats, Shasta Valley, Siskiyou County, Calif.

**13. Poa atropurpúrea** Scribn. (Fig. 127.) Culms erect, 30 to 40 cm. tall;

FIGURE 126.—*Poa fibrata*. Plant, × 1; floret, × 10. (Type.)

FIGURE 127.—*Poa atropurpurea*. Plant, × 1; floret, × 10. (Type.)

blades mostly basal, the uppermost culm leaf below the middle of the culm, folded or involute, firm; panicle contracted, almost spikelike, purple-tinged, 3 to 5 cm. long; spikelets 3 to 4 mm. long, rather thick; glumes broad, less than 2 mm. long; lemmas about 2.5 mm. long, broad, glabrous, not webbed at base, the nerves faint. ♃ —Known only from Bear Valley, San Bernardino Mountains, Calif.

**14. Poa cúrta** Rydb. (Fig. 128.) Culms few in a loose tuft, 40 to 80 cm. tall, rather lax; sheaths glabrous or minutely roughened; ligule truncate, about 1 mm. long; blades 3 to 6 mm. wide; panicle open, 5 to 15 cm. long, nodding, the rather distant branches spreading or reflexed, naked below; spikelets 5 to 10 mm. long, 2- to 6-flowered; lemmas lanceolate, subacute, slightly scaberulous, sometimes slightly pubescent on the back at base, without a web, 4 to 5.5 mm. long, rather strongly nerved or intermediate nerves faint. ♃ —Moist shady places at medium altitudes, western Wyoming, southern Idaho, and Utah.

FIGURE 128.—*Poa curta*. Panicle, × 1; floret, × 10. (Jones 5573, Utah.)

**15. Poa nervósa** (Hook.) Vasey. WHEELER BLUEGRASS. (Fig. 129.) Culms erect, 30 to 60 cm. tall; sheaths glabrous or the lower retrorsely pubescent, often purple, the collar often puberulent; ligule 1 to 2 mm. long; blades sometimes folded; panicle open, usually 5 to 10 cm. long, the apex nodding, the branches mostly in twos or threes, naked below; lemmas rather strongly nerved, glabrous or pubescent on the lower part of the nerves. ♃ (*P. wheeleri* Vasey; *P. olneyae* Piper.)—Open woods at medium altitudes, Alberta and British Columbia, south in the mountains to Colorado, New Mexico, and California. Typical *P. nervosa* (including *P. olneyae*) found mostly in Washington and Oregon, has glabrous to scaberulous strongly nerved lemmas, glabrous sheaths, and a loose open panicle, the capillary lower branches in whorls of 3 or 4, drooping, as much as 8 cm. long; typical *P. wheeleri*, originally described from Colorado, has firmer, less strongly nerved lemmas, more or less pubescent on the

FIGURE 129.—*Poa nervosa*. *A*, Plant, × 1. (Suksdorf 10364, Wash.) *B*, Floret, × 10. (Type of *P. wheeleri*.) *C*, Floret, × 10. (Type of *P. nervosa*.)

FIGURE 130.—*Poa kelloggii*. Plant, × 1; floret, × 10. (Kellogg and Bolander 14, Calif.)

lower part of the keel and marginal nerves, and purplish retrorsely pubescent lower sheaths. These characters are not coordinated, and the forms grade into each other, both as to characters and range.

**16. Poa kellóggii** Vasey. (Fig. 130.) Culms 30 to 60 cm. tall; sheaths slightly scabrous; blades flat or folded, 2 to 4 mm. wide; panicle pyramidal, open, 7 to 15 cm. long, the branches mostly solitary or in twos, spreading or reflexed, bearing a few spikelets toward the ends; spikelets rather loosely flowered, 4 to 6 mm. long; glumes 3 and 4 mm. long; lemmas acute or almost cuspidate, 4 to 5 mm. long, glabrous, rather obscurely nerved, conspicuously webbed at base. ♃ —Moist woods and shady places, Coast Ranges from Corvallis, Oreg., to Santa Cruz County, Calif.

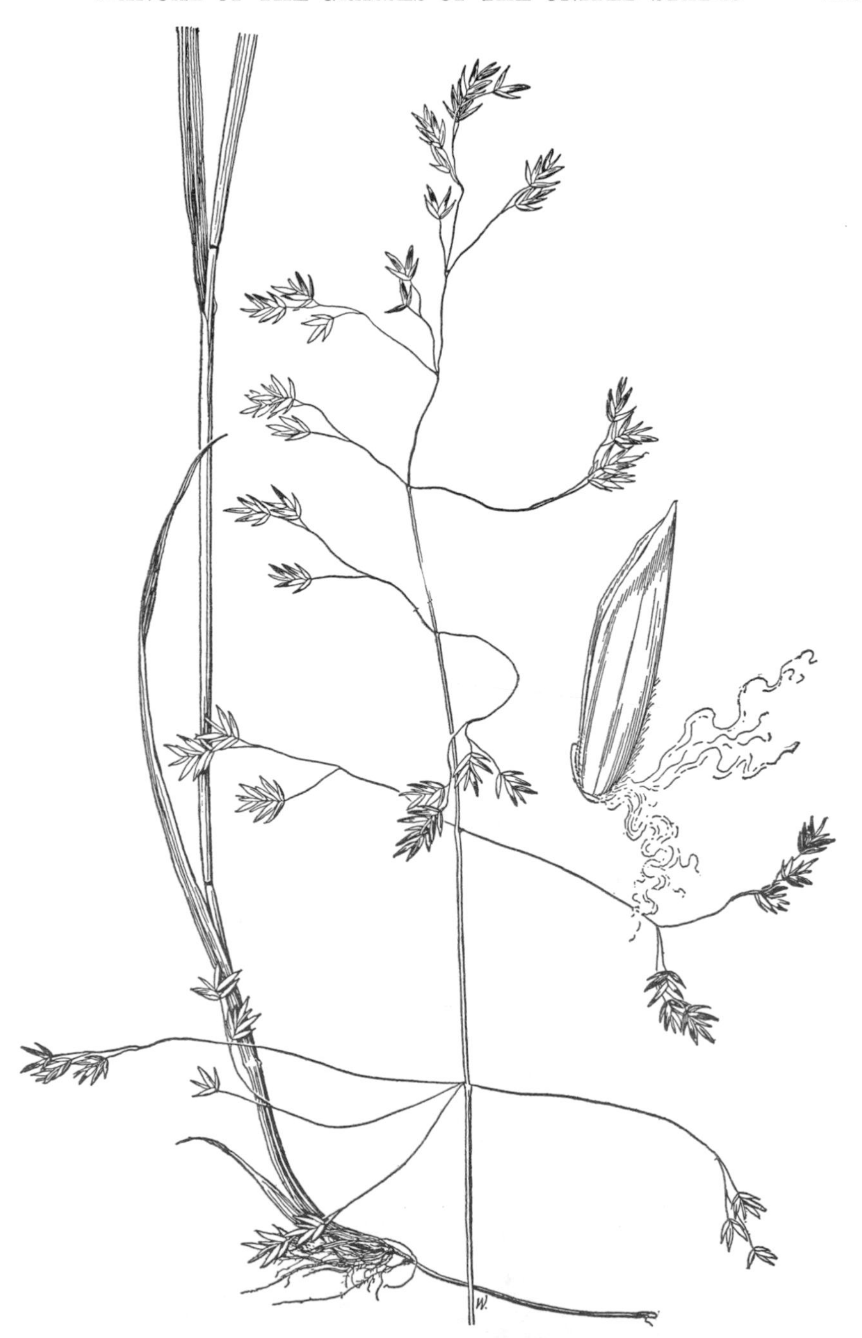

FIGURE 131.—*Poa laxiflora.* Plant, × 1; floret, × 10. (Hitchcock 23468, Wash.)

**17. Poa laxiflóra** Buckl. (Fig. 131.) Culms retrorsely scabrous, 100 to 120 cm. tall; sheaths slightly retrorse-scabrous; ligule 3 to 5 mm. long; blades lax, 2 to 4 mm. wide; panicle loose, open, nodding or drooping, 10

FIGURE 132.—*Poa pratensis*. Plant, × ½; spikelet, × 5; floret, × 10. (Williams, S. Dak.)

to 15 cm. long, the lower branches in whorls of 3 or 4; spikelets 3- or 4-flowered, 5 to 6 mm. long; lemmas about 4 mm. long, webbed at base, rather sparsely pubescent on lower part of the nerves. ♃ —Moist woods, southeastern Alaska (Cape Fox, Hot Springs), Sol Duc Hot Springs, Olympic Mountains, Wash. Sauvies Island (near Portland), Oreg.

**18. Poa praténsis** L. KENTUCKY BLUEGRASS. (Fig. 132.) Culms tufted, erect, slightly compressed, 30 to 100 cm. tall; sheaths somewhat keeled; ligule about 2 mm. long; blades soft, flat or folded, mostly 2 to 4 mm. wide, the basal often elongated; panicle pyramidal or oblong-pyramidal, open, the lowermost branches usually in a whorl of 5, ascending or spreading, naked below, normally 1 central long one, 2 shorter lateral ones and 2 short intermediate ones; spikelets crowded, 3- to 5-flowered, 3 to 6 mm. long; lemmas copiously webbed at base, silky-pubescent on lower half or two-thirds of the keel and marginal nerves, the intermediate nerves distinct, glabrous. ♃ —Open woods, meadows, and open ground, widely distributed throughout the United States and northward, except in arid regions, found in all the States (but not common in the Gulf States) and at all altitudes below alpine regions; introduced from Europe. Bluegrass is commonly cultivated for lawns and pasture in the humid northern parts of the United States.

**19. Poa cuspidáta** Nutt. (Fig. 133.) Culms in large lax tufts, 30 to 50 cm. tall, scarcely longer than the basal blades; blades lax, 2 to 3 mm. wide, abruptly cuspidate-pointed; panicle 7 to 12 cm. long, open, the branches mostly in pairs, distant, spreading, spikelet-bearing near the ends; spikelets 3- or 4-flowered; lemmas 4 to 6 mm. long, tapering to an acute apex, webbed at base, sparingly pubescent on the keel and marginal nerves, the intermediate nerves distinct, glabrous. ♃ (*P. brachyphylla* Schult.) —Rocky woods, New York, New Jersey to Ohio, south to Georgia and Alabama.

**20. Poa árida** Vasey. PLAINS BLUEGRASS. (Fig. 134.) Culms erect, 20 to 60 cm. tall; blades mostly basal, firm, folded, usually 2 to 3 mm. wide, a single culm leaf usually below the middle of the culm, its blade short;

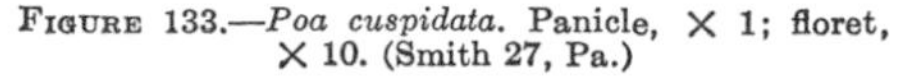
FIGURE 133.—*Poa cuspidata*. Panicle, × 1; floret, × 10. (Smith 27, Pa.)

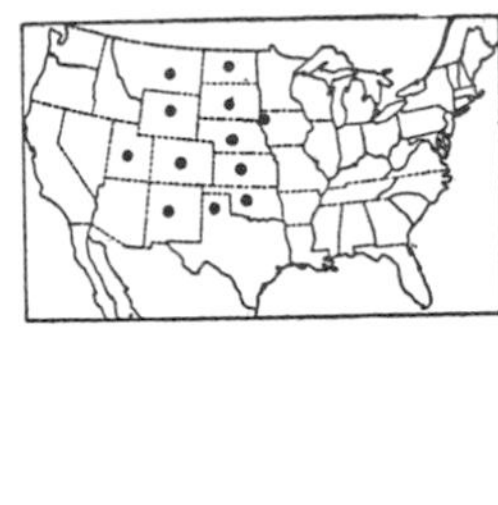

FIGURE 134.—*Poa arida*. Panicle, × 1; floret, × 10. (Jones, Colo.)

panicle narrow, somewhat contracted, 2 to 10 cm. long, the branches appressed or ascending; spikelets rather thick, 5 to 7 mm. long, 4- to 8-flowered; lemmas 3 to 4 mm. long, densely villous on the keel and marginal nerves and more or less villous on the lower part of the intermediate nerves. ♃ (*P. sheldoni* Vasey.)—Prairies, plains, and alkali meadows, up to 3,000 m., Manitoba to Alberta, south to western Iowa, Texas, and New Mexico.

**21. Poa glaucifólia** Scribn. and Will. (Fig. 135.) Plants glaucous; culms in loose tufts, 60 to 100 cm. tall; blades 2 to 3 mm. wide; panicle narrow, open, mostly 10 to 20 cm. long, the branches usually in somewhat distant whorls, mostly in threes, ascending, very scabrous, naked below; spikelets 2- to 4-flowered; glumes

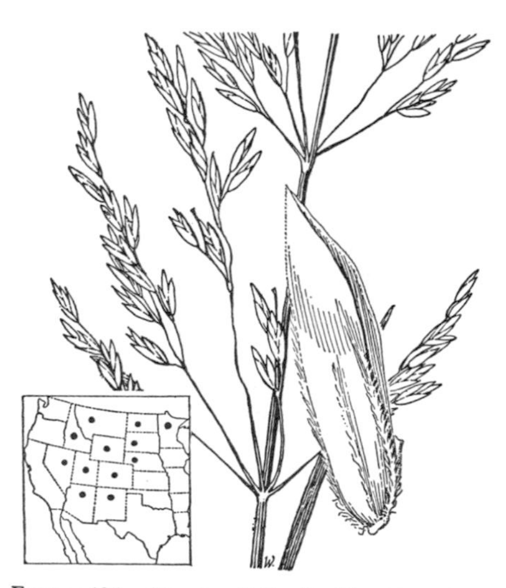

FIGURE 135.—*Poa glaucifolia*. Panicle, × 1; floret, × 10. (Rydberg 3288, Mont.)

4 to 5 mm. long; lemmas about 4 mm. long, villous on the lower half of the keel and marginal nerves and more or less so on the intermediate nerves below. ♃ —Moist places, ditches, and open woods at medium altitudes, British Columbia and Alberta through Idaho to Minnesota, Nebraska, New Mexico, Arizona, and Nevada.

**22. Poa árctica** R. Br. ARCTIC BLUEGRASS. (Fig. 136.) Culms loosely tufted, erect from a decumbent base,

FIGURE 136.—*Poa arctica*. Panicle, × 1; floret, × 10. (Sewall 244, Baffin Land.)

10 to 30 cm. tall; ligule pointed, up to 4 mm. long; blades mostly basal, flat or folded, mostly 2 to 3 mm. wide, one short blade about the middle of the culm; panicle open, pyramidal, 5 to 10 cm. long, the lower branches usually 2, spreading, sometimes reflexed, bearing a few spikelets toward the tip; spikelets 5 to 8 mm. long, 3- or 4-flowered; lemmas densely villous on the keel and marginal nerves and pubescent on the lower part of the internerves, the base often webbed. ♃ (*P. grayana* Vasey; *P. aperta* Scribn. and Merr., a form with pale, rather lax panicles longer than wide.)—Meadows, mostly above timber line, Arctic regions, south to Nova Scotia, in the Rocky Mountains to Nevada and northern New Mexico and in the Cascades to Oregon; California (Inyo County).

**3. Palústres.**—Perennials without creeping rhizomes; lemmas webbed at base, glabrous, or pubescent on the nerves.

**23. Poa triviális** L. ROUGH BLUEGRASS. (Fig. 137.) Culms erect from a decumbent base, often rather lax, scabrous below the panicle, 30 to 100 cm. tall; sheaths retrorsely scabrous or scaberulous, at least toward the summit; ligule 4 to 6 mm. long; blades scabrous, 2 to 4 mm. wide; panicle oblong, 6 to 15 mm. long, the lower branches about 5 in a whorl; spikelets usually 2- or 3-flowered, about 3 mm. long; lemma 2.5 to 3 mm. long, glabrous except the slightly pubescent keel, or lateral nerves rarely pubescent, the web at base conspicuous, the nerves prominent. ♃ —Moist places, Newfoundland and Ontario to North Carolina, Minnesota, South Dakota, and Colorado; on the Pacific coast from southern Alaska to northern California; on ballast, Louisiana; introduced from Europe. Sometimes used in mixtures for meadows and pastures under the name rough-stalked meadow grass.

FIGURE 137.—*Poa trivialis*. Panicle, × 1; floret, × 10. (Coville, N. Y.)

**24. Poa márcida** Hitchc. (Fig. 138.) Culms erect, in small tufts, 40 to 100 cm. tall; ligule very short; blades thin, 1 to 3 mm. wide; panicle drooping, narrow, 10 to 18 cm. long, the capillary branches somewhat distant, solitary or in pairs, ascending or appressed; spikelets mostly 2-flowered; glumes about 3 mm. long; lemmas narrowly lanceolate, acuminate, 4 to 5 mm. long, glabrous, long-webbed at base. ♃ —Bogs and wet shady places, Vancouver Island to the coast mountains of Oregon.

**25. Poa alsódes** A. Gray. (Fig. 139.) Culms in lax tufts, 30 to 60 cm. tall; blades thin, lax, 2 to 5 mm. wide; panicle 10 to 20 cm. long, very open, the slender branches in distant whorls of threes to fives, finally widely spreading, naked below, few-flowered; spikelets 2- or 3-flowered, about 5 mm. long; lemmas gradually acute, webbed at base, pubescent on the lower part of the keel, otherwise glabrous, faintly nerved. ♃ —Rich or moist woods, Ontario and Maine to Minnesota, south to Delaware and the mountains of North Carolina and Tennessee.

**26. Poa lánguida** Hitchc. (Fig. 140.) Culms weak, in loose tufts, 30 to 60 or even 100 cm. tall; ligule about 1 mm. long; blades lax, 2 to 4 mm. wide; panicle nodding, 5 to 10 cm. long, the few slender branches mostly in twos or threes, ascending, few-flowered toward the ends; spikelets 2- to 4-flowered, 3 to 4 mm. long; lemmas 2 to 3 mm. long, glabrous except the webbed base, oblong, rather obtuse, at maturity firm. ♃ (*P. debilis* Torr., not Thuill.)—Dry or rocky woods, Newfoundland and Quebec to Wisconsin, south to Pennsylvania, Kentucky, and Iowa.

FIGURE 138.—*Poa marcida*. Panicle, × 1; floret, × 10. (Type.)

**27. Poa saltuénsis** Fern. and Wieg. (Fig. 141.) Resembling *P. languida;* differing in the thinner, acute, somewhat longer lemmas. ♃ —Woodland thickets, Quebec and Newfoundland to Minnesota, south to Connecticut and Virginia.

**28. Poa occidentális** Vasey. NEW MEXICAN BLUEGRASS. (Fig. 142.) Culms erect, few in a tuft, usually rather stout, scabrous, as much as 1 to 1.5 m. tall; sheaths somewhat keeled, retrorsely scabrous (sometimes faintly so); ligule 2 to 8 mm. long; blades scabrous, 10 to 20 cm. long, 3 to 6 mm. wide; panicle open, 15 to 30 cm. long, the branches in distant whorls of threes to fives, spreading to reflexed, the lower as much as 10 cm. long, spikelet-bearing toward the ends; spikelets 3- to 6-flowered; lemmas 4.5 to 5 mm. long, conspicuously webbed at base, villous on the lower part of the keel and the marginal nerves and sometimes sparingly pubescent on the internerves below. ♃ —Open woods and moist banks at medium altitudes, Wyoming to New Mexico.

**29. Poa trácyi** Vasey. (Fig. 143.) Culms erect, 60 to 80 cm. tall; sheaths glabrous, keeled; ligule truncate, about 2 mm. long; blades 3 to 5 mm. wide; panicle narrowly pyram-

FIGURE 139.—*Poa alsodes*. Panicle, × 1; floret, × 10. (Wilson, N. Y.)

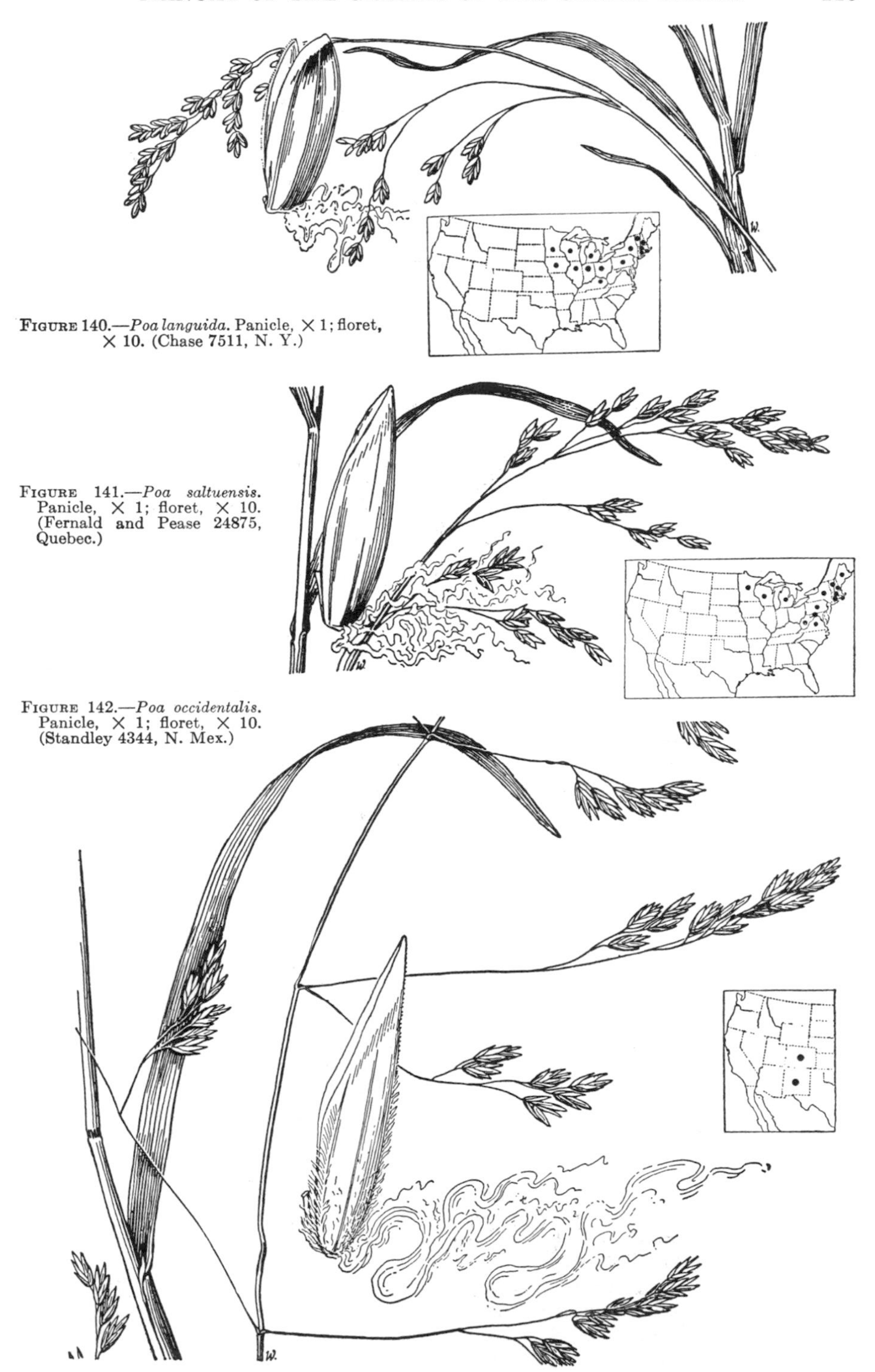


FIGURE 140.—*Poa languida*. Panicle, × 1; floret, × 10. (Chase 7511, N. Y.)

FIGURE 141.—*Poa saltuensis*. Panicle, × 1; floret, × 10. (Fernald and Pease 24875, Quebec.)

FIGURE 142.—*Poa occidentalis*. Panicle, × 1; floret, × 10. (Standley 4344, N. Mex.)

idal, 15 to 20 cm. long, the branches in distant whorls of 2 to 5, spreading, naked on the lower half or two-thirds; spikelets 2- or 3-flowered; lemmas about 3.5 mm. long, oblong-lanceolate or the upper lanceolate, webbed at base, villous on keel and marginal nerves, and more or less so on the internerves below, the intermediate nerves distinct. ♃ —Known only from Raton, N. Mex. May be a form of *P. occidentalis.*

FIGURE 143.—*Poa tracyi.* Panicle, × 1; floret, × 10. (Type.)

**30. Poa sylvéstris** A. Gray. (Fig. 144.) Culms tufted, erect, 30 to 100 cm. tall; sheaths glabrous or rarely pubescent, the lower usually antrorsely scabrous; ligule about 1 mm. long; blades lax, 2 to 6 mm. wide; panicle erect, 10 to 20 cm. long, much longer than wide, the slender flexuous branches spreading, usually 3 to 6 at a node, the lower usually reflexed; spikelets 2- to 4-flowered, 3 to 4 mm. long; lemmas 2.5 to 3 mm. long, webbed at base, pubescent on the keel and marginal nerves and more or less pubescent on the internerves. ♃ —Rich, moist, or rocky woods, New York to Wisconsin and Nebraska, south to Florida and Texas. Sheaths pubescent in a specimen from St. Louis, Mo.

**31. Poa refléxa** Vasey and Scribn. NODDING BLUEGRASS. (Fig. 145.) Culms solitary or in small tufts, erect, 20 to 40 cm. tall; blades rather short, 1 to 4 mm. wide; panicle nodding, 5 to 15 cm. long, the branches naked below, solitary, in pairs or in threes, the lower usually reflexed, sometimes strongly so; spikelets 2- to 4-flowered; lemmas about

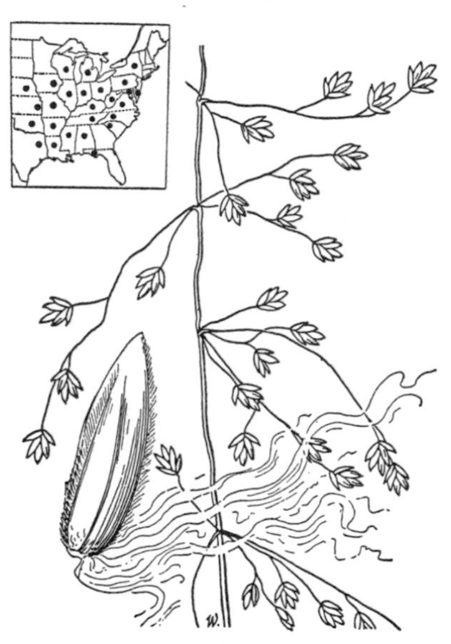

FIGURE 144.—*Poa sylvestris.* Panicle, × 1; floret × 10. (Wheeler 6, Mich.)

3 mm. long, oblong-elliptic, webbed at base, villous on keel and marginal nerves, sometimes on intermediate nerves. ♃ —Open slopes and alpine meadows, 2,000 to 4,000 m., Montana to eastern British Columbia, south in the mountains to New Mexico and Arizona.

**32. Poa wólfii** Scribn. (Fig. 146.) Culms tufted, erect, 40 to 80 cm. tall; sheaths slightly scabrous; blades crowded toward the base of the culms, mostly 1 to 2 mm. wide; panicle drooping, 8 to 15 cm. long, the branches ascending, bearing a few spikelets toward the ends, the lower mostly in pairs; spikelets 2- to 4-flowered, 5 to 6 mm. long; lemmas

3.5 to 4.5 mm. long, acute, webbed at base, pubescent on the keel and marginal nerves, the intermediate nerves distinct. ♃ —Moist woods, Ohio to Minnesota, Nebraska, and Missouri.

**33. Poa paucispícula** Scribn. and Merr. (Fig. 147.) Culms tufted, leafy, rather lax, 10 to 30 cm. tall, the base often decumbent; blades 1 to 2 mm. wide; panicle lax, few-flowered, 2 to 8 cm. long, the branches in pairs or solitary, naked below; spikelets ovate, purple, 4 to 6 mm. long, 2- to 5-flowered; glumes rather broad, acute, 3 to 4 mm. long; lemmas 3 to 4 mm. long, oblong, obtuse, webbed at base (the web sometimes scant), pubescent on the keel and marginal nerves below. ♃ —Rocky slopes, Alaska to Washington (alpine slopes, Mount Rainier, Mount Baker); Glacier National Park, Mont. More leafy than *P. leptocoma*, more tufted, the panicle branches not so long; spikelets broader.

**34. Poa leptocóma** Trin. BOG BLUEGRASS. (Fig. 148.) Culms slender, solitary, or few in a tuft, 20 to 50 cm. tall, often decumbent at base; sheaths usually slightly scabrous; ligule acute, the uppermost 3 to 4 mm. long; blades short, lax, mostly 2 to 4 mm. wide; panicle nodding, delicate, few-flowered, the branches capillary, ascending or spreading, subflexuous, the lower mostly in pairs; spikelets narrow, 2- to 4-flowered; glumes narrow, acuminate; lemmas 3.5 to 4.5 mm. long, acuminate, webbed at base, pubescent on the keel and marginal nerves or sometimes nearly glabrous, the intermediate nerves distinct. ♃ —Bogs, Alaska, south in the mountains to northern New Mexico, Colorado, and California (Mount Dana).

**35. Poa paludígena** Fern. and Wieg. (Fig. 149.) Culms slender, solitary or in small tufts, 15 to 70 cm. tall; sheaths minutely scabrous; ligule short, truncate, the uppermost as much as 1.5 mm. long; blades rather lax, mostly erect, 0.3 to 2 mm. wide; panicle loose and open, mostly 5 to 10 cm. long, the branches long and slender, distant, the lower mostly in twos, spikelet-bearing above the middle; spikelets mostly 4 to 5 mm. long, narrow, 2- to 5-flowered; lemmas 2.5 to 3.5 mm. long, webbed at base with a few long hairs, the keel and lateral nerves pubescent on the lower half or two-thirds, the intermediate nerves glabrous, obscure. ♃ —Bogs and springy places, New York and Pennsylvania to Illinois and Wisconsin.

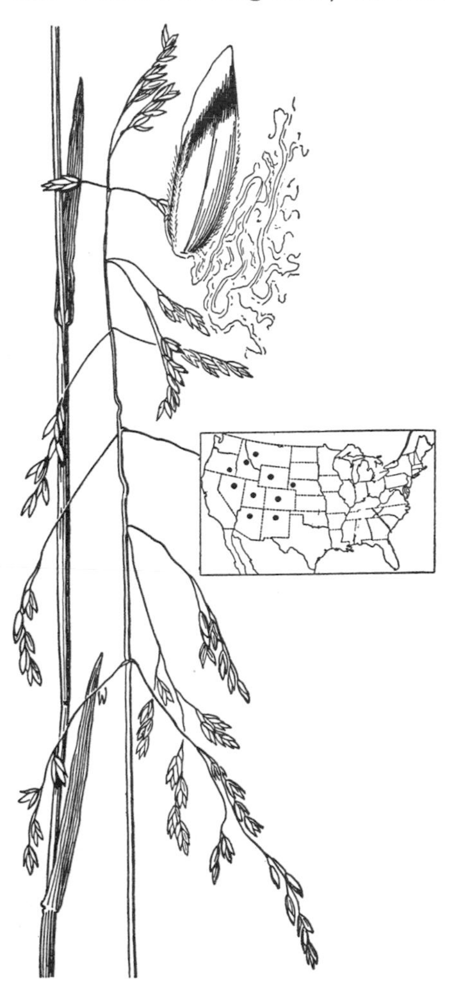

FIGURE 145.—*Poa reflexa*. Panicle, × 1; floret, × 10. (Clokey 11330, Colo.)

FIGURE 146.—*Poa wolfii*. Panicle, × 1; floret, × 10. (Deam 33821, Ind.)

**36. Poa bulbósa** L. BULBOUS BLUEGRASS. (Fig. 150.) Culms densely tufted, more or less bulbous at base, 30 to 60 cm. tall; blades flat or loosely involute, 1 to 2 mm. wide; panicle ovoid, mostly 5 to 8 cm. long, somewhat contracted, the branches ascending or appressed, some floriferous to base; spikelets mostly proliferous, the florets converted into bulblets with a dark purple base (about 2 mm. long), the bracts extending into slender green tips 5 to 15 mm. long; unaltered spikelets about 5-flowered, apparently not perfecting seed; lemmas 2.5 mm. long, webbed at base, densely silky on the keel and marginal nerves, the intermediate nerves faint. ♃ —Fields and meadows, New York to North Carolina; North Dakota to British Columbia and California; Utah; Colorado and Oklahoma; introduced from Europe, propagated by bulblets.

**37. Poa nemorális** L. WOOD BLUEGRASS. (Fig. 151.) Culms tufted, 30 to 70 cm. tall; ligule very short;

blades rather lax, about 2 mm. wide; panicle 4 to 10 cm. long, the branches spreading; spikelets 2- to 5-flowered, 3 to 5 mm. long; glumes narrow, sharply acuminate, about as long as the first floret; lemmas 2 to 3 mm. long, sparsely webbed at base, pubescent on the keel and marginal nerves, the intermediate nerves obscure. ♃ —Labrador to Alaska and British Columbia; occasional in meadows,

FIGURE 147.—*Poa paucispicula*. Panicle, × 1; floret, × 10. (Hitchcock 11711, Wash.)

FIGURE 149.—*Poa paludigena*. Panicle, × 1; floret, × 10. (Eames and Wiegand 9250, N. Y.)

FIGURE 148.—*Poa leptocoma*. Panicle, × 1; floret, × 10. (Arsène and Benedict 15562, N. Mex.)

Maine to Pennsylvania, Michigan, and Minnesota; Wyoming; Washington; Delaware and Virginia; introduced from Europe. Differing from *P. palustris* and *P. interior* in the very short ligule and the narrow acuminate glumes.

**38. Poa macrocláda** Rydb. (Fig. 152.) Culms 50 to 80 cm. tall, glabrous; ligule prominent, 2 to 3 mm. long; blades 2 to 3 mm. wide; panicle open, 10 to 20 cm. long, pyramidal, the branches spreading, distant, in twos or threes, as much as 8 cm. long, naked on the lower half or

FIGURE 150.—*Poa bulbosa*, × 1. (Henderson 6136, Idaho.)

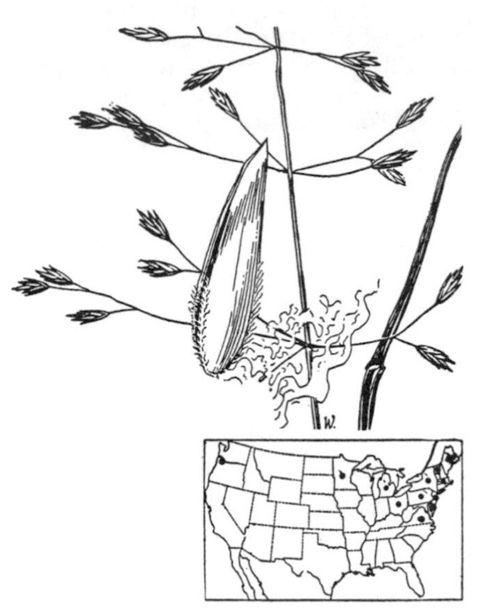

FIGURE 151.—*Poa nemoralis*. Panicle, × 1; floret, × 10. (Hitchcock 23662, Newfoundland.)

two-thirds; spikelets about 6 mm. long, 2- or 3-flowered, purple; glumes 3.5 to 4 mm. long; lemmas 4 to 4.5 mm. long, pubescent on the keel and marginal nerves, the web scant or wanting. ♃ —Moist places, at medium altitudes, Colorado, Montana, and Idaho. A little-known species, allied to *P. palustris*, but with larger spikelets.

**39. Poa palústris** L. FOWL BLUEGRASS. (Fig. 153.) Culms loosely tufted, glabrous, decumbent at the

FIGURE 152.—*Poa macroclada*. Panicle, × 1; floret, × 10. (Duplicate type.)

flattened purplish base, 30 to 150 cm. tall; sheaths keeled, sometimes scaberulous; ligule 3 to 5 mm. long, or only 1 mm. on the innovations; blades 1 to 2 mm. wide; panicle pyramidal or oblong, nodding, yellowish green or purplish, 10 to 30 cm. long, the branches in rather distant fascicles, naked below; spikelets 2- to 4-flowered, about 4 mm. long; glumes lanceolate, acute, shorter than the first floret; lemmas 2.5 to 3 mm. long, usually bronzed at the tip, webbed at base, villous on the keel and marginal nerves. ♃ —Meadows and moist open ground, at low and medium altitudes, Newfoundland and Quebec to Alaska, south to Virginia, Missouri, Nebraska, New Mexico, and California (Sierra Valley, Siskiyou County); Eurasia.

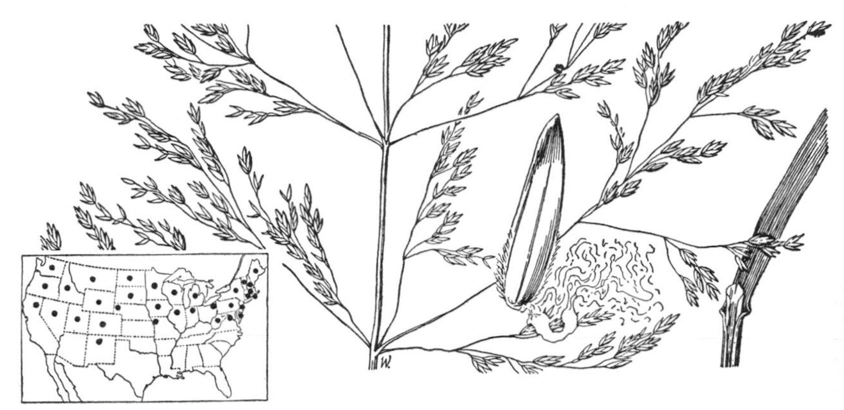

FIGURE 153.—*Poa palustris*. Panicle, × 1; floret, × 10. (Suksdorf 7022, Wash.)

**40. Poa intérior** Rydb. INLAND BLUEGRASS. (Fig. 154.) Culms erect from a usually densely tufted erect base, commonly rather stiff, often scabrous below the panicle, 20 to 50 cm. tall; sheaths slightly keeled or terete; ligule usually less than 1 mm. long; blades 1 to 2 mm. wide; panicle narrowly pyramidal, 5 to 10 cm. long, the branches ascending, the lower 2 or 3 spikelets about 4 mm. long, 2- to 4-flowered; glumes relatively broad, acute to acuminate; lemmas 3 to 3.5 mm. long, webbed at base (the web sometimes scant or obscure), villous on the lower half of the keel and marginal nerves, the intermediate nerves faint. ♃ —Grassy slopes and open woods at medium altitudes, usually not extending much above timber line, Quebec to British Columbia and Washington, south to Vermont, Michigan, Minnesota, western Nebraska, Texas, and Arizona.

FIGURE 154.—*Poa interior*. Panicle, × 1; floret, × 10. (Clements 297, Colo.)

**4. Homalopóae.**—Culms flattened; blades flat or conduplicate, with a conspicuous boat-shaped tip, often splitting at the apex.

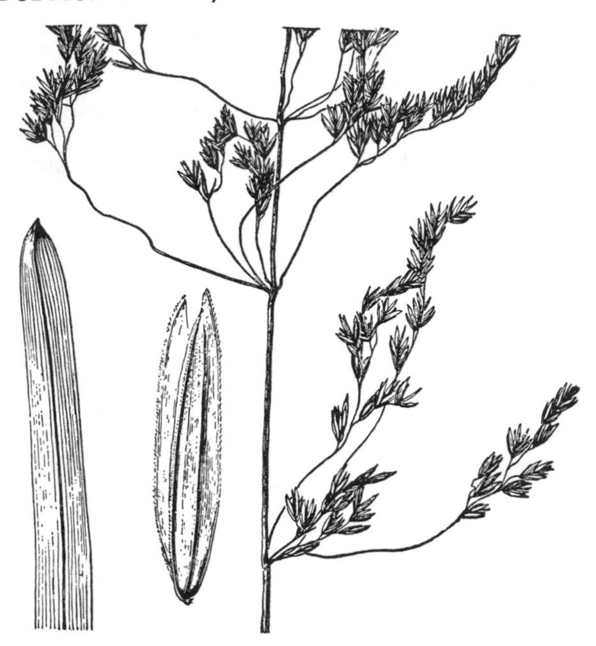

FIGURE 155.—*Poa chaixii.* Panicle, × 1; floret, × 10. (Lakela 2012, Minn.)

41. **Poa cháixii** Vill. (Fig. 155.) Culms erect or ascending, as much as 1 m. tall, soft, flattened, smooth and shining; sheaths compressed, keeled, glabrous, the lower somewhat crowded; blades mostly 10 to 20 cm. long, 4 to 8 mm. wide, flat or conduplicate, glabrous with scabrous margins; panicles about 15 cm. long, the slender spreading branches in whorls of 5, spikelet-bearing above the middle; spikelets 4 to 6 mm. long, 2- to 4-flowered, short-pediceled; lemmas 3.5 to 4 mm. long, acute, glabrous, or scabrous on the keel, distinctly 5-nerved. ♃ —Rich woods, Minnesota (Hunters Hill near Duluth, apparently indigenous); northern Europe.

5. **Alpínae.**—Perennials without creeping rhizomes; lemmas not webbed at base, pubescent on the keel or on the marginal nerves, or both, sometimes also pubescent on internerves.

42. **Poa fendleriána** (Steud.) Vasey. MUTTON GRASS. (Fig. 156.) Incompletely dioecious; culms erect, tufted, scabrous below the panicle, 30 to 50 cm. tall; sheaths somewhat scabrous; ligule less than 1 mm. long, not noticeable viewed from the side of the sheath; blades mostly basal, folded or involute, firm and stiff; panicle long-exserted, oblong, contracted, pale, 2 to 7 cm. long; spikelets 4- to 6-flowered, about 8 mm. long; glumes broad, 3 to 4 mm. long; lemmas 4 to 5 mm. long, villous on lower part of

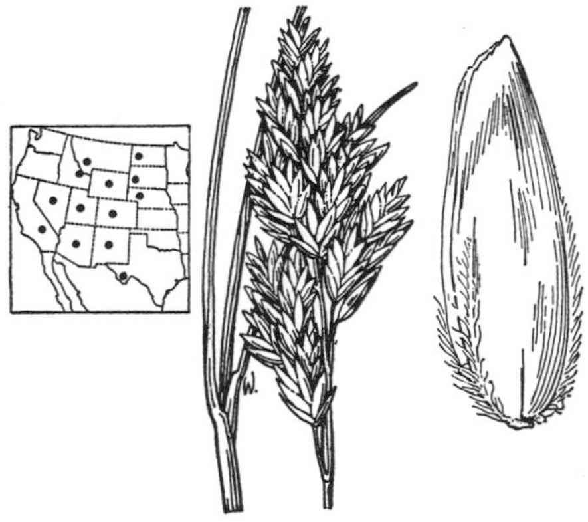

FIGURE 156.—*Poa fendleriana.* Panicle, × 1; floret, × 10. (Eggleston 6463, N. Mex.)

keel and marginal nerves, the intermediate nerves obscure; pistillate spikelets with minute stamens, the anthers about 0.2 mm. long. ♃ —Mesas, open dry woods, and rocky hills at medium altitudes, Manitoba to British Columbia, south through western South Dakota (Black Hills), Nebraska, and Idaho to western Texas (Chisos Mountains) and California; northern Mexico. A very small proportion of specimens have been found with well-developed stamens having large anthers, the pistil also developed.

**43. Poa longilígula** Scribn. and Will. LONGTONGUE MUTTON GRASS. (Fig. 157.) Differing from *P. fendleriana* in the prominent ligule, as much as 5 to 7 mm. long and in the looser, often longer, usually greenish panicle. ♃ —North Dakota to Oregon, south to New Mexico and California.

FIGURE 157.—*Poa longiligula.* Ligule, × 1. (Jones 5149, Utah.)

**44. Poa autumnális** Muhl. ex Ell. (Fig. 158.) Culms in rather large lax tufts, 30 to 60 cm. tall; blades 2 to 3 mm. wide, numerous at base; panicle 10 to 20 cm. long, about as broad, very open, the capillary flexuous branches spreading, bearing a few spikelets near the ends; spikelets 4- to 6-flowered, about 6 mm. long; lemmas oblong, obtusely rounded at the scarious compressed apex, villous on the keel and marginal nerves, pubescent on the internerves below or sometimes

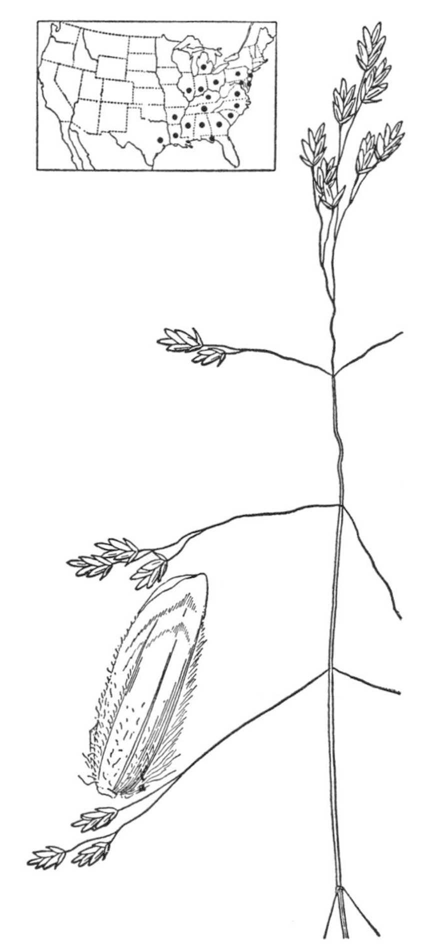

FIGURE 158.—*Poa autumnalis.* Panicle, × 1; floret, × 10. (Curtiss 6787, Ga.)

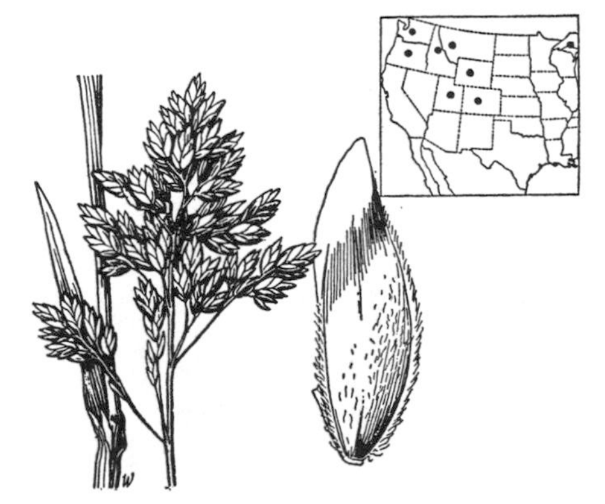

FIGURE 159.—*Poa alpina.* Panicle, × 1; floret, × 10. (Eggleston 11824, Colo.)

nearly to apex. ♃ —Moist woods, New Jersey to Michigan and Illinois, south to Florida and Texas.

**45. Poa alpína** L. ALPINE BLUEGRASS. (Fig. 159.) Culms erect from a rather thick vertical crown, rather stout, 10 to 30 cm. tall; blades short, 2 to 5 mm. wide, the uppermost about the middle of the culm; panicle ovoid or short-pyramidal, rather compact, 1 to 8 cm. long, the lower branches often reflexed; spikelets broad, purple or purplish; glumes broad, abruptly acute; lemmas 3 to 4 mm. long, strongly villous on the keel and marginal nerves, pubescent on the internerves below, the intermediate nerves faint. ♃ —Mountain meadows, Arctic regions of the Northern Hemisphere, extending south to Quebec, northern Michigan (Keweenaw Point), and the alpine summits of Colorado, Utah, Washington, and Oregon (Wallowa Mountains); Mexico.

**46. Poa stenántha** Trin. (Fig. 160.) Culms tufted, 30 to 50 cm. tall; ligule prominent, as much as 5 mm. long; blades flat or loosely involute, rather lax, mostly basal, 1 to 2 mm. wide, the uppermost culm leaf below the middle of the culm; panicle nodd ng, 5 to 15 cm. long, the branches in twos or threes, arcuate-drooping, naked below, with a few spikelets at the ends; spikelets 3- to 5-flowered, 6 to 8 mm. long; lemmas about 5 mm. long, pubescent on the lower part of keel and marginal nerves, sparsely pubescent on the internerves below. ♃ —Moist open ground, Alaska, Alberta, and British Columbia, extending into Montana, Colorado (White River Forest), Idaho, Washington (Nooksack River), and Oregon (Crater Lake).

FIGURE 160.—*Poa stenantha.* Panicle, × 1; floret, × 10. (Blankinship, Mont.)

FIGURE 161.—*A*, *Poa glauca.* Panicle, × 1; floret, × 10. (Hitchcock 16053, N. H.) *B*, *P. glaucantha.* Panicle, × 1; floret, × 10. (Butters, Abbe, and Abbe 258, Minn.)

**47. Poa glaúca** Vahl. (Fig. 161, *A*.) Plants glaucous, in close or loose tufts; culms compressed, stiff, 10 to 30 cm. tall, sometimes taller, naked above, the uppermost leaf usually much below the middle, its ligule about 2 mm. long; blades mostly basal, 3 to 5 cm. long, 1 to 2 mm. wide; panicle 3 to 7 cm. long, narrow, sometimes rather compact, the branches erect or ascending, few-flowered; spikelets mostly 2- or 3-flowered, 5 to 6 mm. long; lemmas 3 to 4 mm. long, strongly pubescent on the lower half of the keel and marginal nerves and often slightly pubescent on the faint intermediate nerves. ♃ —Rocky slopes, Arctic regions south to the alpine summits of New Hampshire; Wisconsin; Minnesota; Colorado. Common in Greenland; Eurasia.

**48. Poa glaucántha** Gaudin. (Fig. 161, *B*.) Plants mostly glaucous, culms compressed, in tufts, usually 30 to 70 cm. tall, leafy throughout; blades to 12 cm. long; panicle 6 to 16 cm. long, loose, but branches mostly ascending; spikelets 5 to 7 mm. long, 3- to 6-flowered; lemmas pubescent on keel and lateral nerves, sometimes with an obscure web at base. ♃ —Mountain meadows, slopes, and cliffs, Newfoundland to Quebec, Minnesota, Montana, and Wyoming; Europe. Resembles both *Poa nemoralis* and *P. interior*, distinguished from both by the florets without web at base or with very obscure web, from *P. nemoralis* by the flat culms, and from *P. interior* by the more strongly keeled sheaths and larger spikelets. A variable and puzzling species, apparently intermediate between *P. nemoralis* and *P. glauca*. *Poa scopulorum* Butters and Abbe is an unusually slender lax form.

Figure 162.—*Poa fernaldiana*. Panicle, × 1; floret, × 10. (Fernald, Maine.)

**49. Poa fernáldiana** Nannf. (Fig. 162.) Plants in loose lax bunches; culms weak and slender, 10 to 20 or sometimes 30 cm. tall; ligule truncate, about 1 mm. long; blades mostly basal, lax, mostly about 1 mm. wide; panicle narrow but loose, few-flowered, 2 to 6 cm. long, the branches ascending, naked below; spikelets 2- to 4-flowered, about 5 mm. long; lemmas 3 to 3.5 mm. long, densely villous on the lower half of the keel and marginal nerves, sometimes sparsely webbed at base. (Has been confused with *P. laxa* Haenke, a European species.) ♃ —Rocky slopes, Newfoundland and Quebec to the alpine summits of Maine, New Hampshire, Vermont, and New York. Common on the upper cone of Mount Washington.

Figure 163.—*Poa pattersoni*. Plant, × 1; floret, × 10. (Patterson 154, Colo.)

**50. Poa pattersóni** Vasey. Patterson bluegrass. (Fig. 163.) Culms loosely tufted with numerous basal leaves, 10 to 20 cm. tall; blades usually folded, rather lax, mostly less than 10 cm. long, about 1 mm. wide; panicle narrow, condensed, purplish,

1 to 4 cm. long; spikelets 2- or 3-flowered, 5 to 6 mm. long; lemmas about 4 mm. long, strongly pubescent on the keel and marginal nerves, short-pubescent on the internerves, sometimes sparsely webbed at base. ♃ —Alpine regions, Montana to Oregon (Mount Hood), Colorado, and Utah.

**51. Poa rupícola** Nash. TIMBERLINE BLUEGRASS. (Fig. 164.) Culms densely tufted, erect, rather stiff, often scaberulous below the panicle, 10 to 20 cm. tall; blades short, 1 to 1.5 mm. wide; panicle narrow, purplish, 2 to 4 cm. long, the short branches ascending or appressed; spikelets usually purple, about 3-flowered; lemmas villous below on keel and marginal nerves and some-

FIGURE 164.—*Poa rupicola*. Plant, × 1; floret, × 10. (Swallen 1348, Colo.)

FIGURE 165.—*Poa involuta*. Plant, × 1; floret, × 10. (Swallen 1110, Tex.)

times pubescent on the internerves below. ♃ —Rocky slopes, British Columbia, south in the mountains, at high altitudes, South Dakota (Black Hills) and Montana to Oregon (Mount Hood and Wallowa Mountains); New Mexico, and California (Mono Pass, Sheep Mountain). Small specimens of *P. interior*, which resemble this, differ in having a small web at the base of the lemma.

**6. Épiles.**—Perennials without rhizomes; lemmas not webbed at base, glabrous or scabrous (minutely pubescent in *P. unilateralis*).

**52. Poa involúta** Hitchc. (Fig. 165.) In dense pale tufts; culms slender, 30 to 40 cm. tall; ligule very short; blades involute, slender, 15 to 25 cm. long, glabrous or slightly scabrous; panicle open, 10 to 15 cm. long, the branches in pairs, few-flowered near the ends; spikelets mostly 3- or 4-flowered, 5 to 6 mm. long; lemmas 3 to 4 mm. long, scabrous. ♃ —Known only from the Chisos Mountains, Tex.

**53. Poa cusíckii** Vasey. Cusick bluegrass. (Fig. 166.) Culms in dense often large tufts, erect, 20 to 60 cm. tall; ligule very short; blades filiform, erect, scabrous, mostly basal; panicle usually pale, tawny, or purple-tinged, narrow, oblong, contracted or somewhat open at anthesis, 3 to 8 cm. long; spikelets 7 to 9 mm. long; lemmas 4.5 to 6 mm. long, smooth or scabrous. ♃ —Dry or rocky slopes at medium and high altitudes, Alberta to British Columbia, south to North Dakota, Colorado, and the central Sierras of California. A form with elongate blades and laxer panicle has been differentiated as *P. filifolia* Vasey; Idaho and Washington.

Figure 166.—*Poa cusickii.* Panicle, × 1; floret, × 10. (Howell 183, Oreg.)

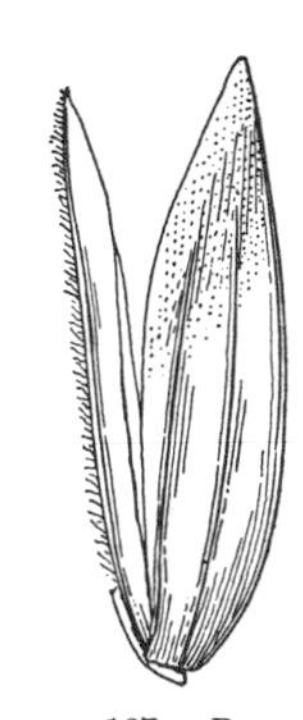

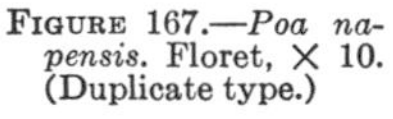

Figure 167.—*Poa napensis.* Floret, × 10. (Duplicate type.)

Figure 168.—*Poa unilateralis.* Panicle, × 1; floret, × 10. (Chase 5653, Calif.)

**54. Poa napénsis** Beetle. (Fig. 167.) Resembling *P. cusickii;* ligule about 1 mm. long, decurrent in young leaves; basal blades filiform, the culm blades 1.5 to 2.5 mm. wide; panicle as in *P. cusickii*, the spikelets slightly smaller; glumes 3 and 3.5 mm. long; lemmas about 4 mm. long, slightly to rather strongly scabrous. ♃ —Known only from Myrtledale Hot Springs, Napa County, Calif.

**55. Poa unilaterális** Scribn. (Fig. 168.) Culms in dense tufts, 10 to 40 cm. tall, sometimes decumbent at base; sheaths tawny, papery; blades flat or folded, shorter than the culms; panicle oblong, dense and spikelike or somewhat interrupted below, 2 to 6 cm. long; spikelets 6 to 8 mm. long; glumes broad, acute; lemmas 3 to 4 mm. long, glabrous except for a few short hairs on the nerves below. ♃ (*P. pachypholis* Piper.)—Cliffs, bluffs, and rocky meadows near the seashore, Washington (Ilwaco), Oregon, and California (Humboldt Bay to Monterey).

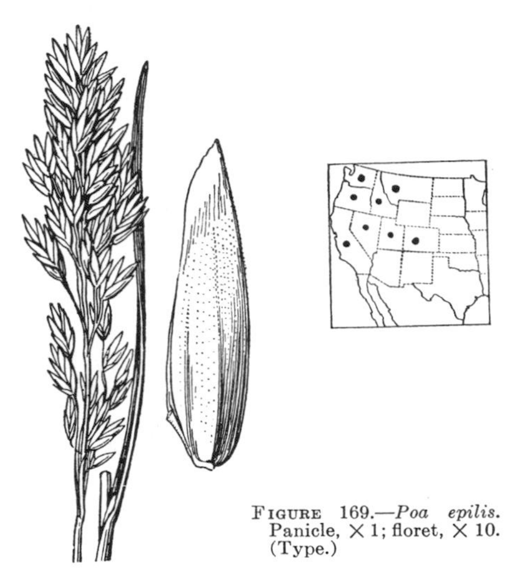

FIGURE 169.—*Poa epilis.* Panicle, × 1; floret, × 10. (Type.)

**56. Poa épilis** Scribn. SKYLINE BLUEGRASS. (Fig. 169.) Culms compressed, erect in rather loose to dense tufts, 20 to 40 cm. tall; ligule about 3 mm. long; blades of the culm about 3, flat, 3 to 6 cm. long, 2 to 3 mm. wide, of the innovations narrow, longer and usually folded or involute; panicle usually condensed, ovoid, 2 to 6 cm. long, long-exserted, usually purple, the lower branches naked below, ascending or appressed; spikelets 3-flowered, about 5 mm. long; lemmas 4 to 5 or even 6 mm. long, glabrous or minutely scabrous. ♃ —Mountain meadows, mostly above timber line, Alberta to British Columbia, south to Colorado and California.

**57. Poa vaseyóchloa** Scribn. (Fig. 170.) In small dense soft lax tufts;

FIGURE 170.—*Poa vaseyochloa.* Plant, × 1; floret, × 10. (Type.)

culms erect, 10 to 20 cm. tall; ligule acute, about 3 mm. long; blades lax, mostly folded or involute, in a basal tuft, mostly less than 5 cm. long, with one or two short ones on the culm, narrow or filiform; panicle ovate, 2 to 4 cm. long, few-flowered, open, the slender branches spreading, bearing 1 or 2 spikelets; spikelets purple, 3- to 6-flowered; glumes 2 to 3 mm. long, rather broad; lemmas smooth or minutely scabrous, 3 mm. long. ♃ —Rocky slopes, Cascade Mountains of Washington and Oregon in the vicinity of Columbia River, and the Wallowa Mountains of Oregon.

**58. Poa prínglei** Scribn. (Fig. 171.) Densely tufted; culms 10 to 20 cm. tall; lower sheaths loose, papery; blades mostly basal, involute, mostly 2 to 5 cm. long, sometimes longer, glabrous on the exposed surface, puberulent on inner surface; panicle narrow, condensed, usually pale or silvery, few- to several-flowered, 1 to 5 cm. long; spikelets 3- to 5-flowered, 6 to 8 mm. long; glumes equal, broad, 4 to 5 or rarely 7 mm. long; lemmas 5 to 6, rarely to 8 mm. long, smooth or scabrous. ♃ —Rocky alpine summits, Montana to Washington, south to Nevada (Mount Rose) and California.

FIGURE 171.—*Poa pringlei.* Plant, × 1; floret, × 10. (Henderson 3080, Idaho.)

**59. Poa lettermáni** Vasey. (Fig. 172.) In low lax tufts; culms mostly less than 10 cm. tall, usually scarcely exceeding the blades; ligule 1 to 2 mm. long; blades lax, usually not more than 1 mm. wide; panicle narrow, contracted, 1 to 3 cm. long; spikelets 3- or 4-flowered, 4 to 5 mm. long; glumes equal, somewhat acuminate, about as long as the first and second florets; lemmas erose at summit, 2.5 to 3 mm. long. ♃ —Rocky alpine summits, British Columbia, Washington, Wyoming, and Colorado to California.

FIGURE 172.—*Poa lettermani.* Plant, × 1; floret, × 10. (Letterman, Colo.)

**60. Poa montevánsi** Kelso. (Fig. 173.) Similar to *P. lettermani*, the culms (in type specimen) only 4.5 cm. tall, differing chiefly in the spikelets, with scabrous glumes and lemmas, the lemmas more acute and scarcely erose. ♃ —Known only from Mount Evans, 14,260 feet altitude, Colo.

FIGURE 173.—*Poa montevansi.* Plant, × 1; spikelet and floret, × 10. (Type.)

**61. Poa leibérgii** Scribn. LEIBERG BLUEGRASS. (Fig. 174.) Usually densely tufted; culms 5 to 30 cm. tall, erect; ligule 1 to 2 mm. long; blades mostly basal, firm, involute, usually less than 10 cm. long; panicle narrow,

FIGURE 174.—*Poa leibergii.* Plant, × 1; floret, × 10 (Type.)

2 to 5 cm. long, often purple, the branches short, appressed or ascending; spikelets 2- to 4-flowered, 4 to 6 mm. long; lemmas 3 to 4 mm. long, smooth or scaberulous. ♃ —Alpine meadows and sterile gravelly alpine flats, Idaho, eastern Oregon, and the Sierras of California.

**7. Scabréllae.**—Perennials, without rhizomes, tufted, with numerous basal leaves; spikelets little compressed, narrow, much longer than wide; lemmas convex, crisp-puberulent on the back towards the base, the keels obscure, the marginal and intermediate nerves usually faint. The whole group of Scabrellae is made up of closely related species which appear to intergrade.

**62. Poa scabrélla** (Thurb.) Benth. ex Vasey. PINE BLUEGRASS. (Fig. 175.) Culms erect, 50 to 100 cm. tall, usually scabrous, at least below the panicle; sheaths scaberulous; ligule 3 to 5 mm. long; blades mostly basal, 1 to 2 mm. wide, lax, more or less scabrous; panicle narrow, usually contracted, sometimes rather open at base, 5 to 12 cm. long; spikelets 6 to 10 mm. long; glumes 3 mm. long, scabrous; lemmas 4 to 5 mm. long, crisp-puberulent on the back toward base. ♃ —Meadows, open woods, rocks, and hills, at low and medium altitudes, western Montana and Colorado to Washington and California; Baja California. A form like *P. scabrella* in other respects but with smooth lemmas has been differentiated as *P. limosa* Scribn. and Will.—California (Mono Lake and Truckee).

**63. Poa gracíllima** Vasey. PACIFIC BLUEGRASS. (Fig. 176.) Culms rather loosely tufted, 30 to 60 cm. tall, usually decumbent at base; ligule 2 to 5 mm. long, shorter on the innovations; blades flat or folded, lax, from filiform to 1.5 mm. wide; panicle pyramidal, loose, rather open, 5 to 10 cm. long, the branches in whorls, the lower in twos to sixes, spreading or sometimes reflexed, naked below; spikelets 4 to 6 mm. long; second glume 3 to 4 mm. long; lemmas minutely scabrous, crisp-pubescent near base, especially on the nerves. ♃ —Cliffs and rocky slopes, Alberta to Alaska, south to Colorado and the southern Sierras of California. *Poa tenerrima* Scribn. is a form with open few-flowered panicles; southern Coast Ranges, California; *P. multnomae* Piper is a loose lax form in which the ligules on the innovations are short and truncate; wet cliffs, Multnomah Falls, Oreg.

**64. Poa secúnda** Presl. SANDBERG BLUEGRASS. (Fig. 177.) Culms erect from a dense, often extensive, tuft of short basal foliage, commonly not more than 30 cm., but sometimes up to 60 cm. tall; ligule acute, rather prominent; blades rather short, soft, flat, folded, or involute; panicle narrow, 2 to 10 cm. long, the branches short, appressed, or somewhat spreading in anthesis; spikelets about as in *P. gracillima.* ♃ (*P. sandbergii* Vasey.)—Plains, dry woods, rocky slopes, at medium and upper altitudes, but not strictly alpine, North Dakota to Yukon Territory, south to Nebraska, New Mexico, and southern California; Chile.

**65. Poa cánbyi** (Scribn.) Piper. CANBY BLUEGRASS. (Fig. 178.) Green or glaucous; culms 50 to 120 cm. tall; ligule 2 to 5 mm. long; blades flat or folded; panicle narrow, compact or rather loose, 10 to 15 cm. long, sometimes as much as 20 cm., the branches short, appressed; spikelets 3- to 5-flowered; lemmas more or less crisp-pubescent on lower part of back. ♃ (*P. lucida* Vasey; *P. laevigata* Scribn.)—Sandy or dry ground, Michigan (Isle Royale) and Minnesota to Yukon Territory, south to Colorado and eastern Washington to northern California; Quebec. *Poa lucida* has a slender but somewhat loose pale or shining panicle; *P. canbyi* has a denser, compact, dull green panicle, but the two forms grade into each other. *Poa lucida* is

FIGURE 175.—*Poa scabrella*. Plant, × ½; spikelet, × 5; floret, × 10. (Chase 5697, Calif.)

more common in Colorado and Wyoming; *P. canbyi* more common in Montana. The pubescence on the lemma may be obvious or obscure.

rather stiff; panicle narrow, 10 to 15 cm. long, pale, rather loose, the branches short-appressed; spikelets 3- to 5-flowered, 6 to 8 mm. long; glumes narrow, the second about as long as the lowest floret; lemmas 4 to 5 mm. long, rather obtuse at the scarious tip. ♃ —Low meadows and wet places, Montana to eastern Washington and Yukon Territory, south to Colorado, Arizona, and the Sierras and San Bernardino Mountains, California; on wool waste in Maine (North Berwick).

FIGURE 176.—*Poa gracillima*. Plant, × 1; floret, × 10. (Sandberg and Leiberg 747, Wash.)

FIGURE 177.—*Poa secunda*. Plant, × 1; floret, × 10. (Hitchcock 23202, Wyo.)

**7. Nevadénses.**—Perennials, without rhizomes, tufted; spikelets little compressed, narrow, much longer than wide; lemmas convex on the back, glabrous or minutely scabrous, not crisp-puberulent; keels obscure, marginal and intermediate nerves usually faint.

**66. Poa nevadénsis** Vasey ex Scribn. NEVADA BLUEGRASS. (Fig. 179.) Culms erect, 50 to 100 cm. tall; sheaths scabrous, sometimes only slightly so; ligule about 4 mm. long, shorter on the innovations, decurrent; blades usually elongate, narrow, involute, sometimes almost capillary,

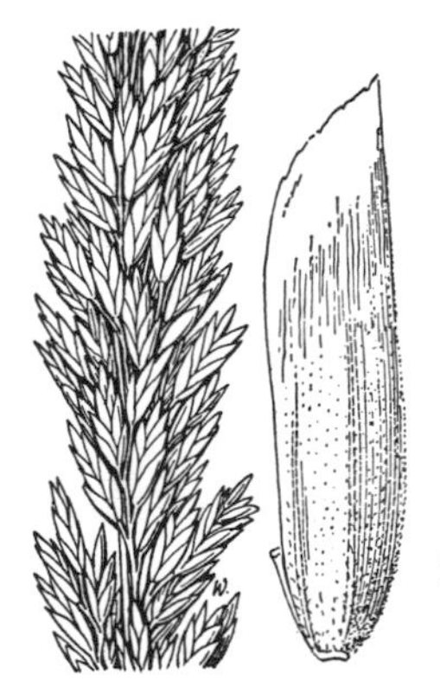

FIGURE 178.—*Poa canbyi*. Panicle, × 1; floret, × 10. (Williams 2787, Wyo.)

**67. Poa curtifólia** Scribn. (Fig. 180.) Culms several in a tuft from firm branched crowns, 10 to 20 cm. tall; ligule prominent, the uppermost as much as 5 mm. long; blades short, the lower 1.5 to 2 cm. long, 2 to 3 mm. wide, the upper successively smaller, the uppermost near the panicle, much reduced; panicle narrow, 3 to 6 cm. long; spikelets about 3-flowered; glumes equal, 5 mm. long, the first acuminate, the second broad, rather obtuse; lemmas 5 to 5.5 mm. long. ♃ —Known only from central Washington.

FIGURE 180.—*Poa curtifolia.* Panicle, × 1; floret, × 10. (Duplicate type.)

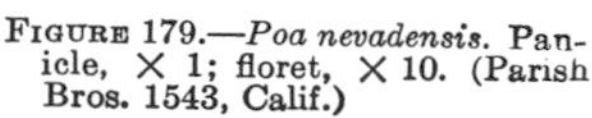

FIGURE 179.—*Poa nevadensis.* Panicle, × 1; floret, × 10. (Parish Bros. 1543, Calif.)

FIGURE 181.—*Poa juncifolia.* Panicle, × 1; floret, × 10. (Type.)

**68. Poa juncifólia** Scribn. Alkali bluegrass. (Fig. 181.) Pale; culms erect, 50 to 100 cm. tall; ligules short, those of the innovations not visible from the sides; blades involute, smooth, rather stiff; panicle narrow, 10 to 20 cm. long, the branches appressed; spikelets 3- to 6-flowered, 7 to 10 mm. long; glumes about equal; lemmas about 4 mm. long. ♃ (*P. brachyglossa* Piper.)—Alkaline meadows, Montana to British Columbia, south to South Dakota, Colorado, and east of the Cascades to northeastern California.

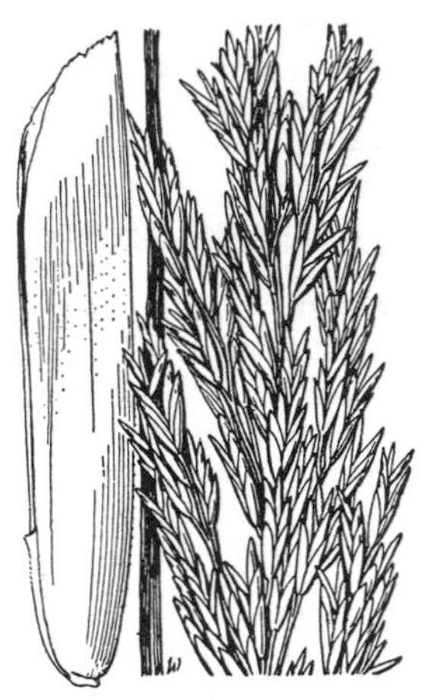

Figure 182.—*Poa ampla.* Panicle, × 1; floret, × 10. (Crandall 205, Colo.)

**69. Poa ámpla** Merr. Big bluegrass. (Fig. 182.) Green or glaucous; culms 80 to 120 cm. tall; sheaths smooth, rarely scaberulous; ligule short, rounded; blades 1 to 3 mm. wide; panicle narrow, 10 to 15 cm. long, usually rather dense; spikelets 4- to 7-flowered, 8 to 10 mm. long; lemmas 4 to 6 mm. long. ♃ —Meadows and moist open ground or dry or rocky slopes, North Dakota to Yukon Territory, south to Nebraska, New Mexico, and California. The typical form is robust and more or less glaucous; this grades into a smaller green form, more common in the eastern part of the range (*P. confusa* Rydb.). Occasional specimens of the typical form have short rhizomes.

## 13. BRÍZA L. Quaking grass

Spikelets several-flowered, broad, often cordate, the florets crowded and spreading horizontally, the rachilla disarticulating above the glumes and between the florets; glumes about equal, broad, papery-chartaceous, with scarious margins; lemmas papery, broad, with scarious spreading margins, cordate at base, several-nerved, the nerves often obscure, the apex in our species obtuse or acutish; palea much shorter than the lemma. Low annuals or perennials, with erect culms, flat blades, and usually open, showy panicles, the pedicels in our species capillary, allowing the spikelets to vibrate in the wind. Standard species, *Briza media.* Name from Greek, *Briza,* a kind of grain, from *brizein,* to nod.

The three species found in this country are introduced from Europe. They are of no importance agriculturally except insofar as *B. minor* occasionally forms an appreciable part of the spring forage in some parts of California. *B. maxima* is sometimes cultivated for ornament, because of the large showy spikelets.

Panicle drooping; spikelets 10 mm. wide .......... 1. B. maxima.
Panicle erect; spikelets 4 to 5 mm. wide.
  Plants perennial; upper ligule 1 mm. long; spikelets about 5 mm. long .... 3. B. media.
  Plants annual; upper ligule 5 mm. or more long; spikelets about 3 mm. long. 2. B. minor.

**1. Briza máxima** L. Big quaking grass. (Fig. 183.) Annual; culms erect or decumbent at base, 30 to 60 cm. tall; panicle drooping, few-flowered; spikelets ovate, 12 mm. long or more, 10 mm. broad, the pedicels slender, drooping; glumes and lemmas usually purple- or brown-margined. ⊙ —Sometimes cultivated for ornament; sparingly escaped in California (Monterey County) and Texas.

**2. Briza mínor** L. LITTLE QUAKING GRASS. (Fig. 184.) Annual; culms erect, 10 to 40 cm. tall; ligule of the upper leaf 5 mm. long or more, acute; blades 2 to 10 mm. wide; panicle 5

FIGURE 183.—*Briza maxima*. × ½. (Baenitz, Dalmatia.)

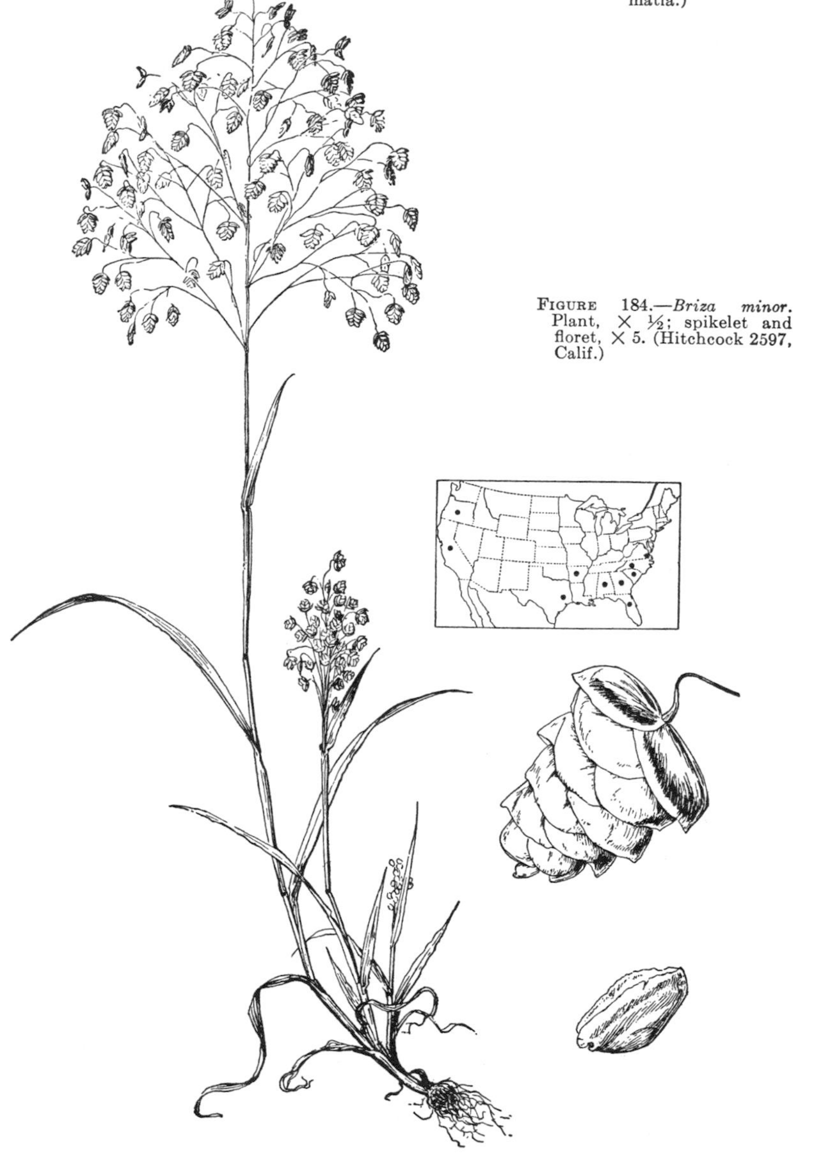

FIGURE 184.—*Briza minor*. Plant, × ½; spikelet and floret, × 5. (Hitchcock 2597, Calif.)

to 12 cm. long, the branches stiffly ascending, the spikelets pendent, triangular-ovate, 3- to 6-flowered, about 3 mm. long. ⊙ —Introduced at several localities in the Eastern States from Canada to Florida, Arkansas, and Texas, becoming common on the Pacific coast, especially in California.

**3. Briza média** L. (Fig. 185.) Perennial; culms 15 to 60 cm. tall; ligule of the upper leaf about 1 mm. long, truncate; blades 2 to 5 mm. wide; panicle erect, 5 to 10 cm. long, the branches rather stiff, ascending, naked below; spikelets 5- to 12-flowered, orbicular, about 5 mm. long. ♃ —Fields and waste places, sparingly introduced, Ontario to Connecticut and Michigan.

Desmazéria sicula (Jacq.) Dum. Low annual; culms spreading with ascending ends; panicles simple, 3 to 5 cm. long, with large flat 2-ranked spikelets. ⊙ —Occasionally cultivated for ornament. Europe. (Name sometimes spelled *Demazeria.*)

Figure 185.—*Briza media.* Panicle, × ½. (Oakes, Mass.)

## 14. ERAGRÓSTIS Beauv. Lovegrass

Spikelets few- to many-flowered, the florets usually closely imbricate, the rachilla disarticulating above the glumes and between the florets, or continuous, the lemmas deciduous, the paleas persistent; glumes somewhat unequal, shorter than the first lemma, acute or acuminate, 1-nerved, or the second rarely 3-nerved; lemmas acute or acuminate, keeled or rounded on the back, 3-nerved, the lateral nerves sometimes obscure; palea usually about as long as the lemma, the keels sometimes ciliate. Annuals or perennials of various habit, the inflorescence an open or contracted panicle. Type species, *Eragrostis eragrostis* Beauv. (*E. poaeoides*). Name from the Greek *eros,* love, and *agrostis,* a kind of grass.

Although the species are numerous, they in general appear to have little forage value. *Eragrostis intermedia* is said to furnish forage on the grazing lands of Arizona and New Mexico. Three introduced African species, *E. curvula, E. lehmanniana,* and *E. chloromelas,* show promise of being valuable in erosion control in the Southwest.

1a. Plants annual.
  2a. Plants creeping, rooting at the nodes, forming mats.
    Plants with perfect flowers; anthers 0.2 mm. long .......... 11. E. hypnoides.
    Plants dioecious; anthers 2 mm. long .......... 10. E. reptans.
  2b. Plants often decumbent at base but not creeping and forming mats.
    3a. Palea prominently ciliate on the keels, the cilia usually as long as the width of the lemma.
      Panicle interruptedly spikelike, rarely somewhat open; spikelets usually 3 to 4 mm. long .......... 7. E. ciliaris.
      Panicle narrow but open, the pedicels ascending or spreading; spikelets 2 mm. long. 8. E. amabilis.
    3b. Palea scabrous to short-ciliate.
      4a. Panicle long, narrow, rather dense, tawny or stramineous; spikelets 2 to 3 mm. long .......... 9. E. glomerata.

4b. Panicle more or less open; spikelets usually more than 3 mm. long.
5a. Spikelets sessile or nearly so........ 12. E. SIMPLEX.
5b. Spikelets pediceled.
6a. Spikelets mostly less than 5-flowered; lemmas obscurely nerved, scarcely keeled.
Panicles two-thirds the entire length of the plant or more, diffuse; pedicels more than 5 mm. long; culms erect, closely tufted.... 14. E. CAPILLARIS.
Panicles less than half the entire length of the plant, oblong, open but scarcely diffuse; pedicels mostly less than 5 mm. long; culms spreading or decumbent at base........ 15. E. FRANKII.
6b. Spikelets mostly more than 5-flowered.
7a. Spikelets ovate to oblong, flat, the florets spreading, closely imbricate. 13. E. UNIOLOIDES.
7b. Spikelets oblong to linear, the florets appressed.
8a. Plants with glandular depressions on the panicle branches, the keel of the lemmas, or on margins of blades or keel of sheaths.
Spikelets 2.5 mm. wide; glands prominent on keel of lemmas. Anthers 0.5 mm. long........ 24. E. CILIANENSIS.
Spikelets not more than 2 mm. wide, mostly less; glandular depressions mostly on panicle branches and leaves.
Panicle narrow, rather dense........ 23. E. LUTESCENS.
Panicle open, at least one-fourth as wide as long.
Spikelets 1.5 to 2 mm. wide, dark drab; panicle branches relatively stout and stiff........ 25. E. POAEOIDES.
Spikelets about 1 mm. wide, pale; panicle branches slender, spreading........ 17. E. PERPLEXA.
8b. Plants not glandular on the branches nor lemmas, sometimes glandular on the sheaths (*E. neomexicana*) and below the nodes (*E. barrelieri*).
Spikelets about 1 mm. wide, linear, slender.
Plant delicate; spikelets 3 to 5 mm. long; lemmas 1 to 1.5 mm. long. 16. E. PILOSA.
Plant rather stout; spikelets 5 to 7 mm. long; lemmas about 2 mm. long. 22. E. ORCUTTIANA.
Spikelets 1.5 mm. wide or wider, ovate to linear.
Panicle narrow, the branches ascending, spikelet-bearing nearly to base, few-flowered; spikelets linear, mostly 10- to 15-flowered. 26. E. BARRELIERI.
Panicle open, often diffuse.
Spikelets linear, mostly 8- to 15-flowered, on slender spreading pedicels mostly longer than the spikelets........ 29. E. ARIDA.
Spikelets ovate to linear, if linear not on spreading pedicels.
Spikelets linear at maturity, appressed along the primary panicle branches, these naked at the base for usually 5 to 10 mm. Lower lemmas 1.5 mm. long.
Primary panicle branches simple or the lower with a branchlet bearing 2 or 3 spikelets; spikelets loosely imbricate or sometimes not overlapping; plants slender, mostly less than 30 cm. tall, the culms slender at base. Chiefly east of the 100th meridian........ 18. E. PECTINACEA.
Primary panicle branches usually bearing appressed branchlets with few to several spikelets, the spikelets thus appearing imbricate or crowded along the primary branches; plants more robust, mostly more than 30 cm. tall, the culms stouter at the base. Chiefly from Texas to southern California........ 19. E. DIFFUSA.
Spikelets ovate to ovate-oblong, rarely linear, if linear not appressed along the primary panicle branches.
Plants comparatively robust, usually more than 25 cm. tall. Texas to southern California.
Panicle large, the branches many-flowered, ascending or drooping. Plant as much as 1 m. tall, with blades as much as 1 cm. wide, but often smaller. 27. E. NEOMEXICANA.
Panicle smaller and more open, the spreading branches few-flowered. Plant usually less than 30 cm. tall. 28. E. MEXICANA

Plants delicate, mostly less than 25 cm. tall; blades mostly not more than 2 mm. wide (see also *E. frankii* var. *brevipes*).
Panicle lax, the branches usually naked at base; spikelets 4 to 7 mm. long........................ 20. E. TEPHROSANTHOS.
Panicle rather stiff, the branches often floriferous nearly to the base; spikelets mostly not more than 3 mm. long. 21. E. MULTICAULIS.

1b. Plants perennial.
9a. Panicle elongate, slender, dense, spikelike........................ 6. E. SPICATA.
9b. Panicle open or contracted, not spikelike.
10a. Plants with stout scaly rhizomes........................ 1. E. OBTUSIFLORA.
10b. Plants without rhizomes.
11a. Spikelets subsessile or nearly so, the lateral pedicels not more than 1 mm. long.
Spikelets subsessile, distant along the few stout panicle branches. 2. E. SESSILISPICA.
Spikelets short-pediceled.
Panicle large, becoming a tumbleweed, the axis and branches viscid. 3. E. CURTIPEDICELLATA.
Panicle narrow (sometimes open in *E. oxylepis*), not a tumbleweed nor viscid; keels of palea forming a thick white band; grain 1 to 1.2 mm. long.
Lemmas 3 mm. long, somewhat abruptly narrowed to the acute apex; panicle usually red brown; anthers 0.2 to 0.3 mm. long............ 4. E. OXYLEPIS.
Lemmas 3.5 mm. long, tapering to the acuminate apex; panicle pale or slightly pinkish; anthers 0.4 to 0.5 mm. long........................ 5. E. BEYRICHII.
11b. Spikelets with pedicels more than 1 mm. long (appressed along the branches in *E. refracta;* sometimes scarcely more than 1 mm. long in *E. chariis* and *E. bahiensis*). Panicles large and open (sometimes condensed in *E. bahiensis*).
12a. Nerves of lemma obscure; lemma rounded on back, sometimes slightly keeled toward apex.
Axils of main panicle branches usually strongly pilose (rarely glabrous in *E. intermedia*).
Sheaths pilose or hirsute (sometimes glabrous in *E. hirsuta*).
Culms mostly more than 50 cm. tall; blades elongate, flat, not crowded at base of culm........................ 30. E. HIRSUTA.
Culms mostly less than 50 cm. tall; blades rather short and crowded at base of culm........................ 32. E. TRICHOCOLEA.
Sheaths glabrous or nearly so, except the pilose summit.
Spikelets about 1 mm. wide, 3- to 7-flowered, 3 to 5 mm. long; lemmas 1.3 to 1.5 mm. long........................ 31. E. LUGENS.
Spikelets about 1.5 mm. wide; 3- to 8-flowered, 3 to 10 mm. long; lemmas 1.8 to 2 mm. long........................ 35. E. INTERMEDIA.
Axils of main panicle branches glabrous or the lower sparsely pilose.
Pedicels bearing above the middle a glandular band or spot; axils glabrous. 36. E. SWALLENI.
Pedicels without glandular band; lower axils sparsely pilose to glabrous.
Lemmas about 3 mm. long........................ 33. E. EROSA.
Lemmas about 2 mm. long........................ 34. E. PALMERI.
12b. Nerves of lemma evident, usually prominent; lemmas keeled.
Spikelets approximate in a somewhat condensed panicle, or along the main branches of a somewhat spreading panicle; florets mostly 15 to 30.
Panicle branches distant, glabrous or nearly so in the axils.
Paleas readily deciduous........................ 45. E. CHARIIS.
Paleas persistent........................ 46. E. BAHIENSIS.
Panicle branches approximate, villous in the axils. Culms densely cespitose with arcuate blades attentuate to long filiform flexuous tips. 47. E. CURVULA.
Spikelets in an open panicle.
Panicle longer than broad, the branches not horizontally spreading.
Culms not more than 60 cm. tall.
Spikelets 9- to 15-flowered; panicle less than one-third the entire length of culm, the branches not viscid........................ 37. E. TRACYI.
Spikelets 4- to 8-flowered; panicle more than half the entire length of culm, the branches viscid........................ 38. E. SILVEANA.
Culms usually 1 m. or more tall.
Spikelets mostly not more than 6-flowered, purplish. 39. E. TRICHODES.

Spikelets mostly 8- to 15-flowered, stramineous to bronze. 40. E. PILIFERA.

Panicle at maturity about as broad as long.
Panicle purple, the branches slender but rigid............ 41. E. SPECTABILIS.
Panicle green to leaden, the branches capillary, fragile.
Spikelets appressed and distant along the nearly simple panicle branches. 44. E. REFRACTA.
Spikelets on long pedicels.
Lemmas 2 mm. long.............................................. 42. E. ELLIOTTII.
Lemmas 3 mm. long.............................................. 43. E. ACUTA.

FIGURE 186.—*Eragrostis obtusiflora*. Plant, × ½, two views of floret, × 10. (Toumey, Ariz.)

SECTION 1. CATACLÁSTOS Doell

Rachilla of spikelets disarticulating between the florets at maturity.

**1. Eragrostis obtusiflóra** (Fourn.) Scribn. (Fig. 186.) Culms erect or ascending, firm, wiry, 30 to 50 cm. tall, from stout creeping rhizomes with closely imbricate hard spiny-pointed scales; sheaths pubescent or pilose at the throat; blades firm, glaucous, flat, becoming involute at least toward the spiny-pointed tip, 5 to 10 cm. long, 2 to 3 mm. wide at base; panicle 5 to 15 cm. long, the rigid simple branches ascending, loosely flowered, 5 to 8 cm. long; spikelets pale or purplish, 6- to 12-flowered, 8 to 12 mm. long, the pedicels about 1 mm. long; glumes acute, 3 and 5 mm. long; lemmas rounded on the back, rather loosely imbricate, obtuse, somewhat lacerate, about 4 mm. long. ♃ — Alkali soil, Arizona, New Mexico (Las Playas); Mexico. "This species is one of the most abundant grasses in the extreme alkaline portions of Sulphur Springs Valley, Arizona, where the large rootstocks in many places bind the shifting sands. It rarely flowers, and its superficial appearance, without flowers, is much the same as our common salt grass (*Distichlis spicata*). It is a hard, rigid grass, but furnishes a large part of the forage of Sulphur Springs Valley, when other grasses are eaten off or are cut short by drought."—Toumey in letter.

**2. Eragrostis sessilispíca** Buckl. (Fig. 187.) Perennial; culms tufted, erect, 20 to 40 cm. tall, with 1 node above the basal cluster of leaves; sheaths glabrous, strongly pilose at the throat; blades flat to rather loosely involute, 1 to 2 mm. wide; panicle loose, open, pilose in the axils, at first about half the entire length of the culm, elongating toward maturity, the axis curving or loosely spiral, as much as 40 cm. long, the distant branches stiffly spreading, 5 to 15 cm. long, floriferous to base, sometimes bearing below a few secondary branches, the whole panicle finally breaking away and tumbling before the wind; spikelets distant, nearly sessile, appressed, linear, 5- to 12-flowered, 8 to 12 mm. long; glumes acute, about 3 mm. long; lemmas loosely imbricate, acuminate, becoming somewhat indurate, 3 to 3.5 mm. long, the lateral nerves prominent; palea prominently bowed out below. ♃ (*Acamptoclados sessilispica* Nash.)—Plains and sandy prairies, Kansas to Texas, New Mexico, and northern Mexico.

FIGURE 187.—*Eragrostis sessilispica*. Panicle, × 1; floret, × 10. (Swallen 1791, Tex.)

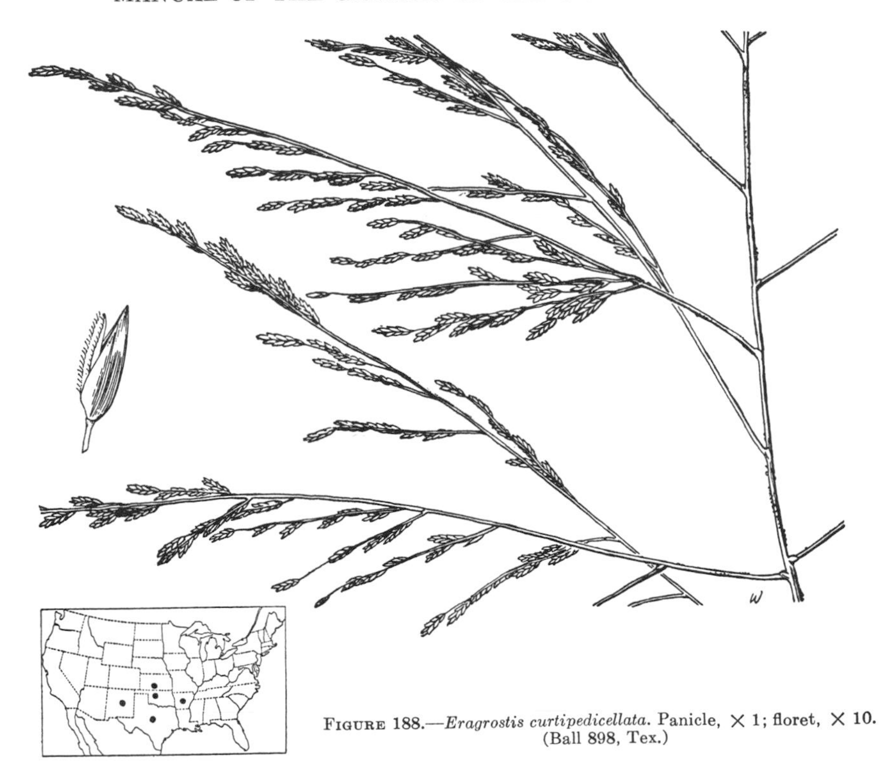


FIGURE 188.—*Eragrostis curtipedicellata*. Panicle, × 1; floret, × 10. (Ball 898, Tex.)

**3. Eragrostis curtipedicelláta** Buckl. (Fig. 188.) Perennial; culms tufted, erect, 20 to 40 cm. tall; sheaths pilose at the throat; blades flat or loosely involute, 1 to 3 mm. wide; panicle open, spreading, at first 15 to 20 cm. long, the axis and branches viscid, rather sparingly pilose in the axils, finally elongating, breaking away and tumbling before the wind, the branches stiffly ascending or spreading; spikelets oblong or linear, short-pediceled, somewhat appressed on the primary and secondary branches, 6- to 12-flowered, 3 to 6 mm. long; glumes about 1.5 mm. long; lemmas rather closely imbricate, oblong, acute, about 1.5 mm. long; palea ciliate on the keels, not bowed out; grain 0.7 mm. long. ♃ —Plains, open woods, and dry slopes, Colorado and Kansas to Arkansas, Texas, and New Mexico.

**4. Eragrostis oxylépis** (Torr.) Torr. (Fig. 189.) Perennial; culms tufted, suberect, 20 to 70 cm. tall;

FIGURE 189.—*Eragrostis oxylepis*. Panicle, × 1; floret, × 10. (Reverchon 3501A, Tex.)

sheaths long-pilose at the throat, the foliage otherwise glabrous, the blades flat, more or less involute in drying, 1 to 4, rarely to 5, mm. wide, tapering to a fine point; panicle 5 to 25 cm. long (mostly 10 to 15 cm.) of several to numerous stiff, ascending or spreading densely flowered branches, approximate to distant, the spikelets mostly aggregate on very short branchlets; spikelets usually red brown, strongly compressed, subsessile, linear at maturity, mostly 10- to 40-flowered, 8 to 15 mm. long; lemmas closely imbricate, 3 mm. long, abruptly narrowed to an acute apex, the tip slightly spreading; palea bowed out below, the keels prominent; anthers 0.2 to 0.3 mm. long; grain 1 to 1.2 mm. long. ♃ —Sandy soil, northern Florida to Colorado, New Mexico, and California (San Diego); eastern Mexico to Vera Cruz. Has been confused with *E. secundiflora* Presl, a rather rare species of Mexico, which it closely resembles, but the latter has less strongly compressed spikelets and grains only 0.4 to 0.5 mm. long.

**5. Eragrostis beyríchii** J. G. Smith. (Fig. 190.) Resembling *E. oxylepis* and possibly only a variety of that species; differing in the softer foliage and panicle, the plant on the average smaller, the panicle mostly smaller, pale or slightly pinkish; spikelets slightly larger; lemmas 3.5 to 4 mm. long (the lower shorter), less firm, tapering to an acuminate apex; palea broader, less bowed out; anthers 0.4 to 0.5 mm. long, yellowish, grain 1 mm. long. ♃ —Sandy soil, Texas and Oklahoma (Wichita Mountains); Mexico.

**6. Eragrostis spicáta** Vasey. (Fig. 191.) Perennial; culms tufted, erect, about 1 m. tall; blades flat, elongate, more or less involute in drying, tapering to a slender point; panicle pale, slender, dense, spikelike, 10 to 30 cm. long, 3 to 4 mm. thick; spikelets strongly compressed, 2- or 3-flowered, 2 mm. long, the somewhat pubescent pedicels less than 1 mm. long; glumes rather broad, obtuse, unequal, the second about 1 mm. long; lemmas about 2 mm. long, all rising to about the same height, the lateral pair of nerves faint. ♃ —Dry ground, Laredo and Brownsville, Tex.; Baja California; Paraguay, Argentina.

FIGURE 190.—*Eragrostis beyrichii*. Panicle, × 1; floret, × 10. (Tracy 7924, Tex.)

FIGURE 191.—*Eragrostis spicata*. Panicle, × 1; spikelet, × 10. (Swallen 1086, Tex.)

**7. Eragrostis ciliáris** (L.) R. Br. (Fig. 192.) Annual; culms branching, erect to spreading, slender, wiry, 15 to 30 cm. tall; blades flat to subinvolute, mostly less than 10 cm. long, 1 to 3 mm. wide; panicle often purplish, condensed, interruptedly spikelike, 3 to 10 cm. long, sometimes looser with stiffly ascending short branches; spikelets 6- to 12-flowered, 2 to 4 mm. long; glumes about 1 mm. long; lemmas oblong, 1 to 1.5 mm. long, obtuse, the midnerve slightly excurrent; keels of the palea conspicuously stiffly long-ciliate, the hairs 0.5 to 0.7 mm. long; grain 0.5 mm. long. ⊙ —Sandy shores, rocky soil, and open ground, South

FIGURE 192.—*Eragrostis ciliaris*. Plant, × ½; spikelet, × 5; floret, × 10. (Nash 2104, Fla.)

Carolina to Florida and Mississippi; Texas; New Jersey (ballast); West Indies and Mexico to Brazil and Peru; Africa; Asia. Specimens with laxer panicles of more spreading loosely flowered branches have been differentiated as *E. ciliaris* var. *laxa* Kuntze.

**8. Eragrostis amábilis** (L.) Wight and Arn. ex Nees. (Fig. 193.) Annual, resembling *E. ciliaris;* blades as much as 5 mm. wide; panicle oblong or oblong-lanceolate, 2 to 4 cm. wide, rather open; spikelets 4- to 8-flowered, about 2 mm. long; glumes less than 1 mm. long; lemmas ovate, obtuse, 1 mm. long; keels of palea long-ciliate, the hairs about 0.3 mm. long. ⊙ (*E. plumosa* Link.)—Gardens and waste places, Georgia and Florida; Texas; tropical America; apparently introduced from the Old World.

**9. Eragrostis glomeráta** (Walt.) L. H. Dewey. (Fig. 194.) Annual; culms erect, 20 to 100 cm. tall, branching below, the branches erect; blades flat, 3 to 8 mm. wide, tapering to a fine point; panicle narrow, erect, densely flowered, somewhat interrupted, 5 to 50 cm. long, greenish or tawny, the branches ascending or appressed, floriferous to base, many-flowered; spikelets short-pediceled, mostly 6- to 8-flowered, 2 to 3 mm. long; glumes minute; lemmas very thin, about 1 mm. long; grain about 0.3 to 0.4 mm. long. ⊙ (*E. conferta* Trin.)—Banks of ponds and streams, and low ground, South Carolina to Florida, Missouri, and eastern Texas, south through Mexico and the West Indies to Uruguay.

FIGURE 193.—*Eragrostis amabilis*. Panicle, × ½; spikelet, × 10. (Meislahn 10, Fla.)

FIGURE 194.—*Eragrostis glomerata*. Panicle, × ½; spikelet and floret, × 10. (Eggert, Ark.)

SECTION 2. PTEROÉSSA Doell

Rachilla of spikelet continuous, not disarticulating at maturity; palea usually persistent for a short time after the fall of the lemma (sometimes falling with it in *E. unioloides* and *E. chariis*).

**10. Eragrostis réptans** (Michx.) Nees. (Fig. 195.) Annual, dioecious; culms branching, creeping, rooting at the nodes, forming mats; blades flat, usually pubescent, mostly 1 to 3 cm. long; panicles numerous, ovoid, usually rather dense or capitate, few- to several-flowered, rarely many-flowered, mostly 1 to 2 cm. long; spikelets several- to many-flowered, linear, at length elongate and more or less curved; lemmas closely imbricate, often sparsely villous, acuminate, about 3 mm. long; palea of pistillate floret about half as long as the lemma,

FIGURE 195.—*Eragrostis reptans.* Pistillate (♀) and staminate (♂) plants, × ½; floret, × 10. (Bush 1306 (♀) and 1307 (♂), Tex.)

FIGURE 196.—*Eragrostis hypnoides.* Plant, × ½; floret, × 10. (Mearns 741, Minn.)

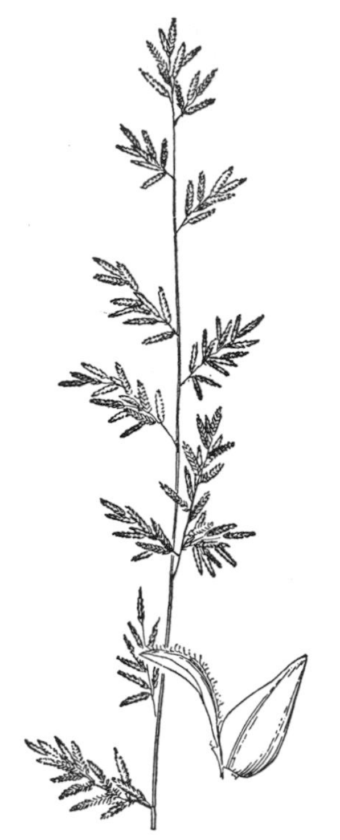

FIGURE 197.—*Eragrostis simplex.* Panicle, × ½; floret, × 10. (Curtiss, Fla.)

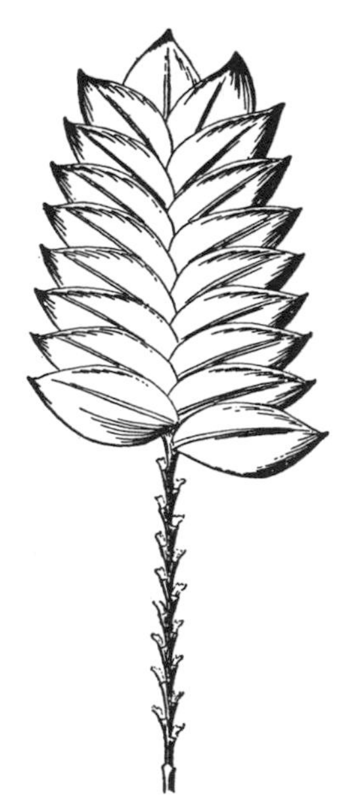

FIGURE 198.—*Eragrostis unioloides.* Spikelet, × 10. (Curtiss 6898, Fla.)

of the staminate floret as long as the lemma; grain ovoid, about 0.5 mm. long; anthers before dehiscing, 1.5 to 2 mm. long. ⊙ (*E. capitata* Nash.)—River banks, sandy land, and open ground, Kentucky to South Dakota and Texas; Florida.

**11. Eragrostis hypnoídes** (Lam.) B. S. P. (Fig. 196.) Annual, branching, creeping, and matlike as in the preceding; blades scabrous or pubescent on the upper surface; panicles elliptic, loosely few-flowered, 1 to 5 cm. long, sometimes somewhat capitate; spikelets several- to many-flowered, linear, mostly 5 to 10 mm. long, sometimes as much as 2 cm. long in a dense cluster; flowers perfect; lemmas glabrous, acute, 1.5 to 2 mm. long; palea about half as long as the lemma; grain 0.5 mm. long; anthers about 0.2 mm. long. ⊙ —Sandy river banks and wet ground, Quebec to Washington, south through Mexico and the West Indies to Argentina; not found in the Rocky Mountains.

**12. Eragrostis símplex** Scribn. (Fig. 197.) Annual; culms spreading to suberect, 10 to 30 cm. tall; blades flat, 1 to 3 mm. wide; panicle narrow, 5 to 20 cm. long, the main axis often curved, the branches solitary, distant, ascending or spreading, sometimes reflexed, floriferous to base, short, with a few crowded spikelets or as much as 5 cm. long, with short branchlets; spikelets nearly sessile, linear, mostly 20- to 50-flowered, 5 to 20 mm. long; lemmas closely imbricate, ovate, acute, 1.5 to 2 mm. long, the lateral nerves near the margin; grain about 0.5 mm. long, anthers about 0.1 mm. long. ⊙ —Sandy woods, dooryards, and waste places, southern Georgia, Florida, and Alabama.

**13. Eragrostis unioloídes** (Retz.) Nees. (Fig. 198.) Annual; culms erect or ascending, 20 to 40 cm. tall; blades flat, 2 to 4 mm. wide; panicle elliptic, open, 10 to 15 cm. long, about half as wide, the branches ascending;

spikelets ovate-oblong, strongly compressed, truncate at base, obtuse, 15- to 30-flowered, 5 to 10 mm. long, 3 mm. wide, often pink or purplish; lemmas closely imbricate, nearly horizontally spreading, strongly keeled, acute, 2 mm. long, the lateral nerves prominent; palea falling with the lemma or soon thereafter; grain about 0.7 mm. long. ⊙ —Waste ground, Georgia and Florida; introduced from southern Asia.

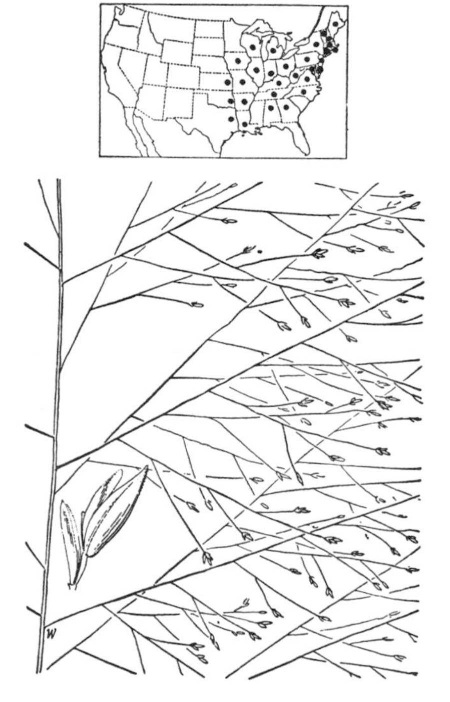

FIGURE 199.—*Eragrostis capillaris*. Panicle, × 1; floret, × 10. (Dewey 35, D. C.)

**14. Eragrostis capilláris** (L.) Nees. LACEGRASS. (Fig. 199.) Annual; culms erect, 20 to 50 cm. tall, much-branched at base, the branches erect; sheaths pilose, at least on the margin, long-pilose at the throat; blades flat, erect, pilose on upper surface near the base, 1 to 3 mm. wide; panicle oblong or elliptic, open, diffuse, usually two-thirds the entire height of the plant, the branches and branchlets capillary; spikelets long-pediceled, 2- to 4-flowered, 2 to 3 mm. long; glumes acute, 1 mm. long; lemmas acute, about 1.5 mm. long, obscurely nerved, rounded on the back, minutely scabrous toward the tip; grain 0.5 mm. long, somewhat roughened. ⊙ —Dry open ground, open woods, and fields, Maine to Wisconsin, south to Georgia, Kansas, and eastern Texas.

**15. Eragrostis fránkii** C. A. Meyer. (Fig. 200.) Resembling *E. capillaris;* culms usually lower, spreading to erect; sheaths glabrous except the pilose throat; blades glabrous; panicle less than half the entire height of the plant, open but not diffuse, mostly less than half as wide as long, the branches ascending, the shorter pedicels not much longer than the spikelets; spikelets 3- to 5-flowered, 2 to 3 mm. long. ⊙ —Sandbars, river banks, and moist open ground, New Hampshire to Minnesota, south to Florida and Oklahoma. ERAGROSTIS FRANKII var. BRÉVIPES Fassett. Spikelets 5- to 7-flowered, 3 to 4 mm. long. ⊙ —Wisconsin (Glenhaven) and Illinois.

FIGURE 200.—*Eragrostis frankii*. Panicle, × 1; floret, × 10. (Chase 2005, Ill.)

**16. Eragrostis pilósa** (L.) Beauv. INDIA LOVEGRASS. (Fig. 201.) Weedy annual; culms slender, erect or ascending from a decumbent base, 10 to 50 cm. tall; blades flat, 1 to 3 mm. wide; panicle delicate, open, becoming somewhat diffuse, 5 to 20 cm. long, the branches capillary, flexuous, ascending or spreading, finally somewhat implicate, the lower fascicled, sparsely long-pilose in the axils; spikelets gray to nearly black, linear, scarcely compressed, 3- to 9-flowered, 3 to 5 mm. long, about 1 mm. wide, the pedicels spreading, mostly longer than the spikelets; glumes acute, the first a little less than, the second a little more than, 1 mm. long; lemmas loosely imbricate, the rachilla more or less exposed, rounded on the back, acute, 1.2 to 1.5 mm. long, 0.5 mm. wide from keel to margin, the nerves obscure; grain 0.6 mm. long. ⊙ —Moist open ground and waste places, Maine to Colorado, south to Florida and Texas, south through Mexico and the West Indies to Argentina; California; introduced from Europe.

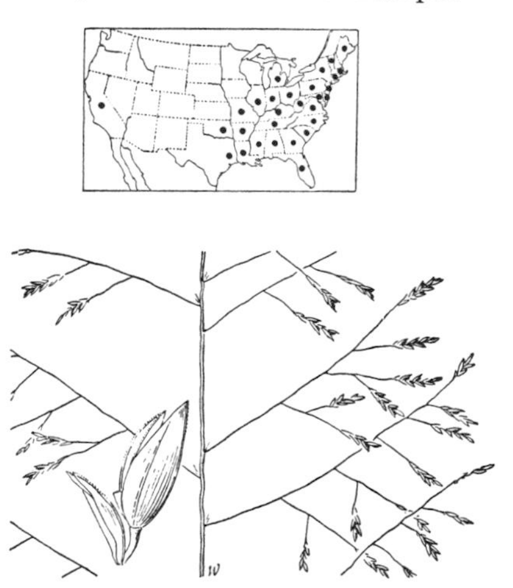

FIGURE 201.—*Eragrostis pilosa.* Panicle, × 1; floret, and palea, × 10. (Ruth 514, Tex.)

**17. Eragrostis perpléxa** L. H. Harvey. (Fig. 202.) Annual resembling *E. pilosa,* but mostly less slender; 20 to 50 cm. tall, or in dry ground 12 to 15 cm. tall; blades mostly 3 to 4 mm. wide; panicle less delicate than in *E. pilosa* and without hairs in the axils; culms below the nodes, keels of the sheaths, and panicle branches bearing small glandular depressions, these often obscure in immature plants. ⊙ —Low alkaline areas and buffalo wallows, North Dakota to Kansas; Texas; Colorado.

**Eragrostis viréscens** Presl. Annual; culms slender, 50 to 60 cm. tall; blades 3 to 6 mm. wide; panicle open, about one third the entire height of the culm, the lower branches mostly solitary, the axils glabrous or nearly so; branchlets and spikelets somewhat appressed along the primary branches; spikelets linear, mostly 7- to 9-flowered, 4 to 5 mm. long, pale or greenish, about 1 mm. wide; lower lemmas scarcely 1.5 mm. long. ⊙ Adventive, Maryland; ballast, Apalachicola, Fla.; Chile. Resembling *E. diffusa;* spikelets smaller.

FIGURE 202.—*Eragrostis perplexa.* Sheath, × 2; panicle, × 1. (Type.)

**18. Eragrostis pectinácea** (Michx.) Nees. (Fig. 203.) Resembling *E. pilosa;* panicles less delicate, the axils glabrous or obscurely pilose, the somewhat larger spikelets appressed along the branches and branchlets, often longer than the pedicels; spikelets at maturity mostly linear, 5 to 8 mm. long; lemmas 1.5 to 1.6 mm. long, the rachilla not or scarcely ex-

FIGURE 203.—*Eragrostis pectinacea*. Panicle, × 1; floret, × 10. (V. H. Chase 84, Ill.)

posed, the nerves evident; grain 0.8 mm. long. ⊙ (*E. caroliniana* (Spreng.) Scribn.; *E. purshii* Schrad.) —Fields, waste places, open ground, moist places, Maine to Washington, south to Florida and Arizona, rare in the Western States. The name *E. pectinacea* has been misapplied to *E. spectabilis*.

**19. Eragrostis diffúsa** Buckl. (Fig. 204.) More robust than *E. pectinacea*, usually 30 to 50 cm. tall, sometimes taller; panicle larger, the primary branches bearing appressed secondary branchlets with few to several spikelets, the main panicle branches thus more densely flowered. ⊙ —A common weed in fields and open ground, Wyoming, Idaho, Oklahoma, and Texas to Nevada and southern California; introduced occasionally in the Eastern States; Mexico. In some specimens the spikelets are ascending rather than appressed, thus making the panicle more open.

FIGURE 204.—*Eragrostis diffusa*. Panicle, × 1; floret, × 10. (Reverchon 1614, Tex.)

**20. Eragrostis tephrosánthos** Schult. (Fig. 205.) Annual, rather soft and lax; culms branching at base, erect to decumbent-spreading, 5 to 20 cm. tall, sometimes taller; blades flat, usually 5 to 10 cm. long, 1 to 2 mm. wide; panicle open, mostly 4 to 10 cm. long, about half as wide, the branches ascending or spreading, naked below, the spikelets appressed or ascending along the upper part, the lower axils pilose; spikelets 6- to 12-flowered, 4 to 7 mm. long, about 1.5 mm. wide; glumes about 1 and 1.3 mm. long; lemmas 1.5 to 2 mm. long, the lateral nerves distinct. ⊙ —Open ground, fields, and waste places, Florida to southern Texas and south through the lowland Tropics to Brazil.

FIGURE 205.—*Eragrostis tephrosanthos*. Panicle, × 1; floret, × 10. (Curtiss 5930, Fla.)

**21. Eragrostis multicaúlis** Steud. (Fig. 206.) Annual; resembling *E. tephrosanthos*, but the axils of the panicle glabrous; panicle branches spikelet-bearing nearly to base; spikelets mostly 4- to 8-flowered, mostly 3 to 4 mm. long. ⊙ (*E. peregrina* Wiegand.)—Waste places, Maine to Wisconsin, south to Pennsylvania and Virginia; ballast, Portland, Oreg.; introduced from Eurasia.

**22. Eragrostis orcuttiána** Vasey. (Fig. 207.) Annual; culms ascending from a decumbent base, rather stout, 60 to 100 cm. tall; blades flat, 2 to 6 mm. wide; panicle open, 15 to 30 cm. long, the branches, branchlets, and pedicels slender, spreading, flexuous, finally implicate, the axils glabrous; spikelets linear, 6- to 10-flowered, sometimes a little falcate, 5 to 7 mm. long, about 1 mm. wide; second glume a little more than 1 mm. long; lemmas loosely imbricate (the rachilla often exposed), narrow, acutish, the

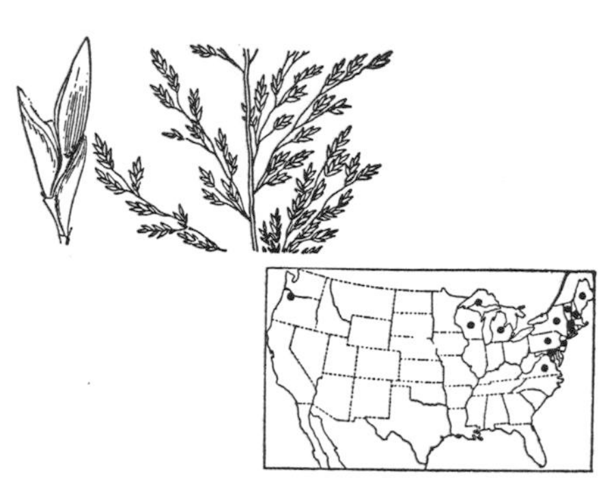

FIGURE 206.—*Eragrostis multicaulis*. Panicle, × 1; floret, × 10. (Hotchkiss 1708, N. Y.)

FIGURE 207.—*Eragrostis orcuttiana*. Panicle, × 1; floret, × 10. (Hitchcock 3063, Calif.)

lower 1.8 mm. long; grain 0.8 mm. long. ☉ —Fields, waste places, and sandy river banks, Oregon (ballast, Portland); Colorado to Arizona and California.

**23. Eragrostis lutéscens** Scribn. (Fig. 208.) Annual; culms freely branching at base, erect or ascending, 5 to 20 cm. tall; sheaths and blades with numerous glandular depressions; blades flat; panicles numerous, narrow, erect, pale or yellowish green, 2 to 10 cm. long, the branches ascending or appressed, beset with glandular depressions; spikelets 6- to 10-flowered, 5 to 7 mm. long, compressed; glumes acute, 1.5 and 2 mm. long; lemmas about 2 mm. long, acute, the nerves prominent; palea 1.5 mm. long. ☉ —Sandy shores, Idaho to Washington, south to Colorado, Arizona, and California; Mexico.

**24. Eragrostis cilianénsis** (All.) Lutati. STINKGRASS. (Fig. 209.) Weedy annual with disagreeable odor when fresh; culms ascending or spreading, 10 to 50 cm. tall, with a ring of glands below the nodes; foliage sparsely beset with glandular depressions, the sheaths pilose at the throat; blades flat, 2 to 7 mm. wide; panicle erect, dark gray green to tawny, usually rather condensed, sometimes, especially in the Southwest, open, 5 to 20 cm. long, the branches ascending; spikelets oblong, compressed, 10- to 40-flowered, 5 to 15 mm. long, 2.5 to 3 mm. wide; lemmas in side view ovate, acutish, about 2.5 mm. long, 1 mm. wide from keel to margin, the keel scabrous toward apex and beset with a few glands, the lateral nerves prominent; palea about two-thirds as long as the lemma, minutely ciliate on the keels; grain ovoid, plump, 0.7 mm. long; anthers 0.5 mm. long. ☉ (*E. major* Host; *E. megastachya* Link.)—Cultivated ground, fields, and waste places, Maine to Washington, south throughout the United States, sparingly in the Northwest, absent from the higher mountains; Mexico and West Indies, south to Argentina; introduced from the Old World.

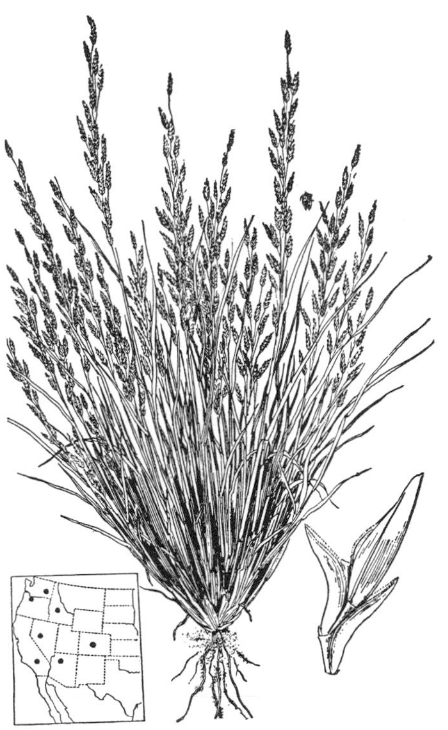

FIGURE 208.—*Eragrostis lutescens*. Plant, × ½; floret, × 10. (Type.)

**25. Eragrostis poaeoídes** Beauv. ex Roem. and Schult. (Fig. 210.) Annual; resembling *E. cilianensis*, mostly more slender; panicles rather more open, the spikelets smaller, 1.5 to 2 mm. wide, the lemmas about 2 mm. long, the glands sometimes obscure; anthers about 0.2 mm. long. ☉ (*E. minor* Host; *E. eragrostis* Beauv.)—Waste places, sparingly introduced from Europe, Maine to Wisconsin and Iowa, south to Georgia, Oklahoma, and Texas; California.

**26. Eragrostis barreliéri** Daveau. (Fig. 211.) Annual; culms erect or decumbent at base, 20 to 50 cm. tall, branching at base, sometimes with a glandular band below the nodes; sheaths pilose at the summit; blades flat, rather short, 2 to 4 mm. wide; panicle erect, open but narrow, 8 to 15 cm. long, the branches ascending or stiffly spreading, few-flowered, spikelet-bearing nearly to base, the

FIGURE 209.—*Eragrostis cilianensis.* Plant, × ½; spikelet, × 5; floret, × 10. (Schuette 155, Wis.)

axils glabrous; spikelets linear, usually 12- to 15-flowered, mostly about 1 cm. long and 1.5 mm. wide; lemmas 2 mm. long or slightly longer. ☉ —Waste places, Colorado and Kansas to Texas and California; Mexico; introduced from southern Europe.

**27. Eragrostis neomexicána** Vasey. (Fig. 212.) Annual; culms usually rather stout, often widely spreading, as much as 1 m. tall; sheaths glabrous, pilose at the throat, often with glandular depressions along the keel or nerves; blades flat, often elongate, 5 to 10 mm. wide; panicle 20 to 40 cm. long, smaller in depauperate specimens, open, the branches ascending or spreading but not divaricate, the branchlets at first appressed along the main branches, finally usually spreading, the axils glabrous; spikelets mostly dark grayish green, ovate to ovate-oblong, or rarely linear, mostly 8- to 12-flowered, 5 to 8 mm. long, about 2 mm. wide; lemmas 2 to 2.3 mm. long. ☉ —Fields, waste places, and wet ground, Texas to southern California, south through Mexico; introduced in Maryland, Indiana, Wisconsin, Iowa, North Dakota, and Missouri.

FIGURE 210.—*Eragrostis poaeoides*. Panicle, × 1 floret, × 10. (Dutton 2235, Vt.)

FIGURE 211.—*Eragrostis barrelieri*. Panicle, × 1; floret, × 10. (Hitchcock 5280, Tex.)

FIGURE 212.—*Eragrostis neomexicana*. Panicle, × 1; floret, × 10. (Type.)

FIGURE 213.—*Eragrostis mexicana.* Panicle, × 1; floret, × 10. (Smith, N. Mex.)

**28. Eragrostis mexicána** (Hornem.) Link. MEXICAN LOVEGRASS. (Fig. 213.) Resembling *E. neomexicana,* but lower, erect or spreading, often simple; panicle erect, comparatively small and few-flowered, less compound, the branches and pedicels spreading; spikelets usually not more than 7-flowered. ⊙ —Open ground, Texas to California; Mexico.

**29. Eragrostis árida** Hitchc. (Fig. 214.) Annual; culms branching at base, erect or more or less decumbent at base, 20 to 40 cm. tall; sheaths not glandular, the hairs at summit in a dense line part way along the collar; blades mostly flat, glabrous, tapering to a fine point, mostly 4 to 8 cm. long, 1 to 2 mm. wide; panicle mostly one-third to half the entire length of the plant, open, the branches, branchlets, and pedicels flexuous, spreading, the lower axils sparsely pilose, the branches solitary or the lower in pairs; spikelets oblong to linear, stramineous or drab, mostly 8- to 15-flowered, 5 to 10 mm. long, 1.5 to 2 mm. wide, somewhat compressed, the lateral pedicels 2 to 3 mm. long; glumes acute, the first narrow, scarcely 1 mm. long, the second a little longer and wider; lemmas 1.6 to 1.8 mm. long, acutish. ⊙ —Dry soil, Missouri; Texas to California and central Mexico.

FIGURE 214.—*Eragrostis arida.* Panicle, × 1; floret, × 10. (Type.)

**30. Eragrostis hirsúta** (Michx.) Nees. (Fig. 215.) Perennial; culms erect, tufted, 50 to 120 cm. tall; sheaths hirsute to glabrous, pilose at the throat and especially along the collar at each side; blades flat, elongate, 5 to 10 mm. wide, becoming more or less involute, tapering to a fine point, scabrous on the upper surface; panicle diffuse, more than half the entire height of the plant, pilose in the axils, branching 4 or 5

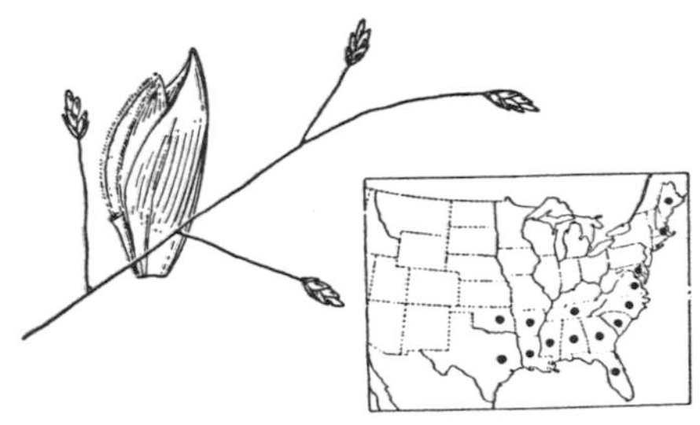

FIGURE 215.—*Eragrostis hirsuta.* Panicle, × 1; floret, × 10. (Curtiss 3499, Fla.)

FIGURE 216.—*Eragrostis lugens*. Plant, × 1; floret, × 10. (Reverchon 16, Tex.)

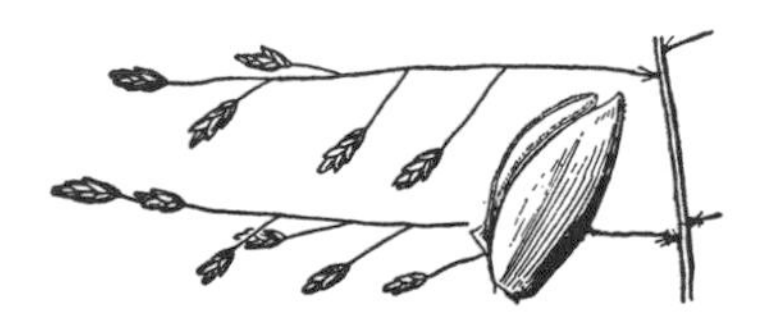

FIGURE 217.—*Eragrostis trichocolea*. Panicle, × 1; floret, × 10. (Curtiss, Fla.)

times; spikelets on long flexuous pedicels, ovate to ovate-oblong, 2- to 6-flowered (rarely to 8-flowered), 3 to 4 mm. long; glumes acuminate, 1.5 and 2 mm. long; lemmas rather turgid, 2 mm. long, acute, the nerves obscure; grain oblong, 1 mm. long, minutely striate and pitted. ♃ —Dry soil, fields and open woods, Maryland to Oklahoma, south to Florida and eastern Texas; British Honduras; introduced in Maine and Massachusetts. Plants with glabrous sheaths have been segregated as *E. hirsuta* var. *laevivaginata* Fern.

**31. Eragrostis lúgens** Nees. (Fig. 216.) Perennial; culms tufted, rather wiry, sometimes geniculate below, sparingly branching; sheaths pilose in the throat, sometimes along the margin and on sides at summit; blades subinvolute, 10 to 25 cm. long, 1.5 to 3 mm. wide, pilose on the upper surface toward base, rarely beneath; panicle rather diffuse, 15 to 30 cm. long, about two-thirds as wide, the axis and ascending to spreading branches capillary, flexuous, the lower branches in pairs or verticils, the axils, except upper, conspicuously long pilose; spikelets on long pedicels, mostly glossy drab, 3- to 7-flowered, 3 to 5 mm. long, 1 to 1.2 mm. wide; glumes thin, 0.7 and 1.2 mm. long, falling early; lemmas closely imbricate, 1.3 to 1.5 mm. long, abruptly acute; grain about 0.7 mm. long. ♃ —Dry prairie, Florida, Louisiana, and Texas; also on ballast, Mobile, Ala.; Mexico and Venezuela to Argentina.

**32. Eragrostis trichocólea** Hack. and Arech. (Fig. 217.) Perennial; culms erect, 30 to 60 cm. tall, the leaves rather short, mostly crowded at the base; sheaths, at least the lower, spreading, pilose; blades spreading, flat or, especially on the innovations, involute, mostly 8 to 12 cm. long, 2 to 4 mm. wide, pilose; panicle diffuse, 15 to 20 cm. long, nearly or quite as wide, the branches stiffly and widely spreading, pilose in the axils; pedicels 2 or 3 times as long as the spikelets; spikelets 3- to 5-flowered, 3 to 4 mm. long, about 1.5 mm. wide; glumes 1 to 1.2 and 1.3 to 1.5 mm. long; lemmas about 1.5 mm. long. ♃ —Sandy woods, Florida (Tampa, Lakeland) and Texas; Mexico to Uruguay.

**33. Eragrostis erósa** Scribn. (Fig. 218.) Perennial; culms tufted, erect, 50 to 90 cm. tall; blades mostly involute; panicle diffuse, less than half the entire height of the plant, usually about one-third, mostly more than half as wide as long, branching 2 or 3 times, sparsely pilose or glabrous in the axils; spikelets mostly 8- to 9-flowered, 5 to 10 mm. long, 1.8 to 2 mm. wide; lemmas 2.5 to 3 mm. long, hyaline-margined toward summit, the tip erose. ♃ —Rocky

FIGURE 218.—*Eragrostis erosa.* Panicle, × 1 (Skehan 58, N. Mex.); floret, × 10. (Type.)

hills, western Texas to New Mexico and northern Mexico.

**34. Eragrostis palméri** S. Wats. (Fig. 219.) Perennial; culms tufted, erect, about 70 cm. tall; blades involute, elongate, erect; panicle open, oblong, 15 to 20 cm. long, 5 to 7 cm. wide, glabrous in the axils; spikelets 5 to 7 mm. long, mostly 7- to 9-flowered, brownish; first glume about 1 mm. long; second glume 1.5 to 2 mm. long; lemmas rounded on the back, bronze-tipped, about 2 mm. long. ♃ —Alkaline banks, Texas; Mexico (Juárez, Coahuila). Differs from *E. erosa* in the oblong panicle and smaller spikelets and lemmas.

FIGURE 219.—*Eragrostis palmeri.* Panicle, × 1; floret, × 10. (Silveus 851, Tex.)

**35. Eragrostis intermédia** Hitchc. PLAINS LOVEGRASS. (Fig. 220.) Perennial; culms erect, tufted, mostly 40 to 80 cm. tall; sheaths glabrous or the lowermost sparsely pilose, conspicuously pilose at the throat, the hairs extending in a line across the collar; blades flat to subinvolute, pilose on the upper surface near the base, otherwise glabrous or with a few scattered hairs, 10 to 25 cm. long, 1 to 3 mm. wide; panicle erect, open, often diffuse, 15 to 35 cm. long, at maturity mostly about three-fourths as wide, the axils pilose, sometimes sparsely so or rarely glabrous, the branches slender but rather stiff, the lower in pairs or verticils, all spreading, often horizontal; spikelets usually 3- to 8-flowered, 3 to 10 mm. long, about 1.5 mm. wide, grayish or brownish green, the pedicels somewhat flexuous, 1 to 3 times as long as the spikelet; glumes acute, 1 to 1.2 and 1.2 to 1.4 mm. long; lemmas turgid, obscurely nerved, 1.8 to 2 mm. long, usually bronze-tipped, not hyaline-margined; grain oblong, about 0.7 mm. long. ♃ —Dry or sandy prairies, Georgia; Louisiana and Missouri to Arizona and south

FIGURE 220.—*Eragrostis intermedia*. Panicle, × 1; floret, × 10. (Type.)

FIGURE 221.—*Eragrostis swalleni*. Plant and panicle, × 1; floret and glandular band, × 10. (Type.)

to Central America. A few specimens from New Mexico have long spikelets (as much as 13-flowered) and glabrous axils.

**36. Eragrostis swalléni** Hitchc. (Fig. 221.) Perennial; culms in dense tufts, erect, 20 to 50 cm. tall, an obscure glandular band below the nodes; sheaths sparingly pilose at the throat; blades involute, glabrous, arching-recurved, 10 to 30 cm. long; panicle erect, open, 10 to 20 cm. long, the branches ascending or spreading, glabrous, stiffly flexuous; spikelets oblong to linear, stramineous or grayish green, 7 to 10 mm. long, about 2 mm. wide, mostly 8- to 12-flowered, the slender pedicels bearing above the middle a glandular band or spot; glumes acutish, rather broad, about 1.2 and 1.8 mm. long; lemmas rather closely imbricate, acutish, about 2 mm. long; palea minutely scabrous on the keels; grain nearly smooth, slightly narrowed toward the summit, 1 mm. long. ♃ —Sandy prairies, southern Texas; northern Mexico.

**37. Eragrostis trácyi** Hitchc. (Fig. 222.) Apparently perennial; culms erect, tufted, 30 to 80 cm. tall; sheaths rather sparsely pilose at the throat; blades flat or, especially of the innovations, involute, 5 to 25

FIGURE 222.—*Eragrostis tracyi*. Panicle, × 1; floret, × 10. (Type.)

FIGURE 223.—*Eragrostis silveana*. Panicle, × 1; spikelet, × 10. (Type.)

cm. long, 1 to 3 mm. wide; panicle erect, open, 10 to 15 cm. long, 5 to 8 cm. wide, the axils glabrous or nearly so, the branches ascending to spreading, flexuous; spikelets linear, mostly 9- to 15-flowered, 5 to 10 mm. long, about 1.5 mm. wide, pinkish or purplish, the flexuous pedicels spreading, 2 to 5 mm. long; glumes acutish, about 1 mm. and 1.5 mm. long; lemmas 1.5 to 2 mm. long, rather soft, loosely imbricate, the lateral nerves distinct; palea somewhat persistent; grain about 0.7 mm. long. ♃ —Sandy soil, known only from Sanibel Island, Fla.

**38. Eragrostis silveána** Swallen. (Fig. 223.) Perennial; culms densely tufted, erect from a knotty base, 40 to 50 cm. tall; sheaths glabrous; blades flat or loosely involute in drying, elongate, 3 mm. wide, attenuate to a fine point, glabrous; panicle 25 to 35 cm. long, 10 to 15 cm. wide, the viscid scabrous branches stiffly ascending or spreading, naked at base, sparsely pilose in the axils; spikelets purplish, 4- to 8-flowered, 2.5 to 4 mm. long, the ultimate pedicels short, usually appressed; glumes about 1 mm. long; lemmas acute, about 1.3 mm. long, the lateral nerves prominent. ♃ —Open ground, southern Texas.

**39. Eragrostis trichódes** (Nutt.) Wood. (Fig. 224.) Perennial; culms tufted, erect, 60 to 120 cm. tall; sheaths pilose at the summit, sometimes on the upper half; blades flat to subinvolute, elongate, 2 to 6 mm. wide, tapering to a slender point, scabrous on the upper surface; panicle usually purplish, diffuse, oblong, usually about half the entire height of the culm, branching 3 or 4 times, the branches capillary, loosely ascending, sparsely pilose in the axils; spikelets long-pediceled, lanceolate to ovate-oblong, mostly 4- to 6-flowered, 4 to 7 mm. long; glumes acuminate, nearly

FIGURE 224.—*Eragrostis trichodes*. Panicle, × 1; floret, × 10. (Reverchon, Tex.)

equal, 2.5 to 3 mm. long, about as long as the first floret; lemmas 2.5 to 3 mm. long, acute, subcompressed, the keel and lateral nerves strong; grain 1 mm. long, minutely pitted; anthers a little more than 1 mm. long. ♃ —Sand barrens and open sandy woods, Illinois to Colorado and Texas.

**40. Eragrostis pilífera** Scheele. (Fig. 225.) Resembling *E. trichodes*, often in smaller tufts and taller; panicle stramineous or golden bronze; spikelets linear, 8- to 15-flowered, 8 to 12 mm. long; glumes and lemmas about 3 mm. long. ♃ (*E. grandiflora* Smith and Bush.)—Sand hills and sand barrens, Illinois and Nebraska to Louisiana and Texas.

**41. Eragrostis spectábilis** (Pursh) Steud. PURPLE LOVEGRASS. (Fig. 226.) Perennial, in dense tufts, rarely producing short or slender rhizomes; culms stiffly erect to spreading, 20 to 60 cm. tall; sheaths glabrous or pilose, conspicuously hairy at the throat; blades flat or folded, rather firm, stiffly ascending, tapering to a fine point, glabrous or rarely pilose, mostly 3 to 8 mm. wide; panicle at first included at base, two thirds the entire height of the culm, diffuse, bright purple, rarely pale, branching 3 or 4 times, the axis stiff, the branches stiffly spreading toward maturity, rarely pilose, strongly pilose in the axils, the lower shorter than the middle ones, finally reflexed, the whole panicle finally breaking away and tumbling before the wind; spikelets long-pediceled, short-pediceled toward the ends of the branches, oblong to linear, 6- to 12-flowered, 4 to 8 mm. long; glumes acute, a little more than 1 mm. long; lemmas acute, about 1.5 mm. long, slightly scabrous toward the tip, the lateral nerves prominent toward the base; palea somewhat bowed out, exposing the rather prominently short-ciliate keels; grain oval, dark-brown, 0.6 mm. long. ♃ —Sandy soil, Maine to Minnesota, south to Florida and Arizona; Mexico (San Luis Potosí). This species was formerly generally called *E. pectinacea.*

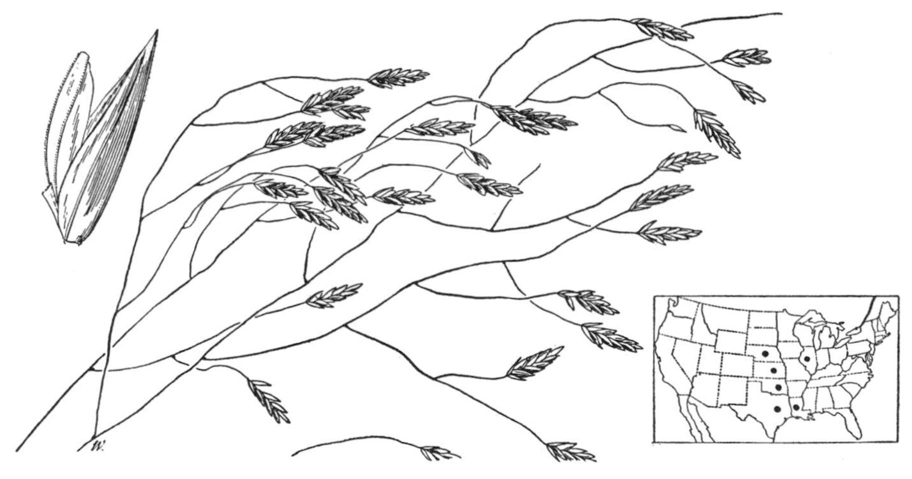

FIGURE 225.—*Eragrostis pilifera.* Panicle, × 1; floret, × 10. (Rydberg 1831, Nebr.)

**42. Eragrostis ellióttii** S. Wats. (Fig. 227.) Perennial; culms tufted. stiffly erect or spreading, 40 to 80 cm. tall; sheaths glabrous, pilose at the throat; blades flat, elongate, scabrous on the upper surface, 2 to 4 mm. wide; panicle diffuse, fragile, usually more than half the entire height of the plant, branching 3 or 4 times, the branches capillary, spreading; spikelets on long capillary spreading pedicels, linear, mostly 8- to 15-flowered, 5 to 12 mm. long, about 2 mm. wide, pale or gray; glumes acute, 1 and 1.5 mm. long; lemmas closely imbricate, acute, about 2 mm. long, bowed out below, fitting into the angles of the zigzag rachilla; grain oval, 0.7 mm. long. ♃ —Low ground, wet meadows, and low pine woods, Coastal Plain, North Carolina to Florida and eastern Texas; West Indies and eastern Mexico.

**43. Eragrostis acúta** Hitchc. (Fig. 228.) Perennial; culms erect, 40 to 60 cm. tall; sheaths glabrous, pilose at the throat; blades flat, becoming more or less involute, 2 to 4 mm. wide; panicle diffuse, more than half

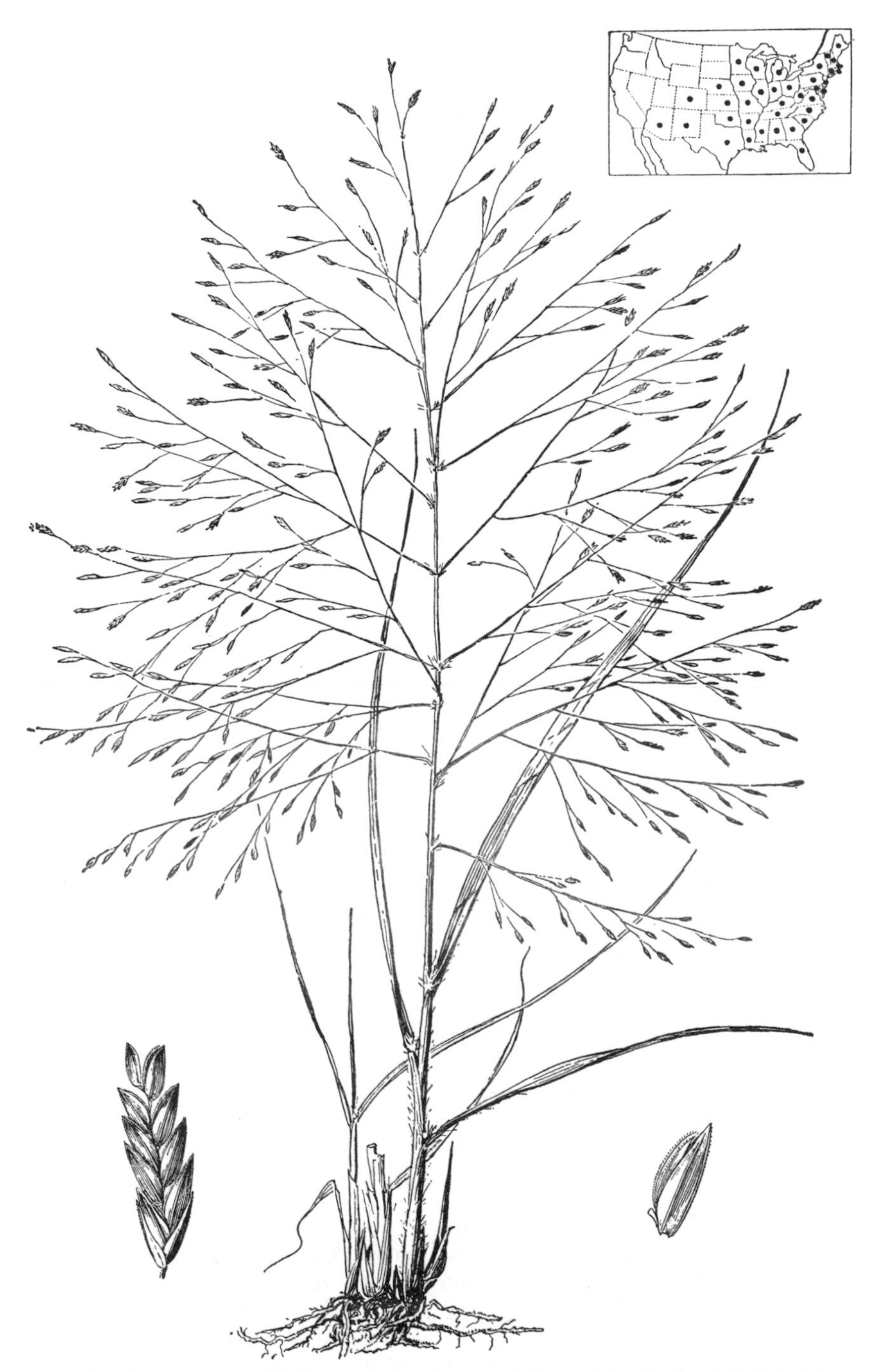

FIGURE 226.—*Eragrostis spectabilis*. Plant, × ½; spikelet, × 5; floret, × 10. (Hitchcock 7849, Md.)

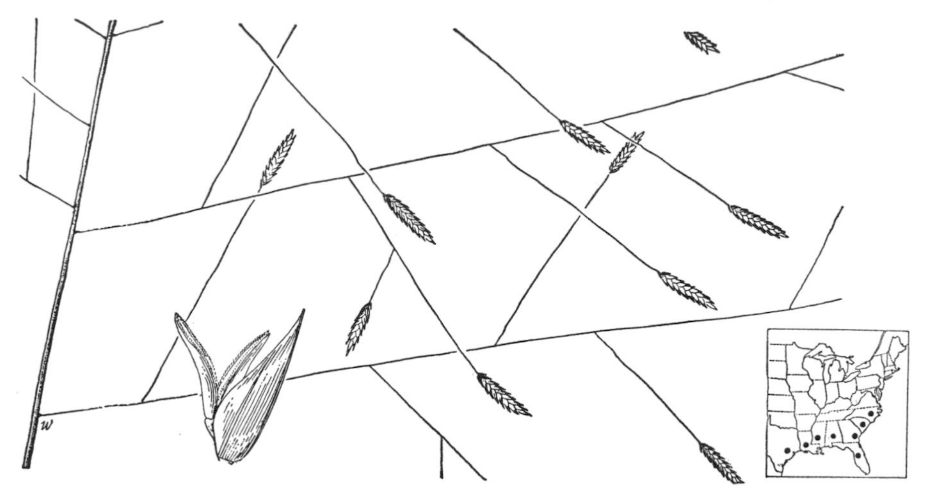

FIGURE 227.—*Eragrostis elliottii*. Panicle, × 1; floret, × 10. (Tracy 7384, Fla.)

FIGURE 228.—*Eragrostis acuta*. Panicle, × 1; floret, × 10. (Type.)

the entire height of the plant, branching 3 or 4 times, the branches less fragile than in *E. elliottii;* spikelets on long spreading pedicels, oblong-elliptic, 10- to 20-flowered, 8 to 14 mm. long, 3 mm. wide, pale or stramineous; glumes acuminate, 2.5 and 3 mm. long; lemmas acuminate, 3 mm. long; grain 0.8 mm. long. ♃ —Low pine woods and moist sandy soil, peninsular Florida.

**44. Eragrostis refrácta** (Muhl.) Scribn. (Fig. 229.) Resembling *E. elliottii;* blades more or less pilose on

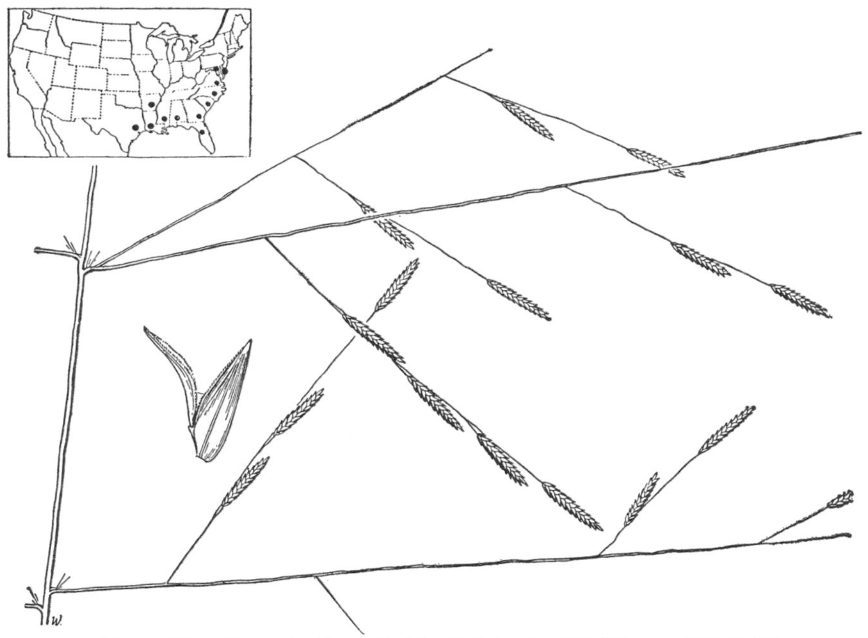

FIGURE 229.—*Eragrostis refracta.* Panicle, × 1; floret, × 10. (Kearney 1922, N. C.)

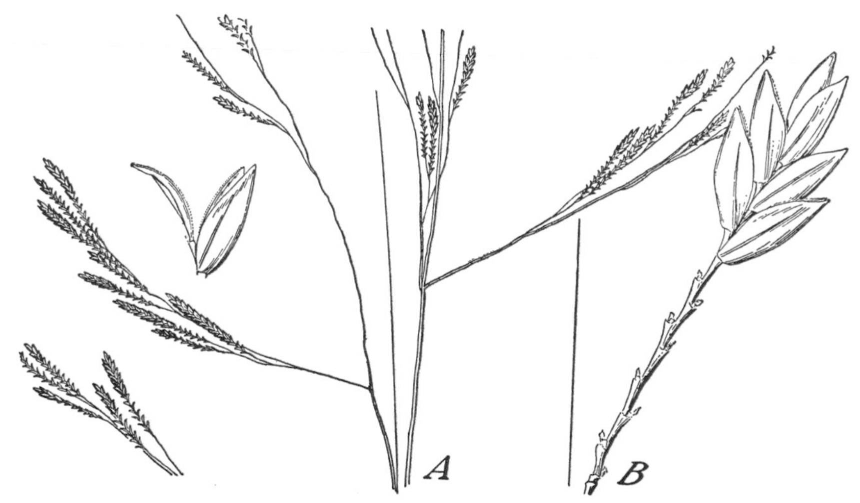


FIGURE 230.—*A*, *Eragrostis bahiensis.* Panicle, × 1; floret, × 10. (Hitchcock 19862, La.) *B*, *E. chariis*, × 10. (Weatherwax 822, Fla.)

the upper surface near base; lower panicle branches usually finally reflexed, long-pilose in the axils; spikelets short-pediceled, appressed and distant along the nearly simple panicle branches, the lemmas on the average shorter than in *E. elliottii.* ♃ —Low sandy soil, Coastal Plain, Delaware to Florida, Arkansas, and eastern Texas.

**45. Eragrostis cháriis** (Schult.) Hitchc. (Fig. 230, *B.*) Perennial; culms erect or ascending at base, 60 to 120 cm. tall; panicle open, 7 to 15 cm. long, nodding, the branches glabrous or with a few hairs in the axils, ascending, solitary, rather distant, naked below, rather closely flowered with ascending or appressed branchlets; spikelets linear, 5 to 10 mm. long, 8- to 20-flowered; glumes about 1.3 and 1.7 mm. long; lemmas 1.5 to 2 mm. long, imbricate; palea persistent only a short time after the fall of the lemma, the naked rachilla persisting. ♃ —Sandy roadsides, Florida (St. Petersburg); introduced from southeastern Asia.

**46. Eragrostis bahiénsis** Schrad. (Fig. 230, *A*.) Resembling *E. chariis;* panicle often more or less condensed; spikelets as much as 30-flowered; lemmas about 2 mm. long; palea persistent. ♃ —Introduced, Florida (Milton, Pensacola), Alabama (Mobile), and Louisiana (Avery Island); Brazil.

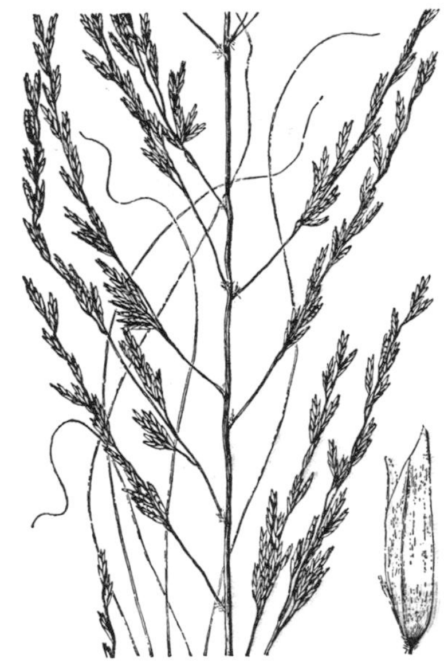

FIGURE 231.—*Eragrostis curvula*. Panicle, × 1; floret, × 10. (Silveus 2156, cult., Tex.)

**47. Eragrostis cúrvula** (Schrad.) Nees. WEEPING LOVEGRASS. (Fig. 231.) Culms 60 to 120 cm. tall, densely tufted, erect, simple or sometimes branching at the lower nodes; sheaths narrow, keeled, glabrous or sparsely hispid, the lower densely hairy toward the base; blades elongate, involute, attenuate to a fine point, arcuate spreading, scabrous; panicles 20 to 30 cm. long, the branches solitary or in pairs, ascending, naked at the base, at least the lower densely pilose in the axils; spikelets 7- to 11-flowered, 8 to 10 mm. long, gray green; lemmas about 2.5 mm. long, obtuse or subacute, the nerves prominent. ♃ —Cultivated for ornament; spontaneous in Florida, Texas, and Arizona. Useful in erosion control and showing promise of being valuable in revegetation of grasslands in the Southern States.

**Eragrostis lehmanniána** Nees. LEHMANN LOVEGRASS. Perennial; culms finally prostrate, 30 to 80 cm. long, the nodes rooting and producing tufts of branches; panicles 10 to 15 cm. long, open; spikelets linear, 10 to 12 mm. long. ♃ —Introduced from Africa, drought-resistant and proving effective in erosion control, Texas, Oklahoma, and Arizona (well established near Tucson).

**Eragrostis stenophýlla** Hochst. Erect smooth annual, 30 to 40 cm. tall, with loosely involute blades and rather loose panicle with ascending branches, the linear spikelets several-flowered, the lemmas 1.3 mm. long. ☉ —Florida, Mississippi (Biloxi), probably escaped from grass garden; India.

**Eragrostis cyperoídes** (Thunb.) Beauv. Stiff stout stoloniferous perennial with sharp-pointed blades and narrow elongate interrupted panicles, the distant branches with naked thornlike tips; spikelets coriaceous, crowded. ♃ —Oregon (Linnton), on ballast; South Africa.

ERAGROSTIS TEF (Zuccagni) Trotter. TEFF. Annual; culms branching and spreading, 30 to 100 cm. tall; panicle large and open; spikelets 5- to 9-flowered, 6 to 8 mm. long. ☉ (*E. abyssinica* (Jacq.) Link.)—Occasionally cultivated for ornament. Africa, where the seed is used for food.

ERAGROSTIS OBTÚSA Munro. Low branching perennial; panicles open, 5 to 10 cm. long; spikelets gray olivaceous, broadly ovate, the lemmas almost horizontally spreading. ♃ —Occasionally cultivated for ornament. South Africa.

ERAGROSTIS CHLOROMÉLAS Steud. BOER LOVEGRASS. Erect branching perennial, 40 to 90 cm. tall, forming dense clumps; blades elongate, subinvolute; panicle 10 to 20 cm. long, loose; spikelets dark olivaceous. ♃ —Introduced from Africa, drought-resistant and promising in erosion control in the Southwest.

## 15. CATABRÓSA Beauv.

Spikelets mostly 2-flowered, the florets rather distant, the rachilla disarticulating above the glumes and between the florets; glumes unequal, shorter than the lower floret, flat, nerveless, irregularly toothed at the broad truncate apex; lemmas broad, prominently 3-nerved, the nerves parallel, the broad apex scarious; palea about as long as the lemma, broad, scarious at apex. Aquatic perennials, with creeping bases, flat soft blades, and open panicles. Type species, *Catabrosa aquatica*. Name from Greek *katabrosis*, an eating up or devouring, referring to the toothed or erose glumes.

**1. Catabrosa aquática** (L.) Beauv. BROOKGRASS (Fig. 232.) Glabrous throughout; culms 10 to 40 cm. long;

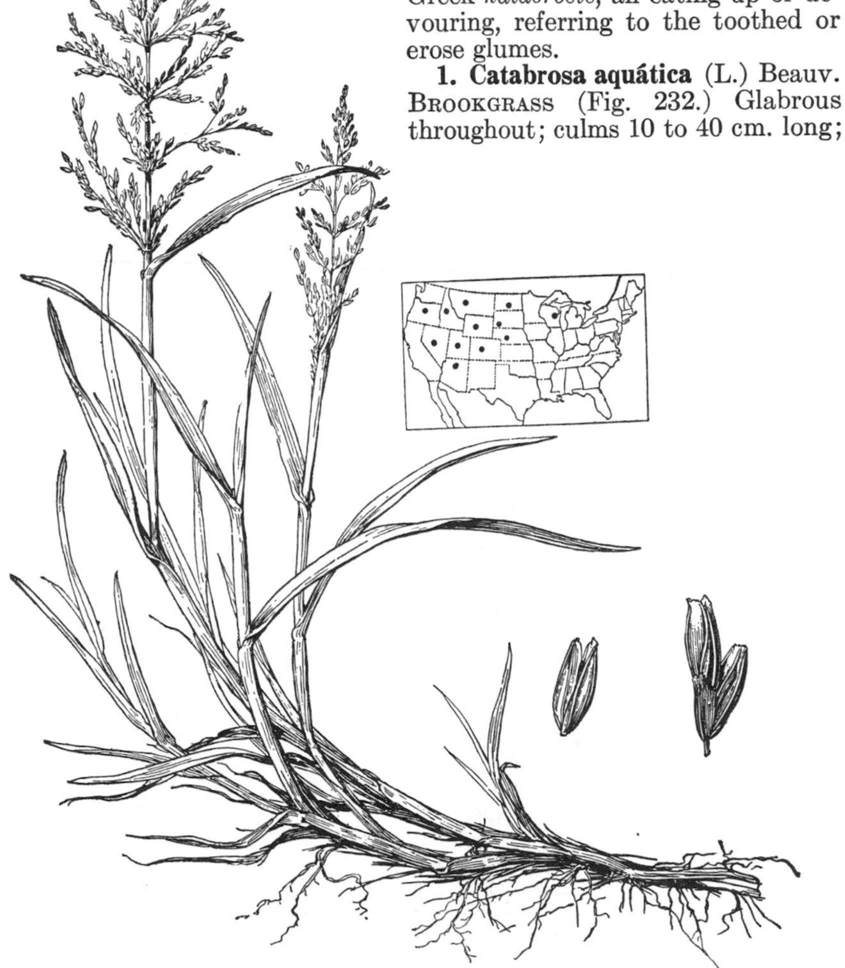

FIGURE 232.—*Catabrosa aquatica*. Plant, × ½; spikelet and floret, × 5. (Williams and Fernald, Quebec.)

FIGURE 233.—*Molinia caerulea*. Plant, × ½; spikelet and floret, × 5. (Kirk 157, Vt.)

blades mostly less than 10 cm. long, 2 to 8 mm. wide; panicle erect, 10 to 20 cm. long, oblong or pyramidal, yellow to brown, the branches spreading in somewhat distant whorls; spikelets short-pediceled, about 3 mm. long; glumes about 1.5 and 2 mm. long; lemmas 2.5 to 3 mm. long. ♃ —Mountain meadows, around springs and along streams, Newfoundland and Labrador to Alberta, south through Wisconsin, North Dakota, South Dakota, and eastern Oregon to northern Arizona; Eurasia. Sometimes 1-flowered spikelets occur in panicles with 2-flowered ones.

---

**Cutándia memphítica** (Spreng.) Richt. Low annual; blades flat; panicle few-flowered; spikelets on short pedicels, finally divergent on the zigzag branches. ☉ —San Bernardino Mountains, Calif.; introduced from the Mediterranean region.

## 16. MOLÍNIA Schrank

Spikelets 2- to 4-flowered, the florets distant, the rachilla disarticulating above the glumes, slender, prolonged beyond the upper floret and bearing a rudimentary floret; glumes somewhat unequal, acute, shorter than the first lemma, 1-nerved; lemmas membranaceous, narrowed to an obtuse point, 3-nerved; palea bowed out below, equaling or slightly exceeding the lemma. Slender tufted perennials, with flat blades and narrow, rather open panicles. Type species, *Molinia caerulea.* Named for J. I. Molina.

**1. Molinia caerúlea** (L.) Moench. (Fig. 233.) Culms erect, 50 to 100 cm. tall; blades 2 to 7 mm. wide, erect, tapering to a fine point; panicle 10 to 20 cm. long, purplish, the branches ascending, rather densely flowered, mostly floriferous to the base; spikelets short-pediceled, 4 to 7 mm. long; lemmas about 3 mm. long. ♃ —Meadows and fields, introduced in a few localities, Maine to Pennsylvania; Eurasia.

## 17. DIARRHÉNA Beauv.

(*Diarina* Raf.)

Spikelets few-flowered, the rachilla disarticulating above the glumes and between the florets; glumes unequal, acute, shorter than the lemmas, the first 1-nerved, the second 3- to 5-nerved; lemmas chartaceous, pointed, 3-nerved, the nerves converging in the point, the upper floret reduced; palea chartaceous, obtuse, at maturity the lemma and palea widely spread by the large turgid beaked caryopsis with hard shining pericarp; stamens 2 or 3. Perennials, with slender rhizomes, broadly linear, flat blades, long-tapering below, and narrow, few-flowered panicles. Type species, *Diarrhena americana.* Name from Greek *dis,* twice, and *arren,* male, alluding to the two stamens.

**1. Diarrhena americána** Beauv. (Fig. 234.) Culms slender, about 1 m.tall, arched-l eaning, leaves approximate below the middle of the culm; sheaths pubescent toward the summit; blades elongate, 1 to 2 cm. wide, scabrous to pubescent beneath; panicle long-exserted, drooping, 10 to 30 cm. long, the branches few, appressed, the lower distant; spikelets 10 to 18 mm. long, at first narrow, the florets expanded at maturity; lemmas 6 to 10 mm. long. ♃ (*Diarina festucoides* Raf.)—Rich or moist woods, Virginia to Michigan and South Dakota, south to Tennessee, Arkansas, Oklahoma, and eastern Texas.

## 18. DISSANTHÉLIUM Trin.

Spikelets mostly 2-flowered, the rachilla slender, disarticulating above the glumes and between the florets; glumes firm, nearly equal, acuminate, much longer than the lower floret, mostly exceeding all the florets, the first 1-nerved, the second 3-nerved; lemmas strongly compressed, oval or elliptic, acute, 3-nerved, the lateral nerves near the margin; palea some-

FIGURE 234.—*Diarrhena americana*. Plant, × ½; spikelet and floret, × 5. (Wilcox 66, Ill.)

what shorter than the lemma. Annuals or perennials with narrow panicles. Type species, *Dissanthelium supinum* Trin. Name from Greek, *dissos*, double, and *anthelion*, a small flower, alluding to the two small florets.

**1. Dissanthelium califórnicum** (Nutt.) Benth. (Fig. 235.) Annual, lax; culms more or less decumbent or spreading, about 30 cm. tall; blades flat, 10 to 15 cm. long, 2 to 4 mm. wide; panicle 10 to 15 cm. long, narrow but rather loose, the branches in fascicles, ascending, slender, flexuous, some of them floriferous to base; glumes narrow, acute, nearly equal, about 3 mm. long; lemmas pubescent, nearly 2 mm. long. ☉ —Open ground, islands off the southern coast of California and of Baja California.

## 19. REDFIÉLDIA Vasey

Spikelets compressed, mostly 3- or 4-flowered, the rachilla disarticulating above the glumes and between the florets; glumes somewhat unequal, 1-nerved, acuminate; lemmas chartaceous, 3-nerved, the nerves parallel, densely villous at base; palea as long as the lemma; grain free. A rather tall perennial, with extensive rhizomes, and a large panicle with diffuse capillary branches. Type species, *Redfieldia flexuosa*. Named for J. H. Redfield.

**1. Redfieldia flexuósa** (Thurb.) Vasey. BLOWOUT GRASS. (Fig. 236.) Culms tough, 60 to 100 cm. tall, the rhizomes long, slender; blades glabrous, involute, elongate, flexuous, tapering to a fine point; panicle oblong, one-third to half the entire length of the culm; spikelets 5 to 7 mm. long, broadly V-shaped, the glumes acuminate, about half as long as the spikelet; lemmas acute, sometimes mucronate, 4 to 5 mm. long. ♃ —Sand hills, North Dakota to Oklahoma, west to Utah and Arizona (Moki Reservation). A sand-binding grass.

FIGURE 235.—*Dissanthelium californicum*. Plant, × ½; spikelet and floret, × 10. (Trask 324, Calif.)

FIGURE 236.—*Redfieldia flexuosa*. Plant, × ½; spikelet and floret, × 5. (Over 2429, S. Dak.)

## 20. MONANTHÓCHLOË Engelm.

Plants dioecious; spikelets 3- to 5-flowered, the uppermost florets rudimentary, the rachilla disarticulating tardily in pistillate spikelets; glumes wanting; lemmas rounded on the back, convolute, narrowed above, several-nerved, those of the pistillate spikelets like the blades in texture; palea narrow, 2-nerved, in the pistillate spikelets convolute around the pistil, the rudimentary uppermost floret enclosed between the keels of the floret next below. Creeping wiry perennial, with clustered short subulate blades, the spikelets inconspicuous at the ends of the short branches, only a little exceeding the leaves. Type species, *Monanthochloë littoralis*. Name from Greek *monos*, single, *anthos*, flower, and *chloe*, grass, alluding to the unisexual flowers.

**1. Monanthochloë littorális** Engelm. (Fig. 237.) Culms tufted, extensively creeping, the short branches erect; blades falcate, mostly less than 1 cm. long, conspicuously distichous in distant to approximate clusters; spikelets 1 to few, nearly concealed in the leaves. ♃ —Muddy seashores and tidal flats, southern Florida, especially on the keys; Texas (Galveston and southward); southern California (Santa Barbara and southward); Mexico, Cuba.

## 21. DISTÍCHLIS Raf. SALTGRASS.

Plants dioecious; spikelets several to many-flowered, the rachilla of the pistillate spikelets disarticulating above the glumes and between the florets; glumes unequal, broad, acute, keeled, 3- to 7-nerved, the lateral nerves sometimes faint; lemmas closely imbricate, firm, the pistillate coriaceous, acute or subacute, with 9 to 11 mostly faint nerves (nerves fewer in *D. texana*); palea as long as the lemma or shorter, the margins bowed out near the base, the pistillate coriaceous, enclosing the grain. Low perennials, with extensively creeping scaly rhizomes, sometimes stolons, erect, rather rigid culms, and dense, rather few-flowered panicles. Type species, *Distichlis spicata*. Name from Greek *distichos*, 2-ranked, alluding to the distichous leaves.

The species of *Distichlis* in general have little value for forage, but in the interior basins, such as the vicinity of Great Salt Lake, *D. stricta* is grazed when better grasses are not available.

Plants mostly more than 30 cm. tall; blades not conspicuously distichous, mostly 20 to 40 cm. long; panicle more than 10 cm. long; stolons present, long and stout .......... 3. D. TEXANA.
Plants mostly less than 30 cm. tall; blades conspicuously distichous, mostly less than 10 cm. long; panicle rarely more than 5 cm. long.
  Panicles condensed, the spikelets imbricate, mostly 5- to 9-flowered; keels of pistillate paleas with narrow entire wings .......... 1. D. SPICATA.
  Panicles looser, the spikelets less imbricate, the individual spikelets plainly visible; keels of pistillate paleas with broader serrate-erose wings .......... 2. D. STRICTA.

**1. Distichlis spicáta** (L.) Greene. SEASHORE SALTGRASS. (Fig. 238.) Culms 10 to 40 cm. tall, sometimes taller; leaves numerous, the sheaths closely overlapping, the spreading blades conspicuously distichous, flat to involute, sharp-pointed, mostly less than 10 cm. long; panicle usually pale or greenish, 1 to 6 cm. long, rarely longer; spikelets mostly 5- to 9-flowered, mostly 6 to 10 mm. long, compressed; lemmas 3 to 6 mm. long, the pistillate more coriaceous and more closely imbricate than the staminate; palea rather soft, narrow, the keels narrowly winged, entire; anthers about 2 mm. long. ♃ —Seashores, forming dense colonies, Nova Scotia to Florida and Texas; British Columbia to California, Mexico, and Cuba; Pacific slope of South

FIGURE 237.—*Monanthochloë littoralis.* Plant, × ½; pistillate spikelet and floret, × 5. (Hitchcock 623, Fla.)

America. Occasional plants produce runners above ground as well as below. Such specimens have been segregated as *D. spicata* var. *stolonifera* Beetle. DISTICHLIS SPICATA var. NÁNA Beetle. Culms slender from slender rhizomes; blades 1 to 8 cm. long, subinvolute, slender; panicles of 2 to 5 spikelets, the spikelets slightly narrower than in the species; keels of the palea densely short-ciliate. ♃ —Alkaline boggy or sandy soil, Stanislaus and Kern Counties, Calif. Insufficiently known.

**2. Distichlis strícta** (Torr.) Rydb. DESERT SALTGRASS. (Fig. 239.) Resembling *D. spicata;* panicles less congested, the individual spikelets easily distinguished; staminate panicles stramineous, the spikelets 8- to 15-flowered; pistillate spikelets greenish leaden, mostly 7- or 9-flowered, broader; lemmas firm, the palea a little shorter, much broader below, the keels with wide serrulate erose or lacerate wings. ♃ (*D. dentata* Rydb., the pistillate plant.) — Alkaline soil of the interior, Saskatchewan to eastern Washington, south to Texas and California; Mexico. This and *D. spicata* appear to be distinct for the most part, but the staminate plants are sometimes difficult to distinguish.[10]

**3. Distichlis texána** (Vasey) Scribn. (Fig. 240.) Culms erect from a decumbent base, 30 to 60 cm. tall, producing extensively creeping rhizomes and long stout stolons; blades flat, firm, glabrous beneath, scabrous on the upper surface, mostly 20 to 40 cm. long, 2 to 6 mm. wide; panicle narrow, pale, 10 to 25 cm. long, somewhat interrupted, the branches appressed; spikelets somewhat compressed, 4- to 8-flowered, 1 to 1.5 cm. long; glumes 5 and 7 mm. long, acute; lemmas of pistillate spikelets closely imbricate and appressed, about 8 mm. long with 3 strong nerves, the intermediate nerves obscure, acute, the margins broad, hyaline; palea of pistillate spikelets shorter than the lemma, strongly bowed out below, closely convolute around the pistil, the keels with narrow erose or toothed wings; lemmas of staminate spikelets more spreading, about 6 mm. long, 3-nerved; palea about as long as the lemma, not bowed out, not convolute, the keels minutely scabrous, not winged; anthers 3 mm. long. ♃ —Sand flats, Presidio and Brewster Counties, Tex., and northern Mexico.

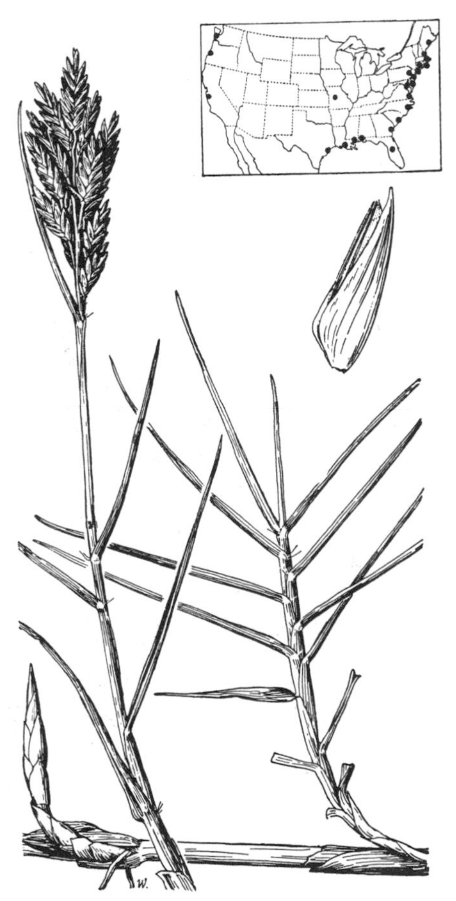

FIGURE 238.—*Distichlis spicata.* Plant, × 1; floret, × 5. (Hitchcock 2826, Oreg.)

[10] REEDER, J. R. STATUS OF DISTICHLIS DENTATA. Torrey Bot. Club Bul. 70: 53–57. 1943.

FIGURE 239.—*Distichlis stricta.* Staminate plant, × ½; staminate spikelet and floret, × 5 (Mearns 3132, Calif.); pistillate panicle, × 1; pistillate floret, × 5 (Sandberg and Leiberg 463, Wash.).

## 22. UNÍOLA L.

Spikelets 3- to many-flowered, the lower 1 to 6 lemmas empty, the rachilla disarticulating above the glumes and between the florets; glumes compressed-keeled, rigid, usually narrow, 3- to 7-nerved, acute or acuminate, rarely mucronate; lemmas compressed, sometimes conspicuously flattened, chartaceous, many-nerved, the nerves sometimes obscure, acute or acuminate, the empty ones at the base and the uppermost usually reduced; palea rigid, strongly keeled, bowed out at base, weakly so in *Uniola paniculata;* stamen 1. Rather tall, erect perennials, with flat or sometimes convolute blades and narrow or open panicles of compressed, sometimes very broad and flat spikelets. Type species, *Uniola paniculata.* Ancient Latin name of a plant.

The inland species are not abundant enough to be of value for forage. *Uniola latifolia* is worthy of cultivation as an ornamental; *U. paniculata* is a sand binder along the southern seacoast; the seeds of *U. palmeri* Vasey of Mexico are used for food by the Cocopa Indians.

Rhizomes extensively creeping; blades firm, flat at base, tapering into a long flexuous involute point; empty lemmas about 4; coastal dunes........................ 1. U. PANICULATA.
Rhizomes wanting or short and knotty; blades thin, flat; empty lemma 1 (2 or 3 in *U. ornithorhyncha*); rich or moist woods.
  Spikelets 8- to 12-flowered on slender pedicels; panicle nodding or drooping. 2. U. LATIFOLIA.
  Spikelets 3- to 7-flowered, nearly sessile; panicle erect, nearly simple, the branches stiff.
    Spikelets more than 10 mm. (usually more than 12 mm.) wide, with 5 to 7 fertile florets.
      Sterile lemma 1; panicle 10 to 15 cm. long, the lower branches with 2 to 5 rather distant spikelets........................ 3. U. NITIDA.
      Sterile lemmas 2 or 3; panicle 3 to 8 cm. long, the branches very short with approximate spikelets........................ 4. U. ORNITHORHYNCHA.
    Spikelets rarely as much as 8 mm. wide at maturity, V-shaped, with 1 to 4 fertile florets (rarely more), and 1 sterile lemma.
      Collar of sheath pubescent, the sheaths commonly loosely long-pubescent, rarely glabrous........................ 5. U. SESSILIFLORA.
      Collar and sheaths glabrous or nearly so........................ 6. U. LAXA.

**1. Uniola paniculáta** L. SEA OATS. (Fig. 241.) Culms stout, about 1 m. tall, from extensively creeping rhizomes; blades flat, firm, elongate,

FIGURE 240.—*Distichlis texana.* Panicle, × 1; lemma and palea, × 5. (Nealley, Tex.)

FIGURE 241.—*Uniola paniculata.* Plant, × 1/10; spikelets, × 1. (Kearney 2134, Va.)

FIGURE 242.—*Uniola latifolia*. Plant, × ½; spikelet and floret, × 3. (Chase 5874, Md.)

becoming involute toward the long, fine flexuous point; panicle pale, narrow, condensed, heavy and nodding, 20 to 40 cm. long, the branches arching and drooping, as much as 12 cm. long; spikelets very flat, 10- to 20-flowered, mostly 2 to 2.5 cm. long, 1 cm. wide, the first 4 to 6 lemmas empty, the slender pedicels shorter than the spikelets; lemmas about 9-nerved, strongly compressed-keeled about 1 cm. long, acute; palea acute, as long as the lemma, the strong wings of the keels ciliate. ♃ —Sand dunes of the seacoast, Northampton County, Va., to Florida and Texas; northern West Indies; eastern Mexico. Spikelets apparently sterile, no caryopses nor stamens found.

**2. Uniola latifólia** Michx. BROADLEAF UNIOLA. (Fig. 242.) Culms 1 to 1.4 m. tall, with short strong rhizomes, forming colonies; blades flat, narrowly lanceolate, 10 to 20 cm. long, mostly 1 to 2 cm. wide; panicle open, drooping, 10 to 20 cm. long, the branches bearing a few large, very flat spikelets, the pedicels capillary; spikelets 8- to 12-flowered, 2 to 3.5 cm. long, 1 to 1.5 cm. wide, green or finally tawny, the first lemma empty; lemmas lanceolate, strongly compressed-keeled, acute, about 1 cm. long, striate-nerved, the keel ciliate with soft ascending hairs, the callus pilose; palea shorter than the lemma, wing-keeled; anther minute, the flower cleistogamous; caryopsis flat, oval, black, 5 mm. long. ♃ —Rich woods, Pennsylvania and New Jersey to Illinois and Kansas, south to Florida and Texas; Arizona (Pinal County).

**3. Uniola nítida** Baldw. (Fig. 243.) Culms slender, 50 to 75 cm. tall, erect, loosely tufted, with short rhizomes; blades flat, spreading, mostly less than 15 cm. long, 4 to 8 mm. wide; panicle open, few-flowered, 10 to 15 cm. long, with a few spreading branches 3 to 8 cm. long, bearing 2 to 5 nearly sessile spikelets; spikelets 4- to 7-flowered, 1 to 1.5 cm. long, about 1 cm. wide, the first lemma empty; lemmas spreading, 7 to 10 mm. long, compressed-keeled, gradually acuminate, striate-nerved; palea equaling the lemma, acuminate, 2-toothed, the keels prominently winged; anther 1.5 mm. long. ♃ —Moist woods, South Carolina to Florida.

**4. Uniola ornithorhyncha** Steud. (Fig. 244.) Culms slender, 30 to 50 cm. tall, loosely tufted with short rhizomes; sheaths pubescent on the collar; blades flat, thin, mostly less than 15 cm. long, 3 to 6 mm. wide; panicle narrow, 3 to 9 cm. long, the short approximate branches with 1 to 3 nearly sessile spikelets or the lower somewhat distant with 4 to 6 spikelets, pubescent in the axils; spikelets very flat, with 3 or 4 widely spreading fertile florets, the 2 or 3 lower lemmas empty, appressed; fertile lemmas about 8 mm. long, narrow, gradually acuminate, striate-nerved; palea as long as or longer than the lemma, acuminate, 2-toothed, strongly bowed out below, the keels rather narrowly winged; anther 1 to 1.8 mm. long. ♃ —Low woods or hummocks in swamps, Alabama to Louisiana.

**5. Uniola sessiliflóra** Poir. (Fig. 245.) Culms erect, 0.5 to 1.5 m. tall, in loose tufts with short rhizomes; sheaths pilose, at least toward the summit; blades elongate, firm, mostly sparsely pilose on the upper surface toward the base, 5 to 10 mm. wide, tapering to base; panicle long-exserted, 20 to 50 cm. long, narrow, the branches distant, stiffly ascending or appressed, the lower as much as 7 cm. long, the upper short, somewhat capitate; spikelets nearly sessile, aggregate in clusters, flat, usually 3- to 5-flowered, broadly V-shaped at maturity, the first lemma empty; glumes about 2 mm. long; lemmas spreading, about 5 mm. long, acuminate, beaked, especially before maturity, striate-nerved; palea shorter than the lemma, acute, broad, the keels narrowly winged; grain black, 3 mm. long, at maturity spreading the lemma and

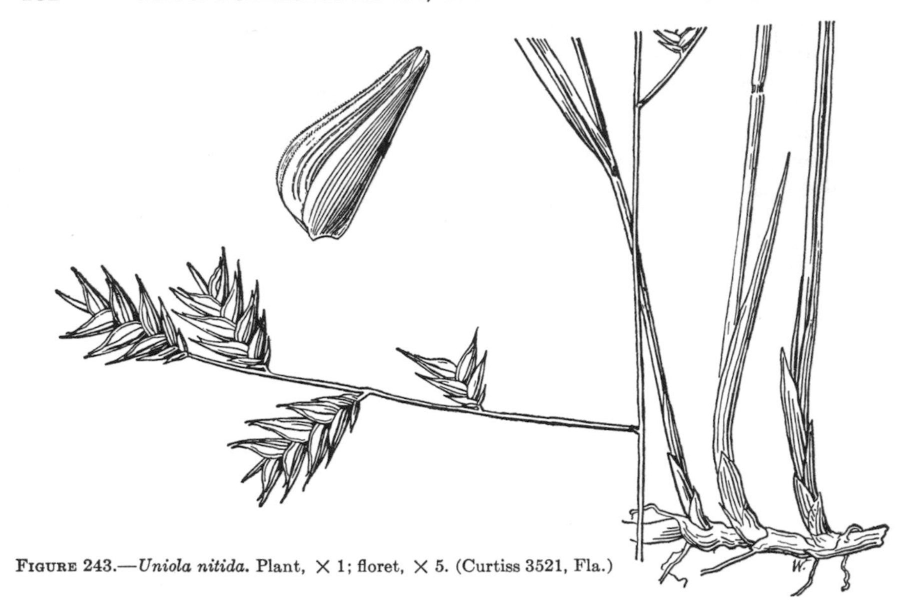


FIGURE 243.—*Uniola nitida*. Plant, × 1; floret, × 5. (Curtiss 3521, Fla.)

palea; anther 1.3 mm. long. ♃ (*U. longifolia* Scribn.)—Rich woods, southeastern Virginia to Tennessee and Oklahoma, south to Florida and eastern Texas.

**6. Uniola láxa** (L.) B. S. P. (Fig. 246.) Culms slender, 60 to 100 cm. tall, erect to nodding from a loosely tufted sometimes knotty base; blades elongate, flat to sometimes loosely involute, 3 to 6 mm. wide; panicle narrow, slender, 15 to 30 cm. long, the branches short, appressed, approximate, the lower sometimes 3 cm. long and distant; spikelets nearly sessile, approximate, flat, usually 3- to 4-flowered, the first lemma empty; lemmas spreading, 4 to 5 mm. long, gradually acuminate, striate-nerved; palea broad, the keels narrowly winged; grain black, 2.5 mm. long, at maturity spreading the lemma and palea; anther 1.2 mm. long. ♃ —Moist woods, Coastal Plain, Long Island to Florida and Texas, extending to western North Carolina, Kentucky, Arkansas, and Oklahoma.

## 23. DÁCTYLIS L. ORCHARD GRASS

Spikelets few-flowered, compressed, finally disarticulating between the florets, nearly sessile in dense 1-sided fascicles, these borne at the ends of the few branches of a panicle; glumes unequal, carinate, acute, hispid-ciliate on the keel; lemmas compressed-keeled, mucronate, 5-nerved, ciliate on the keel. Perennials, with flat blades and fascicled spikelets. Type species, *Dactylis glomerata*. Name from Greek *dactulos*, a finger, alluding to the stiff branches of the panicle.

**1. Dactylis glomeráta** L. ORCHARD GRASS. (Fig. 247.) Culms in large tussocks, 60 to 120 cm. tall; blades elongate, 2 to 8 mm. wide; panicles 5 to 20 cm. long, the few distant stiff solitary branches ascending, or spreading at anthesis, appressed at maturity, the lowermost sometimes as much as 10 cm. long; lemmas about 8 mm. long, mucronate or short-awned. ♃ —Fields, meadows, and waste places, Newfoundland to southeastern Alaska; south to Florida and central California; Eurasia. Commonly cultivated as a meadow and pasture grass. In England called cocksfoot. A variegated form (called by gardeners var. *variegata*) is occasionally cultivated for borders.

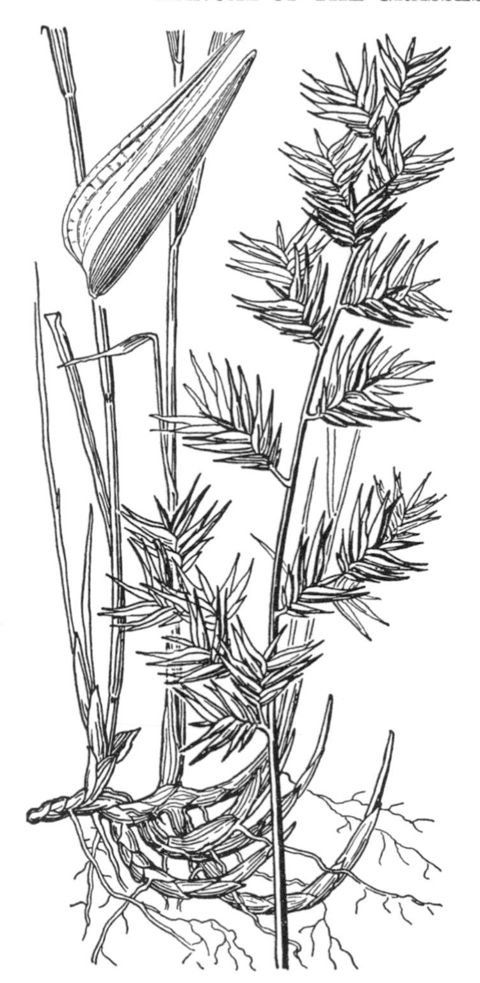

FIGURE 245.—*Uniola sessiliflora*. Plant, × 1; floret, × 5. (Tracy, Miss.)

FIGURE 244.—*Uniola ornithorhyncha*. Plant, × 1; floret, × 5. (Tracy and Lloyd 448, Miss.)

## 24. CYNOSÚRUS L. DOGTAIL

Spikelets of two kinds, sterile and fertile together, the fertile sessile, nearly covered by the short-pediceled sterile one, these pairs imbricate in a dense 1-sided spikelike panicle; sterile spikelets consisting of 2 glumes and several narrow, acuminate, 1-nerved lemmas on a continuous rachilla; fertile spikelets 2- or 3-flowered, the glumes narrow, the lemmas broader, rounded on the back, awn-tipped, the rachilla disarticulating above the glumes. Annuals or perennials with narrow flat blades and dense spikelike or subcapitate panicles. Type species, *Cynosurus cristatus*. Name from Greek *kuon* (*kun-*) dog, and *oura*, tail.

Plants perennial; panicles narrow, spikelike; awns inconspicuous.......... 1. C. CRISTATUS.
Plants annual; panicles subcapitate; awns conspicuous.......................... 2. C. ECHINATUS.

**1. Cynosurus cristátus** L. CRESTED DOGTAIL. (Fig. 248.) Perennial; culms tufted or geniculate at base, erect, 30 to 60 cm. tall; panicle spikelike, linear, more or less curved, 3 to 8 cm.

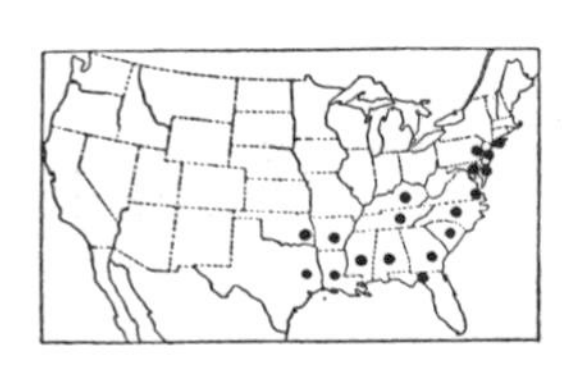

FIGURE 246.—*Uniola laxa*. Plant, × 1; floret, × 5. (Van Eseltine and Moseley 178, D. C.)

long; pairs of spikelets about 5 mm. long; lemmas with awns mostly not more than 1 mm. long. ♃ —Fields and waste places, Newfoundland to Michigan and North Carolina; Idaho, Washington to California; introduced from Europe. Occasionally cultivated in mixtures for meadows, but of little value.

**2. Cynosurus echinátus** L. (Fig. 249.) Annual; culms 20 to 40 cm. tall; blades short; panicle subcapitate, 1 to 4 cm. long, bristly; pairs of spikelets 7 to 10 mm. long; lemmas with awns 5 to 10 mm. long. ⊙ —Open ground, British Columbia; Oregon to central California; Maryland; North Carolina; Arkansas and Oklahoma; introduced from Europe.

## 25. LAMÁRCKIA Moench

(*Achyrodes* Boehmer)

Spikelets of two kinds, in fascicles, the terminal one of each fascicle fertile, the others sterile; fertile spikelet with 1 perfect floret on a slender stipe and a rudimentary floret on a long rachilla-joint, both awned, the glumes narrow, acuminate or short-awned, 1-nerved; lemma broader, scarcely nerved, bearing just below the apex a delicate awn; sterile spikelets linear, 1 to 3 in each fascicle, consisting of 2 glumes similar to those of the fertile spikelet, and numerous imbricate, obtuse, awnless, empty lemmas, a reduced spikelet similar to the fertile one borne on the pedicel with one of the sterile ones.—Low annual with flat blades and oblong, 1-sided, dense panicles, the crowded fascicles drooping, the fertile being hidden, except the awns, by the numerous sterile ones; fascicles falling entire. Type species, *Lamarckia aurea*. Named for J. B. Lamarck.

**1. Lamarckia aúrea** (L.) Moench. GOLDENTOP. (Fig. 250.) Culms erect or decumbent at base, 10 to 40 cm. tall; blades soft, 3 to 7 mm. wide; panicle dense, 2 to 7 cm. long, 1 to 2 cm. wide, shining, golden yellow to purplish, the branches short, erect, the branchlets capillary, flexuous; pedicels fascicled, pubescent, with a tuft of long whitish hairs at the base; fertile spikelet about 2 mm. long, the awn of lemma about twice as long as the spikelet; sterile spikelet 6 to 8 mm. long. ⊙ —Open ground and waste places, Texas, Arizona, southern California, and northern Mexico; introduced from the Mediterranean region. Sometimes cultivated for ornament.

## 26. ARÚNDO L.

Spikelets several-flowered, the florets successively smaller, the summits of all about equal, the rachilla glabrous, disarticulating above the glumes and between the florets; glumes somewhat unequal, membranaceous, 3-nerved, narrow, tapering into a slender point, about as long as the spikelet; lemmas thin, 3-nerved, densely and softly long-pilose, gradually narrowed at the summit, the nerves ending in slender teeth, the middle one extending into a straight awn. Tall perennial reeds, with broad linear blades and large plumelike terminal panicles. Type species, *Arundo donax*. *Arundo*, the ancient Latin name.

**1. Arundo dónax** L. GIANT REED. (Fig. 251.) Culms stout, in large

FIGURE 247.—*Dactylis glomerata*. Plant, × ½; spikelet and floret, × 5. (Wilson 1334, Conn.)

FIGURE 248.—*Cynosurus cristatus*. Plant, × ½; fertile spikelet and floret, × 5. (Waghorne 23, Newf.)

clumps, 2 to 6 m. tall, sparingly branching, from thick knotty rhizomes; blades numerous, elongate, 5 to 7 cm. wide on the main culm, conspicuously distichous, spaced rather evenly along the culm, the margin scabrous; panicle dense, erect, 30 to 60 cm. long; spikelets 12 mm. long. ♃ —Along irrigation ditches, Arkansas and Texas to southern California, occasionally established eastward from Maryland south; tropical America; introduced from the warm regions of the Old World. Frequently cultivated for ornament, including var. VERSÍCOLOR (Miller) Stokes, with white-striped blades. In the Southwest the culms are used for lattices, mats, and screens, and in the construction of adobe huts. In Europe the culms are used for making the reeds of clarinets and organ pipes. If kept cut down the culms branch; in this form used for hedges. Planted in southeastern Texas to prevent wind erosion.

FIGURE 249.—*Cynosurus echinatus*. Panicle, × 1; fertile floret, × 5. (Macoun 80976, Vancouver Island.)

## GYNÉRIUM Willd. ex Beauv.

Plants dioecious; spikelets several-flowered, the pistillate with long-attenuate glumes and smaller long-silky lemmas, the staminate with shorter glumes and glabrous lemmas.

FIGURE 250.—*Lamarckia aurea*. Plant, × ½; fertile spikelet and floret, × 5. (Baker 5275, Calif.)

Tall perennial reeds with plumelike panicles. Type species, *Gynerium saccharoides* (*G. sagittatum*). Name from Greek *gune*, female, and *erion*, wool, referring to the woolly pistillate spikelets.

**Gynerium sagittátum** (Aubl.) Beauv. UVA GRASS. Culms as much as 10 or 12 m. tall, clothed below with the overlapping old sheaths, the blades fallen; blades sharply serrulate, commonly 2 m. long, 4 to 6 cm. wide, forming a great fan-shaped summit to the sterile culms, panicle pale, plumelike, densely flowered, 1 m. or more long, the main axis erect, the branches

FIGURE 251.—*Arundo donax*. Plant, × 1/3; spikelet and floret, × 3. (Biltmore Herb. 7514, N. C.)

drooping. ♃ —Occasionally cultivated for ornament in greenhouses. River banks and wet ground, tropical America; soil binder.

## 27. CORTADÉRIA Stapf

### Pampasgrass

Spikelets several-flowered; rachilla internodes jointed, the lower part glabrous, the upper bearded, forming a stipe to the floret; glumes longer than the lower florets; lemmas of pistillate spikelets clothed with long hairs. Large tussock grasses, with leaves crowded at the base, the blades elongate, narrow, attenuate, the margins usually serrulate; panicle large, plumelike. Type species, *Cortaderia argentea* (*C. selloana*). Name from the Argentine native name *cortadera*, cutting, because of the cutting edges of the blades.

**1. Cortaderia selloána** (Schult.) Aschers. and Graebn. Pampasgrass. (Fig. 252.) Dioecious perennial reed, in large bunches; culms stout, erect 2 to 3 or more m. tall; panicle feathery, silvery white to pink, 30 to 100 cm. long; spikelets 2- to 3-flowered, the pistillate silky with long hairs, the staminate naked; glumes white, papery, long, slender; lemmas bearing a long slender awn. ♃ (*Gynerium argenteum* Nees.)—Plains and open slopes, Brazil to Argentina and Chile. Cultivated as a lawn ornamental in the warmer parts of the United States; in southern California grown commercially for the plumes which are used for decorative purposes, the culms here being sometimes as much as 7 m. tall. Recently planted by Soil Conservation Service for supplementary dry-land pasture in Ventura and Los Angeles Counties, Calif., cattle reported to be thriving on it.

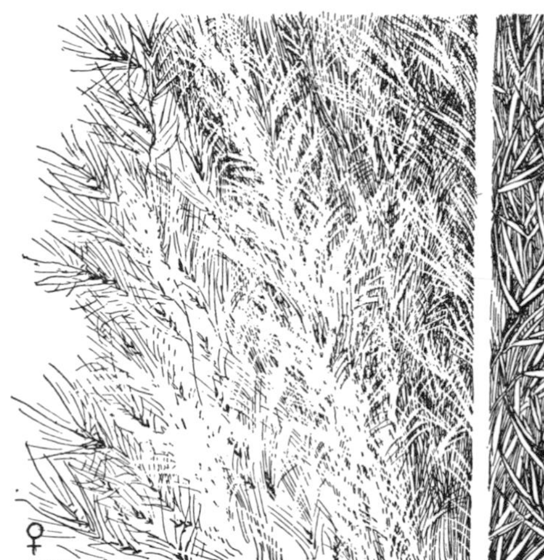

Figure 252.—*Cortaderia selloana.* Pistillate (♀) and staminate (♂) panicles, × 1. (Silveus 308, Tex.)

Cortaderia rudiúscula Stapf. Differing from *C. selloana* in the looser yellowish or purplish panicle; spikelets somewhat smaller. ♃ —Occasionally cultivated for ornament; Argentina. Has been called *C. quila* Stapf, but that name is ultimately based on *Arundo quila* Molina, which is a bamboo, *Chusquea quila* (Molina) Kunth.

---

**Ampelodésmos mauritánicus** (Poir.) Dur. and Schinz. Robust perennial in large clumps, culms solid, 2 to 3 m. tall; blades elongate, wiry, curved at base, bending forward across the culm, the upper surface downward; panicle 20 to 50 cm. long, many-flowered, the slender, flexuous, very scabrous branches naked at base, drooping, the spikelets crowded toward the ends, 2- to 5-flowered, 12 to 15 mm. long, the lower part of lemma and rachilla joints densely pilose with white hairs. ♃ —Occasionally cultivated as an ornamental; escaped and established in Napa County, Calif. Mediterranean region. Generic name often incorrectly spelled *Ampelodesma*.

## 28. PHRAGMÍTES Trin.

Spikelets several-flowered, the rachilla clothed with long silky hairs, disarticulating above the glumes and at the base of each segment between the florets, the lowest floret staminate or neuter; glumes 3-nerved, or the upper 5-nerved, lanceolate, acute, unequal, the first about half as long as the upper, the second shorter than the florets; lemmas narrow, long-acuminate, glabrous, 3-nerved, the florets successively smaller, the summits of all about equal; palea much shorter than the lemma. Perennial reeds, with broad, flat, linear blades and large terminal panicles. Type species, *Arundo phragmites* L. (*Phragmites communis*). Name from the Greek, in reference to its growth like a fence (*phragma*) along streams.

**1. Phragmites commúnis** Trin. Common reed. (Fig. 253.) Culms erect, 2 to 4 m. tall, with stout creeping rhizomes and often also with stolons; blades flat, 1 to 5 cm. wide; panicle tawny or purplish, 15 to 40 cm. long, the branches ascending, rather densely flowered; spikelets 12 to 15 mm. long, the florets exceeded by the hairs of the rachilla. ♃ (*P. phragmites* Karst.)—Marshes, banks of lakes and streams, and around springs, Nova Scotia to British Columbia, south to Maryland, North Carolina, Illinois, Louisiana, and California; Florida; Mexico and West Indies to Chile and Argentina; Eurasia, Africa, Australia.

In the Southwest this, in common with *Arundo donax*, is called by the Mexican name carrizo and is used for lattices in the construction of adobe huts. The stems were used by the Indians for shafts of arrows and in Mexico and Arizona for mats and screens, for thatching, cordage, and carrying nets.

## 29. NEYRAÚDIA Hook. f.

Spikelets 4- to 8-flowered; rachilla jointed about half way between the florets, the part below the joint glabrous, the part above bearded, forming a stipe below the mature floret; glumes unequal, 1-nerved; lemmas narrow, 3-nerved, acuminate, conspicuously long-pilose on the margins, awned from between 2 fine teeth, the awn recurved. Tall perennial with large open many-flowered panicles. Type species, *Neyraudia madagascariensis* (Kunth) Hook. f. (*N. arundinacea* (L.) Henr.) Name an anagram of *Reynaudia*, a genus of Cuban grasses.

**1. Neyraudia reynaudiána** (Kunth) Keng. (Fig. 254.) Reedlike perennial, 1 to 3 m. tall, resembling *Phragmites communis;* sheaths woolly at the throat and on the collar; blades flat, 1 to 2 cm. wide or sometimes narrow and subinvolute; panicle nodding, 30 to 60 cm. long, rather densely flowered; spikelets 4- to 8-flowered, the lowest 1 or 2 lemmas empty, 6 to 8 mm. long, rather short-pediceled along the numerous panicle branches; lemmas somewhat curved, slender, the awn flat, recurved. ♃ —Planted in testing garden at Coconut Grove, Fla., and occasionally escaped; native of southern Asia.

## 30. MÉLICA L. Melicgrass

Spikelets 2- to several-flowered (rarely with 1 perfect floret), the rachilla disarticulating above the glumes and between the fertile florets (in some species spikelets falling entire), prolonged beyond the perfect florets and bearing 2 or 3 approximate gradually smaller empty lemmas, each enclosing the

FIGURE 253.—*Phrag mites communis.* Plant, × 1/3; spikelet and floret, × 3. (Hitchcock 5078, N. Dak.)

one above; glumes somewhat unequal, thin, often papery, scarious-margined, obtuse or acute, sometimes nearly as long as the lower floret, 3- to 5-nerved, the nerves usually prominent; lemmas convex, several-nerved, membranaceous or rather firm, scarious-margined, sometimes conspicuously so, awnless or sometimes awned from between the teeth of the bifid apex, the callus not bearded. Rather tall perennials, the base of the culm often swollen into a corm, with closed sheaths, usually flat blades, narrow or sometimes open, usually simple panicles of relatively large spikelets. Type species, *Melica nutans* L. *Melica*, an Italian name for a kind of sorghum, probably from the sweet juice (mel, honey).

The species are in general palatable grasses but, not being gregarious, do not furnish much forage. Important species are *M. porteri*, *M. imperfecta*, and *M. subulata*.

Spikelets narrow; lemmas acute (obtuse in *M. harfordii*) or awned.
SECTION 1. BROMELICA.
Spikelets broad; lemmas obtuse, awnless...... SECTION 2. EUMELICA.

*Section 1. Bromelica*

Lemmas long-awned from a bifid apex.
Branches of panicle few, distant, spreading, naked on the lower half...... 1. M. SMITHII.
Branches of panicle short, appressed, spikelet-bearing from near the base.
2. M. ARISTATA.
Lemmas awnless or minutely awned.
Culms not bulbous at base; lemmas obtuse, mucronate or awn-tipped. 3. M. HARFORDII.
Culms bulbous at base; lemmas acute or acuminate.
Lemmas acuminate, usually pilose; panicle narrow, the branches short, usually appressed...... 4. M. SUBULATA.
Lemmas acute; panicle broad, the branches long and spreading...... 5. M. GEYERI.

*Section 2. Eumelica*

1a. Culms bulbous at base (see also *M. californica*).
Pedicels capillary, flexuous or recurved; panicle narrow...... 6. M. SPECTABILIS.
Pedicels stouter, appressed.
Rachilla soft, enlarged, wrinkled in drying, usually brownish...... 8. M. FUGAX.
Rachilla firm, whitish, not wrinkled.
Panicle rather dense, the branches short, appressed, usually imbricate; glumes thin, indistinctly nerved...... 7. M. BULBOSA.
Panicle loosely flowered, the branches, or some of them, stiffly ascending-spreading in anthesis, usually somewhat distant, scarcely imbricate; glumes firm, distinctly nerved...... 9. M. INFLATA.
1b. Culms not distinctly bulbous at base (somewhat swollen in *M. californica*.)
2a. Spikelets falling entire, nodding to pendulous on capillary pedicels.
Spikelets 4- or 5-flowered, reflexed; panicle narrow (open in *M. porteri* var. *laxa*).
Spikelets V-shaped; glumes 10 to 15 mm. long...... 10. M. STRICTA.
Spikelets narrow; glumes not more than 7 mm. long...... 11. M. PORTERI.
Spikelets 1- to 3-flowered, nodding; panicle open, the lower branches spreading.
Spikelets with 1 perfect floret; lemma with a few flat, twisted golden hairs on the back about the middle...... 14. M. MONTEZUMAE.
Spikelets with 2 perfect florets, lemmas without hairs.
Glumes nearly as long as the usually 2-flowered spikelet; apexes of the 2 florets about the same height; panicle simple or nearly so...... 12. M. MUTICA.
Glumes shorter than the usually 3-flowered spikelet; apex of second floret a little higher than that of the first; panicle compound...... 13. M. NITENS.
2b. Spikelets not falling entire, not pendulous.
Spikelets 4 to 6 mm. long; fertile florets 1 or 2.
Fertile lemmas pubescent; fertile florets often 2...... 15. M. TORREYANA.
Fertile lemmas glabrous; fertile floret usually 1...... 16. M. IMPERFECTA.
Spikelets 8 to 15 mm. long; fertile florets 2 to several.
Spikelets silvery white; glumes about as long as the spikelet; plant tall, somewhat woody...... 17. M. FRUTESCENS.
Spikelets tawny to purplish; glumes shorter than the spikelet; plant lower, herbaceous...... 18. M. CALIFORNICA.

Section 1. Bromélica Thurb.

Spikelets narrow; glumes usually narrow, scarious-margined (papery in *M. geyeri*); sterile lemmas similar to the acute (obtuse in *M. harfordii*) or awned fertile lemmas.

**1. Melica smíthii** (Porter) Vasey. Smith melic. (Fig. 255.) Culms slender, 60 to 120 cm. tall; sheaths retrorsely scabrous; blades lax, scabrous, 10 to 20 cm. long, 6 to 12 mm. wide; panicle 12 to 25 cm. long, the branches solitary, distant, spreading, naked below, sometimes reflexed, as much as 10 cm. long; spikelets 3- to 6-flowered, 18 to 20 mm. long, sometimes purplish; glumes acute; lemmas about 10 mm. long, with an awn 3 to 5 mm. long. ♃ (*Avena smithii* Porter.)—Moist woodlands, western Ontario and northern Michigan to British Columbia, south to

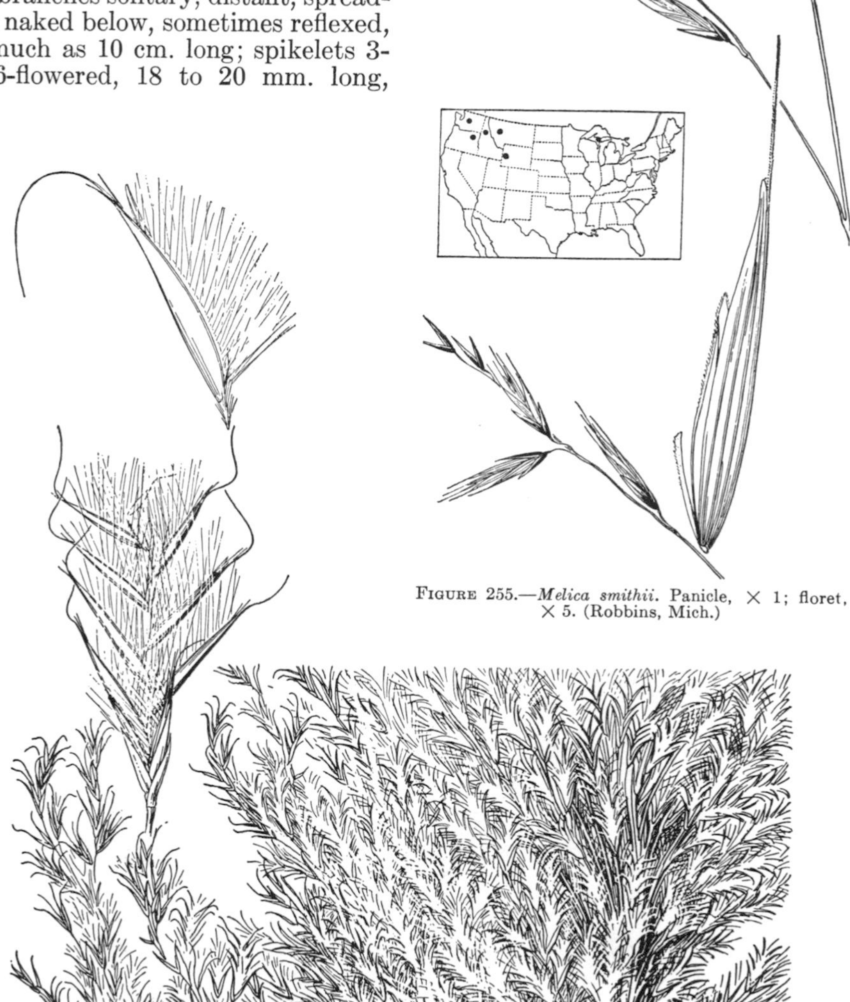

Figure 255.—*Melica smithii*. Panicle, × 1; floret, × 5. (Robbins, Mich.)

Figure 254.—*Neyraudia reynaudiana*. Panicle, × 1; spikelet, × 5; floret, × 10. (Moldenke 432, Fla.)

FIGURE 256.—*Melica aristata*. Plant, × ½; spikelet and floret, × 5. (Cusick 2888, Oreg.)

Wyoming (Teton Mountains) and Oregon (Wallowa Mountains).

**2. Melica aristáta** Thurb. ex Boland. (Fig. 256.) Culms erect or decumbent below, 60 to 100 cm. tall; sheaths scabrous to pubescent; blades 3 to 5 mm. wide, more or less pubescent; panicle narrow, 10 to 15 cm. long, the branches short, mostly appressed or ascending; spikelets, excluding awns, about 15 mm. long; glumes 10 to 12 mm. long; lemmas 7-nerved, scabrous, awned, the awn 6 to 10 mm. long. ♃ —Dry woods, meadows, and open slopes, Montana and Washington to the central Sierras of California.

**3. Melica harfórdii** Boland. HARFORD MELIC. (Fig. 257.) Culms tufted,

FIGURE 257.—*Melica harfordii*. Panicle, × 1; floret, × 5. (Yates 457, Calif.)

60 to 120 cm. tall, often decumbent below; sheaths scabrous to villous; blades scabrous, firm, flat to subinvolute, 1 to 4 mm. wide; panicle narrow, 10 to 15 cm. long, the branches appressed; spikelets 1 to 1.5 cm. long, short-pediceled; glumes 7 to 9 mm. long, obtuse; lemmas rather faintly 7-nerved, hispidulous below, pilose on the lower part of the margin, the apex emarginate, mucronate, or with an awn less than 2 mm. long. ♃ —Open dry woods and slopes, British Columbia to the Cascade Mountains of Oregon, south to Monterey County and Yosemite National Park, Calif. A smaller form with narrow involute blades has been segregated as *M. harfordii* var. *minor* Vasey.

FIGURE 258.—*Melica subulata*. Panicle, × 1; floret, × 5. (Hitchcock 11631, Wash.)

**4. Melica subuláta** (Griseb.) Scribn. ALASKA ONIONGRASS. (Fig. 258.) Culms 60 to 125 cm. tall, mostly bulbous at base; sheaths retrorsely scabrous, often pilose; blades thin,

FIGURE 259.—*Melica geyeri*. Plant, × 1; floret, × 5. (Heller 11932, Calif.)

usually 2 to 5 mm. wide, sometimes wider; panicle usually narrow, mostly 10 to 20 cm. long, the branches appressed or sometimes spreading; spikelets narrow, 1.5 to 2 cm. long, loosely flowered; glumes narrow, obscurely nerved, the second about 8 mm. long; lemmas prominently 7-nerved, tapering to an acuminate point, awnless, the nerves more or less pilose-ciliate. ♃ —Meadows, banks, and shady slopes, western Wyoming and Montana to Alaska, south in the mountains to Mount Tamalpais and Lake Tahoe, Calif.

5. **Melica géyeri** Munro. GEYER ONIONGRASS. (Fig. 259.) Culms 1 to 1.5 m. tall, bulbous at base; sheaths usually glabrous, sometimes slightly scabrous or pubescent; blades scabrous (rarely puberulent), mostly less than 5 mm. wide; panicle 10 to 20 cm. long, open, the branches slender, rather distant, spreading, bearing a few spikelets above the middle; spikelets 12 to 20 mm. long; glumes broad, smooth, papery, the second about 6 mm. long; lemmas 7-nerved, scaberulous or nearly glabrous, narrowed to an obtuse point, awnless. ♃ —Open dry woods and rocky slopes, at medium altitudes, western Oregon to central California in the Coast Range; infrequent in the Sierras to Placer County; Nevada; Yellowstone Park, Wyo.

MELICA GEYERI var. ARISTULÁTA J. T. Howell. Lemma with an awn 0.5 to 2 mm. long from a toothed apex. ♃ —Known only from Marin County, Calif.

SECTION 2. EUMÉLICA Aschers.

Spikelets broad; glumes broad, papery; lemmas awnless; sterile lemmas small, aggregate in a rudiment more or less hidden in the upper fertile lemmas.

6. **Melica spectábilis** Scribn. PURPLE ONIONGRASS. (Fig. 260.) Culms 30 to 100 cm. tall, bulbous at base, rarely with a short rhizome; sheaths pubescent; blades flat to subinvolute, 2 to 4 mm. wide; panicle mostly 10 to 15 cm. long, narrow, the branches appressed; spikelets purple-tinged, rather turgid, 10 to 15 mm. long, the pedicels capillary, flexuous; glumes broad, papery; lemmas strongly 7-nerved, obtuse, scarious-margined, imbricate. ♃ —Rocky or open woods and thickets, Montana to British Columbia, south to Colorado and northern California.

7. **Melica bulbósa** Geyer ex Port. and Coult. ONIONGRASS. (Fig. 261.) Culms 30 to 60 cm. tall, bulbous at base, resembling *M. spectabilis;* sheaths and blades flat to involute, 2

to 4 mm. wide, glabrous, scabrous, or pubescent; panicle narrow, rather densely flowered, the branches short, appressed, rather stiff, mostly imbricate; spikelets papery with age, mostly 7 to 15 mm. long, the short pedicels stiff, erect; lemmas obscurely nerved, obtuse or slightly emarginate. ♃ (*M. bella* Piper.)—Rocky woods and hills, Montana to British Columbia, south to Colorado and California; western Texas (Jeff Davis County). Specimens with pubescent foliage have been differentiated as *M. bella intonsa* Piper.

**8. Melica fúgax** Boland. LITTLE ONIONGRASS. (Fig. 262.) Culms mostly 20 to 60 cm. tall, in loose tufts, the bulbs prominent; sheaths retrorsely scabrous; blades 1.5 to 4 mm. wide, scabrous, usually pubescent on the upper surface; panicle 8 to 15 cm. long, the branches stiffly spreading or

FIGURE 261.—*Melica bulbosa*. Plant, × 1; floret, × 5. (Tidestrom 1252, Utah.)

FIGURE 260.—*Melica spectabilis*. Plant, × 1; floret, × 5. (Tweedy 85, Wyo.)

reflexed at anthesis, the lower 2 to 4 cm. long; spikelets 8 to 14 mm. long, the florets somewhat distant, usually purple-tinged, the rachilla soft, wrinkled in drying, often brownish; second glume nearly as long as the lower lemma; lemmas obscurely nerved, obtuse or emarginate. ♃ —Dry hills and open woods, Washington to Nevada and central California.

**9. Melica infláta** (Boland.) Vasey. (Fig. 263.) Culms 60 to 100 cm. tall, bulbous at base; sheaths glabrous or pubescent; blades flat, 2 to 4 mm.

FIGURE 262.—*Melica fugax*. Plant, × 1; floret, × 5. (Vasey 9, Wash.)

wide; panicle 15 to 20 cm. long, narrow, the rather distant branches, or some of them, stiffly ascending-spreading in anthesis, the lower as much as 5 cm. long; spikelets somewhat inflated, 12 to 20 mm. long, pale green; glumes scabrous on the strong nerves; lemmas strongly nerved, scabrous, acutish. ♃ —California (Yosemite National Park and Mount Shasta), Washington (Chelan County, the sheaths and blades pubescent).

**10. Melica strícta** Boland. ROCK MELIC. (Fig. 264.) Culms 15 to 60 cm. tall, densely tufted, the base somewhat thickened but not bulbous; sheaths scaberulous, sometimes pu-

FIGURE 263.—*Melica inflata*. Plant, × 1; floret, × 5. (Hall and Babcock 3334, Calif.)

FIGURE 264.—*Melica stricta*. Plant, × 1; floret, × 5. (Swallen 720, Calif.)

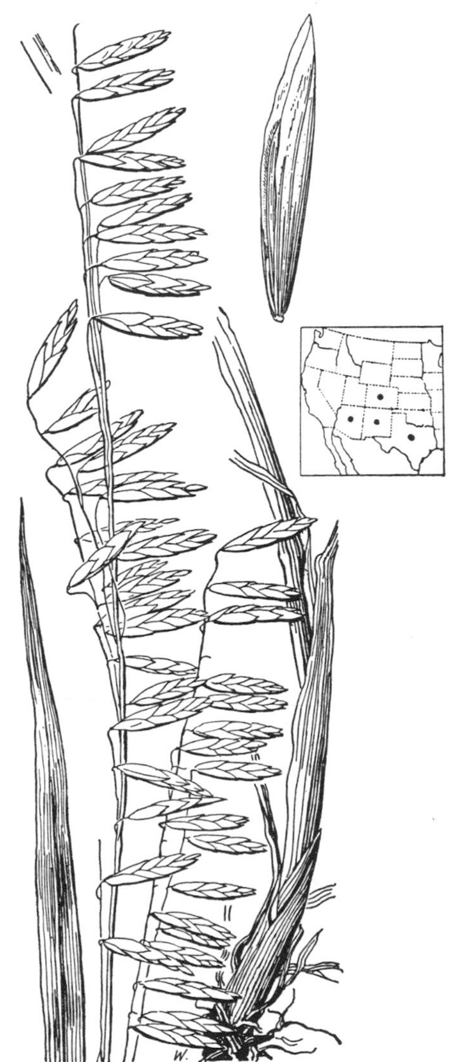

FIGURE 265.—*Melica porteri*. Plant, × 1; floret, × 5. (Shear 726, Colo.)

bescent; blades mostly 1 to 3 mm. wide, scabrous, pubescent on the upper surface; panicle narrow, simple or with 1 or 2 short branches at base; spikelets 12 to 16 mm. long, 4- or 5-flowered, broadly V-shaped, reflexed on capillary pedicels, falling entire; glumes thin, shining, nearly as long as the spikelet; lemmas faintly nerved, scabrous, and obtuse. ♃ —Rocky slopes and banks, at medium alti-

FIGURE 266.—*Melica mutica*. Plant, × ½; spikelet and floret, × 5. (Chase 3695, Va.)

FIGURE 267.—*Melica nitens*. Plant, × 1; floret, × 5. (McDonald 15, Ill.)

tudes, Utah and Nevada to Oregon (Steins Mountains), the Sierras, and the mountains of southern California.

**11. Melica portéri** Scribn. PORTER MELIC. (Fig. 265.) Culms 50 to 100 cm. tall, tufted; sheaths smooth or scabrous; blades 2 to 5 mm. wide; panicle green or tawny, narrow, 1-sided, 15 to 20 cm. long, the branches short, appressed, few-flowered; spikelets 10 to 15 mm. long, 4- or 5-flowered, narrow, reflexed on capillary pubescent pedicels, falling entire; glumes half to two-thirds as long as the spikelet; lemmas with 5 strong nerves and several faint ones, scaberulous. ♃ —Canyons, open woods, and moist places, mostly at 2,000 to 3,000 m., Colorado and Texas to Arizona; Mexico.

MELICA PORTERI var. LÁXA Boyle. Panicles open, the branches 4 to 9 cm. long, spreading to ascending, the glumes often purplish. ♃ —Rocky slopes, Chisos Mountains, Tex., to Arizona. Resembles *M. nitens*, but blades narrower, spikelets 4- or 5-flowered, and rudiment slender.

**12. Melica mútica** Walt. TWO-FLOWER MELIC. (Fig. 266.) Culms 60 to 100 cm. tall, erect, loosely tufted; sheaths scabrous or somewhat pubescent; blades flat, 2 to 5 mm. wide; panicle 10 to 20 cm. long, nearly simple, with 1 to few short, spreading, few-flowered branches below; spikelets broad, pale, 7 to 10 mm. long, usually 2-flowered, the florets spreading, pendulous on slender pedicels, pubescent at the summit, the spikelets falling entire; glumes nearly as long as the spikelet; lemmas scaberulous, strongly nerved, the two florets about the same height; rudiment obconic. ♃ —Rich or rocky woods, Maryland to Iowa, south to Florida and Texas.

**13. Melica nítens** (Scribn.) Nutt. THREE-FLOWER MELIC. (Fig. 267.) Resembling *M. mutica;* on the average culms taller; sheaths glabrous or scabrous; blades 7 to 15 mm. wide; panicle more compound with several spreading branches; glumes shorter than the usually 3-flowered narrower spikelet; apex of the second floret a little higher than that of the first; lemmas acute; rudiment mostly minute. ♃ —Rocky woods, Pennsylvania to Iowa and Kansas, south to Virginia, Arkansas, Oklahoma, and Texas.

**14. Melica montezúmae** Piper. (Fig. 268.) Culms 50 to 100 cm. tall, erect, tufted; sheaths scaberulous; ligule thin, 5 to 10 mm. long; blades flat or subinvolute, 2 to 3 mm. wide; panicle 10 to 20 cm. long, the branches simple or nearly so, distant, the lower 5 to 8 cm. long, spreading to ascend-

FIGURE 268.—*Melica montezumae*. Panicle, × 1; spikelet, × 5. (Pringle 430, Mexico.)

ing; spikelets pale; falling entire, 7 to 8 mm. long, more or less pendulous on filiform pedicels; glumes exceeding the florets, hyaline toward the summit, the first 4 mm. broad, expanded at maturity, the second slightly shorter and narrower; fertile floret 1, the lemma scabrous, strongly nerved and with a few flat twisted golden hairs about the middle; rudiment obconic. ♃ —Shaded mountain slopes and canyons, Pecos and Brewster Counties, Tex., and northern Mexico.

**15. Melica torreyána** Scribn. TORREY MELIC. (Fig. 269.) Culms 30 to 100 cm. tall, ascending from a loose decumbent not bulbous base; blades lax, 1 to 3 mm. wide; panicle narrow, rather loose, 8 to 20 cm. long, the branches more or less fascicled, appressed or ascending, the lower fascicles distant; spikelets 4 to 6 mm. long, with 1 or 2 perfect florets and a minute obovoid, long-stiped rudiment; glumes strongly nerved, as long as the spikelet or nearly so; lemmas pubescent, subacute. ♃ —Thickets and banks at low altitudes, central California, especially in the bay region.

FIGURE 269.—*Melica torreyana.* Panicle, × 1; floret, × 5. (Chase 5686, Calif.)

**16. Melica imperfécta** Trin. CALIFORNIA MELIC. (Fig. 270.) Resembling *M. torreyana;* culms erect or ascending; the base sometimes decumbent or stoloniferous; panicle 5 to 30 cm. long, the lower branches commonly ascending to spreading; spikelets usually with 1 perfect floret and an oblong, short-stiped rudiment appressed to the palea; glumes indistinctly nerved; lemma a little longer than the glumes, glabrous, indistinctly nerved, obtuse. ♃ —Dry open woods and rocky hillsides, at low and medium altitudes, central and southern California, especially in the Coast Ranges; Baja California.

FIGURE 270.—*Melica imperfecta.* Panicle, × 1; spikelet, × 5. (Elmer 4710, Calif.)

A few forms have been distinguished as varieties.

MELICA IMPERFECTA var. REFRÁCTA Thurb. Lower branches of panicle spreading or reflexed; blades pubescent. ♃ —Southern California. MELICA IMPERFECTA var. FLEXUÓSA Boland. Like the preceding but blades glabrous. ♃ —Central and southern California. MELICA IMPERFECTA var. MÍNOR Scribn. Culms less than 30 cm. tall; blades glabrous, 1 to 2 mm. wide. ♃ —Southern California.

**17. Melica frutéscens** Scribn. (Fig. 271.) Culms 0.75 to 2 m. tall, sparingly branching, rather woody below, not bulbous at base; sheaths retrorsely scabrous; blades rather firm, 2 to 4 mm. wide, those of the innova-

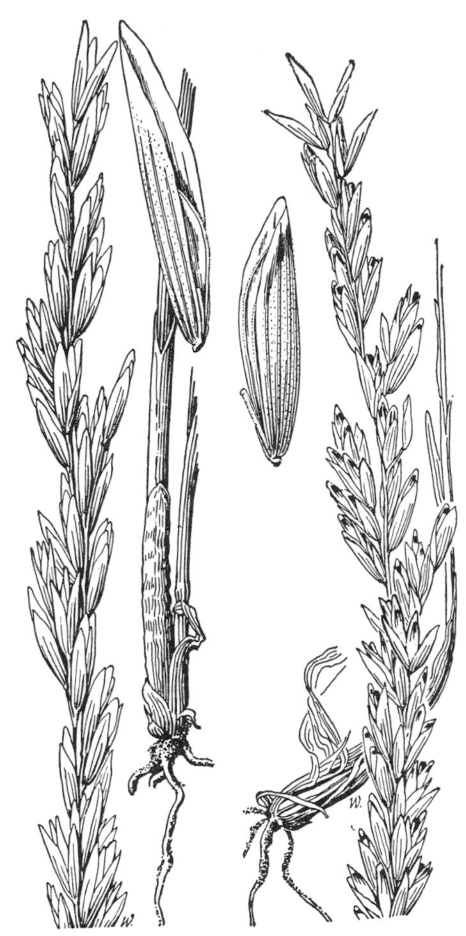

FIGURE 271.—*Melica frutescens*. Plant, × 1; floret, × 5. (Munz, Johnston, and Harwood 4143, Calif.)

FIGURE 272.—*Melica californica*. Plant, × 1; floret, × 5. (Hoffman 37, Calif.)

tions, 1 to 2 mm. wide, subinvolute; panicle silvery-shining, narrow, rather dense, 10 to 30 cm. long, the branches short, appressed; spikelets short-pediceled, 12 to 15 mm. long; glumes nearly as long as the spikelet, prominently 5-nerved; lemmas subacute, faintly 7-nerved. ♃ —Hills and canyons, at low and medium altitudes, Arizona and southern California (Inyo County and southward); Baja California.

**18. Melica califórnica** Scribn. (Fig. 272.) Culms 60 to 120 cm. tall, the base usually decumbent, often more or less bulbous; sheaths glabrous or pubescent, the lower persistent, brown and shredded; blades 1 to 4 mm. wide; panicle narrow, rather dense, 10 to 20 cm. long, tawny to purplish, not silvery; spikelets short-pediceled, 10 to 12 mm. long (rarely shorter) with 2 to 4 florets besides the rudiment; glumes scaberulous, a little shorter than the spikelets; lemmas rather prominently 7-nerved, scaberulous, subacute to obtuse, often emarginate. ♃ (*M. bulbosa* Geyer ex Thurb., not *M. bulbosa* of this work.)—Mountain meadows and rocky woods, at low and medium altitudes, Oregon (Malheur County) and California.

MELICA CALIFORNICA var. NEVADÉNSIS Boyle. Spikelets mostly 2-flowered, 7 to 8 mm. long, the glumes about equaling the upper floret. ♃ —In the lower Sierra Nevada, California.

MELICA ALTÍSSIMA L. Tall perennial; blades 15 to 20 cm. long, 5 to 10 mm. wide; panicle narrow, dense, tawny to purple; spikelets about 12 mm. long; glumes and lemmas broad, papery. ♃ —Sometimes cultivated for ornament. Eurasia.

MELICA CILIÁTA L. Panicle pale, narrow, condensed, silky. ♃ —Occasionally cultivated for ornament. Europe.

## 31. SCHIZÁCHNE Hack.

Spikelets several-flowered, disarticulating above the glumes and between the florets, the rachilla glabrous; glumes unequal, 3- and 5-nerved; lemmas lanceolate, strongly 7-nerved, long-pilose on the callus, awned from just below the teeth of the prominently bifid apex; palea with softly pubescent, thickened submarginal keels, the hairs longer toward the summit. Rather tall perennial with simple culms and open rather few-flowered panicle. Type species, *Schizachne fauriei* Hack. (*S. purpurascens*). Name from Greek *schizein*, to split, and *achne*, chaff, alluding to the b fid lemma.

**1. Schizachne purpuráscens** (Torr). Swallen. FALSE MELIC. (Fig. 273.) Culms erect from a loosely tufted

FIGURE 273.—*Schizachne purpurascens*. Plant, × ½; lemma, palea, and caryopsis, × 5. (Chase 7444, N. Y.)

FIGURE 274.—*Vaseyochloa multinervosa*. Plant, × ½; spikelet and floret, × 5. (Swallen 1854, Tex.)

decumbent base, 50 to 100 cm. tall; sheaths closed; blades flat, narrowed at the base, 1 to 5 mm. wide; panicle about 10 cm. long, the branches single or in pairs, more or less drooping, bearing 1 or 2 spikelets; spikelets 2 to 2.5 cm. long; glumes purplish, less than half as long as the spikelet; lemmas about 1 cm. long, the awn as long as the lemma or longer. ♃ (*Melica striata* Hitchc.; *M. purpurascens* Hitchc.; *Avena torreyi* Nash.)—Rocky woods, Newfoundland to southern Alaska, south to Maryland, Kentucky, South Dakota, and Montana, and in the mountains from British Columbia to New Mexico; Siberia and Japan.

## 32. VASEYÓCHLOA Hitchc.

Spikelets subterete or slightly compressed, several-flowered, the rachilla disarticulating above the glumes and between the florets, the joints very short; glumes rather firm, unequal, much shorter than the lemmas, the first 3- to 5-nerved, the second 7- to 9-nerved; lemmas rounded on the back, firm, closely imbricate, 7- to 9-nerved, broad, narrowed to an obtuse entire apex and with a stipelike hairy callus, pubescent on the lower part of the back and margins; palea shorter than the lemma, splitting at maturity, the arcuate keels strongly wing-margined; caryopsis concavo-convex, oval, black, the base of the styles persistent as a 2-toothed crown. Slender perennial with elongate blades and somewhat open panicles. Type species, *Vaseyochloa multinervosa*. Named from Vasey and Greek, *chloa*, grass.

**1. Vaseyochloa multinervósa** (Vasey) Hitchc. (Fig. 274.) Culms erect, loosely tufted, 40 to 100 cm. tall, with slender rhizomes; sheaths scaberulous, pilose at the throat; blades flat to loosely involute, 1 to 4 mm. wide; panicle narrow, loose, 5 to 20 cm. long, the branches few, at first appressed, later spreading, the lower as much as 8 cm. long, bearing a few spikelets from about the middle; spikelets 12 to 18 mm. long, 6- to 12-flowered, purple-tinged; glumes acute, the first narrow, 4 mm. long, the second broad, 5 mm. long; lemmas narrowed to an obtuse point, about 6 mm. long, the nerves becoming rather obscure toward maturity; grain 2.5 to 3 mm. long, 1.5 to 2 mm. wide, deeply concave on the ventral side. ♃ (*Melica multinervosa* Vasey; *Distichlis multinervosa* Piper.)—Sandy open woods or open ground, southeastern Texas; rare. The rhizomes appear to break off readily, most herbarium specimens being without them.

## 32A. ECTOSPERMA Swallen

(See pp. 860, 995)

## 33. TRÍDENS Roem. and Schult.

(Included in *Triodia* R. Br. in Manual, ed. 1.)

Spikelets several-flowered, the rachilla disarticulating above the glumes and between the florets; glumes membranaceous, often thin, nearly equal in length, the first sometimes narrower, 1-nerved, the second rarely 3- to 5-nerved, acute to acuminate; lemmas broad, rounded on the back, the apex from minutely emarginate or toothed to deeply and obtusely 2-lobed, 3-nerved, the lateral nerves near the margin, the midnerve usually excurrent between the lobes as a minute point or as a short awn, the lateral nerves often excurrent as minute points, all the nerves pubescent below (subglabrous in 1 species), the lateral nerves sometimes conspicuously so throughout; palea broad, the 2 nerves near the margin, sometimes villous; grain concavo-convex. Erect, tufted perennials, rarely rhizomatous or stoloniferous, the blades usually flat, the inflorescence an open to contracted or capitate panicle. Type species,

*T. quinquifidus* Roem. and Schult. (*T. flavus*). Name from Latin, *tria*, thrice, and *dens*, tooth, referring to the 3-toothed lemma.

In general the species of *Tridens* are of little importance economically, *T. grandiflorus*, *T. elongatus*, and *T. pilosus* being the most useful on the range. *Tridens pulchellus* is often abundant, but is not relished by stock, the little dry plants seldom being eaten.

1a, Panicle capitate, exceeded by fascicles of leaves; low creeping plants.
1. T. PULCHELLUS.

1b. Panicle exserted, open or spikelike; plants not creeping.

2a. Panicle open, or loose, not dense or spikelike.

Pedicels of the lateral spikelets less than 1 mm. long........ 8. T. AMBIGUUS.

Pedicels all slender, more than 1 mm. long (some short in *T. buckleyanus*).

Lateral nerves not excurrent.

Spikelets not more than 5 mm. long; lemmas 2 mm. long.
9. T. ERAGROSTOIDES.

Spikelets 6 to 8 mm. long; lemmas 4 to 5 mm. long........ 6. T. BUCKLEYANUS.

Lateral nerves excurrent as short points.

Rhizomes developed, scaly and creeping........ 7. T. CAROLINIANUS.

Rhizomes wanting.

Panicle 5 to 15 cm. long; blades 1 to 3 mm. wide........ 13. T. TEXANUS.

Panicle 15 to 30 cm. long, the branches viscid; blades 3 to 10 mm. wide.

Panicle rather dense, the branches narrowly ascending, floriferous nearly to the base........ 11. T. OKLAHOMENSIS.

Panicle open, the branches widely spreading, loosely flowered, naked at the base.

Panicle erect, the branches stiffly spreading; pulvini hairy, extending entirely around the base of the branches........ 12. T. CHAPMANI.

Panicle drooping; pulvini confined to the upper surface at the base of the branches........ 10. T. FLAVUS.

2b. Panicle narrow, contracted or spikelike, the branches appressed. (See also *T. carolinianus.*)

Panicle dense, oval or oblong, mostly less than 10 cm. long.

Lemmas deeply 2-lobed.

Lobes of lemma 1.5 to 2.5 mm. long, firm, scarcely shining; awn longer than the lobes; panicles mostly oval, not more than 6 cm. long, usually less, often purple tinged........ 2. T. GRANDIFLORUS.

Lobes of lemma 1 to 1.5 mm. long, obtuse, thin, shining; awn scarcely longer than the lobes; panicles oblong, 5 to 8 cm. long, very dense, tawny.
3. T. NEALLEYI.

Lemmas minutely notched, not lobed.

Panicle 1 to 2 cm. long; lemma margins densely long-ciliate; palea half as long as the lemma........ 4. T. PILOSUS.

Panicle 4 to 10 cm. long; lemma margins short-pilose near base; palea about as long as the lemma........ 5. T. CONGESTUS.

Panicle slender, spikelike (long and dense in *T. strictus*).

Lemmas glabrous. Panicle whitish........ 15. T. ALBESCENS.

Lemmas pilose on the margins.

Lemmas mucronate; panicle dense........ 14. T. STRICTUS.

Lemmas not mucronate (rarely lowest lemma obscurely so); panicle not dense.

Glumes acuminate, longer than the lowest floret, the second 3-nerved; blades mostly flat, some of them 2 to 4 mm. wide........ 17. T. ELONGATUS.

Glumes obtuse, short, the second 1-nerved; blades mostly folded or involute, mostly about 1 mm. wide........ 16. T. MUTICUS.

**1. Tridens pulchéllus** (H. B. K.) Hitchc. FLUFFGRASS. (Fig. 275.) Low, tufted, usually not more than 15 cm. high; culms slender, scabrous or puberulent, consisting of 1 long internode, bearing at the top a fascicle of narrow leaves, the fascicle finally bending over to the ground, taking root and producing other culms, the fascicles also producing the inflorescence; sheaths striate, papery-margined, pilose at base; blades involute, short, scabrous, sharp-pointed; panicle capitate, usually not exceeding the blades of the fascicle, consisting of 1 to 5 nearly sessile relatively large white woolly spikelets; glumes

FIGURE 275.—*Tridens pulchellus*. Plant, × ½; spikelet and floret, × 5. (Chase 5511, Ariz.)

FIGURE 276.—*Tridens grandiflorus*. Plant, × ½; floret, × 5. (Eggleston 10973, Ariz.)

glabrous, subequal, broad, acuminate, awn-pointed, 6 to 8 mm. long, nearly as long as the spikelet; lemmas 4 mm. long, conspicuously long-pilose below, cleft about halfway, the awn scarcely exceeding the obtuse lobes, divergent at maturity. ♃ (*Dasyochloa pulchella* Willd.)—Mesas and rocky hills, especially in arid or semiarid regions, Texas to Nevada and southern California to southern Mexico.

**2. Tridens grandiflórus** (Vasey) Woot. and Standl. LARGE-FLOWERED TRIDENS. (Fig. 276.) Culms tufted, erect or geniculate below, 10 to 50 cm. tall, often pubescent at the nodes; blades flat or folded, rather firm, white-margined, appressed-pubescent, 1 to 2 mm. wide, those of the culm less than 10 cm. long; panicle dense, oblong, purplish, 2 to 6 cm. long, cleistogamous spikelets borne in the lower sheaths; spikelets 4- to 8-flowered, 5 to 12 mm. long; glumes acuminate, about as long as the first floret; lemmas 4 to 6 mm. long, conspicuously long-pilose on the margins, densely pilose on the back below, deeply lobed, the awn as long as the lobes, or exceeding them. ♃ —Rocky slopes, western Texas to southern Arizona and northern Mexico. This has been referred to *Triodia avenacea* H. B. K., a Mexican species with stolons and smaller purple panicles.

**3. Tridens néalleyi** (Vasey) Woot. and Standl. (Fig. 277.) Culms erect, 20 to 40 cm., or sometimes as much as 60 cm., tall, glabrous or the lower internodes pilose, at least some of the nodes, especially the lower ones, conspicuously bearded; leaves mostly crowded at the base in a dense cluster, the culm leaves rather distant; blades firm, flat or conduplicate, with thick white midnerve and margins, pilose on both surfaces, 5 to 10 cm. long, about 2 mm. wide, the uppermost usually reduced; panicles 4 to 6 cm. long, pale, very densely flowered, the individual spikelets obscured; spikelets 6 to 8 mm. long; glumes equal, acuminate, as long as or somewhat shorter than the spikelet; lemmas 4 to 6 mm. long, the lobes broad, hyaline, obtuse, more or less erose, spreading at maturity; awn as long as or only slightly exceeding the lobes of the lemma. ♃ —Rocky slopes, southwestern Texas and New Mexico (Las Cruces); northern Mexico.

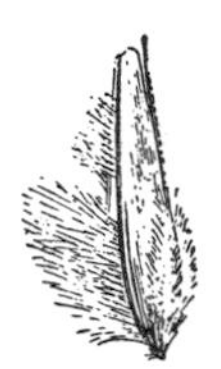

FIGURE 277.—*Tridens nealleyi*. Floret, × 5. (Nealley 153, Tex.)

**4. Tridens pilósus** (Buckl.) Hitchc. HAIRY TRIDENS. (Fig. 278.) Culms erect, densely tufted, 10 to 30 cm. tall, usually only 1 node showing, the tufts easily pulled up; sheaths pilose at the throat; blades 1 to 1.5

mm. wide, flat or folded, mostly in a short basal cluster, somewhat pilose, the margins thick, white, the culm blades 1 to 2 cm. long; panicle long-exserted, ovoid, 1 to 2 cm. long, pale or purplish, of 3 to 10 large short-pediceled spikelets; spikelets 6- to 12-flowered, 1 to 1.5 cm. long, compressed, glumes about two-thirds as long as the lower florets; lemmas about 6 mm. long, densely pilose toward the base, pilose on the margin toward the tip, acute, minutely 2-toothed, the awn 1 to 2 mm. long; palea half as long as the lemma, pilose on the back and margins below. ♃ (*Triodia acuminata* Vasey; *Tricuspis pilosa* Nash; *Erioneuron pilosum* Nash.)—Plains and rocky hills, western Kansas to Nevada, south to Texas, Arizona, and central Mexico.

**5. Tridens congéstus** (L. H. Dewey) Nash. (Fig. 279.) Culms erect, tufted, 30 to 60 cm. tall; blades flat, 2 to 3 mm. wide, tapering to a fine point; panicle mostly dense, pale or pinkish, 4 to 10 cm. long, sometimes interrupted below; spikelets rather turgid, 6- to 12-flowered, 5 to 10 mm. long; lemmas 3 to 4 mm. long, broad, obtuse, short-pilose on the midnerve and margin below, the apex slightly notched, the awn less than 1 mm. long; palea about as long as the lemma, broad, abruptly bowed out below. ♃ —Sandy or dry plains, southern Texas.

FIGURE 279.—*Tridens congestus*. Panicle, × 1; floret, × 5. (Tracy 8879, Tex.)

FIGURE 278.—*Tridens pilosus*. Plant, × ½; floret, × 5. (Griffiths 6427, Tex.)

**6. Tridens buckleyánus** (L. H. Dewey) Nash. (Fig. 280.) Culms erect, tufted, 30 to 60 cm. tall; sheaths scaberulous, sometimes sparsely pilose; blades flat, 1 to 3 mm. wide, tapering to a fine point; panicle 10 to 20 cm. long, the few branches distant, ascending to spreading, as much as 7 cm. long; spikelets pale to dark purple, short-pediceled, appressed, rather few and somewhat distant along the simple branches, 3- to

FIGURE 280.—*Tridens buckleyanus.* Panicle, × 1; floret, × 5. (Tharp 2996, Tex.)

5-flowered, 6 to 8 mm. long; glumes slightly shorter than the lower florets; lemmas 4 to 5 mm. long, pubescent on the callus and on the lower two-thirds of the midnerve and margin, the apex obtuse, entire, the midnerve not or scarcely excurrent; palea a little shorter than the lemma, pubescent along the margins; grain elliptic, 3 mm. long. ♃ —Rocky wooded slopes, southern Texas.

FIGURE 281.—*Tridens carolinianus.* Plant, × 1; floret, × 5. (Bartlett 3224, Ala.)

**7. Tridens caroliniánus** (Steud.) Henr. (Fig. 281.) Culms slender, erect, 1 to 1.5 m. tall, with creeping scaly rhizomes; lower sheaths pubescent; blades flat, elongate, 2 to 7 mm. wide; panicle purplish, narrow, rather loose, nodding, 10 to 20 cm. long, the branches appressed or narrowly ascending; spikelets short-pediceled, 3- to 5-flowered, 7 to 10 mm. long; glumes broad, mucronate from a notched apex; lemmas about 5 mm. long, pilose on the callus and on the lower half of the midnerve and margins, the summit lobed, the 3 nerves excurrent less than 1 mm.; palea glabrous, a little shorter than the lemma, bowed out below. ♃ (*Triodia drummondii* Scribn. and Kearn., *Tridens drummondii* Nash.)—Sandy woods, Coastal Plain, South Carolina to Florida and Louisiana.

**8. Tridens ambíguus** (Ell.) Schult. (Fig. 282.) Culms slender, erect, 60 to 100 cm. tall; lower sheaths glabrous; blades flat or loosely involute, 1 to 5 mm. wide; panicle open, ovoid, pale or purplish, 8 to 20 cm. long, the branches ascending, 3 to 8 cm. long; spikelets on pedicels less than 1 mm. long along the simple branches, 4- to 7-flowered, 4 to 6 mm. long, nearly as broad, the florets crowded; glumes broad, subacute; lemmas 3 to 4 mm. long, mucronate from a minutely lobed apex, the lateral nerves scarcely or barely exserted, pilose on the midnerve and margins on the lower half; palea nearly as long as the lemma, the keels bowed out below. ♃ (*Triodia langloisii* (Nash) Bush.)—Wet pine barrens, on the coast, South Carolina to Florida and Texas.

**9. Tridens eragrostoídes** (Vasey and Scribn.) Nash. (Fig. 283.) Culms slender, erect, densely tufted, 50 to 100 cm. tall; blades flat, 1 to 4 mm. wide, setaceous-tipped; panicle open, 10 to 30 cm. long, the branches rather distant, slender, flexuous, spreading or drooping, 5 to 15 cm. long, nearly simple, rather few-flowered; spikelets on slender pedicels 1 to 10 mm. long, oblong, mostly 6- to 10-flowered, scarcely 5 mm. long; glumes acuminate; lemmas about 2 mm. long, obtuse, obscurely pubescent along the midnerve on the lower half, the margins pubescent, the midnerve minutely excurrent. ♃ —Dry ground among shrubs, Florida Keys, Texas, Arizona, and northern Mexico; Cuba.

**10. Tridens flávus** (L.) Hitchc. PURPLETOP. (Fig. 284.) Culms erect, tufted, 1 to 1.5 m. tall; basal sheaths compressed-keeled; blades elongate, 3 to 10 mm. wide, very smooth; panicle open, 15 to 35 cm. long, usually purple or finally nearly black, rarely yellowish, the branches distant, spreading to drooping, naked below, as much as 15 cm. long, with slender divergent branchlets, the axils pubescent, the axis, branches, branchlets, and pedicels viscid; spikelets oblong, mostly 6- to 8-flowered, 5 to 8 mm. long; glumes subacute, mucronate; lemmas 4 mm. long, obtuse, pubescent on the callus and lower half of keel and margins, the 3 nerves excurrent; palea a little shorter than the lemma, somewhat bowed out below. ♃ (*Tricuspis seslerioides* Torr.)—Old fields and open woods, New Hampshire to Nebraska, south to Florida and Texas. The type specimen is the rare form with yellowish panicle. In some Florida specimens the excurrent nerves of the lemma are as much as 1 mm. long.

**11. Tridens oklahoménsis** (Feath.) Feath. (Fig. 285.) Culms 120 to 150 cm. tall, densely tufted, stout, erect, more or less viscid, especially at and below the nodes; blades to 60 cm. long and 12 mm. wide, flat, glabrous or sparsely pilose on the upper surface at the base; panicles terminal and axillary, purple, the terminal ones 20 to 25 cm. long, the long branches narrowly ascending, floriferous nearly to the base; spikelets 6 to 8 mm. long,

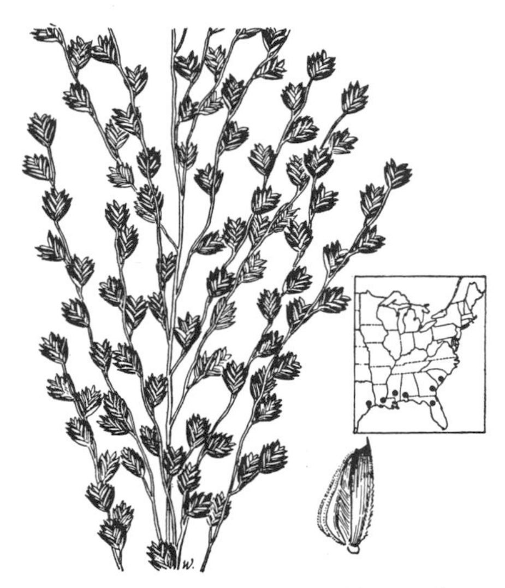

FIGURE 282.—*Tridens ambiguus.* Panicle, × 1; floret, × 5. (Curtiss 5020, Fla.)

7- to 9-flowered, short-pediceled; glumes equal, acute, about 4 mm. long; lowest lemma 4 mm. long. ♃ —Wet meadows, near Stillwater, Okla.

**12. Tridens chapmáni** (Small) Chase. (Fig. 286.) Culms 60 to 160 cm. tall, slender or occasionally rather

FIGURE 283.—*Tridens eragrostoides.* Panicle, × 1; two views of floret, × 5. (Swallen 1471, Tex.)

coarse; lower leaves crowded toward the base, the sheaths narrow, spreading from the culm, keeled, glabrous, densely villous on the collar; blades flat or loosely rolled, elongate, attenuate, 3 to 7 mm. wide, narrowed toward the base; panicles 15 to 25 cm. long, usually erect, the branches

FIGURE 284.—*Tridens flavus*. Plant, × ½; spikelet and floret, × 5. (Dewey 350, Va.)

and branchlets stiffly spreading, the bases of the principal ones surrounded by glandular hairy pulvini; spikelets long-pediceled, divergent, 7 to 10 mm. long, pale or purple-tinged. ♃ —Dry pine and oakwoods, New Jersey, Virginia, Missouri, and Oklahoma, south to Florida and Texas.

**13. Tridens texánus** (S. Wats.) Nash. (Fig. 287.) Culms erect, densely tufted, 20 to 40 cm. tall; sheaths pubescent at throat and on the collar; blades flat or subinvolute, 1 to 4 mm. wide, tapering to a slender point; panicle open, 5 to 15 cm. long, nodding, the branches rather distant, flexuous, drooping, few-flowered; spikelets oblong, 6- to 10-flowered, 6 to 10 mm. long, rather turgid, pink or purplish, more or less nodding on short pedicels; glumes broad, acute to obtuse; lemmas 4 to 5 mm. long, obtuse, minutely lobed, the margins densely pilose near the base, the keel glabrous or sparsely pilose below, the 3 nerves short-excurrent; palea about as long as the lemma, strongly bowed out at base. ♃ —Plains and dry slopes, central and southern Texas, and northern Mexico.

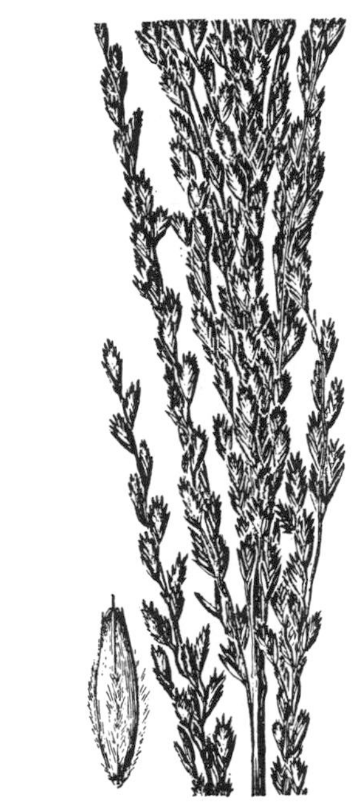

FIGURE 285.—*Tridens oklahomensis*. Panicle, × 1; floret, × 5. (Wade 77, Okla.)

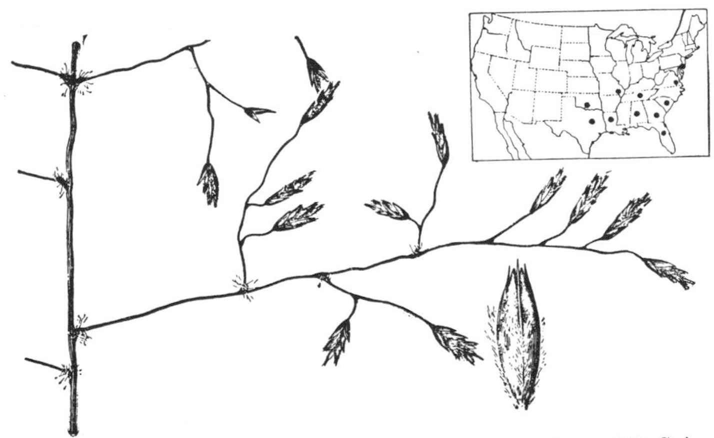

FIGURE 286.—*Tridens chapmani*. Panicle, × 1; floret, × 10. (Harper 1714, Ga.)

**14. Tridens stríctus** (Nutt.) Nash. (Fig. 288.) Culms rather stout, erect, 1 to 1.5 m. tall; blades elongate, flat or loosely involute, 3 to 8 mm. wide; panicle dense, spikelike, more or less interrupted below, narrowed above, 10 to 30 cm. long; spikelets short-pediceled, 4- to 6-flowered, about 5 mm. long, the florets closely imbricate; glumes as long as the spikelet, or nearly so, the apex spreading, the keel glandular viscid toward maturity; lemmas about 3 mm. long, obtuse, the keel and margins pilose on the lower half to two-thirds, the midnerve ex-

FIGURE 287.—*Tridens texanus*. Panicle, × 1; floret, × 5. (Wooton, Tex.)

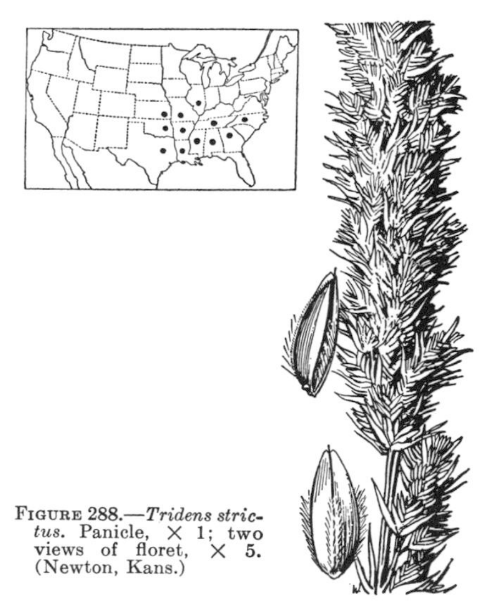

FIGURE 288.—*Tridens strictus*. Panicle, × 1; two views of floret, × 5. (Newton, Kans.)

current as a minute awn; palea about as long as the lemma, short-ciliate on the sharp keels, not strongly bowed out. ♃ (*Tricuspis stricta* A. Gray.) —Low moist ground and low woods, Illinois and Kansas to North Carolina, Alabama, and Texas.

**15. Tridens albéscens** (Vasey) Woot. and Standl. WHITE TRIDENS. (Fig. 289.) Culms erect, tufted, 30 to 80 cm. tall; blades flat to loosely involute, elongate, 2 to 4 mm. wide, tapering to a fine point; panicle narrow, rather dense, greenish to nearly white, 10 to 20 cm. long; spikelets short-pediceled, 8- to 12-flowered, 5 to 7 mm. long, the florets closely imbricate; glumes a little longer than the first lemma, subacute; lemmas 3 mm. long, obscurely pubescent on the callus, otherwise glabrous, obtuse, the midnerve minutely or not at all excurrent; palea a little shorter than the lemma, bowed out below. ♃ (*Rhombolytrum albescens* Nash.)—Plains and open woods, Kansas and Colorado to Texas and New Mexico; northern Mexico.

FIGURE 289.—*Tridens albescens*. Panicle, × 1; two views of floret, × 5. (Ball 1652, Tex.)

**16. Tridens mūticus** (Torr.) Nash. SLIM TRIDENS. (Fig. 290.) Culms slender, densely tufted, 30 to 50 cm. tall; sheaths and blades scaberulous, the sheaths usually loosely pilose, more densely so at the summit; blades flat or subinvolute, 1 to 3 mm. wide, sometimes sparsely pilose; panicle narrow, rather dense, interrupted, the branches short, appressed; spikelets

6- to 8-flowered, about 1 cm. long, pale to purplish, nearly terete; glumes scaberulous, about as long as the lower florets; lemmas about 5 mm. long, densely pilose on the lower half of the nerves and on the callus, obtuse, entire or minutely notched, the midnerve not exserted; palea half or two-thirds as long as the lemma, densely pilose on the keels and puberulent on the back. ♃ —Plains and rocky slopes, Texas to southeastern California, north to Nevada and Utah; Mexico.

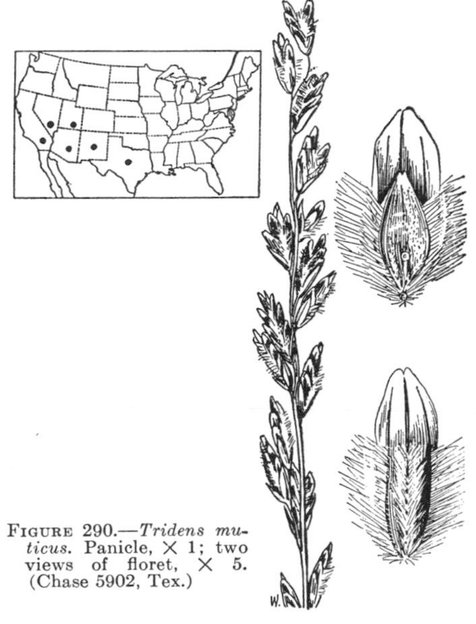

FIGURE 290.—*Tridens muticus.* Panicle, × 1; two views of floret, × 5. (Chase 5902, Tex.)

**17. Tridens elongátus** (Buckl.) Nash. ROUGH TRIDENS. (Fig. 291.) Culms erect, tufted, 40 to 80 cm. tall; sheaths and blades scaberulous, sometimes sparsely pilose, the blades mostly flat, 2 to 4 mm. wide, tapering to a fine point; panicle elongate; erect, pale or purple-tinged, loosely flowered, 10 to 25 cm. long, the branches rather distant, appressed, scarcely or not at all overlapping; spikelets similar to those of *T. muticus*, the glumes longer, the hairs on the florets not so long. ♃ (*Tricuspis elongata* Nash.) —Plains, sandy prairies, and rocky slopes, Missouri to Colorado, Texas, and Arizona.

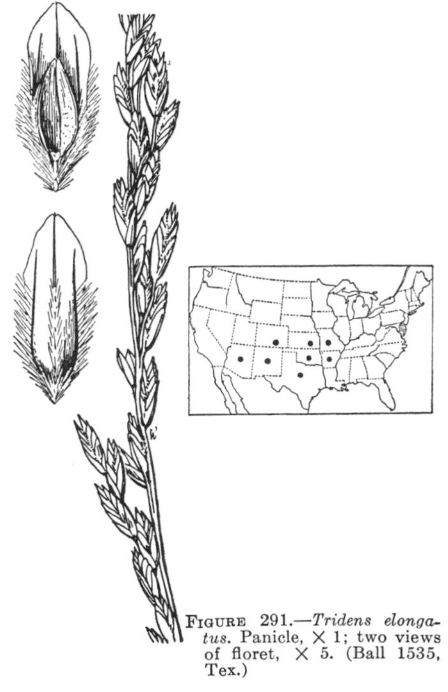

FIGURE 291.—*Tridens elongatus.* Panicle, × 1; two views of floret, × 5. (Ball 1535, Tex.)

## 34. TRÍPLASIS Beauv.

Spikelets few-flowered, V-shaped, the florets remote, the rachilla slender, disarticulating above the glumes and between the florets; glumes nearly equal, smooth, 1-nerved, acute; lemmas narrow, 3-nerved, 2-lobed, the nerves parallel, silky-villous, the lateral pair near the margin, the midnerve excurrent as an awn, as long as the lobes or longer; palea shorter than the lemma, the keels densely long-villous on the upper half. Slender tufted annuals or perennials, with short blades, short, open, few-flowered, purple, terminal panicles and cleistogamous narrow panicles in the axils of the leaves. Both species have, in addition to the small panicles of cleistogamous spikelets in the upper sheaths, additional cleistogamous spikelets, usually reduced to a single large floret, at the bases of the lower sheaths. The culms break at the nodes, the mature cleistogenes remaining within the sheaths. Type species, *Triplasis americana*. Name from Greek *triplasios*, triple, alluding to the awn and the two subulate lobes of the lemma. The species are of no importance except as they tend to hold sandy soil.

Lobes of lemma not subulate-pointed; awn shorter than the lemma; annual. 1. T. PURPUREA.

Lobes of lemma subulate-pointed; awn longer than the lemma; perennial. 2. T. AMERICANA.

**1. Triplasis purpúrea** (Walt.) Chapm. PURPLE SANDGRASS. (Fig. 292.) Annual, often purple; culms ascending to widely spreading, pubescent at the several to many nodes, 30 to 100 cm. tall, rarely taller; blades flat or loosely involute, 1 to 3 mm. wide, mostly 4 to 8 cm. long; panicle 3 to 5 cm. long, with few spreading few-flowered branches, the axillary more or less enclosed in the sheaths; spikelets short-pediceled, 2- to 4-flowered, 6 to 8 mm. long; lemmas 3 to 4 mm. long, the lobes broad, rounded or truncate, the nerves and callus densely short-villous, the awn about as long as the lobes or somewhat exceeding them; palea conspicuously silky-villous on the upper half of the keels; grain about 2 mm. long. ☉ —Dry sand, Ontario, Maine, and New Hampshire to Minnesota and Nebraska, south to Florida and Texas; Colorado (introduced?); Honduras. In autumnal culms the numerous short joints with sheaths swollen at the base, containing cleistogenes, are conspicuous. Plants with awns exceeding the lobes of the lemma have been differentiated as *T. intermedia* Nash.

**2. Triplasis americána** Beauv. (Fig. 293.) Perennial; culms slender, tufted, mostly erect, 30 to 60 cm. tall; blades flat or subinvolute, mostly 15 to 18 cm. long; panicle 2 to 5 cm. long, the few slender ascending branches with 1 or 2 spikelets; spikelets mostly 2- or 3-flowered, about 1 cm. long; lemmas 5 to 6 mm. long, the lobes about half as long as the entire lemma, subulate-pointed, the nerves with a narrow stripe of silky hairs, the awn 5 to 8 mm. long, pubescent below; keels of the palea long-villous, the hairs erect. ♃ —Dry sand, Coastal Plain, North Carolina to Florida and Mississippi.

## 35. NEOSTÁPFIA Davy

(Included in *Anthochloa* Nees in Manual, ed. 1)

Spikelets few-flowered, subsessile, closely imbricate around a simple axis, the rachilla disarticulating between the florets; glumes wanting; lemmas flabellate, prominently many-nerved; palea much narrower and a little shorter than the lemma, obtuse, hyaline. Low annual with loose sheaths merging into rather broad flat blades without definite junction and dense cylindric panicles, the axis prolonged beyond the spikelets, this portion naked or bearing small bracts. Type species, *Neostapfia colusana*. Named for Otto Stapf. (Distinguished from *Anthochloa* Nees, of the Andes, in which the axis is not prolonged, the short-pediceled spikelets have well-developed persistent glumes, the lemmas are not strongly nerved, and the sheaths and blades are distinctly differentiated.)

**1. Neostapfia colusána** (Davy) Davy. (Fig. 294.) Culms 7 to 30 cm. long, ascending from a decumbent base; leaves overlapping, loosely folded around the culm, 5 to 10 cm. long, 6 to 12 mm. wide at the middle, tapering toward both ends, minutely ciliate, with raised viscid glands on the nerves and margins; panicles pale green, at first partly included, later short-exserted, 3 to 7 cm. long, 8 to 12 mm. thick; spikelets usually 5-flowered, 6 to 7 mm. long; lemmas flabellate, very broad, 5 mm. long, ciliolate-fringed, the many nerves viscid-glandular at maturity. ☉ (*Anthochloa colusana* (Davy) Scribn.) —Bordering rain pools on hard alkali

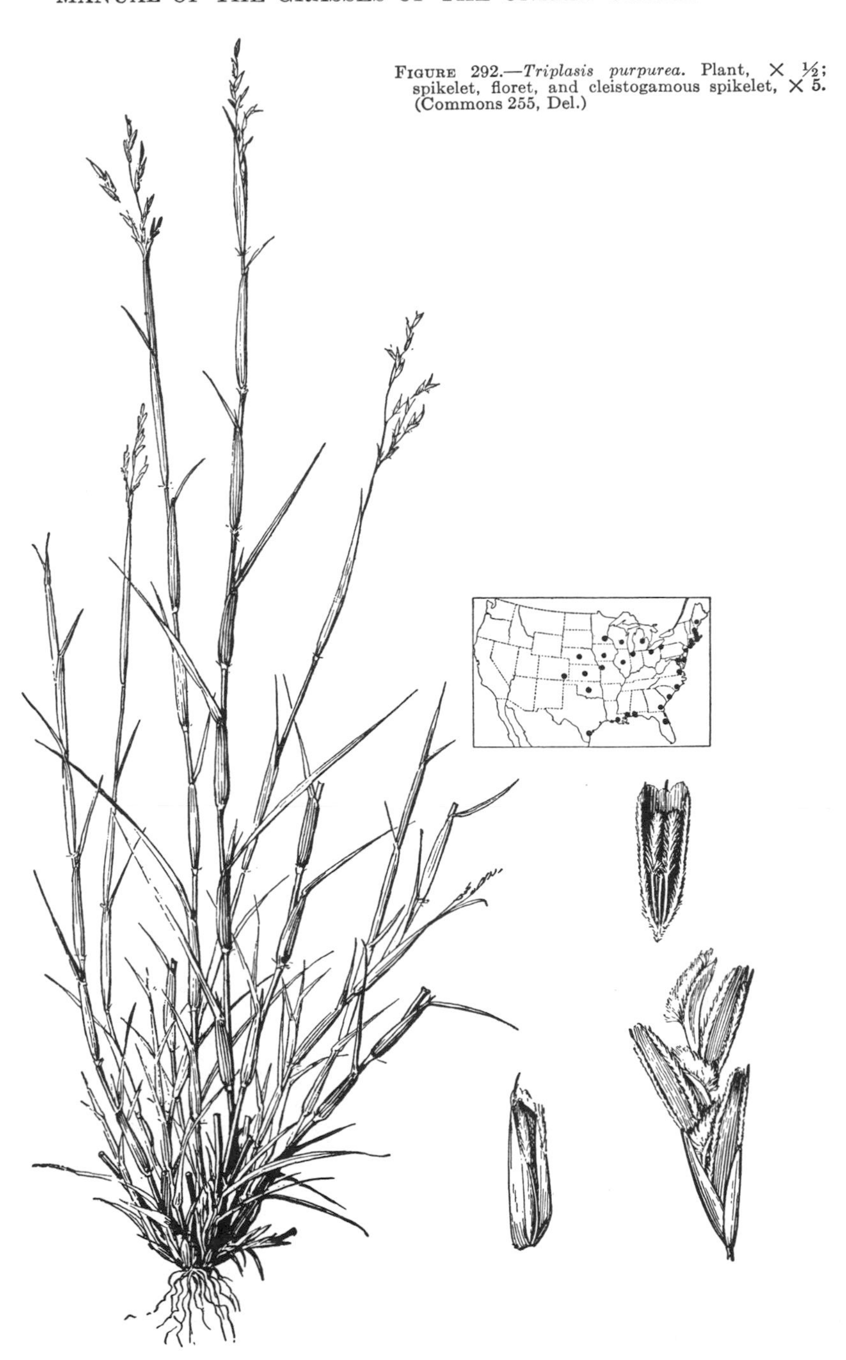

FIGURE 292.—*Triplasis purpurea*. Plant, × ½; spikelet, floret, and cleistogamous spikelet, × 5. (Commons 255, Del.)

soil, Colusa, Stanislaus, and Merced Counties, Calif. At maturity the entire plant is thickly beset with minute raised viscid glands.

## 36. ORCÚTTIA Vasey

Spikelets several-flowered, the upper florets reduced, the rachilla continuous, the spikelets persistent even after maturity; glumes nearly equal, shorter than the lemmas, broad, irregularly 2- to 5-toothed, many-nerved, the nerves extending into the teeth; lemmas firm, prominently 13- to 15-nerved, the broad summit toothed; palea broad, as long as the lemma. Low annuals with short culm blades, solitary spikes or spikelike racemes, the subsessile spikelets relatively large, the upper aggregate, the lower more or less remote. With the exception of *O. greenei*, the young plants produce elongate juvenile leaves before the development of the culms. Type species, *Orcuttia californica*. Named for C. R. Orcutt.

Lemmas with 7 to 11 very short teeth ........ 1. O. GREENEI.
Lemmas with 5 relatively long acuminate or awn-tipped teeth.
  Racemes 2 to 5 cm. long, often capitate, the spikelets usually crowded toward the summit, remote toward the base; teeth of lemma unequal, the middle longer than the lateral ones; nerves of lemma relatively faint ........ 2. O. CALIFORNICA.
  Racemes 5 to 10 cm. long, narrow, not capitate, the spikelets rather evenly distributed (the lower distant in *O. tenuis*); teeth of lemma equal; nerves of lemma prominent.
    Blades 1 to 2 mm. wide; spikelets mostly 2- to 10-flowered, glabrous.... 3. O. TENUIS.
    Blades 2 to 6 mm. wide; spikelets mostly 10- to 40-flowered, pilose........ 4. O. PILOSA.

**1. Orcuttia greénei** Vasey. (Fig. 295.) Culms 15 to 20 cm. tall, suberect; blades 2 to 3 cm. long, subinvolute; raceme 3 to 7 cm. long, pale; spikelets 10 to 15 mm. long, loosely papillose-pilose; glumes 4 to 5 mm. long; lemmas 6 mm. long, the obtuse or truncate tip spreading, 7- to 11-toothed, the teeth mucronate but not awned. ⊙ —Moist open ground, Sacramento and San Joaquin Valleys, Butte and San Joaquin Counties, southeast to Tulare County, Calif. At maturity foliage and spikelets minutely viscid-glandular.

**2. Orcuttia califórnica** Vasey. (Fig. 296.) Culms 5 to 15 cm. long, spreading with ascending ends, forming little mats; foliage thin, pilose, the sheaths loose, the blades 2 to 4 cm. long; raceme loose below, dense or subcapitate at the summit; spikelets 8 to 12 mm. long, densely to sparsely pilose; glumes sharply toothed; lemmas about 6 mm. long, deeply cleft into 5 awn-tipped teeth. The whole plant at maturity more or less viscid-glandular. ⊙ —Drying mud flats, near Murrietta, Riverside County, Calif.; Baja California.

ORCUTTIA CALIFORNICA var. INAEQUÁLIS (Hoover) Hoover. Resembling the species, but differing in having usually shorter capitate inflorescences and unequally toothed lemmas; culms ascending or prostrate. Sacramento and San Joaquin Valleys, Sacramento to Tulare County, Calif.

FIGURE 293.—*Triplasis americana*. Panicle, × 1; floret, × 5. (Curtiss 5570, Fla.)

ORCUTTIA CALIFORNICA var. VÍSCIDA Hoover. Plants very viscid; teeth of lemma awned, giving the capitate inflorescence a distinctly bristly appearance. Near the Sierra Nevada foothills, Sacramento County, Calif.

**3. Orcuttia ténuis** Hitchc. (Fig. 297.) Culms in small tufts, slender,

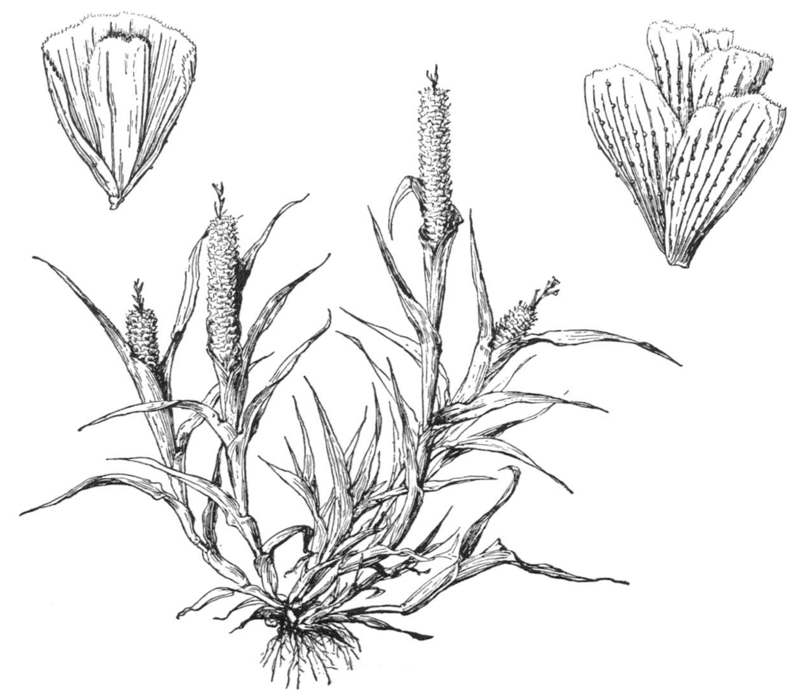

FIGURE 294.—*Neostapfia colusana.* Plant, × ½; spikelet and floret, × 5. (Type.)

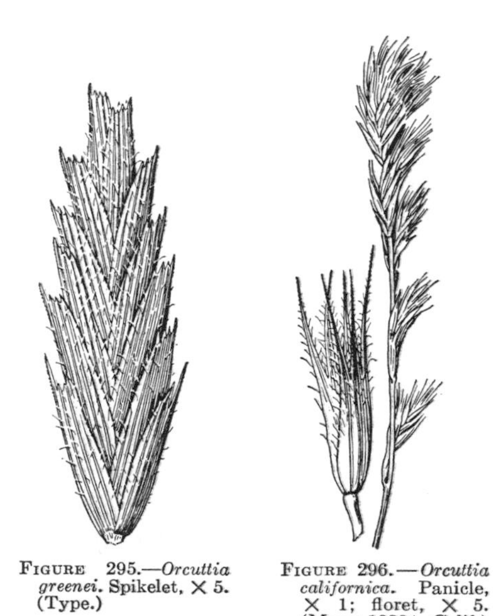

FIGURE 295.—*Orcuttia greenei.* Spikelet, × 5. (Type.)

FIGURE 296.—*Orcuttia californica.* Panicle, × 1; floret, × 5. (Munz 10804, Calif.)

erect, 5 to 12 cm. tall; leaves mostly basal, the blades strongly nerved, 1 to 2 cm. long; raceme more than half the entire height of the plant, the lower spikelets distant, the upper approximate but not crowded; spikelets purple-tinged, 12 to 15 mm. long; glumes and lemmas scabrous, sometimes with a few hairs toward the base of the lemmas; glumes 3 to 4 mm. long, sharply toothed; lemmas 5 mm. long, 5-toothed, the teeth acuminate, awn-tipped, the rigid tips spreading or slightly recurved. ⊙ —Beds of vernal pools, Shasta and Tehama Counties, east of the Sacramento River, Calif.

**4. Orcuttia pilósa** Hoover. (Fig. 298.) Culms densely tufted, 5 to 20 cm. tall, erect or geniculate-decumbent at base, viscid at maturity; sheaths and blades pilose or the blades nearly glabrous beneath; racemes 5 to 10 cm. long; spikelets 10- to 40-flowered, appressed or somewhat spreading, the upper crowded, the lower approximate; glumes about 3 mm. long, irregularly 3-toothed;

FIGURE 297.—*Orcuttia tenuis*. Plant, × ½; spikelet and floret, × 5. (Type.)

FIGURE 298.—*Orcuttia pilosa*. Plant, × ½; floret, × 5. (Hoover 1298, Calif.)

lemmas 4 to 5 mm. long, the teeth equal, acute or awn-tipped, strongly viscid-glandular at maturity; anthers 2.5 to 3 mm. long. ⊙ —San Joaquin Valley, Calif., from Stanislaus County to Madera County.

## 37. BLEPHARIDÁCHNE Hack.

Spikelets compressed, 4-flowered, the rachilla disarticulating above the glumes, but not between the florets; glumes nearly equal, compressed, 1-nerved, thin, smooth; lemmas 3-nerved, the nerves extending into awns, deeply 3-lobed, conspicuously ciliate, the first and second sterile, containing a palea but no flower, the third fertile, the fourth reduced to a 3-awned rudiment. Low annuals or perennials, with short, dense, few-flowered panicles scarcely exserted from the subtending leaves. Type species, *Blepharidachne kingii*. Name from Greek *blepharis* (blepharid-), eyelash, and *achne*, chaff, alluding to the ciliate lemma.

Glumes a little longer than the florets, acuminate; foliage scaberulous.......... 1. B. KINGII.
Glumes a little shorter than the florets, subacute; foliage densely grayish harsh-puberulent. 2. B. BIGELOVII.

**1. Blepharidachne kíngii** (S. Wats.) Hack. (Fig. 299.) Low tufted perennial with the aspect of *Tridens pulchellus*, but not rooting at upper nodes; culms mostly less than 10 cm. tall; sheaths with broad hyaline margins; blades less than 1 mm. wide, involute, curved, sharp-pointed, 1 to 3 cm. long; panicles subcapitate, pale or purplish, 1 to 2 cm. long, exceeded by the upper blades; spikelets flabellate; glumes about 8 mm. long, acuminate, exceeding the florets; sterile lemmas about 6 mm. long, all the lemmas about the same height, long-ciliate on the margins, pilose at the base and on the callus, cleft nearly to the middle, the lateral lobes narrow, obtuse, the nerve at one

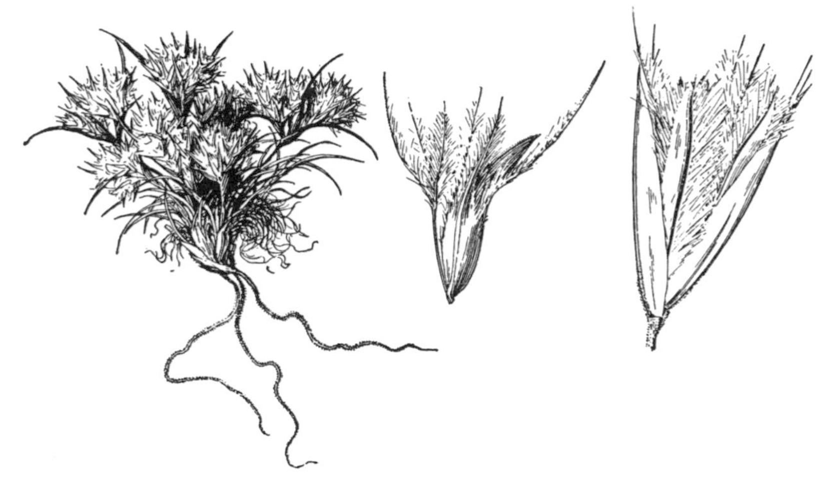

FIGURE 299.—*Blepharidachne kingii.* Plant, × 1; spikelet and perfect floret, × 5. (Jones 4094, Nev.)

margin, awn-tipped, the central lobe consisting of the awn, ciliate below, somewhat exceeding the lateral ones; palea much narrower and somewhat shorter than the lemma; fertile lemma similar to the sterile ones, the palea broad and as long as the lemma; upper sterile lemma on a rachilla segment about 3 mm. long, reduced to 3 plumose awns; grain compressed, 2 mm. long. ♃ —Deserts, Utah, Nevada, and California (Death Valley), apparently rather rare, but reported as common and sometimes the dominant grass in desert regions in Elko and White Pine Counties, Nev.

**2. Blepharidachne bigelóvii** (S. Wats.) Hack. (Fig. 300.) Perennial, culms stiff, 10 to 20 cm. long, the culms and foliage harsh-puberulent; sheaths broad, firm; blades coarser than in *B. kingii;* panicles dense, oblong, 1 to 3 cm. long, the blades not exceeding the panicle; glumes about 6 mm. long, subacute, shorter than the florets; sterile lemmas ciliate and awned as in *B. kingii,* cleft about 1 mm. ♃ —Rocky slopes, Pecos and El Paso Counties, Tex.

FIGURE 300.—*Blepharidachne bigelovii.* Plant, × 1; fertile floret, × 5. (Type.)

## 38. CÓTTEA Kunth

Spikelets several-flowered, the uppermost reduced, the rachilla dis-

FIGURE 301.—*Cottea pappophoroides*. Plant, × ½; spikelet, floret, and cleistogene, × 5. (Griffiths 5946, Ariz.)

articulating above the glumes and between the florets; glumes about equal, nearly equaling the lower lemma, with several parallel nerves; lemmas rounded on the back, villous below, prominently 9- to 11-nerved, some of the nerves extending into awns of irregular size and some into awned teeth; palea a little longer than the body of the lemma, the keels near the margin. An erect tufted branching perennial with flat blades and oblong loose panicle. Type species, *Cottea pappophoroides*. Named for Heinrich Cotta.

**1. Cottea pappophoroídes** Kunth. (Fig. 301.) Softly pubescent throughout; culms 30 to 50 cm. tall; blades 3 to 7 mm. wide; panicle 8 to 15 cm. long, the branches loosely ascending; spikelets 4- to 7-flowered, 5 to 7 mm. long, about 5 mm. wide, green or purplish; glumes 4 to 5 mm. long; lemmas 3 to 4 mm. long, the basal hairs conspicuous, at least the middle awn spreading. ♃ —Plains and dry hills, western Texas to southern Arizona, south to central Mexico; Ecuador to Argentina. Cleistogamous spikelets, usually reduced to a single floret, are found in the lower sheaths, and often large, very turgid ones at the very base. Not abundant enough to have economic importance.

## 39. PAPPÓPHORUM Schreb. PAPPUSGRASS

Spikelets 4- to 6-flowered, the lower 1 to 3 fertile, the upper reduced, the rachilla disarticulating above the glumes, but not or only tardily between the florets, the internodes very short; glumes nearly equal, keeled, thin-membranaceous, as long as the body of the florets, 1-nerved, acute; lemmas rounded on the back, firm, obscurely many-nerved, dissected above into numerous spreading, unequal awns, the florets falling together, the awns of all forming a pappuslike crown; palea as long as the body of the lemma, the nerves near the margin. Erect tufted perennials, with narrow or spikelike whitish to tawny or purplish panicles. Type species, *Pappophorum alopecuroideum* Vahl. Name from Greek *pappos*, pappus, and *phoros*, bearing, alluding to the pappuslike crown of the lemma. Our species are of minor economic importance.

Panicle spikelike, tawny or whitish .......... 1. P. MUCRONULATUM.
Panicle narrow but rather loose, pinkish .......... 2. P. BICOLOR.

**1. Pappophorum mucronulátum** Nees. (Fig. 302.) Culms erect, 60 to 100 cm. tall; blades flat to subinvolute, 2 to 5 mm. wide; panicle spikelike, tawny or whitish, tapering at summit, 10 to 20 cm. long; spikelets short-pediceled with 1 or 2 fertile florets and 2 or 3 sterile reduced ones, the rachilla disarticulating below the fertile floret and tardily above it; glumes 1-nerved; fertile lemma subindurate, the nerves obscure, villous toward base, dissected into numerous unequal awns 2 to 5 mm. long, the body about 3 mm. long. ♃ (*P. apertum* Munro.)—Low places on plains and in valleys, Texas, Arizona, and northern Mexico; South America; wool waste, Maine.

**2. Pappophorum bícolor** Fourn. (Fig. 303.) Culms erect, 30 to 80 cm. tall; blades flat to subinvolute, 1 to 5 mm. wide; panicle mostly 10 to 15 cm. long, usually pink-tinged, rather loose, the branches 1 to 4 cm. long; spikelets on pedicels 1 to 5 mm. long, with 2 or 3 fertile florets and 1 or 2 sterile reduced ones, all about the same height in the spikelet, the rachilla not separating between the florets; glumes 1-nerved; lemmas somewhat indurate, obscurely nerved, pilose on the callus and on the lower half to two-thirds of the midnerve and margins, dissected into about 12 somewhat unequal scabrous awns 2 to 4 mm. long, the body about 3 mm. long, the awns about as long. ♃ —Open valley land, Texas, Arizona (La Noria, near Monument 111), and Mexico.

FIGURE 302.—*Pappophorum mucronulatum*. Plant, × ½; spikelet and perfect floret, × 5. (Pringle, Ariz.)

FIGURE 303.—*Pappophorum bicolor*, × 1. (Griffiths 6291, Tex.)

## 40. ENNEAPÓGON Desv. ex Beauv.

(Included in *Pappophorum* Schreb. in Manual, ed. 1)

Spikelets 3-flowered, the first floret fertile, the second smaller, sterile, the third rudimentary; glumes strongly 7-nerved; lemmas rounded on the back, firm, the truncate summit bearing 9 plumose equal awns; palea a little longer than the body of the lemma, the keels near the margin. Slender tufted perennials, with narrow feathery panicles. Type species *Enneapogon desvauxii* Beauv. Name from *ennea*, nine, and *pogon*, beard, alluding to the 9 plumose or bearded awns. A single species in America.

**1. Enneapogon desvaúxii** Beauv. SPIKE PAPPUSGRASS. (Fig. 304.) Culms numerous, slender, decumbent-spreading, 20 to 40 cm. tall, the nodes pubescent; blades flat to subinvolute, about 1 mm. wide; panicle spikelike, gray green or drab, mostly 2 to 5 cm. long, sometimes interrupted below; glumes longer than the body of the lemmas, 7-nerved, acuminate, pubescent; lemma of first floret (including awns) 4 to 5 mm. long, the body about 1.5 mm. long, villous, 9-nerved, the awns plumose, except at the apex. ♃ (*Pappophorum wrightii* S. Wats.)[11]—Dry plains and stony hills, Utah and Texas to Arizona, south to Oaxaca, Peru, Bolivia, and Argentina. Cleistogamous spikelets are produced in the lower sheaths, the cleistogenes larger than the normal florets, but the awns almost wanting. The culms disarticulate at the lower nodes, carrying the cleistogenes with them. Furnishes a fair proportion of forage on sterile hills.

## 41. SCLEROPÓGON Phil.

Plants monoecious or dioecious. Staminate spikelets several-flowered, pale, the rachilla not disarticulating; glumes about equal, membranaceous, long-acuminate, 1-nerved or obscurely 3-nerved, nearly as long as the first lemma; lemmas similar to the glumes, somewhat distant, 3-nerved or obscurely 5-nerved, mucronate; palea obtuse, shorter than the lemma. Pistillate spikelets subtended by a narrow bract on the pedicel, several-flowered, the upper florets reduced to awns, the rachilla disarticulating above the glumes but not separating between the florets or only tardily so; glumes acuminate, 3-nerved, with a few fine additional nerves, the first about half as long as the second; lemmas narrow, 3-nerved, the nerves extending into slender, scabrous, spreading awns, the florets falling together, forming a cylindric many-awned fruit, the lowest floret with a sharp-bearded callus as in *Aristida;* palea narrow, the 2 nerves near the margin, produced into short awns. Stoloniferous perennial, with short flexuous blades and narrow few-flowered racemes or simple panicles, the staminate and pistillate panicles strikingly different in appearance. Staminate and pistillate panicles may occur on the same plant, or rarely the 2 kinds of spikelets may be found in the same panicle. It may be that the seedlings produce 2 kinds of branches, each kind then re-

[11] For an account of the genus and the identity of this species, see *Chase, A.*, Madroña 7:187–189. 1946.

FIGURE 304.—*Enneapogon desvauxii.* Plant, × ½; spikelet, perfect floret, and cleistogene, × 5. (Purpus 8272, Ariz.)

producing its own sex. This should be investigated. Type species, *Scleropogon brevifolius*. Name from Greek *skleros*, hard, and *pogon*, beard, alluding to the hard awns.

**1. Scleropogon brevifólius** Phil. BURRO GRASS (Fig. 305.) Culms erect, 10 to 20 cm. tall, tufted, producing

FIGURE 305.—*Scleropogon brevifolius*. Pistillate and staminate plants, × ½; pistillate spikelet, × 2; pistillate and staminate florets, × 5. (Zuck, Ariz.)

wiry stolons with internodes 5 to 15 cm. long; leaves crowded at the base, the blades flat, 1 to 2 mm. wide, sharp-pointed; racemes, excluding awns, 1 to 5 cm. long; staminate spikelets 2 to 3 cm. long; body of pistillate spikelets 2.5 to 3 cm. long, the awns 5 to 10 cm. long, loosely twisted. ♃ (*S. karwinskyanus* Benth.)—Semiarid plains and open valley lands, Texas to Colorado, Nevada, and Arizona; south to central Mexico; Argentina. The mature pistillate spikelets break away and with their numerous long spreading awns form "tumbleweeds" that are blown before the wind, the pointed barbed callus readily penetrating clothing or wool, the combined florets acting like the single floret of long-awned aristidas. Spikelets rarely staminate below and pistillate above. On overstocked ranges, where it tends to become established, it is useful in preventing erosion. Often important as a range grass, especially when young.

## TRIBE 3. HORDEAE

### 42. AGROPÝRON Gaertn. Wheatgrass

Spikelets several-flowered, solitary (rarely in pairs), sessile, placed flatwise at each joint of a continuous (rarely disarticulating) rachis, the rachilla disarticulating above the glumes and between the florets; glumes equal, firm, several-nerved, rarely 2-nerved, 1-nerved, or nerveless, usually shorter than the first lemma, acute or awned, rarely obtuse or notched; lemmas convex on the back, rather firm, 5- to 7-nerved, acute or awned from the apex; palea about as long as the lemma. Perennials (our species except *Agropyron triticeum*), often with creeping rhizomes, with usually erect culms and green or purplish, usually erect, spikes. Type species, *Agropyron triticeum* Gaertn. Name from Greek *agrios*, wild, and *puros*, wheat, the two original species being weeds in wheatfields.

Most of the species of *Agropyron* furnish forage, and a few are among the most valuable range grasses of the Western States. In the valleys some species may grow in sufficient abundance to produce hay.

*Agropyron trachycaulum* (*A. tenerum*, *A. pauciflorum*) has been cultivated in the Northwestern States on a commercial scale under the name slender wheatgrass, and the seed has been carried by seedsmen in that region. *A. smithii*, western wheatgrass, sometimes called Colorado bluestem, is a source of hay in alkaline meadows through the Western States. *A. spicatum*, or bluebunch wheatgrass, and *A. dasystachyum* are important range grasses in the Northwestern States. *A. trachycaulum* and *A. subsecundum* (*A. caninum*, so-called), because of their abundance in the mountain grazing regions, are also important. *A. repens*, quackgrass, is a good forage grass, but, because of its creeping rhizomes, is a troublesome weed, especially in the Eastern States where it is widely introduced. The species with strong creeping rhizomes are valuable for holding embankments and sandy soils.

The divisions of the species into those with rhizomes and those without is convenient and usually definite when the entire base is present, but some species normally without rhizomes (as *A. spicatum*) may rarely produce them and species in which rhizomes occur may not show them in herbarium specimens.

1a. Plants with creeping rhizomes.
 Lemmas awned, the awn divergent at maturity.
  Lemmas pubescent .......... 9. A ALBICANS.
  Lemmas glabrous .......... 10. A. GRIFFITHSII.
 Lemmas awnless or with a short straight awn.
  Glumes rigid, gradually tapering into a short awn .......... 5. A. SMITHII.
  Glumes not rigid, acute or abruptly awn-pointed.

Lemmas glabrous (sometimes pubescent in *A. riparium*).
Blades lax, flat.
Glumes shorter than the spikelets; rachilla glabrous.................. 2. A. REPENS.
Glumes nearly as long as the spikelet; rachilla pubescent.
4. A. PSEUDOREPENS.
Blades firm, stiff, often involute.
Spikelets much compressed, closely imbricate, the spike dense.
3. A. PUNGENS.
Spikelets not much compressed, somewhat distant, the spike slender.
8. A. RIPARIUM.
Lemmas pubescent.
Spike 6 to 12 cm. long; spikelets 1 to 1.5 cm. long; glumes 6 to 9 mm. long.
6. A. DASYSTACHYUM.
Spike as much as 25 cm. long; spikelets as much as 2.5 cm. long; glumes to 13 mm. long.................................................................................................. 7. A. ELMERI.
1b. Plants without creeping rhizomes.
Spikelets much compressed, crowded on the rachis............................ 1. A. DESERTORUM.
Spikelets not much compressed nor divergent.
Spikelets awnless or awn-tipped only.
Glumes 2 to 2.5 mm. wide, nearly as long as the spikelet; rachilla villous.
Glumes with a broad subhyaline margin, unsymmetrical at the summit; lemmas commonly pubescent; spike rarely more than 7 cm. long, the spikelets closely imbricate................................................................................ 14. A. LATIGLUME.
Glumes not thin-margined; lemmas glabrous; spike 10 to 25 cm. long, the spikelets mostly scarcely or slightly imbricate............................ 13. A. TRACHYCAULUM.
Glumes narrower, much shorter than the spikelet; rachilla scaberulous.
Blades involute (rarely flat)............................................................ 19. A. INERME.
Blades flat.............................................................................................. 21. A. PARISHII.
Spikelets awned.
Culms prostrate-spreading............................................................ 17. A. SCRIBNERI.
Culms erect (decumbent at base in *A. pringlei*).
Rachis finally disarticulating.
Glumes narrow, 2-nerved; awns of lemmas spreading, out-curved or recurved.
22. A. SAXICOLA.
Glumes broader, with usually 3 to 5 distinct scabrous nerves; awn straight, 2 to 5 cm. long.......................................................................... 23. A. SAUNDERSII.
Rachis continuous.
Awn straight or nearly so.
Spikelets about as long as the internodes of the rachis...... 21. A. PARISHII.
Spikelets imbricate, longer than the internodes of the rachis.
Lemmas coarsely pubescent................................................ 11. A. VULPINUM.
Lemmas glabrous or scabrous toward summit only.
12. A. SUBSECUNDUM.
Awn divergent, when dry.
Spikelets imbricate........................................................................ 15. A. BAKERI.
Spikelets distant.
Spikelets 3 to 7 in a spike, about twice as long as the internode; spike 4 to 7 cm. long............................................................................ 16. A. PRINGLEI.
Spikelets mostly more than 7 in a spike, usually shorter than the internode; spike mostly more than 8 cm. long.
Spike 8 to 15 cm. long; blades 1 to 2 mm. wide.......... 18. A. SPICATUM.
Spike 15 to 30 cm. long; blades 4 to 6 mm. wide...... 20. A. ARIZONICUM.

**1. Agropyron desertórum** (Fisch.) Schult. (Fig. 306.) Culms slender, erect or geniculate at base, in dense tufts, 25 to 100 cm. tall; sheaths glabrous or the lower spreading-hirsute; blades 2 to 4 mm., occasionally to 5 mm. wide; spike 5 to 9 cm. long, 7 to 11 mm. wide, somewhat bristly, the short-jointed rachis pubescent; spikelets closely spaced on the rachis, 8 to 12 mm. long, 5- to 7-flowered, somewhat spreading; glumes and lemmas firm, glabrous to sparsely ciliate on the keel, both abruptly narrowed into an awn 2 to 3 mm. long, the lemma about 6 mm. long, the awn commonly slightly bent to one side. ♃ ("*A. cristatum*" of Manual, ed. 1)—Grown in experiment stations and found here and there in grainfields, Ontario, North Dakota, South Dakota, Montana, Wyoming, Colorado, Utah, Nevada, Arizona, and California; adventive, Albany Port, N. Y. Introduced from Russia, extensively planted in the northern Great Plains area, and spreading readily by reseeding.

**Agropyron cristátum** (L.) Gaertn. CRESTED WHEATGRASS. Spike 2 to 7 cm. long; spikelets more widely spreading, the glumes somewhat contorted, gradually tapering into the awns, these curved, 2 to 5 mm. long. ♃ —Adventive on barrier beach, Fishers Island, N. Y.; Barton, N. Dak. Introduced from Russia, grown in experiment stations, and a valuable dry-land grass for soil conservation and forage in the northern Great Plains. Sometimes found mixed in plantings of *A. desertorum*.

**Agropyron sibíricum** (Willd.) Beauv. Rather smaller with relatively scant foliage; spike 6 to 10 cm. long, the rachis glabrous or nearly so; spikelets somewhat spreading, about as in *A. desertorum*, the glumes and lemmas mucronate or with an awn 1 to 2 mm. long. ♃ —Introduced from Russia, grown in a few experiment stations, spontaneous in Idaho (near Boise) and New Mexico (near Gallup). Better suited to dry soils.

**Agropyron tritíceum** Gaertn. Annual, branching at base; culms slender, erect or usually decumbent, mostly 10 to 30 cm. tall; blades flat, mostly less than 10 cm. long, 2 to 3 mm. wide; spike oval or ovate, 1 to 1.5 cm. long, thick; spikelets crowded, about 7 mm. long; glumes and lemmas acuminate. ⊙ —Absaroka Forest, Mont., Wyoming, Mountain Home, Idaho; Corfu, Wash. Sparingly introduced from southern Russia.

**2. Agropyron rěpens** (L.) Beauv. QUACKGRASS. (Fig. 307, *A*.) Green or glaucous; culms erect or curved at base, 50 to 100 cm. tall, sometimes taller, with creeping yellowish rhizomes; sheaths of the innovations

FIGURE 306.—*Agropyron desertorum*, × 1. (Ball 1768, Colo.)

often pubescent; blades relatively thin, flat, usually sparsely pilose on the upper surface, mostly 6 to 10 mm. wide; spike 5 to 15 cm. long, the rachis scabrous on the angles; spikelets mostly 4- to 6-flowered, 1 to 1.5 cm. long, the rachilla glabrous or scaberulous; glumes 3- to 7-nerved, awn-pointed; lemmas mostly 8 to 10 mm. long, the awn from less than 1 mm. to as long as the lemma; palea obtuse, nearly as long as the lemma, scabrous on the keels. ♃ —Waste places, meadows and pastures, Newfoundland to Alaska (Skagway), south to North Carolina, Arkansas, Utah, and California; Mexico; introduced from Eurasia. Common in the Northern States; a troublesome weed in cultivated ground. Called also quitch grass and couch grass. Awned specimens have been described as *Agropyron leersianum* (Wulf.) Rydb.; also referred to *A. repens* f. *aristatum* (Schum.) Holmb.

**3. Agropyron púngens** (Pers.) Roem. and Schult. (Fig. 307, *B*.) Glaucous, culms 50 to 80 cm. tall,

FIGURE 307.—*A*, *Agropyron repens*. Plant, × ½; spikelet and floret, × 3. *B*, *A. pungens*, × 3. (Scribner, Maine.)

with pale or brownish rhizomes; blades firm, mostly involute, scabrous on the upper surface; spikelets awnless, compressed, often as much as 10-flowered, the florets closely imbricate; glumes firm, acute, obscurely nerved, scabrous on the keel. ♃ —Seacoast, Maine (Cape Elizabeth), Massachusetts (Harwich); ballast, New Jersey and Oregon; introduced from Europe.

**4. Agropyron pseudorépens** Scribn. and Smith. (Fig. 308.) Resembling *A. repens*, often stouter, the rhizomes not yellow; blades commonly narrower; spike 10 to 20 cm. long, the spikelets contracted and appressed, the flat or scarcely keeled glumes 2 to 2.5 mm. wide, nearly equaling the spikelets; lemmas scaberulous to minutely hispidulous, rachilla villous. ♃ —Mostly in bottom lands or valleys, Alberta; Michigan (south shore of Lake Superior); South Dakota and Nebraska to Washington, south to New Mexico and Arizona. Specimens without rhizomes resemble *A. trachycaulum*.

FIGURE 308.—*Agropyron pseudorepens*, × 1. (Chase 5389, Colo.)

**5. Agropyron smíthii** Rydb. WESTERN WHEATGRASS. (Fig. 309.) Usually glaucous; culms erect, 30 to 60 cm. tall, sometimes taller, with creeping rhizomes; sheaths glabrous; blades firm, stiff, mostly flat when fresh, involute in drying, strongly nerved, scabrous or sometimes sparsely villous on the upper surface, mostly 2 to 4 mm. wide, tapering to a sharp point; spike erect, mostly 7 to 15 cm. long, the rachis scabrous on the angles; spikelets rather closely imbricate, occasionally two at a node, 6- to 10-flowered, 1 to 2 cm. long, the rachilla scabrous or scabrous-pubescent; glumes rigid, tapering to a short awn, rather faintly nerved, 10 to 12 mm. long; lemmas about 1 cm. long, firm, glabrous, often pubes-

FIGURE 309.—*Agropyron smithii*, × 1. (Nelson 3918, Wyo.)

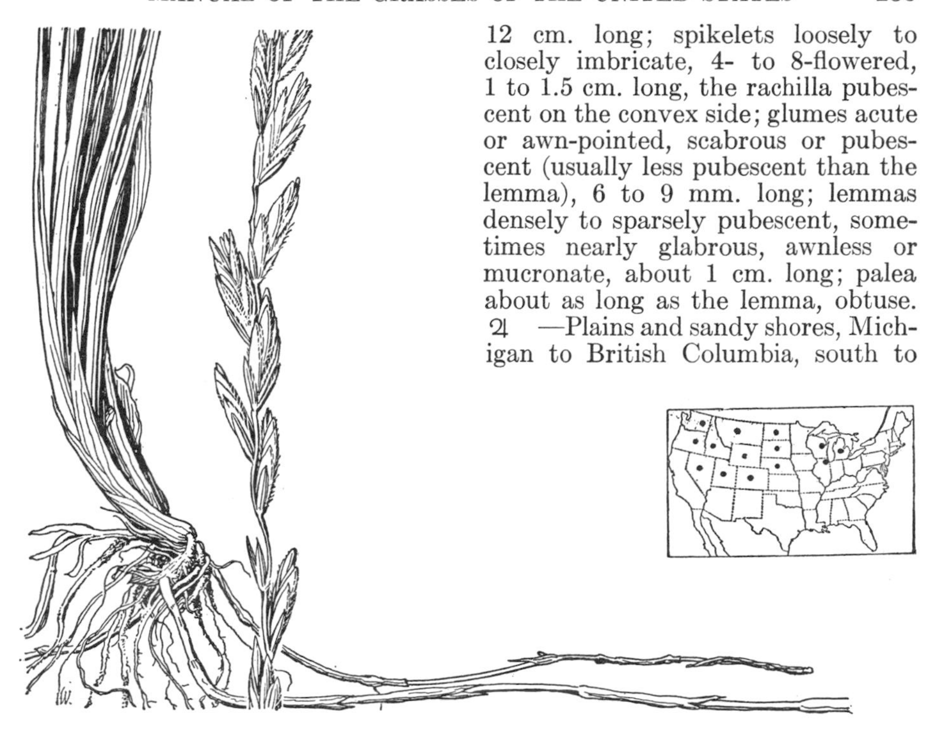

FIGURE 310.—*Agropyron dasystachyum*, × 1. (Griffiths 488, Wash.)

cent near the base, obscurely nerved, acuminate, mucronate, sometimes short-awned; palea scabrous-pubescent on the keels. ♃ —Moist, usually alkaline soil, Ontario to Alberta and British Columbia; New York; Michigan to Washington, south to Tennessee, Texas, Arizona, and northeastern California; mostly introduced east of Iowa and Kansas. Two varieties have been recognized. AGROPYRON SMITHII var. MÓLLE (Scribn. and Smith) Jones. Lemmas and sometimes glumes more or less pubescent. ♃ —About the same range as the species. AGROPYRON SMITHII var. PALMÉRI (Scribn. and Smith) Heller. Lower sheaths pubescent. ♃ —Colorado to Utah, south to New Mexico and Arizona.

**6. Agropyron dasystáchyum** (Hook.) Scribn. THICKSPIKE WHEATGRASS. (Fig. 310.) Often glaucous; culms mostly 40 to 80 cm. tall, with creeping rhizomes; blades flat to involute, 1 to 3 mm. wide; spike mostly 6 to 12 cm. long; spikelets loosely to closely imbricate, 4- to 8-flowered, 1 to 1.5 cm. long, the rachilla pubescent on the convex side; glumes acute or awn-pointed, scabrous or pubescent (usually less pubescent than the lemma), 6 to 9 mm. long; lemmas densely to sparsely pubescent, sometimes nearly glabrous, awnless or mucronate, about 1 cm. long; palea about as long as the lemma, obtuse. ♃ —Plains and sandy shores, Michigan to British Columbia, south to

FIGURE 311.—*Agropyron elmeri*, × 1. (Type.)

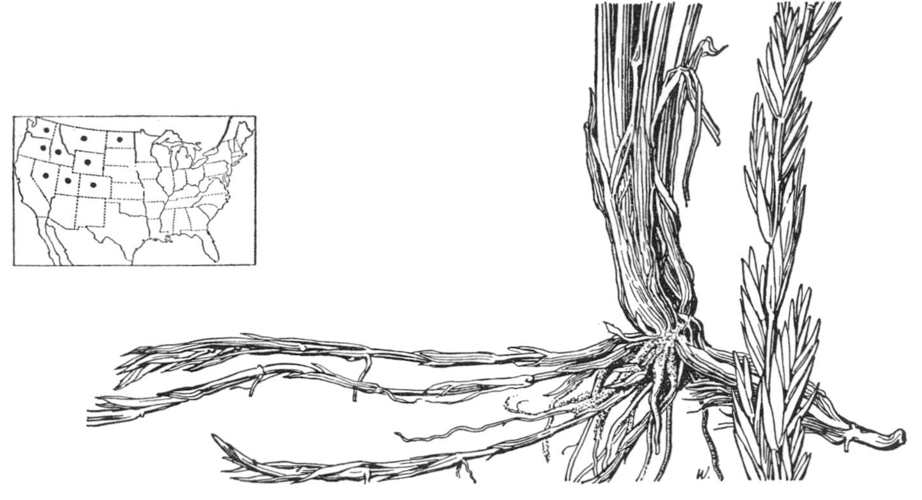

FIGURE 312.—*Agropyron riparium*, × 1. (Nelson 3965, Wyo.)

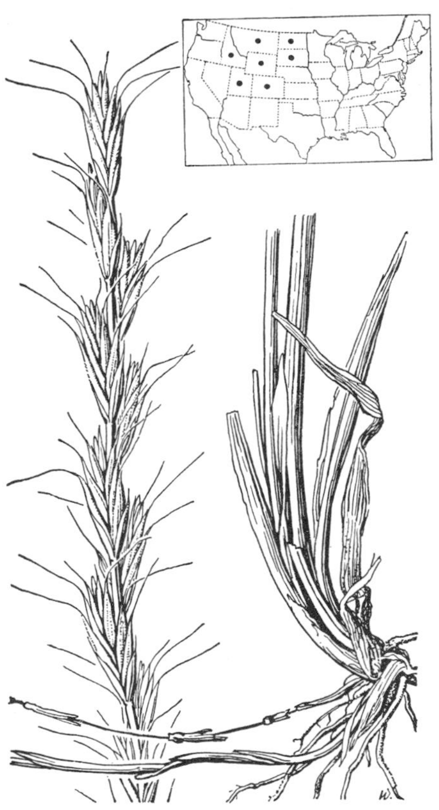

FIGURE 313.—*Agropyron albicans*, × 1. (Griffiths 3013, Wyo.)

Illinois, Nebraska, Colorado, Nevada, and Oregon. In the form growing on the sandy shores of Lake Michigan the lemmas are densely villous, but villous forms occur in other parts of the range of the species.

This and the four following species appear to intergrade, forming a polymorphous group.

**7. Agropyron elméri** Scribn. (Fig. 311.) Resembling *A. dasystachyum*; culms on the average taller, more robust, the spike longer (as much as 25 cm. long), the spikelets larger (as much as 10-flowered and 2.5 cm. long); glumes and lemmas usually longer (as much as 12 mm. and 15 mm., respectively); lemmas pubescent, sometimes sparsely so or scabrous only or pubescent only on the margins at base. ♃ —Dry or sandy soil, British Columbia to Oregon.

**8. Agropyron ripárium** Scribn. and Smith. STREAMBANK WHEATGRASS. (Fig. 312.) Resembling *A. dasystachyum*, with vigorous rhizomes; blades usually narrower; spikelets usually more imbricate; lemmas glabrous or somewhat pubescent along the edges of the lower part of the lemma. ♃ —Dry or moist meadows and hills, North Dakota to Alberta and Washington, south to Oregon and Colorado.

**9. Agropyron álbicans** Scribn. and Smith. (Fig. 313.) Similar to *A. dasystachyum;* glumes awn-pointed, about 1 cm. long; awn of lemma 1 to 1.5 cm. long, divergent when dry. ♃ —Plains and dry hills, South Dakota to Alberta and Idaho, Colorado and Utah.

**10. Agropyron griffíthsi** Scribn. and Smith ex Piper. (Fig. 314.) Resembling *A. albicans*, differing chiefly in having glabrous lemmas, the rachis rarely disarticulating. ♃ —Open, dry, sandy or alkaline soil, western North Dakota to Washington, south to Wyoming and Colorado. In the type specimen the lemmas are smooth, but in several other specimens the lemmas are scabrous. Possibly only a glabrous form of *A. albicans.*

**Agropyron intermédium** (Host) Beauv. Blades short, involute, acutish; glumes about 5-nerved; lemmas awnless. ♃ —Ballast at Camden, N. J.; adventive from Europe. Planted in the Northwest for pastures and for revegetating range lands.

**Agropyron trichóphorum** (Link) Richt. Blades flat; spikelets pubescent, awnless; glumes several-nerved, acutish. ♃ —Lynn, Mass.; adventive from Europe. Planted to some extent in the Northwest.

**Agropyron júnceum** (L.) Beauv. Blades loosely involute; spikelets glabrous; glumes 9-nerved, acutish. ♃ —Ballast near Portland, Oreg.; dunes, San Francisco, Calif.; adventive from Europe.

**11. Agropyron vulpínum** (Rydb.) Hitchc. (Fig. 315.) Culms 50 to 75 cm. tall, somewhat geniculate at base; blades drying loosely involute, 10 to 12 cm. long, 2 to 4 mm. wide; spike nodding, 10 to 15 cm. long, the rachis stiffly scabrous-ciliate on the angles; spikelets imbricate but not appressed, some toward the base two at a node, 3- to 5-flowered, the rachilla appressed-pubescent; glumes scabrous, strongly 5-nerved, awn-tipped; lemmas 5-nerved toward the

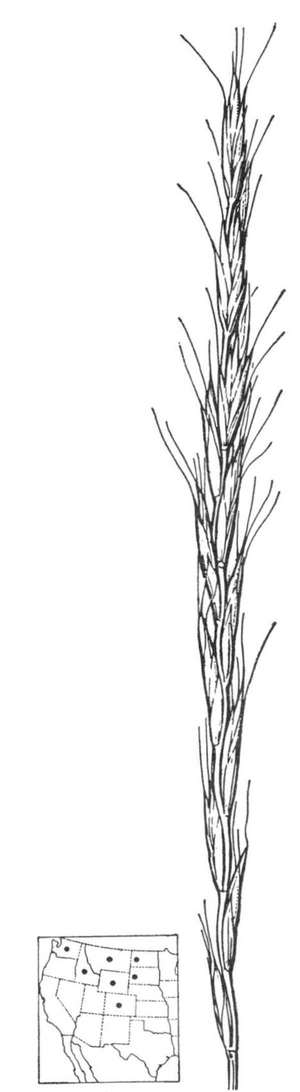

FIGURE 314.—*Agropyron griffithsi*, × 1. (Williams and Griffiths 164, Wyo.)

FIGURE 315.—*Agropyron vulpinum*, × 1. (Type.)

FIGURE 316.—*Agropyron subsecundum*, × 1. (Shear 452, Mont.)

FIGURE 317.—*Agropyron trachycaulum*, × 1. (Shear 404.)

minutely toothed apex, coarsely pubescent, the scabrous awn 8 to 10 mm. long. ♃ (*Elymus vulpinus* Rydb.) —Moist ground, Grant County, Nebr. and Livingston, Mont.

**12. Agropyron subsecúndum** (Link) Hitchc. BEARDED WHEATGRASS. (Fig. 316.) Green or glaucous, without creeping rhizomes; culms erect, tufted, 50 to 100 cm. tall; sheaths glabrous or rarely pubescent; blades flat, 3 to 8 mm. wide; spike erect or slightly nodding, 6 to 15 cm. long, sometimes unilateral from twisting of the spikelets to one side, the rachis scabrous or scabrous-ciliate on the angles, sometimes disarticulating; spikelets rather closely imbricate, few-flowered, the rachilla villous, the callus of the florets short-pilose; glumes broad, rather prominently 4- to 7-nerved, nearly as long as the spikelet, tapering into an awn; lemmas obscurely 5-nerved, the nerves becoming prominent toward the tip, the awn straight or nearly so, usually 1 to 3 cm. long. ♃ —Moist meadows and open woods, Newfoundland to Alaska, south to the mountains of Maryland, west to Washington and California, and south to New Mexico and Arizona. Said by Malte to be self-pollinated. This is the species which has generally been called by American botanists *A. caninum* (L.) Beauv.; that is a European species, differing in having 3-nerved glumes.

AGROPYRON SUBSECUNDUM var. ANDÍNUM (Scribn. and Smith) Hitchc. Culms mostly not more than 50 cm. tall, loosely tufted, usually geniculate at base; lower sheaths pale, usually papery; spike short; awns mostly 5 to 10 mm. long, often curved. An alpine form of mountain meadows. ♃ —Montana to Washington, south to Colorado and Nevada.

**Agropyron canínum** (L.) Beauv. Glumes 3-nerved. ♃ —Ballast near Portland, Oreg.; adventive from Europe.

**13. Agropyron trachycaúlum** (Link) Malte. SLENDER WHEATGRASS. (Fig. 317.) Resembling *A. subsecundum;* sheaths glabrous or rarely pubescent; blades mostly 2 to 4 mm. wide; spike usually more slender, 10 to 25 cm. long, sometimes unilateral; spikelets from rather remote to closely imbricate; glumes and lemmas awnless or nearly so. ♃ (*A. tenerum* Vasey, *A. pauciflorum* (Schwein.) Hitchc.)— Labrador to Alaska, south to the mountains of West Virginia, Missouri, New Mexico, and California; northwestern Mexico. Alpine plants lower, and with shorter denser commonly purplish spikes, resemble *A. subsecundum* var. *andinum*, but the spikelets are awnless. They have been referred to *A. violaceum* (Hornem.) Lange, an Arctic species, and to *A. biflorum* (Brignoli) Roem. and Schult.

FIGURE 318.—*Agropyron latiglume*, × 3. (Type.)

**14. Agropyron latiglúme** (Scribn. and Smith) Rydb. (Fig. 318.) Culms loosely tufted, curved or geniculate below, 20 to 50 cm. tall; blades flat, short, 3 to 5 mm. wide, short-hirsute on both surfaces, rarely glabrous or nearly so beneath; spike mostly 3 to 7 cm. long, rarely longer; spikelets usually closely imbricate; glumes broad, flat, thin-margined, unsymmetrical and slightly notched at summit, awn-tipped; lemmas commonly appressed-pubescent, awnless or awn-tipped. ♃ —Alpine meadows, open slopes, mostly at high altitudes, Montana, Wyoming, and Colorado to Labrador and Alaska.

FIGURE 319.—*Agropyron bakeri*, × 1. (Hitchcock 1686, Colo.)

**15. Agropyron bakéri** E. Nels. BAKER WHEATGRASS. (Fig. 319.) Resembling *A. subsecundum;* culms erect, mostly 50 to 100 cm. tall, rather loosely tufted; spike mostly 5 to 12 cm. long, the spikelets rather loosely imbricate; awns divergently curved when dry, 1 to 4 cm. long. ♃ —Open slopes, upper altitudes, northern Michigan; Alberta to Washington, Oregon, and New Mexico.

**16. Agropyron prínglei** (Scribn. and Smith) Hitchc. (Fig. 320.) Culms tufted, decumbent at base, 30 to 50 cm. tall, the basal sheaths soft and papery; blades flat or loosely involute, mostly less than 10 cm. long, 1 to 3 mm. wide; spike more or less flexuous, 4 to 7 cm. long, the rachis scabrous on the angles, slender, the middle internodes usually 8 to 10 mm. long; spikelets mostly 3 to 7 in each spike, rather distant, the lower and middle ones (excluding awns) about as long as two internodes, mostly 3- to 5-flowered, the rachilla joints minutely

FIGURE 320.—*Agropyron pringlei*, × 1. (Pringle 504, Calif.)

scabrous, about 2 mm. long; glumes rather narrow, about 3-nerved on the exposed side, 7 to 8 mm. long, tapering into a straight awn about 5 mm. long; lemmas tapering into a scabrous, strongly divergent awn 1.5 to 2.5 cm. long; palea 10 to 12 mm. long. ♃ —Stony slopes, 2,500 to 3,500 m., in the Sierra Nevada, Calif.

**17. Agropyron scribnéri** Vasey. SPREADING WHEATGRASS. (Fig. 321) Culms tufted, prostrate or decumbent-spreading, often flexuous, 20 to 40 cm. long; blades flat or, especially on the innovations, loosely involute, more or less pubescent, mostly basal, the 2 or 3 culm blades usually less than 5 cm. long, 1 to 3 mm. wide; spike long-exserted, often nodding or flexuous, dense, 3 to 7 cm. long, the rachis disarticulating at maturity, the internodes glabrous, 3 to 5 mm. long, or the lowermost longer; spikelets 3- to 5-flowered, the rachilla internodes minutely scabrous, about 2 mm. long; glumes narrow, one obscurely nerved, the other with 2 or 3 distinct nerves, tapering into a divergent awn similar to the awns of the lemmas; lemmas nerved toward the tip, tapering to a strongly divergent awn 1.5 to 2.5 cm. long; palea a little longer than the body of the lemma, the apex with 2 short slender teeth. ♃ —Alpine slopes, 3,000 to 4,000 m., Montana and Idaho to New Mexico and California. Characterized by the hard leafy basal tussock with slender spreading flexuous culms.

**18. Agropyron spicâtum** (Pursh) Scribn. and Smith. BLUEBUNCH WHEATGRASS (Fig. 322.) Green or glaucous; culms tufted, often in large bunches, erect, 60 to 100 cm. tall; sheaths glabrous; blades flat to loosely involute, 1 to 2 mm., sometimes to 4 mm., wide, glabrous beneath, pubescent on the upper surface; spike slender, mostly 8 to 15 cm. long, the rachis scaberulous on the angles, the internodes 1 to 2 cm. long, or the

FIGURE 321.—*Agropyron scribneri*, × 1. (Shear 1179, Colo.)

lowermost 2.5 cm.; spikelets distant, not as long (excluding the awns) as the internodes or slightly longer, mostly 6- to 8-flowered, the rachilla joints scaberulous, 1.5 to 2 mm. long; glumes rather narrow, obtuse to acute, rarely short-awned, about 4-nerved, usually about half as long as the spikelet, glabrous or scabrous on the nerves; lemmas about 1 cm. long, the awn strongly divergent, 1 to 2 cm. long; palea about as long as the lemma, obtuse. ♃ —Plains, dry slopes, canyons and dry open woods, northern Michigan to Alaska, south to western South Dakota, New Mexico, and California. A smaller form with smaller spikelets, found in desert regions of the Great Basin has been differentiated as *A. vaseyi* Scribn. and Smith. A. SPICATUM var. PUBÉSCENS Elmer. Culms and foliage pubescent. ♃ —Washington and Idaho.

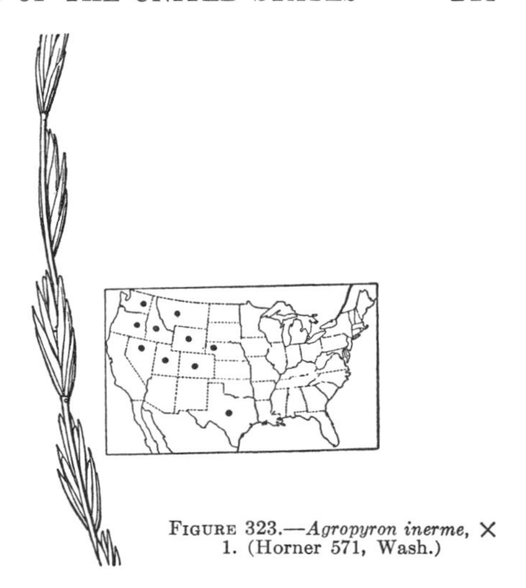

FIGURE 323.—*Agropyron inerme*, × 1. (Horner 571, Wash.)

**19. Agropyron inérme** (Scribn. and Smith) Rydb. BEARDLESS WHEATGRASS. (Fig. 323.) Differing from *A. spicatum* in the awnless spikelets. ♃ (*A. spicatum* var. *inerme* Heller.)—Dry plains and hills, Montana to British Columbia, south to Utah, Wyoming, western Nebraska, and eastern Oregon; Texas. Closely related to *A. spicatum*, but very different in appearance because awnless.

FIGURE 322.—*Agropyron spicatum*, × 1. (Vasey, Wash.)

FIGURE 324.—*Agropyron arizonicum*, × 1. (Type.)

**20. Agropyron arizónicum** Scribn. and Smith. (Fig. 324.) Resembling *A. spicatum*, usually taller and coarser;

FIGURE 325.—*Agropyron parishii*, × 1. (Type.)

FIGURE 326.—*Agropyron saxicola*, × 1. (Type.)

blades commonly 4 to 6 mm. wide; spike 15 to 30 cm. long, flexuous, the rachis more slender; spikelets distant, mostly 3- to 5-flowered; glumes short-awned; awns of the lemmas stouter, mostly 2 to 3 cm. long. ♃ —Rocky slopes, western Texas, New Mexico, Arizona, Nevada, California (Eel Ridge), and Chihuahua, Mexico.

**Agropyron semicostátum** (Steud.) Nees ex Boiss. Blades flat; spike nodding, 10 to 20 cm. long; spikelets several-flowered, imbricate; glumes several-nerved, much shorter than the spikelet, acute but scarcely awned, awn of lemma flexuous or finally divergent, 1.5 to 3 cm. long. ♃ —Ballast near Portland, Oreg. Native of Asia. Cultivated in experiment plots in California, Washington, D. C., and Mississippi in the last century under the unpublished name *Agropyrum japonicum*. Tracy used the name in print in economic notes. (See Synonymy.)

**21. Agropyron paríshii** Scribn. and Smith. (Fig. 325.) Culms 70 to 100 cm. tall, the nodes retrorsely pubescent; blades flat or loosely involute, 2 to 4 mm. wide; spike slender, nodding, 10 to 25 cm. long, the internodes of the rachis 1.5 to 2.5 cm. long; spikelets 4- to 7-flowered, mostly about 2 cm. long, narrow, appressed, the rachilla joints scaberulous, about 2 mm. long; glumes 3- to 5-nerved, 1 to 1.5 cm. long, acute; lemmas acute or with a slender awn 1 to 8 mm. long; palea as long as the lemma, obtuse. ♃ —Canyons and rocky slopes, California (Monterey and San Benito Counties and San Bernardino Mountains); rare. AGROPYRON PARISHII var. LAÉVE Scribn. and Smith. Nodes glabrous; awns usually 1 to 2 cm. long. ♃ —California, more widespread than the species.

**22. Agropyron saxícola** (Scribn. and Smith) Piper. (Fig. 326.) Culms tufted, erect, 30 to 80 cm. tall; sheaths glabrous or sometimes pubescent; blades flat to loosely involute, glabrous or sometimes pubescent, 1 to 4 mm. wide; spike 5 to 12 cm. long, the rachis tardily disarticulating, the internodes more or less scabrous on the angles, 5 to 10 mm. long; spikelets imbricate, sometimes in pairs, about twice as long as the internodes of the rachis, 4- to 6-flowered, the rachilla minutely scabrous; glumes narrow, 2-nerved, the nerves sometimes obscure, sometimes with a third faint nerve, awned, the awn divergent, 5 to 20 mm. long, sometimes with a tooth or short awn at the base of the main awn; lemmas about 8 mm. long, the awn divergent, mostly 2 to 5 cm. long, sometimes with 1 or 2 short additional awns; palea about as long as the lemma, obtuse or truncate. ♃ —Dry or rocky slopes and plains, western South Dakota to Washington, south to Utah, Arizona, and California.

**23. Agropyron saundérsii** (Vasey) Hitchc. (Fig. 327.) Culms erect, 60 to 100 cm. tall; blades flat or loosely involute; spike erect, 8 to 15 cm. long, mostly purplish, the rachis tardily disarticulating; spikelets sometimes in pairs near the middle of the spike, 1 to 1.5 cm. long (excluding awns), 2- to 5-flowered; glumes variable, narrow with 2 nerves or wider with 3 to 5 nerves, the nerves strong and at least the midnerve scabrous, the awn 1 to 5 cm. long, sometimes with a short lateral awn near the base; lemmas scabrous, the awn straight, 2 to 5 cm. long. ♃ (*Elymus saundersii* Vasey.)—Dry slopes, Colorado, Wyoming, Idaho, Utah, Arizona, and California. Only the 5 specimens of the type collection from Veta Pass, Colo., have spikelets with awns to 5 cm. long. In some specimens the awns of the glumes vary from 5 to 16 mm. and those of the lemmas from 7 to 30 mm. (*Elymus saundersii* var. *californicus* Hoover), and in others from 10 to 20 mm. on the glumes and 15 to 35 mm. on the lemmas.

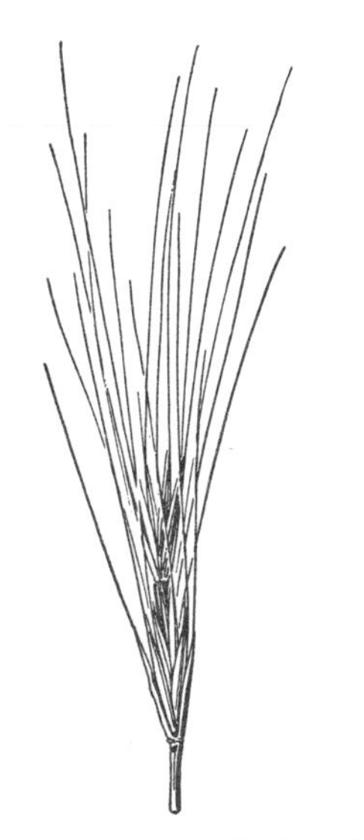

FIGURE 327.—*Agropyron saundersii*, × 1. (Type.)

## 43. TRÍTICUM L.

Spikelets 2- to 5-flowered, solitary, placed flatwise at each joint of a continuous or articulate rachis, the rachilla disarticulating above the glumes and between the florets or continuous; glumes rigid, keeled, 3- to several-nerved, the apex abruptly mucronate or toothed or with one to several awns; lemmas broad, keeled, very asymmetric, many-nerved, abruptly pointed or awned. Low or rather tall annuals, with flat blades and thick spikes. Standard species, *Triticum aestivum*. *Triticum*, the old Latin name for wheat.

**1. Triticum aestívum** L. WHEAT. (Fig. 328.) Culms erect, freely branching at base, 60 to 100 cm. tall; blades 1 to 2 cm. wide; spike mostly 5 to 12 cm. long; internodes of rachis 3 to 6 mm. long; spikelets broad, glabrous or pubescent, long-awned to awnless; glumes usually strongly keeled toward one side, the keel extending into a mucro, the other side usually obtusely angled at apex. ⊙ (*T. vulgare* Vill.; *T. sativum* Lam.)—Commonly cultivated; fields and waste places in the vicinity of cultivated fields or grain elevators, but not established.

Spelt (*T. spelta* L.) and emmer (*T. dicoccum* Schrank) are sometimes cultivated for the grain, used for stock feed, and for forage. In these two species the rachis breaks up, each joint bearing a spikelet which remains entire, each floret permanently enclosing the grain. In spelt the spikelets are somewhat distant, exposing the rachis, in emmer the spikelets are closely imbricate, scarcely exposing the rachis. A large number of varieties of wheat are in cultivation; the lemmas may be glabrous or pubescent, the awns long or nearly or quite wanting.

On the basis of the number of chromosomes the wheats and their

FIGURE 328.—*Triticum aestivum.* Plant with awned spikes (bearded wheat) and a nearly awnless spike (beardless wheat), × ½; spikelet and floret, × 3. (Cult.)

allies may be divided into three groups. The group with 7 chromosomes (probably the most primitive) includes einkorn (*T. monococcum* L.). The group with 14 chromosomes includes durum wheat (*T. durum* Desf.), poulard wheat (*T. turgidum* L.), Polish wheat (*T. polonicum* L.), emmer (*T. dicoccum* Schrank), and also *T. pyramidale* Perciv., *T. orientale* Perciv. (not Biebers. 1806), *T. persicum* Vavilov (not Aitch. and Hemsl. 1888), *T. dicoccoides* Koern. and *T. timopheevi* (Zhukov.) Zhukov.[12] The group with 21 chromosomes includes spelt and the commonly cultivated wheats referred to as *T. vulgare* Vill. and *T. compactum* Host, also *T. macha* Dekap. and Menab. and *T. sphaerococcum* Perciv.[12] Alaska wheat is a variety of poulard wheat with branched heads. It is also known by several other names, such as Egyptian, miracle, and mummy. This variety is considered inferior commercially to standard varieties of wheat. Stories of varieties originating from seed found with mummies 3,000 years old have no basis in fact.

The origin of wheat is not known, as there is no native species like any of the cultivated forms. Some botanists have suggested species of *Aegilops* and others *T. dicoccoides* Koern., a wild species of Palestine, as the possible ancestor.

## 44. AÉGILOPS L. Goatgrass

Spikelets 2- to 5-flowered, solitary, turgid or cylindric, placed flatwise at each joint of the rachis and fitting into it, the joints thickened at the summit, the spikelets usually not reaching the one above on the same side, exposing the rachis; spike usually disarticulating near the base at maturity, falling entire, or finally disarticulating between the spikelets. Annuals with flat blades and usually awned spikes. Type species, *Aegilops ovata*. Name from *Aegilops*, an old Greek name for a kind of grass.

The species of *Aegilops* have been recently introduced into the United States and in some places are becoming troublesome weeds. At maturity the spikes fall entire, the lowest rachis joint serving as a pointed callus to the 2- to several-jointed, strongly barbed fruits, which work their way into the mouths and noses of grazing animals and into the wool of sheep.

Spikelets subovate; rachis not disarticulating .......... 3. A. ovata.
Spikelets cylindric; rachis finally disarticulating.
  Glumes with 1 awn .......... 1. A. cylindrica.
  Glumes with 3 awns .......... 2. A. triuncialis.

**1. Aegilops cylíndrica** Host. Jointed goatgrass. (Fig. 329.) Culms erect, branching at base, 40 to 60 cm. tall; blades 2 to 3 mm. wide; spike cylindric, 5 to 10 cm. long; internodes of rachis 6 to 8 mm. long; spikelets 8 to 10 mm. long, glabrous to hispid; glumes several-nerved, keeled at 1 side, the keel extending into an awn, the main nerve of the other side extending into a short tooth; lemmas mucronate, those of the uppermost spikelets awned like the glumes; awns very scabrous, those of the upper spikelets about 5 cm. long, those of the lower spikelets progressively shorter. ⊙ —Weed in wheatfields, and waste places, New York, and Pennsylvania; Indiana to Wyoming and Utah, south to Texas and New Mexico; Washington; recently introduced from Europe.

**2. Aegilops triunciális** L. Barb goatgrass. (Fig. 330.) Culms branching and spreading at base, 20 to 40 cm. tall; blades rather rigid, sharp-pointed, spreading; spike 3 to 4 cm. long, 2 or 3 of the lower spikelets often reduced, the fertile spikelets 3 to 5; glumes with 3 strong scabrous, somewhat spreading awns, 4 to 8

[12] These names supplied by W. J. Sando, geneticist.

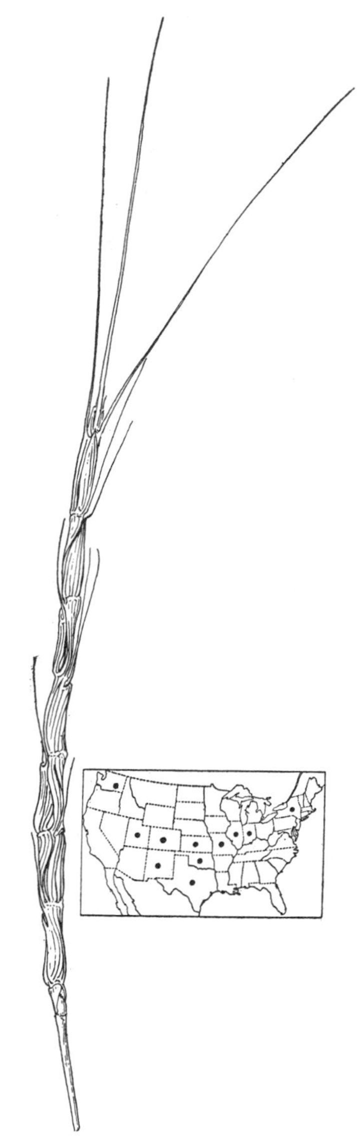

FIGURE 329.—*Aegilops cylindrica*, × ½. (Bush 72148, Mo.)

cm. long; lemmas with three rigid unequal awns. ⊙ —Troublesome weed on range land, California; adventive in Pennsylvania; introduced from Europe.

**3. Aegilops ovâta** L. Culms tufted, geniculate at base, 15 to 25 cm. tall; blades short, sharp-pointed; spike thick, of 2 to 4 subovate spikelets, the upper sterile; glumes with 4 stiff scabrous spreading awns 2 to 3 cm. long; lemmas usually with 1 long and 2 short awns. ⊙ —Weed in fields, California and Virginia; introduced from Europe.

## 45. SECÂLE L. RYE

Spikelets usually 2-flowered, solitary, placed flatwise against the rachis, the rachilla disarticulating above the glumes and produced beyond the upper floret as a minute stipe; glumes narrow, rigid, acuminate or subulate-pointed; lemmas broader, sharply keeled, 5-nerved, ciliate on the keel and exposed margins, tapering into a long awn. Erect, mostly annual grasses, with flat blades and dense spikes. Type species, *Secale cereale*. *Secale*, the old Latin name for rye.

**1. Secale cereâle** L. RYE. (Fig. 331.) In habit resembling wheat, but usually taller, the spike more slender, somewhat nodding, on the average longer. ⊙ —Commonly cultivated; escaped from cultivation, in fields and waste places. This species is thought to be derived from *S. montanum* Guss., a perennial native in the mountains of southwestern Asia.

**Secale montânum** Guss. Culms in rather large dense clumps, erect or geniculate at the base, mostly 100 to 135 cm. tall; blades flat, stiffly spreading; spikes somewhat drooping, 10 to 13 cm. long, the rachis rather readily disarticulating; awns 1 to 2 cm. long, slender, scabrous. ♃ —Persisting along roadsides around the experiment station at Pullman, Wash. Introduced from southwestern Asia.

FIGURE 330.—*Aegilops triuncialis*, × ½. (Cole, Calif.)

## 46. ÉLYMUS L. WILD-RYE

Spikelets 2- to 6-flowered, in pairs (3 or more or solitary sometimes in a few species) at each node of a usually continuous rachis, placed as in *Agropyron* but the rachilla distorted at base, bringing the florets more or less dorsiventral to the rachis; rachilla disarticulating above the glumes and between the florets; glumes equal, somewhat asymmetric, usually rigid, sometimes indurate below, narrow to subulate, 1- to several-nerved, acute to aristate; lemmas rounded on the back or nearly terete, obscurely 5-nerved, acute or usually

FIGURE 331.—*Secale cereale.* Plant, × ½; spikelet, × 3; floret, × 5. (Hill, Ill.)

awned from the tip. Erect, usually rather tall perennials (one annual), with flat or rarely convolute blades and slender or bristly spikes, the spikelets usually crowded, sometimes somewhat distant. Type species, *Elymus sibiricus* L. Name from *Elumos*, an old Greek name for a kind of grain. The species in which the spikelets are mostly solitary can be distinguished from *Agropyron* by the narrow or subulate glumes. The seed of certain species (e.g., *E. mollis* and *E. canadensis*) have been used for food by the Indians.

The species of *Elymus* are for the most part good forage grasses and in some localities form a part of the native hay. In the wooded areas of the Northwest, *E. glaucus* is one of the valuable secondary grasses of the ranges. The species with creeping rhizomes are likely to be of value as soil or sand binders. *E. mollis* is a natural sea-dune grass, and *E. arenicola* and *E. flavescens* are common on inland shifting dunes. *E. triticoides* is to be recommended for holding embankments. On the western ranges *E. cinereus* and *E. triticoides* are important.

1a. Plants annual; spike long-awned, nearly as broad as long........ 1. E. CAPUT-MEDUSAE.
1b. Plants perennial; spike much longer than broad.
  2a. Rhizomes present, slender, creeping.
    Glumes lanceolate, awnless or awn-pointed. Plants of coastal dunes.
      Glumes and lemmas papery, distinctly nerved.......... 2. E. MOLLIS.
      Glumes and lemmas firm, faintly nerved (lemmas nerved at apex).
        3. E. VANCOUVERENSIS.
    Glumes subulate or very narrow.
      Spikelets glabrous.
        Lemmas acute or awn-pointed, brownish or tan-colored; spikelets paired or solitary, crowded.
          Spikelets usually in pairs, or paired and solitary in a single spike; culms 60 to 120 cm. tall.......... 8. E. TRITICOIDES.
          Spikelets solitary in short spikes; culms 10 to 20 cm. tall...... 9. E. PACIFICUS.
        Lemmas awned, the awns 3 to 14 mm. long; spikelets usually solitary, rather distant, pale.......... 10. E. SIMPLEX.
      Spikelets densely villous to coarsely, sometimes sparsely, pubescent.
        Lemmas awned or awn-tipped; spike 5 to 15 cm. long.
          Lemmas copiously villous; awn 1 to 4 mm. long.......... 6. E. INNOVATUS.
          Lemmas hirsute or hirtellous; awn 5 to 10 mm. long........ 7. E. HIRTIFLORUS.
        Lemmas awnless; spike 10 to 25 cm. long.
          Glumes pubescent; lemmas soft, densely villous.......... 4. E. FLAVESCENS.
          Glumes glabrous or nearly so; lemmas relatively firm, coarsely pubescent, sometimes sparsely so.......... 5. E. ARENICOLA.
            Lemmas glabrous to sparsely strigose; culms glabrous; spikes usually compound; blades 15 to 35 mm. wide.......... 13. E. CONDENSATUS.
            Lemmas more or less pubescent; culms harsh-puberulent, at least about the nodes; spikes not or scarcely compound; blades 5 to 15 mm. wide.
              14. E. CINEREUS.
  2b. Rhizomes wanting (or short and stout in *E. condensatus*). Plants tufted.
    3a. Rachis tardily disjointing; glumes and lemmas awned.
      Spike mostly 5 to 7 mm. wide; spikelets mostly in twos; blades subinvolute.
        18. E. MACOUNII.
      Spike 8 to 10 mm. wide; spikelets often in threes; blades flat, 5 to 10 mm. wide.
        19. E. ARISTATUS.
    3b. Rachis continuous.
      4a. Glumes subulate to subsetaceous, not broadened above the base, the nerves obscure except in *E. villosus*.
        Lemmas awnless or awn-tipped, the awn shorter than the body.
          Spike thick, sometimes compound; spikelets commonly in twos to fours.
            Lemmas glabrous to sparsely strigose; culms glabrous; spikes usually compound; blades 15 to 35 mm. wide.......... 13. E. CONDENSATUS.
            Lemmas more or less pubescent; culms harsh-puberulent at least about the nodes; spikes not or scarcely compound; blades 5 to 15 mm. wide.
              14. E. CINEREUS.
          Spike slender; some or most of the spikelets solitary at the nodes, the paired spikelets near the middle.

Culms numerous in a close tuft, the leaves mostly basal; lemmas mostly awnless ........ 12. E. SALINUS.
Culms few, loosely tufted, the leaves scattered along the usually taller culms; lemmas awn-tipped, the awn 2 to 5 mm. long ........ 11. E. AMBIGUUS.
Lemmas awned, the awn as long as the body or longer.
Awns straight; lemmas about 1.2 mm. wide across the back.
20. E. VILLOSUS.
Awns flexuous-divergent; lemmas about 2 mm. wide across the back.
21. E. INTERRUPTUS.
4b. Glumes lanceolate or narrower, broadened above the base, strongly 3- to several-nerved.
Glumes relatively thin, flat, several-nerved, not indurate at base.
Lemmas sparsely long-hirsute on the margins toward the summit.
17. E. HIRSUTUS.
Lemmas glabrous or scabrous.
Lemmas awned ........ 15. E. GLAUCUS.
Lemmas awnless or minutely awn-tipped ........ 16. E. VIRESCENS.
Glumes firm, indurate at base.
Awns divergently curved when dry; base of glumes not terete.
22. E. CANADENSIS.
Awns straight; base of glumes terete.
Glumes about 1 mm. wide about the middle, the bases not bowed out; palea much shorter than the lemma ........ 23. E. RIPARIUS.
Glumes 1.5 to 2 mm. wide about the middle, the bases bowed out; palea as long as the lemma ........ 24. E. VIRGINICUS.

FIGURE 332.—*Elymus caput-medusae*, × 1. (Vasey 3076, Wash.)

**1. Elymus cáput-medúsae** L. (Fig. 332.) Annual; culms ascending from a decumbent, branching base, slender, 20 to 60 cm. tall; blades narrow, short; spike very bristly, 2 to 5 cm. long (excluding the long spreading awns); glumes subulate, smooth, indurate below, tapering into a slender awn 1 to 2.5 cm. long; lemmas lanceolate, 3-nerved, 6 mm. long, very scabrous, tapering into a flat awn 5 to 10 cm. long. ⊙ —Open ground, Idaho and Washington to California; a bad weed, spreading on the ranges in northern California; introduced from Europe.

**2. Elymus móllis** Trin. AMERICAN DUNEGRASS. (Fig. 333.) Culms stout, pubescent below the spike, glaucous, 60 to 120 cm. tall, with numerous overlapping basal leaves, the rhizomes widely creeping; blades firm, 7 to 12 mm. wide, often involute in drying; spike erect, dense, thick, soft, pale, 7 to 25 cm. long; glumes lanceolate, flat, many-nerved, scabrous or pubescent, 12 to 25 mm. long, acuminate, about as long as the spikelet; lemmas scabrous to felty-pubescent, acuminate or mucronate. ♃ —Sand dunes along the coast, Alaska to Greenland, south to Long Island, N.Y., and central California; along Lakes Superior and Michigan; also eastern Siberia to Japan. Closely related to the European *E. arenarius* L. with culm smooth below the spike and glabrous glumes. A form found along the coast of Washington with somewhat compound spikes has been differentiated as *E. arenarius* var. *compositus* (Abrom.) St. John, but the plants are found to be diseased.

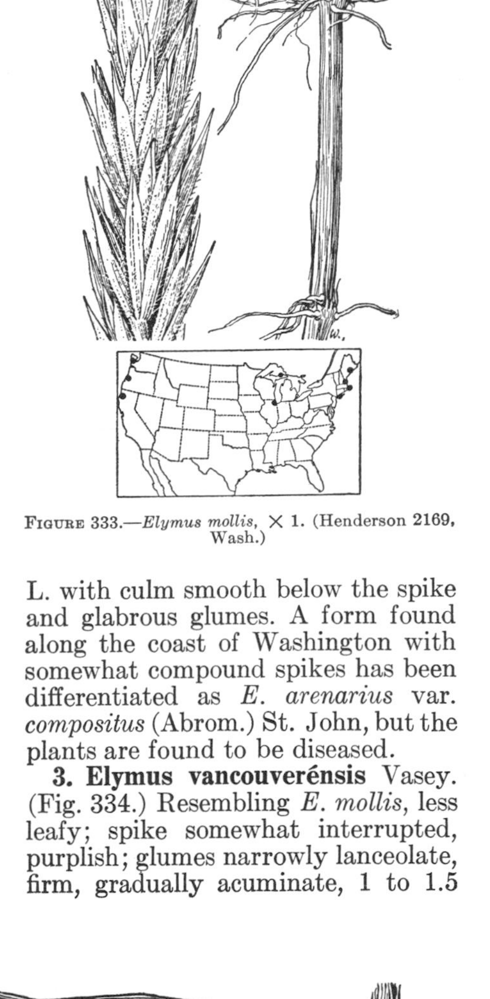

FIGURE 333.—*Elymus mollis*, × 1. (Henderson 2169, Wash.)

**3. Elymus vancouverénsis** Vasey. (Fig. 334.) Resembling *E. mollis*, less leafy; spike somewhat interrupted, purplish; glumes narrowly lanceolate, firm, gradually acuminate, 1 to 1.5

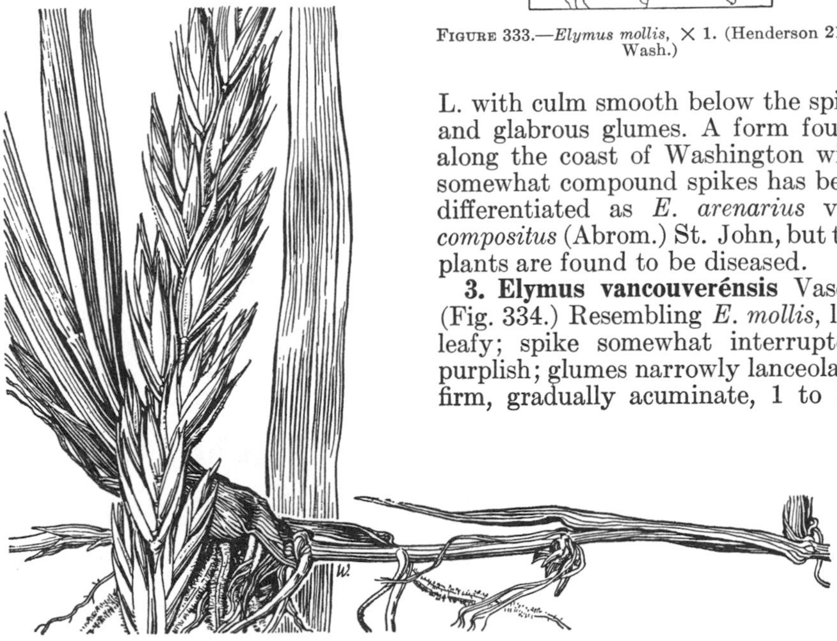

FIGURE 334.—*Elymus vancouverensis*, × 1. (Piper 812, Wash.)

FIGURE 335.—*Elymus flavescens*, × 1. (Merrill and Wilcox 160, Idaho.)

cm. long, sparsely long-villous, especially toward the apex; lemmas firm, 1 to 1.5 cm. long, tapering into a short awn. ♃ —Dunes and sandy shores, British Columbia to northern California.

**4. Elymus flavéscens** Scribn. and Smith. (Fig. 335.) Culms erect, slender, glabrous, 50 to 100 cm. tall, the rhizomes slender, nearly vertical from deep slender horizontal rhizomes with brown scales; sheaths glabrous; blades firm, glabrous beneath, scabrous above, 2 to 5 mm. wide, flat, or involute in drying; spike 10 to 25 cm. long, sometimes with short branches, somewhat nodding; spikelets 2 to 3 cm. long, several-flowered, approximate or somewhat distant; glumes very narrow or subulate, pubescent, nerveless, mostly unequal, 1 to 1.5 cm. long; lemmas awnless, densely silky-villous, the hairs long, yellowish or brownish. ♃ —Sand dunes, eastern Washington and Oregon, Idaho; South Dakota (Black Hills).

**5. Elymus arenícola** Scribn. and Smith. (Fig. 336.) Resembling *E. flavescens* to which it is closely related; glumes glabrous or nearly so; lemmas firmer, coarsely pubescent, sometimes sparsely so, or the pubescence confined to the base or margins, the pubescence grayish rather than yellow. ♃ —Sandy valleys, often in drifting sand, Washington, Oregon, and Idaho.

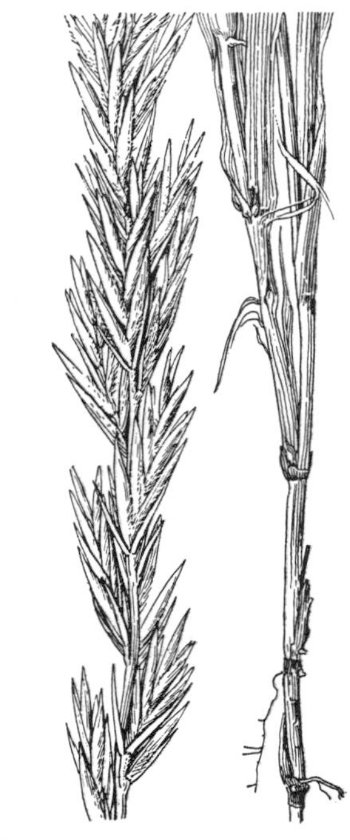

FIGURE 336.—*Elymus arenicola*, × 1. (Palmer 356, Idaho.)

FIGURE 337.—*Elymus innovatus*, × 1. (Hayward 2719, S. Dak.)

**6. Elymus innovátus** Beal. (Fig. 337.) Resembling *E. flavescens;* spike rather dense, 5 to 12 cm. long, the rachis villous; spikelets 1 to 1.5 cm. long, the narrow glumes and the lemmas densely purplish or grayish-villous, the lemmas with an awn mostly 1 to 4 mm. long. ♃ —Open woods and gravelly flats, Alaska to British Columbia; Montana, Wyoming, and South Dakota (Black Hills).

**7. Elymus hirtiflórus** Hitchc. (Fig. 338.) Culms erect, tufted, 40 to 90 cm. tall, with slender creeping rhizomes; blades firm, flat or usually involute, glabrous beneath, 5 to 20 cm. long, 1 to 4 mm. wide when flat; spike erect, 5 to 15 cm. long; spikelets 4- to 6-flowered; glumes firm, hirsute, narrow, tapering into an awn about as long as the body, the entire length 1 to 1.5 cm.; lemmas hirsute, sometimes sparingly so, the lower 8 to 9 mm. long, with an awn 5 to 10 mm. long. ♃ —River banks, Wyoming; Alberta.

**8. Elymus triticoídes** Buckl. Beardless wild-rye. (Fig. 339.) Culms usually glaucous, rarely pubescent below spike, 60 to 120 cm. tall, commonly in large colonies from extensively creeping scaly rhizomes; ligule a truncate rim about 1 mm. long; blades mostly 2 to 6 mm. wide, flat or soon involute; spike erect, slender to rather dense, rarely compound; spikelets mostly 12 to 20 mm. long; glumes very narrow to subulate, firm, nerveless or 1- to 3-nerved, awn-

Figure 338.—*Elymus hirtiflorus*. Spike, × 1; spikelet, × 5. (Type.)

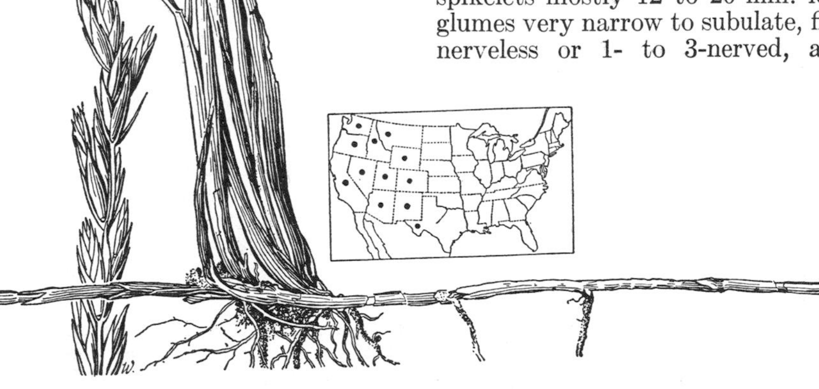

Figure 339.—*Elymus triticoides*, × 1. (Cusick 763, Oreg.)

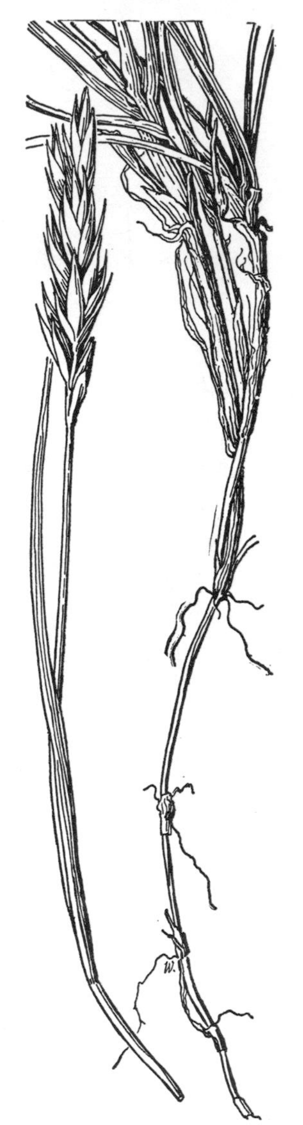

FIGURE 340.—*Elymus pacificus*, × 1. (Davy 6781, Calif.)

tipped, 5 to 15 mm. long, those of the upper spikelets usually reduced or obsolete; lemmas 6 to 10 mm. long, glabrous, firm, brownish, purplish or tawny, awn-tipped. ♃ —Moist or alkaline soil, at low and medium elevations, Montana to Washington, south to western Texas and Baja California. ELYMUS TRITICOIDES var. PUBÉSCENS Hitchc. Sheaths and involute blades pubescent. ♃ —Oregon, California, Nevada; rare.

ELYMUS TRITICOIDES subsp. MULTIFLÓRUS Gould. Plants robust; blades 6 to 12 mm. wide; spike compound, the branches mostly short, congested, but sometimes to 5 cm. long; spikelets 1.5 to 2.5 cm. long. ♃ —Wyoming to Washington, Nevada, and California. Intergrades with the species.

**9. Elymus pacíficus** Gould. (Fig. 340.) Culms low, more or less spreading, 10 to 20 cm. tall, with slender extensively creeping rhizomes; blades involute, mostly longer than the culms, pungent-pointed; spike 2 to 5 cm. long, the rachis glabrous; spikelets solitary, few-flowered, 12 to 15 mm. long; glumes nerveless, firm, tapering into a short awn; lemmas about 1 cm. long, obscurely nerved, pointed or awn-tipped, the margin very narrowly hyaline. (*Agropyron arenicola* Davy, not *Elymus arenicola* Scribn. and Smith.) ♃ —Sandy seacoast, middle California.

**10. Elymus símplex** Scribn. and Williams. (Fig. 341.) More extensively creeping than *E. triticoides*, the rhizomes sometimes as much as 5 m. long; culms ascending, 50 to 90 cm. tall; sheaths crowded, the lower often becoming reddish and papery; blades firm, flat or loosely rolled, strongly nerved; spikes 5 to 20 cm. long; spikelets as much as 2.5 cm. long, usually distant, solitary or sometimes paired; glumes subulate-aristate, 1 to 2 cm. long; rachilla villous; lemmas glabrous, the margins hyaline, awned, the awn 3 to 14 mm. long. ♃ —River banks, alkaline flats, drifting sands, and rocky slopes, southern Wyoming, Colorado, and Utah. Valuable in erosion control.

**11. Elymus ambíguus** Vasey and Scribn. (Fig. 342.) Culms few, loosely tufted, erect, 30 to 70 cm. tall; sheaths glabrous; blades flat to subinvolute, 2 to 5 mm. wide, scabrous; spike erect, rather dense, 5 to 15 cm. long; spikelets solitary toward the base and apex of the spike, mostly 2- to 4-flowered; glumes subulate, scabrous toward the awned tip; lemmas glabrous or scabrous on the back, about 1 cm. long, short-awned, the awn 2 to 5 mm. long. ♃ —Open slopes at medium altitudes in the mountains, Montana, Colorado, and Utah. ELYMUS AMBIGUUS var. STRIGÓSUS (Rydb.) Hitchc. Lemmas strigose or pubescent. ♃ (*E. strigosus* Rydb., lemmas strigose; *E. villiflorus* Rydb.) lemmas pubescent.)—Wyoming, Colorado.

**12. Elymus salínus** Jones. SALINA WILD-RYE. (Fig. 343.) Culms erect, 30 to 80 cm. tall, sometimes scabrous below nodes and below spike; sheaths scabrous; blades firm, involute, scabrous, or rarely softly pubescent; spike slender, erect, 5 to 12 cm. long; spikelets mostly solitary, often rather distant, 1 to 1.5 cm. long; glumes subulate, 4 to 8 mm. long, sometimes reduced, glabrous or scabrous; lemmas about 1 cm. long, awnless or rarely awn-tipped, glabrous or scabrous, rarely sparsely strigose, the nerves obscure. ♃ —Rocky slopes and sagebrush hills, Wyoming and Colorado to Idaho, Nevada, and southern California.

**13. Elymus condensátus** Presl. GIANT WILD-RYE. (Fig. 344.) Culms robust, in large tufts, usually 2 to 3 m. tall, with short thick rhizomes; ligule 2 to 5 mm. long; blades firm, strongly nerved, flat, as much as 3 cm. wide; spike erect, dense, 15 to 50 cm. long, usually more or less compound, the branches erect, 2 to 7 cm. long; spikelets often in threes to fives, commonly distorted by pressure; glumes subulate, awn-pointed, usually 1-nerved or nerveless, about as long as the first lemma, sometimes longer; lemmas glabrous to sparsely strigose, with a hyaline margin, awnless or mucronate. ♃ —Sand dunes, sandy or rocky slopes, moist ravines, mostly near the coast, Alameda County to San Diego County, Calif., and on the adjacent islands off the coast.

FIGURE 341.—*Elymus simplex*, × 1. (Type.)

FIGURE 342.—*Elymus ambiguus*, × 1. (Hitchcock 10990, Colo.)

FIGURE 343.—*Elymus salinus*, × 1. (Rydberg 2041, Wyo.)

FIGURE 344.—*Elymus condensatus*, × 1. (Pringle in 1882, Calif.)

**14. Elymus cinéreus** Scribn. and Merr. (Fig. 345.) Culms robust, but less so than in *E. condensatus*, typically without rhizomes, harsh-puberulent, at least about the nodes; sheaths and blades glabrous to densely harsh-puberulent, the blades mostly less than 15 mm. wide; spikes 10 to 25 cm. long (mostly 12 to 20 cm.), thick and dense but typically not branched, or with 1 of the 3 to 5 spikelets at a node pedicellate; glumes and lemmas like those of *E. condensatus*, but the lemmas more or less pubescent. ♃ (*E. condensatus* var. *pubens* Piper.)—River banks, ravines, moist or dry slopes and plains, mostly at higher altitudes than the preceding, Minnesota to British Columbia, south to Colorado, and California. On the whole this appears to be distinct from *E. condensatus*, but a rather large number of specimens from Wyoming to California have branched spikes, some with blades to 15 mm. wide, a few with rhizomes. These intermediate specimens are more or less harshly puberulent, at least about the nodes. The seeds are sometimes used for food by the Indians.

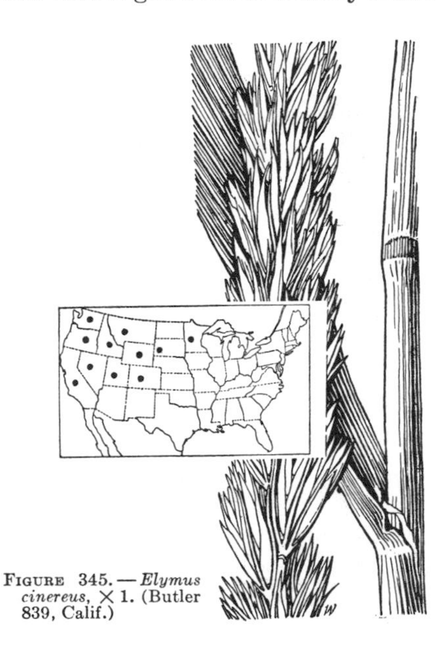

FIGURE 345.—*Elymus cinereus*, × 1. (Butler 839, Calif.)

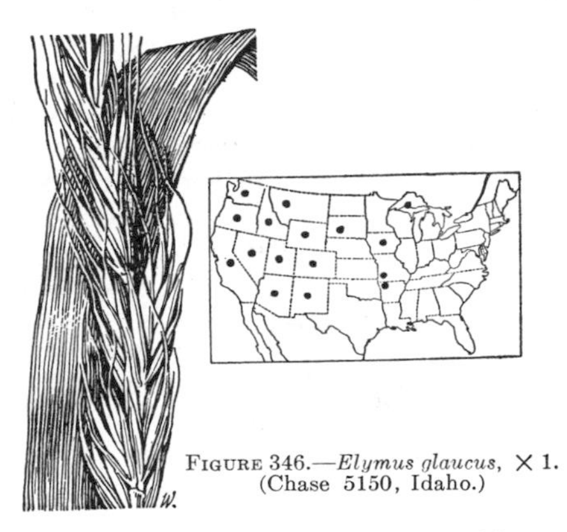

FIGURE 346.—*Elymus glaucus*, × 1. (Chase 5150, Idaho.)

**15. Elymus glaúcus** Buckl. BLUE WILD-RYE. (Fig. 346.) Culms in loose to dense tufts, often bent at base, erect, 60 to 120 cm. tall, without rhizomes, leafy; sheaths smooth or scabrous; blades flat, usually lax, mostly 8 to 15 mm. wide, usually scabrous on both surfaces, sometimes narrow and subinvolute; spike long-exserted, from erect to somewhat nodding, usually dense, commonly 5 to 20 cm. long, occasionally longer; glumes lanceolate at base, 8 to 15 mm. long, with 2 to 5 strong scabrous nerves, acuminate or awn-pointed; lemmas awned, the awn 1 to 2 times as long as the body, erect to spreading. ♃ —Open woods, copses, and dry hills at low and medium altitudes, Ontario and Michigan to southern Alaska, south through South Dakota and Colorado to New Mexico and California; Iowa, Missouri, and Arkansas. Exceedingly variable, the commonest form is loosely tufted, with lax blades 10 to 15 mm. wide and somewhat nodding spike, but plants with narrower blades and stiff spikes are frequent, the extreme form differentiated as *E. angustifolius* Davy. The original specimen described by Buckley is a rather small plant intermediate in blades and spike. Elymus glaucus var. jepsóni Davy. Sheaths and blades pubescent. ♃ —British Columbia to California; Montana and Nevada.

Figure 347.—*Elymus virescens*, × 1. (Flett, Wash.)

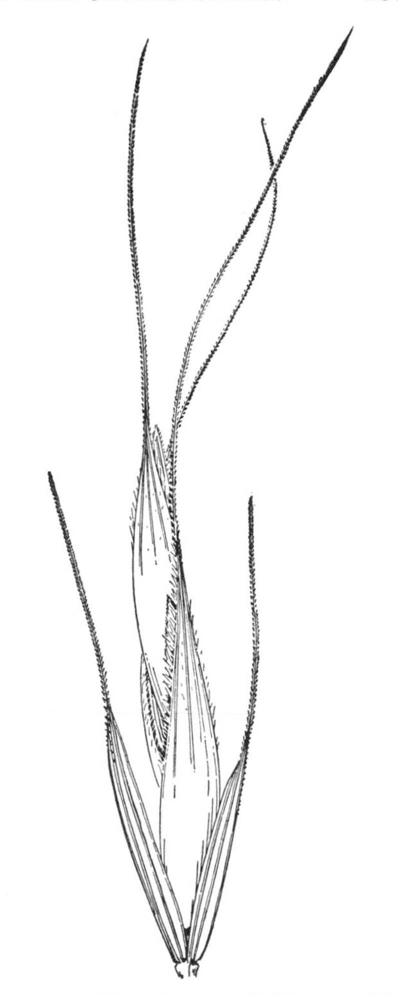

Figure 348.—*Elymus hirsutus*, × 5. (Thompson 7332, Wash.)

**16. Elymus virḗscens** Piper. (Fig. 347.) Resembling *E. glaucus* and nearly as variable in habit, often decumbent at base; sheaths from glabrous to retrorsely pubescent, blades 2 to 12 mm. wide, glabrous to harsh-puberulent; spike 5 to 15 cm. long, dense, spikelets imbricate; glumes flat, 1 to 2 mm. wide, strongly nerved, pointed or awn-tipped; lemmas glabrous to scabrous, barely awn-tipped or with an awn 1 to 4 mm. long. ♃ Moist woods, southern Alaska to California.

**17. Elymus hirsū́tus** Presl. (Fig. 348.) Culms solitary or in small tufts, 50 to 140 cm. tall, rather weak; blades flat, lax, 4 to 10 mm. wide, scabrous; spike drooping, mostly loose, the rachis exposed; spikelets mostly about 15 mm. long; glumes about 1 mm. wide, strongly nerved, awned; lemmas sparsely long-hirsute along the margin toward the summit, sometimes coarsely pubescent on the back, the slender awn flexuous or divergent, 1.5 to 2 cm. long. ♃ —Moist woods or open ground, Alaska to Oregon.

**18. Elymus macoúnii** Vasey. Macoun wild-rye. (Fig. 349.) Culms

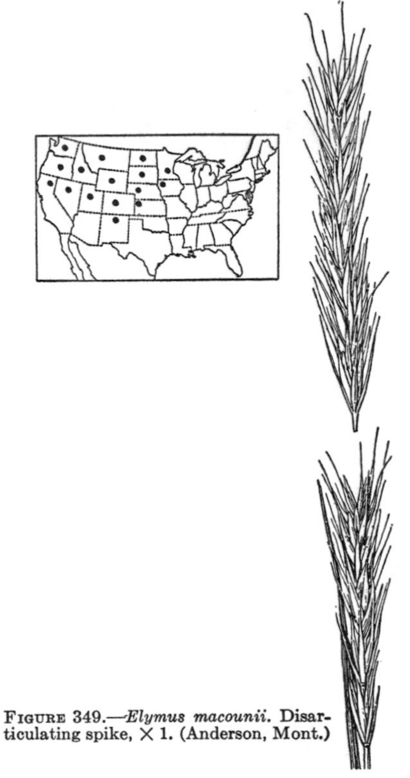

FIGURE 349.—*Elymus macounii*. Disarticulating spike, × 1. (Anderson, Mont.)

densely tufted, erect, slender, 50 to 100 cm. tall; sheaths glabrous or rarely pubescent; blades erect, rather firm, subinvolute, usually scabrous on both surfaces, 10 to 20 cm. long, mostly 2 to 5 mm. wide; spike slender, erect or somewhat nodding, 4 to 12 cm. long, usually about 5 mm. thick (excluding awns), the slender rachis tardily disarticulating; spikelets imbricate, appressed, mostly 2-flowered, about 1 cm. long, excluding the awns; glumes very narrow, scabrous, slightly divergent but not bowed out at base, the midnerve usually distinct; lemmas scabrous toward the apex, extending into slender straight awns 1 to 2 cm. long. ♃ —Meadows and open ground, Minnesota to Alaska and eastern Washington, south to Iowa, Kansas, New Mexico, and California. (Said by Stebbins to be a hybrid between *Agropyron trachycaulum* and species of *Hordeum*.)

**19. Elymus aristátus** Merr. (Fig. 350.) Culms tufted, rather leafy, erect, 70 to 100 cm. tall; sheaths glabrous, blades flat, 5 to 10 mm. wide; spike erect, dense, 6 to 14 cm. long, 5 to 10 mm. thick, the rachis tardily disarticulating; spikelets closely imbricate, often in threes, 1- to 2-flowered, about 1 cm. long, excluding the awns; glumes subsetaceous, scabrous, 10 to 20 mm. long; lemmas slightly wider than in *E. macounii*, sparsely scabrous at least on the upper half, the slender straight awn 10 to 20 mm. long. ♃ —Meadows and open slopes, at middle altitudes, Wyoming to Washington, south to Nevada and California.

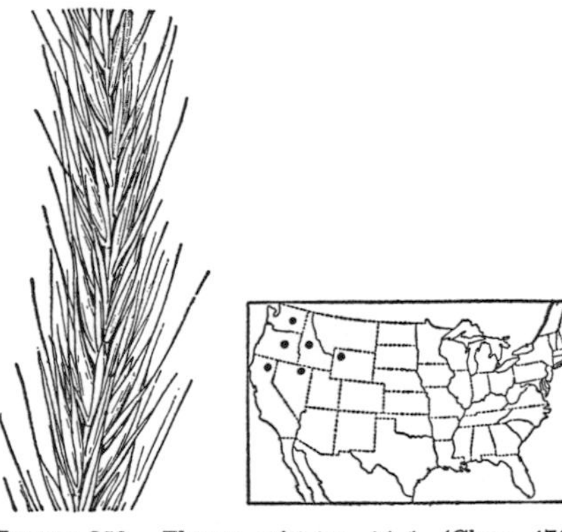

FIGURE 350.—*Elymus aristatus*, × 1. (Chase 4762, Idaho.)

**20. Elymus villósus** Muhl. (Fig. 351.) Culms in small tufts, ascending, slender, 60 to 100 cm. tall; sheaths glabrous to pilose; blades flat, lax, pubescent on upper surface, glabrous and glossy to scabrous beneath; spike drooping, dense, 5 to 12 cm. long; glumes subsetaceous, spreading, distinctly nerved above the firm cylindric nerveless divergent or somewhat bowed-out base, hirsute, 12 to 20 mm. long; lemmas nerved toward the tip, hispidulous to hirsute, 7 to 9 mm. long, about 1.2 mm. across the back, the straight slender awn 1 to 3 cm. long. ♃ (*E. striatus*, American authors, not Willd.) Moist or dry woods and shaded slopes, Canada and Vermont to North Dakota and Wyoming, south to South Carolina, Alabama, and Texas. E. ARKANSÁNUS Scribn. and Ball (*E. villosus* forma *arkansanus* Fernald), a relatively rare form with usually slightly stouter culms, the spikes mostly less drooping, scabrous glumes, and glabrous to scabrous lemmas, is found sparingly in Illinois, North

FIGURE 351.—*Elymus villosus*, × 1. (Commons 163, Del.)

FIGURE 352.—*Elymus interruptus*, × 1. (Grant 3071, Minn.)

Dakota, South Dakota, Nebraska, Missouri, Maryland, Virginia, North Carolina, Kentucky, Tennessee, Arkansas, Texas, and Wyoming. Large specimens resemble *E. riparius*, but the palea reaches the base of the awn.

**21. Elymus interrúptus** Buckl. (Fig. 352.) Culms erect, 70 to 130 cm. tall; sheaths glabrous; blades flat scabrous, 5 to 12 mm. wide; spike flexuous or nodding, 8 to 20 cm. long; glumes setaceous or nearly so, 1 to 3 cm. long, one or both reduced in occasional spikelets, mostly flexuous or spreading, the nerves obscure at least toward the base; lemmas hirsute to scabrous, or glabrous, about 1 cm. long, about 2 mm. across the back, the awn flexuous or divergent, 1 to 3 cm. long. ♃ (*E. diversiglumis* Scribn. and Ball.)—

FIGURE 353.—*Elymus canadensis*. Plant, × ½; spikelet and floret, × 5. (Lansing 3240, Mich.)

Rich, open moist soil, Michigan to North Dakota and Wyoming; Tennessee, Arkansas, Oklahoma, Texas, New Mexico.

**22. Elymus canadénsis** L. CANADA WILD-RYE. (Fig. 353.) Green or often glaucous; culms erect, tufted, mostly 1 to 1.5 m. tall; sheaths glabrous or rarely pubescent; blades flat, scabrous or sparsely hispid on the upper surface, mostly 1 to 2 cm. wide; spike thick and bristly, nodding or drooping, often interrupted below, 10 to 25 cm. long, sometimes glaucous; spikelets commonly in threes or fours, slightly spreading; glumes narrow, mostly 2- to 4-nerved, scabrous, sometimes hispid but less so than the lemmas, the bases somewhat indurate and divergent but scarcely bowed out, the awn about as long as the body; lemmas scabrous-hirsute to hirsute-pubescent, rarely glabrous, strongly nerved above, the awn divergently curved when dry, 2 to 3 cm. long. ♃ —River banks, open ground, and sandy soil, Quebec to southern Alaska, south to North Carolina, Missouri, Texas, Arizona, and northern California. *E. wiegandii* Fernald has been differentiated on lax inflorescence, shorter glumes, and thin flat blades, pilose on the nerves. These characters are found to be rarely coordinated, loose flexuous spikes being not infrequent in humid regions, rarer in dry areas; pilose blades are very rare. ELYMUS CANADENSIS var. ROBÚSTUS (Scribn. and Smith) Mackenz. and Bush. Differing in the stouter and denser only slightly nodding very bristly spikes. ♃ —Prairies, Massachusetts to Montana, south to Kentucky, Missouri, Texas, and Arizona. ELYMUS CANADENSIS var. BRACHÝSTACHYS (Scribn. and Ball) Farwell. Lemmas glabrous or nearly so. ♃ —Moist open or partly shaded ground, Arkansas, Oklahoma, Texas, and New Mexico; Mexico. Grades into *E. canadensis;* many specimens of *E. canadensis* from Kansas to North Dakota have sparingly hirsute lemmas, showing a transition to this variety.

**23. Elymus ripárius** Wiegand. (Fig. 354.) Culms rather slender, erect, 1 to 1.5 m. tall; sheaths glabrous; blades rather thin, flat, 5 to 15 mm. wide, scabrous; spike somewhat nodding, 7 to 20 cm. long; glumes narrow, about 1 mm. wide at the middle, 2- to 4-nerved, somewhat indurate but scarcely bowed out at base; lemmas minutely hispidulous to glabrous, the awn straight, mostly 2 to 3 cm. long. ♃ —River banks and low ground, Quebec and Maine to Wisconsin and Nebraska, south to North Carolina, Arkansas, and Kansas. Differing from *E. virginicus* var. *glabriflorus* in the nodding spike and less indurate glumes; from *E. canadensis* in the straight awns and narrower and somewhat more indurate glumes. When the ranges of *E. riparius* and *E. canadensis* coincide the latter may be distinguished by the hirsute lemmas.

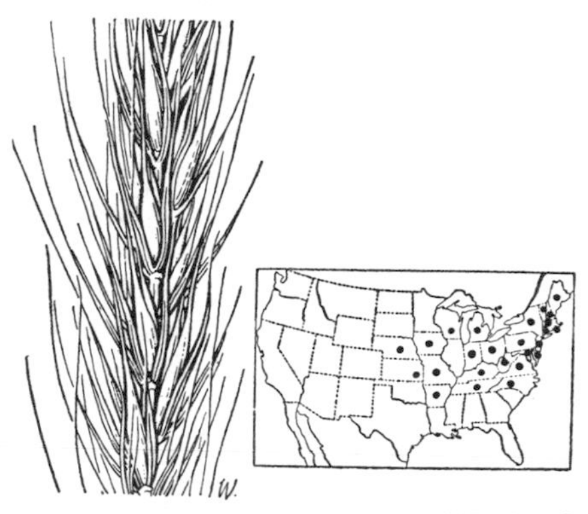

FIGURE 354.—*Elymus riparius*, × 1. (Woodward, Conn.)

**24. Elymus virgínicus** L. VIRGINIA WILD-RYE. (Fig. 355.) Culms tufted, erect, 60 to 120 cm. tall; sheaths glabrous; blades flat, scabrous, mostly 5 to 15 mm. wide; spike usually erect, often partly included, 5 to 15 cm. long; glumes strongly nerved, firm, indurate, yellowish, nerveless and bowed out at base leaving a rounded sinus, broadened above (1.5 to 2 mm. wide), scabrous, the apex somewhat curved, tapering into a straight awn, about as long as the body or shorter; lemmas glabrous and nerveless below, scabrous and nerved above, tapering into a straight awn usually about 1 cm. long. ♃ —

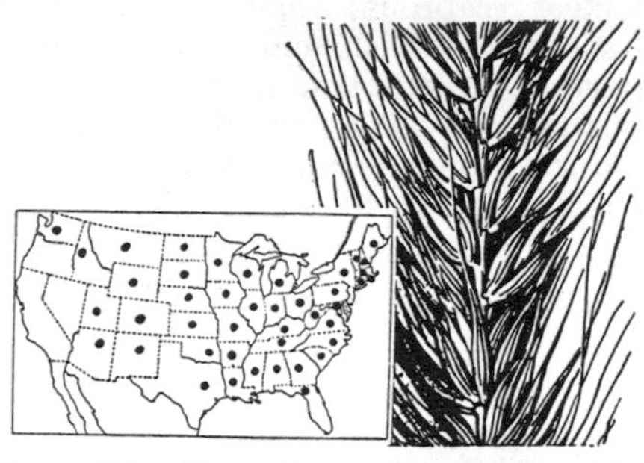

FIGURE 355.—*Elymus virginicus*, × 1. (Hitchcock 79, Va.)

Moist ground, low woods, and along streams, Newfoundland to Alberta, south to Florida and Arizona. Sometimes called Terrell grass. A variable species of which the following intergrading varieties may be distinguished.

ELYMUS VIRGINICUS var. GLABRIFLÓRUS (Vasey) Bush. Glumes mostly less bowed out; lemmas glabrous; awns mostly 2 to 3 cm. long, the spike more bristly. ♃ —Maine to Kansas, south to Florida and New Mexico.

ELYMUS VIRGINICUS var. HALÓPHILUS (Bickn.) Wiegand. More slender, usually glaucous; blades narrower, often becoming involute; spikes and spikelets somewhat smaller. ♃ —Brackish marshes and moist sand along the coast, Maine to Virginia.

ELYMUS VIRGINICUS var. SUBMÚTICUS Hook. Glumes and lemmas awnless or nearly so. ♃ —Woods and open ground, Quebec to Washington, south to Rhode Island; Ohio and Kentucky to Oklahoma and Montana; Utah.

ELYMUS VIRGINICUS var. INTERMÉDIUS (Vasey) Bush. Glumes, lemmas, and rachis more or less hirsute, the awns about as in *E. virginicus*. ♃ (*E. hirsutiglumis* Scribn.)—Thickets and low ground, Maine to Iowa, south to Florida and Texas.

ELYMUS VIRGINICUS var. AUSTRÁLIS (Scribn. and Ball) Hitchc. Differing from *E. virginicus* var. *intermedius* in the stouter, bristly spike and longer awns; differing from *E. virginicus* var. *glabriflorus* in the hirsute or strongly scabrous glumes and lemmas. ♃ —Prairies, rocky hills, and open woods, Vermont to Iowa, south to Florida, Kentucky, and Texas.

ELYMUS GIGANTÉUS Vahl. Robust perennial from stout rhizomes; blades numerous at base, elongate; spike dense, 15 to 20 cm. long, about 2 cm. thick; glumes and lemmas sharp-pointed, the glumes glabrous, the lemmas pubescent below. ♃ —Occasionally cultivated for ornament. Siberia.

## 47. SITÁNION Raf. SQUIRRELTAIL

Spikelets 2- to few-flowered, the uppermost floret reduced, usually 2 at each node of a disarticulating rachis, the rachis breaking at the base of each joint, remaining attached as a pointed stipe to the spikelets above; glumes narrow or setaceous, 1- to 3-nerved, the nerves prominent, extending into one to several awns, these (when more than one) irregular in size, sometimes mere lateral appendages of the long central awn, sometimes equal, the glume being bifid; lemmas firm, convex on the back, nearly terete, 5-nerved, the nerves obscure, the apex slightly 2-toothed, the central nerve extending into a long, slender, finally spreading awn, sometimes one or more of the lateral nerves also extending into short awns; palea firm, nearly as long as the body of the lemma, the two keels serrulate. Low or rather tall tufted perennials, with bristly spikes. Type species, *Sitanion elymoides* Raf. (*S. hystrix*). Name from Greek *sitos*, grain.

The species are exceedingly variable, being glabrous to densely pubescent and green to glaucous; the glumes and lemmas vary in division and length of awns. Some 15 to 25 variations have been recognized as species, but study of extensive collections shows that most of the characters used in differentiating the forms are inconstant and combine in various ways.

The species are widespread in the Western States but do not form complete stands. They have forage value when young but at maturity the disarticulating joints of the spike, with their pointed rachis joints and long-awned spikelets, are blown about by the wind and often cause injury to stock, penetrating the mouth, nose, and ears, working in by means of the forwardly roughened awns, and causing inflammation. Grazed also after the heads are blown off. The commonest species is *S. hystrix*.

Spike much longer than broad; glumes narrowly lanceolate, 2- to 4-nerved.
1. S. HANSENI.

Spike as broad as long or broader; glumes bristlelike, 1- or obscurely 2-nerved.
- Glumes cleft into at least 3 fine divisions ........ 2. S. JUBATUM.
- Glumes entire or 2-cleft ........ 3. S. HYSTRIX.

**1. Sitanion hanséni** (Scribn.) J. G. Smith. HANSEN SQUIRRELTAIL. (Fig. 356.) Culms 60 to 100 cm. tall; sheaths and blades glabrous or scabrous to softly pubescent, the blades flat to subinvolute, 2 to 8 mm. wide; spike somewhat nodding or flexuous, 8 to 20 cm. long; glumes narrowly lanceolate, sometimes bifid, 2- to 3-nerved, long-awned, lower lemmas about 8 mm. long, the awn 4 to 5 cm. long, divergent when dry and mature. ♃ —Open woods and rocky slopes, Wyoming to eastern Washington, Utah, and California. Pubescent plants have been differentiated as *S. anomalum* J. G. Smith. (*S. hanseni* is said by Stebbins to consist of a series of hybrids between *Elymus glaucus* and *Sitanion jubatum* or *S. hystrix*.)

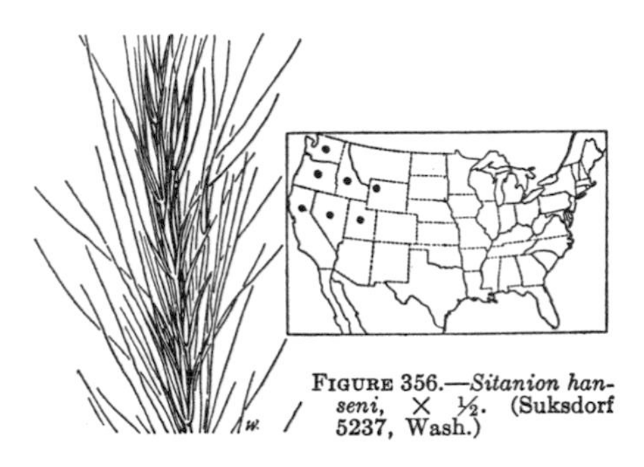

FIGURE 356.—*Sitanion hanseni*, × ½. (Suksdorf 5237, Wash.)

**2. Sitanion jubátum** J. G. Smith. BIG SQUIRRELTAIL. (Fig. 357.) Culms erect to ascending, 20 to 60 cm. tall, rarely taller; foliage glabrous or scabrous to white-villous, the blades flat, often becoming involute, mostly not more than 4 mm. wide; spike erect, dense, 3 to 10 cm. long, thick and bushy from the numerous long slender spreading awns; glumes split into 3 or more long awns; lemmas mostly 8 to 10 mm. long, smooth, or scabrous toward apex, the awns and those of the glumes spreading, 3 to 10 cm. long, rarely shorter. ♃ —Rocky or brushy hillsides and open dry woods and plains, Idaho to eastern Washington, south to Utah, Nevada, Arizona, and Baja California. Occasionally a few of the glumes in a spike are divided into only 2 awns. Short-awned plants have been differentiated as *S. breviaristatum* J. G. Smith and the more densely pubescent plants as *S. villosum* J. G. Smith.

**3. Sitanion hýstrix** (Nutt.) J. G. Smith. SQUIRRELTAIL. (Fig. 358.) Culms erect to spreading, rather stiff, 10 to 50 cm. tall; foliage from glabrous or puberulent to softly and densely white-pubescent, the blades flat to involute, rather stiffly ascending to spreading, 5 to 20 cm. long, 1 to 3 mm. wide, rarely as much as 5 mm. wide; spike mostly short-exserted or partly included, erect, 2 to 7 cm., rarely 10 cm., long or longer, the glumes very narrow, 1- to 2-nerved, the nerves extending into scabrous awns, sometimes bifid to the middle, or bearing a bristle or awn along one margin; lemmas convex, smooth or scabrous to appressed pubescent, sometimes glaucous, the awns of glumes and lemmas widely spreading, 2 to 10 cm. long. ♃ —Dry hills, plains, open woods, and rocky slopes, South Dakota to British Columbia, south to Missouri, Texas, California, and Mexico. At high altitudes plants often dwarf. Softly pubescent plants have been differentiated as *S. cinereum* J. G. Smith (the pubescence whitish) and *S. velutinum* Piper; short-awned plants as *S. insulare* J. G. Smith and *S. marginatum* Scribn. and Merr.; rather small plants with unusually slender awns as *S. minus* J. G. Smith, and tall plants with coarse spikes as *S. brevifolium* J. G. Smith, *S. longifolium* J. G. Smith, and *S. montanum* J. G. Smith.

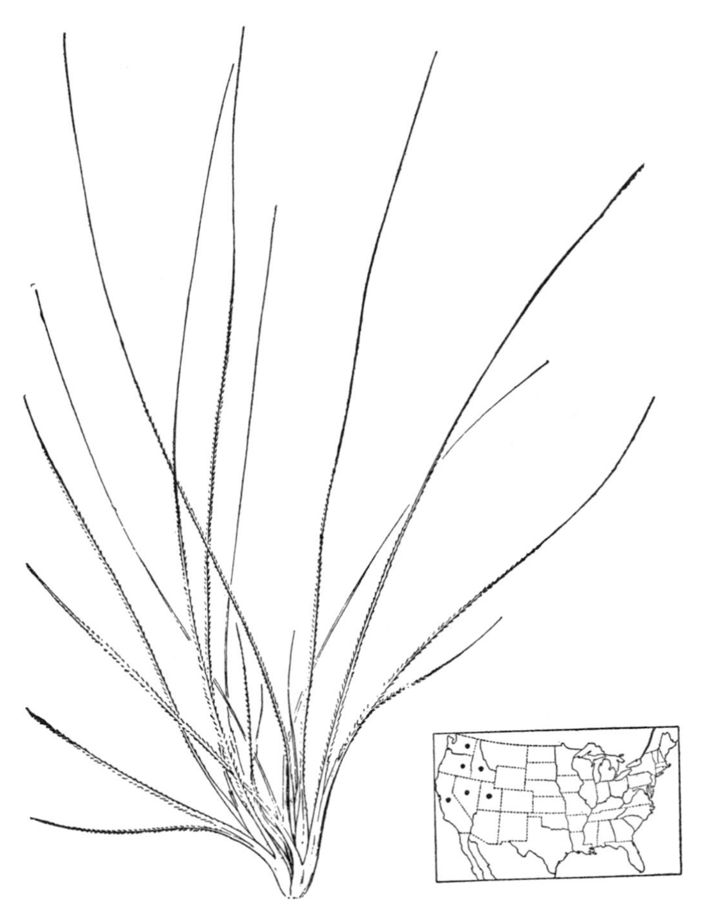

FIGURE 357.—*Sitanion jubatum*. Pair of spikelets, × 2. (Type.)

## 48. HÝSTRIX Moench

Spikelets 2- to 4-flowered, 1 to 4 at each node of a continuous flattened rachis, horizontally spreading or ascending at maturity; glumes reduced to short or minute awns, the first usually obsolete, both often wanting in the upper spikelets; lemmas convex, rigid, tapering into long awns, 5-nerved, the nerves obscure except toward the tip; palea about as long as the body of the lemma. Erect perennials, with flat blades and bristly, loosely flowered spikes. Type species, *Elymus hystrix* L. (*Hystrix patula*). *Hustrix*, Greek name for the porcupine, alluding to the bristly spikes. The species have little forage value, as

FIGURE 358.—*Sitanion hystrix*. Plant, × ½; spikelet and floret, × 3. (Hitchcock 2289, Colo.)

they are nowhere abundant. The first species is worthy of cultivation for ornament.

Spikelets soon divergent; lemmas glabrous or pubescent, not hispid.......... 1. H. PATULA.
Spikelets ascending or appressed; lemmas appressed-hispid................ 2. H. CALIFORNICA.

FIGURE 359.—*Hystrix patula*. Plant, × ½; spikelet and floret, × 3. (Moyer, Minn.)

1. **Hystrix pátula** Moench. BOTTLEBRUSH. (Fig. 359.) Culms slender, 60 to 120 cm. tall; sheaths glabrous or scabrous, rarely retrorsely pubescent; blades mostly 7 to 15 mm. wide; spike nodding, 8 to 15 cm. long, the internodes of the slender rachis 5 to 10 mm. long; spikelets mostly in pairs, 1 to 1.5 cm. long, horizontally spreading toward maturity; lemmas glabrous or

sometimes coarsely pubescent, the awns 1 to 4 cm. long, slender, straight. ♃ (*H. hystrix* Millsp.)—Moist or rocky woods, Nova Scotia to North Dakota, south to Georgia and Arkansas. Plants with pubescent lemmas have been differentiated as *H. patula* var. *bigeloviana* (Fernald) Deam. Such plants occur throughout the range, except from Delaware, Maryland, and southward.

**2. Hystrix califórnica** (Boland.) Kuntze. (Fig. 360.) Culms stout, 1 to 2 m. tall; sheaths hispid or the upper smooth; blades as much as 2 cm. wide; spike 12 to 25 cm. long; spikelets usually 3 or 4 at a node, 1.2 to 1.5 cm. long, thicker than in *H. patula*, ascending at maturity; lemmas hispidulous, the awn about 2 cm. long. ♃ —Woods and shaded banks, near the coast, Sonoma County to Santa Cruz County, Calif. In addition to the sessile spikelets there may be a short branch bearing 1 or 2 spikelets.

FIGURE 360.—*Hystrix californica*. Spike, × ½; floret, × 3. (Vasey, Calif.)

## 49. HÓRDEUM L. BARLEY

Spikelets 1-flowered (rarely 2-flowered), 3 (sometimes 2) together at each node of the articulate rachis (continuous in *Hordeum vulgare*), the back of the lemma turned from the rachis, the middle spikelet sessile, the lateral ones pediceled (except in *H. vulgare* and *H. montanense*); rachilla disarticulating above the glumes and, in the central spikelet, prolonged behind the palea as a bristle and sometimes bearing a rudimentary floret; lateral spikelets usually imperfect, sometimes reduced to bristles; glumes narrow, often subulate and awned, standing in front of the spikelet; lemmas rounded on the back, 5-nerved, usually obscurely so, tapering into a usually long awn. Annual or perennial low or rather tall grasses, with flat blades and dense bristly spikes, disarticulating at the base of the rachis segment, this remaining as a stipe below the attached triad of spikelets. Type species, *Hordeum vulgare*. *Hordeum*, the old Latin name for barley.

Aside from the well-known cultivated barley, *H. vulgare*, the species are of relatively minor value. All furnish forage when young, but many species are aggressive weeds and some (especially *H. jubatum*) at maturity are injurious to stock because of the sharp-pointed joints of the mature spikes, which pierce the nose and mouth parts. The auricle at the base of the blades, characteristic of Hordeae, is wanting in some species of this genus.

Plants perennial; awns slender; auricle wanting.
  Lateral spikelets sessile; central spikelet usually 2-flowered.......... 1. H. MONTANENSE.
  Lateral spikelets pedicellate.
    Spike, including awns, as broad as long or nearly so (narrower in var. *caespitosum*); awns 2 to 5 cm. long.......... 2. H. JUBATUM.
    Spike, including awns, much longer than broad, awns not more than 1 cm. long.
      Floret of lateral spikelet evident, from staminate to reduced and empty; spike 6 to 10 mm. wide; blades 3 to 8 mm. wide.......... 3. H. BRACHYANTHERUM.
      Floret of lateral spikelets scarcely distinct from its awn; spike about 5 mm. wide; blades 2 to 3 mm. wide.......... 4. H. CALIFORNICUM.
Plants annual, branching at base; awns mostly stouter.
  Blades with prominent auricles at base.
    Rachis continuous, the 3 spikelets sessile.......... 11. H. VULGARE.

Rachis disarticulating; lateral spikelets pedicellate.
Floret of lateral spikelets longer and broader than that of central spikelet; rachis internodes mostly 3 mm. long .......... 9. H. LEPORINUM.
Floret of lateral spikelets not larger than that of central spikelet; rachis internodes mostly 2 mm. long .......... 10. H. STEBBINSII.
Blades without auricles.
Glumes of the fertile spikelet dilated above the base .......... 5. H. PUSILLUM.
Glumes of the fertile spikelet not dilated.
Awns slender, 1.5 to 2 cm. long, fragile; one glume of lateral spikelets slightly dilated. 6. H. ARIZONICUM.
Awns relatively stout.
Floret of lateral spikelets awnless; glumes slender, not rigid, not bowed out. 7. H. DEPRESSUM.
Floret of lateral spikelets awned; glumes thickened and slightly bowed out below, rigid .......... 8. H. HYSTRIX.

**1. Hordeum montanénse** Scribn. (Fig. 361.) Culms 60 to 100 cm. tall; sheaths glabrous; blades flat, lax, scabrous, 5 to 8 mm. wide; spike nodding, 8 to 17 cm. long; central spikelets usually 2-flowered, with a rudiment of a third floret; lateral spikelets sessile, usually well developed; glumes slightly broadened above the base, 1 to 3.5 cm. long including awns; lower floret of central spikelet about 8 mm. long, the awn 1.5 to 3.5 cm. long. ♃ (*H. pammeli* Scribn. and Ball.)—Prairies, Illinois, Iowa, South Dakota, Montana, and Wyoming. Variable and somewhat anomalous; lateral spikelets sometimes with 2 florets. Approaches *Elymus;* specimens referred by geneticists to hybrid *Hordeum jubatum* × *Elymus virginicus.*

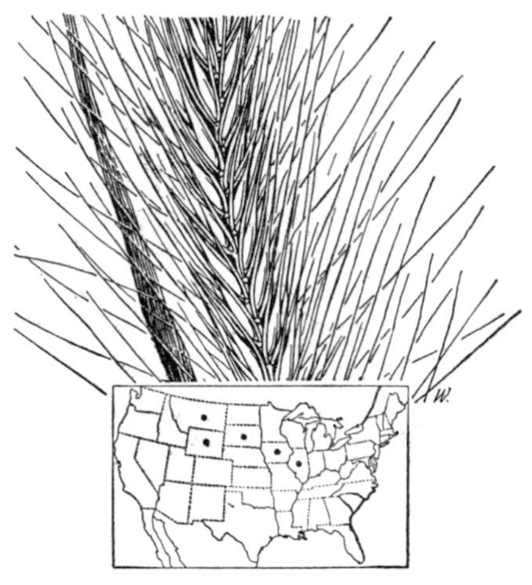

FIGURE 361.—*Hordeum montanense*, × 1. (V. H. Chase 1467, Ill.)

**2. Hordeum jubátum** L. FOXTAIL BARLEY. (Fig. 362.) Perennial, tufted; culms erect, or decumbent at base, 30 to 60 cm. tall; blades 2 to 5 mm. wide, scabrous; spike nodding, 5 to 10 cm. long, about as wide, soft, pale; lateral spikelets reduced to 1 to 3 spreading awns; glumes of perfect spikelet awnlike, 2.5 to 6 cm. long, spreading; lemma 6 to 8 mm. long with an awn as long as the glumes. ♃ —Open ground, meadows and waste places, Newfoundland and Labrador to Alaska, south to Maryland, Missouri, Texas, California, and Mexico; introduced in the Eastern States. A troublesome weed in the Western States, especially in irrigated meadows. HORDEUM JUBATUM var. CAESPITÓSUM (Scribn.) Hitchc. BOBTAIL BARLEY. Awns 1.5 to 3 cm. long. (*H. caespitosum* Scribn.) North Dakota to Alaska, south to California and Arizona; Mexico.

**3. Hordeum brachyántherum** Nevski. MEADOW BARLEY. (Fig. 363.) Perennial, tufted; culms erect or ascending, 20 to 70 cm., sometimes to 100 cm., tall; lower sheaths thin, often shredded, softly retrorse-pubescent to glabrous; blades 3 to 8 mm., mostly 3 to 6 mm., wide, spike erect or slightly nodding, 8 to 10 cm. long, rarely longer, sometimes purplish; floret of central spikelet usually 7 to 10 mm. long, typically 1.5 mm. wide, the awn about 1 cm. long, the glumes slightly shorter; glumes of lateral spikelets usually unequal, somewhat shorter, the floret from well developed and staminate to much reduced and empty (occasionally a staminate and

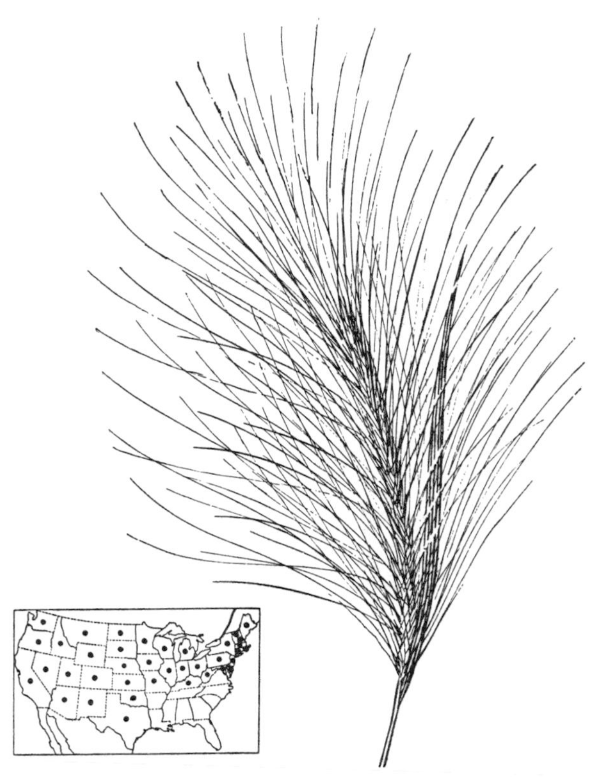

FIGURE 362.—*Hordeum jubatum*, × 1. (Blankinship 189, Mont.)

an empty lateral floret in the same triad), the awn 2 to 5 mm. long; spikelets extremely variable, the spike sometimes slender, the perfect floret 5 to 6 mm. long, the awn 5 to 6 mm. (The name *H. nodosum* L. has been misapplied to this species.) ♃ —Meadows, bottom lands, salt marshes, grassy slopes up to 3,000 m., Aleutian Islands and Alaska to California; Labrador, Newfoundland; Montana to New Mexico and Arizona to California; adventive Maine, Indiana, Mississippi.

**4. Hordeum califórnicum** Covas and Stebbins. Densely tufted perennial; culms slender, 30 to 55 cm. tall; lower sheaths softly retrorse-pubescent to glabrous; blades 2 to 3 mm. wide, the auricle wanting; spike erect, 2.5 to 6 cm. long, mostly purplish; floret of central spikelet 6 to 7 (rarely 8) mm. long, the awn 4 to 10 mm. long, the rachilla behind the palea often wanting; floret of lateral spikelet much reduced, scarcely distinct from the awn. ♃ —Meadows, dried creek beds, and brushy flats and slopes, Oregon and California; scarce, probably depauperate dry ground plants of the preceding.

**5. Hordeum pusíllum** Nutt. LITTLE BARLEY. (Fig. 364.) Annual; culms 10 to 35 cm. tall; blades erect, flat, the auricle wanting; spike erect, 2 to 7 cm. long, 10 to 14 mm. wide; first glume of the lateral spikelets and both glumes of the fertile spikelet dilated above the base, attenuate into a slender awn 8 to 15 mm. long, the glumes very scabrous; lemma of central spikelet awned, of lateral spikelets awn-pointed. ⊙ —Plains and open, especially alkaline, ground, Delaware to Washington, south to Florida, southern California, and northern Mexico;

FIGURE 363.—*Hordeum brachyantherum*. Plant, × ½; group of spikelets and floret, × 3. (Whited 433, Wash.)

adventive in Maine and Pennsylvania; common westward, rare in the Atlantic States; also southern South America. HORDEUM PUSILLUM var. PÚBENS Hitchc. Spike broader; spikelets pubescent; dilated glumes wider. ☉ —Texas to Utah and Arizona.

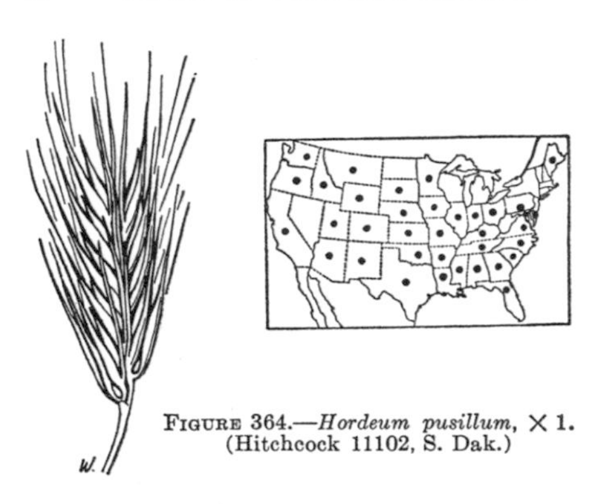

FIGURE 364.—*Hordeum pusillum*, × 1. (Hitchcock 11102, S. Dak.)

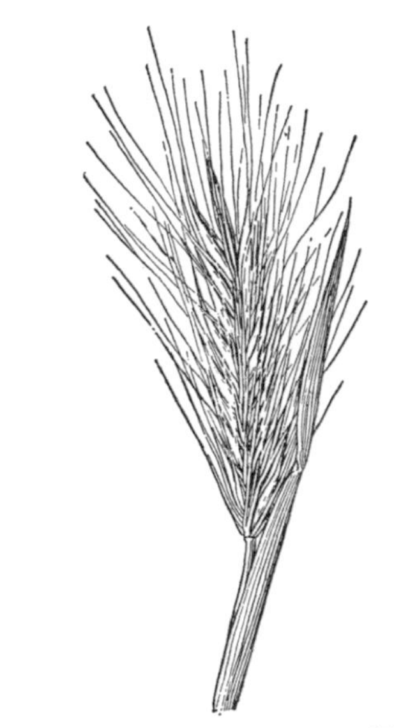

FIGURE 365.—*Hordeum arizonicum*, × 1. (Thornber 536, Ariz.)

**6. Hordeum arizónicum** Covas. (Fig. 365.) Annual; culms geniculate at base, 20 to 60 cm. tall; lower sheaths pubescent, the upper more or less inflated; blades 3 to 5 mm. wide, sparsely pubescent, the auricle wanting; spike erect, 3 to 12 cm. long; floret of central spikelet 8 to 9 mm. long, 1.5 mm. wide, the awn 15 to 22 mm. long, the glumes slightly shorter; glumes of lateral florets nearly as long, one slightly dilated (all awns scabrous, slender, fragile, readily breaking); floret reduced to a small short-awned lemma. (The name *H. adscendens* has been misapplied to this species.) ☉ —Dry open ground (large plants found along irrigation ditches), Arizona and California (Bard).

**7. Hordeum depréssum** (Scribn. and Smith) Rydb. (Fig. 366.) Annual; culms geniculate at base, commonly spreading with ascending ends, 6 to 45 cm. long; upper sheaths often inflated; blades pubescent, mostly not more than 5 cm. long (rarely to 15 cm.), 2 to 4 mm. wide, the auricle wanting; spike erect, 4 to 7 cm. long; floret of central spikelet 7 to 8 mm. long, nearly terete, the awn about 10 mm. long; awns of the glumes and of the glumes of lateral spikelets nearly equal, the whole triad usually about 2 cm. long; floret of lateral spikelet awnless. ☉ —Mostly in moist alkaline soil or along rivers, also in arid or sterile ground, sea level to 600 m., Idaho and Washington to California.

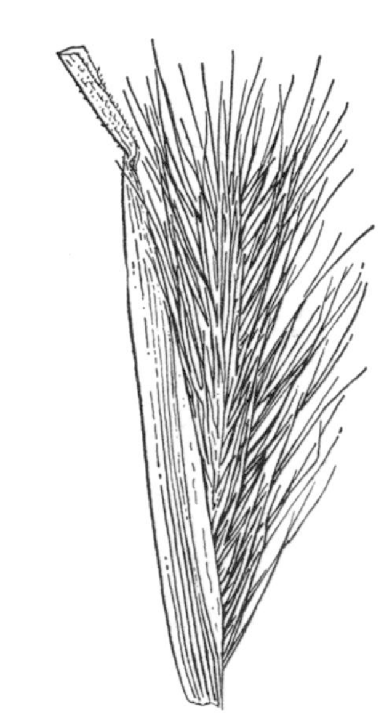

FIGURE 366.—*Hordeum depressum*, × 3. (Type.)

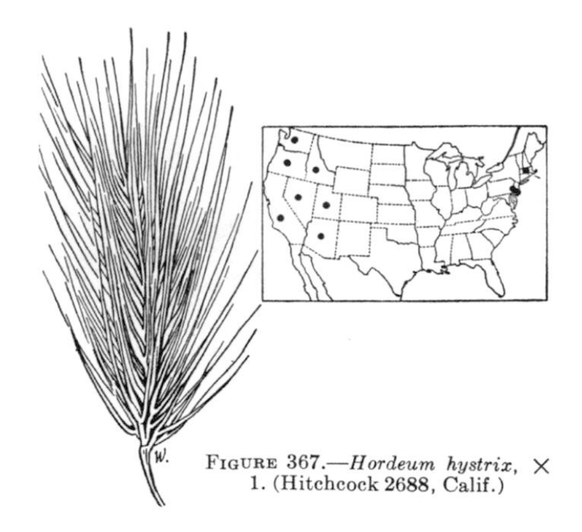

FIGURE 367.—*Hordeum hystrix*, × 1. (Hitchcock 2688, Calif.)

**8. Hordeum hýstrix** Roth. MEDITERRANEAN BARLEY. (Fig. 367.) Annual; culms freely branching and spreading or geniculate at base, 15 to 40 cm. tall; sheaths and blades, especially the lower, more or less pubescent, the auricle wanting; spike erect, 1.5 to 3 cm. long, 10 to 15 mm. wide, the axis usually not readily breaking; glumes setaceous, rigid, nearly glabrous to scabrous, about 12 mm. long; lemma of central spikelet 5 mm. long, the awn somewhat longer than the glumes; floret of lateral spikelets reduced, short-awned. ⊙ (*H. gussonianum* Parl.)—Fields and waste places, Utah to British Columbia, Arizona, and California; adventive in Massachusetts, New Jersey, and Pennsylvania; introduced from Europe.

**Hordeum marínum** Huds. Differing from *H. hystrix* in the glabrous dissimilar glumes of the lateral spikelets, the outer subulate, the inner somewhat broader. ⊙ (*H. marítimum* With.)—On ballast, Camden, N. J.; Europe.

**9. Hordeum leporínum** Link. (Fig. 368.) Annual; branching at base, spreading; sheaths glabrous, blades pilose to glabrous; auricle at base of blade well developed; spike 5 to 9 cm. long, often partly enclosed by the inflated uppermost sheath, the rachis internodes mostly 3 mm. long; glumes of the central spikelet lanceolate, 3-nerved, long-ciliate on both margins, the nerves scabrous, the awn 2 to 2.5 cm. long; floret 1 to 1.2 cm. long, raised on a rachilla segment 1 mm. long, the awn 3 to 4 cm. long; lateral spikelets usually staminate, the glumes much shorter, unlike, the inner similar to those of the central one, the outer setaceous, not ciliate, the lemma broad, 10 to 20 mm. long, the awn 2 to 4 cm. long. ⊙ —Weed, fields, waste places and open ground, introduced from southern Europe; here and there in the Eastern States, Massachusetts to Georgia; Vancouver Island and Washington to California, Utah, and Texas. This and *H. stebbinsii* have been confused with *H. murinum* L., of Europe, not known from America.

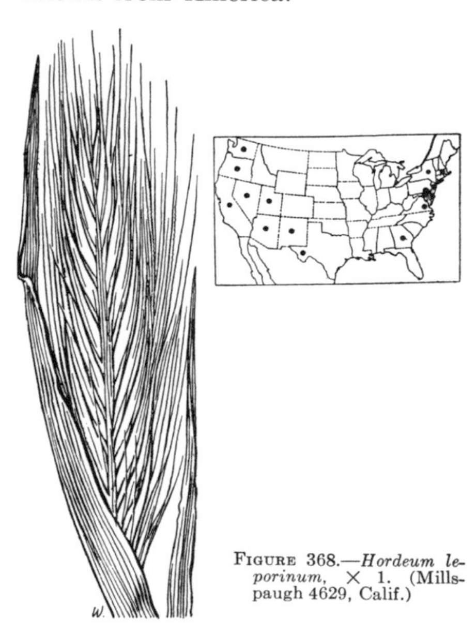

FIGURE 368.—*Hordeum leporinum*, × 1. (Millspaugh 4629, Calif.)

**10. Hordeum stebbínsii** Covas. Similar to the preceding, the culms often shorter and more geniculate; spikes narrower, mostly 9 to 15 mm. wide before beginning to break up, the triads closely ascending and slightly more crowded, the rachis internodes mostly 2 mm. long; florets of lateral spikelets not larger than that of the middle spikelet; all awns mostly shorter and slightly more

FIGURE 369.—*Hordeum vulgare.* Plant, × ½; group of spikelets and floret, × 3; spike of beardless barley (*a*), × ½. (Cult.)

slender. ⊙ —Weed, fields, waste places, and open, mostly arid ground, introduced from the Old World, ballast, Mobile, Ala.; adventive, Oklahoma; Idaho and Washington; New Mexico to California. Often difficult to distinguish from the preceding.

**11. Hordeum vulgáre** L. BARLEY. (Fig. 369.) Annual; culms erect, 60 to 120 cm. tall; blades flat, mostly 5 to 15 mm. wide, the auricle well developed; spike erect or nearly so, 2 to 10 cm. long, excluding awns, the 3 spikelets sessile; glumes divergent at base, narrow, nerveless, gradually passing into a stout awn; awn of lemma straight, erect, mostly 10 to 15 cm. long. ⊙ —Cultivated for the grain, sometimes spontaneous in fields and waste places but not persistent. There are two groups of the cultivated barleys. In the 2-rowed forms (*H. distichon* L.) the lateral spikelets are fairly well developed but sterile. The probable ancestor for at least a part of these is *H. spontaneum* Koch, of Asia. In the second group all the spikelets produce large seed. These are called 6-rowed (*H. hexastichon* L.) or, if the lateral florets overlap, 4-rowed barleys (in European literature). In some varieties the caryopsis is naked. The ancestor of the 6-rowed barleys is not known but probably was similar to some of our cultivated varieties of this group.

HORDEUM VULGARE var. TRIFURCÁTUM (Schlecht.) Alefeld, BEARDLESS BARLEY. Awns suppressed or variously deformed, commonly 3-cleft, the central division converted into a hooded lobe. Adventive or occasional in grainfields and along roads, Connecticut to New Jersey; South Dakota, Montana; Colorado, Utah, New Mexico; California.

## 50. LÓLIUM L. RYEGRASS

Spikelets several-flowered, solitary, placed edgewise to the continuous rachis, one edge fitting to the alternate concavities, the rachilla disarticulating above the glumes and between the florets; first glume wanting (except on the terminal spikelet and rarely in 1 or 2 spikelets in a spike), the second outward, strongly 3- to 5-nerved, equaling or exceeding the second floret; lemmas rounded on the back, 5- to 7-nerved, obtuse, acute, or awned. Annuals or perennials, with flat blades and slender, usually flat spikes. Type species, *Lolium perenne*. *Lolium*, an old Latin name for darnel.

*Lolium perenne*, perennial or English ryegrass, was the first meadow grass to be cultivated in Europe as a distinct segregated species, the meadows and pastures formerly being native species. This and *L. multiflorum*, Italian ryegrass, are probably the most important of the European forage grasses. Both species are used in the United States to a limited extent for meadow, pasture, and lawn. They are of importance in the South for winter forage. In the Eastern States the ryegrasses are often sown in mixtures for parks or public grounds, where a vigorous early growth is required. The young plants can be distinguished from bluegrass by the glossy dark-green foliage. *L. temulentum*, darnel, is occasionally found as a weed in grainfields and waste places. It is in bad repute, because of the presence in the grain of a narcotic poison, said to be due to a fungus. Darnel is supposed to be the plant referred to as the tares sown by the enemy in the parable of Scripture.

Glume shorter than the spikelet.
  Lemmas nearly or quite awnless; culms subcompressed .......... 1. L. PERENNE.
  Lemmas, at least the upper, awned; culms cylindric .......... 2. L. MULTIFLORUM.
Glume as long as or longer than the spikelet. Annuals.
  Spike flat; spikelets much wider than the rachis.
    Florets plump, 6 to 8 mm. long .......... 3. L. TEMULENTUM.
    Florets dorsally compressed, 9 to 10 mm. long .......... 4. L. PERSICUM.
  Spike subcylindric; spikelets scarcely wider than the rachis .......... 5. L. SUBULATUM.

**1. Lolium perénne** L. PERENNIAL RYEGRASS. (Fig. 370, *B*.) Short-lived perennial; culms erect or decumbent at the commonly reddish base, 30 to 60 cm. tall; auricles at summit of sheath, minute or obsolete; foliage glossy, the blades 2 to 4 mm. wide; spike often subfalcate, mostly 15 to 25 cm. long; spikelets mostly 6- to 10-flowered; lemmas 5 to 7 mm. long, awnless or nearly so. ♃ —Meadows and waste places, Newfoundland to Alaska and south to Virginia and California, occasionally farther south; cultivated in meadows, pastures, and lawns, introduced from Europe. Also called English ryegrass. LOLIUM PERENNE var. CRISTÁTUM Pers. Spikes ovate, the spikelets crowded, horizontally spreading. ♃ —Open ground, Wilmington, Del., and Washington, D. C.; ballast, Salem and Eola, Oreg.; adventive from Europe.

**2. Lolium multiflórum** Lam. ITALIAN RYEGRASS. (Fig. 370, *A*.) Differing from *L. perenne* in the more robust habit, to 1 m. tall, pale or yellowish at base; auricles at summit of sheaths prominent; spikelets 10- to 20-flowered, 1.5 to 2.5 cm. long; lemmas 7 to 8 mm. long, at least the upper awned. ♃ (*L. italicum* A. Br.)—About the same range as *L. perenne*, especially common on the Pacific coast where it is often called Australian ryegrass. Introduced from Europe. Closely related to *L. perenne*, but generally recognized as distinct agriculturally. A much reduced form has been called forma *microstachyum* Uechtritz.—California.

LOLIUM MULTIFLORUM var. RAMÓSUM Guss. A peculiar form, the spike transformed into a narrow many-flowered panicle. ♃ —Linn County, Oreg., waif. Europe.

**3. Lolium temuléntum** L. DARNEL. (Fig. 371.) Annual; culms 60 to 90 cm. tall; blades mostly 3 to 6 mm. wide; spike strict, 15 to 25 cm. long; glume about 2.5 cm. long, as long as or longer than the 5- to 7-flowered spikelet, firm, pointed; florets plump, the lemmas as much as 8 mm. long, obtuse, awned, the awn 6 to 12 mm. long. ⊙ —Grainfields and waste places, occasional throughout the eastern United States and rather common on the Pacific coast; introduced from Europe. LOLIUM TEMULENTUM var. LEPTOCHAÉTON A. Br. Lemmas awnless. ⊙ —Washington to California, occasional on the Atlantic coast, Maine to Texas; introduced from Europe.

**4. Lolium pérsicum** Boiss. and Hohen. Annual, resembling small plants of *L. temulentum*, branching at the lower nodes; spike 8 to 12 cm. long; spikelets mostly more distant than in *L. temulentum*, the glume three-fourths to as long as the spikelet, the florets mostly 9 to 10 mm. long, not plump, the awn slender, commonly flexuous, the palea slightly exceeding the lemma. ⊙ —A weed in wheatfields and waste ground, Ontario to Alberta, and in North Dakota, becoming a bad weed. Introduced, probably in wheat seed from Russia.

**5. Lolium subulátum** Vis. (Fig. 372.) Annual; culms freely branching at base, stiffly spreading or prostrate; foliage scant, blades short; spike subcylindric, rigid, often curved; spikelets sunken in the excavations of the rachis, the florets partly hidden by the appressed obtuse strongly nerved glume; lemmas 5 mm. long. ⊙ —On ballast, near Portland, Oreg.; introduced from Europe.

**Lolium stríctum** Presl. Annual; branched and spreading at base, 10 to 30 cm. tall; spike thickish, 5 to 10 cm. long, the rachis thick but flattish and angled. ⊙ —Ballast, Linnton, Oreg., Berkeley, Calif.; Mohave County, Ariz. Introduced from Europe. Resembles *L. subulatum*, but the spikelets not sunken in a cylindric rachis.

LOLIUM REMÓTUM Schrank. Leafy annual; spike slender, spikelets more or less remote; glume half to two-thirds as long as the spikelets; florets 3 to 4 mm. long, plump, awnless. ⊙ —Weed in flax field, North Dakota, the seed from Russia.

FIGURE 370.—*A*, *Lolium multiflorum*. Plant, × ½; spikelet, × 3; floret, × 5. (Suksdorf 5142, Wash.) *B*, *L. perenne*, × ½. (Kimball, D. C.)

FIGURE 371.—*Lolium temulentum*, × ½. (Leiberg 771, Oreg.)

**Nárdus strícta** L. Slender, tufted perennial; sheaths crowded at the base; blades slender, involute, rather stiff; spike slender, 1-sided, 3 to 8 cm. long; spikelets 1-flowered; first glume wanting; second glume minute; lemma narrow, acuminate or short-awned, scabrous. ⊙ —Introduced in Newfoundland and Quebec, and sparingly in dry open ground in New Hampshire, New York, and Michigan; Europe.

## 51. MONÉRMA Beauv.

(Included in *Lepturus* R. Br. in Manual, ed. 1)

Spikelets 1-flowered, embedded in the hard, cylindric articulate rachis and falling attached to the joints; first glume wanting except on the terminal spikelet, the second glume closing the cavity of the rachis and flush with the surface, indurate, nerved, acuminate, longer than the joint of the rachis; lemma with its back to the rachis, hyaline, shorter than the glume, 3-nerved; palea a little shorter than the lemma, hyaline.

Low annual, with slender cylindric spikes. Type species, *M. monandra* Beauv. (*M. cylindrica* (Willd.) Coss. and Dur.) Name from Greek *monos*, one, and *erma*, support, referring to the single spike.

FIGURE 372.—*Lolium subulatum*, × ½. (Sheldon, Oreg.)

**1. Monerma cylíndrica** (Willd.) Coss. and Dur. THINTAIL. (Fig. 373.) Annual; culms bushy-branched, spreading or prostrate, 10 to 30 cm. tall; spike curved, narrowed upward; glume 6 mm. long, acuminate; lemma 5 mm. long, pointed; rachis disarticulating at maturity, the spikelets remaining attached to the joints. ⊙ (*Lepturus cylindricus* Trin.)—Salt marshes, San Francisco Bay, Calif., south to San Diego and Santa Catalina Island; introduced from the Old World.

## 52. PARÁPHOLIS C. E. Hubb.

(Included in *Pholiurus* Trin. in Manual, ed. 1)

Spikelets 1- or 2-flowered, embedded in the cylindric articulate rachis and falling attached to the joints; glumes 2, placed in front of the spikelet and enclosing it, coriaceous, 5-nerved, acute, asymmetric, appearing like halves of a single split glume; lemma with its back to the rachis, smaller than the glumes, hyaline, 1-nerved; palea a little shorter than the lemma, hyaline. Low annuals, with slender cylindric spikes. Type species, *P. incurva* (L.) C. E. Hubb. Name from Greek *para*, beside, and *pholis*, scale, referring to the 2 glumes side by side.

FIGURE 373.—*Monerma cylindrica.* Plant, × ½; rachis joint and spikelet, × 5. (Parish 4446, Calif.)

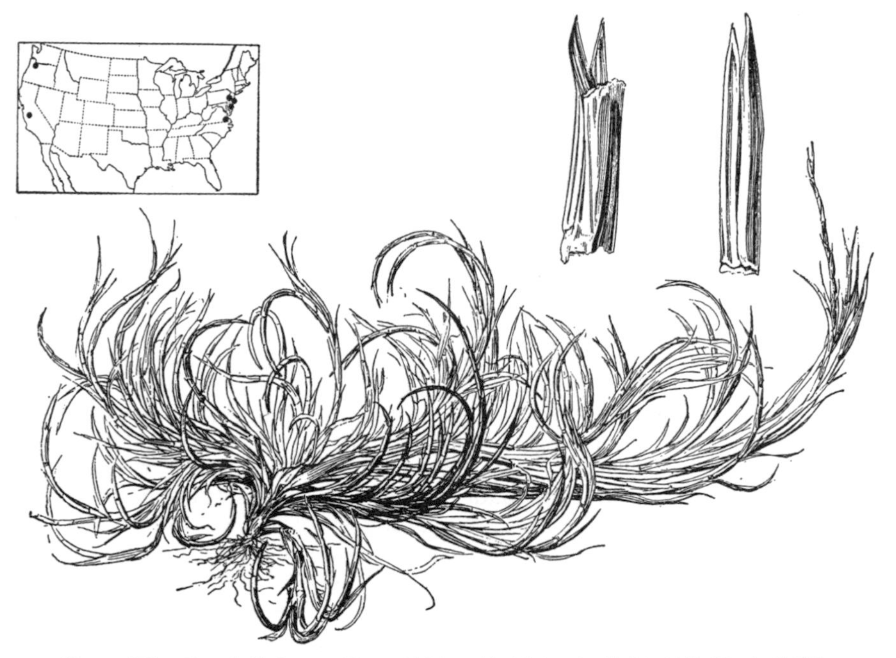

FIGURE 374.—*Parapholis incurva.* Plant, × ½; rachis joint and spikelet, × 5. (Trask, Calif.)

**1. Parapholis incúrva** (L.) C. E. Hubb. SICKLE GRASS. (Fig. 374.) Culms tufted, decumbent at base, 10 to 20 cm. tall; blades short, narrow; spike 7 to 10 cm. long, cylindric, curved; spikelets 7 mm. long, pointed. ⊙ (*Pholiurus incurvus* (L.) Schinz and Thell.)—Mud flats and salt marshes along the coast, New Jersey and Pennsylvania to Virginia; California; Portland, Oreg.; introduced from Europe.

## 53. SCRIBNÉRIA Hack.

Spikelets 1-flowered, solitary, laterally compressed, appressed flatwise against the somewhat thickened continuous rachis, the rachilla disarticulating above the glumes, prolonged as a very minute hairy stipe; glumes equal, narrow, firm, acute, keeled on the outer nerves, the first 2-nerved, the second 4-nerved; lemma shorter than the glumes, membranaceous, obscurely nerved, the apex short-bifid, the faint midnerve extending as a slender awn; palea about as long as

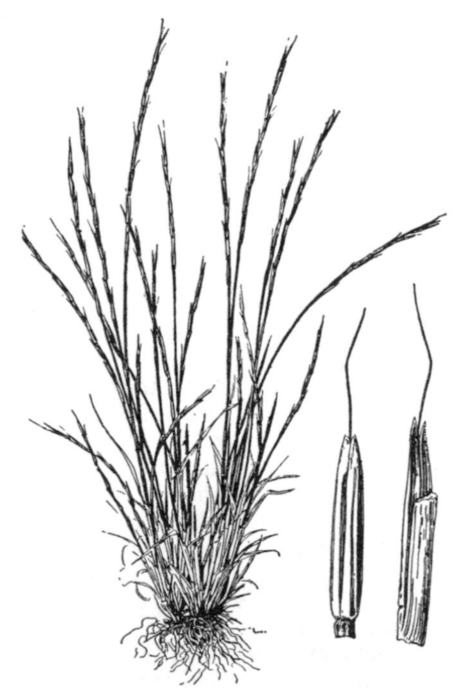

FIGURE 375.—*Scribneria bolanderi.* Plant, × ½; rachis joint and spikelet, × 5. (Suksdorf 217, Wash.)

the lemma; stamen 1. Low annual, with slender cylindric spikes. Type species, *Scribneria bolanderi.* Named for F. Lamson-Scribner.

**1. Scribneria bolandéri** (Thurb.) Hack. (Fig. 375.) Culms branching at base, erect or ascending, 7 to 30 cm. tall; foliage scant, the blades subfiliform; ligule about 3 mm. long; spike about 1 mm. thick, usually one-third to half the entire height of the plant, the internodes 4 to 6 mm. long; spikelets about 7 mm. long; lemmas pubescent at base, the awn erect, 2 to 4 mm. long. ⊙ —Sandy or sterile ground, in the mountains, Washington to California; rare or overlooked, very inconspicuous.

## TRIBE 4. AVENEAE

### 54. SCHÍSMUS Beauv.

Spikelets several-flowered, the rachilla disarticulating above the glumes and between the florets; glumes subequal, longer than the first floret, usually as long as the spikelet, with white membranaceous margins; lemmas broad, rounded on the back, several-nerved, pilose along the lower part of the margin, the summit hyaline, bidentate; palea broad, hyaline, the nerves at the margin. Low tufted annuals with filiform blades and small panicles, the slender pedicels finally disarticulating at the base and falling with the spikelet or with the glumes. Type species, *Schismus marginatus* Beauv. (*S. barbatus*). Name from Greek, *schismos*, a splitting, referring to the bidentate lemmas. This genus has usually been placed in the tribe Festuceae, but its characters place it more naturally in the tribe Aveneae.

Glumes 4 to 5 mm. long; lemmas about 2 mm. long, rounded and emarginate at apex; palea rounded, as long as the lemma ........ 1. S. BARBATUS.
Glumes 5 to 6 mm. long; lemmas 2.5 to 3 mm. long, the apex with 2 acute hyaline lobes; palea acute, shorter than the lemma ........ 2. S. ARABICUS.

**1. Schismus barbátus** (L.) Thell. (Fig. 376.) Culms tufted, erect to prostrate-spreading, 5 to 35 cm. tall; blades usually less than 10 cm. long; panicle oval to linear, 1 to 5 cm. long, usually rather dense, pale or purplish; spikelets about 5-flowered; glumes 4 to 5 mm. long, shorter than the spikelet, 5- to 7-nerved, acute; lemmas about 2 mm. long, 9-nerved, the margin appressed-pilose on the lower half, the teeth minute, sometimes with a mucro between, the rachilla joints slender, flexuous; palea concave, as broad as the lemma and about as long. ⊙ —Open ground in yards, along roadsides, and in dry river beds; Utah to California and southern Arizona; Argentina, Chile. Introduced from the Mediterranean region; India to South Africa.

FIGURE 376.—*Schismus barbatus.* Plant, × ½; spikelet and florets, × 5. (Peebles and Harrison 846, Ariz.)

**2. Schismus arábicus** Nees. (Fig. 377.) Resembling *S. barbatus*, culms widely spreading, the spikelets a little larger, 5- to 7-flowered; lemmas 2.5 to

3 mm. long, longer pilose on the margins and back, the apex cleft into 2 acute lobes, the acute palea reaching the base of the cleft or a little longer. ⊙ —Dry open ground, southern Arizona, Nevada (Clark County), and California; Chile; introduced from southwestern Asia or Africa. Locally dominant in Maricopa County, Ariz., and an excellent forage grass in winter; apparently spreading rapidly.

FIGURE 377.—*Schismus arabicus.* Spikelet, × 10; florets, × 5. (Peebles 9098, Ariz.)

## 55. KOELÉRIA Pers.

Spikelets 2- to 4-flowered, compressed, the rachilla disarticulating above the glumes and between the florets, prolonged beyond the perfect florets as a slender bristle or bearing a reduced floret at the tip; glumes usually about equal in length, unlike in shape, the first narrow, sometimes shorter, 1-nerved, the second wider than the first, broadened above the middle, 3- to 5-nerved; lemmas somewhat scarious, shining, the lowermost a little longer than the glume, obscurely 5-nerved, acute or short-awned, the awn, if present, borne just below the apex. Slender, low or rather tall annuals or perennials, with narrow blades and shining spikelike panicles. Type species, *Koeleria cristata.* Named for G. L. Koeler.

*Koeleria cristata* is a good forage grass and is a constituent of much of the native pasture throughout the Western States. The plants, however, are rather scattering.

Plants perennial........................................ 1. K. CRISTATA.
Plants annual........................................ 2. K. PHLEOIDES.

**1. Koeleria cristáta** (L.) Pers. JUNEGRASS. (Fig. 378.) Tufted perennial; culms erect, puberulent below the panicle, 30 to 60 cm. tall; sheaths, at least the lower, pubescent; blades flat or involute, glabrous or, especially the lower, pubescent, 1 to 3 mm. wide; panicle erect, spikelike, dense (loose in anthesis), often lobed, interrupted, or sometimes branched below, 4 to 15 cm. long, tapering at the summit; spikelets mostly 4 to 5 mm. long; glumes and lemmas scaberulous, 3 to 4 mm. long, sometimes short-awned, the rachilla joints very short. ♃ —Prairie, open woods, and sandy soil, Ontario to British Columbia, south to Delaware, Missouri, Louisiana, California, and Mexico; widely distributed in the temperate regions of the Old World. Variable; several American varieties have been proposed, but the forms are inconstant and intergrading, and it is not practicable to distinguish definite varieties. On the Pacific coast there is a rather large loosely tufted form (*K. cristata* var. *longifolia* Vasey) with long narrow or involute blades and somewhat open panicle.

**2. Koeleria phleoídes** (Vill.) Pers. (Fig. 379.) Annual; culms 15 to 30 cm. tall, smooth throughout; sheaths

FIGURE 378.—*Koeleria cristata*. Plant, × ½; glumes and floret, × 10. (Bebb 2862, Ill.)

and blades sparsely pilose; panicle dense, spikelike, 2 to 7 cm. long, obtuse; spikelets 2 to 4 mm. long; glumes acute; lemmas short-awned from a bifid apex; glumes and lemmas in the typical form papillose-hirsute on the back, but commonly papillose only. ⊙ —Introduced from Europe at Pensacola, Fla., Mobile, Ala., Cameron County, Tex., Portland, Oreg., and at several points in California. Cultivated in nursery plots at Beltsville, Md., and Tucson, Ariz.

## 56. SPHENÓPHOLIS Scribn. WEDGEGRASS

Spikelets 2- or 3-flowered, the pedicel disarticulating below the glumes, the rachilla produced beyond the upper floret as a slender bristle; glumes unlike in shape, the first narrow, usually acute, 1-nerved, the second broadly obovate, 3- to 5-nerved, the nerves sometimes obscure, mostly somewhat coriaceous, the margin scarious; lemmas firm, scarcely nerved, awnless or rarely with an awn from just below the apex, the first a little shorter or a little longer than the second glume; palea hyaline, exposed. Slender perennials (rarely annual) with usually flat blades and narrow shining panicles. Type species, *Sphenopholis obtusata*. Name from Greek *sphen*, wedge, and *pholis*, horny scale, alluding to the hard obovate second glume.

All the species are forage grasses but are usually not abundant. The most important are *S. intermedia* and *S. obtusata*.

Panicle dense, usually spikelike, erect or nearly so; second glume subcucullate. 1. S. OBTUSATA.
Panicle not dense, lax, nodding, from very slender to many-flowered, but not spikelike.
  Spikelets awned ........ 6. S. PALLENS.
  Spikelets awnless (rarely awned in *S. filiformis*).
    Lemmas glabrous; second glume acute or subacute; panicle many-flowered.
      Second glume about 2.5 mm. long ........ 2. S. INTERMEDIA.
      Second glume about 3.5 mm. long ........ 3. S. LONGIFLORA.
    Lemmas scabrous; second glume broadly rounded at the summit; panicle relatively few-flowered.
      Blades rarely more than 10 cm. long, flat, 2 to 5 mm. wide ........ 4. S. NITIDA.
      Blades elongate, flat to subinvolute, mostly less than 2 mm. wide.... 5. S. FILIFORMIS.

**1. Sphenopholis obtusáta** (Michx.) Scribn. PRAIRIE WEDGEGRASS. (Fig. 380.) Culms erect, tufted, 30 to 100 cm. tall; sheaths glabrous to finely retrorsely pubescent; blades flat, glabrous, scabrous, or pubescent, mostly 2 to 5 mm. wide; panicle erect or nearly so, dense, spikelike to interrupted or lobed, rarely slightly looser, 5 to 20 cm. long; spikelets 2.5 to 3.5 mm. long, the two florets closer together than in the other species; second glume very broad, subcucullate, somewhat inflated at maturity, 5-nerved, scabrous; lemmas minutely papillose, rarely mucronate or with a short straight awn, the first about 2.5 mm. long. ♃ —Open woods, old fields, moist ground, and prairies, Maine to British Columbia, south to Florida, Arizona, and California; Mexico; Dominican Republic. Variable in size and in denseness of panicle. Sometimes annual or flowering the first season. Specimens with less dense and lobed panicles may be distinguished from denser panicled specimens of *S. intermedia* by the broader, firmer, subcucullate second glume and more approximate florets.

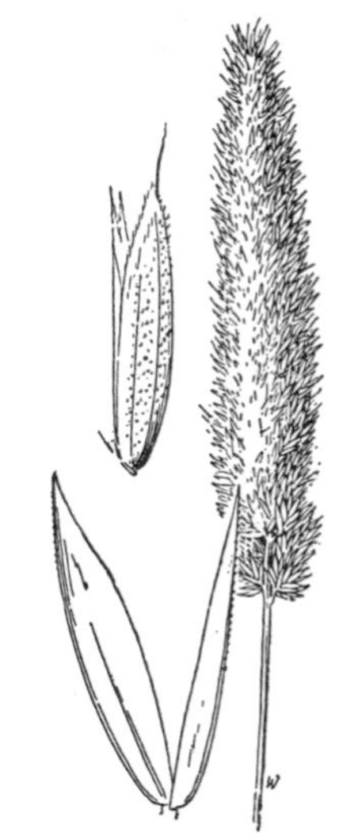

FIGURE 379.—*Koeleria phleoides*. Panicle, × 1; glumes and floret, × 10. (Heller 11417, Calif.)

**2. Sphenopholis intermédia** (Rydb.) Rydb. SLENDER WEDGEGRASS. (Fig. 381.) Culms erect in small tufts, 30 to 120 cm. tall; sheaths glabrous or pubescent; blades flat, often elongate, lax, mostly 2 to 6 mm. wide, sometimes wider, mostly sca-

FIGURE 380.—*Sphenopholis obtusata*. Plant, × ½; glumes and floret, × 10. (Hitchcock 1453, N. C.)

berulous, occasionally sparsely pilose; panicle nodding, from rather dense to open, mostly 10 to 20 cm. long, the branches spikelet-bearing from base; spikelets 3 to 4 mm. long; second glume relatively thin, acute or subacute, about 2.5 mm. long; lemmas subacute, rarely mucronate, smooth or rarely very minutely roughened, mostly 2.5 to 3 mm. long. ♃ —

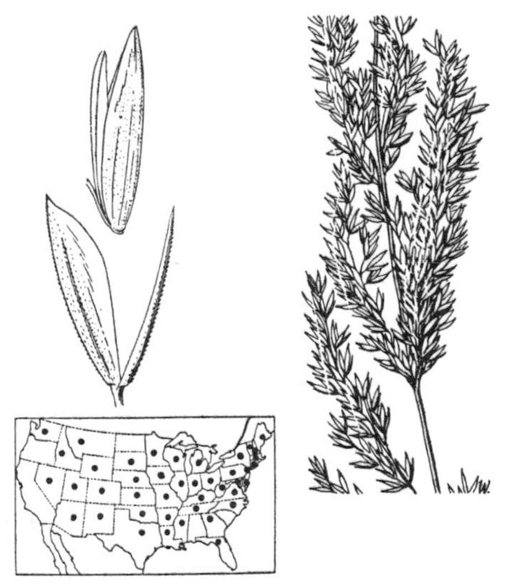

FIGURE 381.—*Sphenopholis intermedia.* Panicle, × 1; glumes and floret, × 10. (Clark 1785, Ind.)

Damp or rocky woods, slopes, and moist places, Newfoundland to British Columbia, south to Florida and Arizona; Tanana Hot Springs, Alaska. Delicate plants with small panicles resembling *S. nitida* may be distinguished by the very narrow first glume, the acute to subacute second glume and lemmas, and usually by the glabrous foliage. Plants with rather dense panicles resembling *S. obtusata* may be distinguished by the thinner, less rounded, more compressed second glume. This is the species called *Sphenopholis pallens* (Spreng.) Scribn. in some manuals. Bieler's description of *Aira pallens* shows that Scribner misapplied the name (see no. 6).

**3. Sphenopholis longiflóra** (Vasey) Hitchc. (Fig. 382.) Culms relatively stout, erect from a decumbent base, 40 to 70 cm. tall; lower sheaths puberulent, the others glabrous; blades thin, flat, scaberulous, 5 to 18 cm. long, 3 to 8 mm. wide; panicle many-flowered, rather loose, slightly nodding, 10 to 18 cm. long; spikelets mostly 2-flowered, the rachilla hispidulous; glumes very scabrous on the green part, the second thin, acute,

FIGURE 382.—*Sphenopholis longiflora.* Panicle, × 1; glumes and floret, × 10. (Nealley, Tex.)

about 3.5 mm. long; lemmas smooth, scaberulous toward the tip, the first about 4 mm. long. ♃ —Wooded banks, Arkansas and Texas. Differing from *S. intermedia* in the larger spikelets, broader blades, and more tapering lemmas.

**4. Sphenopholis nítida** (Bieler) Scribn. (Fig. 383.) Culms tufted, leafy at base, slender, shining, 30 to 70 cm. tall; sheaths and blades mostly softly pubescent, occasionally glabrous, the blades 2 to 5 mm. wide, 3 to 10 cm. long, the basal sometimes longer; panicle rather few-flowered, mostly 8 to 12 cm. long, the filiform branches distant, ascending, spreading in anthesis; spikelets 3 to 3.5 mm. long; glumes about equal in length, usually nearly as long as the first floret, the first glume broader than in the other species, the second broadly rounded at summit, at least the second lemma scabrous-papillose. ♃ —Dry or rocky woods, Massachusetts to North Dakota, south to Florida and Texas.

**5. Sphenopholis filifórmis** (Chapm.) Scribn. (Fig. 384.) Culms erect, very

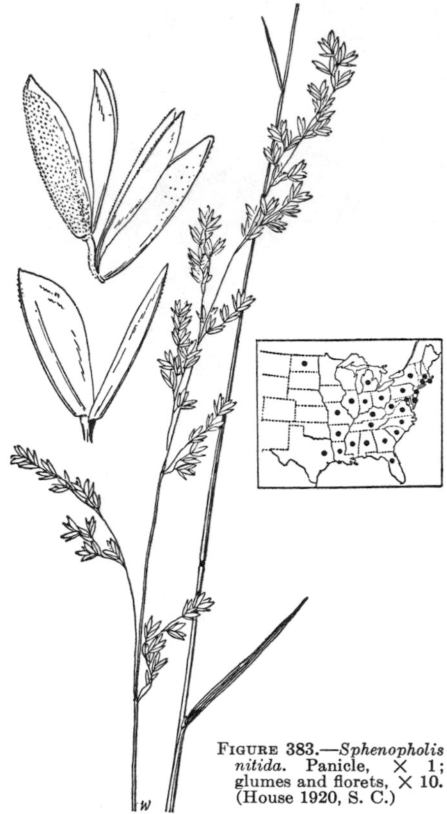
FIGURE 383.—*Sphenopholis nitida*. Panicle, × 1; glumes and florets, × 10. (House 1920, S. C.)

slender, 30 to 60 cm. tall; blades lax, flat to subinvolute, mostly less than 2 mm. wide; panicle slender, often nodding, 5 to 15 cm. long, the short branches rather distant, erect or ascending; spikelets 3 to 4 mm. long, the 2 florets rather distant; second glume broadly rounded at summit, about 2 mm. long; lemmas obtuse to subacute, rarely with a short spreading awn; the first smooth, the second minutely roughened. ♃ —Dry soil, Coastal Plain, southeastern Virginia to Florida, Tennessee, and eastern Texas. Awned lemmas, either the first or second, are occasionally found in some panicles.

**6. Sphenopholis pállens** (Bieler) Scribn. (Fig. 385.) Culms erect, about 60 cm. tall; lower sheaths minutely pubescent, the upper glabrous; blades flat, glabrous, 1 to 2 mm. wide; panicle narrow, nodding, loose or somewhat compact, 15 to 25 cm. long, the branches ascending, the lower distant; spikelets 2- or 3-flowered, 3 to 3.5 mm. long; second floret scaberulous, usually awned just below the apex, the awn scabrous, geniculate, 1 to 2 mm. long. ♃ (*Eatonia aristata* Scribn. and Merr.)—Rich wooded slopes, Southampton County, Va., to South Carolina. The type of *Aira pallens* Bieler has not been examined, but it was received from

FIGURE 384.—*Sphenopholis filiformis*. Panicle, × 1; glumes and florets, × 10. (Hitchcock 1044, Ala.)

FIGURE 385.—*Sphenopholis pallens*. Spikelet, × 10. (Curtiss, S. C.)

Muhlenberg and may be assumed to be the same as the specimen in the Muhlenberg Herbarium described under *Aira pallens* by Muhlenberg.

## 57. TRISÉTUM Pers. TRISETUM

Spikelets usually 2-flowered, sometimes 3- to 5-flowered, the rachilla prolonged behind the upper floret, usually villous; glumes somewhat unequal, acute, the second usually longer than the first floret; lemmas usually short-bearded at base, 2-toothed at apex, the teeth often awned, bearing from the back below the cleft apex a straight and included or usually bent and exserted awn (awnless or nearly so in *Trisetum melicoides* and *T. wolfii*). Tufted perennials (except *Trisetum interruptum*), with flat blades and open or usually contracted or spikelike shining panicles. Type species, *Trisetum flavescens.* Name from Latin *tri,* three, and *setum,* bristle, alluding to the awn and two teeth of the lemma.

Several of the species are valuable for grazing. *Trisetum spicatum* constitutes an important part of the forage on alpine and subalpine slopes and *T. wolfi* at medium altitudes.

Spikelets disarticulating below the glumes.
  Plants perennial; panicle lax, somewhat open .......... 9. T. PENNSYLVANICUM.
  Plants annual; panicle narrow, dense, interrupted .......... 10. T. INTERRUPTUM.
Spikelets disarticulating above the glumes.
  Awn included within the glumes, or wanting.
    Panicle rather lax, nodding .......... 1. T. MELICOIDES.
    Panicle rather dense, erect .......... 2. T. WOLFII.
  Awn exserted.
    Awn straight (see also *T. montanum* var. *shearii*) .......... 3. T. ORTHOCHAETUM.
    Awn geniculate.
      Panicle dense, spikelike, sometimes slightly interrupted below; plants densely tufted .......... 5. T. SPICATUM.
      Panicle loose and open to contracted, but not spikelike; plants in small tufts or solitary.
        Panicle relatively few-flowered, loose, lax or drooping, the filiform branches naked below; florets distant .......... 4. T. CERNUUM.
        Panicle many-flowered, from rather loose to dense and interrupted; florets not distant.
          Panicle yellowish; spikelets mostly 3- or 4-flowered; introduced. .......... 8. T. FLAVESCENS.
          Panicle pale green, sometimes purplish-tinged; spikelets usually 2-flowered.
            Spikelets about 8 mm. long .......... 6. T. CANESCENS.
            Spikelets 5 to 6 mm. long .......... 7. T. MONTANUM.

**1. Trisetum melicoídes** (Michx.) Scribn. (Fig. 386.) Culms 50 to 100 cm. tall; sheaths pubescent or scabrous; blades 2 to 8 mm. wide, scabrous, sometimes pubescent on the upper surface; panicle somewhat open, nodding, 10 to 20 cm. long, the branches slender, ascending, lax or drooping, as much as 7 cm. long, rather closely flowered above the middle; spikelets scaberulous, 6 to 7 mm. long; glumes 4 to 6 mm. long, the second longer and broader; lemmas acute, 5 to 6 mm. long, rarely with a minute awn just below the tip, the rachilla and callus hairs 1 to 2 mm. long. ♃ —River banks, lake shores, mostly in gravelly ground,

FIGURE 386.—*Trisetum melicoides.* Panicle, × 1; glumes and floret, × 5. (Pringle, Vt.)

Newfoundland to Vermont, Michigan, and Wisconsin.

FIGURE 387.—*Trisetum wolfii*. Panicle, × 1; glumes and floret, × 5. (Swallen 809, Calif.)

**2. Trisetum wólfii** Vasey. WOLFS TRISETUM. (Fig. 387.) Culms erect, 50 to 100 cm. tall, loosely tufted, sometimes with short rhizomes; sheaths scabrous, rarely the lower pilose; blades flat, scabrous, rarely pilose on the upper surface, 2 to 4 mm. wide; panicle erect, rather dense but scarcely spikelike, green or pale, sometimes a little purplish, 8 to 15 cm. long; spikelets 5 to 7 mm. long, 2-flowered, sometimes 3-flowered; glumes nearly equal, acuminate, about 5 mm. long; lemmas obtusish, scaberulous, 4 to 5 mm. long, awnless or with a minute awn below the tip, the callus hairs scant, about 0.5 mm. long, the rachilla internode about 2 mm. long, rather sparingly long-villous. ♃ —Meadows and moist ground, at medium altitudes in the mountains, Montana to Washington, south to New Mexico and California.

**3. Trisetum orthochaétum** Hitchc. (Fig. 388.) Culms solitary, erect, slender, 110 cm. tall; sheaths glabrous; blades flat, scabrous, 8 to 20 cm. long, 3 to 7 mm. wide; panicle slightly nodding, lax, pale, about 18 cm. long, the filiform branches loosely ascending, naked below, the lower fascicled, as much as 8 cm. long; spikelets short-pediceled, somewhat appressed, mostly 3-flowered, 8 to 9 mm. long excluding awns, the rachilla appressed-silky; glumes acuminate, about 6 mm. long, the second wider; lemmas rounded on the back, minutely scaberulous on the upper part, obscurely 5-nerved, the callus short-pilose, the apex acute, erose-toothed, awned about 2 mm. below the tip, the awn straight or nearly so, exceeding the lemma about 3 mm. ♃ —Known only from boggy meadows, Lolo Hot Springs, Bitterroot Mountains, Mont.

FIGURE 388.—*Trisetum orthochaetum*. Panicle, × 1; glumes and floret, × 5. (Type.)

**4. Trisetum cérnuum** Trin. NODDING TRISETUM. (Fig. 389.) Culms rather lax, 60 to 120 cm. tall; sheaths glabrous to sparsely pilose; blades thin, flat, lax, scabrous, 6 to 12 mm. wide; panicle open, lax, drooping, 15 to 30 cm. long, the branches verticillate, filiform, flexuous, spikelet-bearing toward the ends; spikelets 6 to 12 mm. long, with usually 3 distant florets, the first longer than the second glume; first glume narrow, acuminate, 1-nerved, 0.5 to 2 mm. long, the second broad, 3-nerved, 3 to 4 mm. long, occasionally reduced; lemma 5 to 6 mm. long, the teeth setaceous, the hairs of the callus 0.5 to 1 mm. long, of the rachilla as much as 2 mm. long, the awns slender, curved, flexuous or loosely spiral, mostly 5 to 10 mm. long, attached 1 to 2 mm. below tip. ♃ —Moist woods, Alberta to southeastern Alaska, south to western Montana and northern California.

**5. Trisetum spicátum** (L.) Richt. SPIKE TRISETUM. (Fig. 390.) Culms densely tufted, erect, 15 to 50 cm. tall, glabrous to puberulent; sheaths and usually the blades puberulent; panicle dense, usually spikelike, often interrupted at base, pale or often dark purple, 5 to 15 cm. long; spikelets 4 to 6 mm. long; glumes somewhat unequal in length, glabrous or scabrous except the keels, or sometimes pilose, the first narrow, acuminate, 1-nerved, the second broader, acute, 3-nerved; lemmas scaberulous, 5 mm. long, the first longer than the glumes, the teeth setaceous; awn attached about one-third below the tip, 5 to 6 mm. long, geniculate, exserted. ♃ —Alpine meadows and slopes, Arctic America, southward to Connecticut, Pennsylvania, northern Michigan and Minnesota, in the mountains to New Mexico and California; also on Roan Mountain, N. C.; high mountains through Mexico to the Antarctic regions of South America; Arctic and alpine regions of the Old World. In northern regions the species descends to low altitudes. Exceedingly variable; several varieties have been proposed, but the characters used to differentiate them are variable and are not correlated. Two rather more outstanding varieties, both intergrading with the species are: *T. spicatum* var. *molle* (Michx.) Beal, with densely pubescent foliage, and *T. spicatum* var. *congdoni* (Scribn. and Merr.) Hitchc., a nearly glabrous alpine form with slightly larger spikelets.

FIGURE 389.—*Trisetum cernuum.* Panicle, × 1; glumes and floret, × 5. (Elmer 1946, Wash.)

**6. Trisetum canéscens** Buckl. TALL TRISETUM. (Fig. 391.) Culms erect, or decumbent at base, 60 to 120 cm. tall; sheaths, at least the lower, sparsely to densely and softly retrorse-pilose, rarely scabrous only; blades flat, scabrous or canescent, sometimes sparsely pilose, mostly 2 to 7 mm. wide; panicle narrow, usually loose, sometimes interrupted and spikelike, 10 to 25 cm. long; spikelets about 8 mm. long, 2- or 3-flowered, the florets not so distant as in *T. cernuum;* glumes smooth, except the keel, the first narrow, acuminate, the second broad, acute, 3-nerved, 5 to 7 mm. long; lemmas rather firm, scaberulous, the upper exceeding the glumes, 5 to 6 mm. long, the teeth aristate, the callus hairs rather scant, the rachilla hairs copious; awn geniculate, spreading, loosely twisted below, attached one-third below the tip, usually about 12 mm. long. ♃ —Mountain meadows, moist ravines and along streams, Montana to British Columbia, south to central California. Plants with less pubescent sheaths and looser panicles resemble *T. cernuum* but in that the spikelets are commonly 3-flowered, the florets distant. Plants with more velvety foliage and narrow panicles with short densely flowered

FIGURE 390.—*Trisetum spicatum*. Plant, × ½; spikelet and floret, × 5. (Rydberg and Bessey 3593, Mont.)

branches, the lower in distant fascicles, have been differentiated as *T. projectum* Louis-Marie. Intergrading specimens are more numerous than the extreme described.

FIGURE 391.—*Trisetum canescens*. Panicle, × 1; glumes and floret, × 5. (Hitchcock 3409, Calif.)

FIGURE 392.—*Trisetum montanum*. Panicle, × 1; glumes and floret, × 5. (Type.)

**7. Trisetum montánum** Vasey. (Fig. 392.) Resembling *T. canescens*, on the average smaller, the blades narrower; sheaths from nearly glabrous to softly retrorsely pubescent; panicles smaller than usual in *T. canescens*, more uniformly rather dense, often purple-tinged; spikelets 5 to 6 mm. long, the glumes and lemmas thinner than in *T. canescens*, the awn more delicate, 5 to 8 mm. long. ♃ —Mountain meadows, gulches and moist places on mountain slopes, between 2,000 and 3,300 m., Colorado, Utah, New Mexico, and Arizona. A form with purplish panicles and erect awns only 2 to 3 mm. long, known from a single collection near Silverton, Colo., has been differentiated as *T. montanum* var. *shearii* Louis-Marie.

**8. Trisetum flavéscens** (L.) Beauv. (Fig. 393.) Resembling *T. canescens;* sheaths glabrous or the lower sparsely pilose; panicle usually yellowish, many-flowered, somewhat condensed; spikelets mostly 3- or 4-flowered; lemmas 4 to 6 mm. long. ♃ —Waste places, Vermont, New York, Missouri, Colorado, Washington, California, and probably other States; introduced from Europe.

**Trisetum aúreum** (Ten.) Ten. Annual; culms 10 to 20 cm. tall; panicle ovate, contracted, 2 to 3 cm. long; spikelets 3 mm. long; awns 2 to 3 mm. long. ⊙ —Ballast, Camden, N. J.; Europe.

**9. Trisetum pennsylvánicum** (L.) Beauv. ex Roem. and Schult. (Fig. 394.) Culms slender, weak, usually subgeniculate at base, 50 to 100 cm. tall; sheaths glabrous or rarely scabrous; blades flat, scabrous, 2 to 5 mm. wide; panicle narrow, loose, nodding, 10 to 20 cm. long; pedicels disarticulating about the middle or toward the base; spikelets 5 to 7 mm.

FIGURE 393.—*Trisetum flavescens*. Panicle, × 1; floret, × 5. (Grant 26, Wash.)

FIGURE 394.—*Trisetum pennsylvanicum.* Panicle, × 1; glumes and florets, × 5. (Heller 4800, Pa.)

FIGURE 395.—*Trisetum interruptum.* Panicle, × 1; glumes and floret, × 5. (Jermy, Tex.)

long, 2-flowered, the long rachilla internodes slightly hairy; glumes mostly 4 to 5 mm. long, acute, the second wider; lemmas acuminate, the first usually awnless, the second awned below the 2 setaceous teeth, the awn horizontally spreading, 4 to 5 mm. long. ♃ —Swamps and wet places, Massachusetts to Ohio and West Virginia, south on the Coastal Plain to Florida and west to Tennessee and Louisiana.

**10. Trisetum interrúptum** Buckl. (Fig. 395.) Annual; culms tufted, sometimes branching, erect or spreading, 10 to 40 cm. tall; sheaths often scabrous or pubescent; blades flat, sometimes pubescent, 1 to 4 mm. wide, mostly 3 to 10 cm. long; panicle narrow, interrupted, from slender to rather dense but scarcely spikelike, 5 to 12 cm. long, sometimes with smaller axillary panicles; pedicels disarticulating a short distance below the summit; spikelets about 5 mm. long, 2-flowered, the second floret sometimes rudimentary; glumes about equal in length, acute, 4 to 5 mm. long, the first 3-nerved, the second a little broader, 5-nerved; lemmas acuminate with 2 setaceous teeth, the awns attached above the middle, flexuous, 4 to 8 mm. long, that of the first lemma often shorter and straight. ⊙ —Open dry ground, Texas to Colorado and Arizona.

## 58. DESCHÂMPSIA Beauv. HAIRGRASS

Spikelets 2-flowered, disarticulating above the glumes and between the florets, the hairy rachilla prolonged beyond the upper floret and sometimes bearing a reduced floret; glumes about equal, acute or acutish, membranaceous; lemmas thin, truncate and 2- to 4-toothed at summit, bearded at base, bearing a slender awn from or below the middle, the awn straight, bent or twisted. Low or moderately tall annuals or usually perennials, with shining pale or purplish spikelets in narrow or open panicles. Standard species, *Deschampsia caespitosa*. Included in *Aira* by some authors. Named for Deschamps.

*Deschampsia caespitosa* is often the dominant grass in mountain meadows, where it furnishes excellent forage.

Plants annual; foliage very scant.......... 1. D. DANTHONIOIDES.
Plants perennial; foliage not scant, one-third to half the entire length of the culm.
  Panicle narrow, the distant branches appressed.
    Glumes 4 to 6 mm. long; lemma smooth, not deeply toothed.......... 2. D. ELONGATA.
    Glumes 7 mm. long; lemma scaberulous, deeply toothed or lacerate.
      3. D. CONGESTIFORMIS.
  Panicle open or contracted, if narrow, not more than one-fourth the length of the culm.
    Blades thin, flat; glumes exceeding the florets.......... 4. D. ATROPURPUREA.

Blades firm or filiform; glumes not exceeding the upper floret.
Blades filiform, flexuous; awn exserted, geniculate, twisted........ 5. D. FLEXUOSA.
Blades flat or folded, stiff; awn included or slightly exserted, straight.
Panicle open, usually nodding or drooping.................................. 6. D. CAESPITOSA.
Panicle narrow, condensed, erect.................................................. 7. D. HOLCIFORMIS.

**1. Deschampsia danthonioídes** (Trin.) Munro ex Benth. ANNUAL HAIRGRASS. (Fig. 396.) Annual; culms slender, erect, 15 to 60 cm. tall; blades few, short, narrow; panicle open, 7 to 25 cm. long, the capillary branches commonly in twos, stiffly ascending, naked below, bearing a few short-pediceled spikelets toward the ends; glumes 4 to 8 mm. long, 3-nerved, acuminate, smooth except the keel, exceeding the florets; lemmas smooth and shining, somewhat indurate, 2 to 3 mm. long, the base of the florets and the rachilla pilose, the awns geniculate, 4 to 6 mm. long. ⊙ —Open ground, Alaska to Montana and Baja California; also Chile.

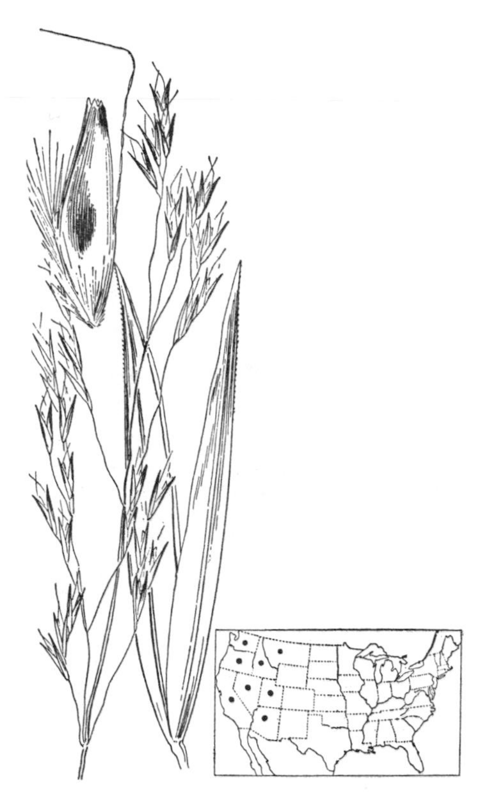

FIGURE 396.—*Deschampsia danthonioides*. Panicle, × 1; glumes and floret, × 10. (Parish 3300, Calif.)

FIGURE 397.—*Deschampsia elongata*. Panicle, × 1 glumes and floret, × 10. (Swallen 780, Calif.)

Variable in the size of the spikelets. A form described from southern California as *D. gracilis* Vasey, with somewhat laxer panicles, the rather more numerous spikelets only 4 to 5 mm. long, grades into the usual form.

**2. Deschampsia elongáta** (Hook.) Munro ex Benth. SLENDER HAIRGRASS. (Fig. 397.) Culms densely tufted, slender, erect, 30 to 120 cm. tall; blades soft, 1 to 1.5 mm. wide, flat or folded, those of the basal tuft filiform; panicle narrow, as much as 30 cm. long, the capillary branches appressed; spikelets on short appressed pedicels; glumes 4 to 6 mm. long, 3-nerved, equaling or slightly exceeding the florets; lemmas 2 to 3 mm. long, similar to those of *D. danthonioides*, the awns shorter, straight. ♃ —Open ground, Alaska to Wyoming, south to Arizona and California; Mexico; Chile.

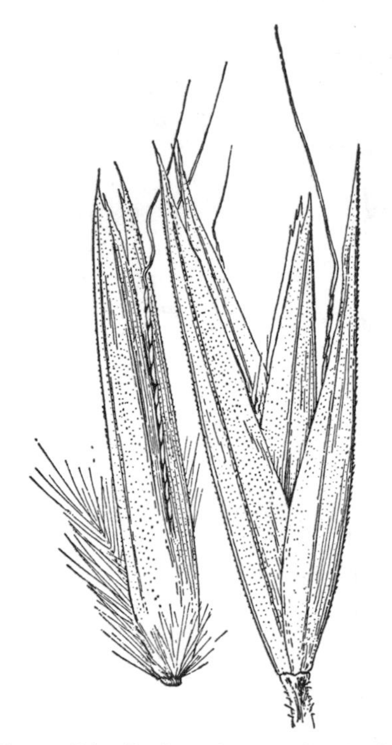

FIGURE 398.—*Deschampsia congestiformis.* Spikelet and floret, × 10. (Type.)

**3. Deschampsia congestifórmis** Booth. (Fig. 398.) Culms in small tufts, slender, 45 to 70 cm. tall, scaberulous above; sheaths scaberulous toward the summit; ligule 1.5 to 3 mm. long; blades flat or folded, scabrous on both surfaces, 2 to 3 mm. wide, the basal 10 to 30 cm. long, those of the culm 3 to 8 cm. long, those of the innovations subfiliform; panicle long-exserted, 6.5 to 10 cm. long, narrow, condensed, the short branches erect, the axis and branches slender, hirtellous; spikelets short-pediceled, appressed, 7 to 10 mm. long; glumes about 7 mm. long, scabrous, especially on the midnerve; lemmas 7 to 8 mm. long, awned from near the base, toothed or lacerate at the apex, sometimes splitting down the back at maturity, the awn twisted and geniculate, exceeding the spikelets 3 to 4 mm., the callus hairs about 0.5 to 1 mm. long, those of the rachilla 1 to 2 mm. long. ♃ —Only known from Gallatin Valley, Bozeman, Gallatin County, and from Cooke, Park County, Mont.

**4. Deschampsia atropurpúrea** (Wahl.) Scheele. MOUNTAIN HAIRGRASS. (Fig. 399.) Culms loosely tufted, erect, purplish at base, 40 to 80 cm. tall; blades flat, rather soft, ascending or appressed, 5 to 10 cm. long, 4 to 6 mm. wide, acute or abruptly acuminate; panicle loose, open, 5 to 10 cm. long, the few capillary drooping branches naked below; spikelets mostly purplish, broad; glumes about 5 mm. long, broad, the second 3-nerved, exceeding the florets; lemmas scabrous, about 2.5 mm. long, the callus hairs one-third to half as long, the awn of the first straight, included, of the second, geniculate, exserted. ♃ —Woods and wet meadows, Newfound-

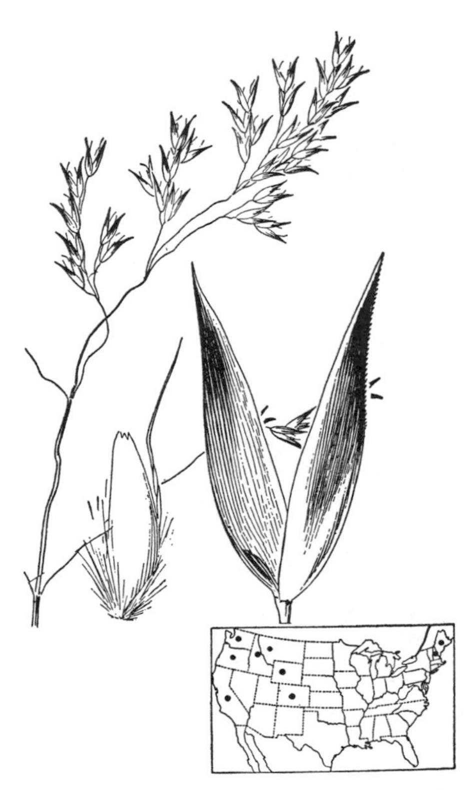

FIGURE 399.—*Deschampsia atropurpurea.* Panicle, × 1; glumes and floret, × 10. (Leiberg 2952, Idaho.)

land and Labrador to Alaska, south to the White Mountains of New Hampshire; Colorado and California; northern Eurasia.

**5. Deschampsia flexuósa** (L.) Trin. CRINKLED HAIRGRASS. (Fig. 400.) Culms densely tufted, erect, slender, 30 to 80 cm. tall; leaves mostly in a basal tuft, numerous, the sheaths scabrous, the blades involute, slender or setaceous, flexuous; panicle loose, open, nodding, 5 to 12 cm. long, the capillary branches naked below, the branchlets spikelet-bearing toward the ends; spikelets 4 to 5 mm. long, purplish or bronze, the florets approximate; glumes 1-nerved, acute, shorter than the florets; lemmas scabrous, the callus hairs about 1 mm. long, the awn attached near the base, geniculate, twisted, 5 to 7 mm. long. ♃ —Dry or rocky woods, slopes, and open ground, Greenland to Alaska, south to Georgia, Michigan, and Wisconsin; Arkansas and Oklahoma (Le Flore County); Mexico; Eurasia. A form with yellow-striped foliage (called by gardeners *Aira foliis variegatis*) is occasionally grown for ornament.

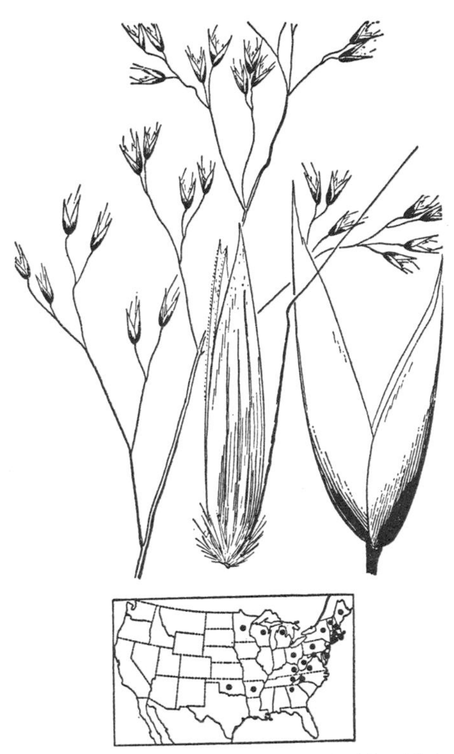

FIGURE 400.—*Deschampsia flexuosa*. Panicle, × 1; glumes and floret, × 10. (Hitchcock 16059, N. H.)

**6. Deschampsia caespitósa** (L.) Beauv. TUFTED HAIRGRASS. (Fig. 401.) Culms in dense tufts, leafy at base, erect, 60 to 120 cm. tall; sheaths smooth; blades 1.5 to 4 mm. wide, often elongate, rather firm, flat or folded, scabrous above; panicle loose, open, nodding, 10 to 25 cm. long, the capillary scabrous branches and branchlets spikelet-bearing toward the ends; spikelets 4 to 5 mm. long, pale or purple-tinged, the florets distant, the rachilla internode half the length of the lower floret; glumes 1-nerved or the second obscurely 3-nerved, acute, about as long as the florets; lemmas smooth, the callus hairs short; awn from near the base, from straight and included in the glumes to weakly geniculate and twice as long as the spikelet. ♃ —Bogs and wet places, Greenland to Alaska, south to New Jersey, West Virginia, North Carolina, Illinois, North Dakota, New Mexico, and California; Arctic and temperate regions of the Old World. Variable in size, in width and texture of blades, in shape of the panicle, and in length of awn. The forms which have been segregated as species and varieties are inconstant, and the characters used to distinguish them are not correlated. Rarely with proliferous spikelets. Large plants from Oregon and California have been described under *Deschampsia caespitosa* subsp. *beringensis* (Hultén) Lawr., but are not *D. beringensis* Hultén, of the Aleutians. Tall plants, with long flat blades, elongate panicles, and spikelets, 3 to 4 mm. long, found in Connecticut, have been referred to D. CAESPITOSA var. PARVIFLÓRA (Thuill.) Coss. and Germ. They agree with

FIGURE 401.—*Deschampsia caespitosa*. Plant, × ½; glumes and floret, × 10. (Nelson 3623, Wyo.)

panicle 10 to 25 cm. long, condensed, many-flowered, the branches appressed to subflexuous-ascending, purplish to brownish; spikelets 6 to 8 mm. long; glumes and lemmas scaberulous, the glumes about equaling the spikelets or shorter, 3-nerved, the lateral nerves of the first often obscure; lemmas awned from below the middle, the awns erect, exceeding the spikelet, the callus hairs short. ♃ —Marshes and sandy soil near the coast, Vancouver Island to central California.

FIGURE 402.—*Deschampsia holciformis*. Panicle, × 1; glumes and floret, × 10. (Bolander, Calif.)

FIGURE 403.—*Aira praecox*. Panicle, × 1; glumes and floret, × 10. (Amer. Gr. Natl. Herb. 375, Del.)

specimens from Germany and are probably introduced.

7. **Deschampsia holcifórmis** Presl. (Fig. 402.) Culms in dense tufts with numerous basal leaves, erect, relatively robust, 50 to 125 cm. tall; blades mostly folded, 20 to 50 cm. long, 2 to 4 mm. wide, rather firm;

## 59. AÍRA L.

(*Aspris* Adans.)

Spikelets 2-flowered, disarticulating above the glumes, the rachilla not prolonged; glumes boat-shaped, about equal, 1-nerved or obscurely 3-nerved, acute, membranaceous or subscarious; lemmas firm, rounded on the back, tapering into 2 slender teeth, bearing on the back below the middle a slender geniculate, twisted, usually exserted, awn, this sometimes wanting in the lower floret or reduced; callus minutely bearded. Delicate annuals with lax, subfiliform blades and open or contracted panicles of small spikelets. Type species, *Aira praecox*. *Aira*, an old Greek name for a weed, probably darnel. Weedy grasses of no economic importance, introduced from Europe.

Panicle dense, spikelike ........ 1. A. PRAECOX.
Panicle open.
  Lower floret with awn as long as that of the upper floret ........ 2. A. CARYOPHYLLEA.
  Lower floret awnless or nearly so ........ 3. A. ELEGANS.

1. **Aira praécox** L. (Fig. 403.) Culms tufted, 10 to 20 cm. tall, usually erect; panicle narrow, dense, 1 to 3 cm. long; spikelets yellowish, shin-

FIGURE 404.—*Aira caryophyllea*. Plant, × ½; spikelet and floret, × 10. (Heller 3889, Wash.)

ing, 3.5 to 4 mm. long; lemmas with awns 2 to 4 mm. long, that of the lower floret the shorter. ☉ —Sandy open ground, along the coast, New Jersey to Virginia; Vancouver to California.

**2. Aira caryophylléa** L. SILVER HAIRGRASS. (Fig. 404.) Culms solitary or in small tufts, erect, 10 to 30 cm. tall; panicle open, the silvery shining spikelets 3 mm. long, clustered toward the ends of the spreading capillary branches; both lemmas with awns about 4 mm. long. ☉ —Open dry ground, Coastal Plain, Massachusetts to Florida and Louisiana; Ohio; common on the Pacific coast from British Columbia to California; southern South America.

**3. Aira élegans** Willd. ex Gaudin. (Fig. 405.) Resembling *A. caryophyllea;* panicle more diffuse; spikelets 2.5 mm. long, scattered at the ends of the branches; lemma of lower floret awnless or with a minute awn just below

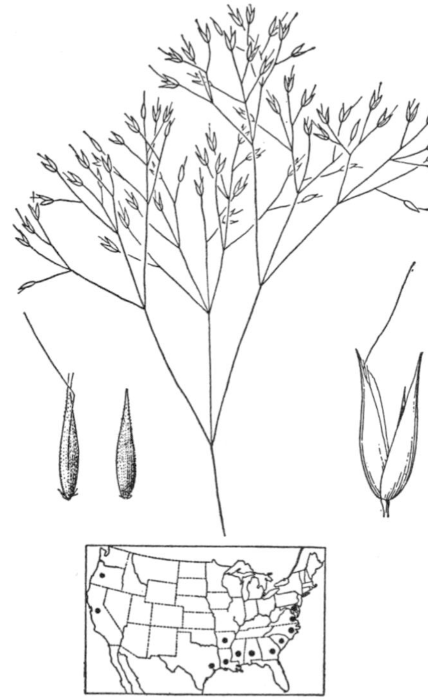

FIGURE 405.—*Aira elegans*. Panicle, × 1; spikelet and florets, × 10. (Davis 2016, S. C.)

the apex, that of the upper floret with an awn 3 mm. long. ☉ (*A. capillaris* Host, not Savi).—Open ground, Coastal Plain, Maryland to Florida; Tennessee; Arkansas and Texas; Oregon and California.

## 60. CORYNÉPHORUS Beauv.

Spikelets 2-flowered, disarticulating above the glumes; glumes nearly equal, 1-nerved, acute, membranaceous; lemmas thin, acute, awned from near the base, the awn jointed about the middle, the joint with a minute ring of hairs, the lower part straight, brown, the upper slender, club-shaped. Slender annuals with subfiliform blades and narrow panicles. Type species *Corynephorus canescens*. Name from Greek *korynephoros*, club-bearing. One species introduced from Europe.

FIGURE 406.—*Corynephorus canescens*. Spikelet and florets, × 10. (Bicknell, Mass.)

**1. Corynephorus canéscens** (L.) Beauv. (Fig. 406.) Culms tufted, 20 to 35 cm. tall, branching and leafy at base; panicle 5 to 10 cm. long, pale or purplish; spikelets about 3.5 mm. long; florets about 1.7 mm. long, faintly nerved, the callus and rachilla softly pilose, the awns equaling or slightly exceeding the glumes. ☉ —Waste ground and ballast, British Columbia. Marthas Vineyard and Long Island, N. Y., New Jersey, and Pennsylvania.

## 61. AVÉNA L. Oats

Spikelets 2- or 3-flowered, the rachilla bearded, disarticulating above the glumes and between the florets; glumes about equal, membranaceous or papery, 7- to 9-nerved, longer than the lower floret, usually exceeding the upper floret; lemmas indurate, except toward the summit, 5- to 9-nerved, bidentate, bearing a dorsal bent and twisted awn (straight and reduced in *Avena sativa*), the awn in age commonly breaking at the bend. Low or moderately tall annuals, with narrow or open, usually rather few-flowered panicles of large spikelets. Type species, *Avena sativa. Avena*, the old Latin name for oats.

The most important species of the genus is *A. sativa*, the familiar cultivated oat. Two other introduced species, *A. fatua* and *A. barbata*, are known as wild oats because of their close resemblance to the cultivated oat. These two species are common on the Pacific coast where they are often utilized for hay. Much of the grain hay of that region is made from either cultivated or wild oats. The varieties of cultivated oat are derived from three species of *Avena*. The common varieties of this country and of temperate and mountain regions in general are derived from *A. fatua*. The Algerian oat grown in North Africa and Italy and the red oat of our Southern States (*A. byzantina* K. Koch) are derived from *A. sterilis*. A few varieties adapted to dry countries are derived from *A. barbata*.

Teeth of lemma setaceous; pedicels curved, capillary .......... 3. A. BARBATA.
Teeth of lemma acute, not setaceous; pedicels stouter.
  Spikelets mostly 2-flowered, the florets not readily separating; awn usually straight or wanting; lemmas glabrous .......... 2. A. SATIVA.
  Spikelets mostly 3-flowered, the florets readily separating; awn stout, geniculate, twisted; lemmas clothed with stiff brown hairs (hairs sometimes white or scant). 1. A. FATUA.

**1. Avena fátua** L. WILD OAT. (Fig. 407, *A*.) Culms 30 to 75 cm. tall, erect, stout; leaves numerous, the blades flat, usually 4 to 8 mm. wide, scabrous; panicle loose and open, the slender branches usually horizontally spreading; spikelets usually 3-flowered; glumes about 2.5 cm. long; rachilla and lower part of the lemma clothed with long stiff brownish, or sometimes whitish, hairs, these sometimes scant; florets readily falling from the glumes; lemmas nerved above, about 2 cm. long, the teeth acuminate, not setaceous; awn stout, geniculate, twisted below, 3 to 4 cm. long. ⊙ —Cultivated soil and waste places; introduced from Europe; rare in the Eastern States; Maine to Pennsylvania, Missouri and westward, a common weed on the Pacific coast. Seed used for food by the Indians.

**Avena stérilis** L. ANIMATED OATS. Resembling *A. fatua*, the spikelets 3.5 to 4.5 cm. long, the awns 5 to 7 cm. long. ⊙ —Sometimes cultivated as a curiosity, occasionally spontaneous. When laid on a moist surface the fruits twist and untwist as the awns lose or absorb moisture. Sometimes used as flies in fishing, the spikelets jerking as the awns untwist.

**2. Avena satíva** L. OAT. (Fig. 407, *B*.) Differing from *A. fatua* in having mostly 2-flowered spikelets, the florets not readily separating from the glumes; lemmas glabrous; awn usually straight, often wanting. ⊙ —Commonly cultivated and occasionally escaped. In *A. nuda* L., NAKED OAT, the caryopsis readily separates from the lemma and palea. *A. brevis* Roth is a form with smaller spikelets, the lemmas plump, awned. *A. strigosa* Schreb. has a 1-sided panicle, the lemmas scabrous toward the apex, both florets awned.

**3. Avena barbáta** Brot. SLENDER OAT. (Fig. 408.) Differing from *A. fatua* in the somewhat smaller, mostly 2-flowered spikelets on curved capillary pedicels; lemmas clothed with stiff red hairs, the teeth ending in fine points 4 mm. long. ⊙ —A common weed in fields and waste places, Washington and Oregon to Arizona and California.

FIGURE 407.—*A*, *Avena fatua*. Plant, × ½; spikelet and floret, × 2. (Umbach, Ill.) *B*, *A. sativa*, × 2. (Deam, Ind.)

Cultivated oats fall into three groups, according to the number of chromosomes. Group 1, 7 chromosomes, *A. brevis*, *A. strigosa*. Group 2, 14 chromosomes, *A. barbata*. Group 3, 21 chromosomes, *A. sativa*, *A. fatua* (including *A. orientalis* Schreb.), *A. nuda*, *A. sterilis*, *A. byzantina* (including *A. sterilis* var. *algeriensis* Trabut).

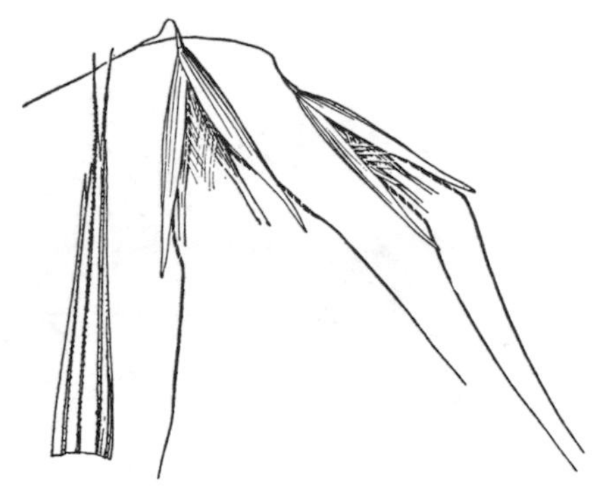

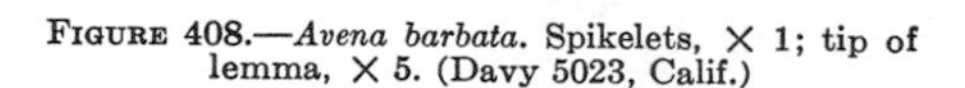

FIGURE 408.—*Avena barbata.* Spikelets, × 1; tip of lemma, × 5. (Davy 5023, Calif.)

## 62. HELICTOTRÍCHON Besser

(*Avena* sec. *Avenastrum* Koch; included in *Avena* L. in Manual, ed. 1)

Spikelets 3- to several-flowered, the rachilla bearded, disarticulating above the glumes and between the florets; glumes about equal, 3- to 5-nerved, subhyaline except toward the base; lemmas convex, the lower half subindurate and several-nerved, the upper part subhyaline, awned from about the middle, the awns twisted and geniculate, much exceeding the spikelets. Tufted perennials with rather narrow panicles of shining spikelets. Type species, *H. sempervirens* (Vill.) Pilger. Name from *helictos*, twisted, and "trichon," apparently referring to the awn, which is twisted. Perennials, numerous in Eurasia, 1 introduced and 2 native in western North America.

Blades involute; panicle 2 to 5 cm. long .......... 3. H. MORTONIANUM.
Blades flat or folded; panicle 5 to 15 cm. long.
  Sheaths and blades glabrous .......... 2. H. HOOKERI.
  Sheaths, at least the lower, and blades pubescent .......... 1. H. PUBESCENS.

**1. Helictotrichon pubéscens** (Huds.) Pilger. (Fig. 409.) Culms erect, 50 to 80 cm. tall; sheaths pubescent; blades flat, pubescent; panicle narrow, open, 10 to 15 cm. long, the flexuous branches ascending; spikelets mostly 3-flowered, 12 to 15 mm. long, glumes and lemmas thin, shining, the rachilla with long white hairs; first glume 1- or 3-nerved, the second 3-nerved; lemmas about 1 cm. long; awn attached about the middle, 1.5 to 2 cm. long. ♃ —Waste places, Connecticut and Vermont; introduced from Europe.

**2. Helictotrichon hookéri** (Scribn.) Henr. SPIKE OAT. (Fig. 410.) Culms densely tufted, 20 to 40 cm. tall; blades firm, flat or folded, 1 to 3 mm. wide, the margins somewhat thickened; panicle long-exserted, narrow, 5 to 10 cm. long, the branches erect or ascending, 1-flowered, or the lower 2-flowered; spikelets 3- to 6-flowered, about 1.5 cm. long; glumes very thin, slightly shorter than the spikelet; lemmas firm, brown, scaberulous, 1 to 1.2 cm. long, the callus short-bearded, the rachilla white-villous; awn 1 to 1.5 cm. long. ♃ Dry slopes and prairies, Manitoba to Alberta, Minnesota, Montana, and New Mexico.

**3. Helictotrichon mortoniánum** (Scribn.) Henr. ALPINE OAT. (Fig. 411.) Culms densely tufted, 10 to 20 cm. tall; blades erect, firm, usually involute; panicle short-exserted, purplish, narrow, 2 to 5 cm. long, the short branches erect, bearing usually a single spikelet, 10 to 12 mm. long, mostly 2-flowered; glumes exceeding the florets; lemmas firm, glabrous, the apex with 4 soft teeth, the callus with a tuft of stiff hairs about 2 mm. long, the rachilla long-villous; awn 1 to 1.5

FIGURE 409.—*Helictotrichon pubescens.* Glumes and floret, × 5. (Weatherby and Harger 4249, Conn.)

cm. long. ♃ —Alpine meadows, Colorado, Utah, and New Mexico.

## 63. ARRHENÁTHERUM Beauv.

Spikelets 2-flowered, the lower floret staminate, the upper perfect, the rachilla disarticulating above the glumes and produced beyond the florets; glumes rather broad and papery, the first 1-nerved, the second a little longer than the first and about as long as the spikelet, 3-nerved; lemmas 5-nerved, hairy on the callus, the lower bearing near the base a twisted, geniculate, exserted awn, the upper bearing a short straight slender awn just below the tip. Rather tall perennials, with flat blades and narrow panicles. Type species, *Arrhenatherum avenaceum* Beauv. (*A. elatius*). Name from Greek *arren*, masculine, and

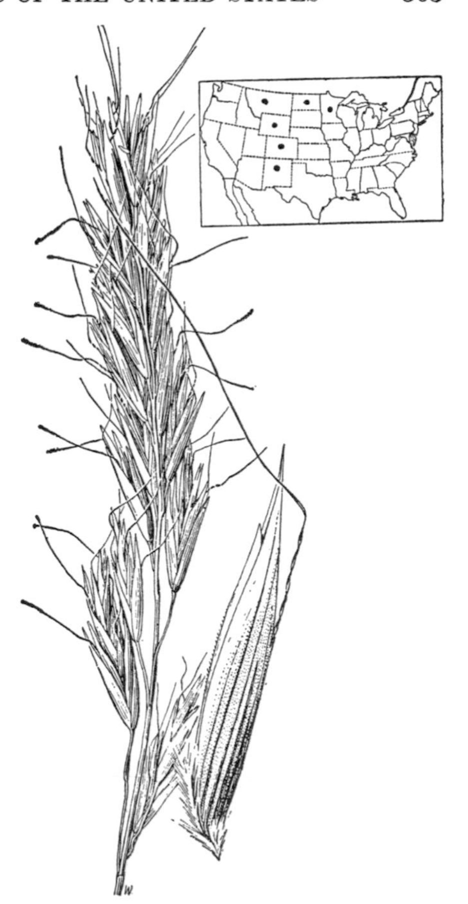

FIGURE 410.—*Helictotrichon hookeri.* Panicle, × 1; floret, × 5. (Scribner 372, Mont.)

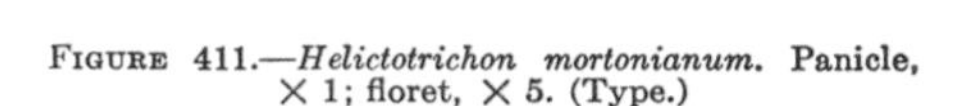

FIGURE 411.—*Helictotrichon mortonianum.* Panicle, × 1; floret, × 5. (Type.)

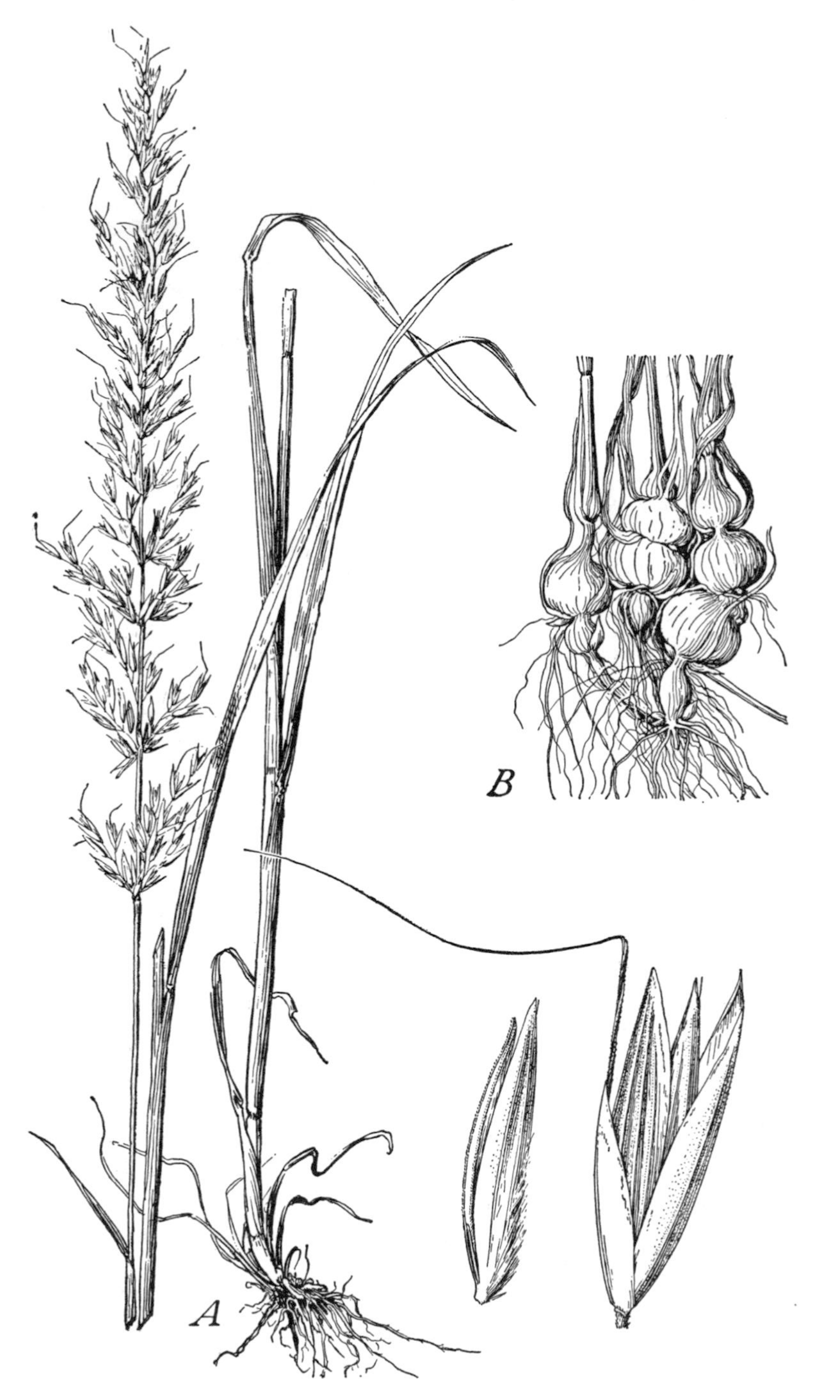


FIGURE 412.—*A*, *Arrhenatherum elatius*. Plant, × ½; spikelet and upper floret, × 5. (McDonald 46, Ill.) *B*, Var. *bulbosum*. Basal corms, × 1. (Harper, Ala.)

*ather*, awn, referring to the awned staminate floret.

**1. Arrhenatherum elátius** (L.) Presl. TALL OATGRASS. (Fig. 412, *A*.) Culms erect, 1 to 1.5 m. tall; blades flat, scabrous, 5 to 10 mm. wide; panicle pale or purplish, shining, 15 to 30 cm. long, the short branches verticillate, spreading in anthesis, usually spikelet-bearing from the base; spikelets 7 to 8 mm. long; glumes minutely scabrous; lemmas scabrous, the awn of the staminate floret about twice as long as its lemma. ♃ —Meadows, open ground, and waste places, Newfoundland to British Columbia, south to Georgia, Tennessee, Iowa, Idaho, Utah, Arizona, and California; frequent in the Northern and Eastern States; introduced from Europe and escaped from cultivation. Cultivated in the northern humid regions as a meadow grass.

ARRHENATHERUM ELATIUS var. BULBÓSUM (Willd.) Spenner. TUBER OATGRASS. (Fig. 412, *B*.) Base of culm consisting of a series of closely approximate corms (short subglobose internodes) 5 to 10 mm. in diameter. ♃ —Occasionally introduced, Michigan, Virginia, and West Virginia to Alabama; California; Europe.

ARRHENATHERUM ELATIUS var. BIARISTÁTUM (Peterm.) Peterm. Both lemmas with well-developed awns. ♃ —Ithaca, N. Y., and Delaware County, Pa.; Europe.

## 64. HÓLCUS L.

(*Notholcus* Nash)

Spikelets 2-flowered, the pedicel disarticulating below the glumes, the rachilla curved and somewhat elongate below the first floret, not prolonged above the second floret; glumes about equal, longer than the 2 florets; first floret perfect, the lemma awnless; second floret staminate, the lemma bearing on the back a short awn. Perennials with flat blades and contracted panicles. Standard species, *Holcus lanatus*. *Holcus*, an old Latin name for a kind of grain.

Rhizomes wanting .......... 1. H. LANATUS.
Rhizomes present .......... 2. H. MOLLIS.

**1. Holcus lanátus** L. VELVET GRASS. (Fig. 413.) Plant grayish, velvety-pubescent; culms erect, 30 to 100 cm. tall, rarely taller; blades 4 to 8 mm. wide; panicles 8 to 15 cm. long, contracted, pale, purple-tinged; spikelets 4 mm. long; glumes villous, hirsute on the nerves, the second broader than the first, 3-nerved; lemmas smooth and shining, the awn of the second hooklike. ♃ —Open ground, meadows, and moist places, Maine to Kansas and Colorado, south to Georgia and Louisiana; common on the Pacific coast, British Columbia, and Montana to Arizona and California; introduced from Europe; occasionally cultivated as a meadow grass on light or sandy land.

**2. Holcus móllis** L. (Fig. 414.) Culms glabrous, 50 to 100 cm. tall, with vigorous slender rhizomes; sheaths, except the lower, glabrous; blades villous or velvety, 4 to 10 mm. wide; panicle ovate or oblong, rather loose, 6 to 10 cm. long; spikelets 4 to 5 mm. long; glumes glabrous; awn of the second floret geniculate, exserted, about 3 mm. long. ♃ —Damp places, recently introduced from Europe and apparently spreading, Washington to California; Lewis County, N. Y.; ballast, Camden, N. J., Delaware County, Pa.

## 65. SIEGLÍNGIA Bernh.

Spikelets 4- to 5-flowered, the rachilla disarticulating above the glumes and between the florets; glumes equal, acute, the first 1- to 3-nerved, the second 3- to 5-nerved; lemmas firm, 7- to 9-nerved, bifid, the midnerve excurrent from between

FIGURE 413.—*Holcus lanatus*. Plant, × ½; spikelet, florets, and mature fertile floret, × 5. (Griffiths 4449 Calif.)

the short teeth in a short flat mucro, the margins densely pilose toward the base. Densely tufted perennial with short narrow blades and narrow, simple, few-flowered panicle. Type species, *Sieglingia decumbens*. Named for Siegling.

**1. Sieglingia decúmbens** (L.) Bernh. (Fig. 415.) Culms 20 to 50 cm. tall, erect, densely tufted; leaves crowded toward the base; blades 5 to 15 cm. long or those of the innovations elongate, 2 to 3 mm. wide; panicles 2 to 7 cm. long, the short few-flowered branches appressed; spikelets 8 to 12 mm. long; lemmas 5 to 6 mm. long. ♃ —Open woods, Long Beach, Wash.; escaped from cultivation, Berkeley, Calif.; Newfoundland and Nova Scotia; Europe. Cleistogamous spikelets sometimes developed in the lower sheaths.

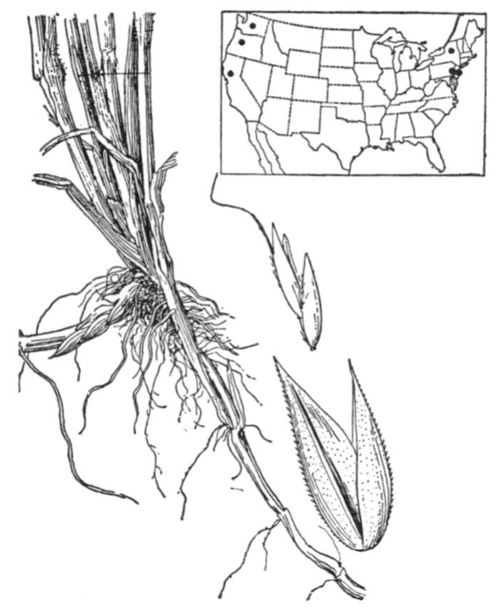

FIGURE 414.—*Holcus mollis*. Plant, × 1; glumes and floret, × 5. (Tracy 2646, Calif.)

## 66. DANTHÓNIA Lam. and DC. OATGRASS

Spikelets several-flowered, the rachilla readily disarticulating above the glumes and between the florets; glumes about equal, broad, papery, acute, mostly exceeding the uppermost floret; lemmas rounded on the back, obscurely several-nerved, the apex bifid, the lobes acute, usually extending into slender awns, a stout flat, twisted, geniculate awn arising from between the lobes. Tufted low or moderately tall perennials, with few-flowered open or spikelike panicles of rather large spikelets. All our species produce cleistogenes (enlarged fertile, 1- or 2-flowered, cleistogamous spikelets) in the lower sheaths, the culms finally disarticulating at the lower nodes. Type species, *Danthonia spicata*. Named for Etienne Danthoine.

The species are found in grassland and contribute somewhat toward the forage value of the range but usually are not abundant. In California *D. californica* is considered a nutritious grass; *D. compressa* is important in the mountains of North Carolina and Tennessee.

Lemmas glabrous on the back, pilose on the margin only.
  Panicle narrow, the pedicels appressed.......... 4. D. INTERMEDIA.
  Panicle open, the slender pedicels spreading or reflexed.
    Panicle usually of a single spikelet.......... 7. D. UNISPICATA.
    Panicle of few to several spikelets.......... 6. D. CALIFORNICA.
Lemmas pilose on the back, sometimes sparsely so.
  Glumes mostly 20 to 22 mm. long.......... 5. D. PARRYI.
  Glumes 10 to 17 mm. long.
    Sheaths pilose (rarely glabrous); glumes 12 to 17 mm. long. Culms 50 to 100 cm. tall. 3. D. SERICEA.
    Sheaths glabrous or nearly so; glumes rarely more than 15 mm. long.
      Panicle simple or nearly so, usually contracted after anthesis; blades rarely more than 15 cm. long, commonly less.......... 1. D. SPICATA.
      Panicle usually compound and somewhat open; blades or some of them more than 15 cm., often as much as 25 cm. long.......... 2. D. COMPRESSA.

**1. Danthonia spicáta** (L.) Beauv. ex Roem. and Schult. POVERTY OATGRASS. (Fig. 416.) Culms 20 to 70 cm. tall, mostly not more than 50

FIGURE 415.—*Sieglingia decumbens*. Panicle, × 1; glumes and floret, × 5. (Robinson and Schrenk 206, Newfoundland.)

cm., slender, terete; leaves numerous in a basal cluster, the blades usually curled or flexuous; sheaths glabrous or pilose above the nodes, with a tuft of long hairs in the throat; blades usually not more than 12 cm. long, filiform, to 2 mm. wide, occasionally a few blades 15 to 20 cm. long, subinvolute or in damp weather flat, glabrous or sparsely pilose; panicle 2 to 5 cm. long, rarely longer, the stiff short branches bearing a single spikelet, or the lower longer with 2 (rarely 3 or 4), usually erect after anthesis; glumes 10 to 12 mm. long (rarely longer); lemmas 3.5 to 5 mm. long, sparsely villous except the 2-toothed summit, the teeth acuminate to subsetaceous; terminal segment of awn about 5 mm. long; palea broad, flat, obtuse, ciliolate, reaching to the base of the awn. ♃ —Dry and sterile or rocky soil, Newfoundland to British Columbia, south to Florida, eastern Texas, and eastern Kansas, in the mountains to New Mexico and Oregon. Variable; tall specimens with longer blades and setaceous teeth resemble *D. compressa*. A rather stiff western form with subsetaceous teeth has been described as *D. thermale* Scribn. Very slender plants with narrow pilose blades and spikelets only 8 to 9 mm. long have been differentiated as var. *longipila* Scribn. and Merr. *D. spicata* var. *pinetorum* (Piper) Piper has been differentiated on variable characters. The basal blades, said to be slightly if at all curling, are closely curled in the type specimen.

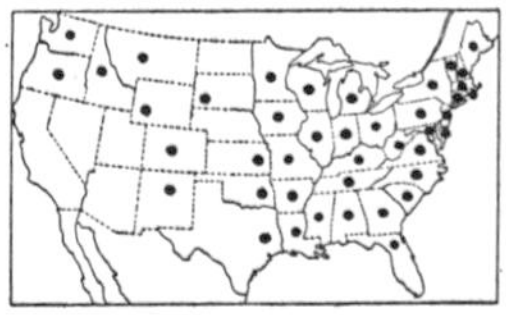

FIGURE 416.—*Danthonia spicata*. Plant, × ½; spikelet, floret, and cleistogene, × 5. (Gayle 787, Maine.)

FIGURE 417.—*Danthonia compressa.* Panicle, × 1; floret, × 5. (Hitchcock 103, Tenn.)

FIGURE 418.—*Danthonia sericea.* Panicle, × 1; floret, × 5. (Kearney 1219, Va.)

**2. Danthonia compréssa** Austin. (Fig. 417.) Culms on the average stouter and taller than in *D. spicata*, compressed, rather loosely tufted, sometimes decumbent or with short rhizomes, 40 to 80 cm. tall; sheaths reddish above the nodes, glabrous, or sparsely pubescent on the collar, a conspicuous tuft of white hairs in the throat; blades elongate, some of them commonly 20 to 25 cm. long, 2 to 3 mm. wide, usually flat, sometimes involute and subfiliform, scabrous; panicle 5 to 8 cm. long (rarely to 10 cm.), the slender branches bearing 2 or 3 spikelets, contracted after anthesis but looser than in *D. spicata;* glumes 10 to 14 mm. (usually about 12 mm.) long; lemma and palea as in *D. spicata* but the teeth of the lemma aristate, 2 to 3 mm. long. ♃ —Meadows, and open woods, Nova Scotia to Quebec, Maine to Ohio and south to the mountains of North Carolina and Georgia. Appears to intergrade with *D. spicata.* Taller stouter plants with panicles of 9 to 20 spikelets with glumes 10 to 13 mm. long have been differentiated as *D. alleni* Austin.

**3. Danthonia serícea** Nutt. DOWNY OATGRASS. (Fig. 418.) Culms erect, densely tufted, 50 to 100 cm. tall; sheaths, especially the lower, villous (rarely glabrous); blades 10 to 25 cm. long, 2 to 4 mm. wide, those of the innovations mostly involute, those of the culm mostly flat; panicle 5 to 10 cm. long, relatively many-flowered, the branches bearing 2 to 6 spikelets, rather open or contracted after anthesis; glumes 12 to 17 mm. long; lemmas densely long-pilose, especially along the margin, about 10 mm. long, including the slender aristate teeth, the teeth about half the entire length; palea concave, narrowed toward the 2-toothed apex. ♃ —Sand barrens, chiefly Coastal Plain, Massachusetts (Sherborn); New Jersey to northern Florida, Kentucky, and Louisiana. A rare form with nearly glabrous foliage has been differentiated as *D. epilis* Scribn. (*D. glabra* Nash, not Phil.) Virginia to Georgia.

**4. Danthonia intermédia** Vasey. TIMBER OATGRASS. (Fig. 419.) Culms 10 to 50 cm. tall; sheaths glabrous (the lower rarely pilose) with long hairs in the throat; blades subinvolute, or those of the culm flat, glabrous or sparsely pilose; panicle purplish, narrow, few-flowered, 2 to 5 cm. long, the branches appressed, bearing a single spikelet; glumes about 15 mm. long; lemmas 7 to 8 mm. long, appressed-pilose along the margin below and on the callus, the summit scaberulous, the teeth acuminate, aristate-tipped; terminal segment of awn 5 to 8 mm. long; palea narrowed above, notched at the apex. ♃ —Meadows and bogs, northern and alpine regions. Newfoundland and Quebec to Alaska, south to northern Michigan, New Mexico, and California.

**5. Danthonia párryi** Scribn. PARRY OATGRASS. (Fig. 420.) Culms rather stout, in tough clumps, 30 to 60 cm. tall, somewhat enlarged at base from the numerous overlapping firm persistent sheaths; sheaths glabrous, somewhat pilose at the throat, a glabrous or pubescent line or ridge on the collar, the lower blades falling from the sheaths; blades erect-flexuous, mostly 15 to 25 cm. long, narrow or filiform, flat or involute, glabrous; panicle 3 to 7 cm. long, usually with 3 to 8 spikelets, the branches more or less pubescent, ascending or appressed, the lowermost 1 to 2 cm. long, with 1 or 2 spikelets; glumes 20 to 22 mm. long, rarely less; lemmas about 1 cm. long, rather densely to sparsely pilose over the back, strongly pilose on the callus at the sides, the rachilla glabrous, the teeth more or less aristate; terminal segment of awn 8 to 12 mm. long; palea narrowed above, nearly as long as the lemma, 2-toothed. ♃ —Open grassland, open woods, and rocky slopes, in the mountains, mostly be-

FIGURE 420.—*Danthonia parryi*. Panicle, × 1; floret, × 5. (Hitchcock 19087, Colo.)

FIGURE 421.—*Danthonia californica*. Panicle, × 1; floret, × 5. (Eastwood 27, Calif.)

FIGURE 419.—*Danthonia intermedia*. Panicle, × 1; floret, × 5. (Hitchcock 11288, Mont.)

low timber line, Alberta and Montana to New Mexico.

**6. Danthonia califórnica** Boland. CALIFORNIA OATGRASS. (Fig. 421.) Culms 30 to 100 cm. tall, glabrous, tending to disarticulate at the nodes; sheaths glabrous, pilose at the throat; blades mostly 10 to 20 cm. long, flat or, especially those of the innovations, involute, glabrous; panicle bearing mostly 2 to 5 spikelets, the pedicels slender, spreading or somewhat reflexed, more or less flexuous, 1 to 2 cm. long, a rather prominent pulvinus at the base of each; glumes 15 to 20 mm. long (rarely less or more); lemmas, excluding awns, 8 to 10 mm. long, pilose on the lower part of the margin and on the callus, otherwise glabrous, the teeth long-aristate; terminal segment of awn 5 to 10 mm. long; palea subacute, usually extending beyond base of awn. ♃ —Meadows and open woods, Montana to British Columbia, south to Colorado, New Mexico, and California.

DANTHONIA CALIFORNICA var. AMERICÁNA (Scribn.) Hitchc. Culms on the average shorter, the tufts usually more spreading; foliage sparsely to conspicuously spreading-pilose; spikelets on the average smaller, but large plants with large spikelets occur, with conspicuously pilose foliage. ♃ —Montana and Wyoming to British Columbia, south to California; Chile. *D. macounii* Hitchc. appears to belong here, differing in having lemmas sparsely pilose on the back. Known only from Nanaimo, Vancouver Island (*Macoun* 78825).

**7. Danthonia unispicáta** (Thurb.) Munro ex Macoun. ONE-SPIKE OATGRASS. (Fig. 422.) Culms 15 to 25 cm. tall, in dense spreading tufts; sheaths and blades pilose, the hairs on the sheaths spreading or reflexed; panicle reduced to a single spikelet or sometimes 2, rarely 3, spikelets, the lower usually reduced, their pedicels appressed or ascending, the long pedicel of the terminal spikelet jointed with the culm; spikelets on the average smaller than in *D. californica;* lemmas usually glabrous, the callus hairy. ♃ —Open or rocky ground, Montana to British Columbia, south to Colorado and California.

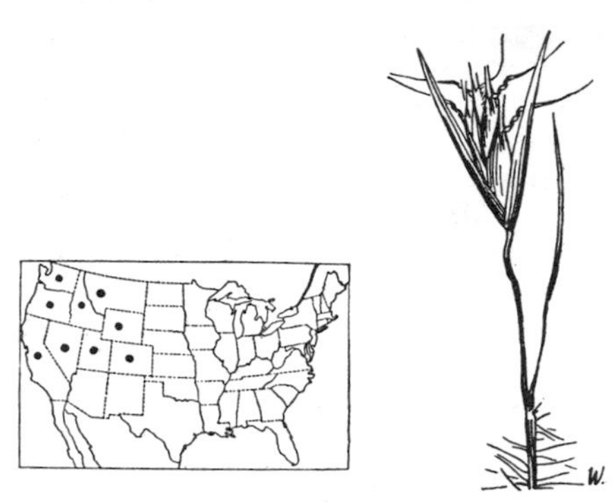

FIGURE 422.—*Danthonia unispicata,* × 2. (Davy, Calif.)

DANTHONIA PILÓSA R. Br. Tufted, 30 to 60 cm. tall, the foliage loosely pilose; panicle narrow, several-flowered; spikelets about 6-flowered; glumes 13 to 14 mm. long; florets disarticulating with a sharp hairy callus, the lemma pilose at base and on the margin, often with a few hairs in the middle of the back; teeth with slender awns 6 to 8 mm. long, the central awn 12 to 15 mm. long. ♃ —Introduced from Australia, escaped in Humboldt, Alameda, and Santa Barbara Counties, Calif.

DANTHONIA SEMIANNULÁRIS (Labill.) R. Br. Tufted, 40 to 100 cm. tall, often rather robust; foliage glabrous or nearly so; panicle many-flowered; glumes mostly 10 to 15 mm. long; florets with a slender hairy callus, the lemma pilose at base and with a conspicuous row of long tufted hairs across the middle; teeth tipped with slender awns, 5 to 8 mm. long, the central awn 10 to 20 cm. long. ♃ —Introduced from Australia; planted on range lands in California and escaped in several localities in the State. Extremely variable with several varieties.

DANTHONIA PURPUREA (Thunb.) Beauv. ex Roem. and Schult. Densely tufted perennial, forming thick mats of filiform curly pilose leaves; culms very slender, 1 to 2 cm. tall, with few short blades; panicle subcapitate, of few to several spikelets on short slender pedicels; spikelets about 8 mm. long; glumes dark purple fading to brown; florets about 4 mm. long, with a slender hairy callus, the lemma pilose at base and with small tufts of white hairs across the middle of the back; awn 2 to 3 mm. long, ♃ —Introduced from South Africa, grown in the grass garden of University of California, Berkeley.

## TRIBE 5. AGROSTIDEAE

### 67. CALAMAGRÓSTIS Adans. Reedgrass

Spikelets 1-flowered, the rachilla disarticulating above the glumes, prolonged behind the palea (in our species, except *Calamagrostis epigeios*) as a short, commonly hairy bristle; glumes about equal, acute or acuminate; lemma shorter and usually more delicate than the glumes, usually 5-nerved, the midnerve exserted as an awn, the callus bearing a tuft of hairs, these often copious and as long as the lemma. Perennial, usually moderately tall grasses, mostly with creeping rhizomes, with small spikelets in open or usually narrow, sometimes spikelike, panicles. Type species, *Arundo calamagrostis* L. Name from Greek *kalamos*, a reed, and *agrostis*, a kind of grass, the type species being a reedy grass. American species belong to the Section Deyeuxia, in which the rachilla is prolonged. In Section Epigeios, of the Old World (one species introduced), the rachilla is not prolonged.

Several species are important native forage grasses. Pinegrass, *C. rubescens*, is a leading range grass in the mountains of Oregon and Washington. Bluejoint, *C. canadensis*, is a source of much of the wild hay of Wisconsin and Minnesota. On the plains and bench lands of Wyoming and northward, *C. montanensis* furnishes forage, especially when young. In low wet lands of the Northern States *C. inexpansa* is grazed especially by horses and cattle.

1a. Awn longer than the glumes, geniculate.
  2a. Panicle open, the branches spreading, naked below.
    Blades scattered, 5 to 9 mm. broad, flat; plant mostly more than 1 m. tall. 1. C. BOLANDERI.
    Blades mostly basal, mostly not more than 2 mm. wide, often involute.
      Awn about 1 cm. long, much longer than the glumes; blades nearly or quite as long as the flowering culms ........ 2. C. HOWELLII.
      Awn only a little exceeding the glumes; blades much shorter than the culms, capillary, sulcate, folded ........ 3. C. BREWERI.
  2b. Panicle compact, the branches appressed, floriferous from base.
    Blades scattered, broad and flat, 6 to 10 mm. wide ........ 4. C. TWEEDYI.
    Blades mostly basal, firm, narrow, becoming involute.
      Glumes about 1 cm. long, gradually long-acuminate; awn nearly 1 cm. long above the bend ........ 5. C. FOLIOSA.
      Glumes 6 to 8 mm. long, abruptly acute or acuminate; awn usually less than 5 mm. long above the bend ........ 6. C. PURPURASCENS.
1b. Awn included or scarcely longer than the glumes, straight or geniculate.
  3a. Awn geniculate, protruding sidewise from the glumes; callus hairs rather sparse, shorter than the lemma.
    Plants tufted, not rhizomatous, less than 40 cm. tall; blades 1 to 2 mm. wide, soon involute, at least toward the tip.
      Panicles compact, spikelike; northwestern ........ **7.** C. MONTANENSIS.
      Panicles loose, open, relatively few-flowered; Tennessee ........ 8. C. CAINII.
    Plants rhizomatous, mostly more than 60 cm. tall; blades mostly more than 4 mm. wide, flat.
      Sheaths, or some of them, pubescent on the collar.
        Callus hairs one-third as long as lemma; western species ........ 9. C. RUBESCENS.
        Callus hairs half to three-fourths as long as lemma; eastern species.
          Palea about as long as the lemma ........ 10. C. PORTERI.
          Palea three-fourths as long as the lemma ........ 11. C. PERPLEXA.
      Sheaths glabrous on the collar.
        Culms stout, mostly more than 1 m. tall.
          Panicles loose, the branches ascending or spreading ........ **17.** C. NUTKAENSIS.
          Panicles compact ........ 18. C. DENSA.
        Culms slender, mostly less than 1 m. tall.
          Hairs on callus and rachilla scant, less than 1 mm. long.
            Spikelets 5 mm. long; panicle spikelike ........ 19. C. KOELERIOIDES.
            Spikelets 4 mm. long; panicles scarcely spikelike, some of the branches naked below ........ 16. C. PICKERINGII.
          Hairs on callus and rachilla rather prominent, at least half as long as the lemma.

FIGURE 423.—*Calamagrostis bolanderi*. Panicle, × 1; glumes and floret, × 10. (Bolander, Calif.)

Callus hairs in 2 tufts, at sides of lemma.
 Plants with creeping rhizomes; spikelets 4 to 5.5 mm. long.
  Blades thin, glabrous on the upper surface, scaberulous beneath; panicle pale, rather loose; glumes relatively thin, 5 to 5.5 mm. long, scaberulous on the keel toward the summit............ 12. C. INSPERATA.
  Blades firm, scabrous; panicle tawny to purplish, rather dense; glumes firm, 4 to 4.5 mm. long, scabrous throughout..... 13. C. LACUSTRIS.
 Plants tufted; spikelets 3.5 to 4 mm. long..................... 14. C. FERNALDII.
Callus hairs surrounding base of lemma.................................... 15. C. NUBILA.

3b. Awn straight (somewhat bent in *C. epigeios* and *C. lactea*), included; callus hairs usually not much shorter than the lemma.
 Sheaths pubescent on the collar (see *C. inexpansa* var. *barbulata*). 20. C. SCRIBNERI.
 Sheaths glabrous on the collar.
  Panicle rather loose and open.
   Callus hairs copious, about as long as the lemma; awn delicate, straight. 21. C. CANADENSIS.
   Callus hairs rather scant, about half as long as the lemma; awn stronger, weakly geniculate........................................................................ 22. C. LACTEA.
  Panicle more or less contracted.
   Blades flat, rather lax.
    Awn attached near the base; rachilla not prolonged................ 29. C. EPIGEIOS.
    Awn attached at or about middle; rachilla prolonged.
     Glumes scabrous; plant green.............................................. 23. C. CINNOIDES.
     Glumes nearly smooth; plant pale.................................. 24. C. SCOPULORUM.

Blades involute or, if flat, rigid and becoming involute.
Blades broad and short, as much as 5 mm. wide, nearly smooth. 28. C. CRASSIGLUMIS.
Blades elongate, smooth or scabrous.
Blades firm, scabrous, rather rigid; ligule 4 to 6 mm. long; panicle firm, dense. 25. C. INEXPANSA.
Blades relatively soft, smooth beneath; ligule 1 to 3 cm. long.
Spikelets 3.9 to 4.2 mm. long; panicle 18 to 22 cm. long. 26. C. CALIFORNICA.
Spikelets 3 to 3.5 mm. long; panicle 5 to 15 cm. long.... 27. C. NEGLECTA.

FIGURE 424.—*Calamagrostis howellii*. Panicle, × 1; glumes and floret, × 10. (Chase 4846, Oreg.)

**1. Calamagrostis bolandéri** Thurb. (Fig. 423.) Culms erect, 1 to 1.5 m. tall, with slender rhizomes; sheaths scabrous; ligule 4 to 5 mm. long; blades flat, 5 to 9 mm. wide, scattered, nearly smooth; panicle open, 10 to 20 cm. long, the branches verticillate, spreading, naked below, the longer 5 to 10 cm. long; glumes 3 to 4 mm. long, purple, scabrous, acute; lemma very scabrous, about as long as the glumes, the awn from near the base, geniculate, exserted, about 2 mm. long above the bend, the callus hairs short; rachilla pilose, 1 to 2 mm. long. ♃ —Bogs and moist ground, prairie or open woods, near the coast, Mendocino and Humboldt Counties, Calif.

**2. Calamagrostis howéllii** Vasey. (Fig. 424.) Culms densely tufted, rather slender, ascending, 30 to 60

FIGURE 425.—*Calamagrostis breweri*. Plant, × 1; glumes and floret, × 10. (Bolander 6098, Calif.)

cm. tall; sheaths smooth or slightly scabrous; ligule 2 to 8 mm. long; blades slender, scabrous on the upper surface, flat or soon involute, especially toward the tip, about as long as the culms, the two cauline shorter, about 1 mm. wide; panicle pyramidal, 5 to 15 cm. long, rather open, the lower branches in whorls, ascending, naked below, 3 to 5 cm. long; spikelets pale or tinged with purple; glumes acuminate, 6 to 7 mm. long; lemma acuminate, a little shorter than the glumes, the awn attached about 2 mm. above the base, geniculate, exserted about 1 cm.; callus hairs and those of the rachilla about half as long as the lemma. ♃ —Perpendicular cliffs, near Columbia River and its tributaries, Washington and Oregon.

**3. Calamagrostis brewéri** Thurb. SHORTHAIR. (Fig. 425.) Culms densely tufted, slender, erect, 15 to 30 cm. tall; leaves mostly basal, usually involute-filiform; panicle ovate, purple, 3 to 8 cm. long, the lower branches slender, spreading, few-flowered, 1 to 2 cm. long; glumes 3 to 4 mm. long, smooth, acute; lemma nearly as long as glumes, cuspidate-toothed, the awn from near the base, geniculate, exserted, twisted below, about 2 mm. long above the bend, the callus hairs short, scant; rachilla long-pilose, about half as long as the lemma. ♃ —Mountain meadows of the high Sierra Nevada, Calif., where it is an important range grass.

**4. Calamagrostis tweédyi** (Scribn.) Scribn. (Fig. 426.) Culms erect, 1 to 1.5 m. tall, smooth, with short rhizomes; sheaths smooth, the lower becoming fibrous; blades flat, somewhat scabrous, the cauline 5 to 15 cm. long, as much as 1 cm. wide, those of the innovations narrower and longer; panicle oblong, rather compact, or interrupted below, about 10 cm. long; glumes abruptly acuminate, purple-tinged, 6 to 7 mm. long; lemma about as long as the glumes, the awn exserted about 5 mm., the callus hairs scant, scarcely 1 mm. long; rachilla pilose, 2 mm. long. ♃ —Moist open alpine slopes, Idaho and Cleland Counties, Idaho, and Kittitas County and Cascade Mountain, Wash.

**5. Calamagrostis foliósa** Kearney. (Fig. 427.) Culms tufted, erect, 30 to 60 cm. tall; leaves numerous, crowded toward the base, the sheaths overlapping, the blades involute, firm, smooth, nearly as long as the culm; panicle pale, dense, spikelike, 5 to 12 cm. long; glumes about 1 cm. long, acuminate; lemma 5 to 7 mm. long, acuminate, the apex with 4 setaceous teeth, the awn from near base, geniculate, about 8 mm. long above the bend, the callus hairs numerous, 3 mm. long; rachilla pilose, nearly as long as lemma. ♃ —Humboldt and Mendocino Counties, Calif.

FIGURE 427.—*Calamagrostis foliosa*. Panicle, × 1; glumes and floret, × 10. (Davy 6602, Calif.)

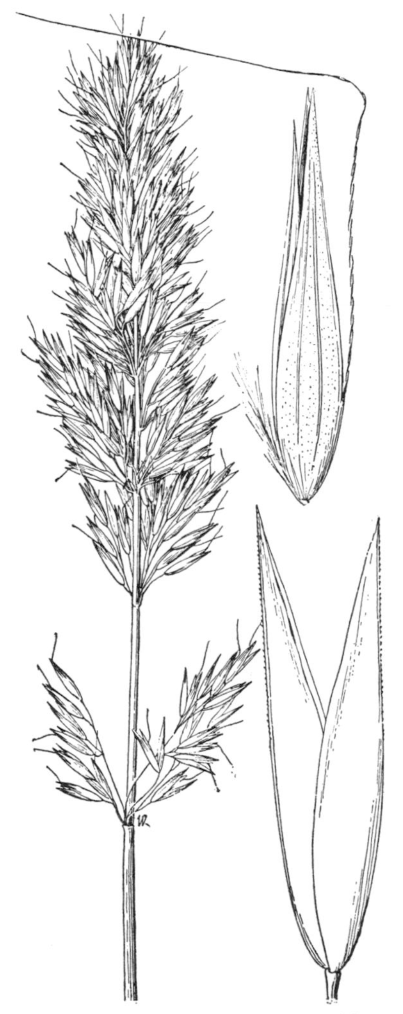

FIGURE 426.—*Calamagrostis tweedyi*. Panicle, × 1; glumes and floret, × 10. (Vasey, Wash.)

**6. Calamagrostis purpuráscens** R. Br. PURPLE REEDGRASS. (Fig. 428.) Culms tufted, sometimes with short rhizomes, erect, 40 to 60 cm. or even 100 cm. tall; sheaths usually scabrous, the old sheaths persistent and fibrous; blades 2 to 4 mm. wide, flat or more or less involute, rather thick, scabrous; panicle dense, usually pinkish or purplish, spikelike, 5 to 12 cm. long, rarely longer; glumes 6 to 8 mm. long, scabrous; lemma nearly as long as glumes, the apex with 4 setaceous teeth, the awn from near base, finally geniculate, exserted about 2 mm.; hairs of callus and rachilla about one-

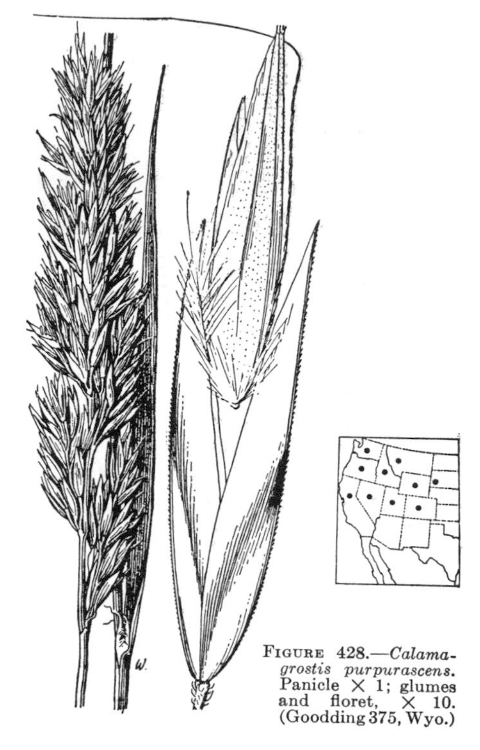

FIGURE 428.—*Calamagrostis purpurascens.* Panicle × 1; glumes and floret, × 10. (Goodding 375, Wyo.)

third as long as the lemma. ♃ (*C. vaseyi* Beal.)—Rocks and cliffs, Greenland to Alaska, south to Quebec, South Dakota (Black Hills), Colorado, and California.

**7. Calamagrostis montanénsis** Scribn. PLAINS REEDGRASS. (Fig. 429.) Culms stiffly erect, scabrous below the panicle, usually 20 to 40 cm. tall, sometimes taller, with slender creeping rhizomes; lower sheaths rather papery, smooth; blades erect, mostly less than 2 mm. wide, more or less involute, scabrous, sharp-pointed; panicle dense, erect, more or less interrupted, usually pale, 5 to 10 cm. long; spikelets 4 to 5 mm. long, the pedicels very scabrous; glumes acuminate, scabrous; lemma nearly as long as the glumes, finely 4-toothed, the awn attached about 1 mm. above base, about equaling the lemma, slightly geniculate and protruding from side of glumes; palea nearly as long as the lemma; hairs of callus and rachilla rather abundant, about half as long as the lemma. ♃ —Plains and dry open ground, Manitoba to Alberta, south to Minnesota, Wyoming, Colorado, and Idaho.

**8. Calamagrostis cáinii** Hitchc. (Fig. 430.) Culms 30 to 60 cm. tall, slender, erect; blades as much as 35 cm. long, 1 to 2 mm. wide, flat or loosely involute, attenuate, scabrous above; panicles 6 to 10 cm. long, pale or purple-tinged, the slender ascending branches 1 to 2 cm. long, few-flowered; glumes narrow, acuminate, 5 to 6 mm. long; lemma acuminate or minutely dentate, the nerves sometimes extending into short mucros, the callus hairs about 1 mm. long; awn attached about 1 mm. above the base, geniculate, a little longer than the glumes; rachilla very short, the hairs 1 to 2 mm. long. ♃ —Shrubby summit and open slopes of Mount LeConte, above 5,000 feet, Tenn.

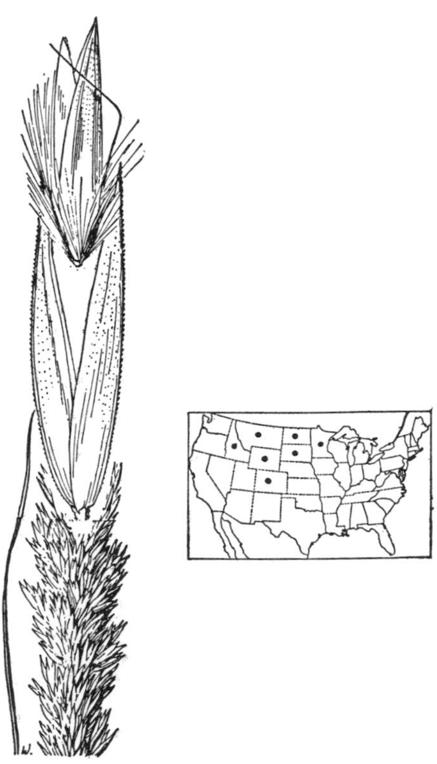

FIGURE 429.—*Calamagrostis montanensis.* Panicle, × 1; glumes and floret, × 10. (Scribner 363, Mont.)

**9. Calamagrostis rubéscens** Buckl. PINEGRASS. (Fig. 431.) Culms slender, tufted, 60 to 100 cm. tall, with creeping rhizomes; sheaths smooth, but pubescent on the collar, sometimes obscurely so; blades erect, 2 to 4 mm. wide, flat or somewhat involute, scabrous; panicle narrow, spikelike or somewhat loose or interrupted, pale or purple, 7 to 15 cm. long; glumes 4 to 5 mm. long, narrow, acuminate; lemma pale, thin, about as long as glumes, smooth, the nerves obscure, the awn from near base, geniculate, exserted from side of glumes, 1 to 2 mm. long above the bend, the callus hairs scant, about one-third as long as the lemma; rachilla 1 mm. long, the sparse hairs extending to 2 mm. ♃ —Open pine woods, prairies, and banks, British Columbia, south to northern Colorado and central California. A valuable range grass. A large form with dense lobed panicle has been differentiated as *C. cusickii* Vasey.

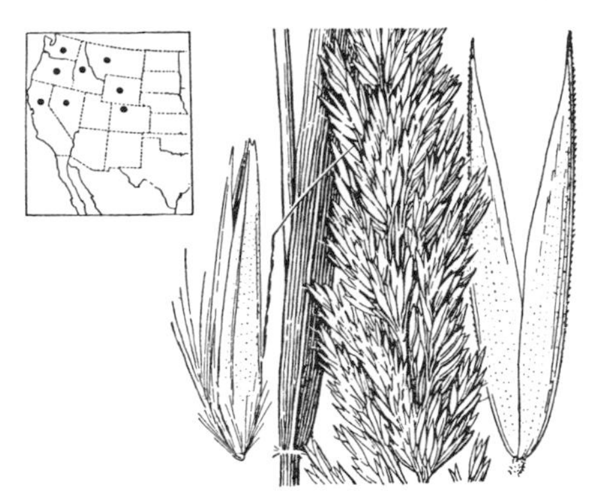

FIGURE 431.—*Calamagrostis rubescens*. Panicle, × 1; glumes and floret, × 10. (Sandberg and Leiberg Wash.)

FIGURE 430.—*Calamagrostis cainii*. Panicle, × 1; glumes and floret, × 10. (Underwood 1210, Tenn.)

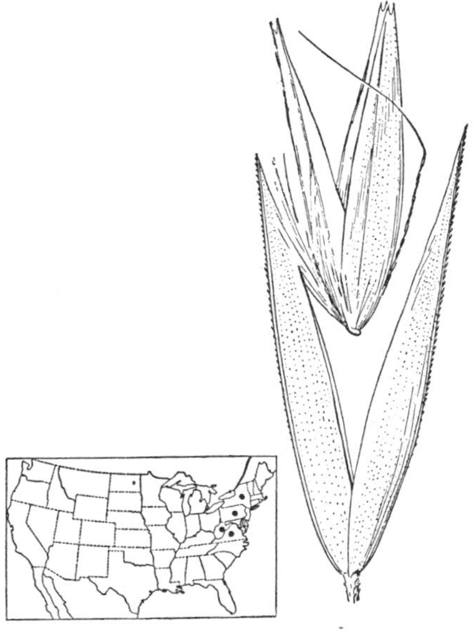

FIGURE 432.—*Calamagrostis porteri*. Glumes and floret, × 10. (Porter, Pa.)

**10. Calamagrostis portéri** A. Gray. (Fig. 432.) Culms slender, 60 to 120 cm. tall, with slender rhizomes; sheaths pubescent on the collar; blades flat, spreading, lax, 4 to 8 mm. wide; panicle narrow but rather loose, erect or somewhat nodding, 10 to 15

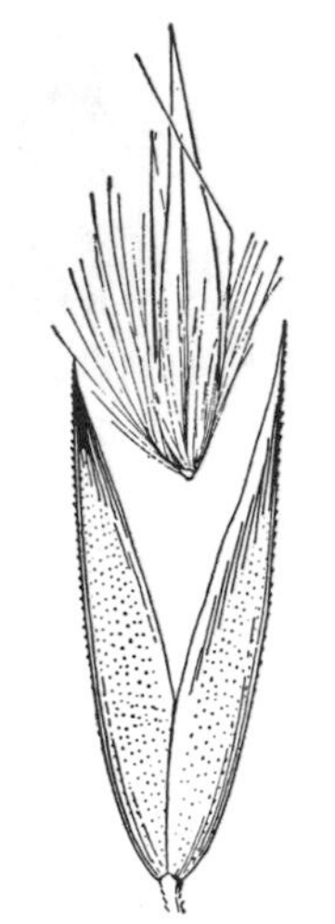

FIGURE 433.—*Calamagrostis perplexa.* Glumes and floret, × 10. (Metcalf 5668, N. Y.)

cm. long; glumes 4 to 6 mm. long, scaberulous; lemma slightly shorter than the glumes, toothed at apex, the awn from near base, about as long as the lemma, bent and protruding from side of glumes; palea about as long as the lemma; callus hairs in tufts at the sides, rather scant, nearly half as long as the lemma; rachilla hairs scant, extending to about 3 mm. ♃ —Dry rocky soil, New York, Pennsylvania, Virginia (Luray), and West Virginia. Apparently flowering irregularly or rarely.

**11. Calamagrostis perpléxa** Scribn. (Fig. 433.) Culms slender, 90 to 100 cm. tall, with slender rhizomes; lower sheaths overlapping and with reduced blades, the others shorter than the internodes, minutely scaberulous, tomentose at the sides of the collar; ligule 3 to 5 mm. long; blades (except the lower) 15 to 35 cm. long, 3 to 6 mm. wide, scabrous; panicle 10 to 15 cm. long, 2 to 3 cm. wide, many-flowered but rather loose, the axis smooth except toward the apex; spikelets 3.5 to 4 mm. long, the glumes nearly equal, acuminate, scaberulous; lemma 3.5 mm. long, acuminate, the awn from near the base, about as long as the lemma; palea and callus hairs about three-fourths as long as the lemma, the hairs in 2 rather dense

FIGURE 434.—*Calamagrostis insperata.* Panicle, × 1; glumes and floret, × 10. (Type.)

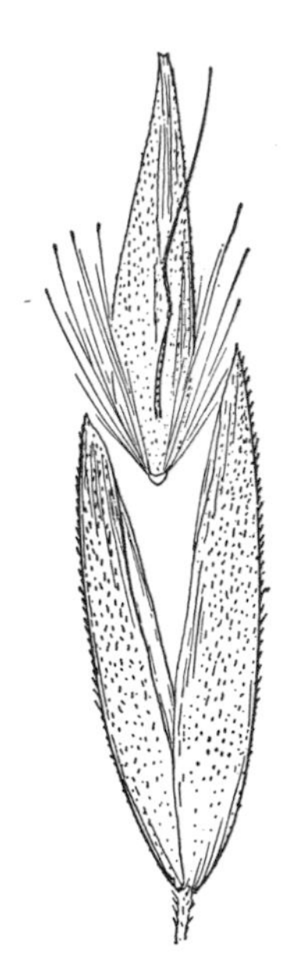

FIGURE 435.—*Calamagrostis lacustris.* Glumes and floret, × 10. (Type.)

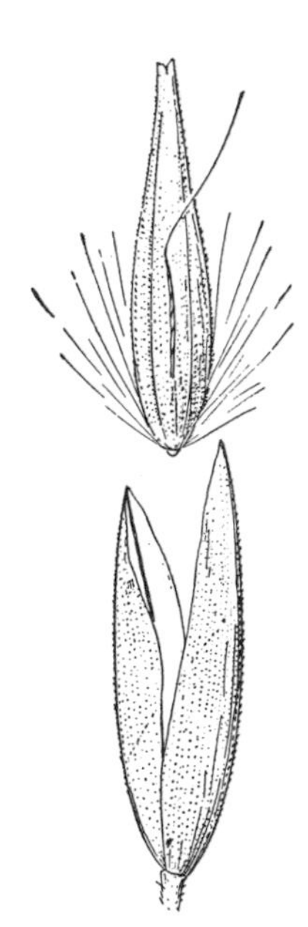

FIGURE 436.—*Calamagrostis fernaldii.* Glumes and floret, × 10. (Fernald 427, Maine.)

tufts at the sides, the hairs of the rachilla scarcely as long, scant. ♃ —Wet rocks, New York (Thatcher's Pinnacle, near Ithaca, type locality).

**12. Calamagrostis insperáta** Swallen. (Fig. 434.) Culms 85 to 95 cm. tall, erect from slender creeping rhizomes; sheaths glabrous on the collar; ligule 5 mm. long; blades flat, rather thin, 4 to 8 mm. wide, 10 to 22 cm. long, acuminate, glabrous, the margins scabrous; panicles 12 to 14 cm. long, the branches narrowly ascending, at least some of them naked at the base; spikelets 5 to 5.5 mm. long; lemma 4 mm. long, scaberulous on the keel, the callus hairs in tufts at the sides, rather dense, some of them half to three-fourths as long as the lemma; rachilla 0.5 mm. long, the hairs as much as 2 mm. long; awn from about one-fourth above the base, about as long as the lemma, geniculate. ♃ —Known only from Ofer Hollow, Jackson County, Ohio.

**13. Calamagrostis lacústris** (Kearn.) Nash. (Fig. 435.) Culms rather slender from short rhizomes, 35 to 100 cm. tall; sheaths and blades scabrous, the blades firm, 2 to 4 mm. wide; panicle 8 to 15 cm. long, 1 to 2.5 cm. wide, relatively dense, or with one of the lower fascicle of branches naked at base, the axis scabrous; spikelets 4 to 4.5 mm. long; glumes firm, rather broad, scabrous; lemma about 3.5 mm. long, scabrous, the awn from near the base, about as long as the lemma, geniculate; callus hairs about half to two-thirds as long as the lemma, in 2 tufts at the sides; rachilla minute, its hairs exceeding those of the callus. ♃ —Mossy rocks, marshy meadows, and sandy shores, Ontario, Vermont, eastern New York, northern Michigan, and eastern Minnesota.

**14. Calamagrostis fernáldii** Louis-Marie. (Fig. 436.) Culms loosely tufted, about 80 cm. tall; sheaths glabrous on the collar; blades elongate, 2 to 4 mm. wide, scabrous on both surfaces; panicle 8 to 9.5 cm. long, narrow, pale; glumes 3.5 to 4 mm. long; lemma 3.2 to 3.6 mm. long, scabrous, minutely toothed, the awn from near the base, scarcely as long as the lemma, the palea about two-thirds as long; callus hairs in tufts at the sides, half to two-thirds as long as the lemma; rachilla hairs two-thirds to three-fourths as long as the lemma. ♃ —Wet cliffs, only known from Boarstone Mountain, Piscataquis County, Maine.

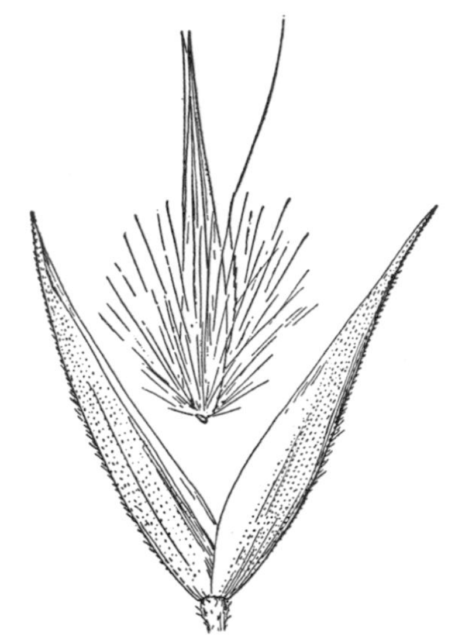

FIGURE 437.—*Calamagrostis nubila.* Glumes and floret, × 10. (Boott, N. H.)

**15. Calamagrostis núbila** Louis-Marie. (Fig. 437.) Culms tufted, erect, 55 cm. tall; sheaths mostly overlapping, scaberulous toward the summit; ligule 3 to 5 mm. long; blades flat, 12 to 18 cm. long, 4 to 5 mm. wide, long-attenuate, scabrous on both surfaces, the upper exceeding the inflorescence; panicle pale, 13 to 14 cm. long, about 4 cm. wide, many-flowered but rather loose, probably nodding, the axis and branches flexuous, scabrous; spikelets on short scabrous pedicels; glumes 4.5 to 5.2 mm. long, scabrous, the second indistinctly 3-nerved; lemma 4.5 mm. long, toothed at the acuminate apex, the awn from near base, about as long as the lemma, bent and protruding from side of glumes; palea about two-thirds as long as the lemma; callus and rachilla hairs rather copious, three-fourths to nearly as long as the lemma. ♃ —Only known from Lake of the Clouds, Mount Washington, N.H.

**16. Calamagrostis pickeríngii** A. Gray. (Fig. 438.) Culms solitary or few in tufts, rather rigid, scabrous below the panicle, 30 to 60 cm. tall, with creeping rhizomes; blades erect, flat, 4 to 5 mm. wide; panicle purplish, erect, contracted and rather

FIGURE 438.—*Calamagrostis pickeringii*. Panicle, × 1; glumes and floret, × 10. (Hubbard 634, Mass.)

dense, 7 to 12 cm. long; glumes acute, about 4 to 4.5 mm. long; lemma a little shorter than the glumes, scaberulous, narrowed to an obtuse point, the awn attached about 1 mm. above the base, about as long as the lemma, slightly bent and protruding somewhat from the side of the glumes; callus hairs in 2 tufts, scant, about 0.5 mm. long; rachilla about 1 mm. long, the hairs short, rather scant. ♃ —Bogs, wet meadows, and sandy beaches, Newfoundland and Labrador to the mountains of Massachusetts and New York; Isle Royale, Mich. Slender plants with slightly smaller spikelets have been differentiated as *C. pickeringii* var. *debilis* (Kearney) Fern. and Wieg.

**17. Calamagrostis nutkaénsis** (Presl) Steud. PACIFIC REEDGRASS. (Fig. 439.) Culms stout, 1 to 1.5 m. tall with short rhizomes (not usually

FIGURE 439.—*Calamagrostis nutkaensis*. Panicle, × 1; glumes and floret, × 10. (Hitchcock 23576, Oreg.)

present in herbarium specimens); ligule 3 to 8 mm. long; blades elongate, 6 to 12 mm. wide, flat, becoming involute, gradually narrowed into a long point, scabrous; panicle usually purplish, narrow, rather loose, 15 to 30 cm. long, the branches rather stiffly ascending; glumes 5 to 7 mm. long, acuminate; lemma about 4 mm. long, indistinctly nerved, the awn rather stout, from near the base, slightly geniculate, about equaling the lemma or shorter; hairs of callus and rachilla scarcely half as long ♃ —Along the coast in moist soil or wet wooded hills, from Alaska to central California.

**18. Calamagrostis dénsa** Vasey. CUYAMACA REEDGRASS. (Fig. 440.) Culms rather stout, densely tufted, smooth or scabrous just below the panicle, mostly more than 1 m. tall, with rather stout rhizomes; sheaths slightly scabrous; ligule 3 to 5 mm. long; blades flat, or subinvolute, scabrous, 15 to 25 cm. long, 3 to 8 mm. wide, the uppermost shorter; panicle spikelike, dense, pale, 10 to 15 cm. long; glumes 4.5 to 5 mm. long, acuminate, scaberulous; lemma 3.5 to 4 mm. long, the awn bent, about as long as the lemma, more or less exserted at the side, the hairs of callus and ra-

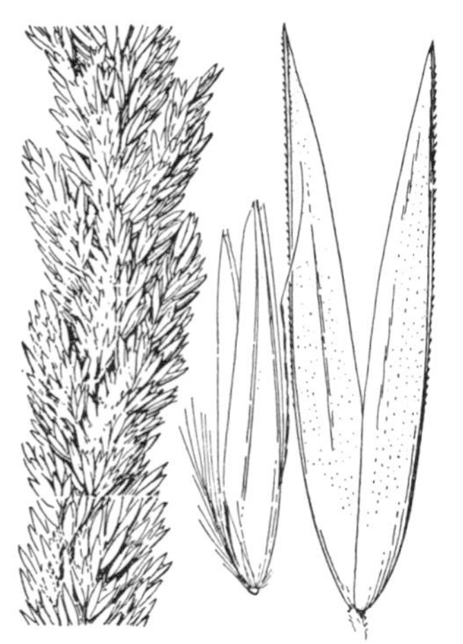

FIGURE 440.—*Calamagrostis densa*. Panicle, × 1; glumes and floret, × 10. (Hitchcock 13163, Calif.)

chilla scant, about 1 mm. long. ♃ —Dry hills, among shrubs, mountains east of San Diego, Calif.

**19. Calamagrostis koelerioídes** Vasey. (Fig. 441.) Differs from *C. densa* in the more slender culms and (often purplish) panicles; lemma nearly as long as the glumes. ♃ —Dry hills, banks, and meadows, Wyoming to Washington, south to southern California. Possibly a form of *C. densa*.

**20. Calamagrostis scribnéri** Beal. SCRIBNER REEDGRASS. (Fig. 442.) Culms tufted, with numerous creeping rhizomes, slender, 60 to 100 cm. tall; lower sheaths loose, thin, upper scabrous, retrorsely pubescent on the collar; ligule about 5 mm. long; blades thin, elongate, 4 to 7 mm. wide, scabrous; panicle pale or purplish, narrow but rather lax, 10 to 15 cm. long (rarely longer); glumes about 4 mm. long, acuminate; lemma a little shorter than the glumes, sharply toothed, the awn about as long as the glumes or a little longer, feebly bent, the callus hairs about half as long as the lemma; rachilla minute, its hairs nearly as long as the lemma. ♃ —Moist meadows, Montana and Washington to New Mexico; infrequent.

**21. Calamagrostis canadénsis** (Michx.) Beauv. BLUEJOINT. (Fig. 443, *A*.) Culms suberect, tufted, 60 to 150 cm. tall, with numerous creeping rhizomes; sheaths glabrous or rarely obscurely pubescent; blades numerous, elongate, flat, rather lax, scabrous, 4 to 8 mm. wide; panicle nodding, from narrow and rather dense to loose and relatively open, especially at base, 10 to 25 cm. long; glumes usually 3 to 4 mm. long, smooth or more commonly scabrous, acute to acuminate; lemma nearly as

FIGURE 441.—*Calamagrostis koelerioides*. Glumes and floret, × 10. (Hitchcock 23558, Oreg.)

FIGURE 442.—*Calamagrostis scribneri*. Panicle, × 1; glumes and floret, × 10. (Rydberg 3083, Mont.)

FIGURE 443.—*A*, *Calamagrostis canadensis*. Plant, × ½; glumes and floret, × 10. (Chase 5077, Mont.) *B*, Var. *scabra*, × 10. (Pringle, N. H.) *C*, Var. *macouniana*, × 10. (Pammel 891, Minn.)

long as the glumes, smooth, thin in texture, the awn delicate, straight, attached just below the middle and extending to or slightly beyond its tip, the callus hairs abundant, about as long as lemma; rachilla delicate, sparsely long-pilose. ♃ —Marshes wet places, open woods, and meadows, Greenland to Alaska, south to West Virginia and North Carolina (Roan Mountain), Missouri, Kansas, to New Mexico and California. A widely distributed and exceedingly variable species. Characters used to differentiate the many proposed varieties are not correlated in the larger proportion of specimens. The panicle varies in density and the glumes in size and scabridity. The following varieties are recognizable but are connected with the species by many intergrading specimens.

Calamagrostis canadensis var. scábra (Presl) Hitchc. (Fig. 443, *B*.) Differing in having spikelets 4.5 to 6 mm. long, the glumes rather firm, hispidly short-ciliate on the keel, strongly scabrous otherwise, but the greater scabridity not constant. ♃ —Mountains of New England, New York, and northward, and along the Pacific coast from Washington to Alaska. This form has been referred to *C. langsdorfii* (Link) Trin., which proves to be an Old World species not found in America.

Calamagrostis canadensis var. macouniána (Vasey) Stebbins. (Fig. 443, *C*.) Differing from *C. canadensis* in the smaller spikelets, about 2 mm. long. Scarcely a distinct variety. ♃ —Saskatchewan (*Macoun* 44, 45), Minnesota (Bemidji), South Dakota (Chamberlain, Redfield), Iowa, Nebraska (Central City), Missouri (Lake City, Little Blue), Montana (Manhattan), Yellowstone Park, Washington (Spokane County), Oregon (Crook County).

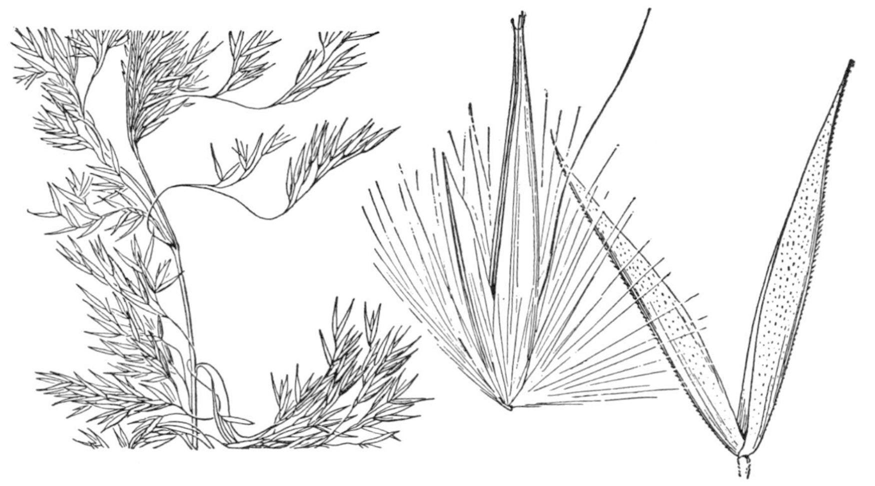

Figure 444.—*Calamagrostis lactea*. Panicle, × 1; glumes and floret, × 10. (Dupl. type.)

**22. Calamagrostis láctea** Beal. (Fig. 444.) Culms ascending, 80 to 150 cm. tall, weak, the nodes subgeniculate, with a short knotty rhizome; sheaths scaberulous; ligule rather firm, 3 to 5 mm. long; blades elongate, flat, lax, scabrous, 6 to 12 mm. wide; panicle pale, narrowly pyramidal, 12 to 20 cm. long, loosely flowered; glumes 5 to 6 mm. long, scabrous, acuminate; lemma shorter than the glumes, scabrous, the apex setaceous-toothed, the awn attached near the base, about equaling the lemma, weakly geniculate; palea slightly exceeding the lemma, the callus hairs about half as long; rachilla minute, sparsely pilose. ♃

—Mountain slopes, Washington to California; apparently rare.

**23. Calamagrostis cinnoídes** (Muhl.) Barton. (Fig. 445.) Glaucous; culms rather stout, erect, 80 to 150 cm. tall, with slender rhizomes readily broken off; sheaths and blades very scabrous, sometimes sparsely hirsute, the blades flat, 5 to 10 mm. wide; panicle erect, dense, more or less lobed (somewhat open at anthesis), 8 to 20 cm. long, purple-tinged; glumes 6 to 7 mm. long, scabrous, long-acuminate or awn-pointed; lemma firm, acuminate, scabrous, shorter than the glumes, the awn attached about one-fourth below the tip, not much exceeding the lemma, the callus hairs copious, about two-thirds as long; rachilla about 1 mm. long, glabrous below, with a brush of long white hairs at the tip about equaling the lemma. ♃ —Bogs and moist ground, Maine to New York, south to Alabama and Louisiana.

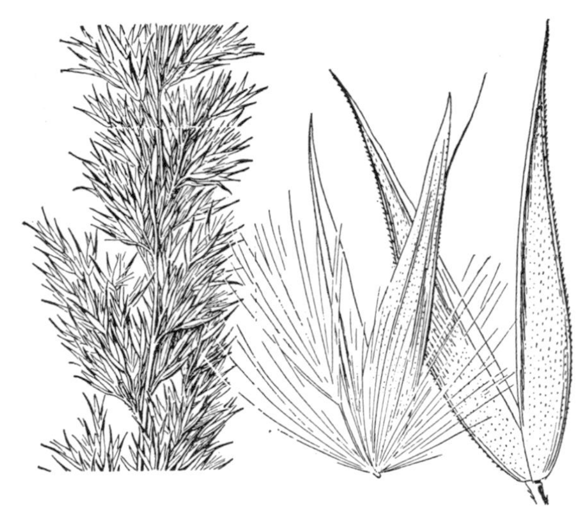

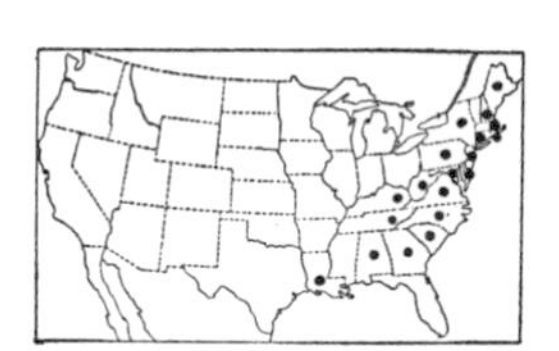

FIGURE 445.—*Calamagrostis cinnoides*. Panicle, × 1; glumes and floret, × 10. (Chase 7518, Md.)

**24. Calamagrostis scopulórum** Jones. (Fig. 446.) Pale, glaucous; culms erect, 50 to 80 cm. tall, with short rhizomes; blades elongate, flat, scabrous, 3 to 7 mm. wide; panicle pale to purplish, contracted, sometimes spikelike, 8 to 15 cm. long; glumes 4 to 6 mm. long, somewhat scabrous, acute or acuminate, not awn-pointed; lemma about as long as the glumes, minutely pilose, the awn attached above the middle, straight, about as long as the lemma, the callus hairs about two-thirds as long; rachilla rather sparsely long-pilose, especially on the upper part. ♃ —Moist soil in gulches, Montana, Wyoming (Wild Cat Peak), Colorado, Utah, New Mexico, and Arizona.

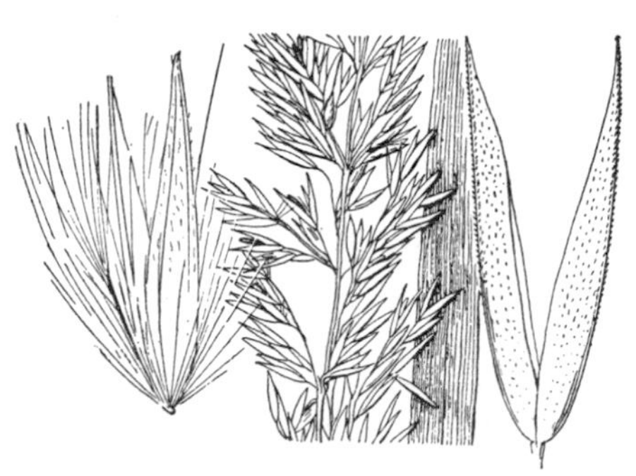

FIGURE 446.—*Calamagrostis scopulorum*. Panicle, × 1; glumes and floret, × 10. (Jones 1145, Utah.)

**25. Calamagrostis inexpánsa** A. Gray. NORTHERN REEDGRASS. (Fig. 447.) Culms tufted, 40 to 120 cm. tall, with rather slender rhizomes, often scabrous below the panicle; sheaths smooth, or somewhat scabrous, the basal ones numerous, withering but persistent; ligule 4 to 6 mm. long; blades firm, rather rigid, flat or loosely involute, very scabrous, 2 to 4 mm. wide; panicle narrow, dense, the branches mostly erect and spikelet-bearing from the base; 5 to 15 cm. long; glumes 3 to 4 mm. long, abruptly acuminate, scaberulous; lemma as long as glumes, scabrous, the awn attached about the middle, straight or nearly so, about as long as glumes, the callus

FIGURE 447.—*Calamagrostis inexpansa.* Panicle, × 1: glumes and floret, × 10. (Ehlers 566, Mich.)

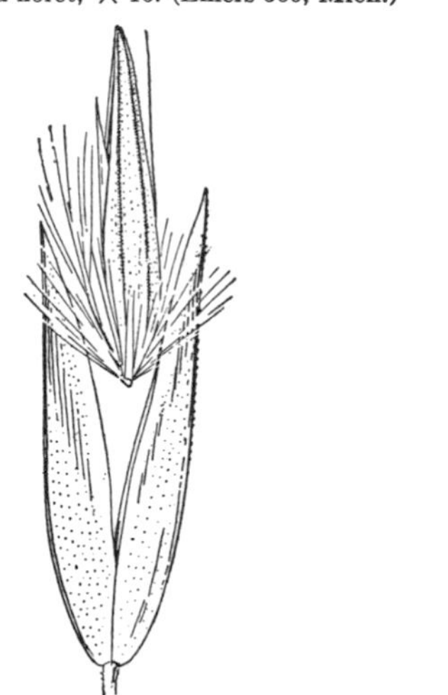

FIGURE 448.—*Calamagrostis californica.* Glumes and floret, × 10. (Type.)

hairs half to three-fourths as long; rachilla 0.5 mm. long, some of the hairs reaching to tip of lemma. ♃ —Meadows, marshes, and wet places, Greenland to Alaska, south to Maine, Virginia (Mountain Lake), Washington, New Mexico, and California. CALAMAGROSTIS INEXPANSA var. NÓVAE-ÁNGLIAE Stebbins. Panicle more loosely flowered, the longer branches naked below. ♃ —Wet granite ledges, Maine to Vermont. CALAMAGROSTIS INEXPANSA var. BARBULÁTA Kearney. Culms robust, puberulent below the nodes; collar of sheaths

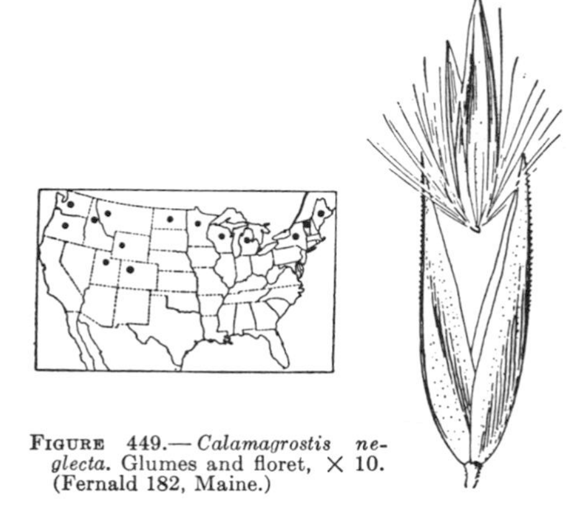

FIGURE 449.—*Calamagrostis neglecta.* Glumes and floret, × 10. (Fernald 182, Maine.)

puberulent; awn minute or obsolete, callus hairs nearly as long as the lemma. ♃ —Known only from Mason County, Wash.

**26. Calamagrostis califórnica** Kearney. (Fig. 448.) Related to *C. inexpansa*, but foliage softer and panicle longer and looser; ligule 2 to 3 mm. long; blades elongate, 1 to 4 mm. wide, mostly involute, scabrous on the upper surface, smooth beneath; panicle 18 to 22 cm. long, the densely flowered branches in rather distant fascicles, some of them naked at base for 1 to 2.5 cm., the axis glabrous; spikelets 3.9 to 4.2 mm. long; glumes acuminate, scabrous; lemma nearly as long as the glumes, strongly nerved, scabrous, the awn attached a little below the middle, straight, scarcely equaling the lemma, the callus hairs scarcely half as long as the lemma, the palea and the hairs of the rachilla about three-fourths as long. ♃ —Only known from a single collection from the Sierra Nevada, particular locality not known.

**27. Calamagrostis neglécta** (Ehrh.) Gaertn. Mey. and Schreb. (Fig. 449.) Resembling *C. inexpansa,* on the average smaller; ligule 1 to 3 mm. long; blades smooth or nearly so, lax and soft, narrow, often filiform; panicles on the average smaller; glumes rather thinner in texture, often smooth ♃ —Marshes, sandy shores, and wet places, Greenland to Alaska, south to Maine, Vermont, New York, Michigan to Washington, Colorado, and Oregon; northern Eurasia.

FIGURE 450.—*Calamagrostis crassiglumis.* Panicle, × 1; glumes and floret, × 10. (Suksdorf 1024, Wash.)

**28. Calamagrostis crassiglúmis** Thurb. (Fig. 450.) Culms rather rigid, 15 to 40 cm. tall, with short rhizomes; lower sheaths overlapping, somewhat papery; blades flat, or somewhat involute, smooth, firm, about 4 to 5 mm. wide; panicle narrow, dense, spikelike, 2 to 5 cm. long, dull purple; glumes 3 to 4 mm. long, ovate, rather abruptly acuminate, purple, scaberulous, firm or almost indurate; lemma about as long as glumes, broad, obtuse or abruptly pointed, the awn attached about the middle, straight, about as long as lemma, the callus hairs abundant, about 3 mm. long; rachilla 1 mm. long, the hairs reaching to apex of lemma. ♃ —Swampy soil, Vancouver Island, Washington (Whatcom Lake), California (Mendocino County). A rare species allied to *C. inexpansa* and *C. neglecta.*

**29. Calamagrostis epigeíos** (L.) Roth. (Fig. 451.) Culms 1 to 1.5 m. tall, with extensively creeping rhizomes; ligule about 4 mm. long, rather firm; blades elongate, 4 to 8, sometimes to 13 mm. wide, scabrous; panicle pale, erect, narrow, rather dense, 25 to 35 cm. long; spikelets crowded; glumes subequal, mostly 5 to 6 mm., sometimes to 8 mm., long, narrowly lanceolate, attenuate; lemma scarcely half as long as the glumes, 2-toothed at the apex, the awn mostly from below the middle, delicate, often obscure, slightly bent, about as long as

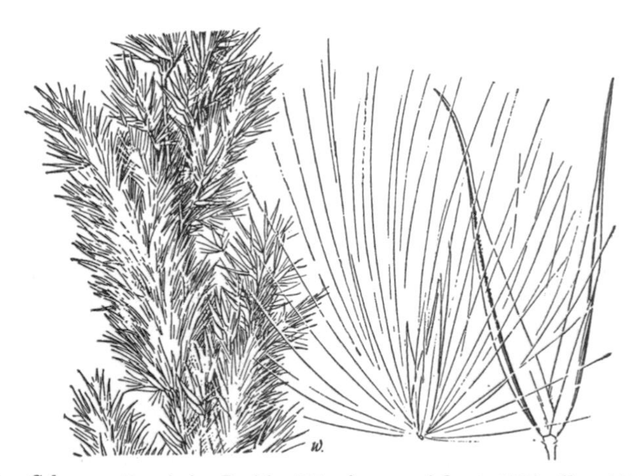

FIGURE 451.—*Calamagrostis epigeios.* Panicle, × 1; glumes and floret, × 10. (Fernald 757, Mass.)

the lemma to equaling the glumes; callus hairs rather copious, about equaling the glumes; rachilla obsolete. ♃ —Sandy woods, salt marshes, fields, and waste ground, near the coast of Massachusetts, Long Island and Saratoga County, N. Y., Montgomery County, Pa., North Dakota to Iowa and Kansas; becoming a weed. Introduced from Eurasia.

## 68. AMMÓPHILA Host. BEACHGRASS

Spikelets 1-flowered, compressed, the rachilla disarticulating above the glumes, produced beyond the palea as a short bristle, hairy above; glumes about equal, chartaceous; lemma similar to and a little shorter than the glumes, the callus bearded; palea nearly as long as the lemma. Tough, rather coarse, erect perennials, with hard, scaly, creeping rhizomes, long, tough, involute blades, and pale, dense spikelike panicles. Type species, *Ammophila arenaria.* Named from the Greek *ammos,* sand, and *philos,* loving, alluding to the habitat.

The species of *Ammophila* are important sand-binding grasses, *A. arenaria* being used in northern Europe to hold the barrier dunes along the coast. In this country it has been tried with success on Cape Cod and at Golden Gate Park, San Francisco. Called also marram, psamma, and sea sandreed.

Ligule thin, 10 to 30 mm. long .......... 2. A. ARENARIA.
Ligule firm, 1 to 3 mm. long .......... 1. A. BREVILIGULATA.

**1. Ammophila breviliguláta** Fernald. AMERICAN BEACHGRASS. (Fig. 452.) Culms in tufts, commonly 70 to 100 cm. tall with deep strong extensively creeping rhizomes, the base of the culms clothed with numerous broad overlapping sheaths; ligule firm, 1 to 3 mm. long; blades elongate, firm, soon involute, curved forward past the culm, the scaberulous upper surface downward; panicle pale, 15 to 30 cm. long, nearly cylindrical; spikelets 11 to 14 mm. long; glumes scaberulous, the first 1-nerved, the second 3-nerved; lemma scabrous, the callus hairs about 2 mm. long, the rachilla about 3 mm. long. ♃ —Sand dunes along the coast from Newfoundland to North Carolina, and on the shores of the Great Lakes from Lake Ontario to Lake Superior and Lake Michigan.

**2. Ammophila arenária** (L.) Link. EUROPEAN BEACHGRASS. (Fig. 453.) Like the preceding in habit, the culms sometimes thicker; ligule thin, 1 to 3 cm. long; panicle often thicker in the middle, tapering to the summit; spikelets 1.2 to 1.5 cm. long; callus hairs about 3 mm. long, the rachilla 2 mm. long. ♃ —Sand dunes along the coast from San Francisco to Washington. Introduced as a sand binder in the vicinity of San Francisco and now established at several places to the north; coast of Europe.

## 69. CALAMOVÍLFA Hack.

Spikelets 1-flowered, the rachilla disarticulating above the glumes, not prolonged behind the palea; glumes unequal, chartaceous, 1-nerved, acute; lemma a little longer than the second glume, chartaceous, 1-nerved, awnless, glabrous or pubescent, the callus bearded; palea about as long as the lemma. Rigid, usually tall perennials, with narrow or open panicles, some species with creeping rhizomes. Type species, *Calamovilfa brevipilis.* Name from Greek *kalamos,* reed, and *Vilfa,* a genus of grasses. *Calamovilfa longifolia* is of some value for forage, but is rather coarse and woody; a variety of this and also *C. gigantea* are inland sand binders.

Rhizomes short and thick.
Panicle narrow, contracted .......... 1. C. CURTISSII.

Panicle subpyramidal, rather open........ 2. C. BREVIPILIS.
Rhizomes extensively creeping.
Lemma glabrous (except for the callus hairs)........ 3. C. LONGIFOLIA.
Lemma villous on the back above the callus hairs........ 4. C. GIGANTEA.

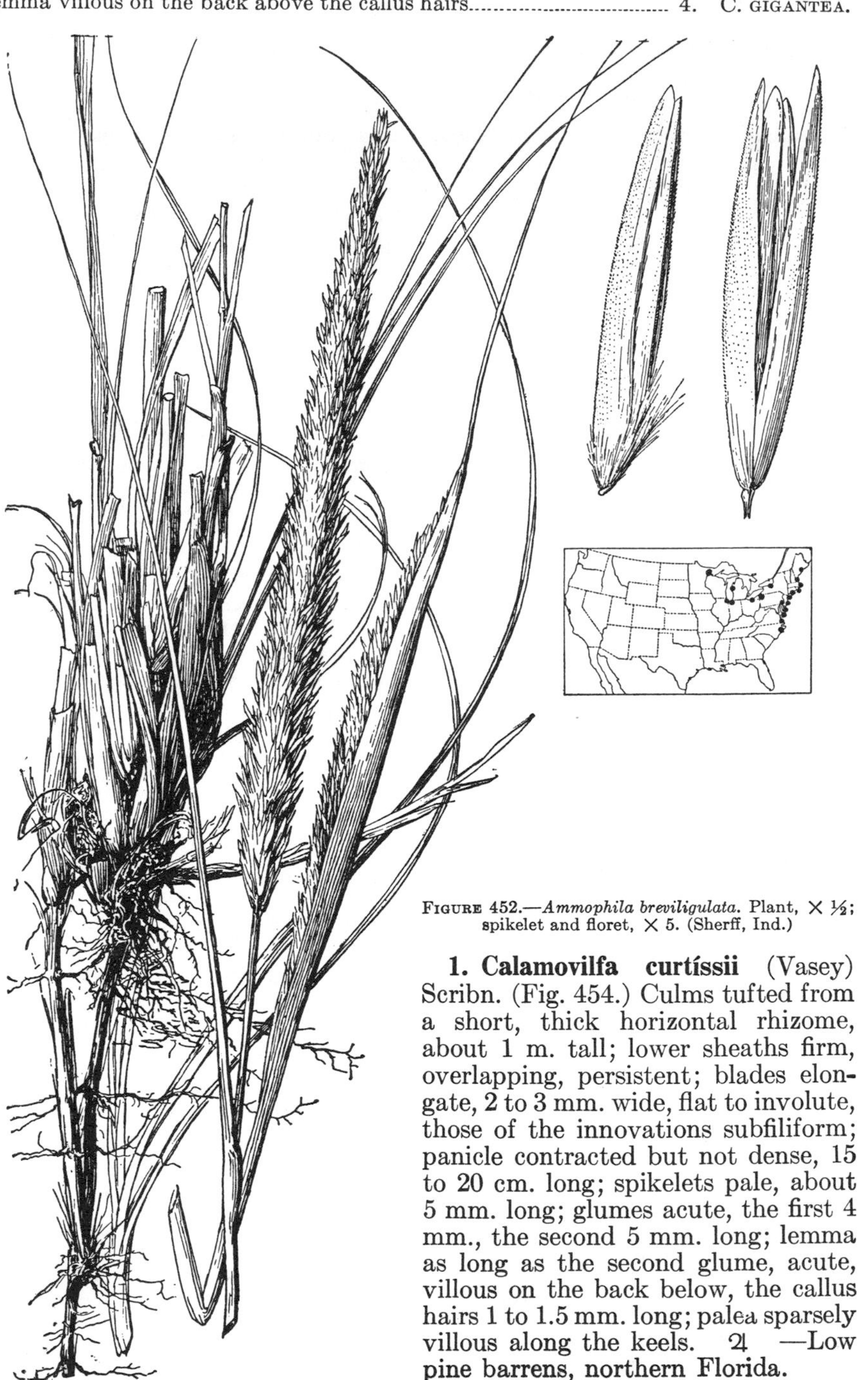

FIGURE 452.—*Ammophila breviligulata*. Plant, × ½; spikelet and floret, × 5. (Sherff, Ind.)

**1. Calamovilfa curtíssii** (Vasey) Scribn. (Fig. 454.) Culms tufted from a short, thick horizontal rhizome, about 1 m. tall; lower sheaths firm, overlapping, persistent; blades elongate, 2 to 3 mm. wide, flat to involute, those of the innovations subfiliform; panicle contracted but not dense, 15 to 20 cm. long; spikelets pale, about 5 mm. long; glumes acute, the first 4 mm., the second 5 mm. long; lemma as long as the second glume, acute, villous on the back below, the callus hairs 1 to 1.5 mm. long; palea sparsely villous along the keels. ♃ —Low **pine barrens, northern Florida.**

**2. Calamovilfa brevípilis** (Torr.) Scribn. (Fig. 455.) Culms solitary or few, compressed, 60 to 120 cm. tall, the base as in *C. curtissii;* blades elongate, 2 to 3 mm. wide, flat to subinvolute; panicle subpyramidal, rather open, 10 to 25 cm. long, the branches ascending, flexuous, naked below; pedicels sparsely pilose at the summit; spikelets brownish, 5 to 6 mm. long; glumes acuminate, the first 2 to 2.5 mm. long, the second about 4 mm. long; lemma villous on the back below, the callus hairs 1.5 mm. long; palea exceeding the lemma, villous on the back. ♃ —Marshes and river banks, New Jersey.

CALAMOVILFA BREVIPILIS var. CÁLVIPES Fernald. Very similar to the species; panicles looser, more open; pedicels glabrous; spikelets 4 to 5 mm. long, the lemma and palea about equal. ♃ —Sphagnous bog, Greensville County, Va.

FIGURE 454.—*Calamovilfa curtissii.* Plant, × ½; glumes and floret, × 5. (Garber, Fla.)

FIGURE 453.—*Ammophila arenaria.* Glumes, floret, and ligule, × 5. (Heller 5670.)

CALAMOVILFA BREVIPILIS var. HETERÓLEPIS Fernald. Panicles somewhat narrower; pedicels with a few short hairs at summit; spikelets more crowded toward the ends of the branches, 5.5 to 6 mm. long, the palea slightly shorter than the lemma. ♃ —Edge of swamps and moist savannas, Virginia to South Carolina.

**3. Calamovilfa longifólia** (Hook.) Scribn. (Fig. 456.) Culms mostly solitary, 50 to 180 cm. tall, with strong scaly creeping rhizomes; sheaths usually more or less appressed-villous, especially near the summit; blades firm, elongate, flat or soon involute, 4 to 8 mm. wide near base, tapering to a long fine point; panicle 15 to 35 cm. long, rather narrow or contracted,

FIGURE 455.—*Calamovilfa brevipilis*. Plant, × ½; glumes and floret, × 5. (Brinton, N. J.)

the branches ascending or appressed, sometimes slightly spreading; spikelets pale, 6 to 7 mm. long; glumes acuminate, the first about 2 mm. shorter than the second; lemma somewhat shorter than the second glume, glabrous, the callus hairs copious, more than half as long as the lemma. ♃ —Sand hills and sandy prairies or open woods, Michigan to Alberta, south to Indiana, Colorado, and Idaho. CALAMOVILFA LONGIFOLIA var. MÁGNA Scribn. and Merr. Panicle more open and spreading. ♃ —Sandy ridges and dunes along Lakes Huron and Michigan.

**4. Calamovilfa gigantéa** (Nutt.) Scribn. and Merr. (Fig. 457.) Culms robust, mostly solitary, usually 1.5 to 2 m. tall, as much as 6 mm. thick at base, with strong creeping rhizomes; sheaths glabrous; blades elongate, 5 to 10 mm. wide at base, tapering to a long involute tip; panicle open, as much as 60 cm. long, the branches rather stiffly spreading, as much as 25 cm. long; spikelets similar to those of *C. longifolia*, but somewhat larger; lemma and palea villous along the back; callus hairs copious, half as long as the lemma. ♃ —Sand dunes, Kansas to Utah, Texas, and Arizona.

## 70. APÉRA Adans.

(Included in *Agrostis* L. in Manual, ed. 1)

Spikelets 1-flowered, disarticulating above the glumes, the rachilla prolonged back of the palea as a naked bristle; glumes subequal, acuminate; lemma firm, subindurate at maturity, acute, bearing a long delicate straight awn just below the tip; palea nearly as long as the lemma, strongly 2-nerved. Annuals with flat blades and loose or narrow panicles. Type species, *Apera spica-venti* (L.) Beauv. Name from Greek *a*, not, and *peros*, maimed, apparently alluding to the long awn, this nearly wanting in *Calamagrostis calamagrostis* (L.) Karst. (*C. lanceolata* Roth), from which Adanson differentiated the genus.

Panicle open, the branches naked below .......... 1. A. SPICA-VENTI.
Panicle narrow, contracted, interrupted, the branches, or some of them floriferous from the base .......... 2. A. INTERRUPTA.

**1. Apera spíca-vénti** (L.) Beauv. (Fig. 458.) Annual; culms branched at base, mostly 40 to 60 cm. tall; ligule as much as 6 mm. long; blades flat, 1 to 3 mm. wide; panicle 10 to

FIGURE 456.—*Calamovilfa longifolia*. Plant, × ½; spikelet and floret, × 5. (Babcock, Ill.)

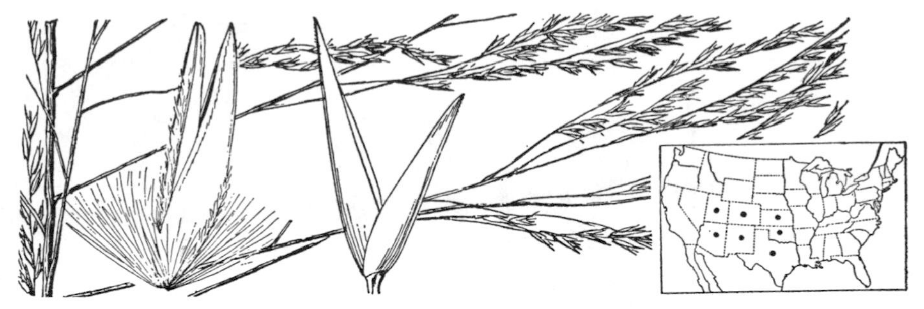

FIGURE 457.—*Calamovilfa gigantea.* Panicle, × ½; glumes and floret, × 5. (White, Okla.)

20 cm. long, usually less than half as broad, the branches capillary, spreading, whorled, naked at base; spikelets 2 to 2.5 mm. long; glumes somewhat unequal, the first shorter and narrower; lemma about as long as the second glume, scaberulous, with a slender awn from below the apex, the awn about twice as long as the glumes; palea about as long as the lemma; rachilla less than 0.5 mm. long. ⊙ —Introduced at a few points from Maine to Maryland; Ohio; Missouri; Portland, Oreg.; Europe.

**2. Apera interrúpta** (L.) Beauv. (Fig. 459.) Similar to *A. spica-venti;* panicle narrower, more condensed, interrupted, the branches or some of them floriferous from the base; awn of lemma about 1 cm. long. ⊙ —Introduced in Missouri (St. Louis), Washington (Spokane), Oregon (Portland), Idaho (Nezperce Forest), and British Columbia (Okanogan); Europe.

## 71. AGRÓSTIS L. BENTGRASS

Spikelets 1-flowered, disarticulating above the glumes, the rachilla usually not prolonged; glumes equal or nearly so, acute, acuminate, or sometimes awn-pointed, usually scabrous on the keel and sometimes on the back; lemma obtuse, usually shorter and thinner than the glumes, mostly 3-nerved, awnless or dorsally awned, often hairy on the callus; palea usually shorter than the lemma, 2-nerved in only a few species, usually small and nerveless or obsolete. Delicate to moderately tall annuals or usually perennials, with flat or sometimes involute, scabrous blades, and open to contracted panicles of small spikelets. Type species, *Agrostis stolonifera.* Name from Greek *agrostis,* a kind of grass, from *agros,* a field; the word agrostology is from the same root. The rachilla is regularly prolonged in a few species and in occasional spikelets of other species.

Most of the species are important forage plants, either under cultivation or in the mountain meadows of the Western States. The three important cultivated species are redtop, *Agrostis alba,* used for meadows, pastures, lawns, and sports turf, Colonial bent, *A. tenuis,* used for pastures, lawns, and sports turf, and creeping bent, *A. palustris,* used for lawns and golf greens. Velvet bent, *A. canina,* is sometimes used for putting greens. Recently forms of *A. palustris,* called Washington bent and Metropolitan bent, have come into use for lawns and especially for golf greens. They are propagated by the stolons. Fiorin is a name applied in England to *A. palustris.*

The native species abundant enough to be of importance as forage plants are *A. exarata,* throughout the western part of the United States, *A. oregonensis* in Oregon, and *A. variabilis* in alpine regions of the Northwest.

1a. Palea evident, 2-nerved, at least half as long as the lemma.

FIGURE 459.—*Apera interrupta*. Panicle, × ½; glumes and floret, × 5. (Bonser 3, Wash.)

FIGURE 458.—*Apera spica-venti*. Plant, × ½; glumes and floret, × 5. (Martindale, N. J.)

2a. Rachilla prolonged behind the palea as a minute bristle.
  Lemma pubescent .......... 1. A. AVENACEA.
  Lemma glabrous.
    Spikelets 2 mm. long .......... 2. A. THURBERIANA.
    Spikelets 3 mm. long .......... 3. A. AEQUIVALVIS.
2b. Rachilla not prolonged.
  Glumes scabrous on the keel and on the back; panicle contracted, lobed, the short branches densely verticillate .......... 4. A. SEMIVERTICILLATA.
  Glumes scabrous on the keel only; panicle open or, if contracted, not lobed nor with densely verticillate branches.
    Plants tufted; dwarf alpine species .......... 10. A. HUMILIS.
    Plants with rhizomes or stolons; taller species of low and medium altitudes.
      Branches of panicle naked at base, the panicle open and delicate; ligule as much as 2 mm. long on culm leaves, less than 1 mm. on the innovations. 9. A. TENUIS.
      Branches of panicle or some of them floriferous from base; ligule as much as 6 mm. long.
        Panicle contracted, the branches appressed; long stolons developed in isolated plants. Culms decumbent at base .......... 6. A. PALUSTRIS.
        Panicle open, the branches ascending, no long stolons developed.
          Culms producing rather stout creeping leafy stolons .......... 7. A. NIGRA.
          Culms decumbent at base; rhizomes wanting .......... 5. A. STOLONIFERA.
          Culms erect; rhizomes developed .......... 8. A. ALBA.
1b. Palea obsolete, or a minute nerveless scale (in *A. exarata* and *A. californica* as much as 0.5 mm. long or more).
  3a. Plants annual, lemma with a slender awn, geniculate or flexuous.
    Lemma awnless .......... 11. A. ROSSAE.
    Lemma with a slender geniculate or flexuous awn.
      Awn flexuous, delicate; Southeastern States .......... 12. A. ELLIOTTIANA.
      Awn geniculate; Pacific coast.
        Spikelets about 1.5 mm. long; lemma awned below the tip .......... 13. A. EXIGUA.
        Spikelets at least 2.5 mm. long; lemma awned from the middle.
          Apex of lemma obscurely toothed or nearly entire; lemma 1.7 to 1.9 mm. long. 16. A. MICROPHYLLA.
          Apex of lemma bearing 2 or 4 delicate awns.
            Lemma pilose; glumes 3.5 to 4 mm. long .......... 15. A. KENNEDYANA.
            Lemma glabrous except on the callus; glumes 5 to 6 mm. long.
              Lemma relatively firm, scabrous, 3.2 to 3.5 mm. long; palea nearly ⅓ as long as the lemma .......... 17. A. ARISTIGLUMIS.
              Lemma thin, glabrous, 3 mm. long; palea obsolete .......... 14. A. HENDERSONI.
  3b. Plants perennial; lemma awned or awnless, the awn when present not much exserted.
    4a. Plants spreading by creeping rhizomes (those of *A. lepida* short).
      Hairs at base of lemma 1 to 2 mm. long .......... 18. A. HALLII.
      Hairs at base of lemma minute or wanting.
        Rhizomes short; alpine tufted plants .......... 19. A. LEPIDA.
        Rhizomes long and slender.
          Panicle spikelike .......... 20. A. PALLENS.
          Panicle open .......... 21. A. DIEGOENSIS.
    4b. Plants without rhizomes, stolons sometimes developed.
      5a. Panicle narrow, contracted, at least some of the lower branches spikelet-bearing from the base.
        Culms slender, not more than 20 cm. tall, in dense tufts with numerous basal leaves; blades not more than 7 cm. long, mostly less, less than 2 mm. wide; panicles seldom more than 5 mm. wide.
          Culms spreading; panicles strict, greenish; lemma with a minute awn or the midnerve ending below the summit .......... 22. A. BLASDALEI.
          Culms erect; panicle narrow but loose, purple; lemma awnless, the midnerve reaching the summit .......... 23. A. VARIABILIS.
        Culms taller, stouter, not in tufts with dense basal foliage; blades or some of them at least 8 to 10 cm. long and 3 to 5 mm. wide, commonly much larger; glumes scabrous on the keel.
          Panicle from loose to dense; lemma acute, not toothed; palea minute.
            Panicle loose, the branches verticillate, not densely flowered at base; awn of lemma twisted, geniculate .......... 25. A. AMPLA.
            Panicle dense to loose, the branches crowded and densely flowered at base; lemma awnless or (in vars. *pacifica* and *monolepis*) awned. 24. A. EXARATA.

Panicle dense and spikelike; lemma minutely 4-toothed; palea ¼ to ⅓ as long as the lemma........ 26. A. CALIFORNICA.

5b. Panicle open, sometimes diffuse; branches very slender, scabrous, the lower branches not spikelet-bearing at the base.

Lemma awned from near the base.

Blades elongate, 3 to 5 mm. wide; panicle branches flexuous; spikelets about 3.5 mm. long........ 28. A. HOWELLII.

Blades about 1 mm. wide or less; panicle branches straight; spikelets 2 to 2.5 mm. long........ 27. A. HOOVERI.

Lemma awnless or awned from the middle or above.

Panicle very diffuse, the capillary branches branching toward the end or (in *A. scabra* var. *geminata*) above the middle.

Spikelets 1.5 to 1.7 mm. long, very densely clustered at the ends of the branchlets; lemma 1 to 1.2 mm. long, scarcely longer than the caryopsis; anthers about 0.2 mm. long........ 29. A. HIEMALIS.

Spikelets 2 to 2.7 mm. long, loosely arranged at the ends of the branchlets; lemma 1.5 to 1.7 mm. long, distinctly longer than the caryopsis; anthers 0.4 to 0.5 mm. long........ 30. A. SCABRA.

Panicle open but not diffuse, the branches branching at or below the middle.

Lemma awnless (occasional plants with awned lemmas).

Spikelets about 2 mm. long; plants of high altitudes, delicate, 10 to 30 cm. tall........ 31. A. IDAHOENSIS.

Spikelets 2 to 3 mm. long; more robust plants of low and medium altitudes.

Panicle rather lax, sometimes delicate and divaricately spreading; blades flat, as much as 6 mm. wide; eastern United States.

Spikelets mostly 2.2 to 2.7 mm. long, not aggregate or but slightly so at the ends of the panicle branches........ 32. A. PERENNANS.

Spikelets mostly 2.7 to 3.5 mm. long, aggregate towards the ends of the panicle branches........ 33. A. ALTISSIMA.

Panicle rather stiff, the branches whorled and rather stiffly ascending; Pacific coast........ 34. A. OREGONENSIS.

Lemma awned.

Spikelets about 2 mm. long; introduced........ 35. A. CANINA.

Spikelets 2.5 to 3 mm. long; native.

Ligule 1 to 2 mm. long........ 36. A. BOREALIS.

Ligule 5 to 8 mm. long........ 37. A. LONGILIGULA.

FIGURE 460.—*Agrostis avenacea*. Panicle, × ½; glumes and floret, × 10. (Tracy and Earle 403, Tex.)

**1. Agrostis avenácea** Gmel. (Fig. 460.) Perennial; culms tufted, erect or decumbent at base, 20 to 60 cm. tall; sheaths smooth; ligule of culm leaves 3 to 5 mm. long; blades flat, scabrous, 1 to 2 mm. wide; panicle diffuse, 15 to 30 cm. long, the branches in distant whorls, capillary, reflexed at maturity, divided above the middle; glumes acuminate, 3 to 4 mm. long; lemma about half as long as the glumes, thin, pubescent, short-bearded on the callus, and bearing about the middle a slender geniculate and twisted awn exserted about the length of the glumes; palea nearly as long as the lemma; rachilla slender, pilose, from half to as long as the lemma. ♃ (*A. retrofracta* Willd.)—Introduced in central California (15 miles south of Stockton), Texas (Kent), and Ohio (Painesville); common in Hawaiian Islands and Polynesia.

FIGURE 461.—*Agrostis thurberiana.* Panicle, × 1; glumes and floret, × 5. (Type.)

FIGURE 462.—*Agrostis aequivalvis.* Panicle, × 1; glumes and floret, × 5. (Howell 1712, Alaska.)

**2. Agrostis thurberiána** Hitchc. THURBER BENT. (Fig. 461.) Culms slender, in small tufts, erect, 20 to 40 cm. tall; leaves somewhat crowded at base, the blades about 2 mm. wide; panicle rather narrow, lax, more or less drooping, 5 to 7 cm. long; spikelets green, pale, or purple, 2 mm. long; lemma nearly as long as the glumes, the palea about two-thirds as long; rachilla hairy, 0.3 mm. long. ♃ —Bogs and moist places, at medium and upper altitudes, Colorado to British Columbia and south in the Sierras to central California.

**3. Agrostis aequiválvis** (Trin.) Trin. (Fig. 462.) Similar to *A. thurberiana;* culms on the average taller, blades longer; panicle usually purple, 5 to 15 cm. long; spikelets about 3 to 4.5 mm. long; palea nearly as long as the lemma; rachilla minutely pubescent, one-fifth to half as long as the lemma. ♃ —Wet meadows and bogs, Alaska, southward (rare) in the Cascade Mountains to Oregon.

**4. Agrostis semiverticilláta** (Forsk.) C. Christ. WATER BENT. (Fig. 463.) Culms usually decumbent at base, sometimes with long creeping and rooting stolons; blades firm, mostly relatively short and broad, but in luxuriant specimens elongate; panicle contracted, 3 to 10 cm. long, densely flowered, lobed, with short verticillate branches, especially at base, the branches spikelet-bearing from the base; spikelets usually falling entire; glumes equal, narrowed to an obtuse tip, scabrous on back and keel, 2 mm. long; lemma 1 mm. long, awnless, truncate and toothed at apex; palea nearly as long as the lemma. ♃ (*A. verticillata* Vill.)—Moist ground at low altitudes, especially along irrigation ditches (in irrigated regions), Texas to California, north to Utah and Washington; on ballast at some Atlantic ports. Introduced in America, south to Argentina; warmer parts of the Eastern Hemisphere.

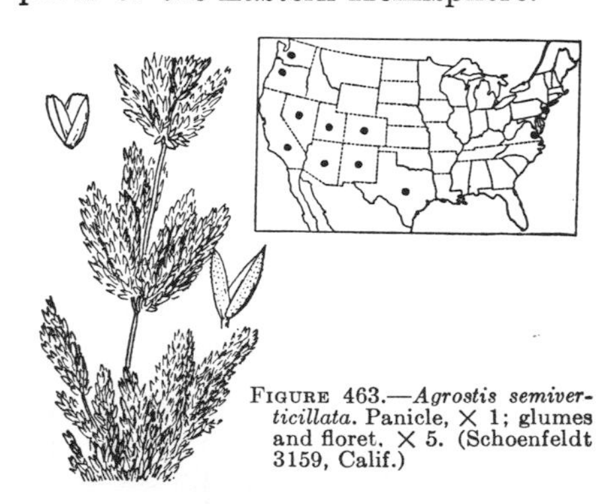

FIGURE 463.—*Agrostis semiverticillata.* Panicle, × 1; glumes and floret, × 5. (Schoenfeldt 3159, Calif.)

**5. Agrostis stolonífera** L. (Fig. 464.) Culms ascending from a spreading base, the decumbent portion rooting in wet soil, 20 to 50 cm. tall; ligule as much as 6 mm. long; blades flat, mostly 1 to 3 mm. wide; panicle oblong, 5 to 15 cm. long, pale or purple, somewhat open, the branches or some of them spikelet-bearing from near the base; spikelets 2 to 2.5 mm. long; glumes acute, glabrous except the scabrous keel; lemma shorter than the glumes, awnless or rarely awned from the back; palea usually half to two-thirds as long as the lemma. ♃ —Moist grassy places, Newfoundland to Alaska, south to Virginia (adventive in South Carolina) in the East and to Washington in the West; northern Europe. This species appears to be native in northern North America.

**6. Agrostis palústris** Huds. CREEPING BENT. (Fig. 465.) Differing from *A. stolonifera* chiefly in the long stolons, the narrow stiff appressed blades, and the condensed (sometimes somewhat open) panicle. ♃ (*A. maritima* Lam.)—Marshes along the

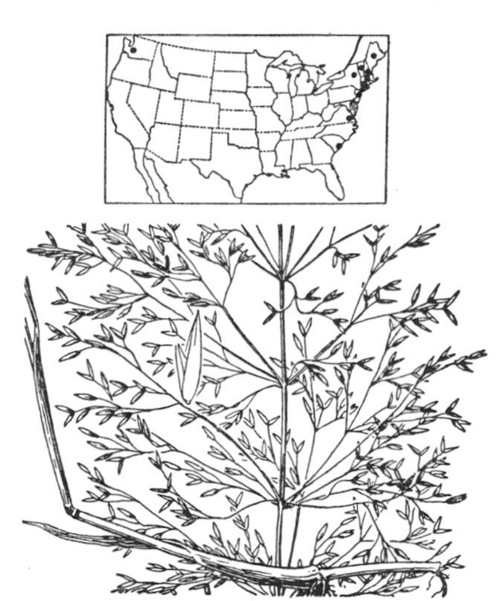

FIGURE 464.—*Agrostis stolonifera*. Panicle, × 1; floret, × 5. (Hitchcock 23899, Newfoundland.)

coast, from Newfoundland to Virginia; British Columbia to northern California; sometimes occupying extensive areas, as at Coos Bay, Oreg.; introduced at various places in the interior of southern Canada and northeastern United States to Virginia and Wisconsin, and occasion-

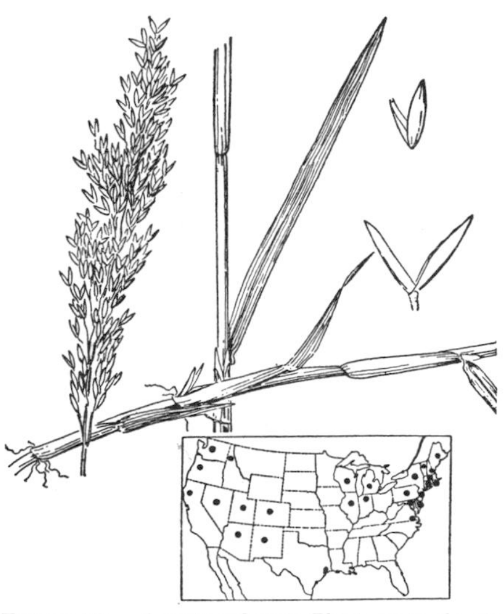

FIGURE 465.—*Agrostis palustris*. Plant, × 1; glumes and floret, × 5. (Hitchcock 11713, Wash.)

FIGURE 466.—*Agrostis nigra*. Plant, × 1; floret, × 5. (Moore 47, cult. Mo. Bot. Gard.)

ally southward, Texas to Arizona, especially along ditches; Idaho and Washington to Colorado and California; Eurasia. Forms of this species, known as seaside, Coos Bay, and Cocoos bents (propagated by seed), and Metropolitan and Washington

FIGURE 467.—*Agrostis alba*. Plant, × ½; 2 spikelets and floret, × 5. (Chase 5191, Mont.)

bents (propagated by stolons and formerly called carpet bent), are used for lawns and extensively for putting greens.

**7. Agrostis nígra** With. Black bent. (Fig. 466.) Culms long-decumbent at base, also with rather stout leafy stolons, the fertile branches ascending or erect, 20 to 30 cm. tall; ligule as in *A. alba;* panicle brownish, open as in *A. alba,* but on the average more condensed along the branches, the base usually partly included. ♃ —Sometimes found mixed with "South German" bent (creeping bent), hence may be a constituent of lawns grown from imported seed; Europe.

**8. Agrostis álba** L. Redtop. (Fig. 467.) Differing from *A. stolonifera* in its usually erect more robust culms, sometimes as much as 1 to 1.5 m. tall, the base erect or decumbent, with strong creeping rhizomes; blades flat, 5 to 10 mm. wide; panicle pyramidal-oblong, reddish, as much as 20 cm. long, the branches spreading in anthesis, sometimes contracting later; lemmas rarely awned ♃ (*A. gigantea* Roth.)—This is the common redtop cultivated for meadows, pastures, and lawns, extensively escaped in all the cooler parts of the United States; Eurasia. This form appears not to be native in America. Plants growing without cultivation often have pale panicles and may tend to take on the aspect of *A. stolonifera.* This and the two preceding are closely allied and appear to intergrade. The name *A. palustris* has been erroneously applied to this species.

Figure 468.—*A, Agrostis tenuis.* Panicle, × 1; glumes, floret, and ligule, × 5. (Waghorne, Newfoundland.) *B,* Var. *aristata.* Floret, × 5. (Gayle 786, Maine.)

**9. Agrostis ténuis** Sibth. Colonial bent. (Fig. 468, *A.*) Culms slender, erect, tufted, usually 20 to 40 cm. tall, with short stolons but no creeping rhizomes; ligule short, less than 1 mm. or on the culm as much as 2 mm. long; blades mostly 5 to 10 cm. long, 1 to 3 mm. wide; panicle mostly 5 to 10 cm. long, open, delicate, the slender branches naked below, the spikelets not crowded. ♃ (*A. vulgaris* With.)—Cultivated for pastures and lawns in the northeastern United States; escaped and well established throughout those regions; Newfoundland south to North Carolina, West Virginia, and Michigan; British Columbia to Montana and California; Europe. This species appears not to be native in America; it has been referred to *A. capillaris* L., a distinct species of Europe. In older works this has been called Rhode Island bent. Forms of this species are sometimes called Prince Edward Island, New Zealand, Rhode Island Colonial, Astoria, and Colonial bent. Highland bent is an aberrant form which may be a distinct species.

Agrostis tenuis var. aristáta (Parnell) Druce. (Fig. 468, *B.*) Differing from *A. tenuis* in having lemma awned from near the base, the awn usually geniculate and exceeding the glumes. ♃ —Fields and open woods, Nova Scotia and Quebec to North Carolina; Alaska to Vancouver Island; northern California; Europe. This form appears to be native, at least in the more northerly part of its range.

**10. Agrostis húmilis** Vasey. (Fig. 469.) Culms low, tufted, mostly not more than 15 cm. tall; leaves mostly basal, the blades flat or folded, usually not more than 1 mm. wide; panicle narrow, purple, 1 to 3 cm. long, the branches appressed to somewhat spreading; spikelets about 2

FIGURE 469.—*Agrostis humilis*. Panicle, × 1; glumes and floret, × 5. (Type.)

mm. long; lemma nearly as long as the glumes, awnless; palea about two-thirds as long as lemma. ♃ —Bogs and alpine meadows at high altitudes, Wyoming and Colorado to Washington, Oregon, and Nevada.

FIGURE 470.—*Agrostis rossae*. Panicle, × 1; glumes and floret, × 5. (V. H. Chase 5740, Yellowstone Natl. Park, Wyo.)

**11. Agrostis róssae** Vasey. (Fig. 470.) Annual, erect, leafy and branching at base, 10 to 19 cm. tall; sheaths rather loose; blades flat, 1 to 2.5 cm. long, 1 to 2 mm. wide; panicle 3.5 to 6 cm. long, usually contracted, the capillary scabrous purplish branches in relatively distant fascicles, narrowly ascending, naked at base; spikelets 2 to 2.5 mm. long; glumes acuminate; lemma 1.5 to 1.6 mm. long, minutely toothed, awnless; palea very minute. ⊙ —Alkali soil near hot springs, Upper Geyser Basin and along Fire Hole River, Yellowstone Park, Wyo.

**12. Agrostis elliottiána** Schult. (Fig. 471.) Annual; culms slender, erect or decumbent at base, 10 to 40 cm. tall; blades flat, about 1 mm. wide; panicle finally diffuse, about half the entire height of the plant, the branches capillary, fascicled, the spikelets toward the ends of the branchlets, the whole panicle breaking away at maturity; spikelets 1.5 to 2 mm. long; glumes acute; lemma

FIGURE 471.—*Agrostis elliottiana*. Panicle, × 1; glumes and floret, × 5. (Johnson, Miss.)

1 to 1.5 mm. long, minutely toothed, awned below the tip, the awn very slender, flexuous, delicately short-pilose, 5 to 10 mm. long, sometimes falling at maturity; palea wanting. ⊙ —Fields, waste places, and open ground, Maryland to Kansas, south to Georgia and eastern Texas; introduced in Maine and Massachusetts; Yucatan.

FIGURE 472.—*Agrostis exigua*. Panicle, × 1; glumes and floret, × 5. (Type.)

**13. Agrostis exígua** Thurb. (Fig. 472.) Annual; culms delicate, 3 to 10 cm. tall, branching from the base; blades 5 to 20 mm. long, subinvolute, scabrous; panicle half the length of the plant, finally open; glumes 1.5 mm. long, scaberulous; lemma equaling the glumes, scaberulous toward the 2-toothed apex, bearing below the tip a delicate bent awn 4 times as long; palea wanting. ⊙ —Foothills and rocky plains, upper Sacramento Valley, and muddy pond border, Howell Mountain, Napa County, Calif.

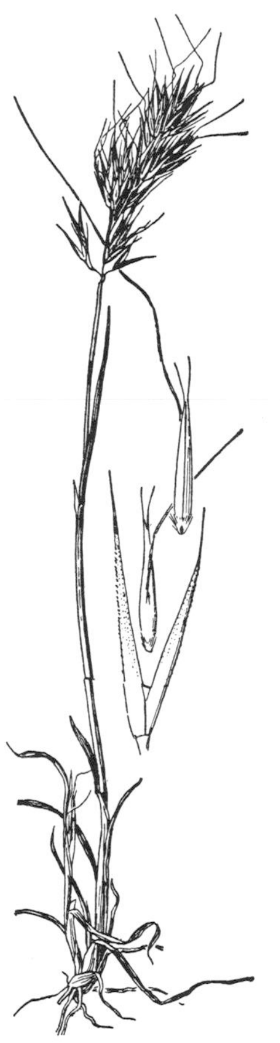

FIGURE 473.—*Agrostis hendersonii*. Plant, × 1; glumes and 2 views of floret, × 5. (Type.)

**14. Agrostis hendersónii** Hitchc. (Fig. 473.) Annual; culms about 10 cm. tall; ligule 2 to 3 mm. long; blades flat or loosely involute, 1 to 3 cm. long, about 1 mm. wide; ligule delicate, about 2 mm. long; panicle condensed, about 2.5 cm. long, purplish; spikelets short-pediceled, 5 to 6 mm. long; glumes subequal, setaceous-tipped; lemma about 3 mm. long, finely 2-toothed, the delicate awns of the teeth readily breaking off, awned from the middle, the awn about 1 cm. long, geniculate, the callus pubescent; palea obsolete. ⊙ —Wet ground, known only from Sams Valley, near Gold Hill, Jackson County, Oreg., and Shasta County, Calif.

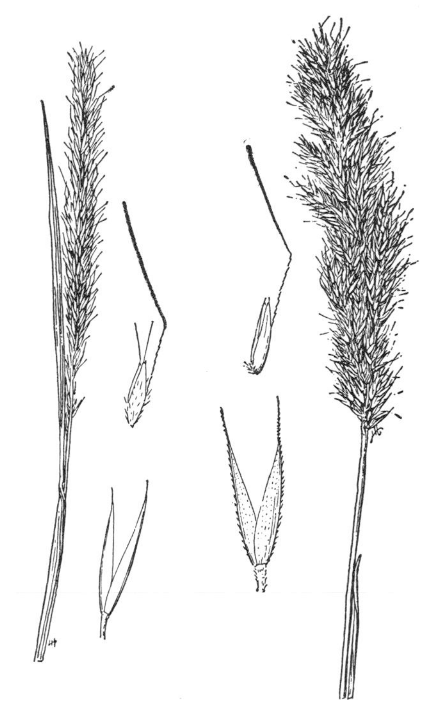

FIGURE 474.—*Agrostis kennedyana*. Panicle, × 1; glumes and floret, × 5. (Type collection.)

FIGURE 475.—*Agrostis microphylla*. Panicle, × 1; glumes and floret, × 5. (Bolander 1512, Calif.)

**15. Agrostis kennedyána** Beetle. (Fig. 474.) Annual; culms very slender, 15 to 23 cm. tall; ligule about 2 mm. long; blades flat or loosely involute, 2 to 4 cm. long, 1 to 1.5 mm. wide, or the basal blades slightly

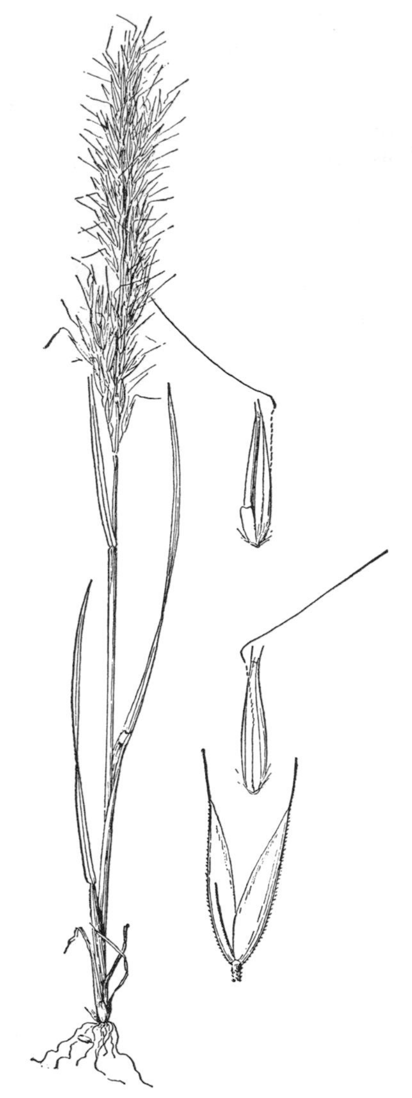

FIGURE 476.—*Agrostis aristigluma*. Plant, × ½; glumes and two views of floret, × 5. (Type.)

longer; panicle spikelike, 2 to 5 cm. long, pale; spikelets short-pediceled; glumes narrow, acuminate, the first 3.5 to 4 mm., the second 3 to 3.2 mm. long; lemma delicate, about 1.7 mm. long, awned from about the middle, the delicate awn about 5 mm. long, geniculate, the lemma loosely pilose except at the 2-toothed summit, the teeth bearing delicate awns about 1 mm. long; palea obsolete. ⊙ —Known only from San Diego County, Calif.

**16. Agrostis microphýlla** Steud. (Fig. 475.) Annual; culms branching at base, slender, erect or ascending, 8 to 40 cm. tall, commonly short and tall culms in the same tuft; blades 2 to 15 cm. long, rarely longer, 1.5 to 3 mm. wide, scabrous; panicle mostly 2 to 8 cm. long (exceptionally less or to 10 cm.), narrow, dense, often lobed; glumes subequal, 3 to 4.4 mm. long, acuminate to awn-tipped; lemma 1.7 to 1.9 mm. long, minutely toothed, awned from about the middle, the awn geniculate, 3.5 to 6 mm. long, rarely longer; palea wanting. ⊙ —Moist open ground, Vancouver Island, Oregon, California, and Baja California. Variable, occasionally small and delicate; glumes rarely only 2 to 2.5 mm. long. A short, densely tufted form with rather thick panicle has been differentiated as *A. inflata* Scribn. The type, from Vancouver Island, is a young plant, the panicles partly included in the slightly inflated upper sheaths. A. MICROPHYLLA var. MÁJOR Vasey is a taller form, 40 to 55 cm. tall, the pale panicles to 15 cm. long; glumes 2.5 to 3 mm. long; lemma 1.6 to 1.7 mm. long. Only known from Humboldt Mountains, Nev.

**17. Agrostis aristiglúmis** Swallen. (Fig. 476.) Annual; culms sparingly branching at base, erect, 5 to 15 cm. tall; ligule 2 to 2.5 mm. long, decurrent; blades flat, 2 to 15 cm. long, rarely longer, 1.5 to 3 mm. wide; panicle mostly 3 to 6 cm. long, 5 to 8 mm. wide, dense; glumes 5 to 6 mm. long, attenuate into an awn 1 to 2 mm. long, the first glume 1-nerved, the second 3-nerved; lemma 3.2 to 3.5 mm. long, relatively firm, scabrous, 5-nerved, awned from the back, the awn geniculate, 6 to 7 mm. long, the lateral nerves excurrent as delicate awns, the inner pair very minute; palea nearly one-third as long as the lemma, nerveless. ⊙ —Only known from a "slope of loose gravelly soil on an outcrop of diatomaceous shale of the Monterey series," west of Mount Vision, Point Reyes Peninsula, Marin County, Calif.

**18. Agrostis hállii** Vasey. (Fig. 477.) Culms erect, 60 to 90 cm. tall, with creeping rhizomes; ligule usually conspicuous, 2 to 7 mm. long; blades flat, 2 to 5 mm. wide; panicle 10 to 15 cm. long, narrow but loose, the branches verticillate; glumes about 4 mm. long; lemma awnless, 3 mm. long, with a tuft of hairs at base about half as long; palea obsolete. ♃ —Mostly in woods near the coast from Oregon to Santa Barbara, Calif. AGROSTIS HALLII var. PRÍNGLEI (Scribn.) Hitchc. Branching, foliage stramineous; blades narrow, usually involute; panicle narrow, compact. ♃ —Near the coast, in sand, Mendocino County, Calif.

FIGURE 477.—*Agrostis hallii.* Panicle, × 1; glumes and floret, × 5. (Bioletti 110, Calif.)

**19. Agrostis lépida** Hitchc. (Fig. 478.) Culms tufted, 30 to 40 cm. tall, erect, with numerous short rhizomes; ligule, at least on the innovations, as much as 4 mm. long; leaves mostly basal, the blades firm, erect, flat or folded, the upper culm leaf below the middle of the culm, the blade 3 cm. long or less; panicle purple, 10 to 15 cm. long, the branches verticillate, becoming divaricately spreading, the lowermost 2 to 5 cm. long; glumes 3 mm. long, smooth or nearly so; lemma 2 mm. long; palea obsolete. ♃ —Meadows and open woods, Sequoia National Park and San Bernardino Mountains, Calif., at upper altitudes.

FIGURE 478.—*Agrostis lepida.* Plant, × ½; glumes and floret, × 5. (Type.)

**20. Agrostis pállens** Trin. DUNE BENT. (Fig. 479.) Culms erect, 20 to 40 cm. tall, with creeping rhizomes; ligule rather firm, 2 to 3 mm. long; blades flat or somewhat involute, 1 to 4 mm. wide; panicle contracted, almost spikelike, 5 to 10 cm. long; glumes 2.5 to 3 mm. long; lemma a little shorter than the glumes, awnless; palea obsolete. ♃ —Sand dunes along the coast, Washington to central California.

FIGURE 479.—*Agrostis pallens*. Plant, × 1; glumes and floret, × 5. (Howell, Oreg.)

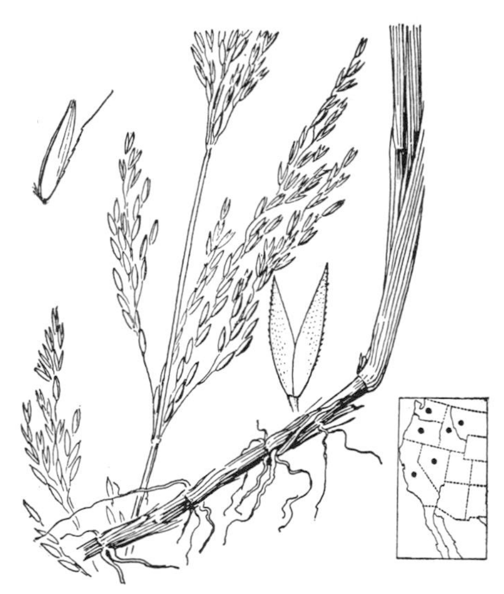

FIGURE 480.—*Agrostis diegoensis*. Plant, × 1; glumes and floret, × 5. (Orcutt, Calif.)

**21. Agrostis diegoénsis** Vasey. THINGRASS. (Fig. 480.) Culms erect, as much as 1 m. tall with creeping rhizomes; blades flat, lax, 2 to 6 mm. wide; panicle narrow, open, 10 to 15 cm. long, the branches ascending, rather stiff, some of them naked below; spikelets about as in *A. pallens*, awned or awnless. ♃ —Meadows and open woods at low and medium altitudes, Montana and British Columbia to southern California and Nevada.

**22. Agrostis blasdálei** Hitchc. (Fig. 481.) Culms 10 to 15 cm. tall, densely tufted; blades narrow or filiform, rigid, involute, 2 to 4 cm. long; panicle strict, narrow, almost spikelike, 2 to 3 cm. long, the short branches closely appressed; spikelets 2.5 to 3 mm. long; lemma about 1.8 mm. long, awnless or with a very short awn just above the middle; palea about 0.3 mm. long, nerveless. ♃ —Cliffs and dunes, Mendocino and Marin Counties, Calif. Previously referred to *A. breviculmis* Hitchc. of Peru.

FIGURE 481.—*Agrostis blasdalei*. Panicle, × 1; glumes and floret, × 5. (Type.)

**23. Agrostis variábilis** Rydb. MOUNTAIN BENT. (Fig. 482.) Culms 10 to 25 cm. tall, densely tufted; blades flat, mostly not more than 1 mm. wide; panicle, 2 to 6 cm. long, the branches ascending; spikelets purple, about 2.5 mm. long; lemma 1.5 mm. long, awnless; palea minute. ♃ —Rocky creeks and mountain slopes at high altitudes; British Columbia and Alberta to Colorado and California. Included in *A. rossae* Vasey in Manual, ed. 1.

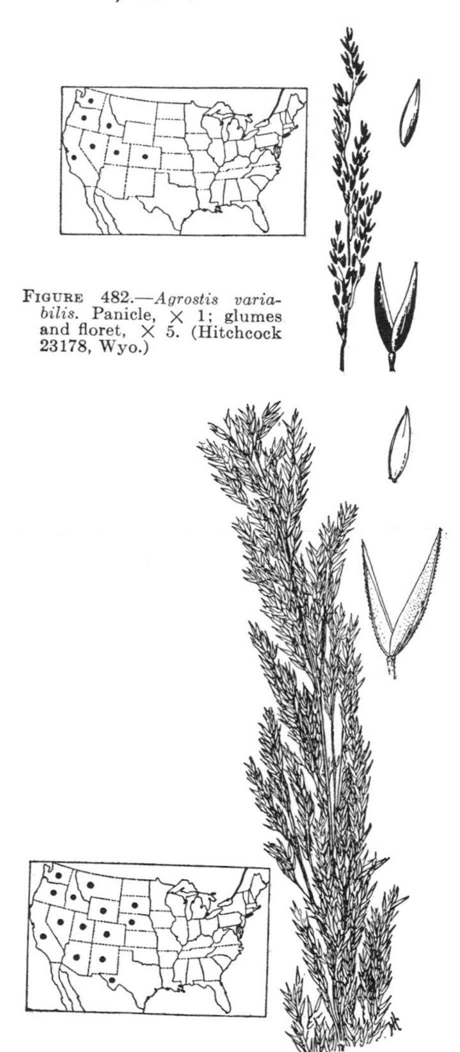

FIGURE 482.—*Agrostis variabilis*. Panicle, × 1; glumes and floret, × 5. (Hitchcock 23178, Wyo.)

FIGURE 483.—*Agrostis exarata*. Panicle, × 1; glumes and floret, × 5. (Davy 4357, Calif.)

**24. Agrostis exaráta** Trin. SPIKE BENT. (Fig. 483.) Culms 20 to 120 cm. tall, slender to relatively stout, mostly tufted; sheaths smooth to somewhat scabrous; ligule to 6 mm. long; blades flat, 2 to 10 mm. wide, usually scabrous; panicle narrow, from somewhat open to dense and interrupted, 5 to 30 cm. long; glumes subequal, 2.5 to 4 mm. long, acuminate to awn-tipped, scabrous on the keel, nearly smooth to scabrous on the back; lemma 1.7 to 2 mm. long, the midnerve ending above the middle or excurrent as a prickle or short awn, sometimes the nearly straight awn exceeding the glumes; palea minute. ♃ —Moist open ground, at low and medium altitudes, South Dakota and Nebraska to Alberta and Alaska, south to Texas, California, and Mexico. Common and extremely variable, ranging from slender plants with narrow blades and few-flowered panicles (*A. scouleri* Trin.) to robust plants a meter or more tall, with dense panicles as much as 30 cm. long (*A. grandis* Trin.). The specimens in the Trinius Herbarium from Unalaska (type) and Sitka, with culms 25 to 60 cm. tall and narrow but not dense panicles, the lemmas awnless, represent about the center of the range of variation. Awnless and awned spikelets are found in the same panicle.

AGROSTIS EXARATA var. PACÍFICA Vasey. Lemma with a straight or weakly geniculate awn exceeding the glumes; habit of the plant, height, and foliage as in the species, the variations similar. ♃ —Frequent from Vancouver Island and Washington to California, rare elsewhere: Canada, the Aleutians, Nebraska, Idaho, Arizona.

AGROSTIS EXARATA var. MONOLÉPIS (Torr.) Hitchc. Panicle narrow, dense, often interrupted; glumes mostly awn-tipped; awn of lemma exceeding the glumes 1.5 to 2 mm. ♃ —Washington to California.

**25. Agrostis ámpla** Hitchc. (Fig. 484.) Resembling *A. exarata* var. *pacifica*, the panicle looser, the branches verticillate, some of them 5 to 9 cm. long, the spikelets less crowded at the base; glumes 3.5 to 4.5 mm. (the first exceeding the second), acuminate to awn-tipped; lemma about 2.5 mm. long, awned from about the middle, the awn twisted, geniculate; anthers 0.8 to 1.8 mm. long. ♃ —Moist or wet places, Pacific slope, Oregon and California; infrequent.

FIGURE 484.—*Agrostis ampla*. Spikelet, and two views of floret, × 5. (Type.)

FIGURE 485.—*Agrostis californica*. Panicle, × 1; glumes and floret, × 5. (Anderson, Calif.)

**26. Agrostis califórnica** Trin. (Fig. 485.) Culms tufted, usually rather stout, erect or somewhat spreading at base, 15 to 60 cm. tall; sheaths sometimes slightly scabrous; ligule truncate, usually shorter than in *A. exarata*, puberulent; blades flat, firm, strongly nerved on the upper surface, usually not more than 10 cm. long, those of the culm comparatively broad and short, often 3 to 5 cm. long and 3 to 5 mm. wide, rarely as much as 10 mm. wide; panicle dense, spikelike, sometimes slightly interrupted, mostly 2 to 10 cm. long and 5 to 15 mm. wide; spikelets about 3 mm. long; glumes acute or acuminate, prominently scabrous on the keel and strongly scabrous on the sides; lemma a little shorter than the glumes, awnless or with a straight awn from minute to somewhat exceeding the glumes; palea one-fourth to one-third as long as the lemma. ♃ (*A. densiflora* Vasey.)—Sandy soil and cliffs near the sea, Mendocino County to Santa Cruz, Calif. This species has been confused with *A. exarata* and with *A. glomerata* (Presl) Kunth of Peru.

**27. Agrostis hoóveri** Swallen. (Fig. 486.) Culms densely tufted, very slender, erect, 55 to 75 cm. tall; ligule 3 to 3.5 mm. long, lacerate, decurrent; blades lax, mostly 10 to 15 cm. long, not or scarcely more than 1 mm. wide; panicle 7 to 17 cm. long, loose, the branches ascending; spikelets slightly purplish, 2 to 2.5 mm. long, the second glume slightly shorter than the first; lemma 2 mm. long, minutely erose, 5-nerved, scaberulous, bearing from near the base a bent awn slightly exceeding the glumes; palea obsolete. ♃ —Dry, mostly sandy open woodland, San Luis Obispo and Santa Barbara Counties, Calif.

FIGURE 486.—*Agrostis hooveri*. Panicle × 1; glumes and floret, × 5. (Type.)

**28. Agrostis howéllii** Scribn. (Fig. 487.) Culms erect or decumbent at base, 40 to 60 cm. tall; ligule 3 to 4 mm. long, lacerate; blades lax, as much as 30 cm. long, 3 to 5 mm. wide; panicle loose and open, 10 to 30 cm. long, the branches flexuous; spikelets pale, clustered toward the ends of the branches; glumes acuminate, rather narrow and firm, somewhat scabrous on the keel, the first about 3.5 mm. long, the second a little shorter; lemma acute, 2.5 mm. long, 4-toothed, faintly 3- to 5-nerved, bearing from

FIGURE 487.—*Agrostis howellii*. Panicle, × 1; glumes and floret, × 5. (Type.)

near the base an exserted bent awn about 6 mm. long; palea obsolete. ♃ —Known only from Oregon (Multnomah and Hood River Counties).

**29. Agrostis hiemális** (Walt.) B. S. P. (Fig. 488.) Culms mostly 30 to 40 cm. tall, erect in small tufts, glabrous; blades crowded toward the base in a dense cluster, 3 to 5 cm. long, less than 1 mm. wide, flat or subfiliform; panicles fragile, the slender filiform branches in rather distant whorls, widely spreading or drooping, unbranched below the middle, spikelet-bearing only at the ends of the branchlets; spikelets 1.5 to 1.7 mm. long, clustered, short-pediceled, appressed; glumes subequal, acute, scabrous on the keels; lemma 1 to 1.2 mm. long, the callus glabrous; anthers 0.2 mm. long. ♃ —Open ground, fields, and waste places, Massachusetts to Florida, west to Wisconsin, Kansas, Oklahoma, and Texas.

**30. Agrostis scábra** Willd. (Fig. 489.) Culms 30 to 85 cm., rarely to 100 cm., tall, erect in small dense tufts; sheaths shorter than the internodes, glabrous; ligule hyaline, 2 to 5 mm. long; blades flat, 8 to 20 cm. long, 1 to 3 mm. wide, scabrous, the basal ones often subfiliform; panicles 15 to 25 cm. long, rarely longer, the brittle scabrous branches in rather distant verticils, ascending or spreading, sometimes drooping, branching above the middle; spikelets 2 to 2.7 mm. long, loosely arranged at the ends of the branchlets; glumes unequal, acuminate, scabrous on the keels; lemma 1.5 to 1.7 mm. long, distinctly longer than the caryopsis, the callus sparsely pilose; anthers 0.4 to 0.5 mm. long. ♃ —Mountain meadows, fields, and open woods, Newfoundland and Alaska, south to Florida, Texas, and California; probably introduced in the Southern States. (Included in *A. hiemalis* in Manual, ed. 1.)

AGROSTIS SCABRA var. GEMINÁTA (Trin.) Swallen. Branches of panicle short and divaricate; lemma awned or awnless. The type specimen, from Alaska, is awned; a large number of specimens over a wide range agree in other respects, but are awnless. ♃ —At high latitudes and altitudes, Newfoundland to Alaska, south to New Hampshire, North Dakota, Colorado, and California.

**31. Agrostis idahoénsis** Nash. IDAHO REDTOP. (Fig. 490.) Culms slender, tufted, 10 to 30 cm. tall; leaves mostly basal, the blades narrow; panicle loosely spreading, 5 to 10 cm. long, the branches capillary,

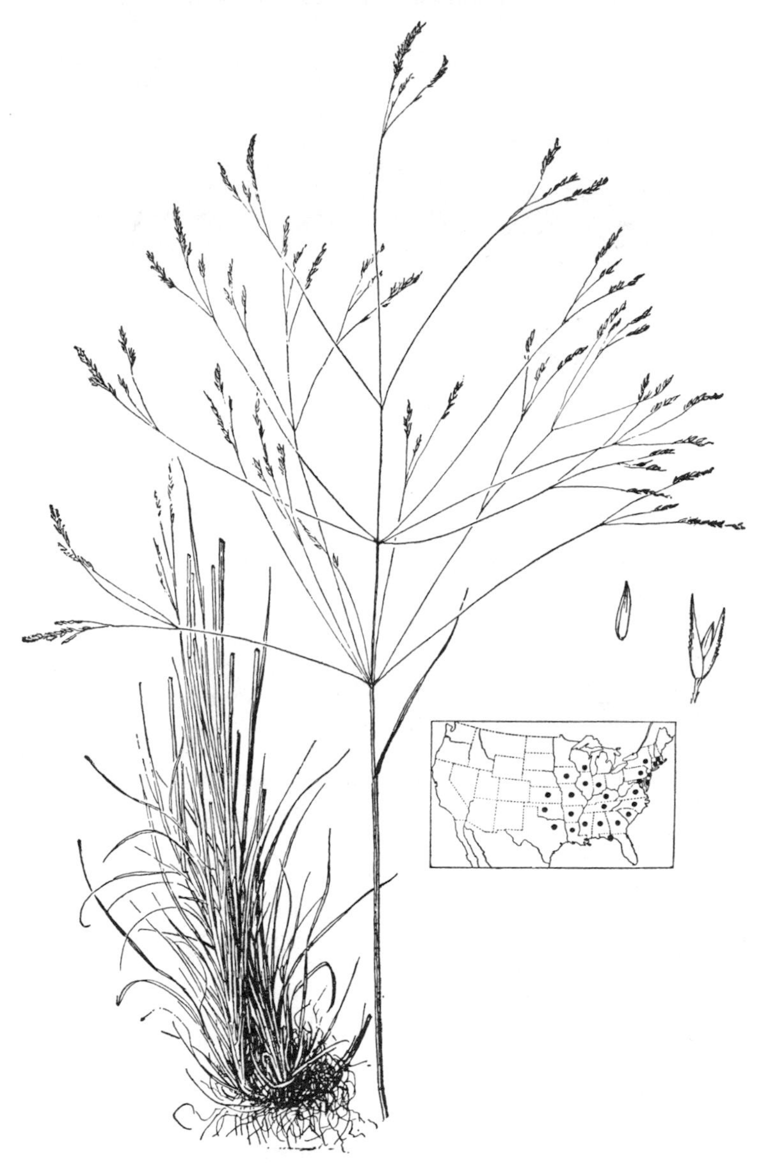

FIGURE 488.—*Agrostis hiemalis*. Plant, × ½; glumes and floret, × 5. (Deam 6514, Ind.)

flexuous, minutely scabrous; spikelets 1.5 to 2.5 mm. long; lemma about 1.3 mm. long, awnless; palea minute. ♃ —Mountain meadows, at medium and high altitudes, western Montana to Washington, south to New Mexico, Arizona, and the high mountains of California; Fairbanks, Alaska. Differs from *A. scabra* in the smaller spikelets and in the narrower panicle with shorter flexuous branches.

**32. Agrostis perénnans** (Walt.) Tuckerm. AUTUMN BENT. (Fig. 491, *A*.) Culms erect to somewhat decumbent at base, varying from weak and lax to relatively stout, 30 to 100 cm. tall, often with lax leafy shoots at base; leaves rather numerous, the

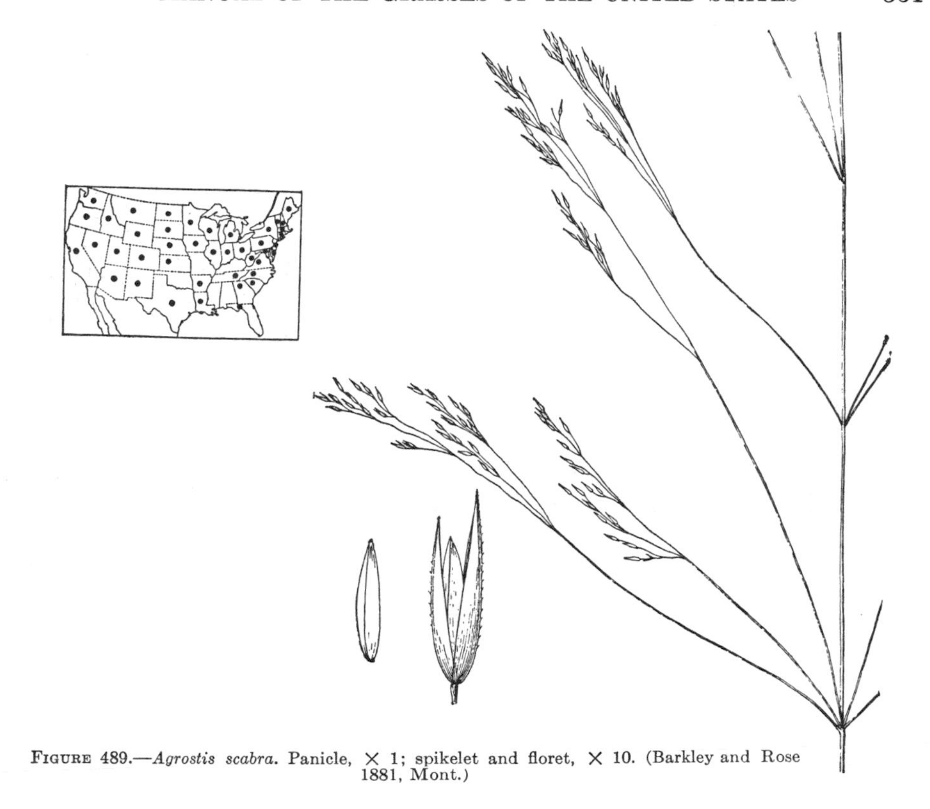

FIGURE 489.—*Agrostis scabra*. Panicle, × 1; spikelet and floret, × 10. (Barkley and Rose 1881, Mont.)

blades from lax to stiffly upright, corresponding to the culms, 10 to 20 cm. long, 1 to 6 mm. wide; panicle pale to tawny, open, oblong, the branches verticillate, mostly lax, ascending, branching about the middle; spikelets 2 to 3.2, mostly 2.2 to 2.7 mm. long, the pedicels spreading, but the spikelets sometimes somewhat aggregate towards the ends of the branchlets; glumes acute or acuminate, the first slightly longer; lemma 1.5 to 2 mm. long, rarely awned (*A. perennans* forma *chaetophora* Fernald); palea obsolete or nearly so. ♃ — Open ground, old fields, open woods, in rather dry soil from sea level to mountain tops, flowering in late summer or autumn, Quebec to Minnesota, south to Florida and eastern Texas; Mexico. Extremely variable, in dry open ground erect and rather stout; in shady places weak, with lax pale panicle and divaricate branchlets and spikelets 2 mm. long (*A. perennans* var. *aestivalis* Vasey). Intergrades with the following, the intermediate specimens (*A. scribneriana* Nash) rather numerous in the Eastern States.

**33. Agrostis altíssima** (Walt.) Tuckerm. (Fig. 491, *B*.) Culms mostly stouter than in the preceding, erect or ascending; panicle branches usually ascending, the spikelets more or less aggregate toward the ends; spikelets 2.3 to 3.7, mostly 2.7 to 3.5 mm. long. ♃ —Mostly in marshy ground, pine barren bogs, and wooded swamps, coastal plain, New Jersey and Maryland to Alabama and Mississippi.

**34. Agrostis oregonénsis** Vasey. OREGON REDTOP. (Fig. 492.) Culms 60 to 90 cm. tall; blades 2 to 4 mm. wide; panicle oblong, 10 to 30 cm. long, open, the branches verticillate, rather stiff and ascending, numerous

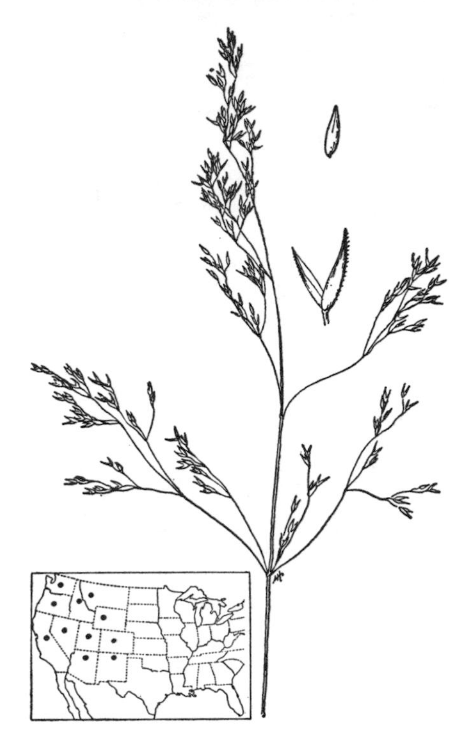

in the lower whorls, the longer 5 to 10 cm. long, branching above the middle; glumes 2.5 to 3 mm. long; lemma 1.5 mm. long, awnless; palea about 0.5 mm. long. ♃ —Marshes, bogs, and wet meadows, Montana to British Columbia, south to Wyoming and California.

**35. Agrostis canína** L. VELVET BENT. (Fig. 493.) Culms tufted, 30 to 50 cm. tall; blades mostly short and narrow, those of the culm 3 to 6 cm. long, usually not more than 2 mm. wide; panicle loose and spreading, mostly 5 to 10 cm. long; glumes equal, acute, 2 mm. long, the lower minutely scabrous on the keel; lemma a little shorter than the glumes, awned about the middle, the awn exserted, bent; callus minutely hairy; palea minute. ♃ —Meadows and open ground, Newfoundland to Quebec, south to Delaware, West Virginia, Tennessee, and Michigan; pos-

FIGURE 490.—*Agrostis idahoensis.* Panicle, × 1; glumes and floret, × 5. (Chase 5040, Idaho.)

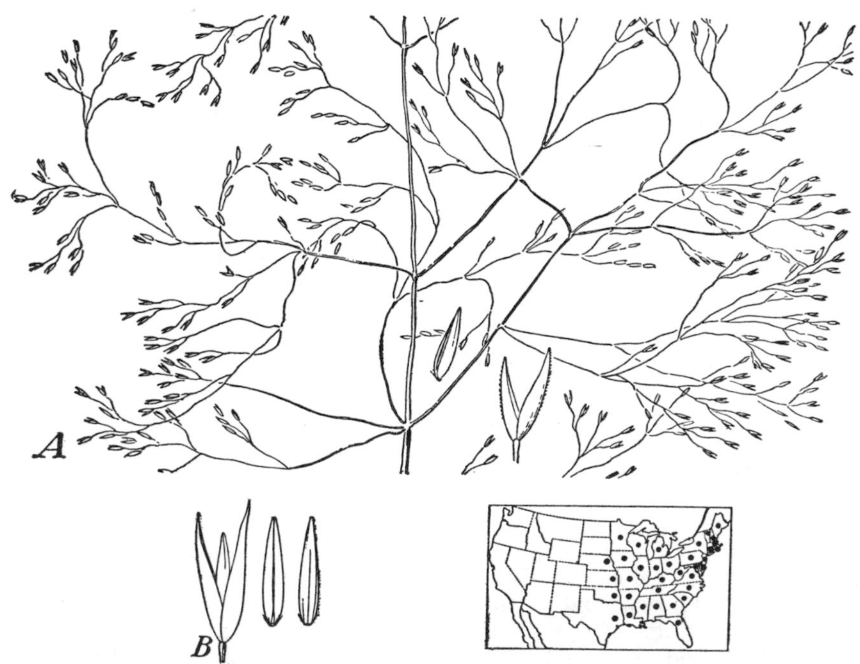


FIGURE 491.—*A, Agrostis perennans.* Panicle, × 1; glumes and floret, × 5. (Millspaugh 53, W. Va.) *B, A. altissima.* Glumes and two views of floret, × 5. (A. Gray, N. J., in Trinius Herb.)

sibly native northward but introduced in the United States; Europe. Sometimes cultivated for putting greens.

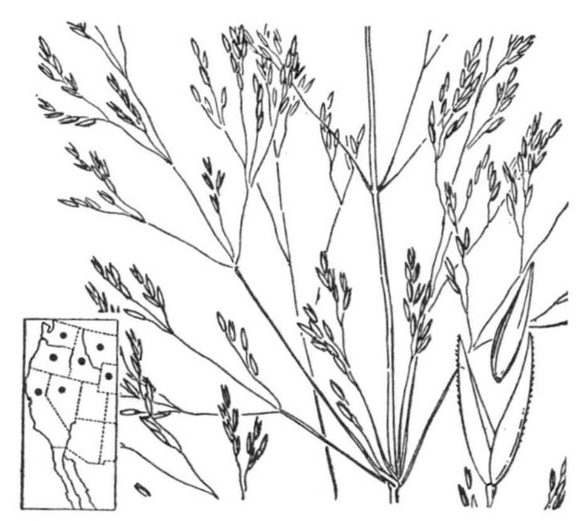

FIGURE 492.—*Agrostis oregonensis*. Panicle, × 1; glumes and floret, × 5. (Hitchcock 23524, Oreg.)

FIGURE 493.—*Agrostis canina*. Panicle, × 1; glumes and floret, × 5. (Commons 99, Del.)

**36. Agrostis boreális** Hartm. (Fig. 494.) Culms tufted, 20 to 40 cm. tall, or, in alpine or high northern plants, dwarf; leaves mostly basal, the blades 5 to 10 cm. long, 1 to 3 mm. wide; panicle pyramidal, 5 to 15 cm. long, the lower branches whorled and spreading; glumes 2.5 to 3 mm. long, acute; lemma a little shorter than the glumes, awned, the awn usually bent and exserted; palea obsolete or nearly so. ♃ (*A. bakeri* Rydb., lemma with a straight awn or awnless.)—Rocky slopes and moist banks at high latitudes and altitudes,

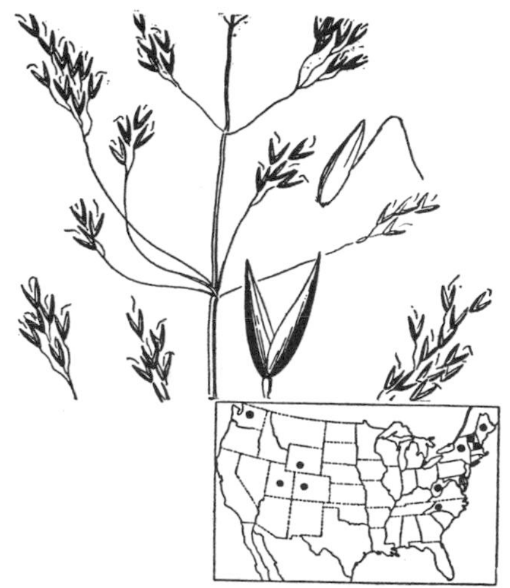

FIGURE 494.—*Agrostis borealis*. Panicle, × 1; glumes and floret, × 5. (Faxon 99, N. H.)

Newfoundland and Greenland to Alaska, south to the high mountains of New England and New York; West Virginia; summit of Roan Mountain, N. C.; Alberta and Washington to Wyoming, Colorado, and Utah; northern Europe.

FIGURE 495.—*Agrostis longiligula*. Panicle, × 1; glumes and floret, × 5. (Type.)

**37. Agrostis longilígula** Hitchc. (Fig. 495.) Culms erect, about 60 cm. tall; ligule 5 to 6 mm. long; blades 10 to 15 cm. long, 3 to 4 mm. wide, scabrous; panicle narrow, but loosely flowered, bronze purple, 10 to 15 cm. long, the branches very scabrous; glumes 4 mm. long; lemma 2.5 mm. long, bearing at the middle a bent exserted awn; palea minute. ♃ —Bogs and marshes at low altitudes, Tillamook County, Oreg., to Marin County, Calif.

AGROSTIS LONGILIGULA var. AUSTRÁLIS J. T. Howell. Ligule 4 to 11 mm. long; awn of lemma straight, about 1 mm. long or obsolete. ♃ —Wet places, Marin, Sonoma, and Mendocino Counties, Calif.

AGROSTIS NEBULÓSA Boiss. and Reut. CLOUDGRASS. Culms slender, branching, about 30 cm. tall; foliage scant; panicle delicate, oblong, half as long as the plant, the branches in verticils; spikelets 1 mm. long. ⊙ (Sometimes called *A. capillaris*, not *A. capillaris* L.)—Cultivated for dry bouquets. Spain.

## 72. PHÍPPSIA (Trin.) R. Br.

Spikelets 1-flowered, the rachilla disarticulating above the glumes, not prolonged; glumes unequal, minute, the first sometimes wanting; lemma thin, somewhat keeled, 3-nerved, abruptly acute; palea a little shorter than the lemma, dentate. Dwarf, tufted perennial, with narrow, few-flowered panicles of small spikelets. Type species, *Phippsia algida*. Named for C. J. Phipps.

**1. Phippsia álgida** (Phipps) R. Br. (Fig. 496.) Culms densely tufted, 2 to 10 cm. tall; blades soft, narrow, with boat-shaped tip; lemma about 1.5 mm. long. ♃ —Summit of Grays Peak, Colo.; Arctic regions of both hemispheres.

## 73. COLEÁNTHUS Seidel

Spikelets 1-flowered; glumes wanting; lemma ovate, hyaline, terminating in a short awn; palea broad, 2-toothed, the keels awn-tipped. Dwarf annual, with short flat blades and small panicles. Type species, *Coleanthus subtilis*. Name from Greek *koleos*, sheath, and *anthos*, flower, alluding to the sheaths enclosing the base of the panicles.

**1. Coleanthus súbtilis** (Tratt.) Seidel. (Fig. 497.) Culms spreading, forming little mats, mostly less than 5 cm. long; panicle 5 to 10 mm. long, the short branches verticillate; lemma about 1 mm. long, the awn about equaling the dark caryopsis. ⊙ —Mud flats along the lower Columbia River, Oregon and Washington, well established but probably introduced; northern Eurasia.

---

**Mibóra mínima** (L.) Desv. Delicate annual, 3 to 10 cm. tall with short narrow blades and slender racemes of 6 to 8 appressed purple spikelets, 2 mm. long, the glumes obtuse, the lemma and palea shorter, pubescent. ⊙ —Plymouth, Mass.; introduced from Europe.

FIGURE 496.—*Phippsia algida*. Plant, × ½; glumes and floret, × 10. (Oldmixon, Alaska.)

FIGURE 497.—*Coleanthus subtilis.* Plant, × 1; lemma and palea and two views of spikelet with ripe caryopsis, × 20. (Howell, Oreg.)

## 74. CÍNNA L. WOODREED

Spikelets 1-flowered, disarticulating below the glumes, the rachilla forming a stipe below the floret and produced behind the palea as a minute bristle; glumes equal or subequal, 1- to 3-nerved; lemma similar to the glumes, nearly as long, 3-nerved, bearing a minute, short, straight awn just below the apex (rarely awnless); palea 1-keeled. Tall perennials with flat blades and close or open panicles. Type species, *Cinna arundinacea. Cinna* (kinna) an old Greek name for a grass.

Our two species furnish highly palatable forage but usually are not abundant enough to be of much importance.

Spikelets 5 mm. long; panicle rather dense, the branches ascending.... 1. C. ARUNDINACEA.
Spikelets 3.5 to 4 mm. long; panicle loose, the branches spreading or drooping.
2. C. LATIFOLIA.

**1. Cinna arundinácea** L. STOUT WOODREED. (Fig. 498.) Culms erect, usually 1 to 1.5 m. tall, often somewhat bulbous at base, solitary or few in a tuft; sheaths glabrous; ligule rather prominent, thin; blades flat, scabrous, mostly less than 1 cm. wide; panicle many-flowered, nodding, grayish, 15 to 30 cm. long, the branches ascending; spikelets 5 to 6 mm. long; glumes somewhat unequal, acute, the second 3-nerved; lemma usually a little longer than the first glume, bearing below the tip a minute straight awn; palea apparently 1-nerved. ♃ —Moist woods, Maine to South Dakota, south to Georgia and eastern Texas. CINNA ARUNDINACEA var. INEXPÁNSA Fern. and Grisc. Panicle narrower, the shorter branches ascending; spikelets 3.7 to 4.2 mm. long. ♃ —Margin of swamps and moist woods, southeast Virginia.

**2. Cinna latifólia** (Trevir.) Griseb. DROOPING WOODREED. (Fig. 499.) Resembling *C. arundinacea;* blades shorter and on the average wider, as much as 1.5 cm. wide; panicle green, looser, the branches fewer, spreading or drooping, naked at base for as much as 5 cm.; spikelets about 4 mm. long; awn of lemma sometimes as much as 1 mm. long (rarely wanting); palea 2-nerved, the nerves very close together. ♃ —Moist woods, Newfoundland and Labrador to Alaska, south to Connecticut, in the mountains to North Carolina and Tennessee, to Michigan, Illinois, South Dakota, in the Rocky Mountains to northern New Mexico, to Utah and central California; northern Eurasia.

## 75. LIMNÓDEA L. H. Dewey

Spikelets 1-flowered, disarticulating below the glumes, the rachilla prolonged behind the palea as a short slender bristle; glumes equal, firm; lemma membranaceous, smooth,

FIGURE 498.—*Cinna arundinacea*. Plant, × ½; glumes and floret, × 10. (Dewey 336, Va.)

FIGURE 499.—*Cinna latifolia*. Panicle, × 1; glumes and floret, × 10. (Sandberg 713, Minn.)

nerveless, 2-toothed at the apex, bearing from between the teeth a slender bent awn, twisted at base; palea a little shorter than the lemma. Slender annual with flat blades and narrow panicles. Type species, *Limnodea arkansana*. Name altered from *Limnas*, a genus of grasses.

FIGURE 500.—*Limnodea arkansana*. Plant, × ½; glumes and floret, × 10. (Orcutt 5910, Tex.)

**1. Limnodea arkansána** (Nutt.) L. H. Dewey. (Fig. 500.) Culms branching at base, 20 to 40 cm. tall; blades more or less pubescent on both surfaces; panicle 5 to 15 cm. long, narrow but loose; spikelets 3.5 to 4 mm. long; glumes hispidulous or pilose; awn 8 to 10 mm. long. ♃ —Dry soil, prairies and river banks, Coastal Plain, Florida to Texas, Arkansas, and Oklahoma. The form with pilose glumes has been called *L. arkansana* var. *pilosa* (Trin.) Scribn.

## 76. ALOPECÚRUS L. Foxtail

Spikelets 1-flowered, disarticulating below the glumes, strongly compressed laterally; glumes equal, usually united at base, ciliate on the keel; lemma about as long as the glumes, 5-nerved, obtuse, the margins united at base, bearing from below the middle a slender dorsal awn, this included or exserted two or three times the length of the spikelet; palea wanting. Low or moderately tall perennials or some annuals, with flat blades and soft, dense, spikelike panicles. Type species, *Alopecurus pratensis*. Name from Greek *alopex*, fox, and *oura*, tail, alluding to the cylindric panicle.

The species of *Alopecurus* are all palatable and nutritious forage grasses, but usually are not found in sufficient abundance to be of great importance. *A. pratensis*, meadow foxtail, is sometimes used as a meadow grass in the eastern United States; *A. aequalis* is the most common on the western ranges.

Spikelets 5 to 6 mm. long. Introduced perennials.
  Panicle slender, tapering at each end; glumes scabrous on the keel. 1. A. MYOSUROIDES.
  Panicle cylindric, dense; glumes conspicuously ciliate on the keel........ 2. A. PRATENSIS.
Spikelets 2 to 4 mm. long (rarely 5 mm. in *A. saccatus*, annual). Native species.
  Plants perennial.
    Spikelets densely woolly all over; panicle oblong, 1 to 5 cm. long, about 1 cm. thick. 3. A. ALPINUS.
    Spikelets not woolly; panicle linear or oblong-linear, less than 1 cm. thick.
      Awn scarcely exceeding the glumes........ 5. A. AEQUALIS.
      Awn exserted 2 mm. or more.
        Awn exserted 2 to 3 mm.; panicle 3 to 4 mm. thick; spikelets 2.5 mm. long. 6. A. GENICULATUS.
        Awn exserted 3 to 5 mm.; panicle 4 to 6 mm. thick; spikelets about 3 mm. long. 4. A. PALLESCENS.
  Plants annual.
    Spikelets 4 to 5 mm. long; panicle relatively loose........ 9. A. SACCATUS.
    Spikelets 2 to 3.5 mm. long; panicle dense.
      Spikelets 2 to 2.5 mm. long; anthers 0.5 mm. long........ 7. A. CAROLINIANUS.
      Spikelets 3 to 3.5 mm. long; anthers about 1 mm. long........ 8. A. HOWELLII.

**1. Alopecurus myosuroídes** Huds. (Fig. 501.) Annual; culms tufted, slightly scabrous, 10 to 50 cm. tall, erect or decumbent at base; blades usually 2 to 3 mm. wide; panicle slender, somewhat tapering at each end, 4 to 10 cm. long, 3 to 5 mm. wide; glumes 6 mm. long, pointed, whitish with 3 green nerves, glabrous, scabrous on the keel, short-ciliate at base; lemma about as long as the glumes, the awn bent, exserted 5 to 8 mm. ⊙ (*A. agrestis* L.)—Fields, waste places, and ballast ground, Maine to North Carolina, Kansas, Texas, Washington, to California; introduced, rare; Eurasia.

**2. Alopecurus praténsis** L. Meadow foxtail. (Fig. 502.) Perennial; culms erect, 30 to 80 cm. tall; blades 2 to 6 mm. wide; panicle 3 to 7 cm. long, 7 to 10 mm. thick; glumes 5 mm. long, villous on the keel and pubescent on the sides; awn exserted 2 to 5 mm. ♃ —Fields and waste places, Newfoundland and Labrador to Alaska, south to Delaware and Missouri; Montana, Idaho, and Oregon.

Introduced; Eurasia. Occasionally cultivated as a meadow grass.

**3. Alopecurus alpínus** J. E. Smith. ALPINE FOXTAIL. (Fig. 503.) Perennial; culms erect or often decumbent at base, rather stiff and rushlike, 10 to 80 cm. tall, with slender rhizomes; sheaths glabrous, often inflated; blades 3 to 5 mm. wide; panicle ovoid or oblong, 1 to 4 cm. long, about 1 cm. wide, woolly; glumes 3 to 4 mm. long, woolly; lemma awned near the base, the awn exserted slightly or as much as 5 mm. ♃ —Mountain meadows and along brooks, Greenland to Alaska, south in the Rocky Mountains to Colorado and Utah; Arctic regions and northern Eurasia.

**4. Alopecurus palléscens** Piper. WASHINGTON FOXTAIL. (Fig. 504.) Perennial, tufted, pale green; culms 30 to 50 cm. tall, erect, or lower nodes geniculate; sheaths somewhat inflated; panicle pale, dense, 2 to 7 cm. long, 4 to 6 mm. thick; glumes about 3 mm. long, ciliate on the keel, appressed-pubescent on the sides; lemma awned near the base, the awn exserted 3 to 5 mm.; anthers about 2 mm. long. ♃ —Edges of ponds and wet places, British Columbia and Montana to Washington and northern California.

**5. Alopecurus aequális** Sobol. SHORT-AWN FOXTAIL. (Fig. 505.) Perennial; culms erect or spreading, usually not rooting at the nodes, 15 to 60 cm. tall; blades 1 to 4 mm. wide; panicle slender, 2 to 7 cm. long, about 4 mm. thick; spikelets 2 mm. long; awn of lemma scarcely exserted; anthers about 0.5 mm. long. ♃ (*A. aristulatus* Michx.)—In water and wet places, Greenland to Alaska, south to Pennsylvania, Illinois, Kansas, New Mexico, and California; Eurasia.

**6. Alopecurus geniculátus** L. WATER FOXTAIL. (Fig. 506.) Differing from *A. aequalis* chiefly in the usually more decumbent culms rooting at the nodes and the longer awn exserted 2 to 3 mm., giving the panicle a softly bristly appearance; spikelets about 2.5 mm. long, the tip dark purple; awn of lemma about as long again as the spikelet; anthers about 1.5 mm. long. ♃ —In water and wet places, Newfoundland to Saskatchewan and British Columbia; Maine to Virginia; Pennsylvania, Michigan, Wisconsin; Kansas and Wyoming to Utah; Montana; Washington to California and Arizona; Eurasia.

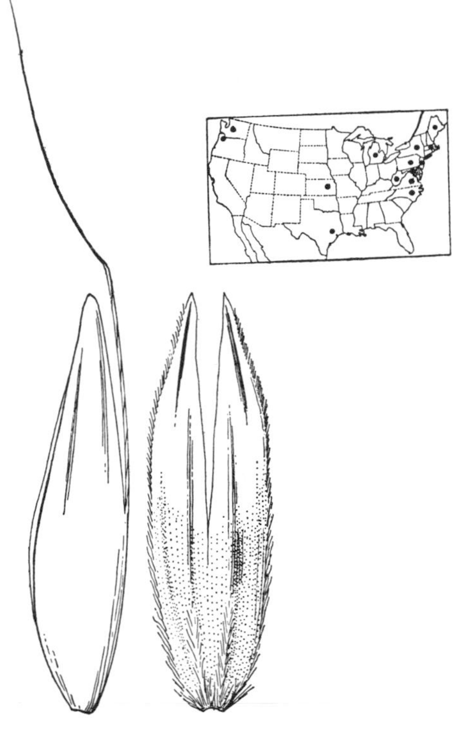

FIGURE 501.—*Alopecurus myosuroides.* Glumes and floret, × 10. (Commons, 14, Del.)

**7. Alopecurus caroliniánus** Walt. (Fig. 507.) Annual; culms tufted, much branched at base, 10 to 50 cm. tall; similar to *A. geniculatus* and *A. aequalis,* but panicle more slender than in the former; spikelets 2 to 2.5 mm. long, pale, the awn as in *A. geniculatus;* anthers about 0.5 mm. long. ⊙ (*A. ramosus* Poir.)—Moist open ground, old fields, and wet places, British Columbia; Long Island, N. Y., to Florida, Washington, and California, except West Virginia, Nevada, and New Mexico.

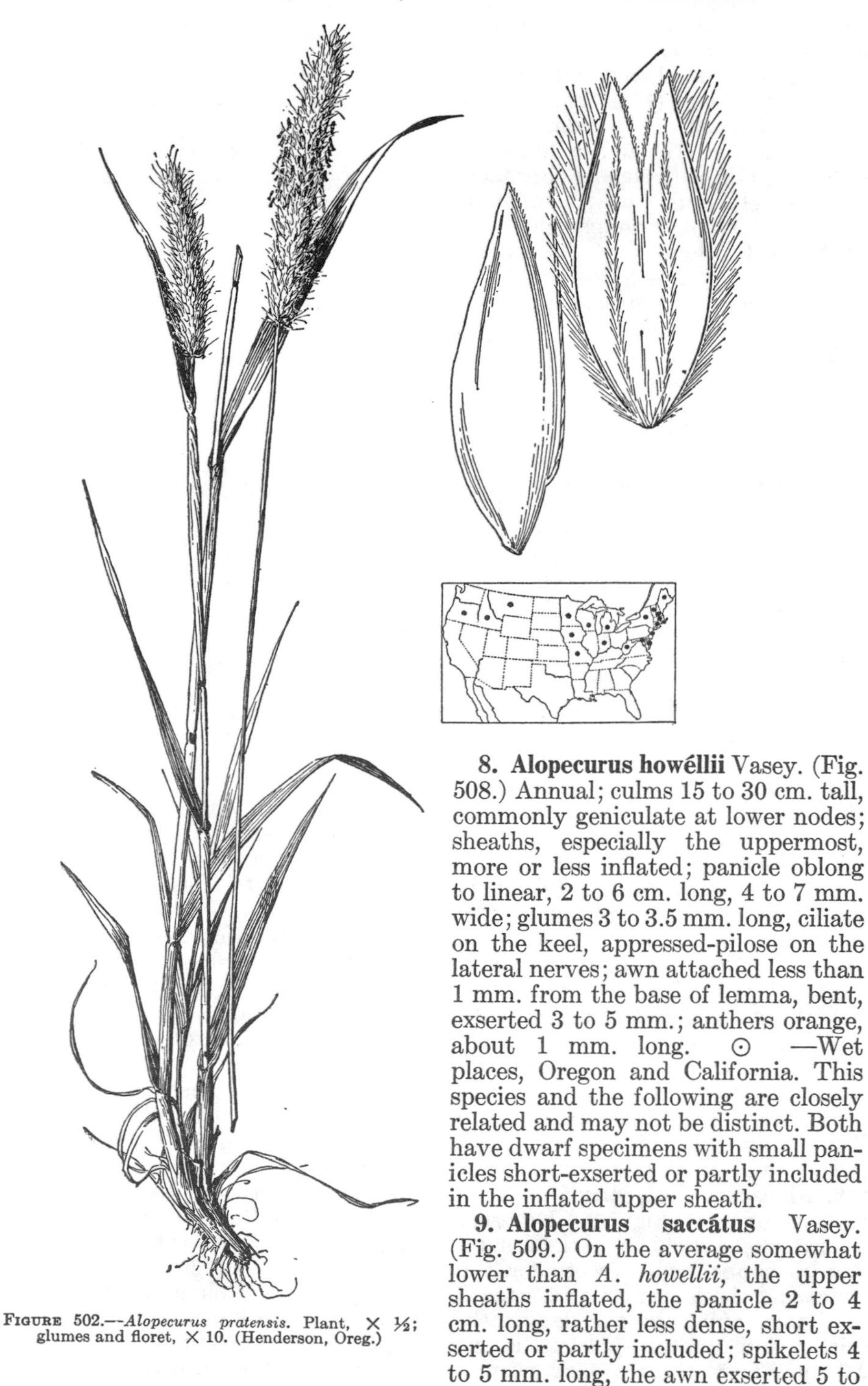

FIGURE 502.—*Alopecurus pratensis*. Plant, × ½; glumes and floret, × 10. (Henderson, Oreg.)

8. **Alopecurus howéllii** Vasey. (Fig. 508.) Annual; culms 15 to 30 cm. tall, commonly geniculate at lower nodes; sheaths, especially the uppermost, more or less inflated; panicle oblong to linear, 2 to 6 cm. long, 4 to 7 mm. wide; glumes 3 to 3.5 mm. long, ciliate on the keel, appressed-pilose on the lateral nerves; awn attached less than 1 mm. from the base of lemma, bent, exserted 3 to 5 mm.; anthers orange, about 1 mm. long. ⊙ —Wet places, Oregon and California. This species and the following are closely related and may not be distinct. Both have dwarf specimens with small panicles short-exserted or partly included in the inflated upper sheath.

9. **Alopecurus saccátus** Vasey. (Fig. 509.) On the average somewhat lower than *A. howellii*, the upper sheaths inflated, the panicle 2 to 4 cm. long, rather less dense, short exserted or partly included; spikelets 4 to 5 mm. long, the awn exserted 5 to

8 mm.; anthers 1 mm. long. ⊙ — Wet places, along the Columbia River, Washington and Oregon; California (Colusa County).

**Alopecurus créticus** Trin. Annual, 10 to 40 cm. tall; panicle dense; spikelets wedge-shaped, 4 mm. long; glumes firm, the keels broadly winged toward the summit, ciliate; lemma truncate, the awn from near the base. Waif, ballast, Philadelphia, Pa.; Europe.

**Alopecurus réndlei** Eig. Annual; culms 15 to 30 cm. tall, geniculate; upper sheaths inflated; panicle 1.5 to 2 cm. long, 7 to 9 mm. wide; spikelets 5 to 6 mm. long, almost diamond-shaped, the glumes inflated-gibbous

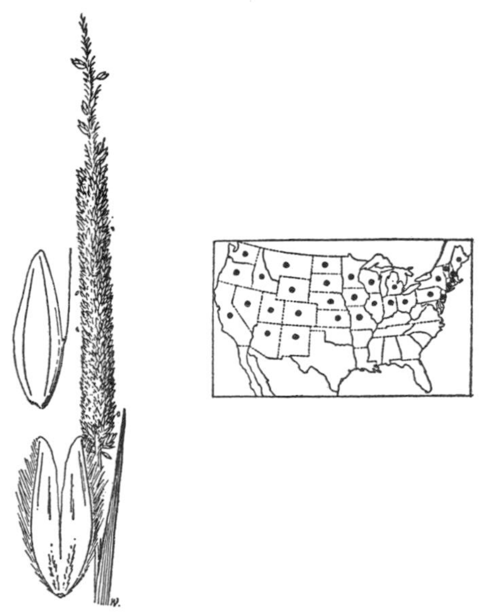

FIGURE 505.—*Alopecurus aequalis*. Panicle, × 1; glumes and floret, × 10. (Fernald, Maine.)

FIGURE 503.—*Alopecurus alpinus*. Panicle, × 1; glumes and floret, × 10. (Hall and Harbour 682, Colo.)

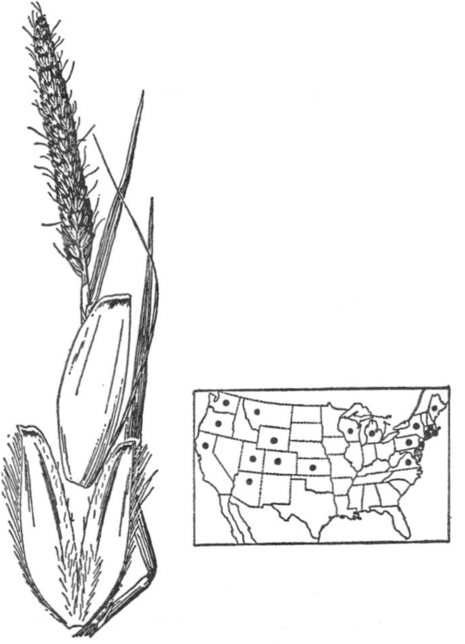

FIGURE 506.—*Alopecurus geniculatus*. Panicle, × 1; glumes and floret, × 10. (Weatherby 3394, Mass.)

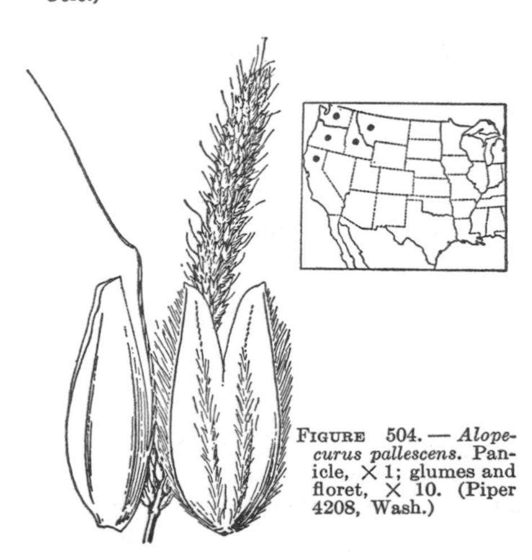

FIGURE 504. — *Alopecurus pallescens*. Panicle, × 1; glumes and floret, × 10. (Piper 4208, Wash.)

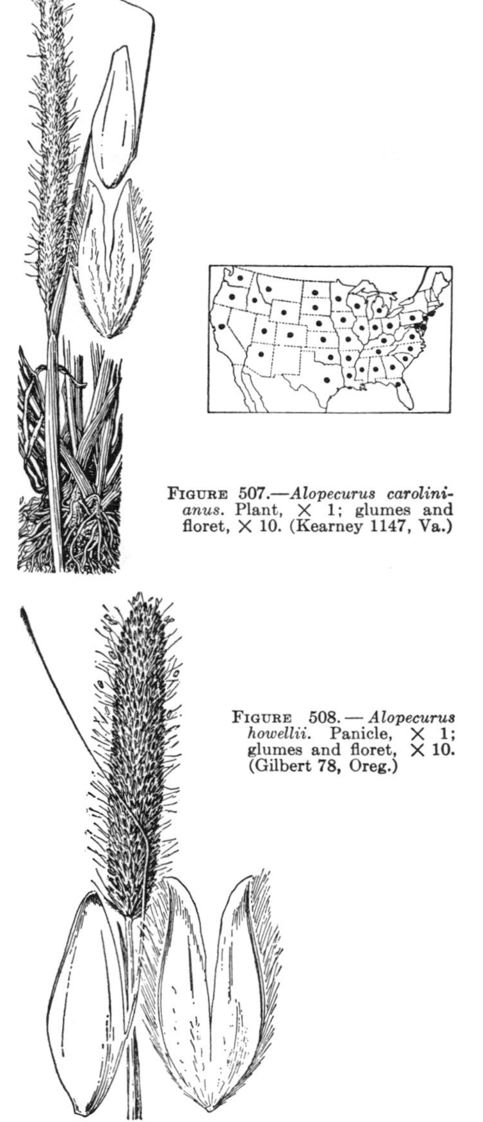

FIGURE 507.—*Alopecurus carolinianus.* Plant, × 1; glumes and floret, × 10. (Kearney 1147, Va.)

FIGURE 508.—*Alopecurus howellii.* Panicle, × 1; glumes and floret, × 10. (Gilbert 78, Oreg.)

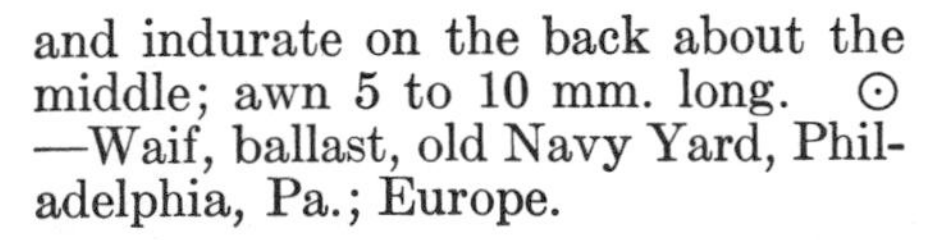

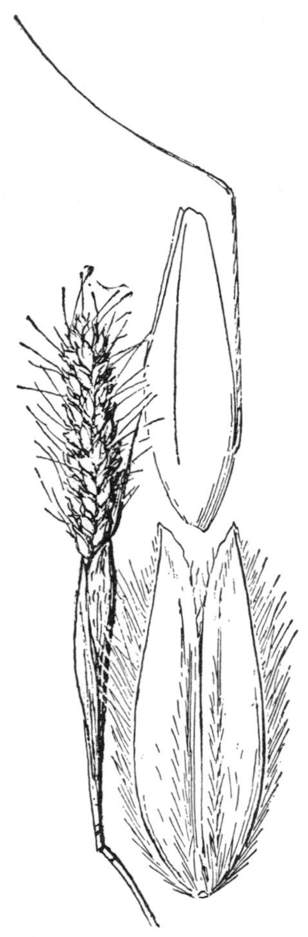

FIGURE 509.—*Alopecurus saccatus.* Panicle, × 1; glumes and floret, × 10. (Suksdorf 188, Wash.)

and indurate on the back about the middle; awn 5 to 10 mm. long. ⊙ —Waif, ballast, old Navy Yard, Philadelphia, Pa.; Europe.

**Alopecurus arundináceus** Poir. Tall rhizomatous perennial; blades 3 to 10 mm. wide; panicle 4 to 10 cm. long, 7 to 8 mm. thick, often purplish; spikelets 4 to 5 mm. long; glumes sparsely pubescent, long-ciliate on the keel; lemma about equaling the glumes, the awn included or exserted 1 to 3 mm. ♃ —Adventive in hay meadows, Labrador; North Dakota; Eurasia.

## 77. POLYPÓGON Desf.

Spikelets 1-flowered, the pedicel disarticulating a short distance below the glumes, leaving a short-pointed callus attached; glumes equal, entire or 2-lobed, awned from the tip or from between the lobes, the awn slender, straight; lemma much shorter than the glumes, hyaline, usually bearing a slender straight awn shorter than the awns of the glumes. Usually decumbent annuals or perennials with flat scabrous blades and dense, bristly, spikelike panicles. Type species, *Polypogon monspeliensis*. Name from Greek *polus*, much, and *pogon*, beard, alluding to the bristly inflorescence.

One species, *P. monspeliensis*, is palatable to stock and is sometimes sufficiently abundant on low meadows to be of importance in the West.

Plants annual.
  Glumes slightly lobed, the lobes not ciliate ........ 1. P. MONSPELIENSIS.
  Glumes prominently lobed, the lobes ciliate-fringed ........ 2. P. MARITIMUS.
Plants perennial.
  Glumes gradually narrowed into the awn ........ 5. P. ELONGATUS.
  Glumes abruptly rounded at summit.
    Awns rather stiff and straight; glumes 2.5 to 3 mm. long ........ 3. P. INTERRUPTUS.
    Awns delicate, flexuous; glumes 1.5 to 2 mm. long ........ 4. P. AUSTRALIS.

**1. Polypogon monspeliénsis** (L.) Desf. RABBITFOOT GRASS. (Fig. 510.) Annual; culms erect or decumbent at base, 15 to 50 cm. tall (sometimes depauperate or as much as 1 m. tall); ligule 5 to 6 mm. long; blades in average plants 4 to 6 mm. wide; panicle dense, spikelike, 2 to 15 cm. long, 1 to 2 cm. wide, tawny yellow when mature; glumes hispidulous, about 2 mm. long, the awns 6 to 8 mm. long, rarely longer; lemma smooth and shining, about half as long as the glumes, the delicate awn slightly exceeding them. ☉ —Ballast and waste places, New Brunswick to Georgia, Oklahoma, and Texas, west to Alaska and California, infrequent in the East, mostly confined to the coastal States, a common weed in the Western States; at low altitudes, south to Argentina; introduced from Europe.

**2. Polypogon marítimus** Willd. (Fig. 511.) Annual; culms 20 to 30 cm. tall, upright or spreading; ligule as much as 6 mm. long; blades usually less than 5 cm. long, 2 to 4 mm. wide; panicle mostly smaller and less dense than in *P. monspeliensis;* glumes about 2.5 mm. long, hispidulous below, the deep lobes ciliate-fringed, the awns 7 to 10 mm. long; lemma awnless. ☉ —Introduced, Georgia (Tybee Island); Nebraska, California (Napa and New York Falls, Amador County); Mediterranean region.

**3. Polypogon interrúptus** H. B. K. DITCH POLYPOGON. (Fig. 512.) Perennial; culms tufted, geniculate at base, 30 to 80 cm. tall; ligule 2 to 5 mm. long or the uppermost longer; blades commonly 4 to 6 mm. wide; panicle oblong, 5 to 15 cm. long, more or less interrupted or lobed; glumes equal, 2.5 to 3 mm. long, scabrous, the awns 3 to 5 mm. long; lemma smooth and shining, 1 mm. long, minutely toothed at the truncate apex, the awn exceeding the glumes. ♃ (*P. lutosus* of Manual, ed. 1, a doubtful species of Europe.)—Ditches and wet places at low altitudes, British Columbia to California, east to Louisiana; Nebraska; Oklahoma; south to Argentina.

**4. Polypogon austrális** Brongn. (Fig. 513.) Perennial; culms as much as 1 m. tall; ligule 2 to 3 mm. long, fragile; blades commonly 5 to 7 mm. wide; panicle soft, lobed or interrupted, mostly 8 to 15 cm. long, the numerous awns purplish; glumes 1.5 to 2 mm. long, hispidulous, the awn flexuous, delicate, 4 to 6 mm. long; lemma about two-thirds as long as the glumes, the awn about 3 mm. long. ♃ (*P. crinitus* Trin., not Nutt.)—Introduced at Bingen, Wash.; Chile and Argentina.

**5. Polypogon elongátus** H. B. K. (Fig. 514.) Perennial; culms rather coarse, as much as 1 m. tall, erect or decumbent at base; sheaths glabrous; ligule prominent, as much as 8 mm. long, lacerate, decurrent; blades 10 to 20 cm. long, 6 to 8 mm. or as much as 10 mm. wide, very scabrous; panicle erect or nodding, loose, interrupted, 15 to 30 cm. long, the branches clustered, densely flowered to the base; glumes about 3 mm. long,

FIGURE 510.—*Polypogon monspeliensis.* Plant, × ½; glumes and floret, × 10. (Chase 5584, Calif.)

hispidulous, gradually narrowed to an awn 2 to 3 mm. long; lemma 1.5 mm. long, the awn 1 to 2 mm. long. ♃ —Wet places, along streams and ditches, Arizona (Santa Rita Mountains); Mexico to Argentina.

## 78. LYCÚRUS H. B. K.

Spikelets 1-flowered; glumes awned, the first usually 2-awned; lemma narrow, firm, longer than the glumes, tapering into a slender awn. Slender perennial, with grayish, bristly spikelike panicles, the spikelets borne in pairs, the lower of the pair sterile, the two falling together. Type species, *Lycurus phleoides*. Name for Greek *lukos*, wolf, and *oura*, tail, alluding to the spikelike panicles.

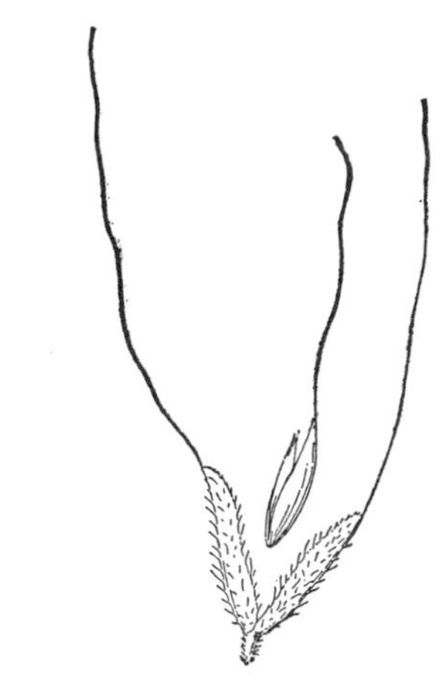

FIGURE 513.—*Polypogon australis*, × 10. (Suksdorf 10091, Wash.)

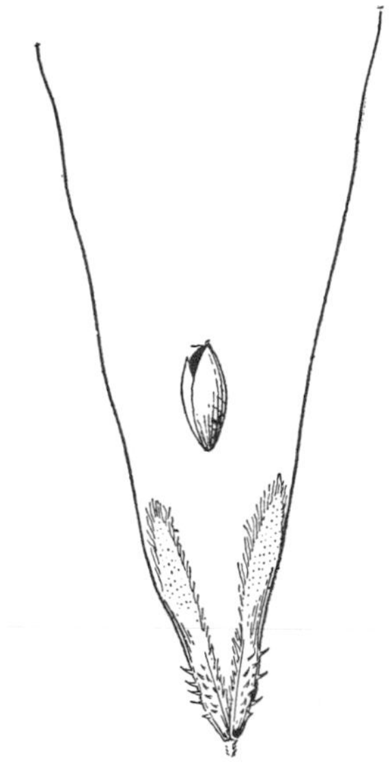

FIGURE 511.—*Polypogon maritimus*, × 10. (Hansen 607, Calif.)

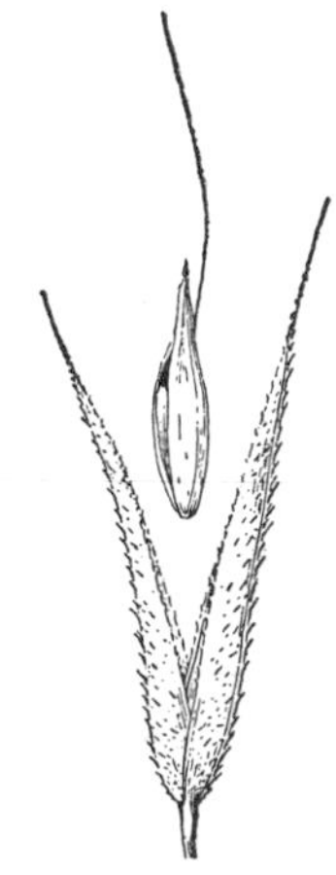

FIGURE 514.—*Polypogon elongatus*. Glumes and floret, × 10. (Silveus 3488, Ariz.)

FIGURE 512.—*Polypogon interruptus*. Panicle, × 1; glumes and floret, × 10. (Hitchcock 2686, Calif.)

**1. Lycurus phleoídes** H. B. K. WOLFTAIL. (Fig. 515.) Culms densely tufted, 20 to 60 cm. tall, compressed, erect or decumbent at base; blades flat or folded, 1 to 2 mm. wide, those of the culm mostly less than 10 cm. long; panicle 3 to 6 cm. long, about 5 mm. thick; spikelets including awns about 5 mm. long, the glumes shorter than the lemma, the first 2- or 3-

FIGURE 515.—*Lycurus phleoides*. Plant, × ½; glumes and floret, × 10. (Rydberg 2363, Colo.)

FIGURE 516.—*Phleum pratense*. Plant, × ½; glumes and floret, × 10. (Mearns 2209, Wyo.)

awned, the second usually 1-awned, the awns slightly spreading; lemma 3-nerved, pubescent on the margins, the awn 2 to 3 mm. long; palea about as long as the lemma, pubescent. ♃ —Plains and rocky hills, Colorado and Utah to Texas and Arizona, south to southern Mexico. Adventive in wool waste, Maine. An important southwestern forage grass.

## 79. PHLÉUM L. Timothy

Spikelets 1-flowered, laterally compressed, disarticulating above the glumes; glumes equal, membranaceous, keeled, abruptly mucronate or awned or gradually acute; lemma shorter than the glumes, hyaline, broadly truncate, 3- to 5-nerved; palea narrow, nearly as long as the lemma. Annuals or perennials, with erect culms, flat blades, and dense, cylindric panicles. Type species, *Phleum pratense*. Name from Greek *phleos*, an old name for a marsh reed.

The common species, *P. pratense*, or timothy, is our most important hay grass. It is cultivated in the humid regions, the Northeastern States, south to the Cotton Belt, and west to the 100th meridian, and also in the humid region of Puget Sound and in mountain districts. The native species, *P. alpinum*, alpine timothy, furnishes forage in mountain meadows of the Western States.

Panicle cylindric, several times longer than wide........ 1. P. pratense.
Panicle ovoid or oblong, usually not more than twice as long as wide........ 2. P. alpinum.

**1. Phleum praténse** L. Timothy. (Fig. 516.) Culms 50 to 100 cm. tall, from a swollen or bulblike base, forming large clumps; blades elongate, mostly 5 to 8 mm. wide; panicle cylindric, commonly 5 to 10 cm. long, often longer, the spikelets crowded, spreading; glumes about 3.5 mm. long, truncate with a stout awn 1 mm. long, pectinate-ciliate on the keel. ♃ —Commonly escaped from cultivation along roadsides and in fields and waste places throughout the United States; Eurasia. In some localities known as herd's grass.

**2. Phleum alpínum** L. Alpine timothy. (Fig. 517.) Culms 20 to 50 cm. tall, from a decumbent, somewhat creeping, densely tufted base; blades mostly less than 10 cm. long, 4 to 6 mm. wide; panicle ellipsoid or short-cylindric, bristly; glumes about 5 mm. long, hispid-ciliate on the keel, the awns 2 mm. long. ♃ —Common in mountain meadows, in bogs and wet places, Greenland to Alaska, south in the mountains of Maine and New Hampshire; northern Michigan; in the mountains of the Western States to New Mexico and California; also on the seacoast at Fort Bragg, Calif., and northward; Mexico; Eurasia and Arctic and alpine regions of the Southern Hemisphere.

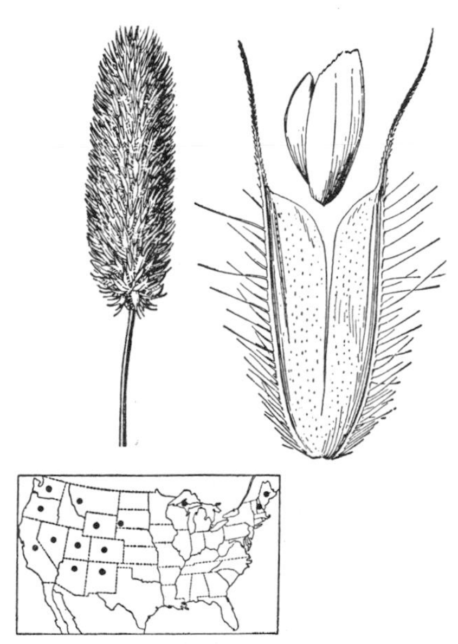

Figure 517.—*Phleum alpinum*. Panicle, × 1; glumes and floret, × 10. (Clements 337, Colo.)

**Phleum arenárium** L. Annual; culms tufted, 5 to 30 cm. tall; foliage scant, mostly basal, the blades 2 to 4 cm. long; panicle 1 to 3 cm. long, somewhat tapering at each end; glumes acuminate, strongly ciliate on

FIGURE 518.—*Gastridium ventricosum*. Plant, × ½; glumes and floret, × 10. (Davy and Blasdale 5340, Calif.)

the keel. ⊙ —Ballast near Portland, Oreg.; coast of Europe and North Africa.

**Phleum subulátum** (Savi) Aschers. and Graebn. Annual; culms 10 to 20 cm. tall; blades 2 to 5 cm. long; panicle linear-oblong, mostly 3 to 8 cm. long, 4 to 5 mm. thick; glumes 2 mm. long, scaberulous, subacute, the tips approaching. ⊙ —Ballast, Philadelphia, Pa., and near Portland, Oreg.; Mediterranean region.

**Phleum paniculátum** Huds. Annual; culms 10 to 30 cm. tall; foliage scabrous; panicle cylindric, 2 to 5 cm. long, 3 to 6 mm. thick; glumes 2 mm. long, glabrous, hard, widened upward to a truncate swollen summit, with a hard awn-point at the tip. ⊙ —Ballast near Portland, Oreg.; Mediterranean region.

## 80. GASTRÍDIUM Beauv.

Spikelets 1-flowered, the rachilla disarticulating above the glumes, prolonged behind the palea as a minute bristle; glumes narrow, unequal, somewhat swollen at the base; lemma much shorter than the glumes, hyaline, broad, truncate, awned or awnless; palea about as long as the lemma. Annual with flat blades and pale, shining, spikelike panicles. Type species, *Milium lendigerum* L. (*G. ventricosum*). Name from Greek *gastridion*, a small pouch, alluding to the slightly saccate glumes.

**1. Gastridium ventricósum** (Gouan) Schinz and Thell. NITGRASS. (Fig. 518.) Culms 20 to 40 cm. tall; foliage scant, blades scabrous; panicle 5 to 8 cm. long, dense, spikelike; spikelets slender, about 5 mm. long; glumes tapering into a long point, the second about one-fourth shorter than the first; floret minute, plump, pubescent, the delicate awn 5 mm. long, somewhat geniculate. ⊙ —Open ground and waste places, Oregon to California; Texas; also Boston, Mass.; introduced from Europe. A common weed on the Pacific coast, of no economic value.

## 81. LAGÚRUS L.

Spikelets 1-flowered, the rachilla disarticulating above the glumes, pilose under the floret, produced beyond the palea as a bristle; glumes subequal, thin, 1-nerved, villous, gradually tapering into a plumose awn-point; lemma shorter than the glumes, thin, glabrous, bearing on the back above the middle a slender, exserted, somewhat geniculate, awn, the summit bifid, the divisions delicately awn-tipped; palea narrow, thin, the two keels ending in minute awns. Annual, with pale, dense, ovoid or oblong woolly heads. Type species, *Lagurus ovatus*. Name from Greek *lagos*, hare, and *oura*, tail, alluding to the woolly heads.

**1. Lagurus ovâtus** L. (Fig. 519.) Culms branching at the base, 10 to 30 cm. tall, slender, pubescent; sheaths and blades pubescent, the sheaths somewhat inflated, the blades flat, lax; panicle 2 to 3 cm. long, nearly as thick, pale and downy, bristling with dark awns; glumes very narrow, 10 mm. long, the awns of the lemmas much exceeding them. ⊙ —Cultivated for ornament and sparingly escaped; New Jersey; Pacific Grove, San Francisco, and Berkeley, Calif.; ballast, Beaufort, N. C.; Mediterranean region.

## 82. MUHLENBÉRGIA Schreb. Muhly

Spikelets 1-flowered (occasionally 2-flowered), the rachilla disarticulating above the glumes; glumes usually shorter than the lemma, sometimes as long, obtuse to acuminate or awned, keeled or convex on the back, the first sometimes small, rarely obsolete; lemma firm-membranaceous, 3-nerved (the nerves sometimes obscure or rarely an obscure additional pair), with a very short callus, rarely long-pilose, usually minutely pilose, the apex acute, awned from the tip or just below it, or from between very short lobes, sometimes only mucronate, the awn straight or flexuous. Perennial, or rarely annual, low or moderately tall or rarely robust grasses, tufted or rhizomatous, the culms simple or much-branched, the inflorescence a narrow (sometimes spikelike) or open panicle. Type species, *Muhlenbergia schreberi*. Named for G. H. E. Muhlenberg.

Many of the western species are important range grasses, forming a considerable proportion of the grass flora of the arid and semiarid regions, and long ago dubbed "muhly" by forest rangers. The most important of these are *M. montana* on mesas and rocky hills of the Western States, *M. pauciflora*, *M. emersleyi*, and *M. wrightii* in the Southwest.

1a. Plants annual.
  2a. Lemma awned.
    Awn of lemma 0.5 to 3 mm. long; glumes acuminate, hirsute.
      Spikelets 1.5 to 1.8 mm. long; relatively long pediceled and spreading along the panicle branches .......... 5. M. TEXANA.
      Spikelets 2 to 2.5 mm. long; short pediceled and mostly appressed along the panicle branches .......... 6. M. ELUDENS.
    Awn of lemma more than 5 mm. long.
      Second glume 3-nerved and often 3-toothed .......... 9. M. PULCHERRIMA.
      Second glume 1-nerved (rarely 2-nerved).
        First glume 2-nerved and usually bidentate.
          Glumes equal to or slightly longer than the floret; lemma about 3 mm. long; awn 2 to 10 mm. long .......... 11. M. DEPAUPERATA.
          Glumes shorter than the floret, sometimes minute, but usually about half as long as the lemma; lemma 4 to 5 mm. long; awn 10 to 20 mm. long. 12. M. BREVIS.
        First glume 1-nerved (rarely 2-nerved), entire or erose, but not bidentate.
          Glumes acuminate or aristate. Lateral nerves of lemma often ciliate. 10. M. PECTINATA.

FIGURE 519.—*Lagurus ovatus*. Plant, × ½; spikelet and floret, × 5. (Heller 5340, Calif.)

Glumes obtuse.

Panicle open, the branches spreading; lemma 2.5 to 3.5 mm. long.
7. M. MICROSPERMA.

Panicle very narrow, the branches appressed; lemma 4.5 to 6 mm. long.
8. M. APPRESSA.

2b, Lemma awnless (see also *M. texana*). Culms branching and panicle-bearing at base.

Pedicels capillary, elongate.
Panicles very diffuse; pedicels straight; glumes glabrous............ 3. M. FRAGILIS.
Panicles open but scarcely diffuse; glumes pilose.
Pedicels sinuous and tangled; glumes long-pilose........................ 4. M. SINUOSA.
Pedicels straight or subflexuous; glumes minutely pilose.... 2. M. MINUTISSIMA.
Pedicels short, appressed; glumes glabrous.
Panicles loose, delicate; spikelets 1 to 1.2 mm. long........................ 1. M. WOLFII.
Panicles narrow, contracted; spikelets 2 mm. long.................. 13. M. FILIFORMIS.
Culms simple, compressed.................................................................. 29. M. UNIFLORA.
1b. Plants perennial.
3a. Rhizomes developed, usually prominent, scaly, creeping, often branching.
4a. Blades 2 mm. wide or less, mostly short and involute.
5a. Panicles open, the spikelets on slender pedicels.
Spikeless awned, 4 to 5 mm. long; blades involute. Panicle branches in stiffly spreading fascicles.................................................................. 50. M. PUNGENS.
Spikelets awnless, acutish or mucronate, 1 to 2 mm. long; blades flat.
Sheaths compressed keeled; panicle oblong; eastern species.
28. M. TORREYANA.
Sheaths rounded; panicle as broad as long; western species.
Ligule 1 to 2 mm. long, auricled............................................ 26. M. ARENACEA.
Ligule minute, not auricled.............................................. 27. M. ASPERIFOLIA.
5b. Panicles narrow, more or less condensed, the spikelets on short pedicels.
Culms tall, stout, somewhat woody at base, as much as 6 mm. thick, 1 to 3 m. tall.
30. M. DUMOSA.
Culms lower, slender.
Lemma and palea glabrous.
Culms smooth, widely creeping, the blades fine, conspicuously recurved, spreading.
Spikelets about 3 mm. long.............................................. 14. M. REPENS.
Spikelets about 2 mm. long.............................................. 15. M. UTILIS.
Culms nodulose-roughened, erect or decumbent at base, sometimes spreading, but not widely creeping........................................ 16. M. RICHARDSONIS.
Lemma and palea pilose or villous on the lower half.
Awns 6 to 10 mm. long.
Panicles densely flowered; glumes as long as the floret.
21. M. POLYCAULIS.
Panicles loosely flowered; glumes about half as long as the floret.
23. M. ARSENEI.
Awns 1 to 3 mm. long or the lemma mucronate only.
Blades 5 to 10 cm., rarely 15 cm. long, flat........................ 20. M. GLAUCA.
Blades 1 to 3 cm. long, involute or pungently pointed.
Glumes about half as long as the floret; lemma 2 to 2.5 mm. long.
17. M. VILLOSA.
Glumes nearly as long as the floret; lemma 3 to 4 mm. long.
Culms glabrous below the nodes; sheaths glabrous (rarely pubescent below the summit); ligule 1 mm. long, short lacerate; lemma loosely villous on the margins on lower half and at the very base, mucronate to short awned.................................. 18. M. THURBERI.
Culms strigose below the nodes; sheaths often strigose to hirsute; ligule 0.5 to 1 mm. long; lemma densely villous on lower half; awn 1 to 3 mm. long.......................................... 19. M. CURTIFOLIA.
4b. Blades flat, at least some of them more than 3 mm., usually 5 mm. wide or more.
6a. Panicles loosely flowered, slender, much exceeding the leaves (see also *M. sylvatica*); glumes broad below, abruptly pointed, shorter than the body of the lemma.
Culms slender, rather weak, becoming much branched, glabrous or slightly scabrous below the nodes. Lemma acuminate, 2.5 to 3.5 mm. long, awned.
37. M. BRACHYPHYLLA.
Culms erect, simple or sparingly branched.
Spikelets 1.5 to 2.5 mm. long; lemma awnless or awn-tipped; blades commonly not more than 5 to 7 mm. wide...................................... 35. M. SOBOLIFERA.
Spikelets 3 to 4 mm. long; lemma with an awn 2 to 5 times as long as the body; blades commonly 8 mm. or more wide........................ 36. M. TENUIFLORA.
6b. Panicles usually densely flowered (sometimes loose in *M. sylvatica*); glumes tapering from base to apex. Culms commonly freely branching (often simple or nearly so in *M. glomerata*).
Hairs at base of floret copious, as long as the body of the lemma.... 31. M. ANDINA.
Hairs at base of floret inconspicuous, not more than half as long as the lemma.

Glumes with stiff scabrous awn-tips, much exceeding the awnless lemma; panicles terminal on the culm or leafy branches, compact, interrupted, bristly.
Culms mostly simple or branching at base; internodes minutely puberulent; sheaths not or scarcely keeled........ 32. M. GLOMERATA.
Culms subcompressed, mostly branching from the middle nodes; internodes smooth and glossy except at summit; sheaths keeled.
33. M. RACEMOSA.
Glumes acuminate, sometimes awn-tipped but not stiff and exceeding the lemma; panicles terminal and axillary, numerous, not bristly.
Culms glabrous below the nodes; panicles not compact, the branches ascending; plants sprawling, top-heavy, the branchlets geniculate-spreading.
38. M. FRONDOSA.
Culms strigose below the nodes; panicles compact or if not the branches erect or nearly so; plants often bushy-branching but not sprawling with geniculate branchlets.
Callus hairs wanting; lemma nearly smooth, awnless.
39. M. GLABRIFLORA.
Callus hairs present; lemma pubescent below.
Panicles not compactly flowered; lemma with awn as much as 10 mm. or more long (nearly awnless in forma *attenuata*); some of the blades 10 to 15 cm. or more long........ 40. M. SYLVATICA.
Panicles compactly flowered or, if not, lemma awnless; blades commonly less than 10 cm. long, but sometimes longer.
Sheaths glabrous........ 41. M. MEXICANA.
Sheaths scabrous........ 34. M. CALIFORNICA.
3b. Rhizomes wanting, the culms tufted, usually erect.
7a. Culms decumbent and rooting at the nodes.
Spikelets awnless; panicles open, diffuse........ 29. M. UNIFLORA.
Spikelets awned; panicles narrow, the branches ascending or appressed.
Glumes minute, the first sometimes wanting........ 42. M. SCHREBERI.
Glumes evident, as much as 3 mm. long (see also *M. schreberi* var. *palustris*).
Awns 1 to 2 mm. long........ 43. M. CURTISETOSA.
Awns 5 to 20 mm. long........ 22. M. PAUCIFLORA.
7b. Culms erect or spreading, but not rooting at the nodes.
Second glume 3-toothed (rarely not toothed in *M. filiculmis*).
Lemma 4 mm. long; culms relatively stout, 15 to 60 cm. tall.... 45. M. MONTANA.
Lemma 2.5 to 3 mm. long; culms filiform, 10 to 20 cm. tall.... 46. M. FILICULMIS.
Second glume usually acute or awned, sometimes erose-toothed, not distinctly 3-toothed.
8a. Panicle narrow or spikelike, the branches floriferous from the base or nearly so (see also *M. metcalfei*).
9a. Lemma acute, acuminate, mucronate or short-awned.
Blades involute.
Panicle elongate and spikelike.
Glumes and lemma or some of them awn-tipped........ 70. M. MARSHII.
Glumes acute to blunt or erose; lemma not awn-tipped.
Ligule 2 to 3 mm. long; lower panicle branches sometimes 5 to 10 cm. long........ 68. M. RIGENS.
Ligule 1 to 2 mm. long; lower panicle branches seldom more than 3 cm. long........ 69. M. MUNDULA.
Panicle narrow but scarcely spikelike, the branches loosely flowered.
Blades mostly in a short basal cluster; panicle 5 to 8 cm. long.
44. M. JONESII.
Blades not in a short basal cluster; panicle 10 to 30 cm. long.
58. M. DUBIA.
Blades flat, folded, or loosely involute.
Panicle more or less spikelike.
Glumes obtuse; culms delicate. Ligule about 2 mm. long.
13. M. FILIFORMIS.
Glumes acute or acuminate; culms wiry.
Glumes gradually acute; culms minutely pubescent; ligule about 0.5 mm. long........ 24. M. CUSPIDATA.
Glumes abruptly acute, usually awn-pointed or awned; culms hispidulous; ligule 1 to 3 mm. long........ 25. M. WRIGHTII.
Panicle narrow, but not spikelike.
Lemma villous below........ 67. M. EMERSLEYI.
Lemma glabrous or obscurely pubescent.

Lower sheaths compressed keeled........................ 65. M. LINDHEIMERI.
Lower sheaths not compressed keeled................. 64. M. LONGILIGULA.

9b. Lemma with an awn usually more than 5 mm. long, or some of the awns less in *M. dubioides* and *M. metcalfei.*

Old sheaths becoming flat and more or less coiled at base of plant. 47. M. VIRESCENS.

Old sheaths not flat and coiled.

Panicle mostly 20 to 40 cm. long.

Ligule 1 to 2 mm. long; glumes acute or awn-pointed. 59. M. DUBIOIDES.

Ligule 4 to 5 mm. long; glumes obtuse to subobtuse. 57. M. METCALFEI.

Panicle mostly 5 to 10 cm. long.

10a. Lemma pilose or villous on lower part.

Culms loosely tufted, hard and wiry at base.

Glumes and floret about equal; lemma 2.5 to 3 mm. long, villous below........................................................ 21. M. POLYCAULIS.

Glumes about half as long as floret; lemma 4 to 5 mm. long, sparsely pilose................................................................ 23. M. ARSENEI.

Culms closely or somewhat loosely tufted, slender but not hard and wiry at base................................................ 48. M. MONTICOLA.

10b. Lemma scaberulous, not pilose.

Glumes less than 1 mm. long................................ 49. M. PARVIGLUMIS.

Glumes 2 to 4 mm. long........................................ 22. M. PAUCIFLORA.

8b. Panicle open, or at least loose, the branches naked at base (sometimes shortly so in *M. metcalfei*).

Plants widely spreading, much branched, wiry, the base knotty. 51. M. PORTERI.

Plants erect, not widely spreading and much branched.

Blades flat, the midnerve and margins white-cartilaginous. 52. M. ARIZONICA.

Blades folded or involute, or occasionally some of them flat.

Blades short in a basal cluster.

Panicle mostly less than 15 cm. long; blades 1 to 3 cm. long, involute, curled or falcate................................................ 53. M. TORREYI.

Panicle mostly more than 20 cm. long; blades commonly 5 to 8 cm. long, flat or usually folded............................................ 54. M. ARENICOLA.

Blades elongate.

11a. Panicle open or diffuse, if narrow, the branches slender; capillary, more or less flexuous.

Awn of lemma less than 5 mm. long; panicle usually not more than twice as long as wide at maturity, the branches and pedicels stiff.

Plants fibrous at the base; lemma awnless or with an awn to 2 mm., rarely to 5 mm., long........................................ 60. M. EXPANSA.

Plants not fibrous at base; lemma with an awn 2 to 5 mm. long. 61. M. REVERCHONI.

Awn of lemma usually more than 10 mm. long (sometimes awnless in *M. emersleyi*); panicle elongate, usually at least 4 times as long as wide at maturity.

Panicle diffuse, the branches more than 10 cm. long; pedicels usually much longer than the spikelets.................... 62. M. CAPILLARIS.

Panicle open but not diffuse.

Panicle deep purple; blades relatively coarse, some of them usually flat.............................................................. 63. M. RIGIDA.

Panicle pale or tawny; blades involute, scabrous.

Ligule 4 to 10 mm. long; glumes obtuse to subacute. 55. M. SETIFOLIA.

Ligule 1 to 3 mm. long; glumes acute to mucronate. 56. M. XEROPHILA.

11b. Panicle narrow, elongate, the branches rather stiffly ascending or appressed.

Lower sheaths rounded................................................ 57. M. METCALFEI.

Lower sheaths compressed-keeled.

Glumes as long as the floret; lemma villous below. 67. M. EMERSLEYI.

Glumes distinctly shorter than the floret; lemma pubescent on the margins toward the base.................................. 66. M. INVOLUTA.

FIGURE 520.—*Muhlenbergia wolfii*. Plant, × 1; spikelet and floret, × 10. (Hitchcock 7661, Mex.)

**1. Muhlenbergia wólfii** (Vasey) Rydb. (Fig. 520.) Annual; culms spreading, branching at base, 6 to 25 cm. tall; blades flat, mostly 1 to 3 cm. long, 1 mm. wide or less; panicle 2 to 6 cm. long, the simple branches ascending, the short, stiff pedicels appressed along the branches; spikelets 1 to 1.2 mm. long; glumes glabrous, about half as long as the spikelet; lemma rather turgid, minutely white-silky along the margins. ⊙ (*Sporobolus ramulosus* of Manual, ed. 1.)—Open or wooded slopes, mostly in thin soil, Colorado to northern Mexico and Arizona.

**2. Muhlenbergia minutíssima** (Steud.) Swallen. (Fig. 521.) Annual; culms erect to spreading, branching at base, 10 to 35 cm. tall; blades flat, mostly less than 10 cm. long, about 1 mm. wide; panicle half to three-fourths the length of the entire plant, the slender pedicels ascending; spikelets 1.2 to 1.5 mm. long, the glumes half to two-thirds as long, minutely pilose; lemma minutely silky-pubescent along the midnerve and margins. ⊙ (*Sporobolus microspermus* of Manual, ed. 1.)—Moist sandy or rocky slopes, Montana to Washington south to Texas, California, and northern Mexico.

**3. Muhlenbergia frágilis** Swallen. (Fig. 522.) Annual; culms geniculate-ascending, freely branching at base, 10 to 30 cm. tall; blades flat, mostly 2 to 6 cm. long, 1 to 1.5 mm. wide; panicle very diffuse, the capillary branches, branchlets, and pedicels widely spreading or reflexed, fragile; spikelets 1 to 1.1 mm. long, the glumes half to two-thirds as long, glabrous; lemma silky-pubescent on the keel and margins, the palea silky-pubescent between the nerves. ⊙ —Moist sandy soil and rocky hills, western Texas to southern Arizona, south to central Mexico.

**4. Muhlenbergia sinuósa** Swallen. (Fig. 523.) Annual; culms geniculate-ascending, freely branching at base; blades flat, mostly 4 to 10 cm. long, 1 to 1.5 mm. wide, minutely pubescent on both surfaces; panicle many-flowered, 14 to 22 cm. long, 2 to 6 cm. wide, the scabrous branches ascending, the elongate, capillary pedicels sinuous and tangled; spikelets often purple-tinged, 1.5 to 2 mm. long, the glumes about half as long, usually conspicuously pilose; lemma obtuse, delicately silky-pubescent below on the midnerve and margins, the broad palea equal. ⊙ —Moist canyon walls and borders of marshes, New Mexico and Arizona.

**5. Muhlenbergia texána** Buckl. (Fig. 524.) Annual, culms delicate, erect or ascending, branching at base, 10 to 30 cm. tall, the culms strongly

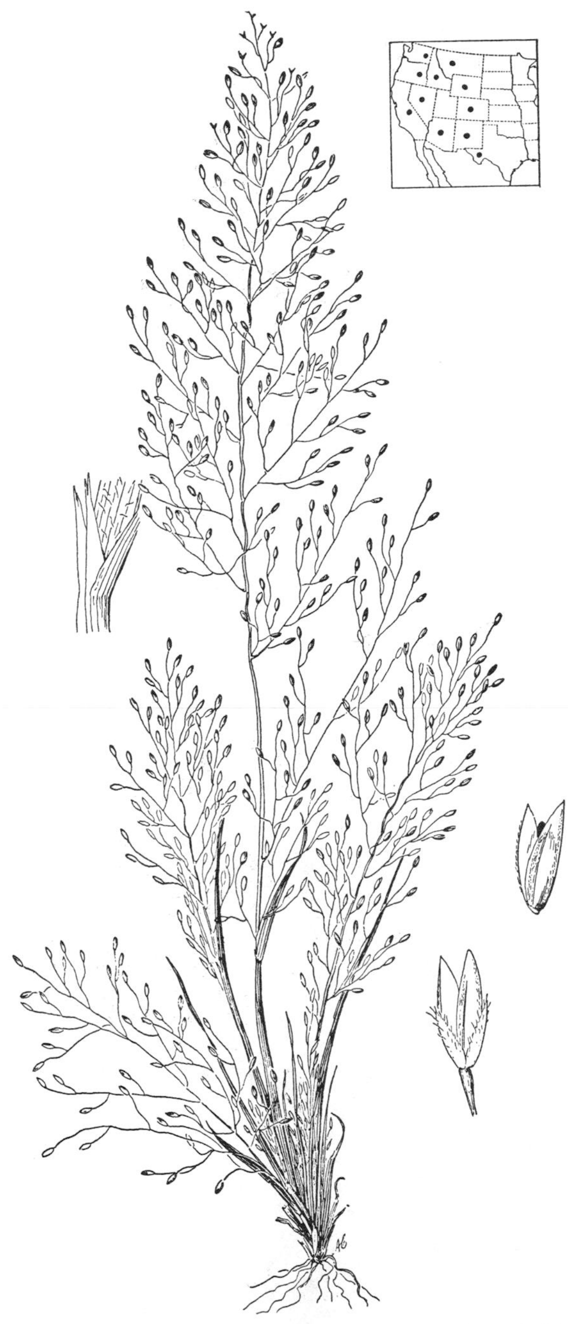

FIGURE 521.—*Muhlenbergia minutissima*. Plant, × 1; spikelet, floret, and ligule, × 10. (Metcalfe 1431, N. Mex.)

unequal; foliage scant, ligule about 2 mm. long, erose, decurrent down the sheath; blades 2 to 5 cm. long, about 1 mm. wide; panicle half to two-thirds

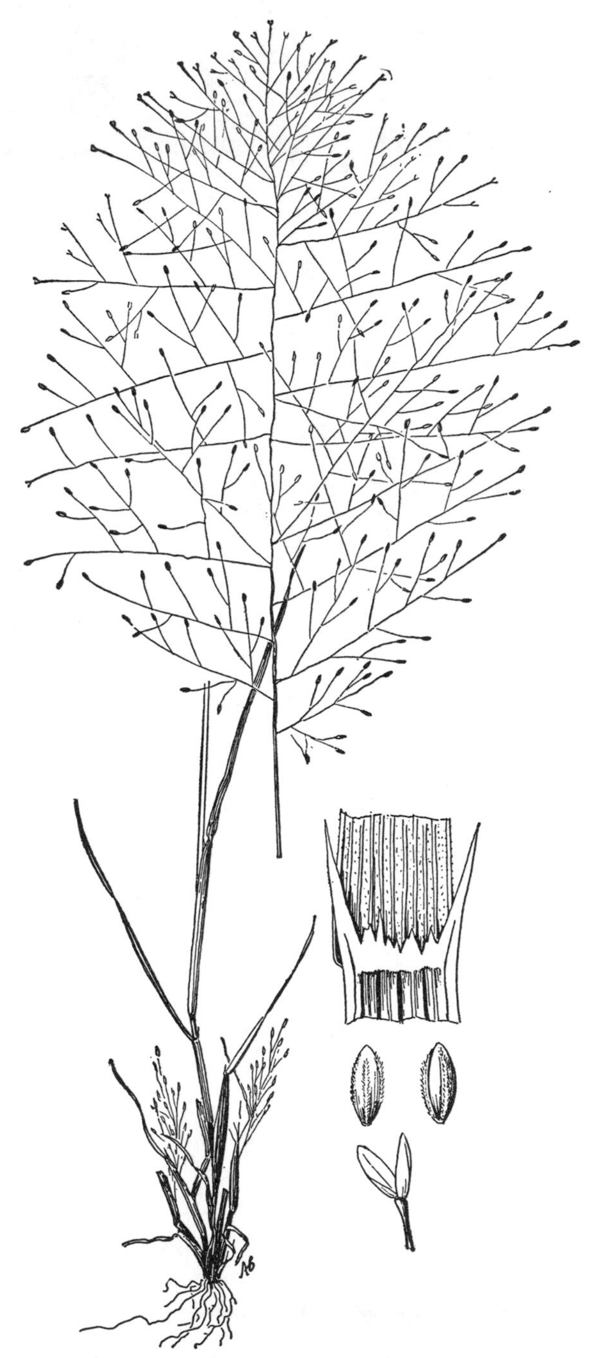

FIGURE 522.—*Muhlenbergia fragilis*. Plant, × 1; glumes, two views of floret, and ligule, × 10. (Type.)

FIGURE 523.—*Muhlenbergia sinuosa*. Plant, × 1; spikelet and ligule, × 10. (Type.)

FIGURE 524.—*Muhlenbergia texana*. Plant, × 1; spikelet and floret, × 10. (Wright 736, western Texas.)

the length of the plant, the delicate branches ascending or spreading; spikelets about 1.5 mm. long, on capillary mostly spreading pedicels 2 to 5 mm. long; glumes 1 and 1.5 mm. long, sparsely hirsute; lemma 1.6 to 1.8 mm. long, minutely silky on the nerves below, slightly notched and with a delicate awn 1 to 1.3 mm. long, the awns sometimes fallen in overmature specimens. ⊙ —Rocky canyons and slopes, western Texas to Arizona, northern Mexico, and Baja California; rare or overlooked.

FIGURE 525.—*Muhlenbergia eludens.* Plant, × 1; glumes and floret, × 8. (Pringle 399, Mex.)

**6. Muhlenbergia elúdens** C. G. Reeder. (Fig. 525.) Annual, branching at base, culms slender, erect, 15 to 35 cm. tall, the culms strongly unequal; foliage scant, scabrous; ligule 2 to 2.5 mm. long; blades mostly 4 to 7 cm. long to 1.5 mm. wide, involute upward; panicle half to three-fourths the length of the plant, the slender branches relatively stiffly spreading; spikelets 2.2 to 2.5 mm. long, on short pedicels, mostly closely appressed to the branches; glumes about 1 to 1.5 mm. long, hirsute; lemma 2.3 mm. long, silky on the midnerve and margins, slightly notched and with an awn 2 to 2.5 mm. long. ⊙ (Included in *M. texana* in Manual, ed. 1.)—Rocky woods and wet ledges and gravel bars, to 2,400 m. altitude, New Mexico, Arizona, and northern Mexico.

**7. Muhlenbergia microspérma** (DC.) Kunth. LITTLESEED MUHLY. (Fig. 526.) Annual; culms densely tufted, branching and spreading at base, often purple, 10 to 30 cm. tall; blades mostly less than 3 cm. long, 1 to 2 mm. wide, scabrous; panicles narrow, 5 to 15 cm. long, the branches rather distant, ascending; spikelets on short thick pedicels; glumes broad, obtuse, subequal, less than 1 mm. long; lemma narrow, 2 to 4 mm. long, scabrous, the slender awn 1 to 3 cm. long. ⊙ —Open dry ground, Nevada, Arizona, and southern California to Peru. Cleistogamous spikelets are developed at the base of lower sheaths, solitary or few in a fascicle in each axil, each spikelet included in an indurate thickened, tightly rolled narrowly conical reduced sheath, which readily disarticulates from the plant at maturity. The glumes are wanting and awn of lemma reduced, but the grain is larger than that of the spikelets in the terminal inflorescence, being about the same length (2 mm.) but much thicker.

FIGURE 526.—*Muhlenbergia microsperma.* Plant, × 1; glumes and floret, × 10. (Mearns 2780, Ariz.)

**8. Muhlenbergia apprėssa** C. O. Goodding. (Fig. 527.) Culms 10 to 40 cm. tall, erect or decumbent at base, much branched below; ligule lacerate, 2 to 3 mm. long; blades flat or folded, 1 to 4 cm. long, scabrous or puberulent; panicles numerous, as much as 20 cm. long, very narrow, loosely flowered, the branches appressed; glumes 1 to 2 mm. long or sometimes less, obtuse; lemma 4.5 to 6 mm. long, scabrous above, densely pilose on the callus and margins at the base; awn 10 to 30 mm. long. ⊙ —Canyons and slopes, southern Arizona. Cleistogamous spikelets similar to those in *M. microsperma* are common in the lower reduced sheaths.

FIGURE 527.—*Muhlenbergia appressa*. Plant, × 1; glumes and floret, × 10. (Type.)

**9. Muhlenbergia pulchérrima** Scribn. (Fig. 528.) Culms 10 to 25 cm. tall, erect, freely branching at the base; sheaths scabrous, longer than the internodes; ligule thin, 2 to 3 mm. long; blades flat, pubescent on the upper surface, mostly less than 5 cm. long and 1 mm. wide; panicles 3 to 5 cm. long, the branches ascending or appressed; first glume 0.5 to 1 mm. long, acute or notched, the second 2 mm. long, 2- or 3-toothed; lemma 3 to 4 mm. long, narrow, acuminate, minutely bifid, scabrous, pubescent on the lower half of the margins; awn slender, flexuous, mostly 10 to 15 mm. long, or sometimes only 5 mm. long. ⊙ —Rocky ledges and open ground, Arizona (Apache County); Chihuahua, Mexico.

FIGURE 528.—*Muhlenbergia pulcherrima*. Plant, × 1; glumes and floret, × 10. (Schroeder, Ariz.)

**10. Muhlenbergia pectináta** C. O. Goodding. (Fig. 529.) Culms 10 to 25 cm. long, erect to decumbent, sometimes rooting at the lower nodes, freely branching, angular; sheath margins often ciliate; ligule erose to ciliate, about 0.5 mm. long; blades flat to involute, 1 to 6 cm. long, 1 to 2 mm. wide, pubescent or sparsely pilose; panicles numerous, narrow, 2 to 12 cm. long; spikelets 3.5 to 4.5 mm. long; glumes abruptly acute or acuminate, commonly aristate, 1.5 to 2 mm. or sometimes 3 mm. long, the awn about half the entire length; lemma 3- to 5-nerved, scabrous to prominently ciliate on the lateral nerves, the callus appressed-pubescent; awn 10 to 30 mm. long. ⊙ —Moist rocky hillsides, southern Arizona; Mexico.

FIGURE 529.—*Muhlenbergia pectinata*. Plant, × 1; glumes and floret, × 10. (Type.)

**11. Muhlenbergia depauperáta** Scribn. (Fig. 530.) Culms 2 to 15 cm. tall, densely tufted, erect, scabrous to hispidulous below the nodes; blades 1 to 1.5 cm. long (rarely 3 cm.), 1 to 1.5 mm. wide, scabrous, puberulent on the upper surface, with white cartilaginous midnerve and margins; panicles narrow, spikelike, often included, 1 to 4 cm. long (rarely to 6 cm.), the branches and pedicels closely appressed; glumes narrow, scabrous, about equal to or slightly longer than the floret, the tips often spreading; first glume 2-nerved, bidentate or entire, 2.5 to 3.5 mm. long; second glume 1-nerved, acuminate-aristate, 3 to 4 mm. long; lemma 3 to 3.5 mm. long, prominently 3-nerved, scabrous above, sparsely pubescent on the internerves, the straight awn 2 to 10 mm. long, rarely less. ⊙ —Open gravelly places, Arizona and New Mexico; northern Mexico.

**12. Muhlenbergia brévis** C. O. Goodding. (Fig. 531.) Culms 3 to 20 cm. tall, erect, tufted, much branched below; ligule 1 to 3 mm. long, lacerate; blades flat to involute, 0.5 to 4 cm. long, scabrous or puberulent above, scabrous below; panicles 1 to 2 cm. long, narrow, rather densely flowered, the branches erect; glumes scabrous, variable, shorter than the floret; first glume 1 to 3 mm. long, 2-nerved, minutely to deeply bifid; second glume 1.5 to 4 mm. long (usu-

FIGURE 530.—*Muhlenbergia depauperata*. Plant, × 1; glumes and floret, × 10. (Hitchcock 13560, N. Mex.)

ally 2 to 3 mm.), 1-nerved, acuminate; lemma 4 to 5 mm. long, 3- to 5-nerved, scabrous, especially on the nerves, sparsely to rather densely appressed-pubescent on the internerves toward the base; awn 10 to 20 mm. long (rarely less). ⊙ (*M. depauperata* of Manual, ed. 1.)—Open ground at higher elevations, Colorado and Texas to Arizona; Mexico.

**13. Muhlenbergia filifórmis** (Thurb.) Rydb. PULL-UP MUHLY. (Fig. 532.) Annual, or sometimes appearing perennial, loosely tufted, rather soft and lax, erect or somewhat spreading; culms filiform, usually 5 to 15 cm. tall, sometimes as much as 30 cm.; ligule about 2 mm. long; blades flat, usually less than 3 cm. long; panicle narrow, interrupted, few-flowered, usually less than 5 cm. long; glumes ovate, 1 mm. long; lemma lanceolate, acute, mucronate, 2 mm. long, minutely pubescent, scaberulous at tip. ⊙ —Open woods and mountain meadows, South Dakota and Kansas to British Columbia, south to New Mexico and California. A somewhat stouter form with thicker panicles has been differentiated as *M. simplex* Rydb.

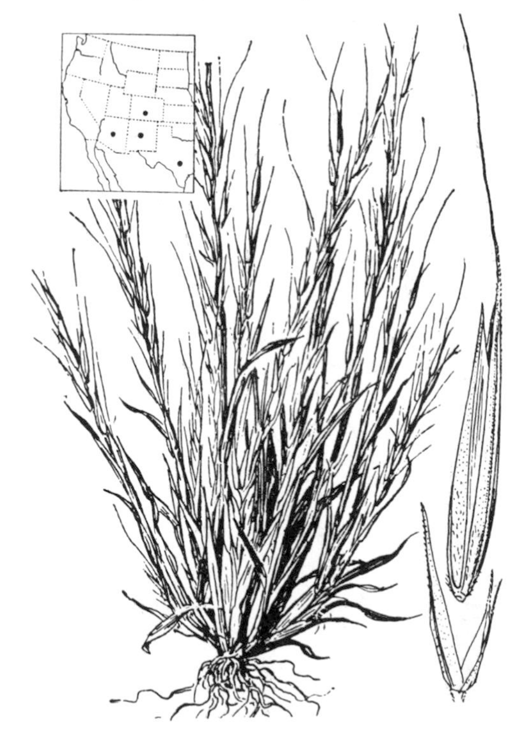

FIGURE 531.—*Muhlenbergia brevis*. Plant, × 1; glumes and floret, × 10. (Type.)

**14. Muhlenbergia répens** (Presl) Hitchc. CREEPING MUHLY. (Fig. 533.) Perennial with widely creeping scaly rhizomes; culms decumbent, branching, spreading, the flowering branches 5 to 20 cm. long; blades mostly 1 to 3 cm. long, flat or soon involute; panicle narrow, 1 to 4 cm. long, sometimes longer, interrupted; spikelets about 3 mm. long; glumes more than half as long as the lemma or a little more, acutish; lemma narrowed to a more or less apiculate summit, minutely roughened, usually darker than the glumes, the lateral nerves obscure. ♃ —Dry rocky or sandy open ground, Texas to Arizona; Mexico.

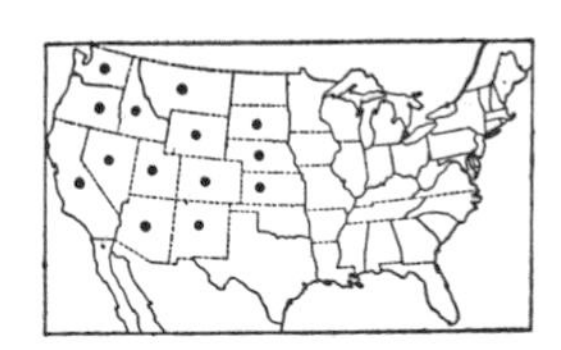

FIGURE 532.—*Muhlenbergia filiformis*. Plant, × 1; glumes and floret, × 10. (Nelson 4011, Wyo.)

**15. Muhlenbergia útilis** (Torr.) Hitchc. APAREJO GRASS. (Fig. 534.) Similar to *M. repens;* usually more delicate and more widely spreading with finer leaves, the blades mostly 1 mm. wide or less; spikelets about 2 mm. long, less pointed, the glumes

FIGURE 533.—*Muhlenbergia repens*. Plant, × 1; glumes and floret, × 10. (Silveus 831, Tex.)

sometimes less than half as long as the paler lemma. ♃ (*Sporobolus utilis* Scribn.)—Wet places, marshy soil, and along ditches and streams, Texas, southern California, Nevada, and Mexico. Used for stuffing pack saddles.

**16. Muhlenbergia richardsónis** (Trin.) Rydb. MAT MUHLY. (Fig. 535.) Perennial from numerous hard creeping rhizomes; culms wiry, nodulose-roughened, erect or decumbent at base, from 5 to 60 cm. tall; ligule 2 to 3 mm. long; blades usually involute, 1 to 5 cm. long, rarely longer; panicle narrow, interrupted, or sometimes rather close and spikelike, 2 to 10 cm. long; spikelets 2 to 3 mm. long, the glumes about half as long, ovate; lemma lanceolate, acute, mucronate. ♃ —Dry or moist open often alkaline soil, New Brunswick and Maine to Alberta, south to Michigan and Nebraska and in the mountains to New Mexico, through eastern Washington to California and Arizona; Baja California. There are two intergrading forms of this species; one with rather stout decumbent or somewhat spreading culms (*M. squarrosa* (Trin.) Rydb.), the other with slender erect culms (*M. richardsonis* (Trin.) Rydb.).

**17. Muhlenbergia villósa** Swallen. (Fig. 536.) Culms 10 to 20 cm. tall, wiry, freely branching, erect from creeping rhizomes, puberulent, obscurely nodulose; blades 2.5 to 3.5 cm. long, firm, involute, glabrous beneath, pubescent above; panicles 2 to 4 cm. long, the branches appressed or spreading, closely flowered; spikelets 2 to 2.5 mm. long, appressed; glumes subequal, 1 to 1.6 mm. long, acute or

FIGURE 534.—*Muhlenbergia utilis*. Plant, × 1; glumes and floret, × 10. (Lindheimer 559, Tex.)

FIGURE 535.—*Muhlenbergia richardsonis.* Plant, × ½; glumes and lemma, × 10. (Jones 5743, Utah.)

subobtuse; lemma and palea villous on the lower half, the lemma acute or mucronate. ♃ —Known only from south of Stanton, Tex. The type of this species was previously referred to *M. thurberi* Rydb.

**18. Muhlenbergia thurbéri** Rydb. (Fig. 537.) Perennial, with creeping rhizomes; culms slender, 10 to 20 cm. tall, branched at base, the branches erect, tufted, the tufts on branches of the rhizome; sheaths glabrous; blades involute, slender, mostly 1 to 3 cm. long; panicle pale, narrow, slender, 3 to 7 cm. long, the branches short, appressed, few-flowered; spikelets 3.5 to 4 mm. long; glumes nearly as long as the lemma, acute; lemma and palea villous on lower half, the lemma mucronate to short-awned. ♃ —Dry hills, New Mexico and Arizona; rare.

**19. Muhlenbergia curtifólia** Scribn. (Fig. 538.) Perennial, with creeping rhizomes; culms 10 to 20 cm. tall, loosely tufted, few from the branches of the rhizome; sheaths glabrous or pubescent; blades 1 to 2.5 cm. long, 2 to 3 mm. wide or less, rigidly spreading, pungently pointed, more or less pubescent; panicle 4 to 8 cm. long, slender, the branches appressed; spikelets 3 to 3.5 mm. long; glumes acute, a little shorter than the floret; lemma and palea villous on the lower half, scabrous above, tapering into an awn 1 to 4 mm. long. ♃ —Rocky soil, southern Utah, southern Nevada, and northern Arizona.

**20. Muhlenbergia glaúca** (Nees) Mez. (Fig. 539.) Perennial, from a slender creeping branching woody rhizome; culms slender, wiry, erect

FIGURE 536.—*Muhlenbergia villosa*. Plant, × 1; glumes and floret, × 10. (Type.)

or ascending, 20 to 60 cm. tall, branching from the lower nodes; blades flat to subinvolute, mostly 5 to 10 cm. long, 1 to 2 mm. wide; panicle 5 to 12 cm. long, narrow, contracted, interrupted, the branches short, appressed; spikelets 3 to 4 mm. long, the glumes nearly as long, acuminate; lemma sparsely pilose on the lower part, acuminate into an awn usually 1 to 3 mm. (rarely as much as 8 mm.) long. ♃ (*M. lemmoni* Scribn.)—Deserts, western Texas to southern California (Jamacha) and northern Mexico.

**21. Muhlenbergia polycaúlis** Scribn. (Fig. 540.) Perennial, from a firm crown; culms numerous, wiry, decumbent and scaly at base, 30 to

FIGURE 537.—*Muhlenbergia thurberi*. Plant, × 1; glumes and floret, × 10. (Standley 7345, Ariz.)

FIGURE 538.—*Muhlenbergia curtifolia*. Plant, × 1; glumes and floret, × 10. (Type.)

50 cm. tall; blades mostly flat and less than 5 cm. long, about 1 mm. wide; panicle narrow, contracted, interrupted, 3 to 8 cm. long; spikelets, excluding awns, 2.5 to 3 mm. long, the glumes a little shorter, tapering to slender awn tips; lemma and palea loosely villous below, the lemma tapering into a delicate awn 1 to 2 cm. long. ♃ —Shaded ledges and grassy slopes, western Texas to southern Arizona and central Mexico.

**22. Muhlenbergia pauciflóra** Buckl. New Mexican muhly. (Fig. 541.) Perennial; culms loosely tufted, wiry, erect, branching at the lower nodes, 30 to 60 cm. tall; blades 1 mm. wide or less; panicle narrow, contracted, interrupted, 5 to 12 cm. long, the branches erect or ascending; spike-

Figure 539.—*Muhlenbergia glauca.* Plant, × 1; glumes and floret, × 10. (Nealley 726, Tex.)

Figure 540.—*Muhlenbergia polycaulis.* Plant, × 1; glumes and floret, × 10. (Type.)

FIGURE 541.—*Muhlenbergia pauciflora*. Plant, × 1; glumes and floret, × 10. (Wright 732, Tex.)

lets, excluding awn, about 4 mm. long, the glumes about half as long, acuminate to awn-tipped; lemma scaberulous only, tapering into a slender flexuous awn, 5 to 20 mm. long. ♃ —Rocky hills and canyons, western Texas and Colorado, Utah, and Arizona, south to northern Mexico.

**23. Muhlenbergia arsénei** Hitchc. (Fig. 542.) Perennial, without rhizomes but the spreading base sometimes rhizomatous in appearance, loosely tufted; culms wiry, 10 to 45 cm. tall, branched below, the branches erect; leaves crowded toward the base, the blades slender, involute, sharp-pointed, 1 to 3 cm. long; panicle narrow, rather loose, purplish, 2 to 10 cm. long, the branches ascending, floriferous from base; spikelets, excluding the awns, 4 to 5 mm. long, the glumes shorter, acute or subacute, awnless; lemma sparsely pubescent below, tapering into a flexuous awn 6 to 10 mm. long. ♃ —Arid slopes, northern New Mexico and southeastern Utah; southern California (Clark Mountains).

**24. Muhlenbergia cuspidáta** (Torr.) Rydb. PLAINS MUHLY. (Fig. 543.) Culms slender, wiry, 20 to 40 cm.

tall, erect, in dense tufts with hard bulblike scaly bases; ligule minute; blades flat or loosely involute, erect or ascending, 1 to 2 mm. wide; panicle narrow, somewhat spikelike, 5 to 10 cm. long, the short branches appressed; spikelets about 3 mm. long; glumes subequal, acuminate-cuspidate, about two-thirds as long as the spikelet; lemma acuminate-cuspidate, minutely pubescent. ♃ —Prairies and gravelly or stony slopes, Michigan and Wisconsin to Alberta, south to Ohio, Kentucky, and New Mexico.

**25. Muhlenbergia wríghtii** Vasey. Spike muhly. (Fig. 544.) Culms closely tufted from a hard crown, erect, wiry, 20 to 60 cm. tall; sheaths compressed-keeled; ligule 1 to 3 mm. long, sometimes longer; blades flat, 1 to 3 mm. wide; panicle spikelike, interrupted below, 5 to 10 cm. long; spikelets about 2.5 mm. long, the

Figure 542.—*Muhlenbergia arsenei*. Plant, × 1; glumes and floret, × 10. (Type.)

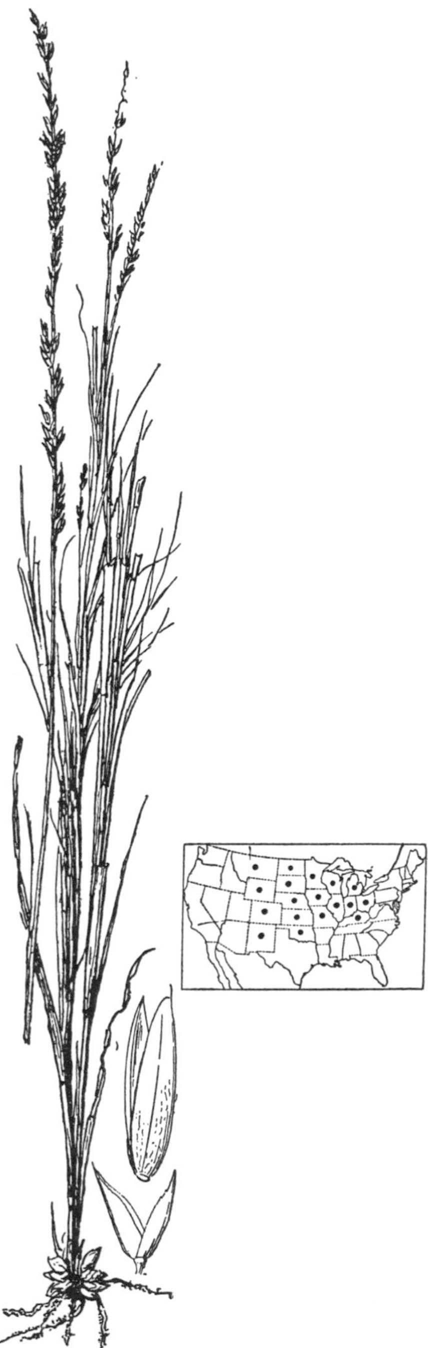

Figure 543.—*Muhlenbergia cuspidata*. Plant, × 1; glumes and floret, × 10. (Crattv, Iowa.)

glumes rather thin, mostly about half as long as the spikelet, broad at base, tapering to an awn point; lemma glabrous, acuminate, awn-tipped. ♃ —Plains and open slopes at medium altitudes, Oklahoma, Colorado, Utah, New Mexico, Arizona, and northern Mexico.

**26. Muhlenbergia arenácea** Buckl. Hitchc. (Fig. 545.) Perennial, with creeping rhizomes; culms tufted from the branches of the rhizomes, sometimes decumbent at base, 10 to 35 cm. tall; ligule prominent, decurrent, 1 to 2 mm. long, the margins usually split away, forming an erect auricle at each side; blades flat, wavy, mostly 1 to 3 cm. long, about 1 mm. wide, sharp-pointed, the margins and midnerve white and cartilaginous; panicle diffuse, 7 to 12 cm. long, about as broad, the branches and pedicels capillary; spikelets about 2 mm. long, rarely 2-flowered; the glumes about half as long, abruptly apiculate or subacute; lemma glabrous, abruptly mucronate. ♃ (*Sporobolus auriculatus* Vasey.)—Low places in mesas, Texas and Colorado to Arizona and Sonora. This species and the next three are placed in *Muhlenbergia* because of the 3-nerved mucronate lemma. The caryopsis does not fall from the lemma and palea as in most species of *Sporobolus*, nor can the pericarp be separated from the grain by moistening it.

FIGURE 544.—*Muhlenbergia wrightii*. Plant, × 1; glumes and floret, × 10. (Standley 8249, N. Mex.)

**27. Muhlenbergia asperifólia** (Nees and Mey.) Parodi. SCRATCHGRASS. (Fig. 546.) Perennial, pale or glaucous, with slender scaly rhizomes; culms branching at base, spreading, slender, compressed, 10 to 50 cm. tall, the branches ascending or erect; sheaths somewhat compressed-keeled, usually overlapping; ligule minute, erose-toothed; blades flat, crowded, scabrous, mostly 2 to 5 cm. long, 1 to 2 mm. wide; panicle diffuse, 5 to 15 cm. long, about as wide, the capillary scabrous branches finally widely spreading, the panicle at maturity breaking away; spikelets 1.5 to 2 mm. long, occasionally 2-flowered, the pedicels capillary; glumes acute, from half to nearly as long as the spikelet; lemma thin, broad, minutely mucronate from an obtuse apex. ♃ (*Sporobolus asperifolius* Nees and Mey.)—Damp or marshy, often alkaline soil, along irrigation ditches and banks of streams, New York, Indiana and Alberta to British Columbia, south to Texas, California, and Mexico; southern South America. The caryopsis is frequently affected by a fungus (*Tilletia asperifolia* Ell. and Everh.) which produces a large globular body.

**28. Muhlenbergia torreyána** (Schult.) Hitchc. (Fig. 547.) Perennial, strongly compressed at base, with short very scaly rhizomes; culms simple, or sparingly branching at base, erect, 30 to 60 cm. tall; blades elongate, rather firm, flat or folded, 1 to 3 mm. wide; panicle oblong, open, 10 to 20 cm. long, the capillary branches and pedicels ascending; spikelets about 2 mm. long, the glumes subequal, slightly shorter; lemma and palea minutely sca-

FIGURE 545.—*Muhlenbergia arenacea.* Plant, × 1; glumes and floret, × 10. (Tracy 7909, Tex.)

FIGURE 546.—*Muhlenbergia asperifolia.* Plant, × 1; glumes and floret, × 10. (Griffiths 212, S. Dak.)

FIGURE 547.—*Muhlenbergia torreyana*. Plant, × 1; glumes and floret, × 10. (Vasey, N. J.)

berulous-puberulent. ♃ (*Sporobolus compressus* Kunth; *S. torreyanus* Nash.)—Moist pine barrens and meadows, New Jersey and Delaware; Georgia (Sumter County), Kentucky, and Tennessee.

**29. Muhlenbergia uniflóra** (Muhl.) Fernald. (Fig. 548.) Perennial, but often appearing like an annual, tufted, often with decumbent bases; culms slender, erect, 20 to 40 cm. tall, the base and lower sheaths compressed; blades flat, crowded along the lower part of the culm, about 1 mm. wide; panicle loose, open, oblong, 7 to 20 cm. long, 2 to 4 cm. wide, the branches and pedicels capillary; spikelets dark purplish, about 1.5 mm. long, rarely 2-flowered; glumes scarcely half as long as the spikelet, subacute; lemma faintly 3-nerved, acutish. ♃ (*Sporobolus serotinus* A. Gray; *S. uniflorus* Scribn. and Merr.)—Bogs and wet meadows, Newfoundland to Michigan and New Jersey.

**30. Muhlenbergia dumósa** Scribn. (Fig. 549.) Perennial, with short, stout creeping scaly rhizomes; culms robust, solid, thick, and scaly at base (here as much as 6 mm. thick), the main culm erect or leaning, 1 to 3 m. tall, the lower part clothed with bladeless sheaths, freely branching at the middle and upper nodes, the branches numerous, fascicled, spreading, decompound, the ultimate branchlets filiform; blades

FIGURE 548.—*Muhlenbergia uniflora*. Plant, × 1; glumes and floret, × 10. (Chamberlain 147, Maine.)

FIGURE 549.—*Muhlenbergia dumosa*. Plant, × 1; glumes and floret, × 10. (Pringle, Ariz.)

flat or soon involute, smooth, those of the branches mostly less than 5 cm. long and 1 mm. wide; panicles numerous on the branches, commonly exceeded by the leaves, 1 to 3 cm. long, narrow, somewhat flexuous; spikelets, excluding the awn, about 3 mm. long, the glumes scarcely half as long, thin, pale with a green midnerve, usually minutely awn-tipped or with an awn as much as 9 mm. long; lemma narrow, pubescent about the base and margin, pale with green nerves, the awn from the slightly notched apex, flexuous, 3 to 5 mm. long. ♃ —Canyons and valley flats, southern Arizona to Jalisco, Mexico. Has the aspect of a miniature bamboo.

**31. Muhlenbergia andína** (Nutt.) Hitchc. FOXTAIL MUHLY. (Fig. 550.) Perennial, with numerous scaly rhizomes; culms erect or sometimes spreading, scabrous-puberulent below the nodes and the panicle, 50 to 100 cm. tall; sheaths smooth or slightly scabrous, keeled; ligule 1 mm. long, membranaceous, short-ciliate; blades flat, 2 to 6 mm. wide, scabrous; panicle narrow, spikelike, usually more or less lobed or interrupted, grayish, silky, often purple-tinged, 7 to 15 cm. long; glumes narrow, acuminate, ciliate-scabrous on the keels, 3 to 4 mm. long; lemma 3 mm. long, tapering into a capillary awn 4 to 8 mm. long, the hairs at base

FIGURE 550.—*Muhlenbergia andina*. Plant, × 1; glumes and floret, × 10. (Elmer 558, Wash.)

FIGURE 551.—*Muhlenbergia glomerata.* Plant, × 1; glumes and floret, × 8. (Macoun 26241, Ontario.)

FIGURE 552.—*Muhlenbergia racemosa.* Panicle, × 1; glumes and floret, × 10. (V. H. Chase 940, Ill.)

of floret copious, nearly as long as the body of the lemma. ♃ (*M. comata* Benth.)—Meadows, moist thickets, gravelly river beds, and open ground, at medium altitudes, Montana to eastern Washington, south to Kansas, New Mexico, and central California.

**32. Muhlenbergia glomeráta** (Willd.) Trin. (Fig. 551.) Perennial from creeping branching scaly rhizomes; culms slender, erect or suberect, 30 to 90 cm. tall, simple or with a few erect branches at base, the internodes minutely puberulent; sheaths rounded on the back; ligule minute; blades flat, 5 to 15 cm. long, lax, 2 to 5 mm. wide, ascending; panicle narrow, compact, lobed, mostly interrupted at base, often purplish, 3 to 10 cm. long; spikelets 5 to 6 mm. long, the narrow, attenuate subequal glumes stiffly awn-tipped; lemma about 3 mm. long, pointed, pilose on the lower part. ♃ —Sphagnum bogs, swamps, and moist ground, Newfoundland to British Columbia, Maine to Wisconsin, Virginia, and Indiana; Nebraska. Has been confused with *M. racemosa*; occasionally difficult to distinguish. Internodes are sometimes glabrous, but are roughish to the fingernail.

**33. Muhlenbergia racemósa** (Michx.) B. S. P. (Fig. 552.) Perennial from creeping scaly branching rhizomes, these and culms usually somewhat stouter than in the preceding; culms erect or ascending, subcompressed, 30 to 100 cm. tall, usually finally branching from the middle nodes, the branches mostly erect, the internodes smooth and shining except toward the summit; sheaths loose, keeled; ligule 1 to 1.5 mm. long; blades flat, 4 to 18 cm. long, 2 to 7 mm. wide, commonly somewhat firmer than those of *M. glomerata*, erect to ascending; panicle 3 to 14 cm. long, narrow, compact, often lobed, less commonly purple and thicker than in *M. glomerata;* spikelets 5 (rarely 4.5) to 7 mm. long, the narrow attenuate subequal glumes stiffly awn-tipped; lemma 2.5 to 3.5 mm. long, acuminate, rarely with a short awn, pilose on the lower part. ♃ Meadows, prairies, alluvial soil along rivers, irrigation ditches, rocky slopes, dry ground and waste places, occasionally in wet meadows, swamps, and moist canyon bottoms, found in a wide range of habitats; Manitoba to Alberta; Michigan and Indiana to Washington, Oklahoma, and Arizona. Specimens from Orono, Maine, and Washington, D. C., were doubtless from cultivated plants.

**34. Muhlenbergia califórnica** Vasey. (Fig. 553.) Perennial, pale, leafy, the base more or less creeping and rhizomatous; culms ascending, somewhat woody below, 30 to 60 cm. tall, branching below; sheaths scaberulous; blades flat, 3 to 6 mm. wide, scabrous, usually short; panicle narrow, dense but interrupted, 7 to 15 cm. long; spikelets 3 to 4 mm. long, the glumes slightly shorter, scabrous, acuminate, awn-tipped; lemma scabrous, acuminate, awn-tipped, with sparse callus hairs about half as long as the lemma. ♃—Stream borders and gullies, foothills and mountain slopes up to 2,000 m., confined to southern California.

FIGURE 553.—*Muhlenbergia californica*. Plant, × 1; glumes and floret, × 10. (Parish 2113, Calif.)

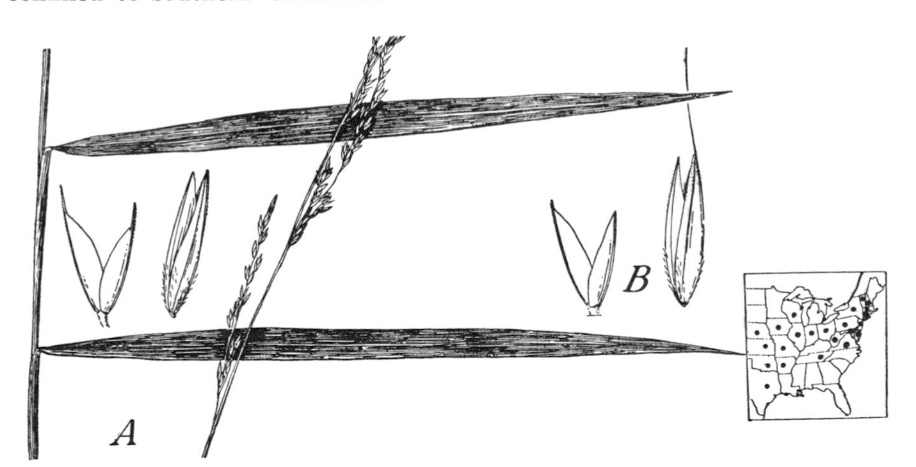


FIGURE 554.—*A*, *Muhlenbergia sobolifera*. Plant, × 1; glumes and floret, × 10. (Metcalf 1589, N. Y.) *B*, Var. *setigera*, × 10. (Reverchon 1049, Tex.)

**35. Muhlenbergia sobolífera** (Muhl.) Trin. (Fig. 554, *A*.) Perennial, with numerous creeping scaly rhizomes 2 to 3 mm. thick; culms erect, slender, solitary or few in a tuft, glabrous, 60 to 100 cm. tall, sparingly branching, the branches erect; blades flat, spreading, scabrous, those of the main culm 5 to 15 cm. long, 3 to 8 mm. wide, occasionally larger, at time of flowering aggregate along the middle part of

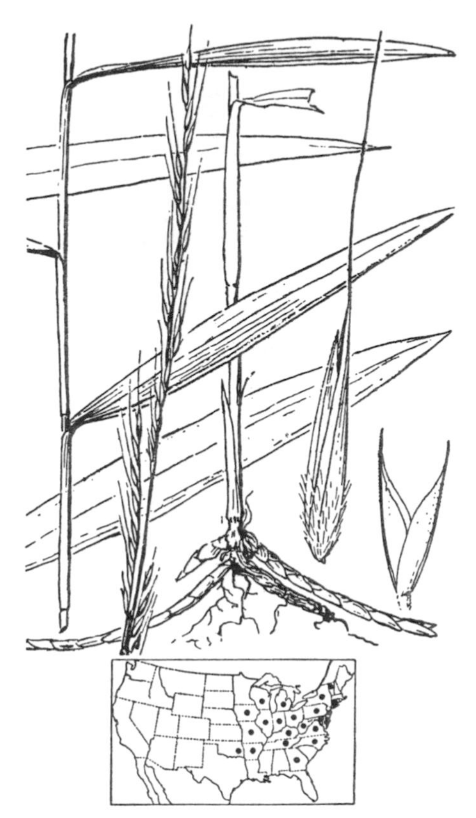

FIGURE 555.—*Muhlenbergia tenuiflora*. Plant, × 1; glumes and floret, × 10. (Mosley, Ohio.)

the culm; panicles slender, somewhat nodding, mostly 5 to 15 cm. long, the distant branches appressed, floriferous from base, overlapping or the lower more distant; spikelets mostly 2 to 2.5 mm. long, the glumes about two-thirds as long, abruptly acuminate or awn-tipped; lemma elliptic, bluntish, pubescent on the lower part, usually apiculate. ♃ —Dry rocky woods and cliffs, New Hampshire to Nebraska, south to Virginia, Tennessee, and Texas.

MUHLENBERGIA SOBOLIFERA var. SETÍGERA Scribn. (Fig. 554, *B*.) Branching more freely in the later stages; lemma with an awn 1 to 3 mm. long. ♃ —Dry woods, Arkansas and Texas.

**36. Muhlenbergia tenuiflóra** (Willd.) B. S. P. (Fig. 555.) Similar to *M. sobolifera* in habit; culms often more robust; blades mostly 10 to 18 cm. long and 6 to 10 mm. wide; panicles on the average longer; culms retrorsely puberulent at least around the nodes; sheaths puberulent or scaberulous toward the summit; spikelets (excluding the awns) 3 to 4 mm. long, the glumes about half as long, broad at base, abruptly acuminate, scaberulous; lemma narrow, pubescent toward the base, tapering into a slender straight awn 3 to 10 mm. long. ♃ —Rocky woods, Ontario and Vermont to Iowa, south to Georgia, Tennessee, and Oklahoma.

**37. Muhlenbergia brachyphýlla** Bush. (Fig. 556.) Perennial, with numerous slender scaly rhizomes; culms slender, suberect, freely branching at the middle nodes, the branches lax, glabrous or obscurely scabrous below the nodes; blades flat, spreading, scaberulous, mostly 7 to 15 cm. long and 3 to 5 mm. wide; panicles on filiform peduncles, very slender, lax, relatively few-flowered, mostly 8 to 15 cm. long; spikelets, excluding the awn, about 3 mm. long, the glumes about two-thirds as long, awn-tipped; lemma minutely pubescent toward the base, tapering into a slender awn 3 to 6 mm. long, rarely shorter. ♃ —Low woods, Maryland to North Carolina; Indiana and Wisconsin to Nebraska, south to Texas. Resembling *M. tenuiflora*, but with numerous filiform branches and more slender panicles.

**38. Muhlenbergia frondósa** (Poir.) Fernald. WIRESTEM MUHLY. (Fig. 557.) Perennial, with creeping scaly rhizomes; culms often relatively stout, glabrous below the nodes, finally decumbent, often rooting at the geniculate lower nodes, freely branching from all the nodes (occasionally simple below), the branches ascending or somewhat spreading, the plants becoming top-heavy and bushy, 40 to 100 cm. long; blades flat, scabrous, usually not more than 10 cm. long, sometimes as much as 15 cm., 3 to 7

FIGURE 556.—*Muhlenbergia brachyphylla.* Plant, × 1; glumes and florets, × 10. (V. H. Chase 3759, Ill.)

mm. wide; panicles numerous, short-exserted or partly included, terminal and axillary, the larger as much as 10 cm. long (the axillary shorter), narrow, sometimes rather loose, the branches ascending, mostly densely flowered from the base; glumes 2 to 3 mm., rarely to 4 mm., long, tapering into an awned tip, subequal or unequal, shorter than the floret, or the second glume exceeding it; lemma 2 to 3 mm. long, pointed, short-pilose at base. ♃ (Described under *M. mexicana* in Manual, ed. 1.)—Thickets, low ground, and waste places, New Brunswick to North Dakota, south to Georgia and Texas.

MUHLENBERGIA FRONDOSA forma COMMUTÁTA (Scribn.) Fernald. Lemmas awned. ♃ —Quebec and Maine to South Dakota, south to Virginia and Missouri. May be distinguished from the awned forms of *M. mexicana* by the culms smooth below the nodes.

**39. Muhlenbergia glabriflóra** Scribn. (Fig. 558.) In habit resem-

FIGURE 557.—*Muhlenbergia frondosa.* Plant, × 1; glumes and floret, × 10. (V. H. Chase 1166, Ill.)

bling *M. frondosa*, freely branching; culms scaberulous below the nodes as in *M. sylvatica;* blades numerous, short, narrow, appressed; panicles on the average shorter and narrower than in *M. frondosa;* spikelets about as in *M. frondosa* but the lemma glabrous. ♃ —Low woods, Maryland to North Carolina; Indiana to Missouri, Arkansas, and Texas.

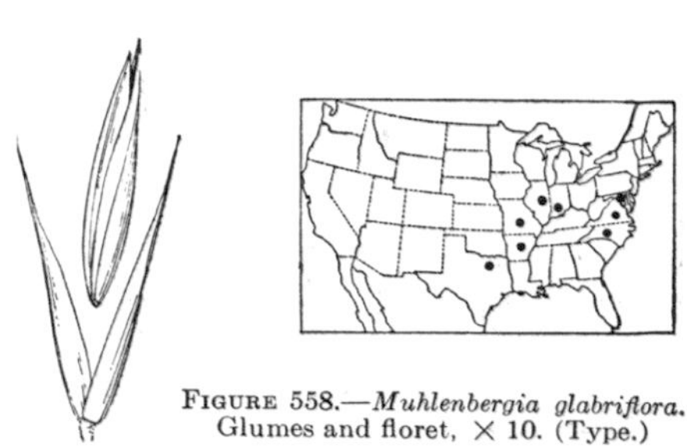

FIGURE 558.—*Muhlenbergia glabriflora.* Glumes and floret, × 10. (Type.)

**40. Muhlenbergia sylvática** (Torr.) Torr. (Fig. 559.) Perennial with creeping scaly rhizomes, culms slender, retrorsely scaberulous below the nodes, rather sparingly branching from the middle and upper nodes, finally leaning, the subfiliform branches often elongate, drooping, the plant 40 to 100 cm. tall; blades flat, lax, ascending to spreading, 0.5 to 18, commonly 8 to 15 cm., long, 2 to 8 mm. wide; panicles slender, nodding, the slender branches appressed, slightly overlapping; glumes lanceolate, acuminate or awn-tipped, 2 to 3 mm. long; lemma slightly exceeding the glumes, pilose below, tapering into a slender awn 5 to 10 mm. long. ♃ (*M. umbrosa* Scribn.)—Moist woods and thickets, Quebec and Maine to South Dakota, south to Alabama and Texas; Arizona.

MUHLENBERGIA SYLVATICA forma ATTENUÁTA (Scribn.) Palmer and Steyermark. Lemmas short-awned or nearly awnless. ♃ —Ontario, Maine, Connecticut, Indiana, Illinois, Michigan, South Dakota, Missouri, District of Columbia, and Oklahoma.

MUHLENBERGIA SYLVATICA var. ROBÚSTA Fernald. Culm stiffer, blades somewhat firmer, some of them 7 to 10 mm. wide; panicles with more densely flowered branches; glumes slightly broader. ♃ —Maine, Connecticut, New York, New Jersey, and Indiana.

**41. Muhlenbergia mexicána** (L.) Trin. (Fig. 560.) Resembling *M. frondosa*, the culms erect or ascending, usually simple below, less freely branching, scaberulous below the nodes; blades lax, often 10 to 20 cm. long, mostly 2 to 4 mm. wide; panicles mostly long-exserted, narrow, the upper often 10 to 15 cm. long, of numerous short appressed densely flowered somewhat aggregate branches; spikelets 2 to 3 mm. long, glumes narrow, attenuate, awn-

FIGURE 559.—*Muhlenbergia sylvatica.* Plant, × 1; glumes and floret, × 10. (Conant, Mass.)

FIGURE 560.—*Muhlenbergia mexicana.* Plant, × 1; glumes and floret, × 10. (Deam 19225, Ind.)

tipped, about equaling the pointed or awn-tipped lemma, the lemma long-pilose below. ♃ (Described under *M. foliosa* in Manual, ed. 1. The name *M. mexicana* had long been misapplied to the recently recognized *M. frondosa* (Poir.) Fernald.)—Moist thickets, low woods, and low open ground, Quebec and Maine to British Columbia and Washington, south to North Carolina, New Mexico, and California.

MUHLENBERGIA MEXICANA forma AMBÍGUA (Torr.) Fernald. Lemmas with an awn 4 to 10 mm. long. ♃ —Range of the species to North Dakota; intergrading with forma *setiglumis* in Indiana and westward.

MUHLENBERGIA MEXICANA forma SETIGLÚMIS (S. Wats.) Fernald. ♃ —Glumes with an awn 1 to 2 mm. long; lemma awned as in the preceding, the two scarcely distinct. ♃ —Iowa and South Dakota to Washington, south to New Mexico and California.

**42. Muhlenbergia schrebéri** Gmel. NIMBLEWILL. (Fig. 561.) Culms slender, branching, spreading and decumbent at base, usually rooting at the lower nodes, but not forming definite creeping rhizomes, the flowering branches ascending, 10 to 30 cm. long; blades flat, mostly less than 5 cm. long, and 2 to 4 mm. wide; panicles terminal and axillary, slender, loosely flowered, lax, nodding, 5 to 15 cm. long; glumes minute, the first often obsolete, the second rounded, 0.1 to 0.2 mm. long; lemma narrow, somewhat pubescent around the base, the body about 2 mm. long, the slender awn 2 to 5 mm. long. ♃ —Damp shady places, New Hampshire to Wisconsin and eastern Nebraska, south to Florida and Texas; eastern Mexico. In spring and early summer the culms are short and erect with spreading blades, the plants being very different in appearance from the flowering phase of fall. MUHLENBERGIA SCHREBERI var. PALÚSTRIS (Scribn.) Scribn. Glumes developed as much as 1 mm. long. ♃ —Washington, D. C.; Bull Run Mountains, Va.

**43. Muhlenbergia curtisetósa** (Scribn.) Bush. (Fig. 562.) A little-known form, differing from *M. schreberi* in having stouter culms, coarser panicles, the glumes evident, rarely as much as 2 mm. long, the lemma 2.5 to 3 mm. long, the awn 1 to 2 mm. long. ♃ —Delaware County, Pa., Illinois (Clinton), Missouri (Eagle Rock).

**44. Muhlenbergia jonésii** (Vasey) Hitchc. (Fig. 563.) Perennial, closely tufted; culms erect, 20 to 40 cm. tall; leaves mostly basal, the numerous lower sheaths finally flattened and loose; ligule 2 to 4 mm. long; blades subfiliform, involute, scabrous; panicle narrow, 5 to 15 cm. long, the branches ascending, rather loosely flowered; spikelets 3 to 4 mm. long; glumes broad, scabrous-puberulent, about one-third as long as the spikelet, obtuse, often erose; lemma obscurely pubescent below, tapering to an acuminate or awned tip. ♃ — Open ground, northeastern California.

FIGURE 562.—*Muhlenbergia curtisetosa*. Glumes and floret, × 10. (Wolf 30, Ill.)

FIGURE 563.—*Muhlenbergia jonesii*. Plant, × 1; glumes and floret, × 10. (Austin 1230, Calif.)

FIGURE 561.—*Muhlenbergia schreberi*. Plant, × ½; glumes and floret, × 10. (Curtiss 3400, Tenn.)

**45. Muhlenbergia montána** (Nutt.) Hitchc. MOUNTAIN MUHLY. (Fig. 564.) Perennial; culms densely tufted, erect, 15 to 60 cm. tall; sheaths glabrous, mostly basal, becoming flat and loose; blades flat to involute, 1 to 2 mm. wide; panicle narrow, rather

FIGURE 564.—*Muhlenbergia montana.* Plant, × 1; glumes and floret, × 10. (Patterson 156, Colo.)

loose, 5 to 15 cm. long, the branches ascending or appressed, floriferous from base; first glume acute, 1.5 mm. long, the second longer, broader, 3-nerved, 3-toothed; lemma 3 to 4 mm. long, pilose below, scaberulous above, the awn slender, flexuous, 1 to 1.5 cm. long, sometimes shorter. ♃ (*M. trifida* Hack., *M. gracilis* of authors, not Kunth.)—Canyons, mesas, and rocky hills, 2,000 to 3,000 m., Montana to Utah and central California, south to western Texas and southern Mexico.

**46. Muhlenbergia filicúlmis** Vasey. SLIMSTEM MUHLY. (Fig. 565.) Culms densely tufted, erect, filiform, 10 to 20 cm. tall, the leaves in a short basal cluster; ligule prominent; blades involute, filiform, mostly less than 5 cm. long; panicle slender, the branches erect, mostly 2 to 5 cm. long, sometimes as much as 10 cm.; spikelets about 2.5 to 3 mm. long, the glumes about half as long, awn-tipped, the first rather narrow, acuminate, the second broader, 3-nerved, sharply 3-toothed, rarely entire or erose only; lemma pubescent on the lower half, tapering to an awned tip, or rarely with an awn as much as 4 mm. long. ♃ —Open sandy or

FIGURE 565.—*Muhlenbergia filiculmis.* Panicle, × 1; glumes and floret, × 10. (Type.)

rocky soil, 2,000 to 3,000 m. altitude, Wyoming, Colorado, New Mexico, and Utah.

**47. Muhlenbergia viriscéns** (H. B. K.) Kunth. SCREWLEAF MUHLY. (Fig. 566.) Perennial; culms densely tufted, erect, 40 to 60 cm. tall, the old basal sheaths flattened and more or less coiled; ligule, except the margin, delicate, 3 to 10 mm. long; blades flat or those of the innovations involute, mostly elongate and flexuous; panicle narrow but rather loose, 5 to 15 cm. long, the branches erect; spikelets, excluding awns, about 5 mm. long, the glumes slightly shorter, acute, the second 3-nerved; lemma and palea pubescent on the lower half, the lemma tapering into a slender flexuous awn 1 to 2 cm. long. ♃ —Canyons, rocky hills, and mesas, New Mexico and Arizona to central Mexico.

**48. Muhlenbergia montícola** Buckl. MESA MUHLY. (Fig. 567.) Perennial; culms tufted, slender, erect or decumbent at base, 30 to 50 cm. tall, branching at the lower and middle nodes, leafy throughout; blades 3 to 7 cm. long, narrow, flat, or soon involute; panicle soft, narrow, contracted, 5 to 10, sometimes to 20 cm. long, the branches appressed or slightly spreading; spikelets, excluding awns, about 3 mm. long, the glumes about two-thirds as long, subacute to obtuse and erose at tip; lemma pubescent at base and on lower half of margin, tapering into a delicate flexuous awn 1 to 2 cm. long. ♃ —Rocky hills and canyons, western Texas to Arizona and central Mexico.

**49. Muhlenbergia parviglúmis** Vasey. (Fig. 568.) Perennial, with the habit of *M. monticola;* blades on the average somewhat longer, 1 to 3 mm. wide; panicle looser, the branches filiform, longer; glumes minute, erose, subacute to truncate; lemma scaberulous only, tapering in-

FIGURE 566.—*Muhlenbergia virescens.* Plant, × 1; glumes and floret, × 10. (Palmer 565, Ariz.)

FIGURE 567.—*Muhlenbergia monticola*. Plant, × 1; glumes and floret, × 10. (Nealley 399, Tex.)

FIGURE 568.—*Muhlenbergia parviglumis*. Panicle, × 1; glumes and floret, × 10. (Vasey, Tex.)

to a delicate awn 2 to 3 cm. long. ♃ —Canyons, Texas, New Mexico, and northern Mexico; Cuba.

**50. Muhlenbergia púngens** Thurb. (Fig. 569.) Perennial, with strong creeping rhizomes; culms tufted, erect from a decumbent leafy base, 20 to 40 cm. tall, sometimes taller; blades short, involute, sharp-pointed; panicle long-exserted, open, oblong, 5 to 15 cm. long; the main branches 3 to 5, these dividing into fascicles of capillary finally spreading or divaricate very scabrous branchlets; spikelets purple to brownish, 4 to 5 mm. long, the glumes about one-third as long, scabrous, often erose or toothed, the midnerve extending into a short awn; lemma terete, tapering into an awn about 1 mm. long; palea about as long as the lemma, the keels awn-tipped. ♃ —Dry hills and sandy plains, South Dakota and Nebraska to Wyoming, New Mexico, and Arizona.

**51. Muhlenbergia portéri** Scribn. BUSH MUHLY. (Fig. 570.) Perennial; culms woody or persistent at base, numerous, wiry, widely spreading or ascending through bushes, scaberulous, mostly branching from all the nodes, 30 to 100 cm. tall or more; sheaths smooth, spreading away from the branches, the prophylla conspicuous; blades mostly about 1 mm. wide, flat, 2 to 8 cm. long, early deciduous from the sheaths; panicle 5 to 10 cm. long, open, the slender branches and branchlets brittle, widely spreading, bearing rather few long-pediceled spikelets; glumes narrow, acuminate, slightly unequal, the second about 2 mm. long; lemma purple, acuminate, sparsely pubescent, 3 to 4 mm. long, with a delicate awn 5 to 12 mm. long. ♃ —Dry mesas and hills, canyons, and rocky deserts, western Texas to Colorado,

FIGURE 569.—*Muhlenbergia pungens*. Plant, × 1; glumes and floret, × 10. (Jones 6046, Utah.)

Nevada, and southern California, south to northern Mexico. Known also as mesquite grass and black grama.

**52. Muhlenbergia arizónica** Scribn. (Fig. 571.) Perennial, in close tufts; culms slender, erect or decumbent at base, 15 to 40 cm. tall; sheaths keeled; ligule thin, 1 to 2 mm. long, decurrent; blades flat or folded, mostly less than 5 cm. long, 1 to 2 mm. wide, the margins and midnerve white, cartilaginous; panicle open, 5 to 12 cm. long 4 to 8 cm. wide, the branches capillary, compound; spikelets long-pedicellate, about 3 mm. long, the glumes about one-third as long, ovate, subacute; lemma narrowly lanceolate, minutely pubescent along the midnerve and margins below, the awn about 1 mm. long, from a minutely notched apex. ♃ —Stony hills, southern Arizona and northwestern Mexico.

**53. Muhlenbergia torréyi** (Kunth) Hitchc. ex Bush. RINGGRASS. (Fig. 572.) Perennial in loose tufts, with numerous innovations, the base decumbent or forming short rhizomes, the plants usually gregarious, sometimes forming large patches or "fairy rings"; culms slender, 10 to 30 cm. tall; leaves in a short basal cluster; blades closely involute, usually 2 to 3 cm. long, falcate or flexuous, forming a crisp curly cushion; panicle open, usually about half the entire length of the culm, commonly purple, the capillary branches finally spreading, the pedicels mostly as long as the spikelets or longer; spikelets about 3 mm. long, the glumes, including the awn-tip, about two-thirds as long; lemma nearly glabrous, tapering into a delicate awn about 3 mm. long. ♃ (*M. gracillima* Torr.)—Plains, mesas, and dry hills, western Kansas and Wyoming to Texas and Arizona.

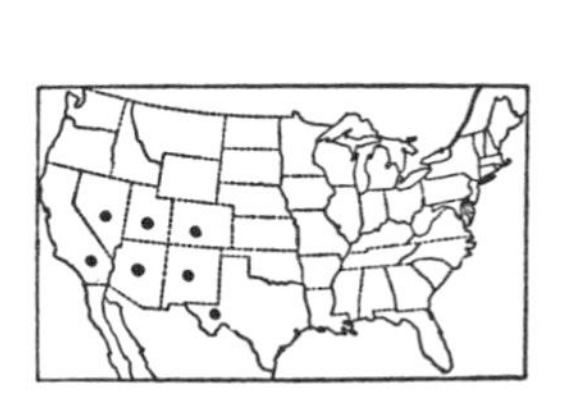

FIGURE 570.—*Muhlenbergia porteri*. Plant, × 1; glumes and floret, × 10. (Chase 5887, Tex.)

**54. Muhlenbergia arenícola** Buckl. (Fig. 573.) Resembling *M. torreyi;* culms mostly 30 to 50 cm. tall; blades usually straight and on the average longer; panicle larger, mostly pale, the branchlets and pedicels appressed; spikelets slightly longer, the lemma scabrous, the awn 1 to 2 mm. long. ♃ —Sandy plains and mesas, western Kansas to Arizona, south to northern Mexico.

**55. Muhlenbergia setifólia** Vasey. (Fig. 574.) Perennial, tufted; culms erect, hard, wiry, 50 to 80 cm. tall; sheaths with erect auricles, 2 to 10 mm. long; blades involute, fine, scarcely 0.5 mm. thick, very scabrous, flexuous, as much as 20 cm. long; panicle narrow, open, 10 to 30 cm. long, the capillary branches ascending, flexuous; spikelets, excluding awns, about 5 mm. long, the glumes one-third to half as long, obtuse to subacute; lemma hairy on the callus, otherwise smooth, tapering into a flexuous awn 1.5 to 2 cm. long. ♃ —Rocky hills, western Texas and northern Mexico.

**56. Muhlenbergia xeróphila** C. O. Goodding. (Fig. 575.) Culms 45 to 90 cm. tall, densely tufted, glabrous or scaberulous; sheaths scaberulous; ligule 2 to 4 mm. long, obtuse; blades involute, 15 to 50 cm. long, 1 to 1.5 mm. wide; panicle open (contracted at maturity), 15 to 35 cm. long, with capillary, flexuous, spreading branches; spikelets about 4 mm. long; glumes equal or subequal, 2 to 2.5 mm. or sometimes as much as 3 mm. long, acute or acuminate, scabrous or pubescent; lemma 4 mm. long, scabrous, the callus appressed-pilose, the hairs about 1 mm. long, the delicate capillary awn 10 to 35 mm. long. ♃ —Canyons and rocky slopes, southern Arizona.

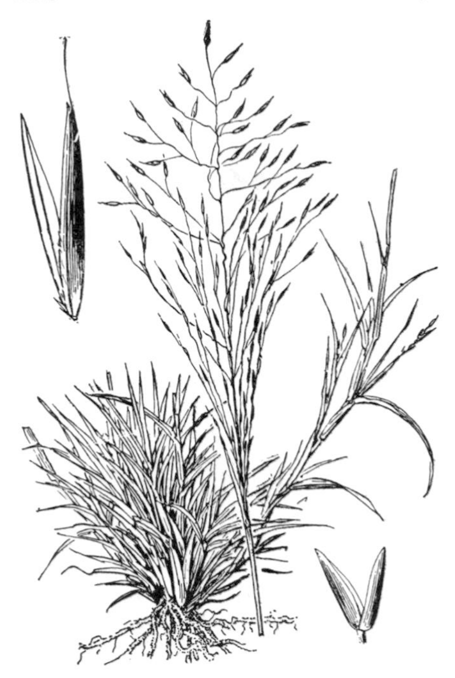

FIGURE 571.—*Muhlenbergia arizonica*. Plant, × 1; glumes and floret, × 10. (Griffiths 3368, Ariz.)

**57. Muhlenbergia metcálfei** Jones. (Fig. 576.) Perennial, in close tufts; culms erect, 50 to 80 cm. tall; ligule 3 to 15 mm. long, sometimes longer; blades involute, slender, flexuous, scabrous, sometimes only slightly so, not crowded at base; panicle narrow but somewhat loose, pale or slightly purplish, 15 to 40 cm. long, the branches usually naked at base; spikelets tapering to summit, about 4 mm. long; glumes nearly equal, obtuse, a little less than half as long as spikelet; lemma scaberulous toward summit, the awn 3 to 10 mm. long. ♃ —Rocky hills, Texas and New Mexico.

**58. Muhlenbergia dúbia** Fourn. PINE MUHLY. (Fig. 577.) Perennial, closely tufted; culms erect, hard and wiry at base, 30 to 100 cm. tall; sheaths with erect firm auricles, 4 to 10 mm. long, rarely longer; blades involute, scabrous; panicle narrow, sometimes almost spikelike, grayish, 10 to 30 cm. long, rarely longer; spikelets about 4 mm. long; glumes about half as long as the spikelet, minutely scaberulous, obtuse; lemma minutely scaberulous, with an awn as much as 4 mm. long, rarely acuminate only. ♃ (*M. acuminata* Vasey; *Sporobolus ligulatus* Vasey and Dewey.)—Canyons and rocky hills, up to 7,000 feet elevation, western Texas, New Mexico, and northern Mexico.

**59. Muhlenbergia dubioídes** C. O. Goodding. (Fig. 578.) Culms 50 to 100 cm. tall, densely tufted, erect;

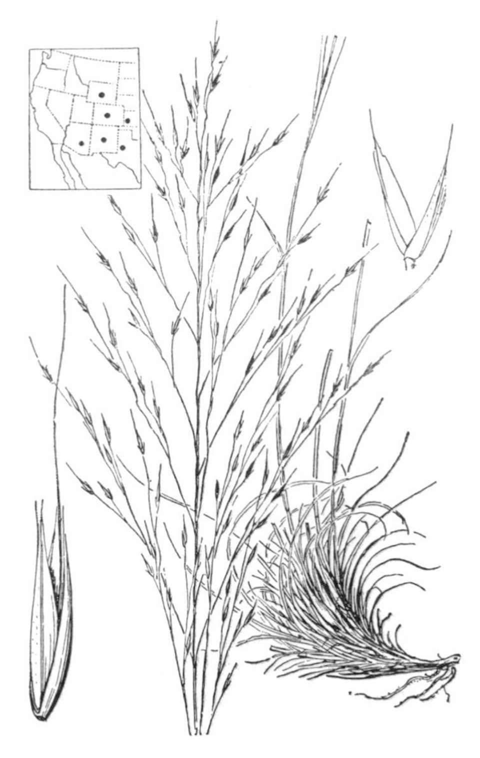

FIGURE 572.—*Muhlenbergia torreyi*. Plant, × 1; glumes and floret, × 10. (Chase 5298, Colo.)

FIGURE 573.—*Muhlenbergia arenicola*. Plant, × 1; glumes and floret, × 10. (Hitchcock 13602, Tex.)

ligule truncate, 1 to 2 mm. long; blades 15 to 50 cm. long, 1 to 2 mm. wide, involute, glabrous, or scaberulous below; panicle 15 to 35 cm. long, 2 to 4 cm. wide, densely flowered, the branches appressed; spikelets about 4 mm. long; glumes subequal, 2 to 3 mm. long, acute, more or less erose, scaberulous; lemma 3.5 to 4 mm. long, the callus appressed-pilose with hairs 1 to 1.5 mm. long; awn straight, scabrous, 3 to 10 mm. long. ♃ —Canyons and rocky slopes, Santa Cruz and Pima Counties, southern Arizona.

**60. Muhlenbergia expánsa** (DC.) Trin. (Fig. 579.) Resembling *M. capillaris*, in denser tufts, the old basal sheaths forming a curly fibrous mass; blades narrow, flat, becoming involute; panicle relatively smaller, narrower, the capillary branches and branchlets mostly straight; spikelets 3.5 to 5 mm. long, the glumes one-third to two-thirds as long, acute to acuminate; lemma scaberulous, nearly glabrous at base, awnless or with an awn 2 to 3 mm. long, rarely longer. ♃ (*M. trichopodes* Chapm.)—Moist pine barrens near the coast, Virginia to Florida and Texas.

**61. Muhlenbergia reverchóni** Vasey and Scribn. (Fig. 580.) Resembling *M. expansa*, culms more slender, foliage finer; glumes less than half as long as the lemma, subacute or erose; lemma with an awn 2 to 5 mm. long. ♃ —Rocky prairies, Texas and Oklahoma.

**62. Muhlenbergia capilláris** (Lam.) Trin. (Fig. 581.) Perennial, in tufts; culms rather slender, erect, 60 to 100 cm. tall; sheaths scaberulous, at least toward the summit, and with auricles mostly 3 to 5 mm. long; blades elongate, flat or involute, 1 to 4 mm. wide, those of the innovations narrower, involute; panicle purple, oblong, diffuse, one-third to half the entire height of the culm, the branches capillary, flex-

FIGURE 574.—*Muhlenbergia setifolia*. Plant, × 1; glumes and floret, × 10. (Hitchcock 13507, N. Mex.)

uous, the branchlets and pedicels finally spreading; spikelets, excluding awns, 3 to 4 mm. long, the glumes one-fourth to two-thirds as long, acute, the second often short-awned; lemma scaberulous, minutely hairy on the callus and with a delicate awn 5 to 15 mm. long. ♃ —Rocky or sandy woods, Massachusetts to Indiana and Kansas, south to Florida and Texas; West Indies, eastern Mexico.

MUHLENBERGIA CAPILLARIS var. FÍLIPES (M. A. Curtis) Chapm. ex Beal. Culms stouter; blades mostly involute; glumes with delicate awns, mostly longer than the lemma; lemma with a delicate setaceous tooth each side of the awn. ♃ (*M. filipes* M. A. Curtis.)—Moist pine barrens near the coast, North Carolina, Florida, Mississippi, and Texas.

**63. Muhlenbergia rígida** (H. B. K.) Kunth. PURPLE MUHLY. (Fig. 582.) Perennial, densely tufted; culms erect, 60 to 100 cm. tall; leaves crowded at base, old sheaths persistent, the sheaths with auricles 2 to 5 mm., rarely longer; blades flat or soon involute, flexuous, those of the innovations involute; panicle dark purple, narrow, finally loose and open, 15 to 30 cm. long, the capillary branches ascending, the lower as much as 10 cm. long; spikelets, excluding awns, about 4 mm. long, the glumes from minute to about one-fourth as long, acute to erose-obtuse; lemma strongly nerved, hairy on the callus and with a

FIGURE 575.—*Muhlenbergia xerophila*. Panicle, × 1; glumes and floret, × 10. (Silveus 3477, Ariz.)

FIGURE 576.—*Muhlenbergia metcalfei*. Panicle, × 1; glumes and floret, × 10. (Metcalfe, N. Mex.)

flexuous awn 1 to 1.5 cm. long. ♃ (*M. berlandieri* Trin.)—Rocky or gravelly soil, Texas to Arizona and northern Mexico.

**64. Muhlenbergia longilígula** Hitchc. (Fig. 583.) Culms erect, about 1 m. tall, the base hard, wiry, cylindric, the lower sheaths expanded; ligule (or auricle of sheath) firm, usually about 1 cm. long; blades as much as 50 cm. long, 2 to 5 mm. wide, flat to subinvolute, very scabrous, usually drying involute; panicle narrow, somewhat loose, erect, 20 to 40 cm. long, the branches ascending or appressed; spikelets 2 to 3 mm. long; glumes subequal, acutish, usually glabrous; lemma usually about as long as the glumes, glabrous, awnless, rarely with a minute awn. ♃ (*Epi-*

FIGURE 577.—*Muhlenbergia dubia*. Plant, × 1; glumes and floret, × 10. (Hitchcock 3775, N. Mex.)

FIGURE 578.—*Muhlenbergia dubioides*. Panicle, × 1; glumes and floret, × 10. (Type.)

FIGURE 579.—*Muhlenbergia expansa.* Panicle, × 1; glumes and floret, × 10. (Tracy 3701, Miss.)

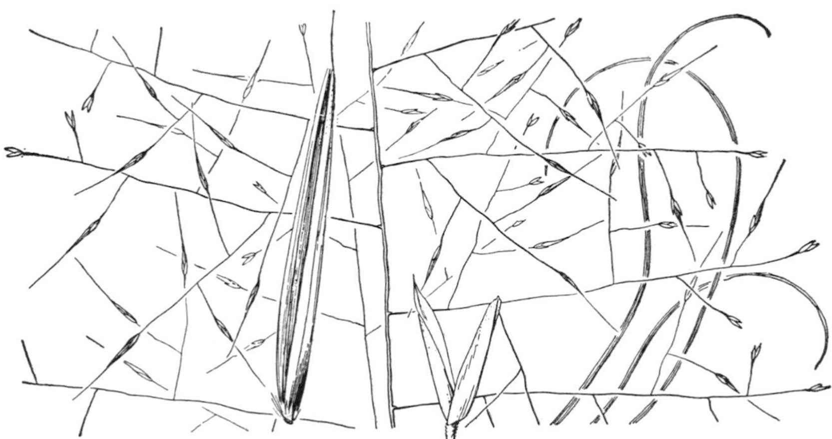

FIGURE 580.—*Muhlenbergia reverchoni.* Panicle, × 1; glumes and floret, × 10. (Reverchon, Tex.)

*campes ligulata* Scribn., not *Muhlenbergia ligulata* Scribn. and Merr.)—Canyons and rocky slopes, western New Mexico, Arizona, southern Nevada, and northern Mexico.

**65. Muhlenbergia lindheímeri** Hitchc. (Fig. 584.) Culms erect, 1 to 1.5 m. tall, the numerous overlapping lower sheaths keeled; ligule rather thin, elongate, mostly hidden in the folded base of the blade; blades elongate, firm, flat or usually folded, about 3 mm. wide, scaberulous or glabrous; panicle narrow, pale, somewhat loose, erect, 20 to 40 cm. long, the branches ascending or appressed; spikelets 2.5 to 3 mm. long; glumes acute to rather obtuse, scabrous-puberulent; lemma a little shorter to a little longer than the glumes, 3-nerved, glabrous or obscurely pubescent, awnless or rarely with an awn as much as 3 mm. long. ♃ —Rocky slopes, Texas.

**66. Muhlenbergia involúta** Swallen. (Fig. 585.) Culms erect, densely tufted, 60 to 135 cm. tall; sheaths compressed-keeled, scabrous; ligule about 10 mm. long; blades elongate, involute, wiry, scabrous; panicle erect, narrow, 30 to 40 cm. long, the subcapillary branches ascending or appressed, naked toward the base, the lower as much as 20 cm. long; spikelets 3 to 4.5 mm. long; glumes acute or somewhat erose, scabrous, 2 to 2.5 mm. long; lemma densely pubescent on the margin toward the very base, the minutely toothed apex awned from just below the teeth, the awn

FIGURE 581.—*Muhlenbergia capillaris*. Plant, × ½; glumes and floret, × 10. (Scribner, Tenn.)

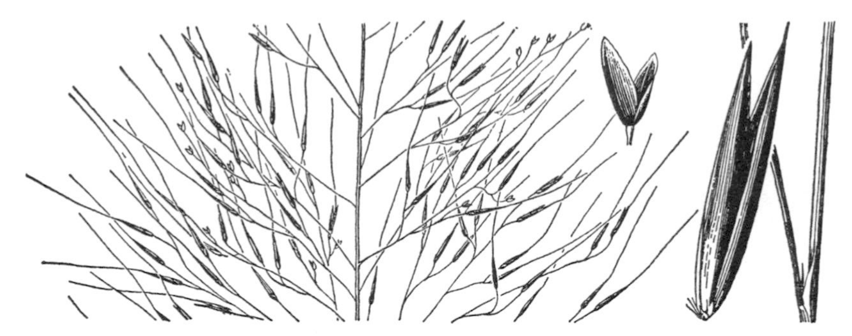

FIGURE 582.—*Muhlenbergia rigida.* Panicle and ligule, × 1; glumes and floret, × 10. (Metcalfe 1447, N. Mex.)

FIGURE 583.—*Muhlenbergia longiligula.* Panicle and ligule, × 1; glumes and floret, × 10. (Jones, Ariz.)

slender, 1.5 to 2 mm. long. ♃ — Canyons and ravines, southern Texas.

**67. Muhlenbergia emersléyi** Vasey. BULLGRASS. (Fig. 586.) Culms in large clumps, erect, 50 to 100 cm. tall; sheaths glabrous, slightly scabrous, compressed-keeled, especially those of the innovations; ligule softly membranaceous, 1 to 2 cm. long; blades flat or folded, scabrous, 1 to 4 mm. wide, the lower as much as 50 cm. long; panicle narrow but rather loose, erect or nodding, mostly 20 to 40 cm. long, the branches ascending, more or less fascicled or whorled, naked below; spikelets 2.5 to 4 mm. long, often purplish; glumes thin, equal, acutish, scabrous; lemma about as long as the glumes, narrowed and scabrous above, villous below, with a delicate flexuous awn, about 1.5 cm. long, or sometimes awnless. ♃ — Rocky woods and ravines, Texas to Arizona and Mexico. A good soil binder on steep slopes.

**68. Muhlenbergia rígens** (Benth.) Hitchc. DEERGRASS. (Fig. 587.) Culms rather slender, stiffly erect, in small bunches, with a hard tough base, 1 to 1.5 m. tall; sheaths smooth

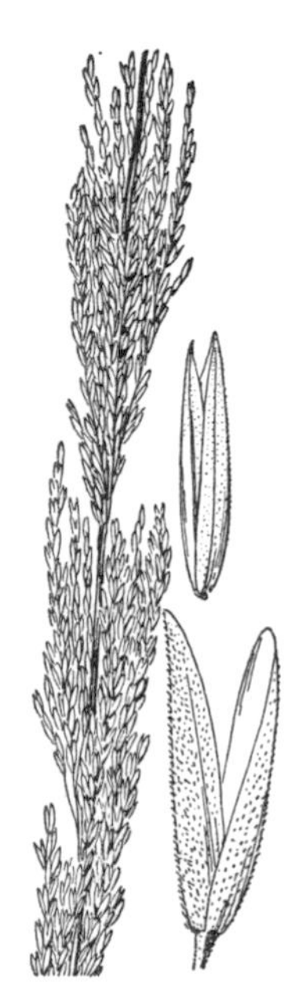

FIGURE 584.—*Muhlenbergia lindheimeri.* Panicle, × 1; glumes and floret, × 10. (Type.)

FIGURE 585.—*Muhlenbergia involuta*. Panicle and ligule, × 1; spikelet and floret, × 10. (Type.)

FIGURE 587.—*Muhlenbergia rigens*. Spikelet and floret, × 10. (Type collection.)

or slightly scabrous, mostly overlapping, the lower crowded, expanded, somewhat papery; ligule firm, truncate, 2 to 3 mm. long; blades scabrous, elongate, involute, tapering into a long slender point; panicle grayish or

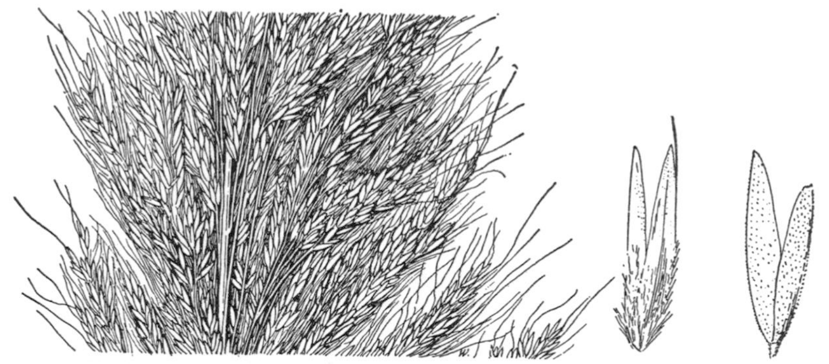

FIGURE 586.—*Muhlenbergia emersleyi*. Panicle, × 1; glumes and floret, × 10. (Wooton and Standley, N. Mex.)

pale, slender, mostly spikelike, 20 to 60 cm. long or more, the lower branches sometimes 5 to 10 cm. long; spikelets 2.5 to 3.5 mm. long, the glumes shorter than the lemma, from acute to obtuse or somewhat erose, scabrous-puberulent, rarely faintly 3-nerved; lemma scaberulous, sparsely pilose at base, 3-nerved toward the narrowed summit, awnless. ♃ (*Epicampes rigens* Benth.)—Dry or open ground, hillsides, gullies, and

FIGURE 588.—*Muhlenbergia mundula*. Plant, × 1; glumes and floret, × 10. (Metcalfe 10, N. Mex.)

open forest, southern California. Used by Indians in basket making.

**69. Muhlenbergia múndula** I. M. Johnston. (Fig. 588.) Similar to the preceding; ligule 1 to 2 mm. long; panicle similar, but lower branches not more than 4 cm. long; spikelets 3 to 4 mm. long, the glumes shorter than the lemma or sometimes about equaling it. ♃ (This and the next species included in *M. rigens* in Manual, ed. 1.)—Rocky canyons and gullies, Nevada, New Mexico, Arizona, and northern Mexico. This and the following doubtfully distinct from *M. rigens*. Many intermediates are found.

**70. Muhlenbergia márshii** I. M. Johnston. (Fig. 589.) Often smaller than *M. rigens*, differing in the minute ligule and narrower, usually awn-tipped glumes and lemma. ♃ — Rocky stream banks and canyons, Texas and northern Mexico.

FIGURE 589.—*Muhlenbergia marshii*. Glumes and lemma, × 10. (Type collection.)

## 83. SPORÓBOLUS R. Br. DROPSEED

Spikelets 1-flowered, the rachilla disarticulating above the glumes; glumes 1-nerved, usually unequal, the second often as long as the spikelet; lemma membranaceous, 1-nerved, awnless; palea usually prominent and as long as the lemma or longer; caryopsis free from the lemma and palea, falling readily from the spikelet at maturity, the pericarp free from the seed, usually thin and closely enveloping it, but readily slipping away when moist. Annuals or perennials, with small spikelets in open or contracted panicles. Type species, *Sporobolus indicus*. Name from Greek *spora*, seed, and *ballein*, to throw, alluding to the free seeds. In some species of this genus the palea splits at maturity, giving the impression of an extra lemma. The first glume is early deciduous in some species. The size of the spikelets is often variable in the same panicle.

Most of the perennial species are palatable forage grasses, but few of them are abundant enough to be of importance. Two species of the Southwest, *S. airoides* and *S. wrightii*, are valuable grasses in the arid and semiarid regions; *S. interruptus* is common on the Arizona Plateau; and the widely distributed *S. cryptandrus* is also important. The seed of *S. flexuosus* and *S. cryptandrus* have been used for food by the Indians.

1a. Plants annual.
  Panicles pyramidal, many-flowered, the lower branches verticillate.
    Spikelets appressed, short-pediceled, 1.5 to 1.7 mm. long; panicle branches densely flowered ........ 1. S. PULVINATUS.
    Spikelets spreading, long-pediceled, 1.8 to 2 mm. long; panicle branches loosely few-flowered ........ 2. S. PATENS.
  Panicles narrow, spikelike, few-flowered, usually included in the sheaths.
    Lemma pubescent ........ 3. S. VAGINIFLORUS.
    Lemma glabrous ........ 4. S. NEGLECTUS.
1b. Plants perennial.
  2a. Plants producing creeping rhizomes. Panicle narrow or spikelike.
    Rhizomes extensively creeping; leaves numerous, crowded, the blades involute, conspicuously distichous; panicle spikelike ........ 10. S. VIRGINICUS.

Rhizomes short; leaves not numerous nor crowded nor involute; panicle narrow but loose ........ 6. S. MACER.

2b. Plants without creeping rhizomes.

3a. Glumes nearly equal, much shorter than the lemma. Panicle narrow or spikelike.

Panicle branches short and appressed, the panicle spikelike ........ 8. S. POIRETII.

Panicle branches slender, ascending, the panicle scarcely spikelike .... 9. S. INDICUS.

3b. Glumes unequal or if equal as long as the spikelet.

4a. Spikelets mostly 3 to 7 mm. long. Plants usually less than 1 m. tall.

Second glume shorter than the lemma; panicle contracted, more or less included in the sheath.

Lemma glabrous, the palea not exceeding it ........ 5. S. ASPER.

Lemma pubescent, the palea acuminate, exceeding it ...... 7. S. CLANDESTINUS.

Second glume about as long as the lemma; panicle open (contracted in *S. purpurascens*), not included.

Branches of the narrow panicle in distinct whorls, usually less than 4 cm. long.

Branches 2 to 3 cm. long, somewhat distant, more or less spreading, the panicle open ........ 17. S. JUNCEUS.

Branches 1 to 2 cm. long, ascending or appressed, overlapping, the panicle contracted ........ 18. S. PURPURASCENS.

Branches of the open panicle not in distinct whorls, usually more than 4 cm. long.

Spikelets short-pediceled and appressed along the main panicle branches.

Spikelets about 4 mm. long, purplish ........ 14. S. CURTISSII.

Spikelets about 3 mm. long, pale ........ 30. S. THARPII.

Spikelets not appressed, the branches and pedicels somewhat spreading.

Blades terete ........ 15. S. TERETIFOLIUS.

Blades flat or folded.

Glumes about equal, as long as the lemma ........ 16. S. FLORIDANUS.

Glumes unequal.

Panicles 30 to 50 cm. long, purple; culms mostly more than 1 m. tall. 13. S. SILVEANUS.

Panicles 10 to 20 cm. long, gray or lead-colored; culms 30 to 70 cm. tall.

Blades elongate ........ 12. S. HETEROLEPIS.

Blades mostly less than half as long as culm .... 11. S. INTERRUPTUS.

4b. Spikelets 1 to 2.5 mm. long (sometimes 3 mm. in *S. giganteus*).

5a. Lower panicle branches in distinct whorls, the mature panicle pyramidal; spikelets about 1 mm. long ........ 19. S. PYRAMIDATUS.

5b. Lower panicle branches not in distinct whorls (occasionally whorled in *S. domingensis*); spikelets 1.5 to 2.5 mm. long.

6a. Basal sheaths compressed-keeled. Panicle branches few, widely spreading, naked for about one-third their length; spikelets 1.5 mm. long. 26. S. BUCKLEYI.

6b. Basal sheaths not compressed-keeled.

7a. Sheaths with a conspicuous tuft of white hairs at summit.

Culms robust, 1 to 2 m. tall; spikelets 2.5 to 3 mm. long. 25. S. GIGANTEUS.

Culms more slender, mostly less than 1 m. tall; spikelets 2 to 2.5 mm. long.

Panicle open, often large, the branches and branchlets flexuous, the spikelets loosely arranged ........ 22. S. FLEXUOSUS.

Panicle open or compact, if open the spikelets crowded on the branchlets.

Panicle, or the exserted portion, somewhat open, the branches naked below (sometimes entirely enclosed).

Base of plant a close tuft ........ 21. S. CRYPTANDRUS.

Base of plant a cluster of knotty rhizomes. Culms erect, slender, mostly less than 30 cm. tall; blades short, involute, spreading. 23. S. NEALLEYI.

Panicle compact, spikelike, usually exserted ........ 24. S. CONTRACTUS.

7b. Sheaths naked or nearly so at the summit.

Pedicels elongate, capillary ........ 29. S. TEXANUS.

Pedicels short.

Panicle 1 to 2 times as long as wide, loose, the branches not crowded; blades mostly involute ........ 27. S. AIROIDES.

Panicle more than 3 times as long as wide, relatively dense; blades mostly flat.

Panicle not more than 20 cm. long, usually smaller. 20. S. DOMINGENSIS.

Panicle commonly 50 cm. long, rarely as small as 25 or 30 cm. 28. S. WRIGHTII.

FIGURE 590.—*Sporobolus pulvinatus.* Panicle, × 1; glumes and floret, × 10. (Type.)

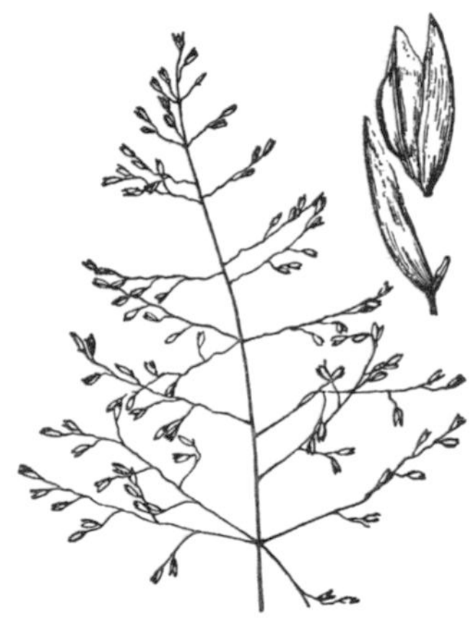

FIGURE 591.—*Sporobolus patens.* Panicle, × 1; glumes and floret, × 10. (Type.)

**1. Sporobolus pulvinátus** Swallen. (Fig. 590.) Culms 5 to 30 cm. tall in small erect or spreading tufts; blades mostly 4 to 7 cm. long, 2 to 5 mm. wide, lanceolate-acuminate, scabrous, the uppermost much reduced; panicles 2 to 5 or rarely to 8 cm. long, pyramidal, the branches erect to spreading, densely flowered, usually naked at the base; spikelets 1.5 to 1.7 mm. long, appressed; first glume minute, the second as long as the spikelet, abruptly acute or subobtuse; lemma acute or subobtuse; palea broad, conspicuous, as long as the lemma. ⊙ —Sandy land, Texas, New Mexico, and Arizona; northern Mexico.

**2. Sporobolus pátens** Swallen. (Fig. 591.) Culms 10 to 25 cm. tall, slender, erect; sheaths glabrous, sparsely hispid at the throat, the uppermost elongate, almost bladeless; blades 1 to 3.5 cm. long, 1 to 2 mm. wide, flat, scabrous on the margins; panicles pyramidal, 2.5 to 5 cm. long, the slender branches spreading or even reflexed, few-flowered, the branchlets abruptly spreading; spikelets 1.8 to 2 mm. long, the pedicels slender, spreading, as much as 3 mm. long; first glume minute; second glume and lemma equal, acute; palea shorter than the lemma, truncate, minutely

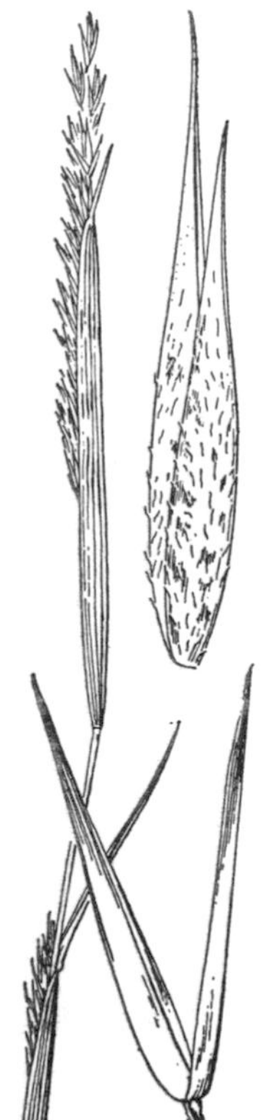

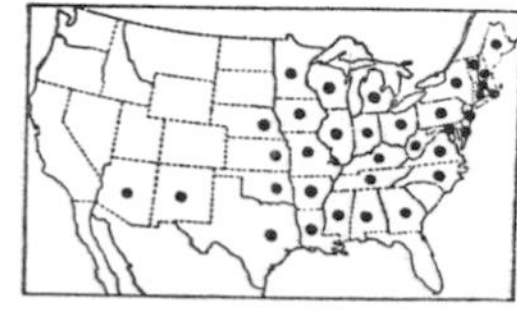

FIGURE 592.—*Sporobolus vaginiflorus.* Plant, × 1; glumes and floret, × 10. (Deam 39615, Ind.)

dentate. ⊙ —Open dry ground, southern Arizona.

**3. Sporobolus vaginiflórus** (Torr.) Wood. (Fig. 592.) Annual, branching from base; culms erect to spreading, mostly 20 to 40 cm. tall, sometimes as much as 75 cm.; blades slender, subinvolute, the lower elongate; panicles terminal and axillary, slender, mostly not more than 3 cm. long, the terminal exserted or partly included, the axillary included in the sheaths or slightly exserted, late in the season the sheaths swollen and containing cleistogamous spikelets; glumes acute, about equal, 3 to 5 mm. long; lemma as long as the glumes or exceeding them, acute or acuminate, rather sparsely pubescent, sometimes mottled with dark spots; palea acuminate, sometimes longer than the lemma. ⊙ —Sandy soil or open waste ground, Maine and Ontario to Minnesota and Nebraska, south to Georgia, Texas, and Arizona.

FIGURE 593.—*Sporobolus neglectus*. Plant, × 1; spikelet and floret, × 10. (Deam 33426, Ind.)

FIGURE 594.—*Sporobolus asper*. Plant × 1; glumes and floret, × 10. (Deam 42707, Ind.)

**4. Sporobolus neglέctus** Nash. (Fig. 593.) Differing from *S. vaginiflorus* chiefly in the smaller, paler, plumper spikelets, 2 to 3 mm. long, and in the glabrous lemma; lower blades often sparsely pilose; panicles usually entirely hidden in the more swollen sheaths. ☉ —Dry open ground and sandy fields, Quebec and Maine to Montana, south to Virginia, Tennessee, and Texas; also Washington and Arizona. A form from Missouri (Ozark Mountains), with rather strongly pilose leaves, has been differentiated as *S. ozarkanus* Fernald.

**5. Sporobolus ásper** (Michx.) Kunth. (Fig. 594.) Perennial; culms erect, often rather stout, solitary or in small tufts, 60 to 120 cm. tall; blades elongate, flat, becoming involute, 1 to 4 mm. wide at base, tapering to a fine point; panicle terminal and axillary, pale or whitish, sometimes purplish, contracted, more or less spikelike, usually enclosed at base or sometimes entirely in the inflated upper sheath, 5 to 15 cm. long; spikelets 4 to 6 mm. long; glumes rather broad, keeled, subacute, the first about half as long as the spikelet, the second two-thirds to three-fourths as long; lemma and palea subequal, glabrous, the tip boat-shaped. ♃ —Prairies and sandy meadows, Vermont to Montana, south to Louisiana and Arizona; eastern Washington.

Sporobolus asper var. pilósus (Vasey) Hitchc. Sheaths and blades more or less pilose. ♃ (*S. pilosus* Vasey.)—Prairies and rocky hills, Kansas (Saline County and westward), Texas (Del Rio), and Montana.

Sporobolus asper var. hookéri (Trin.) Vasey. Less robust, the more slender, fewer flowered, panicle looser; spikelets usually smaller, 3 to 5 mm. long. ♃ (*S. attenuatus* Nash; *S. drummondii* Vasey.)—Plains, Missouri, Kansas, Mississippi, Texas, and Oklahoma. Foliage rarely somewhat villous.

**6. Sporobolus mácer** (Trin.) Hitchc. (Fig. 595.) Perennial, with short scaly rhizomes; culms erect, 50 to 70 cm. tall; blades flat, 10 to 20 cm. long, 1 to 2 mm. wide, sometimes wider, pilose on the upper surface near base and at the throat of the sheath; panicle narrow, often enclosed at base, 5 to 15 cm. long, the branches erect; spikelets 4 to 5 mm.

Figure 595.—*Sporobolus macer*. Plant, × ½; glumes and floret, × 10. (Chase 4341, Miss.)

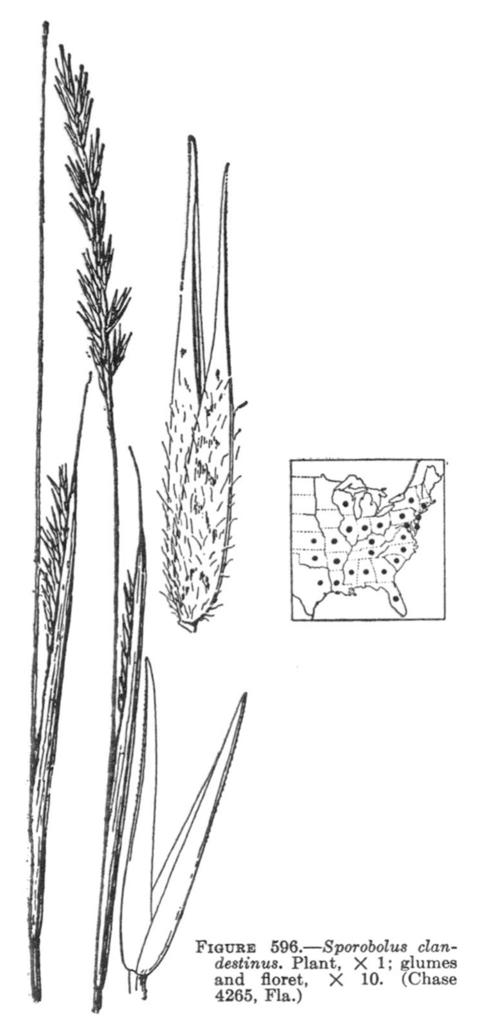

FIGURE 596.—*Sporobolus clandestinus*. Plant, × 1; glumes and floret, × 10. (Chase 4265, Fla.)

long, the glumes keeled, the first about two-thirds as long, the second a little longer than the first; lemma and palea subequal, the tips boat-shaped. ♃ —Wet pineland, Oklahoma, Mississippi, Louisiana, and Texas. Except for the rhizomes this species resembles *S. asper* var. *hookeri*.

**7. Sporobolus clandestínus** (Bieler.) Hitchc. (Fig. 596.) Perennial; culms relatively stout to slender, erect to spreading, 50 to 100 cm. tall; lower sheaths sometimes pilose; blades flat, becoming involute, with a long fine point; panicle narrow, contracted, 5 to 10 cm. long, usually partly enclosed; spikelets 5 to 7 mm. long, the glumes keeled, acute or subacute, the first more than half as long as the spikelet, the second longer than the first; lemma sparsely appressed-pubescent, acuminate, the palea longer, sometimes as much as 10 mm. long. ♃ (*S. canovirens* Nash.)—Sandy fields, pine barrens, hills, and prairies, Connecticut to Wisconsin and Kansas, south to Florida and Texas.

**8. Sporobolus poirétii** (Roem. and Schult.) Hitchc. SMUTGRASS. (Fig. 597.) Perennial; culms erect, solitary or in small tufts, 30 to 100 cm. tall; blades flat to subinvolute, rather firm, 2 to 5 mm. wide at base, elongate, tapering to a fine point; panicle usually spikelike but more or less interrupted, 10 to 40 cm. long, the branches appressed or ascending; spikelets about 2 mm. long; glumes obtuse, somewhat unequal, about half as long as the spikelet or less; lemma acutish. ♃ (*Sporobolus berteroanus* (Trin.) Hitchc. and Chase.)—Open ground and waste places, Virginia to Tennessee and Oklahoma, south to Florida, Texas, and the warmer parts of America to Argentina; on ballast in Oregon and New Jersey; tropical Asia, apparently introduced in America. At maturity the extruded reddish caryopses remain for some time sticking to the panicle by the mucilaginous pericarp. Often affected with a black fungus. This species has been referred to the Australian *S. elongatus* R. Br., which seems to be distinct, differing in its looser panicle.

**9. Sporobolus índicus** (L.) R. Br. (Fig. 598.) Resembling *S. poiretii*, but the blades more slender, especially at base, and the panicle branches longer, more slender, less densely flowered, loosely ascending to somewhat spreading, the panicle not spikelike. ♃—Punta Gorda, Fla.; ballast, Mobile, Ala.; tropical America.

**10. Sporobolus virgínicus** (L.) Kunth. (Fig. 599.) Perennial, with

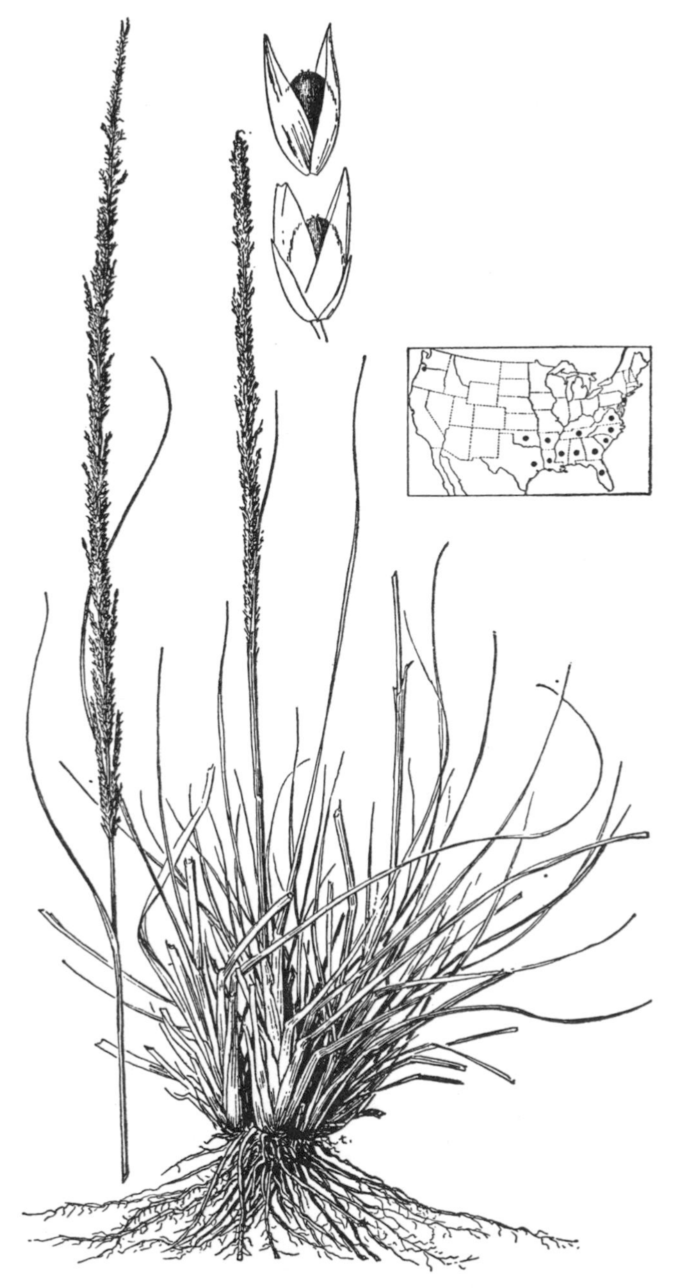

FIGURE 597.—*Sporobolus poiretii*. Plant, × ½; spikelet and floret, × 10. (Chase 7043, Fla.)

numerous branching widely creeping slender rhizomes (yellowish in drying); culms erect, 10 to 40 cm. tall; sheaths overlapping, more or less pilose at the throat; blades flat or becoming involute especially toward the fine point, conspicuously distichous, mostly less than 5 cm. long or

FIGURE 598.—*Sporobolus indicus*. Panicle, × 1; spikelet and floret, × 10. (Léon 867, Cuba.)

on the innovations longer; panicle pale, contracted or spikelike, 2 to 8 cm. long, 5 to 10 mm. thick; spikelets 2 to 2.5 mm. long; glumes and lemma about equal. ♃ —Sandy or muddy seashores and saline marshes, forming extensive colonies, with relatively few flowering culms, southeastern Virginia (Gronovius, Fl. Virg.) to Florida and Texas, south through the West Indies to Brazil. Readily grazed where available. A robust form (called *S. littoralis* (Lam.) Kunth), with culms as much as 1 m. tall and panicles as much as 15 cm. long, is found in the West Indies and extends into Florida. Complete intergradations are found, and the type specimen is not the robust form.

FIGURE 599.—*Sporobolus virginicus*. Plant, × 1; glumes and floret, × 10. (Nash 2467, Fla.)

**11. Sporobolus interrúptus** Vasey. BLACK DROPSEED. (Fig. 600.) Perennial, densely tufted; culms erect, 30 to 60 cm. tall, the leaves crowded at base, about 2 on the culm; sheaths more or less pilose; blades flat or folded, sparsely pilose to glabrous, 1 to 2 mm. wide; panicle 10 to 20 cm. long, brownish-leaden, the branches distant, finally spreading, naked at base; spikelets about 6 mm. long, short-pediceled; glumes acute, the first 2 to 3 mm., the second 4 to 6 mm. long; lemma and palea acute, about equal. ♃ —Grassy plains and hills, Arizona. The second glume and lemma may have wrinkles toward the summit that look like nerves.

**12. Sporobolus heterólepis** (A. Gray) A. Gray. PRAIRIE DROPSEED. (Fig. 601.) Perennial, in dense tufts; culms erect, slender, 30 to 70 cm. tall; sheaths somewhat pilose at the throat, the lower sometimes sparsely pilose on the back; blades elongate, flat, becoming involute at the slender attenuate tip, 2 mm. or less wide; panicle, 5 to 20 cm. long, the branches ascending or spreading, 3 to 6 cm. long, naked below, few-flowered above; spikelets grayish; glumes acuminate, the first 2 to 4 mm. long, the second 4 to 6 mm. long; lemma shorter than the second glume, palea slightly longer than the lemma; caryopsis globose, nutlike, nearly 2 mm. thick, finally splitting the palea. ♃ —

Prairies, Quebec to Saskatchewan, south to Connecticut, eastern Texas, and Colorado.

**13. Sporobolus silveánus** Swallen. (Fig. 602.) Culms 85 to 115 cm. tall, densely tufted, erect, scabrous; sheaths glabrous or scaberulous, pubescent on the collar, the uppermost elongate, the lower shiny, becoming more or less papery with age; blades as much as 45 cm. long, 1 to 2 mm. wide, usually involute, curved or flexuous; panicles 30 to 50 cm. long, the ascending branches rather distant, few-flowered, naked at the base; spikelets 5 to 6 mm. long, purple; first glume 3 to 4.5 mm. long, the second 4.5 to 6 mm. long; lemma subacute; palea as long as the lemma, the keels obscure. ♃ —Open woods, western Louisiana and eastern Texas.

**14. Sporobolus curtíssii** (Vasey) Small ex Scribn. (Fig. 603.) Perennial, in dense tufts; culms slender, 30 to 70 cm. tall; basal sheaths pilose at the throat; blades flat or folded, flexuous, about 1 mm. wide, pilose on the upper surface near the base; panicle pyramidal, open, 7 to 20 cm. long, the branches solitary or in twos, ascending; spikelets appressed along the main branches, bronze or purplish, about 4.5 mm. long; glumes about equal, acuminate, as long as or longer than the lemma and palea. ♃ —Dry pine barrens, North Carolina to Florida.

**15. Sporobolus teretifólius** Harper. (Fig. 604.) Perennial, in tufts; culms erect, wiry, 60 to 80 cm. tall, sheaths pilose at the throat; blades elongate, slender, terete, wiry, flexuous, pilose on the upper surface at base; panicle pyramidal, open, 15 to 20 cm. long, the capillary branches, branchlets, and pedicels ascending to spreading; spikelets purplish brown, 4 to 5 mm. long; glumes acute, the first half as long, the second as long as the equal lemma and palea. ♃ —Moist pine barrens, North Carolina and Georgia.

**16. Sporobolus floridánus** Chapm. (Fig. 605.) Plants more robust than *S. curtissii*, as much as 1 m. tall; sheaths keeled, the basal ones somewhat pilose at throat, the base indurate and shining, blades folded at base, usually flat above, 2 to 5 mm. wide, abruptly narrowed at apex; panicle narrow, open, 15 to 35 cm. long, the branches and branchlets ascending; spikelets 4 to 5 mm. long;

FIGURE 600.—*Sporobolus interruptus*. Plant, × 1; glumes and floret, × 10. (Rusby, Ariz.)

FIGURE 601.—*Sporobolus heterolepis*. Plant, × 1; spikelet and floret with caryopsis and split palea, × 10. (McDonald, Ill.)

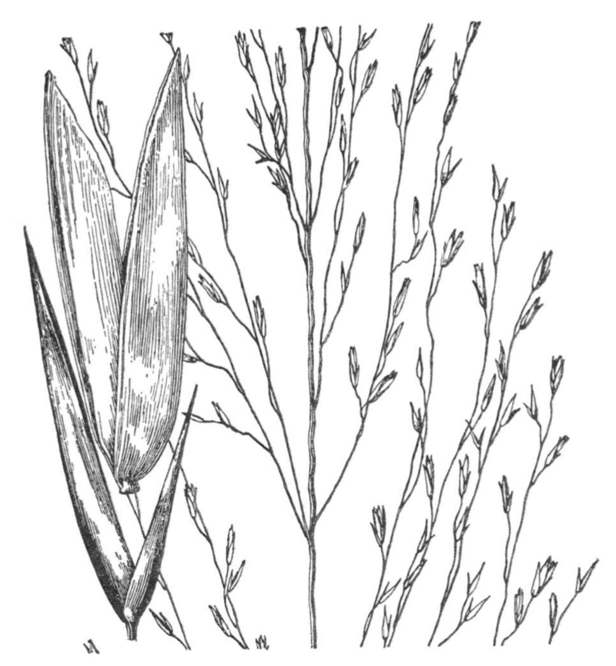

FIGURE 602.—*Sporobolus silveanus*. Panicle, × 1; glumes and floret, × 10. (Type.)

FIGURE 603.—*Sporobolus curtissii*. Panicle, × 1; glumes and floret, × 10. (Bitting 1050, Fla.)

glumes acute, subequal, about as long as the lemma and palea. ♃ —Low pine barrens, North Carolina to Florida.

**17. Sporobolus júnceus** (Michx.) Kunth. (Fig. 606.) Perennial, in dense bunches; culms erect, slender, about 3-noded, 30 to 60 cm. tall; blades folded or involute, slender, glabrous; panicle mostly bronze brown, oblong or narrowly pyramidal, open, 7 to 15 cm. long, 2 to 5 cm. wide, the flexuous branches (2 to 3 cm. long) in rather regular whorls 1 to 3 cm. apart, widely spreading to ascending, naked at base, the short-pediceled spikelets appressed along the upper part; spikelets about 3 mm. long; first glume about half as long, the second glume as long as the acute lemma or a little longer. ♃ (*S. gracilis* (Trin.) Merr.)—Pine barrens of the Coastal Plain, southeastern Virginia to Florida and Texas. Common in the high pineland of Florida.

**18. Sporobolus purpuráscens** (Swartz) Hamilt. (Fig. 607.) Re-

FIGURE 604.—*Sporobolus teretifolius*. Plant, × 1; glumes and floret, × 10. (Harper 677, Ga.)

sembling *S. junceus;* blades flat or folded, 1 to 3 mm. wide; panicle 10 to 15 cm. long, more contracted than in *S. junceus,* the shorter branches numerous in the whorls, ascending or appressed, floriferous nearly to the base; spikelets about as in *S. junceus,* greenish purple. ♃ —Sandy prairies, southern Texas and eastern Mexico; West Indies to Brazil.

**19. Sporobolus pyramidátus** (Lam.) Hitchc. (Fig. 608.) Perennial, in spreading or prostrate tufts; culms 10 to 40 cm. tall; leaves crowded at the base, the sheaths pilose at the throat; blades flat, mostly less than 10 cm. long, 2 to 4 mm. wide, sparsely long-ciliate toward the base; panicle pale, pyramidal, 3 to 7 cm. long, rarely longer, the branches spreading, somewhat viscid, 1 to 3 cm. long, naked below, closely flowered above, the lowermost in a distinct whorl; spikelets a little more than 1 mm.

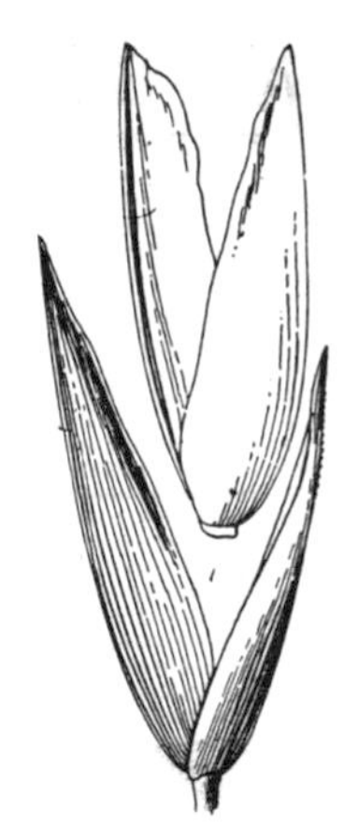

FIGURE 605.—*Sporobolus floridanus*. × 10. (Curtiss 4054, Fla.)

FIGURE 606.—*Sporobolus junceus*. Panicle, × 1; glumes and floret, × 10. (Curtiss 4056, Fla.)

long; first glume minute, the second as long as the lemma and palea. ♃ (*S. argutus* Kunth.) —Sandy or gravelly soil, especially along streets and along the seashore and in the interior in alkaline soil, Kansas and Colorado to Louisiana and Texas; southern Florida; tropical America.

**20. Sporobolus domingénsis** (Trin.) Kunth. (Fig. 609.) Perennial; culms erect, 20 to 100 cm. tall; leafy at base; blades rather firm, mostly 5 to 20 cm. long, 3 to 8 mm. wide, drying subinvolute, panicle pale, mostly 10 to 15 cm. long, the branches ascending or appressed; spikelets about 2 mm. long, the first glume half as long. ♃ —Coral sand and rocks along the coast of southern Florida, mostly on the Keys, north to Sanibel Island; West Indies.

FIGURE 608.—*Sporobolus pyramidatus*. Panicle, × 1; glumes and floret, × 10. (Hitchcock 5343, Tex.)

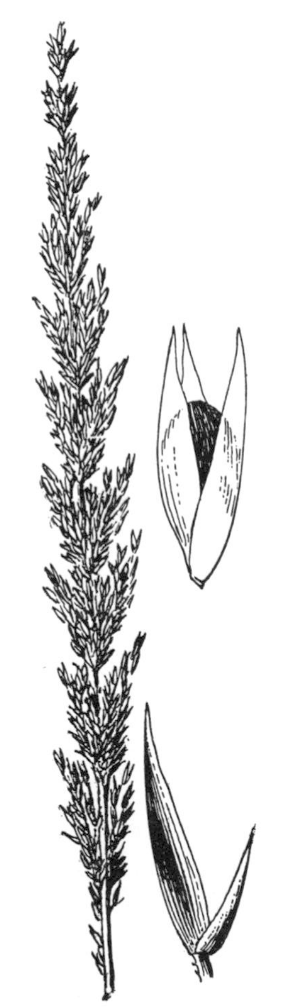

FIGURE 607.—*Sporobolus purpurascens*. Panicle, × 1; glumes and floret, × 10. (Hitchcock, Tex.)

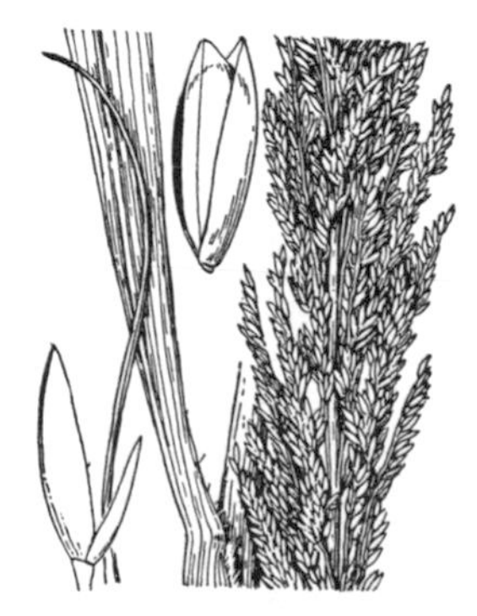

FIGURE 609.—*Sporobolus domingensis*. Plant, × 1; glumes and floret, × 10. (Hitchcock 530, Fla.)

**21. Sporobolus cryptándrus** (Torr.) A. Gray. SAND DROPSEED. (Fig. 610.) Perennial, usually in rather small tufts; culms erect or spreading, sometimes prostrate, 30 to 100 cm. tall; sheaths with a conspicuous tuft of long white hairs at summit; blades flat, 2 to 5 mm. wide, more or less involute in drying, tapering to a fine point; panicles terminal and axillary, usually included at base, sometimes entirely included, the well-

FIGURE 610.—*Sporobolus cryptandrus*. Plant, × ½; glumes and floret, × 10. (Shear 253, Nebr.)

developed terminal panicles open, as much as 25 cm. long, the branches spreading or sometimes reflexed, rather distant, naked at base, as much as 8 cm. long or even more, the spikelets crowded along the upper part of the main branches; spikelets from pale to leaden, 2 to 2.5 mm. long; first glume one-third to half as long, the second about as long as the acute lemma and palea. ♃ —Sandy open ground, Maine and Ontario to Alberta and Washington, south to North Carolina, Indiana, Louisiana, southern California, and northern Mexico.

**22. Sporobolus flexuósus** (Thurb.) Rydb. MESA DROPSEED. (Fig. 611.) Resembling *S. cryptandrus*, differing in the more open often elongate panicles, the slender branches and branchlets spreading or drooping, flexuous, loosely flowered. ♃ —Mesas, western Texas to southern Utah, Nevada, southern California, and northern Mexico.

**23. Sporobolus nealléyi** Vasey. NEALLEY DROPSEED. (Fig. 612.) Resembling dwarf forms of *S. cryptandrus*, but differing in the loose rhizomatous base; culms slender, erect, 15 to 40 cm. tall; blades slender, involute, squarrose-spreading, mostly less than 5 cm. long; panicle delicate, open, 3 to 8 cm. long, sometimes enclosed in the sheaths, the branches and branchlets spreading, the spikelets less crowded than in *S. cryptandrus*. ♃ —Gypsum sands, western Texas, Nevada, New Mexico and Arizona.

FIGURE 612.—*Sporobolus nealleyi*. Panicle, × ½; glumes and floret, × 10. (Nealley, Tex.)

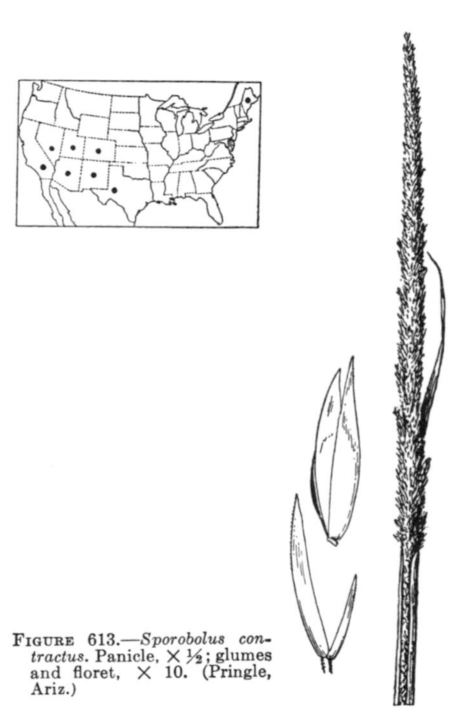

FIGURE 613.—*Sporobolus contractus*. Panicle, × ½; glumes and floret, × 10. (Pringle, Ariz.)

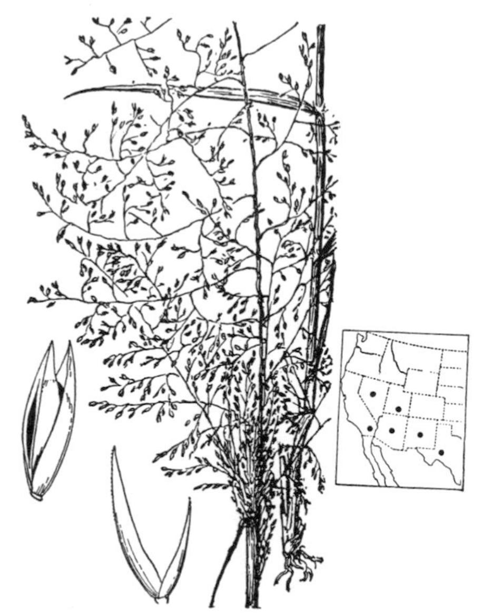

FIGURE 611.—*Sporobolus flexuosus*. Plant, × ½; glumes and floret, × 10. (Vasey, N. Mex.)

**24. Sporobolus contráctus** Hitchc. SPIKE DROPSEED. (Fig. 613.) Differing from *S. cryptandrus* in the spikelike panicle as much as 50 cm. long, usually

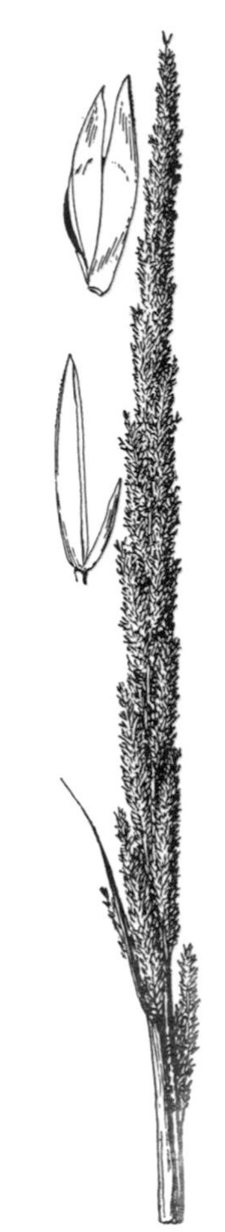

FIGURE 614.—*Sporobolus giganteus*. Panicle, × ½; glumes and floret, × 10. (Nealley, Tex.)

included at the base, rarely entirely included in the sheath. ♃ (*S. strictus* Merr.)—Mesas, dry bluffs, and sandy fields, Arkansas, Colorado to Nevada, south to western Texas, southeastern California, and Sonora; adventive in Maine.

**25. Sporobolus gigantéus** Nash. GIANT DROPSEED. (Fig. 614.) Resembling *S. cryptandrus* and *S. contractus;* culms 1 to 2 m. tall, erect, robust; blades as much as 1 cm. wide; panicle usually thicker than in *S. contractus*, less spikelike; spikelets 2.5 to 3 mm. long. ♃ —Mesas and sandhills, Oklahoma and western Texas to Colorado and Arizona.

**26. Sporobolus buckléyi** Vasey. (Fig. 615.) Perennial, the base strongly compressed; culms erect, slender, 40 to 80 cm. tall; sheaths keeled, pubescent on the margin and collar; blades flat, 4 to 7 mm. wide; panicle open, 10 to 30 cm. long, the slender branches widely spreading, as much as 10 cm. long, solitary, rather distant, naked below, with closely flowered short-appressed branchlets above; spikelets about 1.5 mm. long; glumes narrow, the first a little shorter, the second a little longer than the acute lemma; palea about as long as the lemma, splitting as the grain (1 mm. long) ripens. ♃ —Texas and eastern Mexico.

**27. Sporobolus airoídes** (Torr.) Torr. ALKALI SACATON. (Fig. 616.) Perennial, in large tough bunches; culms erect to spreading, 50 to 100 cm. tall; sheaths pilose at the throat; ligule pilose; blades elongate, flat, soon becoming involute, usually less than 4 mm. wide, often flexuous; panicle nearly half the entire height of the plant, at maturity half to two-thirds as wide as long, the stiff slender branches and branchlets finally widely spreading, naked at base, the spikelets aggregate along the upper half to two-thirds; spikelets 2 to 2.5 mm. long, the first glume about half as long, commonly falling toward

FIGURE 615.—*Sporobolus buckleyi*. Panicle, × ½; glumes and floret, × 10. (Nealley, Tex.)

FIGURE 616.—*Sporobolus airoides*. Plant, × ½; glumes and floret, × 10. (Metcalfe, N. Mex.)

FIGURE 617.—*Sporobolus wrightii*. Panicle, × ½; glumes and floret, × 10. (Hitchcock 3648, Ariz.)

maturity; second glume, lemma, and palea about equal, the palea splitting as the grain ripens. ♃ —Meadows and valleys, especially in moderately alkaline soil, South Dakota and Missouri to eastern Washington, south to Texas and southern California; Mexico. Mature spikelets with the first glume fallen and the palea split to the base are puzzling to the beginner. Less mature complete spikelets will usually be found at the base of the panicle. A good forage grass in alkaline regions; often called bunchgrass.

**28. Sporobolus wríghtii** Munro ex Scribn. SACATON. (Fig. 617.) Perennial, in large dense tufts; culms robust, erect, firm and hard, 1 to 2 m. tall; sheaths sparsely pilose at the throat; ligule pilose; blades elongate, flat, involute in drying, 3 to 6 mm. wide; panicle pale, narrow, open, mostly 30 to 60 cm. long, the branches crowded, straight, stiffly ascending, the branchlets appressed, closely flowered from the base or nearly so; spikelets 2 to 2.5 mm. long, the first glume about one-third as long, the second two-thirds to three-fourths as long, acute; lemma and palea about equal. ♃ —Mesas and valleys, southern and western Texas and Oklahoma to southern California and central Mexico. Useful for grazing when young; also furnishes hay and makes good winter range.

**29. Sporobolus texánus** Vasey. (Fig. 618.) Perennial, in close hemispherical tufts; culms erect to spreading, slender, wiry, 30 to 50 cm. tall; sheaths pilose at the throat, the lower often papillose-pilose on the surface; blades flat, involute in drying, mostly less than 10 cm. long, 1 to 4 mm. wide; panicle open, rather diffuse, breaking away at maturity, 15 to 30 cm. long, about as wide, the capillary scabrous branches, branchlets, and long pedicels stiffly spreading; spikelets about 2.5 mm. long, the first glume acute, one-third to half as long, the second, acuminate, slightly exceeding the acute lemma and palea, the palea early splitting. ♃ —Mesas, valleys, and salt marshes, Kansas and Colorado to Texas and Arizona.

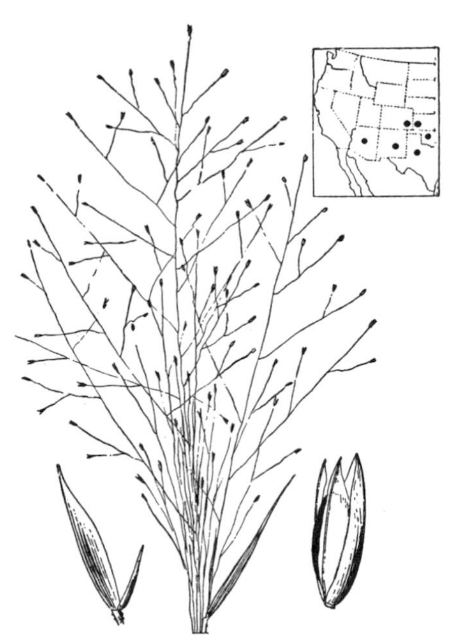

FIGURE 618.—*Sporobolus texanus*. Panicle, × ½; glumes and floret with caryopsis, × 10. (Nealley, Tex.)

**30. Sporobolus thárpii** Hitchc. (Fig. 619.) Perennial, densely tufted; culms 60 to 100 cm. tall; sheaths glabrous, the lower firm, loose, shining;

blades elongate, involute, flexuous, about 1 mm. thick, tapering to a long

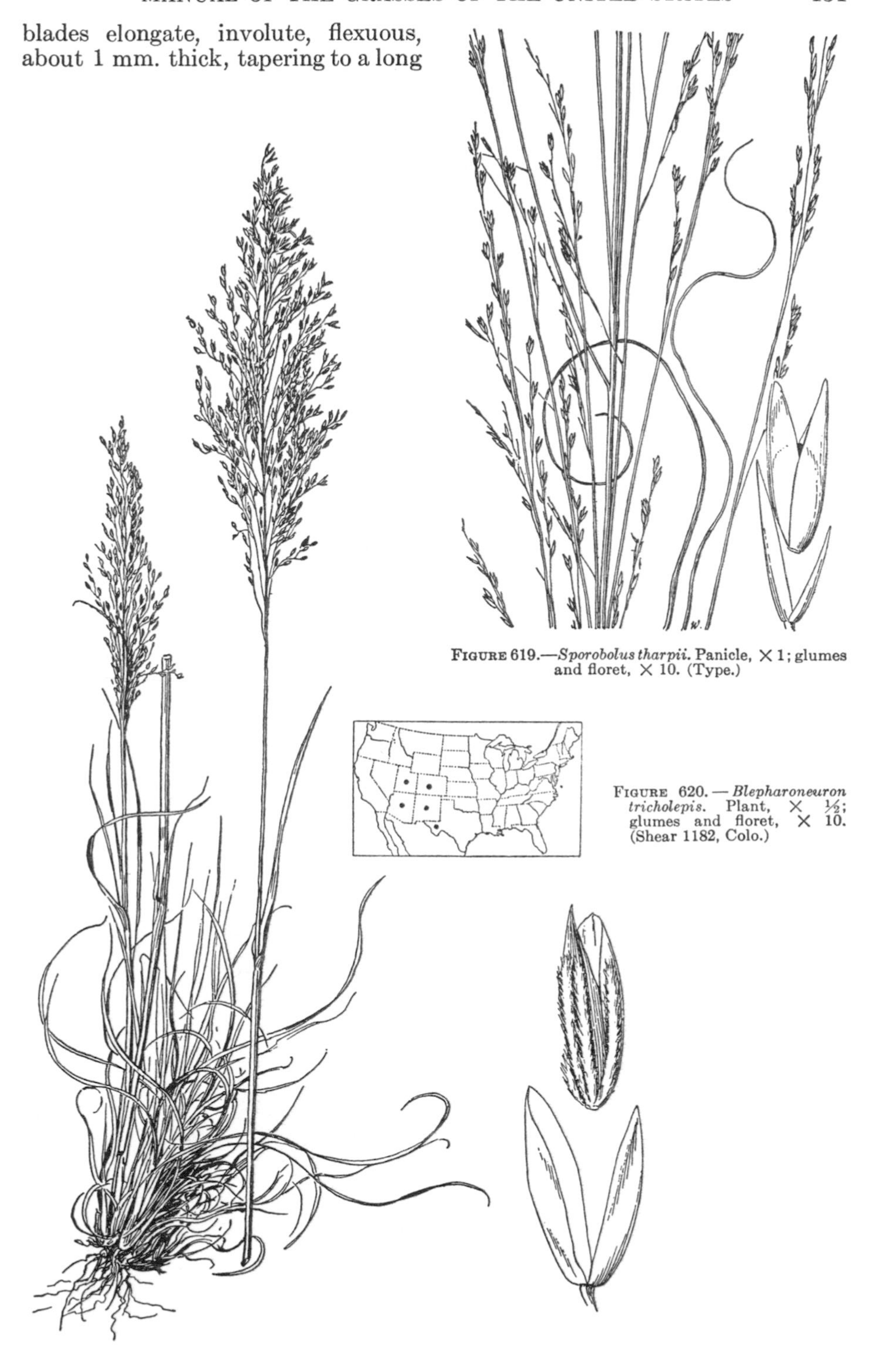

FIGURE 619.—*Sporobolus tharpii.* Panicle, × 1; glumes and floret, × 10. (Type.)

FIGURE 620.—*Blepharoneuron tricholepis.* Plant, × ½; glumes and floret, × 10. (Shear 1182, Colo.)

fine point, long-ciliate at base; panicle open, as much as 30 cm. long, the branches stiffly ascending, the lower as much as 15 cm. long; spikelets appressed along the nearly simple branches and branchlets, about 3 mm. long; first glume narrow, acuminate, about half as long as the spikelet, the second glume, lemma, and palea acute, about equal. ♃ —Known only from Padre Island, Tex.

## 84. BLEPHARONEŪRON Nash

Spikelets 1-flowered, the rachilla disarticulating above the glumes; glumes subequal, rather broad; lemma 3-nerved, the nerves densely silky villous; palea densely villous between the two nerves. Tufted perennial, with open, narrow panicles. Type species, *Blepharoneuron tricholepis*. Name from Greek *blepharis*, eyelash, and *neuron*, nerve, alluding to the villous nerves of the lemma.

**1. Blepharoneuron trichólepis** (Torr.) Nash. Hairy dropseed (Fig. 620.) Culms erect, densely tufted, slender, 20 to 60 cm. tall; leaves crowded on the innovations, mostly less than half as long as the culm, the slender blades flat, soon becoming involute, often flexuous; panicle grayish, elliptic, 5 to 20 cm. long, 2 to 5 cm. wide, many-flowered, the branches ascending, the pedicels capillary, flexuous; spikelets 2.5 to 3 mm. long; glumes obtuse or subacute, a little shorter than the abruptly pointed lemma; palea slightly exceeding the lemma. ♃ —Rocky slopes and dry open woods, 2,000 to 3,500 m., Colorado to Utah, south to Texas, Arizona, and Mexico. Palatable and sufficiently abundant in places to be of importance.

## 85. CRÝPSIS Ait.

Spikélets 1-flowered, disarticulating below the glumes; glumes about equal, narrow, acute; lemma broad, thin, 1-nerved; palea similar to the lemma, about as long, splitting between the nerves; fruit readily falling from the lemma and palea, the seed free from the thin pericarp (easily removed when wet). Spreading annual, with capitate inflorescences in the axils of a pair of broad spathes, these being enlarged sheaths with short rigid blades. Type species, *Crypsis aculeata* (L.) Ait. Name from Greek *krupsis*, concealment, alluding to the partially hidden inflorescence.

Figure 621.—*Crypsis niliaca*. Plant, × ½; glumes and floret, × 10. (Brandegee, Calif.)

**1. Crypsis nilíaca** Fig. and De Not. (Fig. 621.) Freely branching, prostrate, the mats to 30 cm. in diameter, often depauperate, 2 to 3 cm. wide; sheaths tuberculate, the summit bearded; blades flat, involute toward the apex, 2 to 5 cm. long, spreading, readily falling from the sheaths, mature plants mostly bladeless; glumes about 3 mm. long, minutely pilose; lemma and palea about as long as the glumes, the broad palea readily splitting between the nerves. (Described under *C. aculeata* in Manual, ed. 1.) ⊙ —Overflowed land, dried mud flats, sand bars, and wet alkali ground, Sacramento and San Joaquin Valleys and in Humboldt, Santa Clara, and Los Angeles Counties, Calif. Introduced; first found at Norman, Glenn County, and in alkali hollow, Colusa County, in May 1898, the source of the seed not known. The grass is slowly spreading, the latest collection being made in Santa Clara County in 1942. Egypt and southwestern Asia.

## 86. HELEÓCHLOA Host ex Roemer

Spikelets 1-flowered, the rachilla mostly disarticulating above the glumes; glumes about equal, narrow, acute; lemma broader, thin, 1-nerved, a little longer than the glumes; palea nearly as long as the lemma, readily splitting between the nerves. Low spreading tufted annuals with oblong, dense, spikelike panicles, the subtending leaves with inflated sheaths and reduced blades. Type species, *Heleochloa alopecuroides.* Name from Greek *helos,* marsh, and *chloa* grass, alluding to the habitat of the type species.

**1. Heleochloa schoenoídes** (L.) Host. (Fig. 622, *A.*) Culms tufted, branching, erect to spreading and geniculate, 10 to 30 cm. long; sheaths often somewhat inflated; blades flat, with involute slender tips, mostly less than 10 cm. long, 2 to 4 mm. wide; panicle pale, 1 to 4 cm. long, 8 to 10 mm. thick; spikelets about 3 mm. long; pericarp readily separating. ⊙ —Waste places, Massachusetts to Wisconsin, south to Delaware, Ohio, Illinois, and Iowa; California; introduced from Europe.

**Heleochloa alopecuroídes** (Pill. and Mitterp.) Host. (Fig. 622, *B.*) Differing from *H. schoenoides* in the more slender panicles, 4 to 5 mm. thick, exserted at maturity; spikelets about 2 mm. long. ⊙ —Ballast, Philadelphia, Pa., and near Portland, Oreg.; Europe.

## 87. BRACHYÉLYTRUM Beauv.

Spikelets 1-flowered, the rachilla disarticulating above the glumes, prolonged behind the palea as a slender naked bristle; glumes minute, the first often obsolete, the second sometimes awned; lemma firm, narrow, 5-nerved, the base extending into a pronounced oblique callus, the apex terminating in a long straight scabrous awn. Erect, slender perennials with short slender knotty rhizomes, flat blades, and narrow, rather few-flowered panicles. Type species, *Brachyelytrum erectum.* Name from Greek *brachus,* short, and *elutron,* cover or husk, alluding to the short glumes.

**1. Brachyelytrum eréctum** (Schreb.) Beauv. (Fig. 623.) Culms 60 to 100 cm. tall; sheaths sparsely retrorse-hispid, rarely glabrous; blades mostly 7 to 15 cm. long, 1 to 1.5 cm. wide, scabrous, sparingly pilose beneath, at least on the nerves and margin; panicle 5 to 15 cm. long, the short branches appressed; second glume 0.5 to 2 mm. long; lemma subterete, about 1 cm. long, scabrous, the nerves sometimes hispid, the awn 1 to 3 cm. long. ♃ —Moist or rocky woods, Newfoundland to Minnesota, south to Georgia, Louisiana, and Oklahoma. Plants with lemmas scabrous only toward the summit and on the nerves have been named *B. erectum* var. *septentrionale* Babel.

FIGURE 622.—*A*, *Heleochloa schoenoides*. Plant, × ½; spikelet and floret, × 5. (Smith, Pa.) *B*, *H. alopecuroides*, × 5. (Burk, Pa.)

## 88. MÍLIUM L.

Spikelets 1-flowered, disarticulating above the glumes; glumes equal, obtuse, membranaceous, rounded on the back; lemma a little shorter than the glumes, obtuse, obscurely nerved,

FIGURE 623.—*Brachyelytrum erectum*. Plant, × ½; branchlet with glumes of two spikelets, and floret, × 5. (Bissell, Conn.)

rounded on the back, dorsally compressed, in fruit becoming indurate, smooth and shining, the margins enclosing the lemma as in *Panicum*. Moderately tall grasses with flat blades and open panicles. Type species, *Milium effusum*. *Milium*, old Latin name for millet.

**1. Milium effúsum** L. (Fig. 624.) Smooth perennial, somewhat succulent; culms slender, erect from a bent base, 1 to 1.5 m. tall; blades mostly 10 to 20 cm. long, flat, lax, 8 to 15 mm. wide; panicle 10 to 20 cm. long, the slender branches in remote spreading or drooping pairs or fascicles, naked below; spikelets pale, 3 to 3.5 mm. long; glumes scaberulous. ♃ —Damp or rocky woods, Quebec and Nova Scotia to Minnesota, south to Maryland and Illinois; Eurasia. A handsome grass, sometimes cultivated as an annual.

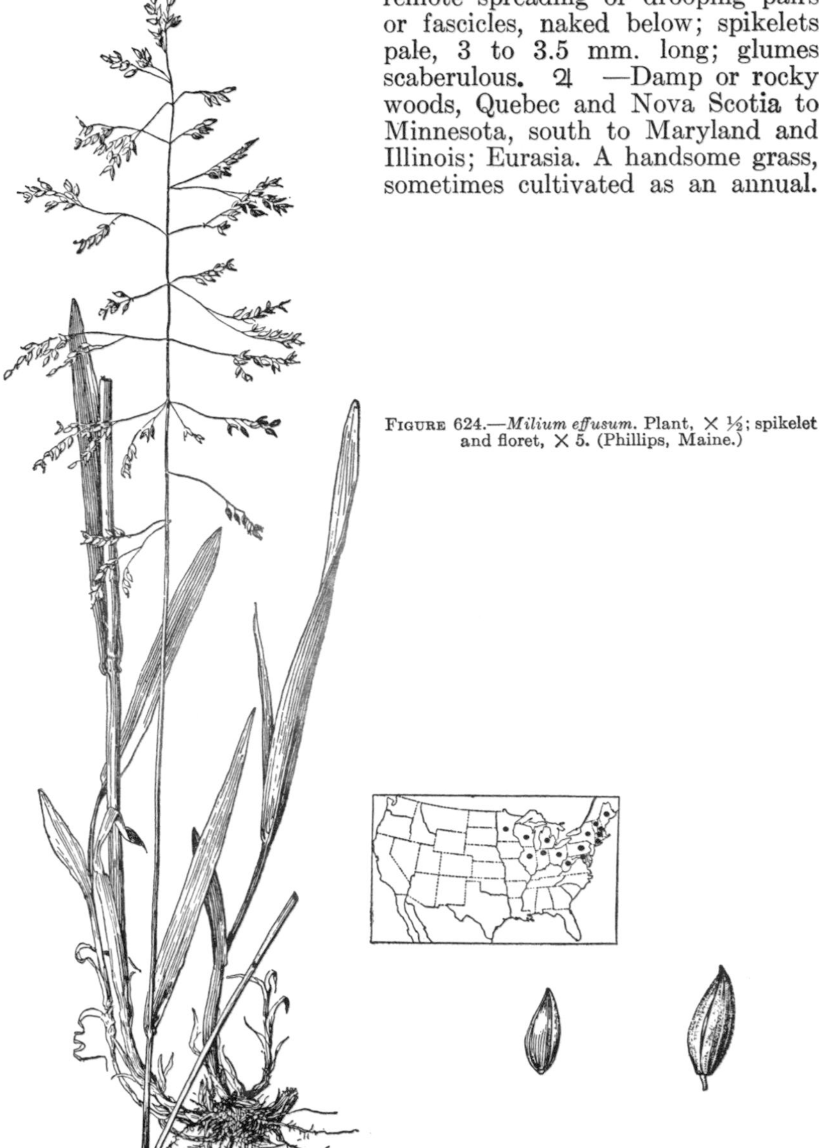

FIGURE 624.—*Milium effusum*. Plant, × ½; spikelet and floret, × 5. (Phillips, Maine.)

## 89. ORYZÓPSIS Michx. RICEGRASS

Spikelets 1-flowered, disarticulating above the glumes; glumes about equal, obtuse to acuminate; lemma indurate, usually about as long as the glumes, broad, oval or oblong, nearly terete, usually pubescent, with a short, blunt, oblique callus, and a short deciduous, sometimes bent and twisted awn; palea enclosed by the edges of the lemma. Mostly slender perennials, with flat or often involute blades and terminal narrow or open panicles. Type species, *Oryzopsis asperifolia*. Name from *oruza*, rice, and *opsis*, appearance, alluding to a fancied resemblance to rice.

Nearly all the species are highly palatable to stock, but are usually not in sufficient abundance to be of importance, except *O. hymenoides* (Indian ricegrass), which is common in the arid and semiarid regions of the West and furnishes much feed. The seed has been used for food by the Indians. Locally important may be *O. micrantha* in the Black Hills region and *O. kingii* in the high Sierras. *O. miliacea* is cultivated for forage in California.

As the result of study of several species of *Oryzopsis* and *Stipa*, Johnson and Rogler[13] conclude that the types of *Oryzopsis caduca* and *O. bloomeri* are hybrids between *O. hymenoides* and *Stipa viridula* and *O. hymenoides* and *S. occidentalis*, respectively. For these the generic name *Stiporyzopsis* is proposed. Other hybrids between *O. hymenoides* and six other species of *Stipa*—*S. elmeri*, *S. thurberiana*, *S. californica*, *S. scribneri*, *S. robusta*, and *S. columbiana*—are described, but not transferred to *Stiporyzopsis*.

Lemma smooth (rarely pubescent in *O. micrantha*).
  Blades flat, 5 mm. wide or more. Spikelets numerous, about 3 mm. long. 1. O. MILIACEA.
  Blades more or less involute, less than 2 mm. wide.
    Panicle branches spreading or reflexed; fruit about 2 mm. long, pale. 2. O. MICRANTHA.
    Panicle branches ascending or appressed; fruit about 4 mm. long, dark brown. 3. O. HENDERSONI.
Lemma pubescent.
  Pubescence on lemma long and silky.
    Panicle branches and the capillary pedicels divaricately spreading. 12. O. HYMENOIDES.
    Panicle branches and pedicels erect or ascending.
      Awn 6 mm. long; culms usually not more than 30 cm. tall. 11. O. WEBBERI.
      Awn 12 mm. long; culms 30 to 60 cm. tall. 10. O. BLOOMERI.
  Pubescence on lemma short, appressed.
    Spikelets, excluding awn, 6 to 9 mm. long; blades flat.
      Basal blades elongate, uppermost not more than 1 cm. long. 8. O. ASPERIFOLIA.
      Basal blades reduced, upper elongate. 9. O. RACEMOSA.
    Spikelets, excluding awn, 5 mm. long or less; blades involute or subinvolute.
      Panicle branches erect or appressed.
        Blades and panicle stiff, erect; awns about 5 mm. long. 4. O. EXIGUA.
        Blades flexuous, the panicle somewhat so; awns at least 10 mm. long. 7. O. KINGII.
      Panicle branches loosely ascending or spreading.
        Awn not more than 2 mm. long, straight or nearly so. 5. O. PUNGENS.
        Awn 10 to 20 mm. long, weakly twice-geniculate. 6. O. CANADENSIS.

**1. Oryzopsis miliácea** (L.) Benth. and Hook. ex Aschers. and Schweinf. SMILO GRASS. (Fig. 625.) Culms relatively stout, sometimes branching, erect from a decumbent base, 60 to 150 cm. tall; ligule about 2 mm. long; blades flat, 8 to 10 mm. wide; panicle 15 to 30 cm. long, loose, the branches spreading with numerous short-pediceled spikelets beyond the middle; glumes acuminate, 3 mm. long; lem-

[13] Amer. Jour. Bot. 30: 49–56. f. 1–40. 1943; JOHNSON, B. L., Amer. Jour. Bot. 32: 599–608. f. 1–71. 1945; Bot. Gaz. 107: 1–32. 1945.

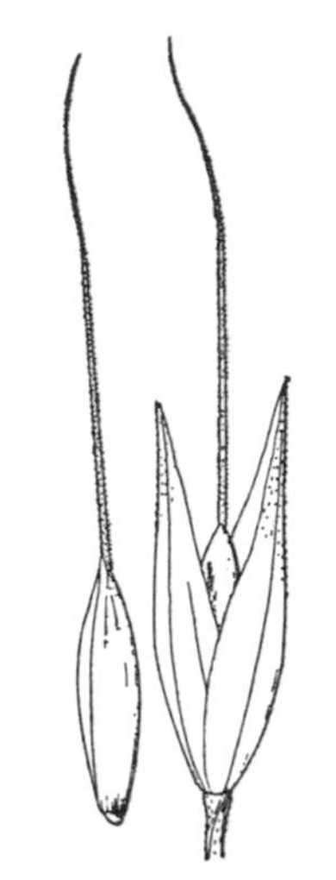

FIGURE 625.—*Oryzopsis miliacea*, × 5. (Kralik, Europe.)

ma smooth, 2 mm. long, the straight awn about 4 mm. long. ♃ —Introduced in California; ballast, Camden, N. J., and Philadelphia, Pa.; Mediterranean region.

**2. Oryzopsis micrántha** (Trin. and Rupr.) Thurb. LITTLESEED RICEGRASS. (Fig. 626.) Culms densely tufted, erect, slender, 30 to 70 cm. tall; ligule about 1 mm. long; blades scabrous, flat or involute, 0.5 to 2 mm. wide; panicle open, 10 to 15 cm. long, the branches distant, single or in pairs, spreading or finally reflexed, 2 to 5 cm. long, with short-pediceled appressed spikelets toward the ends; glumes thin, acuminate, 3 to 4 mm. long; lemma elliptic, glabrous, or rarely appressed-pilose, 2 to 2.5 mm. long, yellow or brown, the straight awn 5 to 10 mm. long. ♃ —Open dry woods and rocky slopes, medium altitudes, Saskatchewan to North Dakota and Montana, south to Nevada, New Mexico, Arizona, and California (Mohave Desert). The form with pilose lemmas is found from Colorado to Arizona.

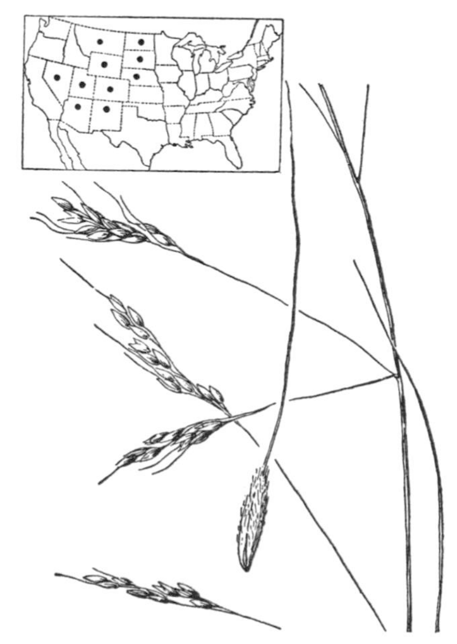

FIGURE 626.—*Oryzopsis micrantha*. Panicle, × 1; floret, × 5. (Hitchcock 22993, N. Mex.)

**3. Oryzopsis hendersóni** Vasey. (Fig. 627.) Culms densely tufted, scabrous, 10 to 40 cm. tall; leaves mostly basal, the sheaths broad, papery, glabrescent; ligule very short; blades subfiliform, involute, scabrous, firm, mostly less than 10 cm. long, the one or two culm blades 4 to 5 cm. long; panicle few-flowered, 5 to 12 cm. long, the few scabrous branches appressed or ascending, spikelet-bearing toward the ends, the lower as much as 8 cm. long; spikelets short-pediceled; glumes abruptly acute, 5 to 6 mm. long; lemma nearly as long as the glumes, glabrous, dark brown at maturity, the awn early deciduous, nearly straight, 6 to 10 mm. long. ♃ —Dry or gravelly soil. Known only from Mount Clements, Wash., and from the Ochoco National Forest, Oreg.

**4. Oryzopsis exígua** Thurb. LITTLE RICEGRASS. (Fig. 628.) Culms densely tufted, stiffly erect, scabrous, 15 to 30 cm. tall; sheaths smooth or somewhat scabrous; ligule 2 to 3 mm. long; blades involute-filiform, stiffly erect, scabrous, 5 to 10 cm. long, the culm blades about 2, shorter; panicle narrow, 3 to 6 cm. long, the branches appressed, the lower 1 to 2 cm. long; spikelets short-pediceled, glumes abruptly acute, 4 mm. long;

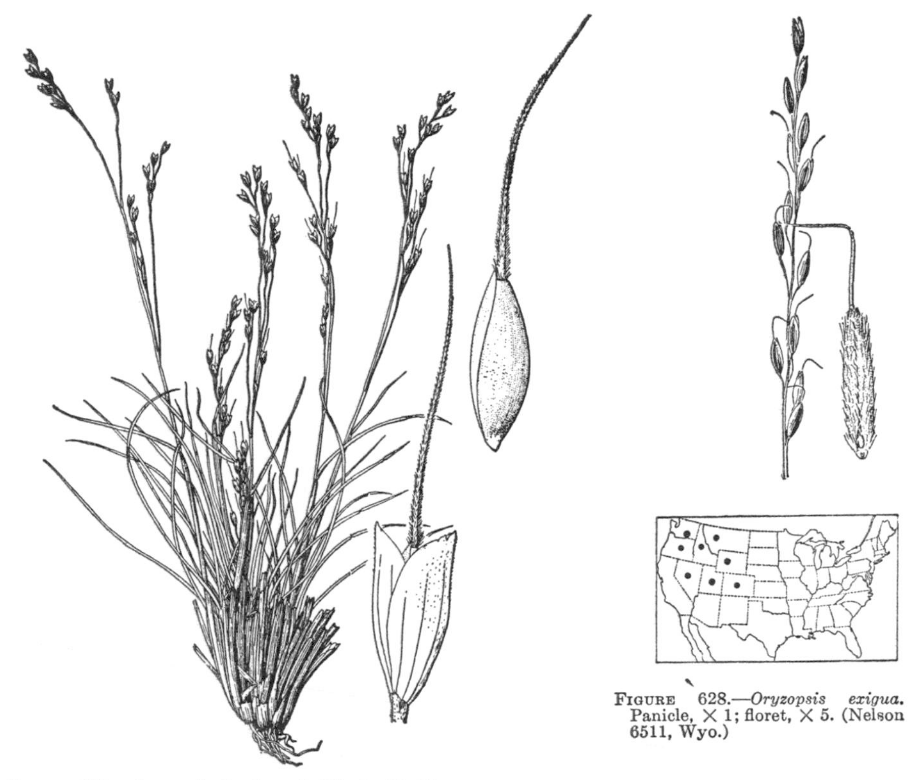

FIGURE 628.—*Oryzopsis exigua.* Panicle, × 1; floret, × 5. (Nelson 6511, Wyo.)

FIGURE 627.—*Oryzopsis hendersoni.* Plant, × ½; spikelet and floret, × 5. (Type.)

lemma appressed-pilose, about as long as the glumes, the awn about 5 mm. long, not twisted, geniculate. ♃ —Dry open ground or open woods, at moderately high altitudes, Montana to Washington, south to Colorado, Nevada, and Oregon.

**5. Oryzopsis púngens** (Torr.) Hitchc. (Fig. 629.) Culms tufted, erect, slender, 20 to 50 cm. tall; blades elongate, slender, flat or involute, less than 2 mm. wide; panicle narrow, 3 to 6 cm. long, the branches erect or ascending or spreading in anthesis; spikelets long-pediceled; glumes 3 to 4 mm. long, obscurely 5-nerved, obtuse; lemma about as long as the glumes, rather densely pubescent, the awn usually 1 to 2 mm. long. ♃ —Sandy or rocky soil, Labrador to British Columbia, south to Connecticut, Indiana, South Dakota, and Colorado.

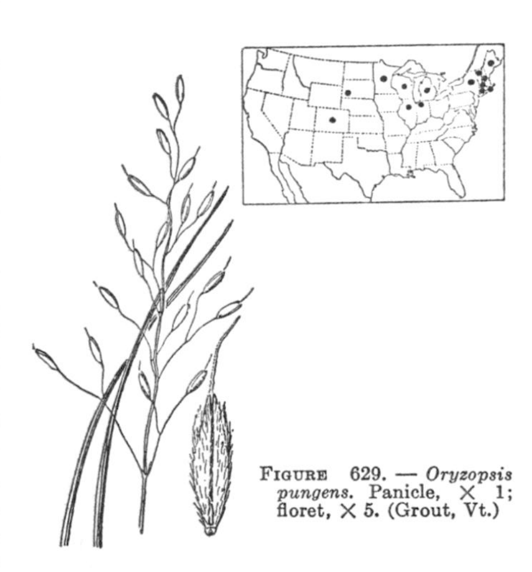

FIGURE 629. — *Oryzopsis pungens.* Panicle, × 1; floret, × 5. (Grout, Vt.)

**6. Oryzopsis canadénsis** (Poir.) Torr. (Fig. 630.) Culms slender, tufted, erect, 30 to 70 cm. tall; ligule about 2 mm. long; blades flat

FIGURE 630.—*Oryzopsis canadensis.* Panicle, × 1; floret, × 5. (Rand, Maine.)

to involute, scabrous; panicle open, 5 to 10 cm. long, the slender flexuous branches ascending or spreading, naked below, few-flowered above; spikelets long-pediceled; glumes 4 to 5 mm. long, abruptly acute; lemma about 3 mm. long, rather sparsely appressed-pilose, the awn 1 to 2 cm. long, weakly twice geniculate. ♃ —Woods and thickets, Newfoundland to Alberta, south to New Hampshire, New York, West Virginia (Panther Knob, Pendleton County), northern Michigan, Wisconsin, northern Minnesota, and Wyoming.

**7. Oryzopsis kíngii** (Boland.) Beal. (Fig. 631.) Culms tufted, slender, 20 to 40 cm. tall; leaves numerous at the base, the blades involute, filiform, flexuous; ligule about 1 mm. long; panicle narrow, loose, the short slender branches appressed or ascending, few-flowered; spikelets rather short-pediceled; glumes broad, papery, nerveless, obtuse, purple at base, the first about 3.5 mm. long, the second a little longer; lemma elliptic, 3 to 3.5 mm. long, rather sparingly appressed-pubescent; awn bent in a wide curve or indistinctly geniculate below the middle, not twisted, minutely pubescent, about 12 mm. long, not readily deciduous. ♃ —Meadows at upper altitudes, central Sierra Nevada, Calif.

**8. Oryzopsis asperifólia** Michx. (Fig. 632.) Culms tufted, the innovations erect, the fertile culms widely spreading or prostrate, 20 to 70 cm. long, nearly naked, the two or three sheaths bearing reduced or obsolete blades; basal blades erect, firm, scabrous, flat to somewhat revolute, elongate, 3 to 8 mm. wide, tapering toward each end, glaucous beneath; panicle nearly simple, rather few-flowered, 5 to 8 cm. long, the branches appressed; spikelets on appressed pedicels 3 to 6 mm. long; glumes 6 to 8 mm. long, somewhat obovate, about 7-nerved, abruptly pointed or apiculate; lemma about as long as the glumes, sparingly pubescent, more densely so on the callus, pale or yellowish at maturity, the awn 5 to 10 mm. long. ♃ —Wooded slopes and dry banks, Newfoundland to British Columbia; Maine to West Virginia (Panther Knob, Pendleton

FIGURE 631.—*Oryzopsis kingii.* Plant, × 1; floret × 5. (Bolander 6097, Calif.)

FIGURE 632.—*Oryzopsis asperifolia.* Plant, × ½; spikelet and floret, × 5. (Amer. Gr. Natl. Herb. 834, N. Y.)

County), Indiana to Idaho, south in the mountains to Utah and New Mexico.

FIGURE 633.—*Oryzopsis racemosa*. Panicle, × ½; floret, × 5. (Sartwell, N. Y.)

**9. Oryzopsis racemósa** (J. E. Smith) Ricker. (Fig. 633.) Culms tufted, from a knotty rhizome, erect, 30 to 100 cm. tall; culm leaves several, the lowermost blades reduced, the others elongate, flat, 5 to 15 mm. wide, tapering at both ends, rather thin, scabrous above, pubescent beneath; panicle 10 to 20 cm. long, the branches distant, the lower spreading or reflexed at maturity, bearing a few spikelets toward the end; glumes 7 to 9 mm. long, about 7-nerved, abruptly acuminate; lemma slightly shorter than the glumes, sparsely pubescent, nearly black at maturity, the awn 1.5 to 2.5 cm. long, slightly flexuous. ♃ —Rocky woods, Quebec to Minnesota and South Dakota, south to Virginia, Kentucky, and Iowa.

**10. Oryzopsis blooméri** (Boland.) Ricker. (Fig. 634.) Culms tufted, 30 to 60 cm. tall; leaves crowded at the base; ligule about 1 mm. long; blades narrow, involute, firm; panicle 7 to 15 cm. long, the branches slender, rather stiffly ascending, the longer 5 to 7 cm. long, spikelet-bearing from about the middle; spikelets rather long-pediceled; glumes broad, indistinctly 3- to 5-nerved, rather abruptly acuminate, 8 to 10 mm. long; lemma elliptic, 5 mm. long, densely long-villous; awn about 12 mm. long, tardily deciduous, slightly twisted and appressed-villous below, weakly geniculate. ♃ —Dry ground, medium altitudes, North Dakota to eastern Washington, south to New Mexico and California, rather rare.

FIGURE 634.—*Oryzopsis bloomeri*. Panicle, × 1; floret, × 5. (Sandberg and Leiberg 231, Wash.)

**11. Oryzopsis webbéri** (Thurb.) Benth. ex Vasey. (Fig. 635.) Culms densely tufted, erect, 15 to 30 cm. tall; blades involute, filiform, scabrous; panicle narrow, 2.5 to 5 cm. long, the branches appressed; glumes about 8 mm. long, narrow, obscurely 5-nerved, minutely scaberulous, acuminate; lemma narrow, 6 mm. long, densely long-pilose, the awn about 6 mm. long, straight or bent, not twisted. ♃ —Deserts and plains, Idaho, Colorado, Nevada, and California.

**12. Oryzopsis hymenoídes** (Roem. and Schult.) Ricker. INDIAN RICEGRASS. (Fig. 636.) Culms densely tufted, 30 to 60 cm. tall; ligule about 6 mm. long, acute; blades slender, involute, nearly as long as the culms; panicle diffuse, 7 to 15 cm. long, the slender branches in pairs, the branchlets dichotomous, all divaricately spreading, the ultimate pedicels capillary, flexuous; glumes about 6 to 7 mm. long, puberulent to glabrous, rarely hirsute, papery, ovate, 3- to 5-nerved, abruptly pointed; lemma fusiform, turgid, about 3 mm. long, nearly black at maturity, densely long-pilose with white hairs 3 mm. long; awn about 4 mm. long, straight, readily deciduous. ♃ —Deserts and plains, medium altitudes, Manitoba to British Columbia, south to Texas, California, and northern Mexico.

ORYZOPSIS HYMENOIDES var. CONTRÁCTA B. L. Johnson. Panicles narrow, the branches ascending; lemmas less turgid and less copiously pilose. ♃ —Dry soil, Wyoming.

---

**Nassélla chilénsis** (Trin. and Rupr.) E. Desv. Slender tufted perennial; blades narrow, flat or loosely involute; panicle narrow, 3 to 5 cm. long, the few branches appressed, 1 to 1.5 cm. long; glumes 4 mm. long, awn-pointed; mature lemma flattish, obovate-oblong, gibbous at apex, smooth and shining, 2 mm. long; awn geniculate, 1 cm. long, soon deciduous. ♃ (*N. major* (Trin. and Rupr.) E. Desv.)—Ballast, Portland, Oreg. Introduced from Chile.

## 90. PIPTOCHAÉTIUM Presl

Spikelets 1-flowered, disarticulating above the glumes, the callus of the floret short, acutish, usually bearded; glumes about equal, broad, ovate, convex on the back, thin, abruptly acuminate; fruit brown or dark gray,

FIGURE 635.—*Oryzopsis webberi.* Panicle, × 1; floret, × 5. (Hillman, Nev.)

FIGURE 636.—*Oryzopsis hymenoides.* Panicle, × 1; floret, × 5. (Mearns 2583, Wyo.)

coriaceous, obovate, shorter than the glumes, glabrous or hispid above the callus, often minutely striate, sometimes tuberculate near the summit, the lemma turgid, usually somewhat compressed and keeled on the back, gibbous near the summit back of the awn, the edges not meeting but clasping the sulcus of the palea, the summit sometimes expanded into a crown; awn deciduous or persistent, curved,

FIGURE 637.—*Piptochaetium fimbriatum*. Plant, × ½; glumes, floret, and palea, × 5. (Hitchcock 13511, N. Mex.)

flexuous or geniculate, often twisted below; palea narrow, indurate, except toward the margins, central keel consisting of two nerves and a narrow channel or sulcus between, the apex of the keel projecting above the summit of the lemma as a minute point. Tufted perennials with narrow usually involute blades and rather few-flowered panicles. Type species, *Piptochaetium setifolium* Presl. Name from Greek *piptein*, to fall, and *chaite*, bristle, alluding to the deciduous awns of the type species.

**1. Piptochaetium fimbriátum** (H. B. K.) Hitchc. PINYON RICEGRASS. (Fig. 637.) Culms densely tufted, erect, slender, 40 to 80 cm. tall; blades involute-filiform, flexuous, elongate; panicle open, 5 to 15 cm. long, the slender branches spreading, few-flowered toward the ends; spikelets long-pediceled; glumes about 5 mm. long, abruptly acuminate, 7-nerved; lemma a little shorter than the glumes, appressed-pubescent, especially on the callus, dark brown at maturity with a circular ridge at the base of the awn; awn weakly twice geniculate, 1 to 2 cm. long. ♃ (*Oryzopsis fimbriata* Hemsl.)—Open rocky woods, Colorado to western Texas, Arizona, and Mexico. A fine forage grass. Specimens from the United States and most of those from northern Mexico have pale glumes (*P. fimbriatum* var. *confine* I. M. Johnston), while those of middle and southern Mexico have purple or brown glumes, as in the type of *P. fimbriatum*. In that, 1 panicle is open and 2 are narrow, as in var. *confine*.

## 91. STÍPA L. NEEDLEGRASS

Spikelets 1-flowered, disarticulating above the glumes, the articulation oblique, leaving a bearded, sharp-pointed callus attached to the base of the floret; glumes membranaceous, often papery, acute, acuminate, or even aristate, usually long and narrow; lemma narrow, terete, firm or indurate, strongly convolute, rarely the margins only meeting, terminating in a prominent awn, the junction of body and awn evident, the awn twisted below, geniculate, usually persistent; palea enclosed in the convolute lemma. Tufted perennials, with usually convolute blades and mostly narrow panicles. Type species, *Stipa pennata* L. Name from Greek *stupe*, tow, alluding to the feathery awns of the type species.

The species are for the most part valuable forage plants. Several, all western, such as *Stipa comata*, *S. occidentalis*, *S. lemmoni*, and *S. neomexicana*, are grazed chiefly when young. *Stipa lettermani* is important at high altitudes, in the mountains of the West; *S. columbiana* at medium altitudes; *S. viridula* in the Rocky Mountains; *S. pulchra*, *S. thurberiana*, and *S. speciosa* in California. Some of the species, when mature, particularly *S. spartea* and *S. comata*, are injurious, especially to sheep, because of the hard sharp points to the fruits which penetrate the skin. Sleepy grass, *S. robusta*, acts as a narcotic (see p. 458). One of the Old World species, *S. tenacissima* L., furnishes a part of the esparto or alfa grass of Spain and Algeria that is used in the manufacture of paper and cordage.

1a. Terminal segment of awn plumose.
  Awn 12 to 18 cm. long ........ 1. S. NEOMEXICANA.
  Awn 1.2 to 1.5 cm. long ........ 16. S. PORTERI.
1b. Terminal segment of awn not plumose, or somewhat plumose in *S. occidentalis*.
  2a. First segment of the once-geniculate awn strongly plumose, the ascending hairs 5 to 8 mm. long ........ 2. S. SPECIOSA.
  2b. First segment of awn sometimes plumose but the hairs not more than 2 mm. long.
    3a. Mature lemma 2 to 3 mm. long. Awn capillary, flexuous, about 5 cm. long. ........ 34. S. TENUISSIMA.
    3b. Mature lemma at least 5 mm. long.

4a. Lemma densely appressed-villous with white hairs 3 to 4 mm. long, rising above the summit in a pappuslike crown.
Culms 1 to 2 m. tall; spikelets about 2 cm. long; awns 4 to 5 cm. long. 5. S. CORONATA.
Culms not more than 50 cm. tall; spikelets less than 1 cm. long; awns about 2 cm. long ... 32. S. PINETORUM.
4b. Lemma often villous but the hairs not more than 1 mm. long, or sometimes those at the summit as much as 2 mm. long.
5a. Summit of mature lemma smooth, cylindric, whitish, forming a ciliate crown 0.5 to 1 mm. long (see also *S. pulchra*) ... 3. S. LEUCOTRICHA.
5b. Summit of mature lemma not forming a crown.
6a. Lemma 2-lobed at summit, the lobes extending into awns 2 to 3 mm. long on each side of the central awn ... 4. S. STILLMANII.
6b. Lemma not lobed at summit or only obscurely so.
7a. Awn plumose below, the hairs ascending or spreading (compare *S. pulchra*, with appressed-hispid awn).
Awns once or obscurely twice-geniculate, hairs at summit of lemma longer. 23. S. CURVIFOLIA.
Awns distinctly twice geniculate.
Ligule 3 to 6 mm. long, hyaline ... 17. S. THURBERIANA.
Ligule minute, mostly hairy.
Lemma 8 to 10 mm. long; glumes firm ... 19. S. LATIGLUMIS.
Lemma 6 to 8 mm. long; glumes thin.
Hairs on upper part of lemma longer than those below; culms 60 to 125 cm. tall.
Sheaths pubescent ... 18. S. ELMERI.
Sheaths glabrous ... 22. S. CALIFORNICA.
Hairs short all over the lemma; culms 25 to 40 cm. tall. 20. S. OCCIDENTALIS.
7b. Awn scabrous or nearly glabrous, rarely appressed-hispid, not plumose.
8a. Lemma more than 7 mm. (often 1 to 2 cm.) long, glabrous or sparsely pubescent above the callus, mostly cylindric (somewhat fusiform in *S pulchra*).
Mature lemma pale or finally brownish, sparsely pubescent to summit, mostly more than 1 cm. long ... 10. S. COMATA.
Mature lemma dark.
Lemma 8 to 10 mm. long.
Glumes 3-nerved; summit of lemma hispidulous-ciliate, the hairs erect, nearly 1 mm long.
Lemma slender, cylindric; basal blades usually numerous, narrow, involute, glaucous, pilose ... 12. S. CERNUA.
Lemma fusiforme; blades green ... 11. S. PULCHRA.
Glumes 5- to 9-nerved.
Lemmas glabrous above the base, minutely roughened at apex; callus with fine sharp point ... 8. S. AVENACEA.
Lemmas sparsely pubescent to apex; callus rather blunt. 13. S. PRINGLEI.
Lemma 12 to 25 mm. long, cylindric.
Mature lemma glabrous above the callus ... 7. S. AVENACIOIDES.
Mature lemma more or less pubescent above the callus. 9. S. SPARTEA.
8b. Lemma less than 7 mm. long, or if as long as 7 to 8 mm., distinctly pubescent on the upper part (see also *S. cernua*).
Panicle open, the branches spreading or ascending, naked at base.
Panicle diffuse, the branches divergent, drooping; lemma about 5 mm. long; awn about 2 cm. long ... 6. S. RICHARDSONI.
Panicle open but not diffuse.
Ligule 3 to 6 mm. long; awn about 5 cm. long, the terminal segment flexuous ... 14. S. EMINENS.
Ligule 1 mm. long or less; awn 2.5 to 4 cm. long ... 15. S. LEPIDA.
Panicle narrow, the branches appressed.
Hairs on lemma copious, at least at summit, 2 mm. long.
Lemmas evenly villous all over; summit with lobes 0.8 to 1.5 mm. long ... 21. S. LOBATA.
Lemmas conspicuously villous above, less so below; summit not lobed or obscurely so.
Lemma about 8 mm. long, villous at summit, pubescent below. 24. S. SCRIBNERI.

Lemma about 5 mm. long, villous all over but more so above.
32. S. PINETORUM.
Hairs not copious, usually not more than 1 mm. long at summit.
Glumes broad, abruptly acuminate, rather firm, the first 5-nerved.
25. S. LEMMONI.
Glumes narrow, gradually acuminate, usually hyaline, the first usually 3-nerved.
Awn 4 to 6 cm. long, obscurely geniculate, the terminal segment flexuous........ 33. S. ARIDA.
Awn mostly less than 5 cm. long, if as much as 4 cm. long, twice-geniculate and the terminal segment straight or nearly so.
Sheaths, at least the lowermost, pubescent.
30. S. WILLIAMSII.
Sheaths glabrous.
Sheaths villous at the throat; fruit rather turgid, the callus broad and short; lower nodes of panicle villous.
Glumes thin, papery; plants rather slender, mostly less than 1 m. tall; panicle rather slender, loose.
26. S. VIRIDULA.
Glumes firm, the nerves inconspicuous; plants robust, mostly more than 1 m. tall; panicle larger, more compact........ 27. S. ROBUSTA.
Sheaths not villous at the throat or only slightly so; fruit slender, the callus narrow, sharp-pointed; nodes of panicle glabrous or nearly so.
Culms densely pubescent below the nodes.
31. S. DIEGOENSIS.
Culms glabrous throughout.
Awn mostly more than 2 cm. long; hairs at summit of lemma about as long as the others.
28. S. COLUMBIANA.
Awn mostly less than 2 cm. long; hairs at summit of lemma longer than those on the body, 1 to 1.5 mm. long........ 29. S. LETTERMANI.

**1. Stipa neomexicána** (Thurb.) Scribn. NEW MEXICAN FEATHER-GRASS. (Fig. 638.) Culms mostly 40 to 80 cm. tall; sheaths glabrous or the lower minutely pubescent; ligule very short, ciliate; blades slender, firm, convolute, glabrous beneath, the basal 10 to 30 cm. long, scarcely 1 mm. wide when unrolled; panicle narrow, 3 to 10 cm. long; spikelets pale, more or less shining; glumes 3 to 5 cm. long, tapering to a fine point; lemma about 15 mm. long including the pilose callus 4 to 5 mm. long; awn readily deciduous, 12 to 18 cm. long, the lower one-fourth to one-third straight, strongly twisted, appressed-villous, the middle segment 1 to 2 cm. long, the terminal segment flexuous, plumose, the hairs about 3 mm. long. ♃ —Mesas, canyons, and rocky slopes, western Texas, Oklahoma, Wyoming, and Colorado to Utah and Arizona.

FIGURE 638.—*Stipa neomexicana*. Plant, × ½; lemma, × 5. (Jones 5377, Utah.)

FIGURE 639.—*Stipa speciosa*. Panicle, × ½; floret, × 5. (Reed 4853, Calif.)

**2. Stipa speciósa** Trin. and Rupr. DESERT NEEDLEGRASS. (Fig. 639. Culms numerous, 30 to 60 cm. tall; sheaths brownish, smooth or the lower pubescent or even felty at the very base, the throat densely short-villous; ligule short; blades elongate, involute-filiform, mostly basal, more or less deciduous from the outer and older persistent sheaths; panicle narrow, dense, 10 to 15 cm. long, not much exceeding the leaves, white or tawny, feathery from the plumose awns; glumes smooth, 14 to 16 mm. long, 3-nerved, long-acuminate, papery; lemma 7 to 9 mm. long, narrow, densely short-pubescent, the callus sharp and smooth below; awn with one sharp bend, the first section 1.5 to 2 cm. long, densely long-pilose on the lower two-thirds or more, the hairs 5 to 8 mm. long, the remaining portion of the awn scabrous, the second segment about 2.5 cm. long. ♃ —Deserts, canyons, and rocky hills, Colorado and Arizona to southern California; southern South America.

**3. Stipa leucótricha** Trin. and Rupr. TEXAS NEEDLEGRASS. (Fig. 640.) Culms 30 to 60 cm. tall, the nodes pubescent; blades 10 to 30 cm. long, flat, often becoming involute, hispidulous beneath, 2 to 4 mm. wide; panicle narrow, mostly not more than 10 cm. long; glumes 12 to 18 mm. long; lemma about 1 cm. long, the slender callus about 4 mm. long, the body oblong, brownish, appressed-pubescent on the lower part, papillose-roughened at least toward the summit, abruptly narrowed into a cylindric smooth neck about 1 mm. long, the crown ciliate with short stiff hairs; awn 6 to 10 cm. long, rather stout, twice-geniculate, the first segment hispidulous, twisted, 2 to 3.5 cm. long. ♃ —Dry, open grassland, Oklahoma and Texas to central Mexico. Cleistogamous spikelets with glumes obsolete and lemma nearly awnless are borne in basal sheaths just after maturity of panicle.

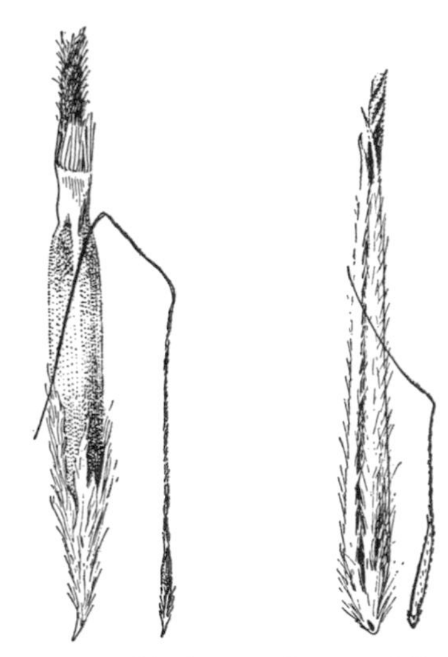

FIGURE 640.—*Stipa leucotricha*. Floret, × 1; lemma, × 5. (Hitchcock 5138, Tex.)

FIGURE 641. — *Stipa stillmanii*. Floret, × 1; lemma, × 5. (Bolander, Calif.)

**4. Stipa stillmánii** Boland. (Fig. 641.) Culms stout, 60 to 100 cm. tall; sheaths smooth, puberulent at the throat and collar; ligule very short; blades elongate, scattered, folded or involute, firm, the uppermost filiform; panicle 10 to 20 cm. long, narrow, dense or interrupted at base, the branches short, fascicled; glumes equal, 14 to 16 mm. long, papery, minutely scabrous, acuminate into a scabrous awn-point, the first 3-nerved, the second 5-nerved; lemma 9 mm. long, short-pilose, bearing 2 slender teeth at the apex, the callus short; awn about 2.5 cm. long, once- or indistinctly twice-geniculate, scabrous. ♃ —Rocky slopes, Sierra Nevada, from Lassen National Forest to Tahoe National Forest, Calif.; apparently rare.

**5. Stipa coronáta** Thurb. (Fig. 642.) Culms stout, 1 to 2 m. tall, as much as 6 mm. thick at base, smooth or pubescent below the nodes; sheaths smooth, the margin and throat villous; ligule about 2 mm. long, ciliate; blades elongate, 4 to 6 mm. wide, flat to subinvolute with a slender involute point; panicle 30 to 40 cm. long, contracted, erect, purplish; glumes gradually acuminate, 3-nerved, the first about 2 cm. long, the second 2 to 4 mm. shorter; lemma about 8 mm. long, densely villous with long appressed hairs 3 to 4 mm. long; awn usually 4 to 5 cm. long, scabrous, twice-geniculate, the first and second segments about 1 cm. long. ♃ —Open ground in the Coast Range, California, from Monterey to Baja California; Grand Canyon, Ariz.

STIPA CORONATA var. DEPAUPERÁTA (Jones) Hitchc. Culms usually 30 to 50 cm. tall; blades 10 to 20 cm. long; panicle 10 to 15 cm. long, rather few-flowered, the spikelets commonly smaller than in the species, the lemma 6 to 7 mm. long, the awn about 2.5 cm. long, once-geniculate, the first segment twisted and scabrous-pubescent, about 1 cm. long, the second segment bent about horizontally. ♃ —Dry or rocky slopes, Utah and Nevada to Arizona and southern California. Many intermediates occur between the variety and the species.

**6. Stipa richardsóni** Link. RICHARDSON NEEDLEGRASS. (Fig. 643.) Culms 50 to 100 cm. tall; blades mostly basal, usually 15 to 25 cm. long, involute, subfiliform, scabrous; panicle 10 to 20 cm. long, open, the

FIGURE 642.—*Stipa coronata.* Floret, × 1; lemma × 5. (Orcutt 1068, Calif.)

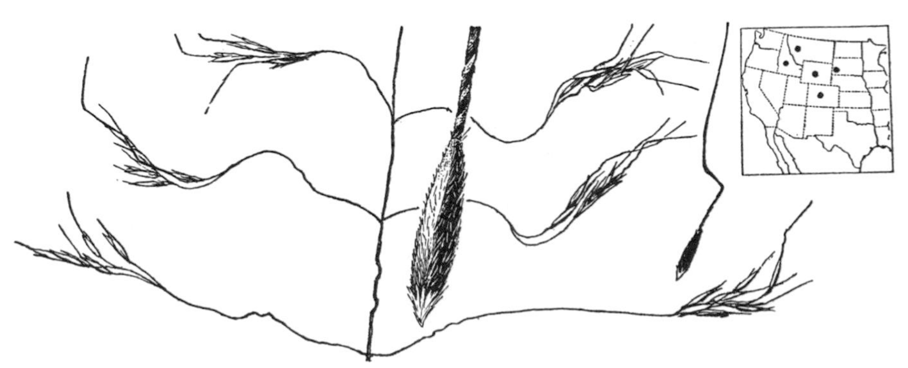

FIGURE 643.—*Stipa richardsoni.* Panicle, × ½; floret, × 1; lemma, × 5. (Hitchcock 11468, Alberta.)

branches slender, distant, spreading or drooping, naked below; glumes 8 to 9 mm. long; lemma about 5 mm. long, subfusiform, brown at maturity; awn 2.5 to 3 cm. long. ♃ — Bottom lands and wooded slopes, Saskatchewan to South Dakota, Colorado, Idaho, and British Columbia.

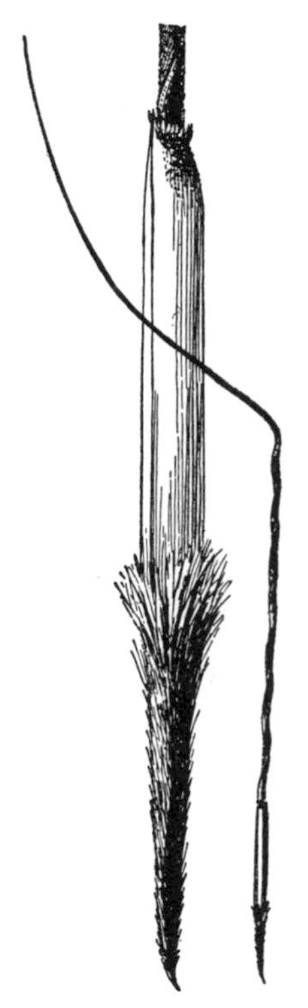

FIGURE 644.—*Stipa avenacioides*. Floret, × 1; lemma, × 5. (Curtiss 5834, Fla.)

**7. Stipa avenacioídes** Nash. (Fig. 644.) Culms about 1 m. tall; ligule 2 to 3 mm. long; blades elongate, involute, subfiliform; panicle 10 to 25 cm. long, open, the branches slender, spreading, naked below; glumes about 2 cm. long; lemma brown, linear, 1.5 to 2 cm. long including the callus 7 mm. long, the body glabrous, minutely papillose at the slightly contracted summit, slightly hispidulous on the crown; awn 8 to 11 cm. long, scabrous, twice geniculate. ♃ — Dry pine woods, peninsular Florida.

**8. Stipa avenácea** L. BLACKSEED NEEDLEGRASS. (Fig. 645.) Culms 60 to 100 cm. tall; ligule about 3 mm. long; blades 20 to 30 cm. long, 1 mm. wide, flat or involute; panicle 10 to 15 cm. long, open, the slender branches 2 to 4 cm. long, bearing 1 or 2 spikelets; glumes 1.5 cm. long; lemma dark brown, 9 to 10 mm. long, the callus 2 mm. long, the body glabrous, papillose-roughened toward the summit, awn scabrous, 4.5 to 6 cm. long, twice-geniculate. ♃ — Dry or rocky open woods, Massachusetts to Michigan south to Florida and Texas, mostly on the Coastal Plain.

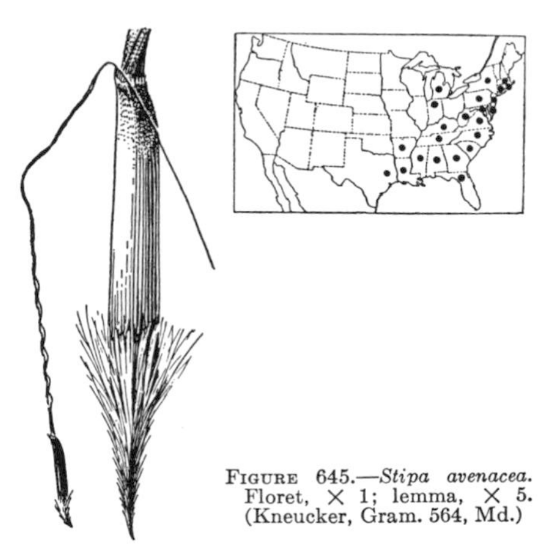

FIGURE 645.—*Stipa avenacea*. Floret, × 1; lemma, × 5. (Kneucker, Gram. 564, Md.)

**9. Stipa spártea** Trin. PORCUPINE GRASS. (Fig. 646.) Culms about 1 m. tall; ligule rather firm, 4 to 5 mm. long; blades 20 to 30 cm. long, 3 to 5 mm. wide, flat, involute in drying; panicle 15 to 20 cm. long, narrow, nodding, the few slender branches bearing 1 or 2 spikelets; glumes 3 to 4 cm. long; lemma subcylindric, brown, 1.6 to 2.5 cm. long, the callus about 7 mm. long, the body pubescent below, glabrous above except for a line of pubescence on one side, the crown erect-ciliate; awn stout, 12 to 20 cm. long, twice geniculate. ♃ —Prairies, Ontario to British Columbia; Pennsylvania to Montana, Missouri, and New Mexico. STIPA SPARTEA var. CURTISÉTA Hitchc. Glumes

FIGURE 646.—*Stipa spartea*. Plant, × ½; glumes and floret, × 2. (McDonald 16, Ill.)

2 to 3 cm. long; lemma 12 to 15 mm. long; awn mostly not more than 7 or 8 cm. long. ♃ —Manitoba to Alberta, Montana, South Dakota, and Wyoming.

**10. Stipa comáta** Trin. and Rupr. NEEDLE-AND-THREAD. (Fig. 647.) Culms 30 to 60 cm. tall, sometimes

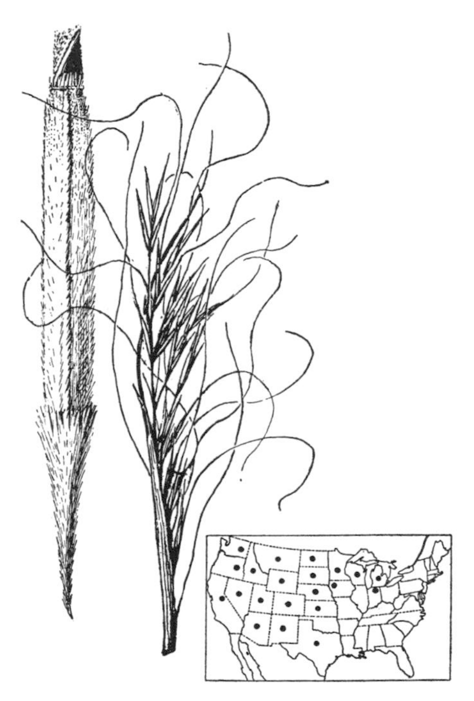

FIGURE 647.—*Stipa comata*. Panicle, × ½; lemma, × 5. (Hitchcock 1700, Colo.)

taller; ligule thin, 3 to 4 mm. long; blades 10 to 30 cm. long, 1 to 2 mm. wide, flat or involute, panicle commonly included at base, narrow, 10 to 20 cm. long; glumes 1.5 to 2 cm. long, the attenuate tips subhyaline; lemma 8 to 12 mm. long, mostly about 1 cm., pale or finally brownish, the callus about 3 mm. long, the body sparsely pubescent or glabrate toward the summit; awn 10 to 15 cm. long, indistinctly twice-geniculate, very slender, loosely twisted below, flexuous above, often deciduous. ♃ —Prairies, plains, and dry hills, Indiana to Yukon Territory, south to Texas and California. A form from Washington with pubescent foliage has been differentiated as *S. comata* var. *intonsa* Piper. STIPA COMATA var. INTERMÉDIA Scribn. and Tweedy. Differing from *S. comata* in the shorter straight third segment of the awn; glumes and lemma on the average a little longer; panicle usually exserted.

FIGURE 648.—*Stipa pulchra*. Panicle, × ½; lemma, × 5. (Chase 5598, Calif.)

—Canada; Montana to Washington, south to New Mexico and California.

**11. Stipa púlchra** Hitchc. PURPLE NEEDLEGRASS. (Fig. 648.) Culms 60 to 100 cm. tall; blades long, narrow, flat or involute; ligule about 1 mm. long; panicle nodding, about 15 to 20 cm. long, loose, the branches spreading, slender, some of the lower 2.5 to 5 cm. long; glumes narrow, long-acuminate, purplish, 3-nerved, the first about 20 mm. long, the second 2 to 4 mm. shorter; lemma 7.5 to 13 mm. long, fusiform, sparingly pilose, sometimes only in lines above, minutely papillose-roughened, the callus about 2 mm. long, the summit sometimes with a smooth neck and a ciliate crown (as in *S. leucotricha*); awn 7 to 9 cm. long, short-pubescent to the second bend, the first segment 1.5 to 2 cm. long, the second shorter, the third 4 to 6 cm. long. ♃ —Open ground, northern California to Baja California, mostly in the Coast Ranges.

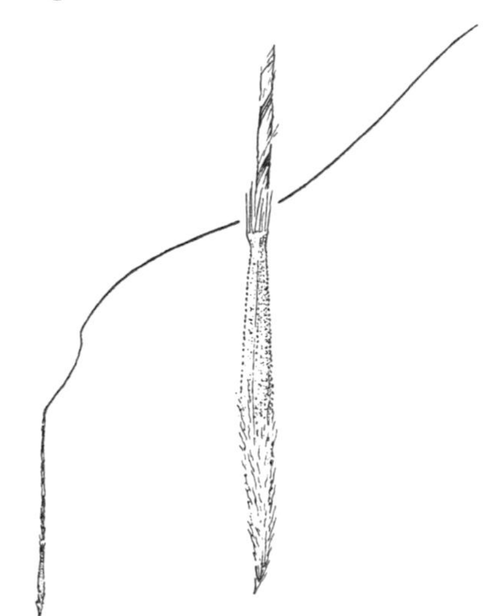

FIGURE 64 .—*Stipa cernua*. Glumes and floret, × 5. (Hall 2921, Calif.)

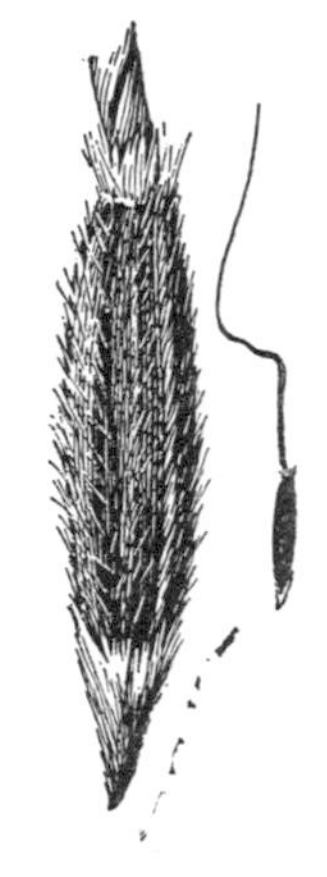

FIGURE 650. — *Stipa pringlei*. Floret, × 1; lemma, × 5. (Hitchcock 7691, Mexico.)

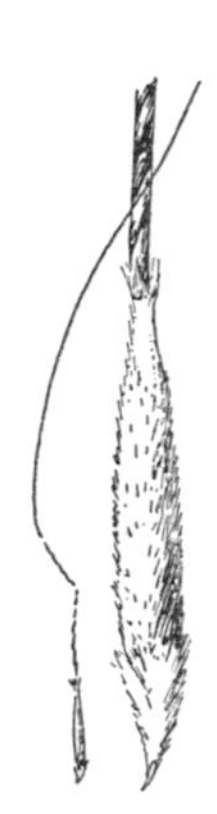

FIGURE 651.—*Stipa eminens*. Floret, × 1; lemma, × 5. (Palmer 523, Mexico.)

**12. Stipa cérnua** Stebbins and Love. (Fig. 649.) Culms mostly 60 to 90 cm. tall, in rather large clumps; basal blades numerous, narrow, glaucous, those of the culm 1.2 to 2.4 mm. wide; panicle open with slender flexuous branches; glumes acuminate, the first 12 to 19 mm. long, the second a little shorter; lemma 5 to 10.5 mm. long, papillose, silky-pilose below and on the nerves, the callus acute, densely bearded; awn 6 to 11 cm. long, the terminal segment flexuous. ♃ —Foothills of Sierra Nevada and Coast Ranges, Calif.

**13. Stipa prínglei** Scribn. PRINGLE NEEDLEGRASS. (Fig. 650.) Culms, about 1 m. tall; ligule about 2 mm. long; blades 10 to 30 cm. long, 1 to 3 mm. wide, flat or those of the innovations involute, firm, erect, scabrous, panicle nodding, 10 to 15 cm. long, the branches ascending, few-flowered, naked below; glumes about 1 cm. long, broad, rather abruptly narrowed into a short point, 7- to 9-nerved; lemma 7 to 8 mm. long, oblong-elliptic, brown, minutely papillose and brownish pubescent, the callus 1 mm. long; awn about 3 cm. long, obscurely twice-geniculate. ♃ —Rocky woods and slopes, Texas, New Mexico, and Arizona to Chihuahua, Mex.

**14. Stipa éminens** Cav. (Fig. 651.) Culms slender, rather wiry, 80 to 120 cm. tall; ligule 3 to 6 mm. long; blades mostly elongate, flat or involute, 1 to 4 mm. wide; panicle nodding, open, 10 to 20 cm. long, usually densely pilose on the lower node, the branches slender, spreading, often flexuous, usually 3 to 4 or even more at the node; glumes about 1.5 cm. long; lemma pale, 5 to 7 mm. long, pubescent; awn 3 to 6 cm. long, obscurely twice-geniculate, the third segment flexuous. ♃ —Rocky hills, Texas to Arizona and central Mexico.

**15. Stipa lépida** Hitchc. FOOTHILL NEEDLEGRASS. (Fig. 652.) Culms slender, puberulent below the nodes, 60

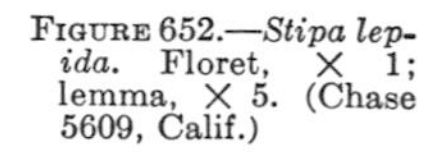

FIGURE 652.—*Stipa lepida*. Floret, × 1; lemma, × 5. (Chase 5609, Calif.)

FIGURE 653.—*Stipa porteri*. Floret, × 1; lemma, × 5. (Wolf 1109, Colo.)

to 100 cm. tall; sheaths smooth, rarely puberulent, sparingly villous at throat; ligule very short; blades 10 to 30 cm. long, flat, 2 to 4 mm. wide, pubescent on upper surface near base; panicle rather loose and open, usually 15 to 20 cm. long, sometimes more than 30 cm. long, the branches distant, slender; glumes 3-nerved, smooth, acuminate, the first 6 to 10 mm. long, the second about 2 mm. shorter; lemma about 6 mm. long, brown, sparingly villous, nearly glabrous toward the hairy-tufted apex; awn indistinctly twice-geniculate, about 2.5 to 4 cm. long, scabrous. ♃ —Dry hills, open woods, and rocky slopes, central California to Baja California, in the Coast Range. STIPA LEPIDA var. ANDERSÓNII (Vasey) Hitchc. Differing only in the more slender culms, the slender involute blades, and in the narrow or reduced panicle.—Same range as the species.

**16. Stipa portéri** Rydb. (Fig. 653.) Culms 20 to 35 cm. tall; ligule 2 to 3 mm. long; blades 2 to 12 cm. long, involute, subfiliform, sulcate, scaberulous; panicle mostly 5 to 10 cm. long, open, the branches distant, capillary, flexuous, few-flowered; glumes 5 to 6 mm. long; lemma about 5 mm. long, oblong-elliptic, softly pilose on the lower half, scaberulous above, lobed at summit; awn 12 to 15 mm. long, plumose with hairs 1 to 2 mm. long, with a single bend one-third from the base, the first segment weakly twisted. ♃ —High mountains of Colorado.

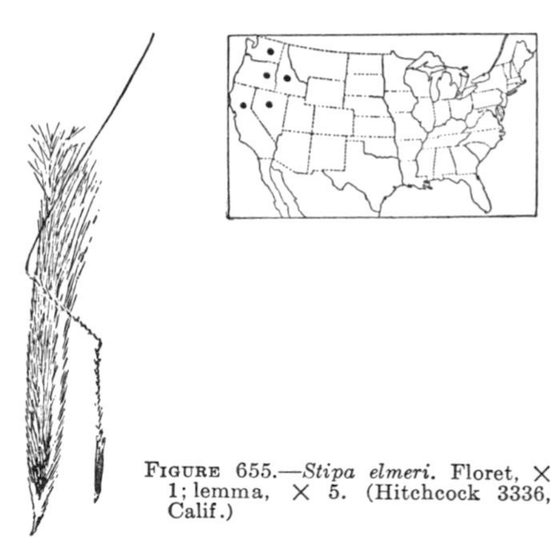

FIGURE 655.—*Stipa elmeri*. Floret, × 1; lemma, × 5. (Hitchcock 3336, Calif.)

**17. Stipa thurberiána** Piper. THURBER NEEDLEGRASS. (Fig. 654.) Culms mostly 30 to 60 cm. tall; sheaths scaberulous or the upper glabrous; ligule hyaline, 3 to 6 mm. long; blades 10 to 25 cm. long, filiform, involute, scabrous to densely soft-pubescent, flexuous; panicle mostly 8 to 15 cm. long, narrow, the ascending branches few flowered; glumes 11 to 13 mm. long, the acuminate summit hyaline; lemma 8 to 9 mm. long, appressed-pubescent, callus about 1 mm. long; awn 4 to 5 cm. long, twice-geniculate, the first and second segments plumose with

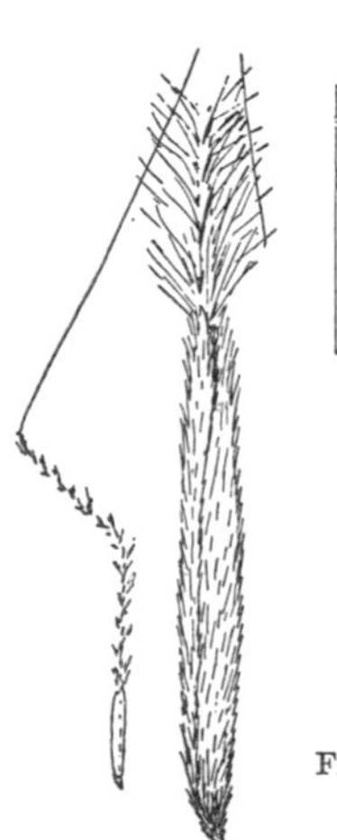

FIGURE 654.—*Stipa thurberiana*. Floret, × 1; lemma, × 5. (Chase 4689, Idaho.)

hairs 1 to 2 mm. long. ♃ —Mesas and rocky slopes, Idaho to Washington and central California.

**18. Stipa elméri** Piper and Brodie ex Scribn. (Fig. 655.) Culms 60 to 100 cm. tall, more or less puberulent, especially at the nodes; sheaths pubescent; ligule very short; blades 15 to 30 cm. long, 2 to 4 mm. wide, flat or becoming involute, pubescent on the upper surface, or those of the innovations also on the lower surface; panicle narrow, 15 to 35 cm. long, rather loose; glumes 12 to 14 mm. long, long-acuminate, hyaline except toward base; lemma about 7 mm. long, appressed-pubescent, the

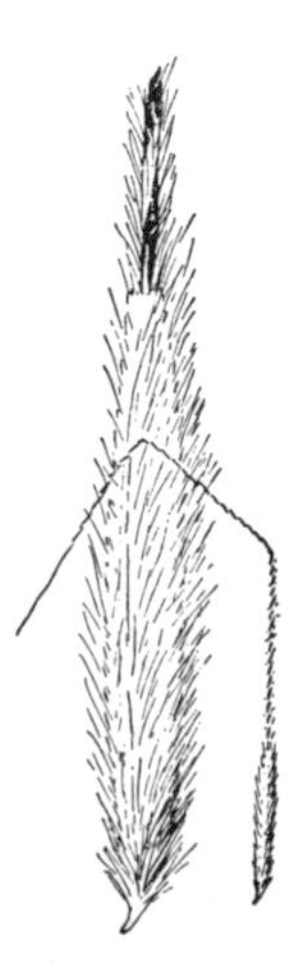

FIGURE 656.—*Stipa latiglumis.* Floret, × 1; lemma, × 5. (Type.)

callus 1 mm. long; awn 4 to 5 cm. long, distinctly twice-geniculate, the segments nearly equal, the first and second finely plumose. ♃ —Dry hills, sandy plains, and open woods, Washington and Idaho to California and Nevada.

**19. Stipa latiglúmis** Swallen. (Fig. 656.) Culms slender, erect, strigose below, 50 to 110 cm. tall; sheaths, at least the lower, pubescent; blades flat or loosely involute, pilose on the upper surface, glabrous beneath; ligule 1 to 4 mm. long; panicle narrow, loosely flowered, 15 to 30 cm. long, the branches distant, slender, the

FIGURE 657.—*Stipa occidentalis.* Panicle, × ½; lemma, × 5. (Hitchcock 11740, Oreg.)

lower as much as 10 cm. long; glumes about equal, firm, rather abruptly acute or acuminate, 3-nerved, tinged with purple, 13 to 15 mm. long, 1.5 mm. wide from keel to margin; lemma densely pubescent, 8 to 9 mm. long, the sharp callus 1 mm. long; awn twice-geniculate, 3.5 to 4.5 cm. long, the first and second segments plumose. ♃ —Sierras of central California at medium altitudes.

**20. Stipa occidentális** Thurb. WESTERN NEEDLEGRASS. (Fig. 657.) Culms mostly 25 to 40 cm. tall;

FIGURE 658.—*Stipa lobata.* Floret, × 1; lemma, × 5; summit of lemma, × 15. (Type.)

FIGURE 659.—*Stipa californica*. Floret, × 1; lemma, × 5. (Hall 2556, Calif.)

sheaths glabrous to pubescent; blades 10 to 20 cm. long, 1 to 2 mm. wide, usually involute, glabrous beneath, white-puberulent on the upper surface; panicle 10 to 20 cm. long, lax, the few slender branches narrowly ascending; glumes about 12 mm. long, the attenuate tips hyaline; lemma pale brown, about 7 mm. long, rather sparsely appressed-pubescent; awn 3 to 4 cm. long, twice-geniculate, plumose, the hairs on first and second segments about 1 mm. long, shorter on third segment. ♃ —Plains, rocky hills, and open woods, Wyoming to Washington, Arizona, and California.

**21. Stipa lobáta** Swallen. (Fig. 658.) Culms densely tufted, erect, scaberulous below the panicle, 35 to 85 cm. tall; blades flat or loosely folded toward the base, tapering into a fine point, as much as 50 cm. long, 1 to 4 mm. wide at the base, scabrous on the upper surface, glabrous beneath; ligule less than 0.5 mm. long; panicle narrow, 10 to 18 cm. long, the branches appressed; glumes about equal, acuminate, 3-nerved, scabrous, 9 to 10 mm. long; lemma brownish, 6 mm. long, densely pubescent with hairs 1 to 2 mm. long, the callus very short, blunt, the summit 2-lobed, the lobes 0.8 to 1.5 mm. long, awned from between the lobes; awn twice-geniculate, 12 to 16 mm. long, the first and second segments appressed-hispid. ♃ —Rocky hills at medium altitudes, western Texas and New Mexico.

**22. Stipa califórnica** Merr. and Davy. (Fig. 659.) Culms 75 to 125 cm. tall; ligule rather firm, 1 to 2 mm. long; blades 10 to 25 cm. long, 1 to 4 mm. wide, flat, becoming involute, those of the innovations slender and involute; panicle 15 to 30 cm., sometimes to 50 cm., long, slender, pale; glumes about 12 mm. long; lemma 6 to 8 mm. long, rather sparsely villous with ascending white hairs, those at the summit about 1.5 mm. long; awn 2.5 to 3.5 cm. long, twice-geniculate, the first and second segments plumose. ♃ —Dry open ground, Idaho and Washington to California and western Nevada.

FIGURE 660.—*Stipa curvifolia*. Floret, × 1; lemma, × 5. (Type.)

**23. Stipa curvifólia** Swallen. (Fig. 660.) Culms densely tufted, erect, about 35 cm. tall; leaves clustered toward the base, the lowermost sheaths pubescent, the blades involute, becoming curved with age; panicle 7 to 8 cm. long, dense, the branches short, appressed; glumes about 10 mm. long; lemma 5.5 mm. long, light brown, evenly white pilose; awn once or obscurely twice-geniculate, 22 to 25 mm. long, twisted and densely plumose below the bend.

♃ —Known only from limestone cliffs, Guadalupe Mountains, N. Mex.

**24. Stipa scribnéri** Vasey. SCRIBNER NEEDLEGRASS. (Fig. 661.) Culms 30 to 70 cm. tall; sheaths villous at the throat; ligule less than 1 mm. long; blades 15 to 25 cm. long, 2 to 4 mm. wide, flat or sometimes involute; panicle 10 to 25 cm. long, contracted, the rather short stiff branches erect; glumes 10 to 15 mm. long, relatively firm, attenuate; lemma about 8 mm. long, pale, narrow-fusiform, villous with white hairs, those at the summit about 2 mm. long, forming a brushlike tip; awn 14 to 20 mm. long, twice-geniculate. ♃ —Mesas and rocky slopes, Colorado, Utah, New Mexico, and Arizona.

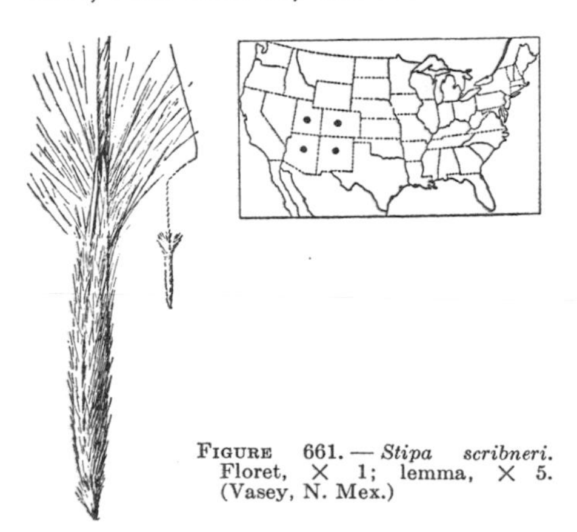

FIGURE 661. — *Stipa scribneri*. Floret, × 1; lemma, × 5. (Vasey, N. Mex.)

**25. Stipa lemmóni** (Vasey) Scribn. LEMMON NEEDLEGRASS. (Fig. 662.) Culms 30 to 80 cm. tall, scaberulous, usually puberulent below the nodes; ligule 1 to 3 mm. long; blades 10 to 20 cm. long, flat or involute, 1 to 2 mm. wide, or those of the innovations very narrow; panicle 5 to 12 cm. long, narrow, pale or purplish; glumes 8 to 10 mm. long, rather broad and firm, somewhat abruptly acuminate, the first 5-nerved, the second 3-nerved; lemma 6 to 7 mm. long, pale or light brown, the callus rather blunt, the body fusiform, 1.2 mm. wide, villous with appressed hairs; awn 20 to 35 mm. long, twice-geniculate, appressed-pubescent to the second bend. ♃ —Dry open ground and open woods, British Columbia to Idaho and California.

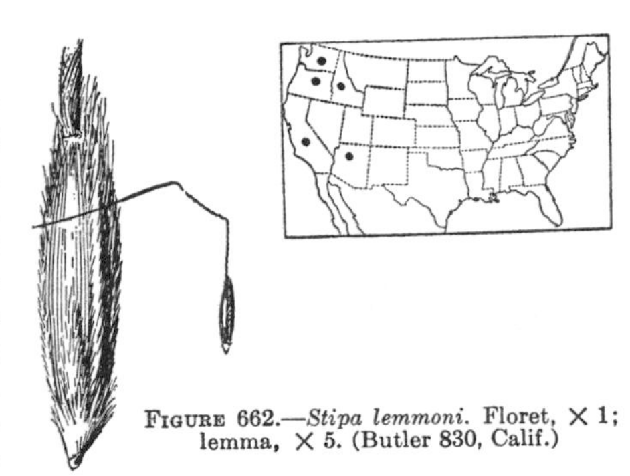

FIGURE 662.—*Stipa lemmoni*. Floret, × 1; lemma, × 5. (Butler 830, Calif.)

**26. Stipa virídula** Trin. GREEN NEEDLEGRASS. (Fig. 663.) Culms 60 to 100 cm. tall; sheaths villous at the throat, often rather sparingly so, more or less hispidulous in a line across the collar; ligule about 1 mm. long; blades 10 to 30 cm. long, 1 to 3 or even 5 mm. wide, flat or, especially on the innovations, involute; panicle 10 to 20 cm. long, narrow, rather closely flowered, greenish or tawny at maturity; glumes 7 to 10 mm. long, hyaline-attenuate; lemma 5 to 6 mm. long, fusiform, at maturity plump, more than 1 mm. wide, the body at maturity brownish, appressed-pubescent, the callus rather blunt; awn 2 to 3 cm. long, twice-geniculate. ♃ —Plains and dry slopes, Alberta and Saskatchewan to Wisconsin and Illinois, west to Montana and Arizona; New York (near Rochester); east of the Mississippi, found near railways.

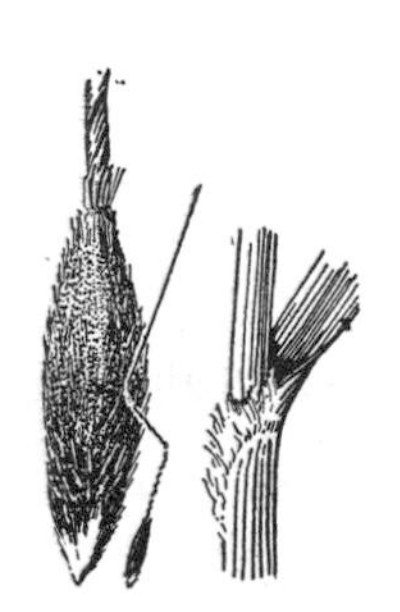

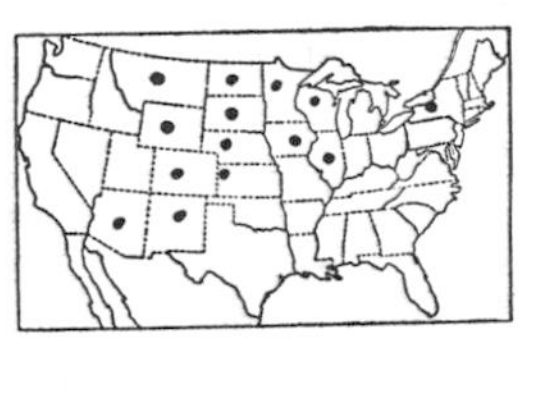

FIGURE 663.—*Stipa viridula*. Floret, × 1; lemma and summit of sheath, × 5. (Griffiths 201, S. Dak.)

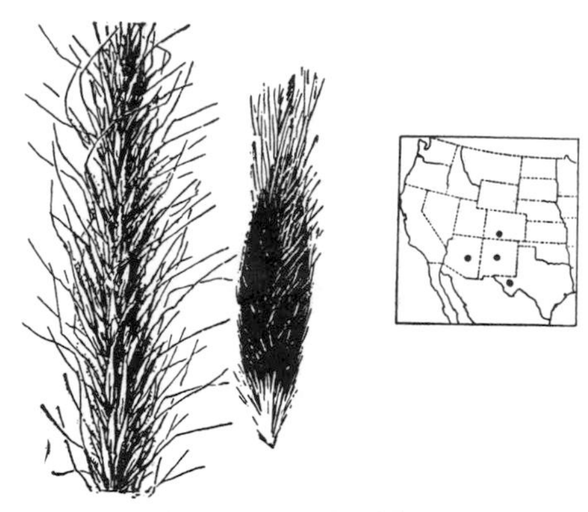

FIGURE 664.—*Stipa robusta*. Panicle, × ½; lemma, × 5. (Hitchcock 13280, N. Mex.)

**27. Stipa robústa** (Vasey) Scribn. SLEEPY GRASS. (Fig. 664.) Culms robust, mostly 1 to 1.5 m. tall; sheaths villous at the throat and on the margin, a strong hispidulous line across the collar; ligule 2 to 4 mm. long; blades elongate, flat or on the innovations involute, those of the culm as much as 8 mm. wide; panicle narrow, compact, often more or less interrupted below, as much as 30 cm. long and 2 cm. thick; glumes about 1 cm. long, attenuate into a fine soft point; lemma 6 to 8 mm. long, about as in *S. viridula;* awn 2 to 3 cm. long, rather obscurely twice-geniculate. ♃ (*S. vaseyi* Scribn.)—Dry plains and hills and dry open woods, Colorado to western Texas, Arizona, and northern Mexico. Said to act as a narcotic on animals that graze upon it, especially affecting horses.

**28. Stipa columbiána** Macoun. COLUMBIA NEEDLEGRASS. (Fig. 665.) Culms mostly 30 to 60 cm. tall, sometimes as much as 1 m.; sheaths naked at the throat; ligule 1 to 2 mm. long; blades 10 to 25 cm. long, 1 to 3 mm. wide, mostly involute, especially on the innovations, those of the culm sometimes flat; panicle 7 to 20 cm. long, narrow, mostly rather dense, often purplish; glumes about 1 cm. long; lemma 6 to 7 mm. long, pubescent as in *S. viridula,* the body narrower, the callus sharper; awn 2 to 2.5 cm. long, twice-geniculate. ♃ (*S. minor* Scribn.)—Dry plains, meadows, and open woods, at medium and high altitudes, South Dakota to Yukon Territory, south to Texas and California. Differing from *S. viridula* in the glabrous throat of the sheath and in the shape of the fruit.

STIPA COLUMBIANA var. NELSÓNI (Scribn.) Hitchc. Differing in its usually larger size, often as much as 1 m. tall, the broader culm blades, and the larger and denser panicle; lemma 6 to 7 mm. long; awn as much as 3.5 cm. long, sometimes longer. ♃ —Alberta to Washington, south to Colorado and Arizona.

FIGURE 665.—*Stipa columbianai.* Panicle, × ½; lemma, × 5. (Nelson 7478, Wyo.)

**29. Stipa lettermáni** Vasey. LETTERMAN NEEDLEGRASS. (Fig. 666.) Resembling small forms of *S. columbiana;* culms often in large tufts, 30 to 60 cm. tall; blades slender, involute; panicle slender, narrow, loose, 10 to 15 cm. long; glumes about 6 mm. long; lemma 4 to 5 mm. long, slender and more copiously hairy than in *S. columbiana;* awn 1.5 to 2 cm. long. ♃ —Open ground or open woods at upper altitudes, Wyoming to Montana and Oregon, south to New Mexico and California.

**30. Stipa williámsii** Scribn. WILLIAMS NEEDLEGRASS. (Fig. 667.) Differing from *S. columbiana* chiefly in having more or less pubescent culms, sheaths, and blades; culms

60 to 100 cm. tall; panicle 10 to 20 cm. long; lemma about 7 mm. long; awn usually 3 to 5 cm. long. ♃ —Dry hills and plains, Montana to Washington, south to Colorado and California.

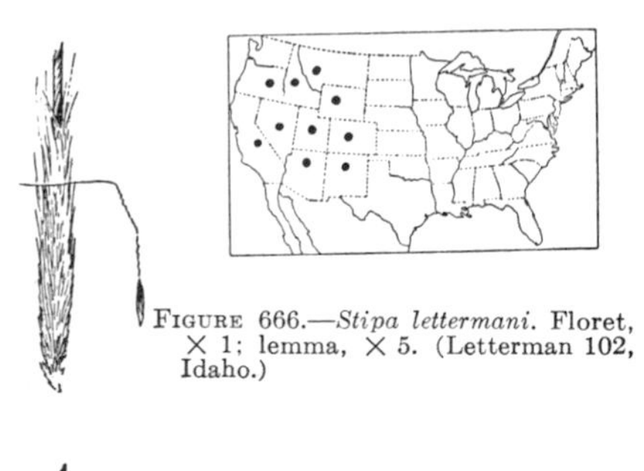

FIGURE 666.—*Stipa lettermani*. Floret, × 1; lemma, × 5. (Letterman 102, Idaho.)

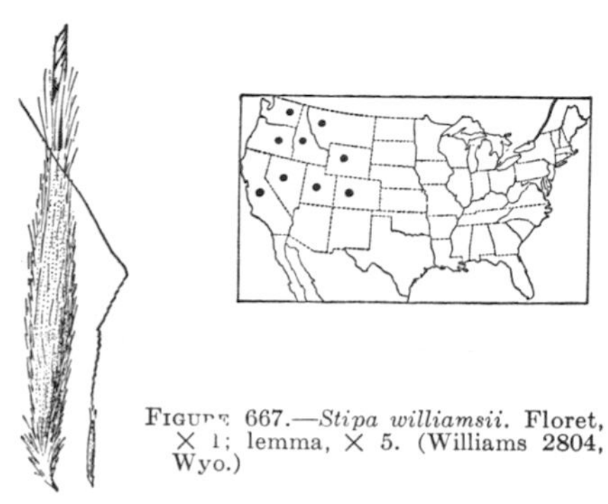

FIGURE 667.—*Stipa williamsii*. Floret, × 1; lemma, × 5. (Williams 2804, Wyo.)

**31. Stipa diegoénsis** Swallen. (Fig. 668.) Culms 70 to 100 cm. tall, scaberulous, densely pubescent below the nodes; sheaths glabrous or scaberulous; ligule 1 to 2 mm. long, obtuse or truncate; blades 15 to 40 cm. long, 2 to 4 mm. wide, flat or involute, scabrous below, pubescent above; panicle 15 to 30 cm. long, dense, the branches appressed; glumes acuminate, the first 9 to 10 mm. long, 1-nerved, the second 8 to 9 mm. long, 3-nerved; lemma 6.5 to 7.5 mm. long, the hairs at the summit 1 to 2 mm. long, the callus 0.5 mm. long, sharp-pointed; awn 2 to 3.3 cm. long, twice-geniculate, scabrous. ♃ —Along streams in chaparral. Known only from San Diego County (Jamul), Calif., and northern Baja California.

**32. Stipa pinetórum** Jones. (Fig. 669.) Culms in large tufts, 30 to 50

FIGURE 668.—*Stipa diegoensis*. Floret, × 1; lemma × 5. (Type.)

cm. tall; ligule very short; leaves mostly basal, the blades 5 to 12 cm. long, involute-filiform, more or less flexuous, slightly scabrous; panicle narrow, 8 to 10 cm. long; glumes about 9 mm. long; lemma 5 mm. long, narrowly fusiform, clothed especially on the upper half with hairs 2 mm. long, forming a conspicuous tuft exceeding the body of the lemma, and bearing 2 hyaline teeth 1 mm. long at the summit; awn about 2 cm. long, twice-geniculate, nearly glabrous. ♃ —Open pine woods at high altitudes, rare, Colorado to Montana, Idaho, and California.

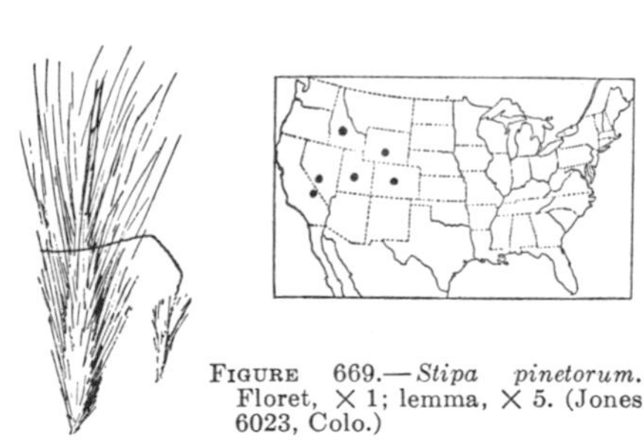

FIGURE 669.—*Stipa pinetorum*. Floret, × 1; lemma, × 5. (Jones 6023, Colo.)

**33. Stipa árida** Jones. (Fig. 670.) Culms 40 to 80 cm. tall; blades 10 to 20 cm. long, 1 to 2 mm. wide, flat or involute, scabrous; panicle 10 to 15 cm. long, narrow, compact, pale or silvery; glumes 8 to 12 mm. long; lemma 4 to 5 mm. long; appressed-pubescent on the lower half and along the margin, slightly roughened toward the summit; awn 4 to 6 cm. long, capillary, scaberulous, loosely twisted for 1 or 2 cm., flexuous beyond. ♃ —Rocky slopes, Texas, Colorado to Arizona and California (Funeral Mountains).

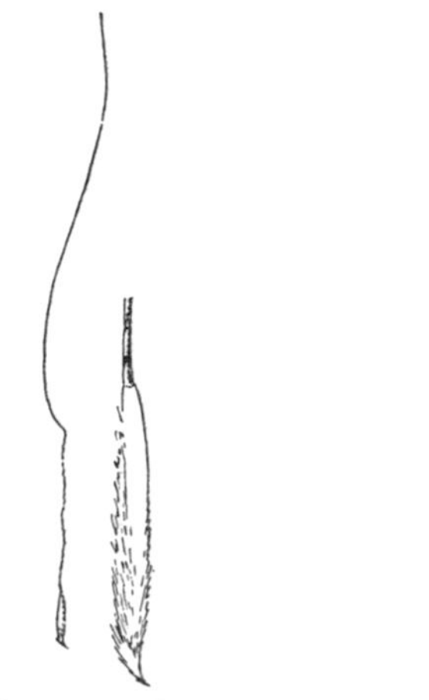

FIGURE 670.—*Stipa arida*. Floret, × 1; lemma, × 5. (Jones 5377, Utah.)

**34. Stipa tenuíssima** Trin. (Fig. 671.) Culms in large tufts, slender, wiry, 30 to 70 cm. tall; ligule 2 mm. long; blades 15 to 30 cm. long, sometimes longer, filiform, wiry, closely involute; panicle 10 to 30 cm. long, narrow, soft, nodding; glumes about 1 cm. long; lemma 2 to 3 mm. long, oblong-elliptic, glabrous, minutely papillose-roughened, the short callus densely pilose; awn about 5 cm. long, capillary, flexuous, obscurely geniculate about the middle. ♃ —Dry open ground, rocky slopes, and open dry woods, Texas and New Mexico to central Mexico; Argentina.

**Stipa neesiána** Trin. and Rupr. Related to *S. leucotricha* but with shorter lemma with thickened erose crown. ♃ —Ballast, Mobile, Ala.; South America.

**Stipa brachychaéta** Godr. Blades firm, flat, or loosely involute; panicle narrow, open, the few spikelets on slender pedicels; glumes 8 mm. long; lemma 5 mm. long, brown, pubescent in lines; awn 12 mm. long. ♃ —Ballast near Portland, Oreg.; Argentina.

Stipa elegantíssima Labill. Tufted perennial; foliage scant; panicle commonly half the height of the plant, the filiform spreading branches conspicuously feathery; spikelets purple, long-awned. ♃ —Sometimes cultivated for ornament; Australia.

Stipa pennáta L. Tufted perennial; blades elongate, involute; panicle few-flowered, the large spikelets with awns 25 to 35 cm. long, conspicuously feathery above the bend. ♃ —Sometimes cultivated for ornament; Europe.

Stipa tenacíssima L. Esparto. Tufted perennial with tough branching base; blades elongate, involute, tomentose at base and with erect auricles 3 to 10 mm. long; panicle narrow, dense; awns 4 to 6 cm. long, feathery below the bend. ♃ —Sometimes cultivated for ornament; Spain and Algeria, where it is gathered for making paper and cordage; also in Portugal and Morocco.

Stipa spléndens Trin. Robust perennial, 1.2 to 2 m. tall; foliage scabrous; panicle 30 to 50 cm. long, many-flowered, but loose; spikelets 5 to 6 mm. long; lemma as long as the glumes, silky; awn weakly geniculate, 10 to 15 mm. long. ♃ —Introduced from Siberia under the name "chee grass," sparingly cultivated. Seed of *Calamagrostis epigeios* was mixed with the first introduction and "chee grass" was erroneously applied to that, which thrived more vigorously than the *Stipa*.

## 92. ARÍSTIDA L. Three-awn

Spikelets 1-flowered, the rachilla disarticulating obliquely above the glumes; glumes equal or unequal, narrow, acute, acuminate, or awn-tipped; lemma indurate, narrow, terete, convolute, with a hard, sharp-pointed, usually minutely bearded callus, terminating above in a usually trifid awn (the lateral divisions reduced or obsolete in Section Streptachne), the base sometimes undivided, forming a column. Annual or perennial, mostly slender tufted grasses, with narrow, frequently convolute blades and narrow or sometimes open panicles. Type species, *Aristida adscensionis* L. Name from Latin *arista*, awn.

The species are of distinctly minor importance for forage except in the Southwest, where several, such as *A. longiseta*, are eaten by stock before the

FIGURE 671.—*Stipa tenuissima*. Plant, × ½; spikelet, × 2; glumes and floret, × 5. (Bailey 694, Tex.)

flowers are produced. The ripe fruits of several species are troublesome to stock on the plains because of the sharp hard points. These fruits are produced sometimes in vast numbers and are carried far and wide by the wind in open country. *Aristida adscensionis* is one of the annuals that make up the "six-weeks" grasses of the Southwest.

Lemma articulate with the column of the awns; awns nearly equal. SECTION 1. ARTHRATHERUM.
Lemma not articulate.
  Lateral awns minute (less than 1 mm. long) or wanting........ SECTION 2. STREPTACHNE.
  Lateral awns more than 1 mm. long (rarely obsolete in *A. ramosissima*), usually well developed........ SECTION 3. CHAETARIA.

*Section 1. Arthratherum*

Plants annual.
  Column very short........ 1. A. DESMANTHA.
  Column 10 to 15 mm. long, twisted........ 2. A. TUBERCULOSA.
Plants perennial.
  Culms pubescent........ 3. A. CALIFORNICA.
  Culms glabrous........ 4. A. GLABRATA.

*Section 2. Streptachne*

Awn (column) twisted at base........ 7. A. ORCUTTIANA.
Awn not twisted.
  Branches of panicle distant, spreading, mostly more than 5 cm. long, naked at base; awn straight or abruptly divergent........ 5. A. TERNIPES.
  Branches of panicle short, approximate, 3 to 5 cm. long, floriferous nearly to base; awn curved and flexuous........ 6. A. FLORIDANA.

*Section 3. Chaetaria*

1a. Central awn spirally coiled at base, the lateral straight. Plants annual. (Group DICHOTOMAE.)
  Lateral awns half to two-thirds as long as the central, somewhat spreading. 8. A. BASIRAMEA.
  Lateral awns much shorter than the central, 1 to 3 mm. long, erect.
    Glumes nearly equal, 6 to 8 mm. long; lemma sparsely appressed-pilose, 5 to 6 mm. long........ 9. A. DICHOTOMA.
    Glumes unequal, the second longer, about 1 cm. long; lemma glabrous except the keel, scabrous toward the apex, about 1 cm. long........ 10. A. CURTISSII.
1b. Central awn not spirally coiled (in a few species all the awns loosely contorted in the lower part).
  2a. Plants annual. (Group ADSCENSIONES.)
    Awns mostly 4 to 7 cm. long, about equal, divergent........ 11. A. OLIGANTHA.
    Awns mostly less than 2 cm. long, often unequal.
      Central awn with a semicircular bend at base, spreading or reflexed.
        Lateral awns much reduced; lemma about 2 cm. long........ 12. A. RAMOSISSIMA.
        Lateral awns one-third to half as long as the central; lemma 4 to 5 mm. long. 13. A. LONGESPICA.
      Central awn not sharply curved, the awns about equally divergent.
        Glumes unequal; awns flat at base, 10 to 15 mm. long........ 14. A. ADSCENSIONIS.
        Glumes about equal; awns terete, 15 to 20 mm. long........ 15. A. INTERMEDIA.
  2b. Plants perennial.
    3a. Panicle open, the branches spreading (in *A. pansa* ascending), naked at base. (Group DIVARICATAE.)
      Panicle branches stiffly and abruptly spreading or reflexed at base.
        Branchlets divaricate and implicate........ 16. A. BARBATA.
        Branchlets appressed.
          Summit of lemma narrowed into a twisted neck 2 to 5 mm. long. 17. A. DIVARICATA.
          Summit of lemma somewhat narrowed but not twisted........ 18. A. HAMULOSA.
      Panicle branches drooping or ascending, not abruptly spreading at base.
        Lateral awns one-fourth to half as long as the central one........ 19. A. PATULA.
        Lateral awns about as long as the central, at least more than half as long. 20. A. PANSA.
    3b. Panicle narrow, the branches ascending or appressed (branches sometimes somewhat spreading in *A. parishii* and *A. purpurea*).
      Column 1 cm. or more long, twisted; glumes awned........ 21. A. SPICIFORMIS.
      Column less than 1 cm. long.

Creeping rhizomes present. Glumes unequal, awned; awns loosely twisted at base, the central a little longer, 18 to 24 mm. long........ 31. A. RHIZOMOPHORA.
Creeping rhizomes wanting (sometimes short ones in *A. stricta*).
4a. First glume about half as long as the second (as much as two-thirds as long in *A. glauca*). (Group PURPUREAE.)
Lemma tapering into a slender somewhat twisted beak 5 to 6 mm. long; awns 1.5 to 2.5 cm. long, widely spreading........ 22. A. GLAUCA.
Lemma beakless or only short-beaked.
Branches of the rather loose and nodding panicle slender and flexuous (see also *A. longiseta* var. *rariflora*).
Lemma about 1 cm. long; awns 3 to 5 cm. long........ 23. A. PURPUREA.
Lemma 7 to 8 mm. long; awns about 2 cm. long.... 24. A. ROEMERIANA.
Branches of the erect panicle stiff and appressed, or the lowermost sometimes somewhat flexuous.
Panicle mostly more than 15 cm. long, the branches several-flowered; awns about 2 cm. long. Sheaths with a villous line across the collar. 25. A. WRIGHTII.
Panicle mostly less than 15 cm. long, the branches few-flowered; awns 2 to several cm. long.
Lemma gradually narrowed above, scaberulous on the upper half; leaves mostly in a short curly cluster at the base of the plant. 27. A. FENDLERIANA.
Lemma scarcely narrowed above, scaberulous only at the tip; leaves not conspicuously basal........ 26. A. LONGISETA.
4b. First glume more than half as long as the second. (Usually the glumes about equal or the first sometimes a little longer.)
Sheaths lanate-pubescent. Panicle branched, somewhat spreading; central awn 1.5 to 2.5 cm. long, spreading or reflexed from a curved base. 28. A. LANOSA.
Sheaths not lanate-pubescent.
Column of awn at maturity 3 to 5 mm. long, distinctly twisted. 29. A. ARIZONICA.
Column of awn less than 3 mm. long, or if so long, not twisted.
Blades villous on upper surface near base, involute...... 30. A. STRICTA.
Blades not involute and villous at base.
Awns at maturity about equally divergent, sometimes slightly twisted but not spirally contorted at base.
Lemma about 7 mm. long; awns horizontally spreading; panicle usually more than 20 cm. long........ 32. A. PURPURASCENS.
Lemma 10 to 12 mm. long; awns somewhat spreading but scarcely horizontal; panicle mostly 10 to 15 cm. long. 33. A. PARISHII.
Awns at maturity unequally divergent or spirally contorted at base.
Awns not spirally contorted at base; central awn more spreading than the others, curved at base, sometimes reflexed.
Lateral awns erect, two-thirds to three-fourths as long as the central.
Glumes about 12 mm. long........ 34. A. AFFINIS.
Glumes about 6 mm. long........ 35. A. VIRGATA.
Lateral awns spreading or reflexed. Panicles nearly simple.
Glumes 6 to 7 mm. long; spikelets mostly in pairs. 36. A. SIMPLICIFLORA.
Glumes about 1 cm. long; spikelets solitary.... 37. A. MOHRII.
Awns spirally contorted at base, spreading.
Blades flat (sometimes subinvolute in *A. condensata*).
Panicle slender, the branches short, rather distant, few-flowered. 38. A. TENUISPICA.
Panicle rather thick, the branches as much as 10 cm. long, rather densely many-flowered........ 39. A. CONDENSATA.
Blades involute........ 40. A. GYRANS.

SECTION 1. ARTHRÁTHERUM (Beauv.) Reichenb.

Lemma articulate with the column of the awns, the latter finally deciduous; glumes 1-nerved; awns nearly equal.

**1. Aristida desmántha** Trin. and Rupr. (Fig. 672.) Annual; branching,

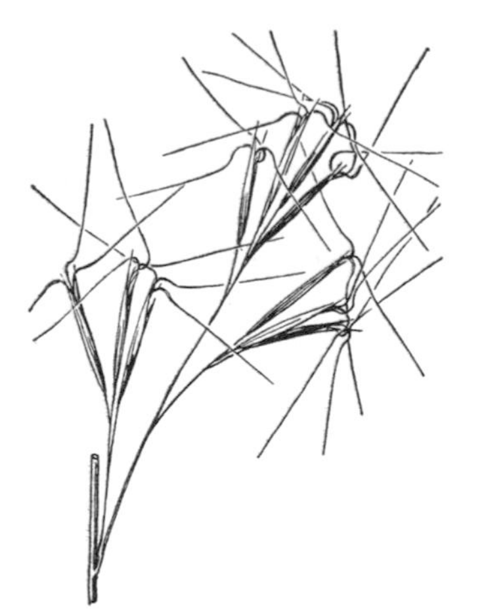

FIGURE 672.—*Aristida desmantha*, × 1. (Reverchon 3428, Tex.)

as much as 80 cm. tall; sheaths often woolly; blades folded or involute, 2 to 3 mm. wide; panicle as much as 20 cm. long, the branches stiffly ascending, very scabrous, bearing 1 to few spikelets; glumes slightly unequal, the body about 1 cm. long, tapering into an awn about half as long; lemma 7 to 8 mm. long, glabrous below,

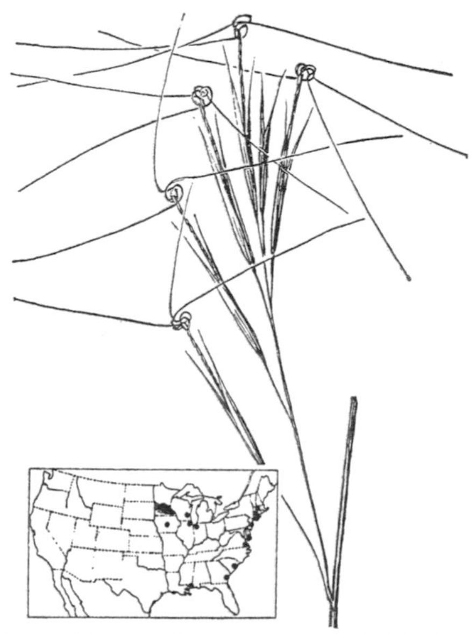

FIGURE 673.—*Aristida tuberculosa*, × 1. (V. H. Chase 322, Ind.)

somewhat laterally compressed and slightly twisted at summit, the densely pubescent callus about 2 mm. long; awns 2 to 2.5 cm. long, united for 1 to 2 mm., the bases curved in a semicircular somewhat contorted bend, the upper part thus usually deflexed. ☉ —Open sandy soil or sandy woods, Illinois, Nebraska, and Texas.

**2. Aristida tuberculósa** Nutt. (Fig. 673.) Annual; culms branching, 30 to 60 cm. or even 1 m. tall; blades involute, 2 to 4 mm. wide when flat; panicle 10 to 20 cm. tall, the branches stiffly ascending; glumes about equal, gradually narrowed into an awn, about 2.5 cm. long, including the awn; lemma 11 to 13 mm. long, glabrous, except for the slightly scabrous summit, extending downward into a densely pubescent callus 3 to 4 mm. long; column of awns twisted, 10 to 15 mm. long, the upper 2 or 3 mm. twisted but not united, above this forming a semicircular bend, the terminal straight part of the awns usually deflexed, 3 to 4 cm. long. ☉ —Open sandy woods, Massachusetts to Georgia and Mississippi near the coast; around the southern end of Lake Michigan and in other localities in Wisconsin, Indiana, Illinois, Iowa, and Minnesota.

**3. Aristida califórnica** Thurb. (Fig. 674.) Perennial, tufted, much branched at base; culms pubescent, 10 to 30 cm. tall; blades mostly involute and less than 5 cm. long; panicles numerous, mostly reduced to few-flowered racemes; first glume about 8 mm. long, the second about 12 mm. long; lemma 5 to 7 mm. long, glabrous below, scaberulous toward the summit, the strongly pubescent callus 1.5 to 2 mm. long; column 15 to 20 mm. long, the awns about equal, 2.5 to 3.5 cm. long, spreading horizontally, the bases arcuate and slightly contorted. ♃ —Dry sandy or gravelly soil, deserts of southern California, southwestern Arizona, and northern Mexico.

**4. Aristida glabráta** (Vasey) Hitchc. (Fig. 675.) Perennial; culms erect, branched, glabrous, 20 to 40 cm. tall; blades mostly involute, those of the culm 1 to 3 cm. long; panicle narrow, 3 to 6 cm. long; first glume 5 to 6 mm., the second 10 to 12 mm. long; lemma 5 to 7 mm. long, the twisted column 6 to 14 mm. long; awns about equal, divergent, 2 to 3 cm. long. ♃ —Open dry ground, southern Arizona to Baja California.

SECTION 2. STREPTÁCHNE (R. Br.) Domin (Sect. *Uniseta* Hitchc.)

Lateral awns minute (less than 1 mm. long) or wanting (see also *A. dichotoma* and *A. ramosissima* of Section Chaetaria); lemma not articulate with the column of the awn.

**5. Aristida térnipes** Cav. SPIDER GRASS. (Fig. 676.) Perennial; culms erect, 50 to 100 cm. tall; blades flat, involute toward the end and tapering into a fine point, as much as 40 cm. long, 2 to 3 mm. wide; panicle open, one-third to half the entire height of the plant, the branches few, distant, spreading, scabrous, mostly naked at base; spikelets appressed at the ends of the branches; glumes about equal, 8 to 10 mm. long; lemma glabrous, often strongly scabrous on the keel, gradually narrowed into a laterally compressed scabrous falcate beak, 1-nerved on each side, this extending into a single straight or divergent scabrous nearly terete awn, the obsolete

FIGURE 675.—*Aristida glabrata*, × 1. (Griffiths 7312, Ariz.)

FIGURE 674.—*Aristida californica*, × 1. (Kearney 3524, Ariz.)

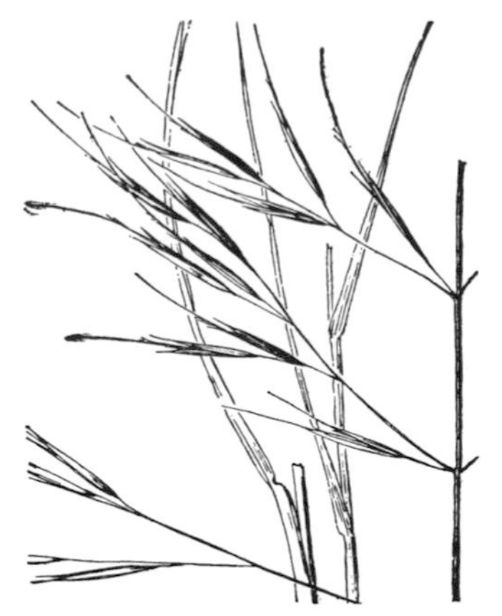

FIGURE 676.—*Aristida ternipes*, × 1. (Griffiths 7271, Ariz.)

or minute lateral awns about 1.7 mm. above the lemma, the central awn 10 to 15 mm. long. ♃ (*A. scabra* Kunth.)—Rocky hills and dry plateaus, Texas, New Mexico, and Arizona to northern South America; Bahamas, Cuba. ARISTIDA TERNIPES var. MÍNOR (Vasey) Hitchc. Smaller and often prostrate or ascending, the panicle usually more than half the length of the entire plant, less diffuse, the shorter branches usually stiffly spreading or somewhat deflexed. ♃ (*A. divergens* Vasey.)—Rocky hills and plains, Texas to Arizona; Nicaragua.

FIGURE 677.—*Aristida floridana*, × 1. (Blodgett, Fla.)

**6. Aristida floridâna** (Chapm.) Vasey. (Fig. 677.) Resembling *A. ternipes*, but differing in having a narrow panicle with ascending branches 3 to 5 cm. long, spikelet-bearing nearly to the base; awns sickle-shaped, the column somewhat twisted. ♃ —Known only from the original collection from Key West, Fla.

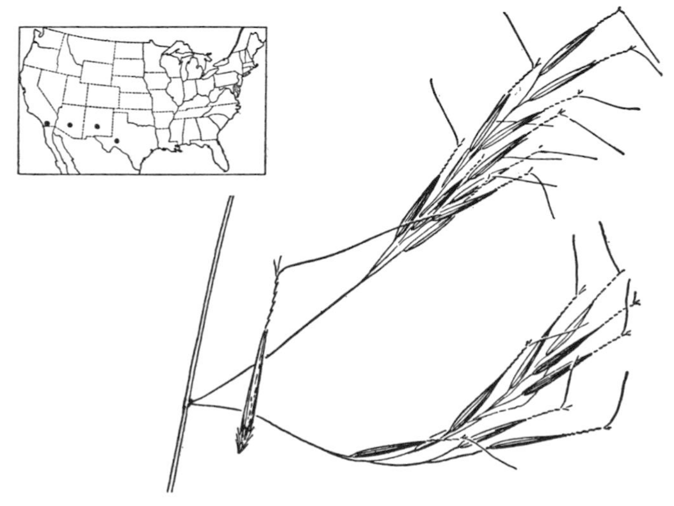

FIGURE 678.—*Aristida orcuttiana*. Panicle, × 1; floret, × 2. (Smith, N. Mex.)

**7. Aristida orcuttiána** Vasey. BEGGARTICK GRASS. (Fig. 678.) Perennial; culms erect, 30 to 60 cm. or even 1 m. tall; blades flat or the upper involute, as much as 3 mm. wide; panicle open, as much as 30 cm. long, nodding or drooping, the branches few, distant, spreading or drooping, as much as 20 cm. long; glumes equal or nearly so, 10 to 15 mm. long; lemma 8 to 10 mm. long, gradually narrowed into a scabrous twisted column, the total length to the bend 10 to 17 mm.; central awn divergent, 5 to 10 mm. long, the lateral awns from obsolete to as much as 1 to 2 mm. long, erect. ♃ —Rocky hills and plains, Texas to southern California (San Diego), and northwestern Mexico.

SECTION 3. CHAETÁRIA (Beauv.) Trin.

Lateral awns more than 1 mm. long, usually well developed; lemma not articulate with the column of the awns.

FIGURE 679.—*Aristida basiramea*, × 1. (Pammel 174, Iowa.)

**8. Aristida basirámea** Engelm. ex Vasey. (Fig. 679.) Annual; branching at base, 30 to 50 cm. tall; blades flat, as much as 15 cm. long and 1.5 mm. wide; panicles terminal and axillary, the terminal 5 to 10 cm. long, the axillary mostly enclosed in the sheaths; glumes somewhat unequal, 12 to 15 mm. long; lemma about 1 cm. long; central awn coiled at base, 10 to 15 mm. long, the lateral awns half to two-thirds as long, somewhat spreading. ⊙ —Open barren or sandy soil, Maine to North Dakota, south to Kentucky, Oklahoma, and Colorado; introduced in Maine.

FIGURE 680.—*Aristida dichotoma*, × 1. (Jackson 1829, Del.)

**9. Aristida dichótoma** Michx. (Fig. 680.) Annual; culms branched at base, 20 to 40 cm. tall; blades short, the lower mostly flat, scarcely 1 mm. wide, the upper involute; panicles terminal and axillary, the terminal usually less than 10 cm. long, the lateral small; glumes about equal, 6 to 8 mm. long; lemma 5 to 6 mm. long; central awn spirally coiled, horizontally bent, 3 to 6 mm. long, the lateral awns erect, about 1 mm. long. ⊙ —Dry open ground, Maine to Wisconsin and eastern Kansas, south to Florida and Texas.

**10. Aristida curtíssii** (A. Gray) Nash. (Fig. 681.) Annual; similar to *A. dichotoma*, differing in the less branching habit, the longer and more

FIGURE 681.—*Aristida curtissii*, × 1. (Waite, Ill.)

conspicuous blades, the looser panicles of larger spikelets, the more unequal glumes, the longer second glume (about 1 cm. long), the longer smooth lemma (about 1 cm. long) and central awn, and the usually longer lateral awns; central awn about 1 cm. long, the lateral awns 2 to 4 mm. long. ⊙ —Open dry ground, Maryland and Virginia to South Dakota, Wyoming, Colorado, and Kentucky to Oklahoma; Florida.

**11. Aristida oligántha** Michx. PRAIRIE THREE-AWN. (Fig. 682.) Annual, much branched; culms 30 to 50 cm. tall; blades flat or loosely involute, usually not more than 1 mm. wide; panicle loose, 10 to 20 cm. long; spikelets short-pediceled, the lower often in pairs; glumes about equal, 2 to 3 cm. long, tapering into an awn, the first 3- to 5-nerved; lemma about 2 cm. long, the awns about equal, divergent, 4 to 7 cm. long, somewhat spirally curved at base. ⊙ —Open dry ground, Massachusetts to South Dakota, south to Florida and Texas; Oregon to Arizona.

**12. Aristida ramosíssima** Engelm. ex A. Gray. (Fig. 683.) Annual, much branched; culms 30 to 50 cm. tall; blades flat or involute, about 1 mm. wide; panicle narrow, 8 to 12 cm. long; glumes 3- to 5-nerved, the first about 15 mm., the second about 2 cm. long, including an awn 3 to 5 mm. long; lemma about 2 cm. long, tapering into a neck about 5 mm. long; central awn with a semicircular bend or part of a coil at base, 15 to 20 mm. long, spreading, the lateral awns reduced or as much as 6 mm. long, rarely longer. ⊙ —Open sterile soil, Indiana to Iowa, south to Tennessee, Louisiana, Oklahoma, and Texas.

**13. Aristida longespíca** Poir. (Fig. 684.) Annual, branched; culms 20 to 40 cm. tall; blades flat or involute, about 1 mm. wide; panicles narrow, slender, the terminal 10 to 15 cm. or even 20 cm. long; glumes about equal, 5 mm. long; lemma 4 to 5 mm. long; central awn sharply curved at base, spreading, 5 to 15 mm. long, the lateral awns erect, one-third to half as long as the central, sometimes only 1 mm. long. ⊙ (*A. gracilis* Ell.)—Sterile or sandy soil, New Hampshire to Michigan and Kansas, south to Florida and Texas, especially on the Coastal Plain. In the typical form the lateral awns are short; in var. *geniculata* Fernald (*A. geniculata* Raf.) the lateral awns are more than one-third as long as the central one.

**14. Aristida adscensiónis** L. SIX-WEEKS THREE-AWN. (Fig. 685.) Annual, branched at base, erect or spreading; culms 10 to 80 cm. tall; panicle narrow and usually rather compact, 5 to 10 cm. long, or longer in large plants; first glume 5 to 7

FIGURE 682.—*Aristida oligantha*. Plant, × ½; glumes and floret, × 2. (Fitzpatrick 21, Iowa.)

FIGURE 683.—*Aristida ramosissima*, × 1. (Deam 18549, Ind.)

mm. long, the second 8 to 10 mm. long; lemma 6 to 9 mm. long, compressed toward the scarcely beaked summit, scabrous on the upper part of the keel; awns about equal (the lateral occasionally shorter), mostly 10 to 15 mm. long, about equally divergent at an angle of as much as 45°, flat and without torsion at base. ☉ —Dry open ground, Missouri (Courtney); southern Kansas to Texas, west to Nevada and southern California, southward to Argentina; a common weed in the American tropics; warmer parts of the Old World. Originally described from Ascension Island. Variable in size from depauperate plants a few centimeters tall with shorter contracted panicle (*A. bromoides* H. B. K.) to tall slender plants with large open panicle (*A. fasciculata* Torr.).

**15. Aristida intermédia** Scribn. and Ball. (Fig. 686.) Annual, simple or branched, 20 to 40 cm. tall; blades flat or involute, mostly less than 10 cm. long and 2 mm. wide; panicle narrow, slender, loosely flowered, 10 to 20 cm. long; glumes about equal, 1 cm. long; lemma 8 mm. long; awns about equal, all somewhat divergent, 1.5 to 2 cm. long. ☉ —Low sandy soil, Indiana and Michigan to Nebraska, south to Florida (Pensacola), Mississippi, and Texas. The measurements of the spikelet are sometimes less than those given, especially in plants attacked by smut.

**16. Aristida barbáta** Fourn. HAVARD THREE-AWN. (Fig. 687.) Perennial, forming hemispherical tufts as much as 30 cm. in diameter, the culms rather stiffly radiating in all directions, 15 to 30 cm. long; blades closely involute, mostly less than 10 cm. long and 0.5 mm. thick; panicles about half the length of the entire plant, open, the branches divaricately spreading or somewhat reflexed, mostly 3 to 6 cm. long, in pairs or with short basal branchlets, but without long naked base, the branchlets and pedicels implicate or flexuous, the whole panicle fragile at maturity, breaking away and rolling before the wind; glumes about equal, 1 cm. long; lemma gradually narrowed into a straight or twisted scaberulous beak, the entire length 8 to 10 mm.; awns somewhat di-

FIGURE 684.—*Aristida longespica*, × 1. (Vasey, D. C.)

FIGURE 685.—*Aristida adscensionis*, × 1. (Earle 559, N. Mex.)

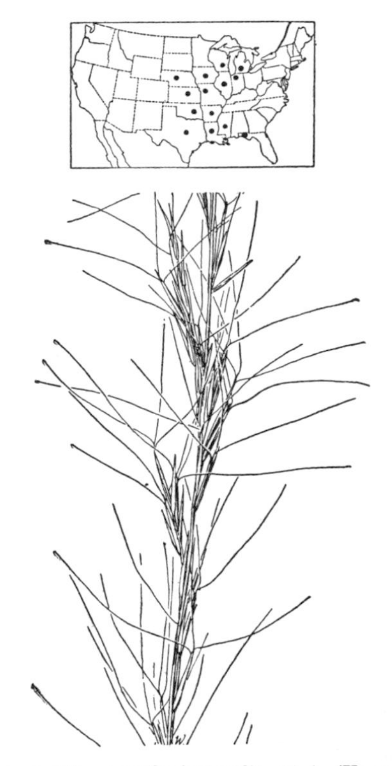

FIGURE 686.—*Aristida intermedia*, × 1. (Kearney 236, Miss.)

vergent, nearly equal, 15 to 20 mm. long. ♃ (*A. havardii* Vasey.)—Hills and plains, western Texas to Arizona and central Mexico.

**17. Aristida divaricáta** Humb. and Bonpl. ex Willd. POVERTY THREE-AWN. (Fig. 688.) Perennial; culms erect or prostrate-spreading, usually 30 to 60 cm. long, sometimes longer; blades flat or usually loosely involute, or the basal closely involute, mostly less than 3 mm. wide; panicle large, diffuse, usually as much as half the entire length of the culm, the branches spreading or reflexed, naked below; glumes nearly equal, 1 cm. long; lemma 1 cm. long, narrowed into a twisted beak 2 to 5 mm. long; awns about equal, 10 to 15 mm. long. ♃ —Dry hills and plains, Kansas to southern California, south to Texas and Guatemala.

**18. Aristida hamulósa** Henr. (Fig. 689.) Resembling *A. divaricata;* lemma somewhat narrowed at summit but not twisted, central awn a little longer than the two lateral ones. ♃ —Dry hills and plains, western Texas to southern California, south

FIGURE 687.—*Aristida barbata*, × 1. (Wooton, N. Mex.)

FIGURE 688.—*Aristida divaricata*, × 1. (Talbot, N. Mex.)

to Guatemala. In Arizona more common than *A. divaricata*.

**19. Aristida pátula** Chapm. ex Nash. (Fig. 690.) Perennial, erect, as much as 1 m. tall; blades flat, becoming involute especially at the slender tip, elongate, 2 to 4 mm. wide; panicle loose and open, one-third to half the entire length of the culm, the branches drooping, naked below, as much as 20 cm. long; glumes 12 to 15 mm. long, nearly equal; lemma 10 to 12 mm. long; central awn straight, 2 to 2.5 cm. long, the lateral scarcely diverging, 5 to 10 mm. long. ♃ —Moist sandy pine barrens and low open ground, peninsular Florida.

**20. Aristida pánsa** Woot. and Standl. WOOTON THREE-AWN. (Fig. 691.) Perennial; culms stiffly erect, slender, wiry, 20 to 40 cm. tall; blades closely involute, 0.5 mm. thick, often flexuous; panicle rather nar-

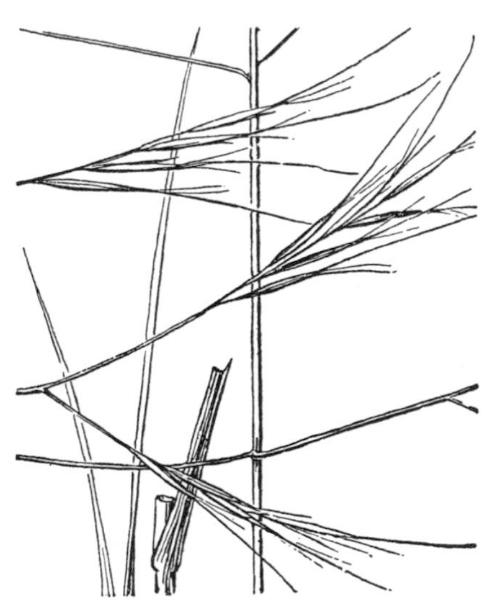

FIGURE 690.—*Aristida patula*, × 1. (Hitchcock, Fla.)

FIGURE 689.—*Aristida hamulosa*. Panicle, × 1; floret, × 3. (Type.)

row, open, rather stiffly upright, 10 to 20 cm. long, the branches stiffly ascending, 4 to 8 cm. long; spikelets erect or narrowly ascending on the branchlets; first glume 5 to 7 mm. long, the second 7 to 10 mm. long; lemma about as long as the second glume, or slightly longer, tapering into a scabrous slightly twisted beak about 2 mm. long; awns about equal,

FIGURE 691.—*Aristida pansa*, × 1. (Wooton, N. Mex.)

divergent or finally nearly horizontally spreading, 10 to 20 mm. long, the bases finally somewhat curved or warped. ♃ —Plains and open ground, western Texas to Arizona; northern Mexico.

**21. Aristida spicifórmis** Ell. (Fig. 692.) Perennial; culms strictly erect, 50 to 100 cm. tall; blades erect, flat or usually involute, elongate, 1 to 3 mm. wide; panicle erect, dense and spikelike, 10 to 15 cm. long, more or less spirally twisted; glumes unequal, abruptly long-awned, the first 4 mm. long, the second 8 to 10 mm. long, the awns usually 10 to 12 mm. long; lemma 5 to 6 mm. long, extending into a slender twisted column 1 to 3 cm. long; awns about equal, 2 to 3 cm. long, divergent or horizontally spreading, more or less curved or warped at base. ♃ —Pine barrens along the coast, South Carolina to

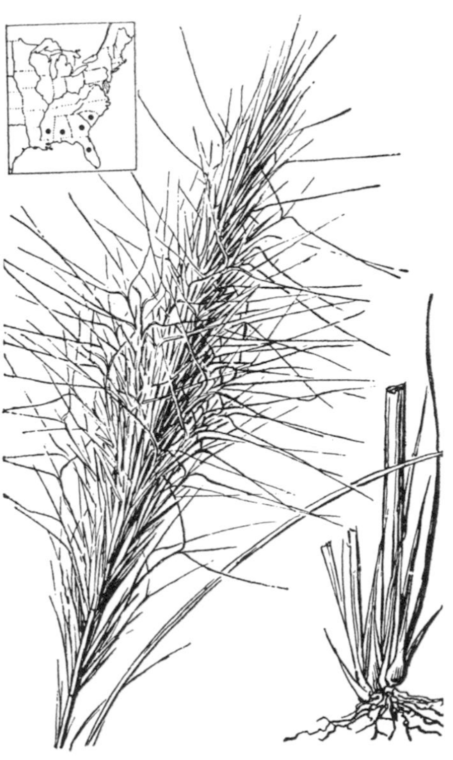

FIGURE 692.—*Aristida spiciformis*, × 1. (Combs and Baker 1115, Fla.)

Florida and Mississippi; Cuba, Puerto Rico.

**22. Aristida glaúca** (Nees) Walp. REVERCHON THREE-AWN. (Fig. 693.) Perennial; culms erect, 20 to 40 cm. tall; blades involute, mostly curved or flexuous, 5 to 10 cm. long, about 1 mm. thick; panicle narrow, erect, rather few-flowered, mostly 8 to 15 cm. long, the branches stiffly appressed; first glume 5 to 8 mm. long, the second about twice as long; lemma 10 to 12 mm. long, tapering into a minutely scabrous, slender, somewhat twisted beak about half the total length of the lemma; awns equal, divergent or horizontally spreading, 1.5 to 2.5 cm. long. ♃ (*A. reverchoni* Vasey.)—Dry or rocky hills and plains, Texas to Utah, Nevada, and southern California, south to Puebla, Mexico.

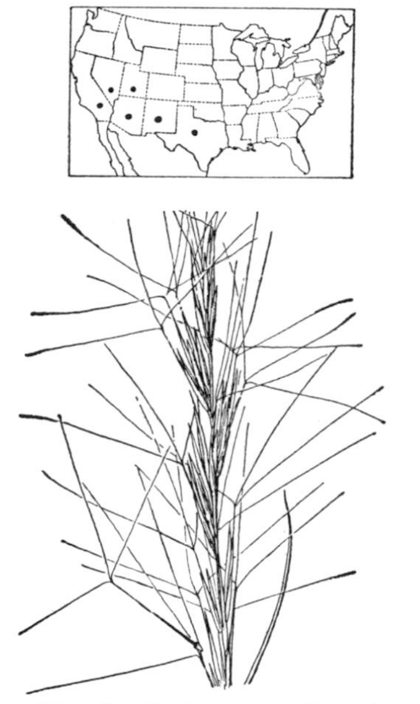

FIGURE 693.—*Aristida glauca*, × 1. (Reverchon 1237, Tex.)

**23. Aristida purpúrea** Nutt. PURPLE THREE-AWN. (Fig. 694.) Perennial, often in large tufts; culms 30 to 50 cm. tall; blades usually involute and less than 10 cm. long, 1 to 1.5 mm. wide when unrolled; panicle narrow, nodding, rather lax and loose, usually purplish, 10 to 20 cm. long, the branches and longer pedicels capillary, more or less curved or flexuous; first glume 6 to 8 mm. long, the second about twice as long; lemma about 1 cm. long, the body tapering to a scarcely beaked summit, tuberculate-scabrous in lines from below the middle to the summit; awns nearly equal, spreading, 3 to 5 cm. long. ♃ —Dry hills and plains, Arkansas and Kansas to Utah and Texas to southern California; northern Mexico. ARISTIDA PURPUREA var. LAXIFLÓRA Merr. Panicle few-flowered, the capillary branches bearing 1 or 2 spikelets. ♃ —Texas to Arizona.

**24. Aristida roemeriána** Scheele. (Fig. 695.) Differing from *A. purpurea* chiefly in the smaller spikelets; first glume 4 to 5 mm. long; lemma 7 to 8 mm. long, the awns about 2 cm. long. ♃ (*A. micrantha* Nash.) —Texas, New Mexico, and northern Mexico.

**25. Aristida wríghtii** Nash. (Fig. 696.) Perennial; culms tufted, erect, 30 to 60 cm. tall; sheaths villous at the throat and with a more or less hispid or villous line across the collar; blades involute, curved or flexuous; panicle erect, narrow, 15 to 20 cm. long; first glume 6 to 7 mm. long, the second about twice as long; lemma 10 to 12 mm. long; awns nearly equal, about 2 cm. long, divergent. ♃ —Dry plains and hills, Oklahoma, Texas, Colorado, and Utah to southern California and central Mexico.

**26. Aristida longiséta** Steud. RED THREE-AWN. (Fig. 697.) Perennial, often in large bunches; culms 20 to 30 cm. tall; blades involute, curved or flexuous, usually less than 15 cm. long; panicle narrow, erect but not stiff, few-flowered, the axis only a few cm. long, the branches ascending or appressed, or the lower more or less curved or flexuous; first glume 8 to 10 mm. long, the second about twice as long; lemma terete, 12 to 15 mm. long, only slightly narrowed above, glabrous or the upper part scaberulous but scarcely tuberculate-scabrous in lines as in *A. purpurea;* awns about equal, divergent, 6 to 8 cm. long. ♃ —Plains and foothills, North Dakota and Iowa to Montana and British Columbia, south to Texas, Arizona, and northern Mexico. ARISTIDA LONGISETA var. RARIFLÓRA Hitchc. Differing in the few-flowered panicles with capillary flexuous branches bearing 1 or 2 spikelets.

FIGURE 694.—*Aristida purpurea*, × 1. (Bush 665, Tex.)

FIGURE 695.—*Aristida roemeriana*, × 1. (Swallen 1585, Tex.)

FIGURE 696.—*Aristida wrightii*, × 1. (Ball 1511, Tex.)

♃ —Texas to Colorado and Arizona.

ARISTIDA LONGISETA var. ROBÚSTA Merr. Taller and more robust, 30 to 50 cm. tall, the blades longer and not in conspicuous basal tufts, the panicle longer, stiffer, and the branches more stiffly ascending, the awns mostly 4 to 5 cm. long. ♃ —Same range but more common northward, extending east to Minnesota and west to Washington and California.

**27. Aristida fendleriána** Steud. FENDLER THREE-AWN. (Fig. 698.) Resembling *A. longiseta;* differing in the numerous short curly blades at the base of the plant, the shorter glumes (the first about 7 mm. long), the gradually narrowed lemma, scaberulous on the upper half, and the shorter awns (2 to 5 cm. long). ♃ —Dry plains and hills, North Dakota and Montana, south to Texas, Nevada, and southern California; Mexico.

**28. Aristida lanósa** Muhl. ex Ell. (Fig. 699.) Perennial; culms solitary or few in a tuft, rather robust, 1 to 1.5 m. tall; sheaths lanate-pubescent or rarely glabrous; blades flat, elon-

FIGURE 697.—*Aristida longiseta*, × 1. (Thompson 63, Kans.)

gate, as much as 4 mm. wide; panicle narrow, rather loose, as much as 40 cm. long; first glume 12 to 14 mm. long, the second about 10 mm.; lemma 8 to 9 mm. long; central awn horizontally spreading or reflexed from a curved base, 1.5 to 3 cm. long, the lateral half to two-thirds as long, erect or spreading. ♃ —Dry sandy soil of the Coastal Plain, New Jersey and West Virginia to Florida and Texas; Tennessee; Oklahoma and Missouri. A slender form 65 cm. tall, with fewer-flowered panicle, the lemma 10 mm. long, the central awn 2.5 to 3 cm. long, has been differentiated as *A. lanosa* var. *macera* Fern. and Grisc. ♃ —Cape Henry, Va.

FIGURE 699.—*Aristida lanosa*, × 1. (Canby, Md.)

FIGURE 698.—*Aristida fendleriana*, × 1. (Coville 1089, Ariz.)

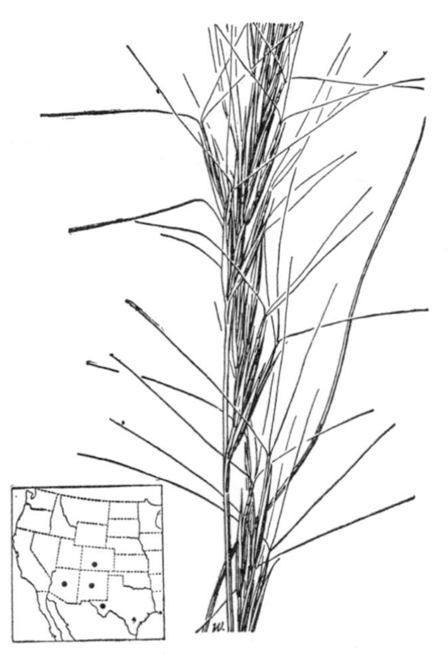

FIGURE 700.—*Aristida arizonica*, × 1. (Rusby, Ariz.)

**29. Aristida arizónica** Vasey. ARIZONA THREE-AWN. (Fig. 700.) Perennial; culms erect, 30 to 120 cm. tall; blades flat, narrowed to a fine involute point or some of them involute throughout, 1 to 4 mm. wide, the old ones usually curled or flexuous; panicle narrow, erect, closely flowered or more or less interrupted at base, 10 to 25 cm. long; glumes equal or nearly so, awn-pointed, 10 to 15 mm. long; lemma 1 to 1.5 cm. long, including the more or less twisted beak of about 3 to 5 mm.; awns about equal, ascending, 1 to 2 cm. long. ♃ —Dry plains, stony hillsides, and open forest, mostly at 1,500 to 2,500 m. altitude, southern Colorado and western Texas to Arizona.

FIGURE 701.—*Aristida stricta*, × 1. (Chase 4565, N. C.)

**30. Aristida strícta** Michx. PINELAND THREE-AWN. (Fig. 701.) Perennial; culms erect, 50 to 100 cm. tall; blades closely involute, villous on the upper surface above the base (the hairs visible without unrolling the blade), elongate, 1 mm. thick; panicle slender, as much as 30 cm. long; glumes about equal, 7 to 10 mm. long; lemma 6 to 8 mm. long, scarcely beaked; awns divergent, the central 1 to 1.5 cm. long, the lateral a little shorter. ♃ —Common in pine barrens, North Carolina to Florida, west to Mississippi.

**31. Aristida rhizomóphora** Swallen. (Fig. 702.) Perennial; culms tufted, erect, 65 to 80 cm. tall, producing well-developed scaly rhizomes; blades firm, flat or folded, 7 to 10 cm. long, 1 to 2 mm. wide, those of the innovations flexuous, as much as 30 cm. long; panicle flexuous, 20 to 30 cm. long, the distant branches somewhat spreading, few-flowered, spikelet-bearing from near the base; glumes acuminate, usually awned, the first 8 to 14 mm. long, the second 12 to 17 mm. long (including the awn); lemma 9 to 12 mm. long, the callus 1 mm. long, the awns flexuous, curved or loosely twisted at base, spreading, the central often reflexed by a semicircular bend, 18 to 28 mm. long, the lateral 15 to 20 mm. long. ♃ —Prairies, peninsular Florida.

**32. Aristida purpuráscens** Poir. ARROWFEATHER. (Fig. 703.) Perennial; culms tufted from a rather thin, weak, sometimes decumbent base, slender, 40 to 70 cm. or even 1 m. tall; blades flat, rather lax and flexuous (especially the old ones), usually less than 2 mm. wide; panicle narrow, rather lax and nodding, one-third to half the entire length of the plant; glumes about equal, mostly 8 to 12 mm. long; lemma about 7 mm. long; awns about equal, divergent or somewhat reflexed, 1.5 to 2.5 cm. long. ♃ —Dry sandy soil, Massachusetts to Wisconsin and Kansas, south to Florida and Texas; British Honduras.

**33. Aristida paríshii** Hitchc. (Fig. 704.) Perennial; culms erect, 30 to 50 cm. tall; blades more or less involute, sometimes flat, 1 to 2 mm. wide; panicle narrow, 15 to 30 cm. long; glumes short-awned, the first 12 mm. long, the second 1 or 2 mm. longer; lemma about 12 mm. long, tapering into a short, straight or obscurely twisted beak; awns about equal, divergent, about 2.5 cm. long. ♃ —

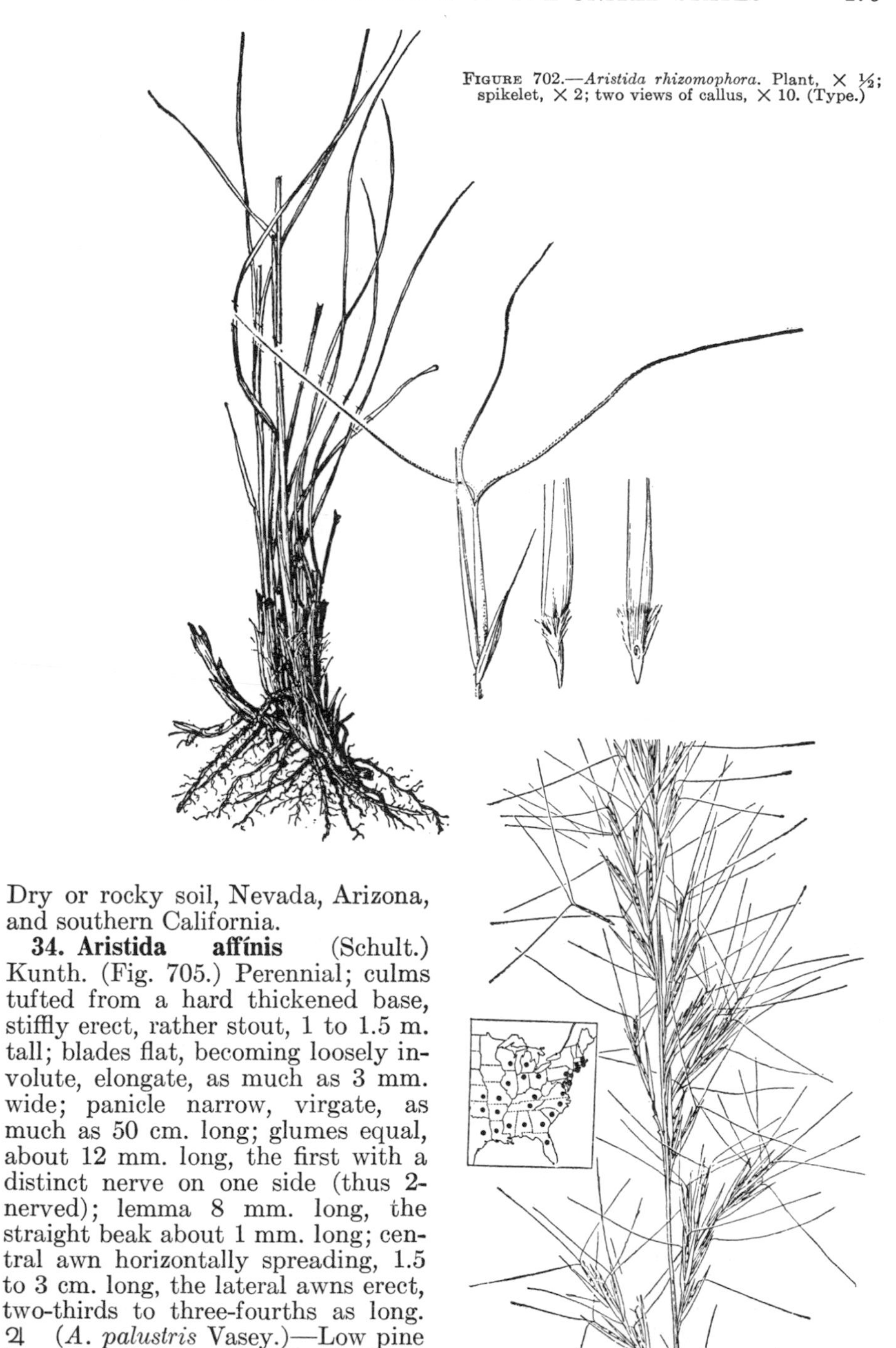

FIGURE 702.—*Aristida rhizomophora.* Plant, × ½; spikelet, × 2; two views of callus, × 10. (Type.)

FIGURE 703.—*Aristida purpurascens,* × 1. (Chase 4563, N. C.)

Dry or rocky soil, Nevada, Arizona, and southern California.

**34. Aristida affínis** (Schult.) Kunth. (Fig. 705.) Perennial; culms tufted from a hard thickened base, stiffly erect, rather stout, 1 to 1.5 m. tall; blades flat, becoming loosely involute, elongate, as much as 3 mm. wide; panicle narrow, virgate, as much as 50 cm. long; glumes equal, about 12 mm. long, the first with a distinct nerve on one side (thus 2-nerved); lemma 8 mm. long, the straight beak about 1 mm. long; central awn horizontally spreading, 1.5 to 3 cm. long, the lateral awns erect, two-thirds to three-fourths as long. ♃ (*A. palustris* Vasey.)—Low pine barrens and flatwoods, North Carolina and Kentucky to Florida and Texas, mostly on the Coastal Plain.

FIGURE 704.—*Aristida parishii*, × 1. (Parish 1029A, Calif.)

**35. Aristida virgáta** Trin. (Fig. 706.) Perennial; culms tufted from a rather slender soft base, erect, 50 to 80 cm. tall; blades flat, rather lax, usually not more than 2 mm. wide; panicle slender, erect, though not very stiff, rather loosely flowered, one-third to half the entire length of the culm; glumes about equal, 6 to 7 mm. long; lemma 4 to 5 mm. long; central awn horizontally spreading or somewhat reflexed, 1.5 to 2 cm. long, the lateral awns erect, about two-thirds as long as the central. ♃ (*A. chapmaniana* Nash.)—Moist sandy soil of the Coastal Plain, New Jersey to Florida and Texas.

**36. Aristida simpliciflóra** Chapm. (Fig. 707.) Perennial; culms erect from a rather delicate base, slender, 30 to 60 cm. tall; blades flat, 5 to 15 cm. long, 1 mm. wide; panicle slender, somewhat nodding, 10 to 20 cm. long, few-flowered, the spikelets mostly in pairs; glumes equal, 6 to 7 mm. long; lemma a little shorter than the glumes; central awn finally reflexed by a semicircular bend, 1 to 1.5 cm. long, the lateral awns horizontally spreading, a little shorter than the central one. ♃ —Moist pine woods, rare, western Florida; Mississippi (McNeill).

**37. Aristida móhrii** Nash. (Fig. 708.) Perennial; culms erect, 40 to 60 cm. tall; blades flat or those of the innovations involute, 10 to 15 cm. long, 1 to 2 mm. wide, the uppermost reduced; panicle slender, strict, as much as 30 cm. long; spikelets solitary, appressed, distant, even the upper not overlapping; glumes equal, firm, rather broad toward the mucronate apex, 1 cm. long; lemma terete, a little shorter than the glumes; awns divergent, the central one reflexed by a semicircular bend near the base, 1.5 to 2 cm. long, the lateral ones scarcely shorter than the central, horizontally spreading or re-

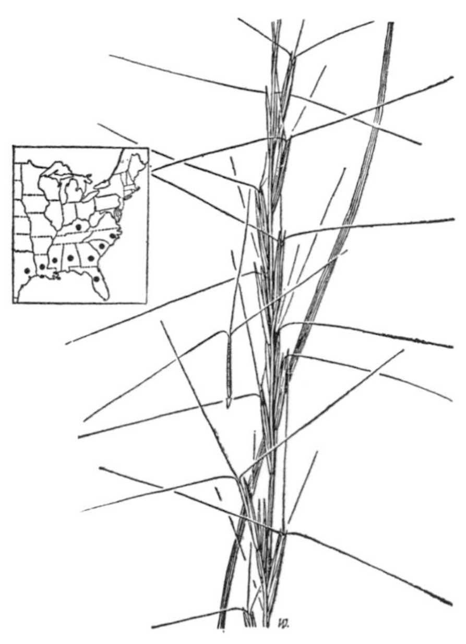

FIGURE 705.—*Aristida affinis*, × 1. (Combs 688, Fla.)

flexed. ♃ —Sterile soil, South Carolina, Florida, and Alabama.

**38. Aristida tenuispíca** Hitchc. (Fig. 709.) Perennial; culms slender, 60 to 100 cm. tall; blades flat, 10 to 20 cm. long, 1 to 2 mm. wide, bearing scattered long hairs on the upper surface; panicle slender, about half the entire length of the culm; glumes nearly equal, about 8 mm. long; lemma 7 mm. long including a 1-mm. long beak; awns equal, 12 to 15 mm. long, spreading or reflexed, somewhat spirally contorted at base. ♃ —Low pine barrens, peninsular Florida; British Honduras.

**39. Aristida condensáta** Chapm. (Fig. 710.) Perennial; culms rather robust, a meter or more tall; lower sheaths usually appressed pubescent; blades firm, flat, becoming involute, elongate, 2 to 3 mm. wide; panicle narrow, as much as 30 cm. long, the branches 5 to 12 cm. long, ascending, closely flowered; glumes equal, 8 to 9 mm. long; awns equal, divergent, 10 to 15 mm. long, the base more or less contorted, finally forming a loose spiral. ♃ —Sandy pine or oak barrens, North Carolina, Georgia, Florida, and Alabama, on the Coastal Plain. Specimens with glabrous lower sheaths have been differentiated as *A. condensata* var. *combsii* (Scribn. and Ball) Heur.

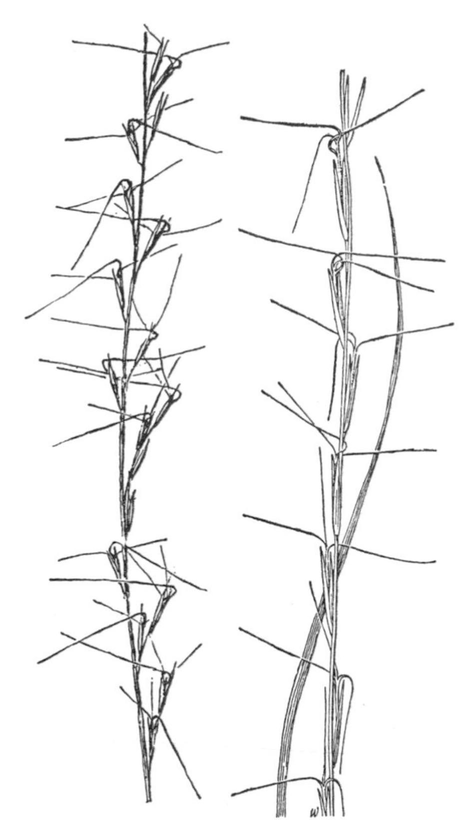

FIGURE 707.—*Aristida simpliciflora*, × 1. (Chapman, Fla.)

FIGURE 708.—*Aristida mohrii*, × 1. (Mohr 53, Ala.)

FIGURE 706.—*Aristida virgata*, × 1. (Tracy 4667, Miss.)

**40. Aristida gýrans** Chapm. (Fig. 711.) Perennial; culms erect, slender, 40 to 70 cm. tall; blades involute, 10 to 15 cm. long, 1 mm. wide; panicle slender, rather lax, 15 to 30 cm. long, the branches appressed, not at all or only slightly overlapping, bearing mostly 1 to 3 spikelets; first glume 7 to 8 mm. long, the second 10 to 11 mm. long; lemma about 6 mm. long, the callus 1.5 mm. long, sharp; awns equal, divergent, 1 to 1.5 cm. long, about equally contorted at base in a loose spiral. ♃ —Dry sandy soil, Georgia and Florida.

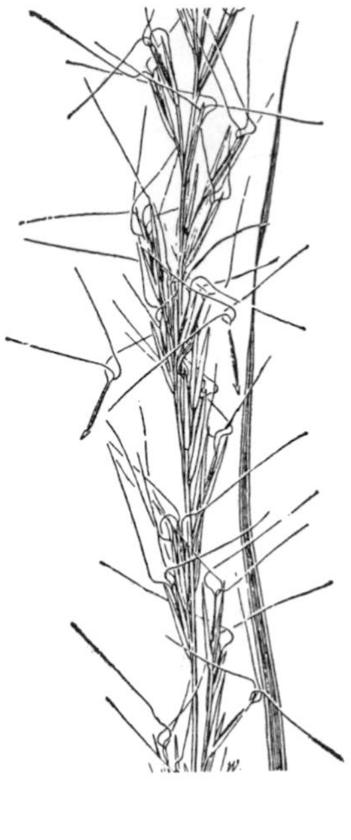

FIGURE 709.—*Aristida tenuispica*, × 1. (Tracy 7104, Fla.)

FIGURE 711.—*Aristida gyrans*, × 1. (Combs 1289, Fla.)

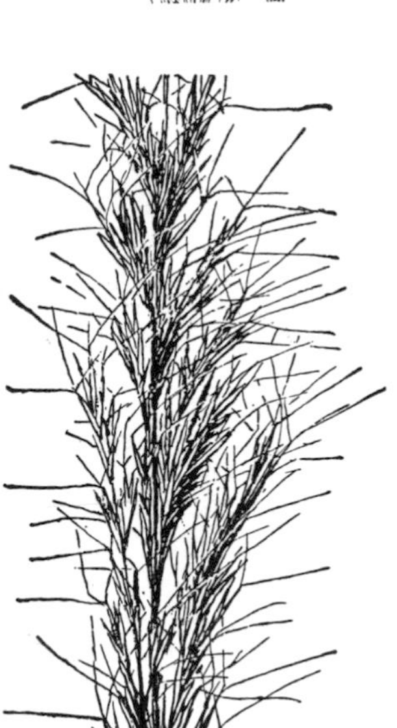

FIGURE 710.—*Aristida condensata*, × 1. (Chapman, Fla.)

## TRIBE 6. ZOYSIEAE

### 93. TRÁGUS Hall.

(*Nazia* Adans.)

Spikelets 1-flowered, in small spikes of 2 to 5, the spikes subsessile, falling entire, the spikelets sessile on a very short zigzag rachis, the first glumes small, thin, or wanting, appressed to the rachis, the second glumes of the 2 lower spikelets strongly convex with 3 thick nerves bearing a row of squarrose, stout hooked prickles along each side, the 2 second glumes forming the halves of a little bur, the upper 1 to 3 spikelets reduced and sterile; lemma and palea thin, the lemma flat, the palea strongly convex. Low annuals, with flat blades and terminal inflorescence, the burs or spikes rather closely arranged along an elongate, slender axis. Type species, *Tragus racemosus*. Name from Greek *tragos*, he-goat, applied by Plinius to a plant.

Spikelets 2 to 3 mm. long, the apex scarcely projecting beyond the spines, the bur nearly sessile........ 1. T. BERTERONIANUS.
Spikelets 4 to 4.5 mm. long, the acuminate apex projecting beyond the spines, the bur pediceled........ 2. T. RACEMOSUS.

**1. Tragus berteroniánus** Schult. (Fig. 712.) Culms branched at base, spreading, 10 to 40 cm. long; blades firm, mostly less than 5 cm. long, 2 to 4 mm. wide, the cartilaginous margin bearing stiff white hairs or short slender teeth; raceme dense, 4 to 10 cm. long, 4 to 5 mm. thick; burs 2 to 3 mm. long, nearly sessile, the apex scarcely exceeding the spines. ⊙ (The name *Nazia aliena* Scribn. has been erroneously applied to the species.)—Dry open ground, probably introduced, Texas to Arizona, south to Argentina; also in the warmer parts of the Old World; on ballast at Boston and on wool waste in Maine.

FIGURE 712.—*Tragus berteronianus*. Plant, × ½; bur and spikelet, × 5. (Hitchcock 3745, N. Mex.)

FIGURE 713.—*Tragus racemosus*, × 1. (Griffiths 1529, Ariz.)

**2. Tragus racemósus** (L.) All. (Fig. 713.) Differing from *T. berteronianus* in the larger burs, the spikelets 4 to 4.5 mm. long, in the acuminate apex projecting beyond the spines, and in the pediceled burs. ⊙ (*Nazia racemosa* Kuntze.)—Waste ground and on ballast at a few places from Maine to North Carolina; Texas to Arizona; introduced from the Old World.

## ANTHÉPHORA Schreb.

Spikelets with 1 perfect floret and a sterile lemma below, in clusters of 4, the indurate first glumes united at base, forming a pitcher-shaped pseudo-involucre, the clusters subsessile and erect on a slender, flexuous, continuous axis, deciduous at maturity. Type species, *Anthephora elegans* Schreb. (*A. hermaphrodita*). Name from *anthe*, blossom, and *pherein*, to bear.

**Anthephora hermaphrodíta** (L.) Kuntze. Leafy ascending or decumbent annual; culms mostly 20 to 50 cm. tall; blades flat, thin, 5 to 10 mm. wide; spikes erect, 5 to 10 cm. long; first glume 5 to 7 mm. long, about 9-nerved; second glume narrow, acuminate, shorter than the first, pubescent; sterile lemma 5-nerved, about as long as the fertile floret. ⊙ —Escaped from experiment station plots, Florida (Gainesville); a common weed in tropical America.

## 94. ZOÝSIA Willd.

(*Osterdamia* Neck.)

Spikelets 1-flowered, laterally compressed, appressed flatwise against the slender rachis, glabrous, disarticulating below the glumes; first glume wanting; second glume coriaceous, mucronate, or short-awned, completely infolding the thin lemma and palea, the palea sometimes obsolete. Low perennials, with creeping rhizomes, short, pungently pointed blades, and terminal spikelike racemes, the spikelets on short appressed pedicels. Type species, *Zoysia pungens* Willd. Named for Karl von Zois.

Several years ago a species of this genus was introduced into the United States as a lawngrass under the names Korean lawngrass and Japanese lawngrass. It was recommended for the Southern States and was said to be hardy as far north as Connecticut. The species then introduced appears to be *Zoysia japonica* Steud. Recently a fine-leaved species, *Zoysia tenuifolia* Willd. ex Trin. (Mascarene grass), has been introduced in Florida and southern California (called in the latter region Korean velvet grass) and has given favorable results. These species may escape from cultivation. The original species, *Z. matrella* (L.) Merr. (*Z. pungens* Willd.), Manila grass (fig. 714.) common in the Philippine Islands, has been used in recent years for lawns from the Gulf States to Long Island, propagated by cuttings. The spikelets are about 2.5 mm. long and 0.8 mm. wide. But little seed is produced. Sometimes called "Flawn."

In *Z. japonica* (Japanese lawngrass) the blades are flat and rather stiff, 2 to 4 mm. wide, the spikelets about 3 mm. long and a little more

FIGURE 714.—*Zoysia matrella.* Plant, × ½; spikelet and floret, × 10. (Whitford 1303, P. I.)

than 1 mm. wide. The rhizomes are underground. In *Z. tenuifolia* the blades are involute-capillary, the spikelets much narrower than in *Z. japonica,* and the stolons are at or near the surface of the soil.

## 95. HILÁRIA H. B. K.

Spikelets sessile, in groups of 3, the groups falling from the axis entire, the central spikelet (next the axis) fertile, 1-flowered (occasionally 2-flowered), the 2 lateral spikelets staminate, 2-flowered (occasionally 3-flowered); glumes coriaceous, those of the 3 spikelets forming a false involucre, in some species connate at the base, more or less asymmetric, usually bearing an awn on one side from about the middle (extension of the midnerve of the asymmetric glume); lemma and palea hyaline, about equal in length. Perennials, with stiff, solid culms and narrow blades, the groups of spikelets appressed to the axis, in terminal spikes. Type species, *Hilaria cenchroides* H. B. K. Named for Auguste St. Hilaire.

All the species are important range grasses and resist close grazing. Curly mesquite is the dominant "short grass" of the Texas plains. The larger species are well known on the range in the arid and semiarid regions of the Southwest.

Culms white felty-pubescent ........ 5. H. RIGIDA.
Culms not felty-pubescent.
Cluster of spikelets not flabellate; glumes of lateral spikelets narrowed toward summit. 4. H. JAMESII.
Cluster of spikelets flabellate; glumes (at least the outer one) of lateral spikelets broadest toward summit.
Glumes subhyaline and fimbriate at summit; plants tufted, not stoloniferous. 3. H. MUTICA.
Glumes firm, not fimbriate; plants stoloniferous (except in *H. belangeri* var. *longifolia*).
Glumes of lateral spikelets much shorter than the florets, pale; group of spikelets mostly 5 mm. long ........ 1. H. BELANGERI.
Glumes of lateral spikelets about equaling the florets, blackish; group of spikelets 7 to 8 mm. long ........ 2. H. SWALLENI.

**1. Hilaria belangéri** (Steud.) Nash. CURLY MESQUITE. (Fig. 715.) Plants in tufts, sending out slender stolons, these producing new tufts, the internodes of the stolons wiry, 5 to 20

FIGURE 715.—*Hilaria belangeri*. Plant, × ½; two views of group of spikelets, × 5; fertile spikelet, staminate spikelet, and fertile floret, × 5. (Hitchcock, Tex.)

cm. long; culms erect, slender, 10 to 30 cm. tall, villous at the nodes; blades flat, 1 to 2 mm. wide, scabrous, more or less pilose, usually short, crowded at base, often forming a curly tuft, but sometimes longer and erect; spike usually 2 to 3 cm. long, with mostly 4 to 8 clusters of spikelets, the axis flat, the internodes alternately curved, 3 to 5 mm. long; group of spikelets 5 to 6 mm. long; lateral spikelets attenuate at base, the glumes united below, firm, scabrous, the outer lobe broadened upward, 2- to 3-nerved, the inner much reduced, the midnerve of both glumes extending into short awns, the first glume smaller, the lateral nerves sometimes excurrent into awns or teeth (the glumes variable in a single spike); fertile spikelet usually shorter than the sterile, rounded at base; glumes firm with deeply lobed thinner upper part, the midnerves extending into awns mostly exceeding the staminate spikelets; lemma compressed, narrowed above, awnless ♃ (*H. texana* Nash.)—Mesas and plains, Texas to Arizona and northern Mexico. *H. cenchroides* H. B. K., to which this species has commonly been referred, is confined to Mexico. H. BELANGERI var. LONGIFÓLIA (Vasey) Hitchc. Stolons wanting; blades elongate. ♃ —Arizona and Sonora.

**2. Hilaria swálleni** Cory. (Fig. 716.) Resembling *H. belangeri*, culms to 35 cm. tall; blades usually 2 mm. wide, scarcely curled; spike 2 to 4.5 cm. long, with 3 to 8 clusters of spikelets, the internodes of the flat axis 4 to 6 mm. long; glumes of lateral spikelets similar, oblong, narrowed at base, about equaling the florets, firm and strongly pigmented except toward the summit, the nerves often rather obscure; awns of all glumes slightly longer than those of the preceding; fertile spikelet about equaling the sterile, the fertile floret slightly larger than in *H. belangeri*. ♃ —Mesas and rocky plains, western Texas and northern Mexico. Said to be better forage than *H. belangeri*.

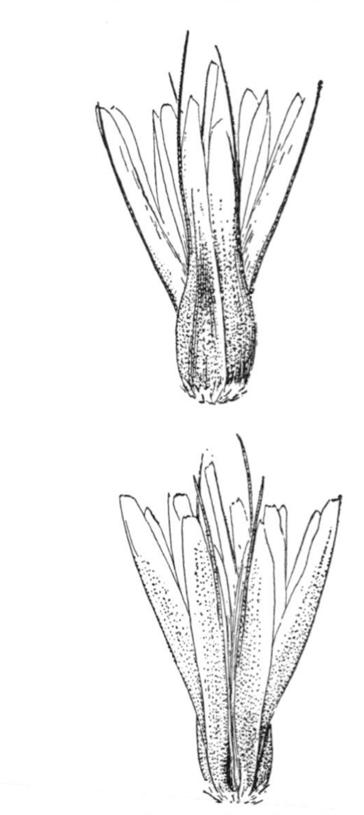

FIGURE 716.—*Hilaria swalleni*. Two views of group of spikelets, × 5. (Young 46, Tex.)

FIGURE 717.—*Hilaria mutica*, × 1. (Toumey, Ariz.)

**3. Hilaria mútica** (Buckl.) Benth. TOBOSA GRASS. (Fig. 717.) Culms from a tough rhizomatous base, 30

FIGURE 718.—*Hilaria jamesii*. Plant, × ½; single spike, × 1; group of spikelets, two views (*A*), × 5; fertile spikelet (*B*), staminate spikelet (*C*), and fertile floret (*D*), × 5. (Tidestrom 1449, Utah.)

to 60 cm. tall, glabrous, the nodes pubescent; blades flat or somewhat involute, rather rigid, 2 to 3 mm. wide; spikes 4 to 6 cm. long; group of spikelets about 7 mm. long; bearded at base; glumes of lateral spikelets very unsymmetrical, widened toward the ciliate summit, the nerves flabellate, not excurrent or barely so; fertile spikelet about equaling the lateral ones, its glumes strongly keeled, cleft into few to several narrow ciliate lobes and slender awns; lemma exceeding the glumes, mucronate between 2 rounded lobes. ♃ (*Pleuraphis mutica* Buckl.)—Dry plains and hills, Texas to Arizona and northern Mexico.

**4. Hilaria jamésii** (Torr.) Benth. GALLETA. (Fig. 718.) Plants erect, the base often decumbent or rhizomatous, bearing also tough scaly rhizomes; culms glabrous, the nodes villous; sheaths glabrous or slightly scabrous, sparingly villous around the short membranaceous ligule; blades mostly 2 to 5 cm. long, 2 to 4 mm. wide, rigid, soon involute, the upper reduced; group of spikelets 6 to 8 mm. long, long-villous at base, similar to those of *H. rigida*, but the glumes of lateral spikelets acute, usually with a single awn; lemma of the fertile spikelet exceeding its glumes. ♃ (*Pleuraphis jamesii* Torr.)—Deserts, canyons, and dry plains, Wyoming and Utah to Texas and Inyo County, Calif.

**5. Hilaria rígida** (Thurb.) Benth. ex Scribn. BIG GALLETA. (Fig. 719.) Plants rather robust at base, branching, the branches mostly erect or ascending, the base rather woody, decumbent or rhizomatous; culms numerous, rigid, felty-pubescent, glabrate and scabrous above, 50 to 100 cm. tall; leaves felty or glabrous, usually woolly at the top of the sheath; blades spreading, 2 to 5 cm. long, or longer on sterile shoots, 2 to 4 mm. wide, more or less involute, acuminate into a rigid coriaceous point; group of spikelets about 8 mm. long, densely bearded at base; glumes of lateral spikelets thin, long-ciliate, about 7-nerved, usually 2- to 4-lobed at the broad summit and with 1 to 3 nerves excurrent into slender awns, nerves sometimes obscure and scarcely excurrent (variable in a single spike); fertile spikelet about equaling the lateral ones, its narrow glumes deeply cleft into few to several acuminate ciliate lobes and slender awns; lemma scarcely exceeding the glumes, thin, ciliate, 2-lobed, the midnerve excurrent into a short awn. ♃ (*Pleuraphis rigida* Thurb.)—Deserts, southern Utah and Nevada to Arizona, southern California, and Sonora.

FIGURE 719.—*Hilaria rigida*, × 1. (Palmer 494, Utah.)

## 96. AEGOPÓGON Humb. and Bonpl. ex Willd.

Spikelets on short flat pedicels, in groups of 3, the group short-pedunculate, spreading, the peduncle disarticulating from the axis and forming a pointed stipe below the group, this falling entire; central spikelet shorter pedicellate, fertile, the two lateral ones longer pedicellate and staminate or neuter; glumes membranaceous, notched at the apex, the

FIGURE 720.—*Aegopogon tenellus.* Plant, × ½; group of spikelets, × 5; lateral spikelets and central spikelet, × 10. (Pringle 1407, Mexico.)

midnerve extending into a delicate awn; lemma and palea thinner than the glumes, extending beyond them, the lemma 3-nerved, the central nerve and sometimes also the lateral ones extending into awns, the palea 2-awned. Low, lax annuals, with short, narrow, flat blades and loose racemes of delicate groups of spikelets. Type species, *Aegopogon cenchroides* Humb. and Bonpl. Name from Greek *aix*, goat, and *pogon*, beard, alluding to the fascicle of awns of the spikelets.

**1. Aegopogon tenéllus** (DC.) Trin. (Fig. 720.) Culms 10 to 20 cm. long, usually spreading or decumbent; blades 1 to 2 mm. wide; racemes 3 to 5 cm. long; spikelets, excluding awns, about 2 mm. long; lemma and palea of lateral spikelets broad and rounded at summit with a single delicate awn, those of the fertile spikelet narrower, with one long and 2 short awns. ☉ —Open ground, mountains of southern Arizona, south to northern South America. Lateral spikelets sometimes reduced or rudimentary (var. *abortivus* (Fourn.) Beetle), but such spikelets and also central spikelets with reduced awns are found in plants with normal spikelets.

## TRIBE 7. CHLORIDEAE

### 97. LEPTÓCHLOA Beauv. SPRANGLETOP

Spikelets 2- to several-flowered, sessile or short-pediceled, approximate or somewhat distant along one side of a slender rachis, the rachilla disarticulating above the glumes and between the florets; glumes unequal or nearly equal, awnless or mucronate, 1-nerved, usually shorter than the first lemma; lemmas obtuse or acute, sometimes 2-toothed and mucronate or short-awned from between the teeth, 3-nerved, the nerves sometimes pubescent. Annuals or perennials, with flat blades and numerous usually slender spikes or racemes borne on a common axis forming a long or sometimes short panicle. Type species, *Leptochloa virgata*. Name from Greek *leptos*, slender, and *chloa*, grass, alluding to the slender spikes.

The only species of *Leptochloa* important as a forage grass is *L. dubia*, or sprangletop, of the Southwest, useful for grazing and for hay.

Plants perennial.
  Lemmas broad, notched at apex, the lateral nerves glabrous........ 1. L. DUBIA.
  Lemmas acute or awned, the lateral nerves pubescent.
    Lemmas about 3 mm. long; panicle flabellate, the axis short.... 2. L. CHLORIDIFORMIS.
    Lemmas about 1.5 mm. long; panicle oblong, the axis relatively long.
      Sheaths and blades glabrous; lemmas awnless or nearly so........ 3. L. VIRGATA.
      Sheaths and blades sparsely pilose; lemmas awned........ 4. L. DOMINGENSIS.
Plants annual.
  Sheaths papillose-pilose; first floret not longer than the second glume; spikelets mostly 1 to 2 mm. long........ 5. L. FILIFORMIS.
  Sheaths smooth or scabrous, not pilose; spikelets more than 2 mm. long.
    Lemmas awned, awns sometimes minute. Culms freely branching.
      Lemmas viscid on the back; panicle oval, usually less than 10 cm. long, the longer branches usually less than 5 cm. long; second glume 1.5 mm. long. 6. L. VISCIDA.
      Lemmas not viscid; panicle more than 10 cm. long, the longer branches usually as much as 10 cm. long; second glume 3 mm. long........ 7. L. FASCICULARIS.
    Lemmas awnless or mucronate only.
      Lemmas obtuse, sometimes mucronate.
        Spikelets 5 to 7 mm. long, 6- to 9-flowered, lead color........ 8. L. UNINERVIA.
        Spikelets 2 to 3 mm. long, 3- to 4-flowered, pale........ 9. L. NEALLEYI.
      Lemmas acuminate.
        Sheaths scabrous, keeled and compressed........ 10. L. SCABRA.
        Sheaths smooth or slightly scabrous near apex, scarcely keeled or compressed. 11. L. PANICOIDES.

**1. Leptochloa dúbia** (H. B. K.) Nees. GREEN SPRANGLETOP. (Fig. 721.) Perennial; culms wiry, erect, 50 to 100 cm. tall; sheaths glabrous; blades flat or sometimes folded or loosely involute, scabrous, as much as 1 cm. wide, usually narrower; panicle of few to many spreading or ascending racemes 3 to 12 cm. long, approximate or somewhat distant on an axis as much as 15 cm. long; spikelets 5- to 8-flowered (or in reduced specimens only 2-flowered), 5 to 10 mm. long; lemmas broad, glabrous on the internerves, obtuse or emarginate, the midnerve sometimes extending into a short point, the florets at maturity widely spreading, very different in appearance from their early phase. ♃ —Rocky hills and canyons and sandy soil, southern Florida; Oklahoma and Texas to Arizona, south through Mexico; Ar-

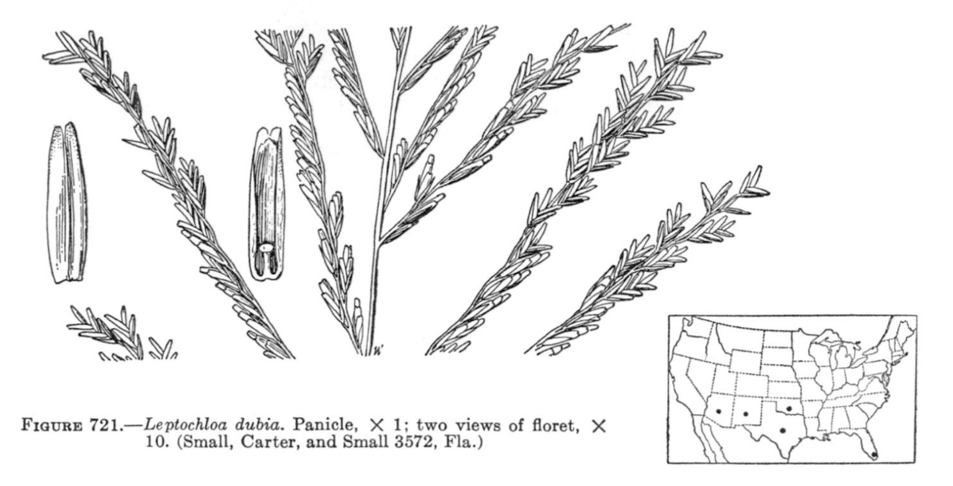

FIGURE 721.—*Leptochloa dubia*. Panicle, × 1; two views of floret, × 10. (Small, Carter, and Small 3572, Fla.)

gentina. Racemes of cleistogamous spikelets are often found in the sheaths.

**2. Leptochloa chloridifórmis** (Hack.) Parodi. (Fig. 722.) Robust tufted perennial, somewhat glaucous; culms erect, 80 to 150 cm. tall; sheaths scaberulous; ligule a dense line of white hairs, 1 to 2 mm. long; blades erect, elongate, flat, rather firm, 3 to 4 mm. wide, villous on the upper surface near the base, the margins scabrous, long-attenuate; panicle long-exserted; spikes numerous (usually 10 to 15), pale or stramineous, erect at base, flabellate or outcurved above, 10 to 15 cm. long, aggregate in 2 or 3 whorls on an axis 3 to 4 cm. long; spikelets closely imbricate on a rachis 0.5 mm. wide, 4-flowered, about 4 mm. long; glumes acute, the first 1.5 mm. long, the second 2.5 to 3 mm. long; lemmas keeled, pilose on the margins nearly to apex, the midnerve extending beyond the obtuse tip as a minute mucro, the first and second florets about 3 mm. long, the other shorter, not extending much beyond the first two. ♃ —Dry open ground, Cameron County, Tex.; Paraguay and Argentina.

**3. Leptochloa virgáta** (L.) Beauv. (Fig. 723.) Perennial; culms wiry, erect, 50 to 100 cm. tall; blades flat; racemes several to many, slender, laxly ascending, 5 to 10 cm. long, the lower distant, the others often aggregate; spikelets nearly sessile, mostly 3- to 5-flowered; lemmas 1.5 to 2 mm. long, awnless or the lower with a short awn. ♃ —Open ground and grassy slopes, southern Florida and southern Texas; tropical America.

**4. Leptochloa domingénsis** (Jacq.) Trin. (Fig. 724.) Resembling *L. virgata;* sheaths and blades sparsely pilose; panicle more elongate, the racemes shorter and more numerous; lemmas appressed-pubescent on the internerves, awned, the awn of the lower florets 1 to 3 mm. long. ♃ —Open ground and grassy slopes, southern Florida; Texas; tropical America.

**5. Leptochloa filifórmis** (Lam.) Beauv. RED SPRANGLETOP. (Fig. 725.) Annual; the foliage and panicles often reddish or purple; culms erect or branching and geniculate below, 40 to 70 cm. tall, or often dwarf; sheaths papillose-pilose, sometimes sparsely so; blades flat, thin, as much as 1 cm. wide; panicle somewhat viscid, of numerous approximate slender racemes 5 to 15 cm. long, on an axis mostly about half the entire length of the culm; spikelets 3- to 4-flowered, 1 to 2 mm. long, rather distant on the rachis; glumes acuminate, longer than the first floret, often as long as the spikelet; lemmas awnless, pubescent on the nerves, 1.5 mm. long. ⊙

FIGURE 722.—*Leptochloa chloridiformis*. Panicle, × 1; floret, × 10. (Silveus 622, Tex.)

(*L. mucronata* Kunth.)—Open or shady ground, a common weed in gardens and fields, Virginia to southern Indiana and eastern Kansas, south to Florida and Texas, west to southern California; Massachusetts; throughout tropical America. Much of the material from the Southwest has shorter racemes. Smaller forms occur throughout. These have been called *L. attenuata* (Nutt.) Steud.

**6. Leptochloa víscida** (Scribn.) Beal. (Fig. 726.) Annual, freely branching at base and from all the nodes, spreading or prostrate, the foliage and panicles somewhat viscid; culms 10 to 30 cm. tall; blades flat; panicles ovoid, rather dense, 1 to 8 cm. long, tinged with purple, included at base; spikelets 3 to 5 mm. long, 5- to 7-flowered; lemmas pubescent on the nerves, about 2 mm. long, short-awned. ⊙ —Open ground and waste places, New Mexico, Arizona, and northern Mexico.

**7. Leptochloa fasciculáris** (Lam). A. Gray. (Fig. 727.) Annual, somewhat succulent; culms erect to spreading or prostrate, freely branching, 30 to 100 cm. tall; blades flat to loosely involute; panicles more or less included, mostly 10 to 20 cm. long, often smaller, occasionally longer, the racemes several to numerous, as much

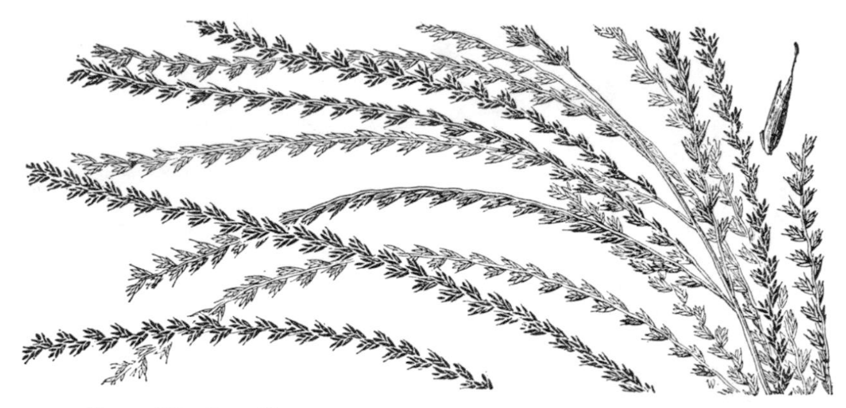

FIGURE 723.—*Leptochloa virgata.* Panicle, × 1; floret, × 10. (Wilson 9402, Cuba.)

as 10 cm. long, usually ascending or appressed, or at maturity spreading; spikelets usually overlapping, 7 to 12 mm. long, 6- to 12-flowered; lemmas 4 to 5 mm. long, the lateral nerves pubescent below, acuminate, the awn from short to as long as the body. ☉ (*Diplachne fascicularis* Beauv.) —Brackish marshes along the coast, New Hampshire and New York to Florida and Texas and in alkali flats, ditches, and marshes, Ohio to North Dakota; Washington and Colorado to New Mexico, Arizona, and California; south through tropical America to Argentina. A prostrate form has been called *Diplachne procumbens* (Muhl.) Nash and *D. maritima* Bickn.

**8. Leptochloa uninérvia** (Presl) Hitchc. and Chase. (Fig. 728.) Resembling *L. fascicularis*, rather sparingly branching, usually strictly erect, the panicle more oblong in outline, with shorter, denser-flowered racemes; spikelets 5 to 7 mm. long, 6- to 9-flowered, lead-color; glumes broader, more obtuse; lemmas scarcely narrowed toward tip, apiculate but not awned, the lateral nerves more or less excurrent. ☉ (*L. imbricata* Thurb.)—Ditches and moist places, North Carolina; Mississippi to Texas; Colorado and New Mexico to Oregon and California, south to Mexico; Peru to Argentina; introduced in Maine, Massachusetts, and New Jersey.

FIGURE 724.—*Leptochloa domingensis.* Panicle, × 1; floret, × 10. (Hitchcock 10055, Trinidad.)

**9. Leptochloa nealléyi** Vasey. (Fig. 729.) Annual, usually erect and rather robust; culms mostly 1 to 1.5 m. tall, simple or sparingly branching at base; sheaths glabrous or slightly scabrous, mostly keeled; blades elongate, flat to loosely involute; panicle commonly 25

FIGURE 725.—*Leptochloa filiformis.* Plant, × ½; spikelet and floret, × 10. (Ruth 51, Tenn.)

FIGURE 726.—*Leptochloa viscida.* Panicle, × 1; floret, × 10. (Mearns 833, Ariz.)

to 50 cm. long, not more than 4 cm. wide, the racemes subverticillate, overlapping, 2 to 4 cm. long, appressed or ascending; spikelets crowded, 3- or 4-flowered, 2 to 3 mm. long; lemmas about 1.5 mm. long, obtuse, the nerves sparingly pubescent, the lateral close to the margin. ⊙ —Marshes, mostly near the coast, Louisiana (Cameron) and Texas; also eastern Mexico.

**10. Leptochloa scábra** Nees. (Fig. 730.) Annual; culms erect, about 1 m. tall, somewhat robust and succulent,

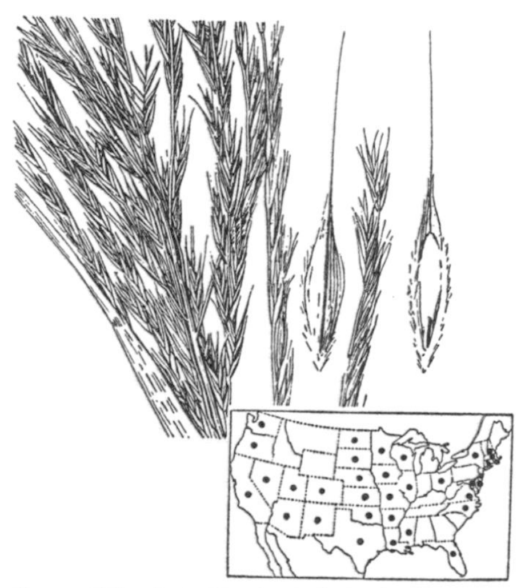

FIGURE 727.—*Leptochloa fascicularis.* Panicle, × 1; two views of floret, × 10. (Hitchcock 7876, Md.)

sparingly branching; sheaths and blades scabrous, the blades elongate, 8 to 12 mm. wide; panicle 20 to 40 cm. long, not more than 7 cm. wide, usually less, the slender racemes crowded, 4 to 8 cm. long, ascending or somewhat drooping, usually curved or flexuous; spikelets crowded, mostly 3-flowered, about 3 mm. long; lemmas acute, awnless, the nerves pubescent. ⊙ —Marshes and ditches, Louisiana (near New Orleans) and tropical America.

**11. Leptochloa panicoídes** (Presl) Hitchc. (Fig. 731.) Annual; culms erect or spreading, 50 to 100 cm. tall, branching; sheaths glabrous; blades

FIGURE 728.—*Leptochloa uninervia.* Panicle, × 1; two views of floret, × 10. (Tharp 3123, Tex.)

thin, 5 to 10 mm. wide, scaberulous; panicle oblong, 10 to 20 cm. long, 3 to 5 cm. wide, the racemes approximate, 3 to 5 cm. long, ascending, rather lax; spikelets 5- to 7-flowered, 4 to 5 mm. long; lemmas 2.5 mm. long, apiculate, the lateral nerves minutely pubescent at base. ⊙ (*L. floribunda* Doell.)—Indiana (Posey County) and Missouri to Mississippi (Holmes County), Arkansas, and Texas; Brazil.

## 98. TRICHONEÚRA Anderss.

Spikelets few-flowered, the rachilla disarticulating above the glumes, the internodes pilose at base, disarticulating near their summit, the upper part forming a short callus below the floret; glumes about equal, 1-nerved, long-acuminate, mostly as long as the spikelet or longer; lemmas bidentate, 3-nerved, the lateral nerves near the margin, the midnerve usually excurrent as a short awn, the margins long-ciliate; palea broad, the nerves near the margin. Annuals or perennials with simple panicles, the spikelets short-pediceled along one side of the main branches. Type species, *Trichoneura hookeri* Anderss. Name from Greek *thrix*, hair, and *neuron*, nerve, alluding to the ciliate nerves of the lemma.

FIGURE 730.—*Leptochloa scabra*. Panicle, × 1; two views of floret, × 10. (Tracy 8388, La.)

FIGURE 729.—*Leptochloa nealleyi*. Panicle, × 1; two views of floret, × 10. (Fisher 25, Tex.)

**1. Trichoneura élegans** Swallen. (Fig. 732.) Annual, branching at base; culms erect, rather robust, or ascending, 40 to 110 cm. tall, several-noded; sheaths scaberulous; blades flat, or subinvolute toward the tip, scabrous, elongate, 3 to 7 mm. wide; panicle erect, 10 to 18 cm. long, the axis angled, scabrous; branches numerous, stiffly ascending, the lower 5 to 8 cm. long, rather densely flowered; spikelets mostly 5- to 8-flowered, 9 to 10 mm. long; glumes about equaling the spikelet, the setaceous tips slightly spreading; lemmas scaberulous toward the obtuse minutely lobed summit, the awn minute, the margins conspicuously ciliate on the lower half to two-thirds, the hairs as much as 1 mm. long. ⊙ —Sandy soil, southern Texas.

## 99. TRIPÓGON Roth

Spikelets several-flowered, subsessile, appressed in 2 rows along one

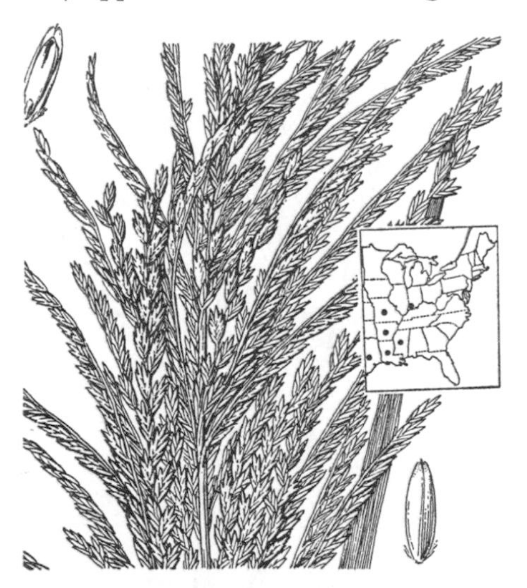

FIGURE 731.—*Leptochloa panicoides*. Panicle, × 1; two views of floret, × 10. (Tracy 7451, Miss.)

FIGURE 732.—*Trichoneura elegans*. Plant, × 1; spikelet and floret, × 10. (Type.)

side of a slender rachis, the rachilla disarticulating above the glumes and between the florets; glumes somewhat unequal, acute or acuminate, narrow, 1-nerved; lemmas narrow, 3-nerved, bearing at base a tuft of long hairs, the apex bifid, the midnerve extending as a short awn. Our species a low, tufted perennial, with capillary blades and slender solitary spikes, the spikelets somewhat distant. Type species, *Tripogon bromoides* Roth. Name from Greek *treis*, three, and *pogon*, beard, alluding to the hairs at the base of the three nerves of the lemma.

**1. Tripogon spicátus** (Nees) Ekman. (Fig. 733.) Culms 10 to 20 cm. tall; spike from one-fourth to half the entire height of the plant; spikelets 5 to 8 mm. long. ♃ —Rocky hills, central Texas, Mexico; Cuba; South America.

## 100. ELEUSÍNE Gaertn.

Spikelets few to several-flowered, compressed, sessile and closely imbricate, in 2 rows along one side of a rather broad rachis, not prolonged beyond the spikelets; rachilla disarticulating above the glumes and between the florets; glumes unequal, rather broad, acute, 1-nerved, shorter than the first lemma; lemmas acute, with 3 strong green nerves close together, forming a keel, the uppermost somewhat reduced; seed dark brown, roughened by fine ridges, loosely enclosed in the thin pericarp. Annuals, with 2 to several rather stout spikes, digitate at the summit of the culms, sometimes with 1 or 2 a short distance below, or rarely with a single spike. Type species, *Eleusine coracana*. Name from Eleusis, the town where Demeter was worshipped.

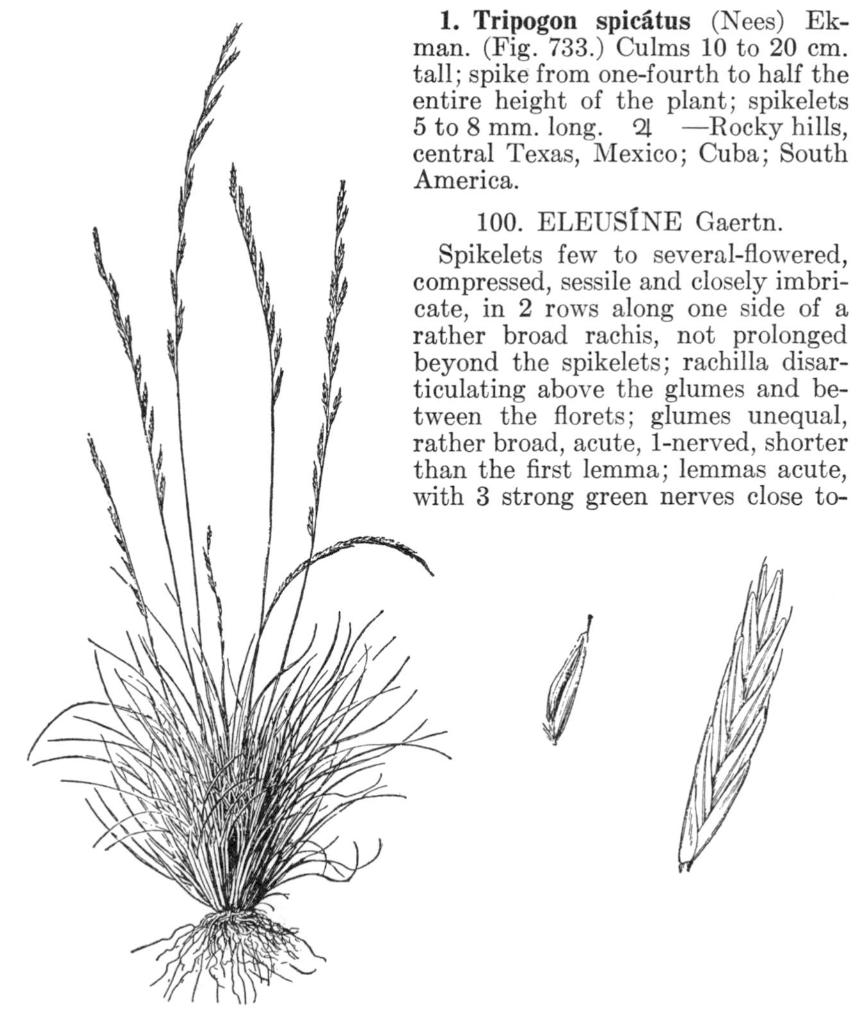

FIGURE 733.—*Tripogon spicatus*. Plant, × ½; spikelet and floret, × 5. (Nealley 78, Tex.)

**1. Eleusine índica** (L.) Gaertn. GOOSEGRASS. (Fig. 734.) Branching at base, ascending to prostrate, very smooth; culms compressed, usually

FIGURE 734.—*Eleusine indica*. Plant, × ½; spikelet, floret, and seed (without pericarp), × 5. (Fredholm 5331, Fla.)

less than 50 cm. long, but sometimes as much as 1 m.; blades flat or folded, 3 to 8 mm. wide; spikes mostly 2 to 6, rarely more, or but 1 in depauperate plants, flat, 4 to 15 cm. long. ⊙ —Waste places, fields, and open ground, Massachusetts to South Dakota and Kansas, south to Florida and Texas; occasional in Oregon, Utah, Arizona, and California; introduced; a common weed in the warmer regions of both hemispheres.

**Eleusine tristáchya** (Lam.) Lam. Spikes 1 to 3, rarely more, 1 to 2.5 cm. long, 8 to 10 mm. thick; resembling *E. indica*, but the spikes short and thick. ⊙ —On ballast, Camden, N. J. and Mobile, Ala.; Portland, Oreg. and elsewhere; tropical Africa; introduced in tropical South America.

**Eleusine coracána** (L.) Gaertn. African millet. More robust than *E. indica;* spikes thicker, heavier, sometimes incurved at the tip, brownish at maturity. A cultivated form of *E. indica*; the seed used for food among primitive peoples in Africa and southern Asia. ⊙ —Occasionally grown at experiment stations. Called also ragi, coracan millet, and finger millet.

## 101. DACTYLOCTÉNIUM Willd.

Spikelets 3- to 5-flowered, compressed, sessile and closely imbricate, in two rows along one side of the rather narrow flat rachis, the end projecting in a point beyond the spikelets; rachilla disarticulating above the first glume and between the florets; glumes somewhat unequal, broad, 1-nerved, the first persistent upon the rachis, the second mucronate or short-awned below the tip, deciduous; lemmas firm, broad, keeled, acuminate or short-awned, 3-nerved, the lateral nerves indistinct, the upper floret reduced; palea about as long as the lemma; seed subglobose, ridged or wrinkled, enclosed in a thin, early-disappearing pericarp. Annuals or perennials with flat blades and 2 to several short thick spikes, digitate and widely spreading at the summit of the culms. Type species, *Dactyloctenium aegyptium.* Name from Greek *daktulos*, finger, and *ktenion*, a little comb, alluding to the pectinate arrangement of the spikelets.

**1. Dactyloctenium aegýptium** (L.) Beauv. (Fig. 735.) Culms compressed, spreading with ascending ends, rooting at the nodes, branching, commonly forming radiate mats, usually 20 to 40 cm. long, sometimes as much as 1 m.; blades flat, ciliate; spikes 1 to 5 cm. long. ⊙ —Open ground, waste places, and fields, Coastal Plain, North Carolina to Florida and Texas; also occasional at more northern points (Maine to New Jersey; Illinois); Colorado, Arizona, and California; tropical America; introduced from Old World Tropics.

## 102. MICROCHLÓA R. Br.

Spikelets 1-flowered, awnless, sessile in 2 rows along one side of a narrow flattened rachis, the rachilla disarticulating above the glumes; glumes subequal, longer than the floret, acute, 1-nerved; floret with a soft, pointed callus; lemma thin, 3-nerved, flabellate; palea narrow, a little shorter than the lemma. Slender perennials with simple culms and slender solitary falcate spikes. Type species, *Microchloa setacea* R. Br. Name from the Greek *micros*, small, and *chloe*, grass.

**1. Microchloa kúnthii** Desv. (Fig. 736.) Perennial; culms very slender, erect in small dense tufts, 10 to 30 cm. tall; sheaths, except the lowermost, much shorter than the internodes, scaberulous; ligule ciliate, 1 to 1.5 mm. long; blades firm, flat or usually folded, with thick white scabrous margins, those of the culm 1 to 2.5 cm. long, those of the innovations to 6 cm. long, 1 to 1.5 mm. wide; spike 6 to 15 cm. long, falcate, the rachis ciliate; spikelets 2.5 to 3.5 mm. long; lemma 2 to 2.5 mm. long, pilose on the midnerve, the margins densely ciliate with hairs about 1 mm. long. ♃ —Granitic outcrop on rocky slope, Carr Canyon, Huachuca Mountains,

FIGURE 735.—*Dactyloctenium aegyptium*. Plant, × ½; spikelet, floret, and seed (without pericarp), × 5. (Small and Heller 378, N. C.)

southern Arizona; Mexico and Guatemala.

## 103. CÝNODON L. Rich.

(*Capriola* Adans.)

Spikelets 1-flowered, awnless, sessile in 2 rows along one side of a slender continuous rachis and appressed to it, the rachilla disarticulating above the glumes and prolonged behind the palea as a slender naked bristle, sometimes bearing a rudimentary lemma; glumes narrow, acuminate, 1-nerved, about equal, shorter than the floret; lemma firm, strongly compressed, pubescent on the keel, 3-nerved, the lateral nerves close to the margins. Perennial, usually low grasses, with creeping stolons or rhizomes, short blades, and several slender spikes digitate at the summit of the upright culms. Type species, *Cynodon dactylon*. Name from *kuon* (*kun-*), dog, and *odous*, tooth, alluding to the sharp hard scales of the rhizome.

**1. Cynodon dáctylon** (L.) Pers. Bermuda grass. (Fig. 737.) Extensively creeping by scaly rhizomes or by strong flat stolons, the old bladeless sheaths of the stolon and the lowest one of the branches often forming conspicuous pairs of "dog's teeth"; flowering culms flattened, usually erect or ascending, 10 to 40 cm. tall; ligule a conspicuous ring of white hairs; blades flat, glabrous or pilose on the upper surface, those of the innovations often conspicuously distichous; spikes usually 4 or 5, 2.5 to 5 cm. long; spikelets imbricate, 2 mm. long, the lemma boat-shaped, acute. ♃ (*Capriola dactylon* Kuntze.)—Open ground, grassland, fields, and waste places, common, Maryland to Oklahoma, south to Florida and Texas, west to California; also occasional north of this region (Massachusetts to Michigan, Oregon); warm regions of both hemispheres, introduced in America. Bermuda grass is the most important pasture grass of the Southern States, and is also widely utilized there as a lawngrass. On alluvial ground it may grow sufficiently rank to be cut for hay. It propagates readily by its rhizomes and stolons, and on this account may become a troublesome weed in cultivated fields. This grass is known also as wire-grass (especially the weedy form in fields). A more robust form,

Figure 736.—*Microchloa kunthii*. Plant, × ½. (Conzatti 3605, Mexico.)

found along the seacoast of Florida, has been called *C. maritimus* H. B. K., though the type of that (from Peru) is characteristic *C. dactylon*. There are large areas of Bermuda grass around the Roosevelt Dam,

FIGURE 737.—*Cynodon dactylon*. Plant, × ½; spikelet and two views of floret, × 5. (Kearney, Tenn.)

Ariz., where it survives submergence and furnishes grazing at low water.

CYNODON TRANSVAALÉNSIS Davy. Extensively creeping with fine foliage, the blades rarely more than 1 mm. wide; spikes mostly 2 or 3, the spikelets a little narrower and the glumes shorter than in *C. dactylon*. ♃ —Coming into cultivation as a lawn-grass, escaped, Ames, Iowa, and Bard, Calif. Introduced from South Africa.

## 104. WILLKÓMMIA Hack.

Spikelets 1-flowered, dorsally compressed, sessile in 2 rows on one side of a slender rachis and appressed to it, the rachilla somewhat lengthened below and above the second glume, disarticulating just above it, not prolonged above the floret; glumes thin,

unequal, the first narrow, nerveless, the second 1-nerved; lemma awnless, 3-nerved, the lateral nerves near the margin, the back of the lemma sparingly pubescent between the nerves, the margins densely covered with silky hairs; nerves of the palea densely silky hairy. Annuals or perennials, with several short spikes racemose on a slender axis; our species a low tufted perennial. Type species, *Willkommia sarmentosa* Hack. Named for H. M. Willkomm.

**1. Willkommia texána** Hitchc. (Fig. 738.) Culms erect to spreading, 20 to 40 cm. tall; blades flat or more or less involute, short; spikes few to several, 2 to 5 cm. long, somewhat overlapping or the lower distant, appressed, the axis 4 to 15 cm. long; spikelets about 4 mm. long, narrow, acute; first glume about two-thirds as long as the second, obtuse; second glume subacute; lemma about as long as the second glume. ♃ — Spots of hardpan, central and southern Texas. A stoloniferous form has been found in Argentina.

## 105. SCHEDONNÁRDUS Steud.

Spikelets 1-flowered, sessile and somewhat distant in 2 rows on one side of a slender, continuous 3-angled rachis, appressed to its slightly concave sides, the rachilla disarticulating above the glumes, not prolonged; glumes narrow, stiff, somewhat unequal, acuminate, 1-nerved; lemmas narrow, acuminate, a little longer than the glumes, 3-nerved. Low,

FIGURE 738.—*Willkommia texana*. Plant, × ½; two views of spikelet and floret, × 5. (Tracy 8903, Tex.)

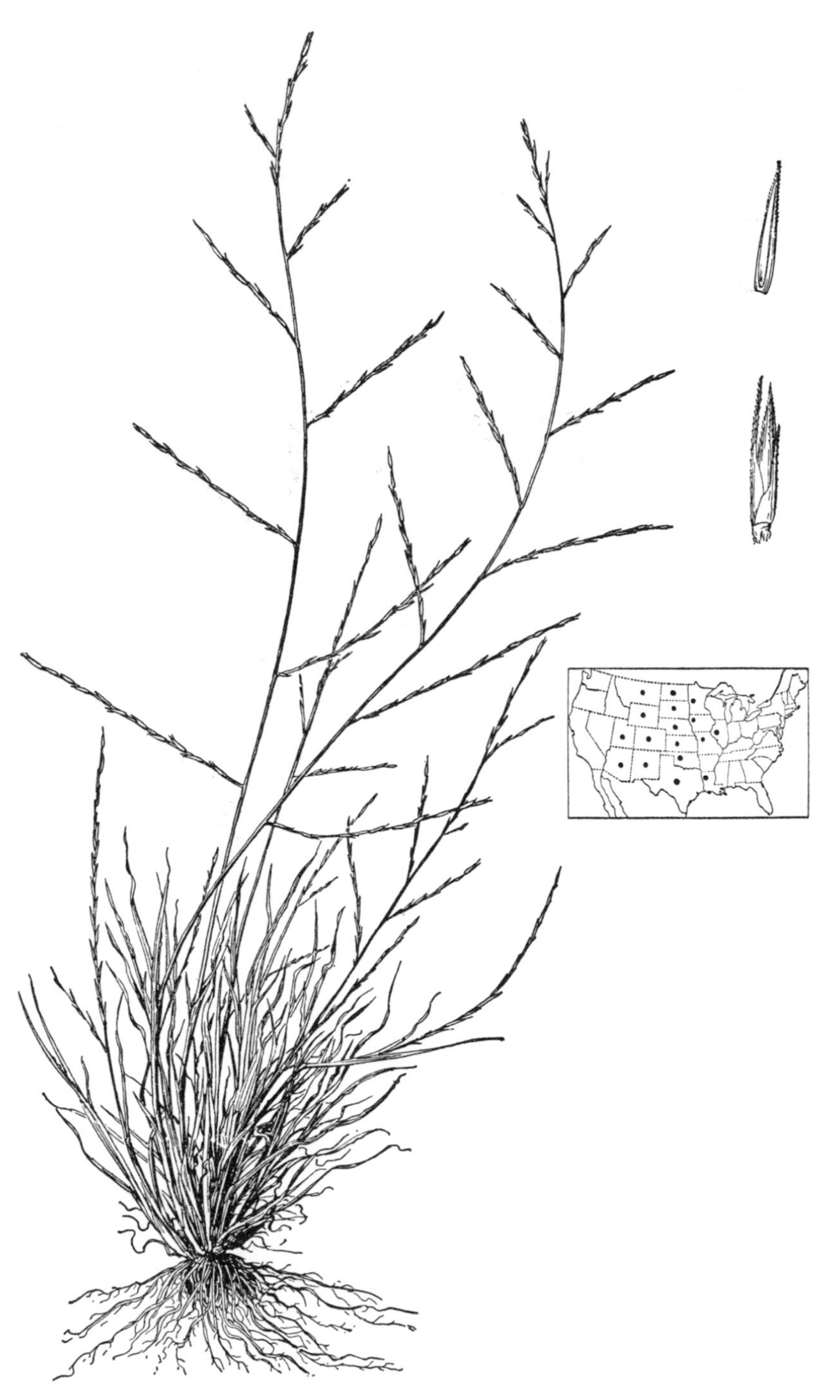

FIGURE 739.—*Schedonnardus paniculatus*. Plant, × ½; spikelet and floret, × 5. (Hall 797, Tex.)

FIGURE 740.—*Beckmannia syzigachne*. Plant, × ½; spikelet and floret, × 5. (Hitchcock 4668, Alaska.)

tufted perennial, with stiff, slender, divergent spikes rather remote along a common axis. Type species, *Schedonnardus texanus* Steud. (*S. paniculatus*). Name from Greek *schedon*, near, and *Nardus*, a genus of grasses (Steudel places *Schedonnardus* next to *Nardus* in his classification).

**1. Schedonnardus paniculátus** (Nutt.) Trel. TUMBLEGRASS. (Fig. 739.) Culms 20 to 40 cm. tall; leaves crowded at the base; blades flat, mostly 2 to 5 cm. long, about 1 mm. wide, wavy; spikes 2 to 10 cm. long; spikelets narrow, acuminate, about 4 mm. long. The axis of the inflorescence elongates after flowering, becoming 30 to 60 cm. long, curved in a loose spiral; the whole breaks away at maturity and rolls before the wind as a tumbleweed. ♃ —Prairies and plains, Illinois to Saskatchewan and Montana, south to Louisiana and Arizona; Argentina. This species forms an inconsiderable part of the forage on the Great Plains.

### 106. BECKMÁNNIA Host
SLOUGHGRASS

Spikelets 1- or 2-flowered, laterally compressed, subcircular, nearly sessile and closely imbricate, in 2 rows along one side of a slender continuous rachis, disarticulating below the glumes, falling entire; glumes equal, inflated, obovate, 3-nerved, rounded above but the apex apiculate; lemma narrow, 5-nerved, acuminate, about as long as the glumes; palea nearly as long as the lemma. Erect, rather stout annuals with flat blades and numerous short appressed or ascending spikes in a narrow more or less interrupted panicle. Type species, *Beckmannia erucaeformis* (L.) Host, to which our species was formerly referred. Named for Johann Beckmann.

**1. Beckmannia syzigáchne** (Steud.) Fernald. AMERICAN SLOUGHGRASS. (Fig. 740.) Light green; culms 30 to 100 cm. tall; panicle 10 to 25 cm. long, the erect branches 1 to 5 cm. long; spikes crowded, 1 to 2 cm. long; spikelets 1-flowered, 3 mm. long; glumes transversely wrinkled and with a deep keel, the acuminate apex of the lemma protruding. ⊙ —Marshes and ditches, Manitoba to Alaska; New York and Ohio to the Pacific coast, south to Kansas and New Mexico; Asia. The European *B. erucaeformis* (L.) Host has 2-flowered spikelets. Our species is palatable to stock, sometimes sufficiently abundant locally to be an important forage grass, and is frequently cut for hay.

### 107. SPARTÍNA Schreb. CORDGRASS

Spikelets 1-flowered, much flattened laterally, sessile and usually closely imbricate on one side of a continuous rachis, disarticulating below the glumes, the rachilla not produced beyond the floret; glumes keeled, 1-nerved, or the second with a second nerve on one side, acute or short-awned, the first shorter, the second often exceeding the lemma; lemma firm, keeled, the lateral nerves obscure, narrowed to a rather obtuse point; palea 2-nerved, keeled and flattened, the keel between or at one side of the nerves. Erect, often stout tall perennials, with usually extensively creeping, firm, scaly rhizomes (wanting in *Spartina spartinae*, *S. bakeri*, and sometimes in *S. caespitosa*), long tough blades, and 2 to many appressed or sometimes spreading spikes racemose on the main axis, the slender tips of the rachises naked, often prolonged. Type species, *Spartina schreberi* Gmel. Name from Greek *spartine*, a cord made from *spartes* (*Spartium junceum*), probably applied to *Spartina* because of the tough leaves.

The species with rhizomes often form extensive colonies to the exclusion of other plants. They are important soil binders and soil builders in coastal and interior marshes. A European species, *S. townsendi* H. and J. Groves, has in recent years assumed much importance, especially in southern England, the Netherlands, and northern France, as a soil builder along the coast where it is reclaiming extensive areas of marsh land. The marsh hay of the Atlantic coast, much used for packing and formerly for bedding, often consists largely of *S. patens*.

Blades usually more than 5 mm. wide, flat when fresh, at least at base, the tip involute; plants mostly robust and more than 1 m. tall.
  First glume nearly as long as the floret, slender-acuminate, the second with an awn as much as 7 mm. long; spikes somewhat distant, mostly more or less spreading. 1. S. PECTINATA.
  First glume shorter than the floret, acute, the second acute or mucronate but not slender-awned; spikes approximate, usually appressed.
    Blades very scabrous on the margins; glumes strongly hispid-scabrous on the keels. 2. S. CYNOSUROIDES.
    Blades glabrous throughout or minutely scabrous on the margins; glumes glabrous or usually softly hispidulous or ciliate on the keels.
      Inflorescence dense and spikelike, the spikes closely imbricate; the spikelets mostly somewhat curved, giving a slightly twisted effect; blades mostly comparatively short.......... 3. S. FOLIOSA.
      Inflorescence less dense, the spikes more slender, less crowded, the spikelets not curved, the inflorescence with no suggestion of a twist...... 4. S. ALTERNIFLORA.
Blades less than 5 mm. wide (rarely more in *S. gracilis*); involute (sometimes flat in *S. gracilis*); plants mostly slender and less than 1 m. tall (taller in *S. bakeri*).
  Inflorescence dense, cylindric; spikes numerous.......... 5. S. SPARTINAE.
  Inflorescence not cylindric; spikes not more than 10, usually fewer.
    Creeping rhizomes absent (see also *S. caespitosa*); plants in large hard tufts with culms 1.5 to 2 m. tall and long slender blades.......... 6. S. BAKERI.
    Creeping rhizomes present (except occasionally in *S. caespitosa*); plants usually less than 1 m. tall.
      Second glume 12 to 16 mm. long, aristate.......... 7. S. CAESPITOSA.
      Second glume less than 10 mm. long, acute.
        Blades usually flat; glumes conspicuously hispid-ciliate on the keels; spikes several, appressed.......... 8. S. GRACILIS.
        Blades usually involute; glumes scabrous on the keels; spikes few, ascending to spreading.......... 9. S. PATENS.

**1. Spartina pectináta** Link. PRAIRIE CORDGRASS. (Fig. 741.) Culms 1 to 2 m. tall, firm or wiry; blades elongate, flat when fresh, soon involute in drying, as much as 1.5 cm. wide, very scabrous on the margins; spikes mostly 10 to 20, sometimes fewer or as many as 30, mostly 4 to 8 cm. long, ascending, sometimes appressed, rarely spreading, on rather slender peduncles; glumes hispid-scabrous on the keel, the first acuminate or short-awned, nearly as long as the floret, the second exceeding the floret, tapering into an awn as much as 7 mm. long; lemma glabrous except the scabrous keel, 7 to 9 mm. long, the apex with 2 rounded teeth; palea usually a little longer than the lemma. ♃ (*S. michauxiana* Hitchc.)—Fresh-water marshes, Newfoundland and Quebec to eastern Washington and Oregon, south to North Carolina, Arkansas, Texas, and New Mexico; in the Eastern States extending into brackish marshes along the coast.

**2. Spartina cynosuroídes** (L.) Roth. BIG CORDGRASS. (Fig. 742.) Culms 1 to 3 m. tall, stout, the base sometimes as much as 2 cm. thick; blades flat, 1 to 2.5 cm. wide, very scabrous on the margins; spikes numerous, ascending, approximate, often dark-colored, usually more or less peduncled, mostly 3 to 8 cm. long; spikelets about 12 mm. long; glumes acute, hispid-scabrous on the keel, the first much shorter than the floret, the second longer than the floret, sometimes rather long-acuminate; lemma not toothed at apex; palea a little longer than the lemma. ♃ (*S. polystachya* (Michx.) Beauv. (*S. cynosuroides* var. *polystachya* Beal) has been differentiated on its strictly maritime habitat, but morphological

FIGURE 741.—*Spartina pectinata*. Plant, × ½; spikelet and floret, × 5. (Worthern, Mass.)

characters are not coordinated with habitat.)—Salt or brackish marshes along the coast, and margins of tidal streams, Massachusetts to Florida and Texas.

**3. Spartina foliósa** Trin. (Fig. 743.) Culms 30 to 120 cm. tall, stout, as much as 1 cm. thick at base, somewhat spongy, usually rooting at the lower nodes; blades 8 to 12 mm. wide at the flat base, gradually narrowed to a long involute tip, smooth throughout; inflorescence dense, spikelike, about 15 cm. long; spikes numerous, approximate, closely appressed, 3 to 5 cm. long; spikelets very flat, 9 to 12 mm. long, occasionally longer; glumes firm, glabrous or hispid-ciliate on the keel, acute, the first narrow, half to two-thirds as long as the second, smooth, the second sparingly hispidulous and striate-nerved; lemma hispidulous on the sides, mostly smooth on the keel, shorter than the second glume; palea thin, longer than the lemma. ♃ (*S. leiantha* Benth.)—Salt marshes along the coast from San Francisco Bay, Calif., to Baja California.

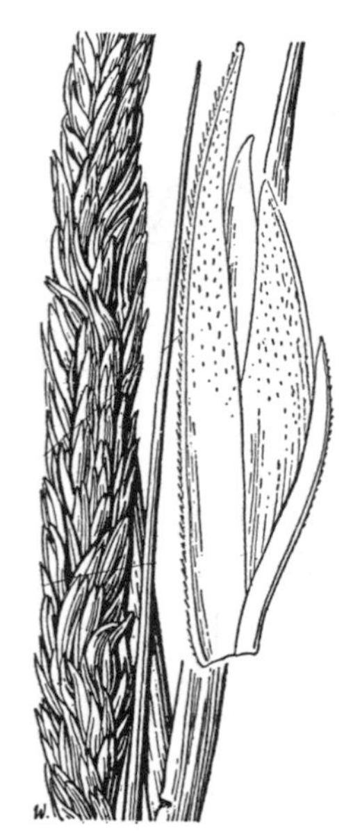

FIGURE 743.—*Spartina foliosa*. Panicle, × 1; spikelet × 5. (Heller 13871, Calif.)

FIGURE 742.—*Spartina cynosuroides*. Panicle, × 1; spikelet, × 5. (Boettcher 444, Va.)

**4. Spartina alterniflóra** Loisel. SMOOTH CORDGRASS. (Fig. 744.) Smooth throughout or the margins of the blades minutely scabrous, 0.5 to 2.5 m. tall; culms soft and spongy or succulent at base, often 1 cm. or more thick; blades flat, tapering to a long involute tip, 0.5 to 1.5 cm. wide; spikes appressed, 5 to 15 cm. long; spikelets somewhat remote, barely overlapping or sometimes more imbricate, mostly 10 to 11 mm. long; glumes glabrous or hispid on the keel, the first acute, narrow, shorter than the lemma, the second obtusish,

FIGURE 744.—*Spartina alterniflora*. Panicle, × 1; spikelet, × 5. (Scribner 155, Maine.)

a little longer than the lemma; floret sparingly pilose or glabrous. ♃ —Salt marshes along the coast, often growing in the water, Quebec and Newfoundland to Florida and Texas; recently introduced in oyster culture, Pacific County, Wash., and spreading; Atlantic coast of Europe. Through the southern part of the range of the species the spikelets are often more imbricate. The imbricate form with glabrous spikelets has been differentiated as *S. alterniflora* var. *glabra* (Muhl.) Fernald; that with sparsely pilose spikelets as *S. alterniflora* var. *pilosa* (Merr.) Fernald.

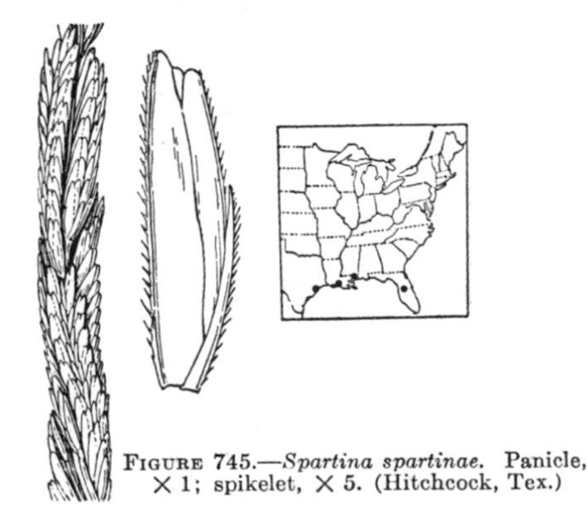

FIGURE 745.—*Spartina spartinae*. Panicle, × 1; spikelet, × 5. (Hitchcock, Tex.)

**5. Spartina spartínae** (Trin.) Merr. (Fig. 745.) In large dense tufts without rhizomes; culms stout, 1 to 2 m. tall; blades narrow, firm, strongly involute; spikes short and appressed, closely imbricate, forming a dense cylindric inflorescence 10 to 30 cm. long; spikelets closely imbricate, 6 to 8 mm. long; glumes hispid-ciliate on the keel, the first shorter than the lemma, the second usually a little longer. ♃ (*S. junciformis* Engelm. and Gray.)—Marshes, swamps, and moist prairies near the coast, Florida to Texas and eastern Mexico.

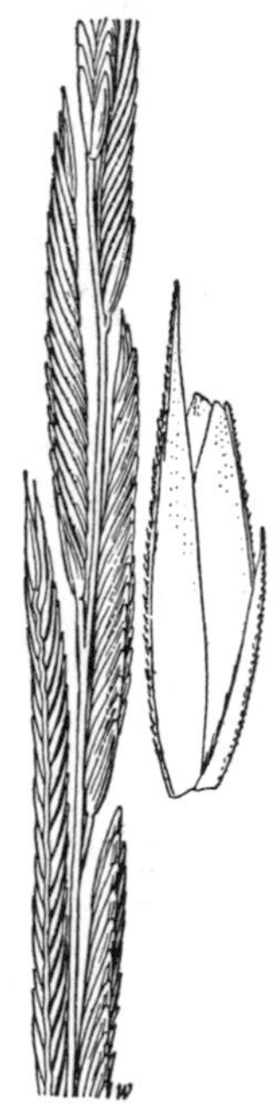

FIGURE 746.—*Spartina bakeri*. Panicle, × 1; spikelet, × 5. (Type.)

**6. Spartina bakéri** Merr. (Fig. 746.) In large dense tufts without rhizomes; culms stout, 1 to 2 m. tall; blades 4 to 8 mm. wide, involute or occasionally flat; inflorescence 12 to 18 cm. long, the spikes 5 to 12, 3 to 6 cm. long, appressed; spikelets closely appressed, 6 to 8 mm. long; glumes scabrous, hispid-ciliate on the keel, the first about half as long as the lemma, the second longer, acuminate. ♃ —Sandy soil, South Carolina, Georgia, and Florida.

**7. Spartina caespitósa** A. A. Eaton. (Fig. 747.) Culms 70 to 100 cm. tall, erect, from coarse widely spreading rhizomes or tufted, the rhizomes nearly wanting; blades 10 to 40 cm. long, 3 to 7 mm. wide, flat or becoming involute, scabrous on the upper surface and margins; spikes 2 to 7, 3 to 9 cm. long, finally spreading, rather distant; glumes acuminate, aristate, conspicuously hispid-ciliate on the keels, the second 12 to 16 mm. long; lemma about 8 mm. long, minutely lobed. ♃ —Salt marshes near the coast, New Hampshire to Maryland.

FIGURE 747.—*Spartina caespitosa*. Panicle, × 1; spikelet, × 5. (Type collection.)

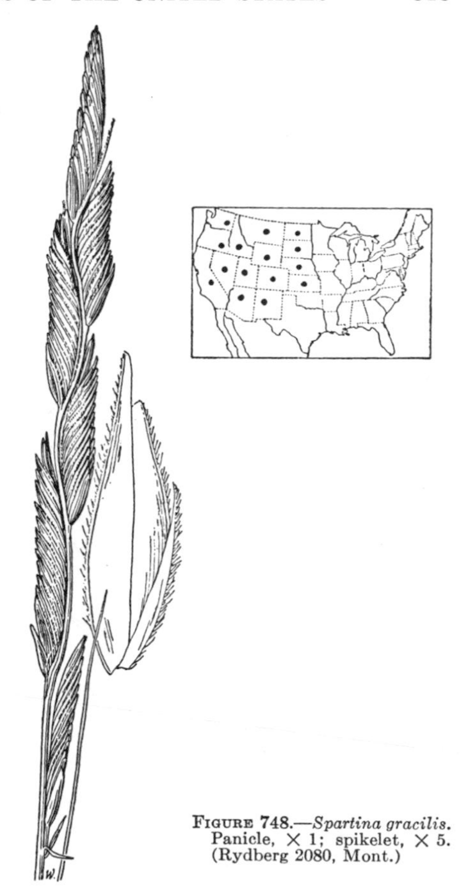

FIGURE 748.—*Spartina gracilis*. Panicle, × 1; spikelet, × 5. (Rydberg 2080, Mont.)

**8. Spartina grácilis** Trin. ALKALI CORDGRASS. (Fig. 748.) Culms 60 to 100 cm. tall; blades flat, becoming involute, 15 to 20 cm. long, very scabrous above, mostly less than 5 mm. wide; spikes 4 to 8, closely appressed, 2 to 4 cm. long; spikelets 6 to 8 mm. long; glumes ciliate on the keel, acute, the first about half as long as the second; lemma nearly as long as second glume, ciliate on the keel; palea as long as lemma, obtuse. ♃ —Alkaline meadows and plains, Saskatchewan to British Columbia, south to Kansas and New Mexico,

FIGURE 749.—*Spartina patens*. Panicle, × 1; spikelet, × 5. (Killip 6359, Md.)

and through eastern Washington to Arizona.

**9. Spartina pátens** (Ait.) Muhl. SALTMEADOW CORDGRASS. (Fig. 749.) Culms slender, mostly less than 1 m. tall, with long slender rhizomes; blades sometimes flat but mostly involute, less than 3 mm. wide; spikes 2 to several, appressed to somewhat spreading, 2 to 5 cm. long, rather remote on the axis; spikelets 8 to 12 mm. long; first glume about half as long as the floret, the second longer than the lemma; lemma 5 to 7 mm. long, emarginate at apex; palea a little longer than the lemma. ♃ —Salt marshes and sandy meadows along the coast, Quebec to Florida and Texas, and in saline marshes inland, New York and Michigan. SPARTINA PATENS var. MONÓGYNA (M. A. Curtis) Fernald. Often taller and coarser, commonly with 4 to 8 spikes, the spikelets slightly smaller and more closely imbricate. Intermediate specimens rather frequent. ♃ (*S. juncea* Willd., *S. patens* var. *juncea* Hitchc.)—Along the coast, New Jersey to Texas.

## 108. CTÉNIUM Panzer

(*Campulosus* Desv.)

Spikelets several-flowered but with only 1 perfect floret, sessile and pectinately arranged on one side of a continuous rachis, the rachilla disarticulating above the glumes; first glume small, hyaline, 1-nerved, the second about as long as the lemmas, firm, 3- to 4-nerved, bearing on the back a strong divergent awn; lemmas rather papery, 3-nerved, with long hairs on the lateral nerves and a short straight or curved awn on the back just below the apex, the first and second lemmas empty, the third enclosing a perfect flower, the upper 1 to 3 empty and successively smaller. Erect, slender, rather tall perennials, with usually solitary, often curved spikes. Type species, *Ctenium carolinianum* Panzer. (*C. aromaticum*). Name from Greek *ktenion*, a little comb, alluding to the pectinate arrangement of the spikelets.

Plants forming dense tussocks; second glume with a row of prominent glands on each side of the midnerve; awn stout, at maturity horizontal or nearly so; ligule about 1 mm. long. 1. C. AROMATICUM.

Plants with slender scaly rhizomes; second glume glandless or with obscure glands; awn rather slender, not horizontally spreading; ligule 2 to 3 mm. long.... 2. C. FLORIDANUM.

FIGURE 750.—*Ctenium aromaticum*. Plant, × ½; spikelet and fertile floret, × 5. (McCarthy, N. C.)

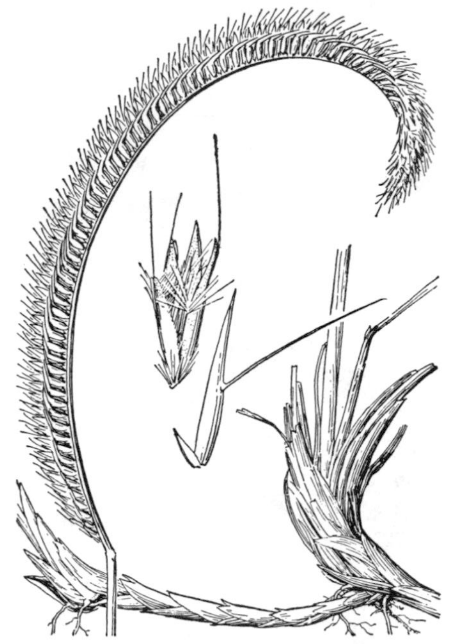

FIGURE 751.—*Ctenium floridanum.* Plant, × 1; glumes and florets, × 5. (Combs 702a, Fla.)

**1. Ctenium aromáticum** (Walt.) Wood. TOOTHACHE GRASS. (Fig. 750.) Culms 1 to 1.5 m. tall, the old sheaths persistent and fibrillose at base; ligule about 1 mm. long; blades flat or involute, stiff; spike 5 to 15 cm. long; spikelets 5 to 7 mm. long. ♃ (*Ctenium carolinianum* Panzer.)—Wet pine barrens, Coastal Plain, Virginia to Florida and Louisiana. The roots spicy when freshly dug. Furnishes fair cattle forage in moist pine barrens of Florida.

**2. Ctenium floridánum** (Hitchc.) Hitchc. (Fig. 751.) Differs from *C. aromaticum* in having creeping scaly rhizomes, ligule 2 to 3 mm. long, second glumes with longer, more slender awns and without glands or with only obscure ones. ♃ (Erroneously referred by American authors to *Campulosus chapadensis* Trin.)—Moist pine barrens, Florida.

## 109. GYMNOPÓGON Beauv.

Spikelets 1- or rarely 2- or 3-flowered, nearly sessile, appressed and usually remote in 2 rows along one side of a slender continuous rachis, the rachilla disarticulating above the glumes and prolonged behind the 1 or more fertile florets as a slender stipe, bearing a rudiment of a floret, this sometimes with 1 or 2 slender awns; glumes narrow, acuminate, 1-nerved, usually longer than the floret; lemmas narrow, 3-nerved, the lateral nerves near the margin, the apex minutely bifid, bearing between the teeth a slender awn, rarely awnless. Perennials or rarely annuals (ours perennial), with short, stiff, flat blades, often folded in drying, numerous long slender divergent or reflexed spikes, approximate on a slender stiff axis. Type species, *Gymnopogon racemosus* Beauv. (*G. ambiguus*). Name from Greek *gumnos*, naked, and *pogon*, beard, alluding to the naked prolongation of the rachilla.

Awn 4 to 6 mm. long, longer than the lemma .......... 1. G. AMBIGUUS.
Awn 1 to 3 mm. long, usually shorter than the lemma.
  Spikelets 1-flowered; spikes floriferous only in the upper half .......... 2. G. BREVIFOLIUS.
  Spikelets 2- to 3-flowered; spikes floriferous to the base.
    Spikes stiffly ascending, usually more than 20; glumes widely spreading even on young spikelets .......... 3. G. CHAPMANIANUS.
    Spikes spreading or reflexed, usually fewer than 15; glumes not spreading, even in mature spikelets .......... 4. G. FLORIDANUS.

**1. Gymnopogon ambíguus** (Michx.) B. S. P. (Fig. 752.) Culms 30 to 60 cm. tall in small clumps with short scaly rhizomes, suberect to spreading, rigid, sparingly branching; leaves numerous, approximate with overlapping sheaths, or the lower rather distant; blades spreading, 5 to 15 mm., mostly about 10 mm. wide, the base

FIGURE 752.—*Gymnopogon ambiguus*. Plant, × ½; spikelet and floret, × 5. (Tracy 8292, Tex.)

rounded-truncate; spikes 10 to 20 cm. long, floriferous from base, the lower spikelets often remote; glumes 4 to 6 mm. long; lemma with an awn 4 to 6 mm. long, the rudiment bearing a delicate shorter awn. ♃ —Dry pinelands, Coastal Plain, New Jersey to Florida and Texas; dry woods, Ohio to Kansas and south.

**2. Gymnopogon brevifólius** Trin. (Fig. 753.) Differing from *G. ambiguus* in the longer, more slender, somewhat straggling culms, narrower, less crowded blades, and in the subcapillary spikes, floriferous only on the upper half or third; lemma awnless or with a minute awn. ♃ —Dry ground, Coastal Plain, New Jersey to Florida and Louisiana.

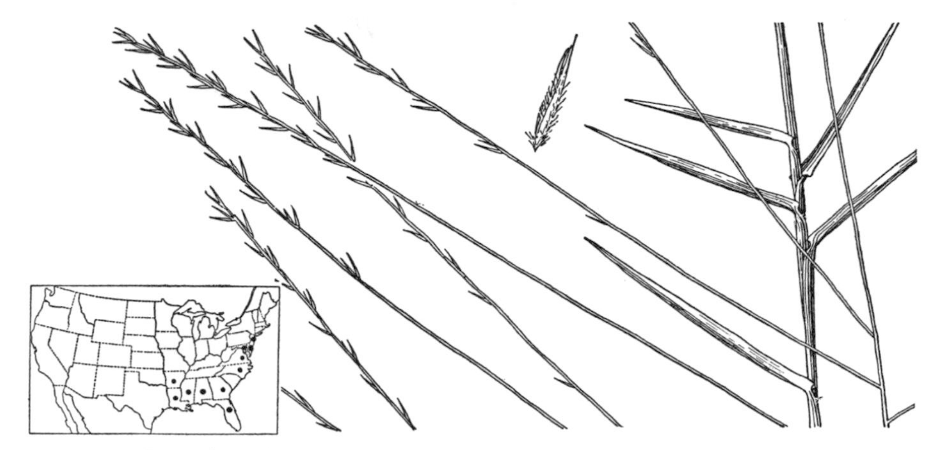

FIGURE 753.—*Gymnopogon brevifolius*. Plant, × 1; floret, × 5. (Chase 3669, Va.)

FIGURE 754.—*Gymnopogon chapmanianus*. Plant, × 1; florets, × 5. (Tracy 7102, Fla.)

**3. Gymnopogon chapmaniánus** Hitchc. (Fig. 754.) Culms 30 to 40 cm. tall, in small tufts, ascending, sparingly branching from lower nodes, rigid; leaves approximate toward the base, the blades 5 to 6 cm. long, about 5 mm. wide, sharp-pointed, often subinvolute in drying; spikes ascending to spreading (not reflexed), floriferous from base, spikelets not remote, 2- or 3-flowered, the florets somewhat spreading; lemmas pubescent, with a minute awn or awnless; palea very narrow, arched. ♃ —Sandy pinelands, Florida.

**4. Gymnopogon floridánus** Swallen. (Fig. 755.) Plants in small tufts, commonly purple below; culms 15 to 45 cm. tall; sheaths glabrous, overlapping, and crowded toward the base, minutely hairy in the throat, the uppermost elongate; blades firm, mostly about 3 cm. long, 2 to 4 mm. wide, sometimes to 6 cm. long and 6 mm. wide, flat, stiffly spreading; spikes 5 to 20, very slender, 10 to 15 cm. long, spreading or reflexed, spikelet-bearing to the base or nearly so; spikelets 2- or 3-flowered, 3 to 5 mm. long; glumes about equal, acuminate, as long as the florets, not spreading; lemma 2 to 2.2 mm. long. ♃ —Sandy prairies and pine barrens, peninsular Florida.

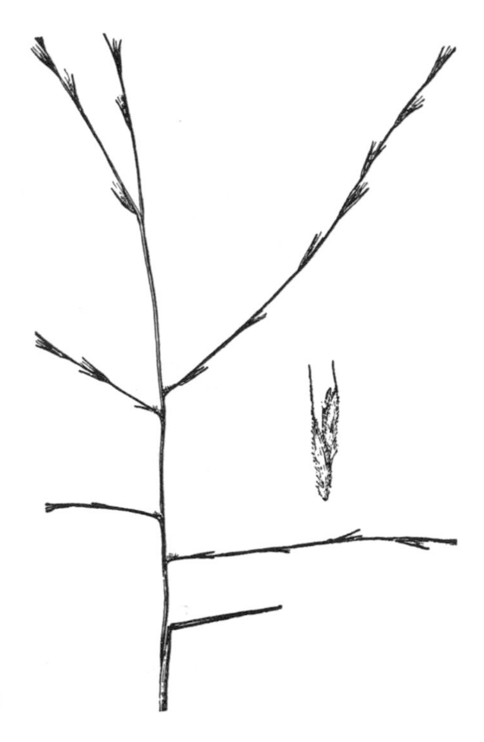

FIGURE 755.—*Gymnopogon floridanus*. Panicle, × 1; florets, × 5. (Type.)

## 110. CHLÓRIS Swartz. FINGERGRASS

Spikelets with 1 perfect floret, sessile, in 2 rows along one side of a continuous rachis, the rachilla disarticulating above the glumes, produced beyond the perfect floret and bearing 1 to several reduced florets consisting of empty lemmas (a few species occasionally with a second fertile floret), these often truncate, and, if more than 1, the smaller ones usually enclosed in the lower, forming a somewhat club-shaped rudiment; glumes somewhat unequal, the first shorter, narrow, acute; lemma keeled, usually broad, 1- to 5-nerved, often villous on the callus and villous or long-ciliate on the keel or marginal nerves, awned from between the short teeth of a bifid apex, the awn slender or sometimes reduced to a mucro, the sterile lemmas awned or awnless. Tufted perennials or sometimes annuals with flat or folded scabrous blades and 2 to several, sometimes showy and feathery, spikes aggregate at the summit of the culms. Type species, *Chloris cruciata* (L.) Swartz. Named for Greek *Chloris*, the goddess of flowers.

Several species are found on the plains of Texas, where they form part of the forage for grazing animals. *C. virgata* is a rather common annual weed in the Southwest, especially in alfalfa fields. It may be locally abundant and then furnishes considerable forage. *C. gayana*, Rhodes grass, is cultivated in the irrigated regions of the Southwest, where it is valuable as a meadow grass. It is also used in the Hawaiian Islands on some ranches in the drier regions. In a few species 2 or 3 internodes of the culm may be greatly reduced, bringing the nodes and sheaths close together.

Lemmas firm, dark brown, awnless or mucronate. Perennials with strongly compressed

culms and sheaths, and firm flat or folded blades abruptly rounded at the tip.
SECTION 1. EUSTACHYS.
Lemmas distinctly awned (awn very short in *C. cucullata*), pale or fuscous.
SECTION 2. EUCHLORIS.

*Section 1. Eustachys*

Spikes numerous, usually more than 10 ........ 1. C. GLAUCA.
Spikes usually not more than 6.
Spikelets 2 mm. long; lemmas dark ........ 2. C. PETRAEA.
Spikelets 3 mm. long; lemmas pale to golden brown until maturity.
Spikes 2, sometimes 1 or 3 ........ 3. C. FLORIDANA.
Spikes 4 to 6 ........ 4. C. NEGLECTA.

*Section 2. Euchloris*

Rudiment narrow, oblong, acute, often inconspicuous. (Second rudiment truncate in *C. gayana*).
Plant producing long, stout stolons ........ 5. C. GAYANA.
Plant not stoloniferous (occasionally with short stolons in *C. andropogonoides*).
Fertile lemma about 2.5 mm. long; plants mostly less than 50 cm. tall; spikes mostly less than 10 cm. long ........ 7. C. ANDROPOGONOIDES.
Fertile lemma 4 to 7 mm. long; plants 40 to 100 cm. or more tall; spikes mostly more than 10 cm. long.
Blades folded, abruptly acute or rounded; spikes whorled, naked at base.
8. C. TEXENSIS.
Blades flat, long-acuminate; spikes racemose on a short axis, solitary or in small fascicles ........ 6. C. CHLORIDEA.
Rudiment truncate-broadened at apex, usually conspicuous (rather narrow in *C. virgata*).
Lemma conspiculusly ciliate-villous, the spikes feathery.
Plants annual. Lemma long-ciliate on the lateral nerves near apex ........ 9. C. VIRGATA.
Plants perennial.
Spikes flexuous, nodding, mostly 10 to 15 cm. long; hairs much exceeding the spikelets ........ 10. C. POLYDACTYLA.
Spikes straight or subflexuous, 5 to 7 cm. long; hairs about equaling the spikelets.
11. C. CILIATA.
Lemma minutely ciliate on the nerves or glabrous, the spikes not feathery.
Awn of fertile lemma usually 3 to 8 mm. long; spikes mostly 7 to 12 cm. long, the spikelets not closely crowded ........ 12. C. VERTICILLATA.
Awn of fertile lemma usually less than 3 mm. long; spikes usually less than 6 cm. long, the spikelets crowded.
Awns about 1 mm. long; rudiment prominent, inflated, broadly triangular-truncate, about 1.5 mm. wide as folded at summit ........ 15. C. CUCULLATA.
Awns 2 to 3 mm. long; rudiment not inflated, not more than 1 mm. wide as folded at summit.
Rudiment oblong-cuneate, about 0.6 mm. wide as folded at summit.
13. C. SUBDOLICHOSTACHYA.
Rudiment triangular-truncate, about 1 mm. wide as folded at summit.
14. C. LATISQUAMEA.

SECTION 1. EÚSTACHYS (Desv.) Reichenb.

Lemmas firm, brown to blackish, awnless or mucronate only; glumes scabrous, the second mucronate from a notched or truncate summit. Perennials.

**1. Chloris glaúca** (Chapm.) Wood. (Fig. 756.) Glaucous; culms erect, compressed, stout, 70 to 150 cm. tall; basal sheaths several, broad, compressed, keeled, overlapping and equitant, those of the succeeding 1 or 2 distant nodes similar, 2 to 4 leaves aggregate; blades flat or folded, as much as 1 cm. wide, the tip abruptly rounded; spikes several to many (as many as 20), ascending, 7 to 12 cm. long; spikelets about 2 mm. long; lemma glabrous or scaberulous on the nerves. ♃ (*Eustachys glauca* Chapm.)—Brackish marshes, wet prairies, and swamps, North Carolina (Wilmington), Georgia (Baker County), and Florida.

**2. Chloris petraéa** Swartz. (Fig. 757.) Often glaucous, sometimes purplish; culms slender, 50 to 100 cm. tall, more or less decumbent and root-

FIGURE 756.—*Chloris glauca*. Plant, × 1; florets, × 5. (Combs and Baker 1143, Fla

ing or producing distinct stolons; sheaths compressed, strongly keeled, usually 2 to 4 aggregate below; blades 3 to 8 mm. wide, often short and numerous on the stolons; spikes mostly 4 to 6, 4 to 10 cm. long; spikelets 2 mm. long; lemma mucronate, short-ciliate on the nerves. ♃ (*Eustachys petraea* Desv.)—Strands, sandy fields, and open pine woods, Coastal Plain, North Carolina to Florida and Texas; tropical America.

**3. Chloris floridána** (Chapm.) Wood. (Fig. 758.) Culms slender, 40 to 80 cm. tall; sheaths compressed, crowded at base but not paired or aggregate at succeeding nodes; blades 3 to 7 mm. wide, somewhat narrowed toward the acutish tip; spikes mostly 2, sometimes 1 or 3, 5 to 10 cm. long; spikelets 3 mm. long; second glume with an awn about 1 mm. long; lemma with a slender mucro 0.5 to 1 mm. long, stiffly ciliate on keel and lateral nerves. ♃ (*Eustachys floridana* Chapm.)—Dry sandy woods and open ground, Georgia and Florida.

**4. Chloris neglécta** Nash. (Fig. 759.) Differing from *C. floridana* in having usually taller, stouter culms, the leaves sometimes paired at the lower nodes; spikes 3 to 8, mostly 4 to 6. ♃ (*Eustachys neglecta* Nash.) —Open sandy woods and swamps, Florida.

**Chloris distichophýlla** Lag. Culms about 1 m. tall; spikes several (as many as 20), drooping, feathery; lemma ciliate with silky hairs 1 mm. long. ♃ —Escaped from cultivation in southern California. A specimen from Bastrop, Tex., is probably also an escape from cultivation; South America.

CHLORIS ARGENTÍNA (Hack.) Lillo and Parodi. Culms erect, compressed, 30 to 90

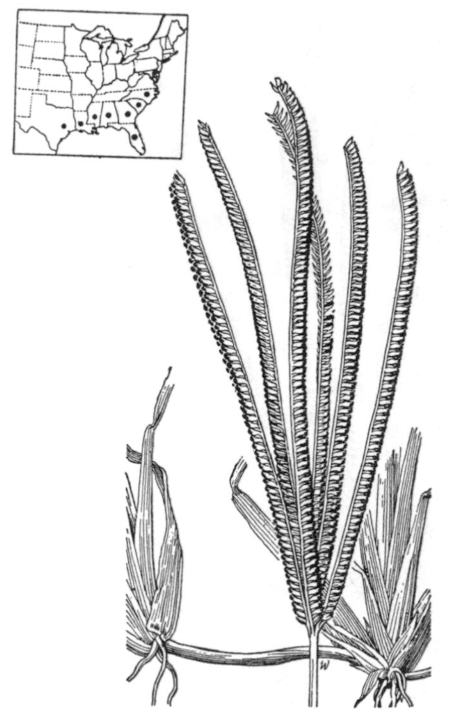

FIGURE 757.—*Chloris petraea.* Plant, × 1; florets, × 5. (Curtiss, Fla.)

FIGURE 758.—*Chloris floridana.* Panicle, × 1; florets, × 5. (Nash 2198, Fla.)

cm. tall; leaves mostly crowded toward the base, the sheaths compressed, keeled, the blades short, 4 to 10 mm. wide; racemes 7 to 12, mostly 5 to 10 cm. long, erect or ascending, crowded, brown, appearing feathery from the cilia on the margins of the lemma; spikelets about 2 mm. long. ♃ —Introduced from Argentina. Roadsides near Tifton, Ga. Probably escaped from cultivation.

CHLORIS CAPÉNSIS (Houtt.) Thell. Stoloniferous perennial; culms 40 to 75 cm. tall; blades obtuse; spikes few to several, finally

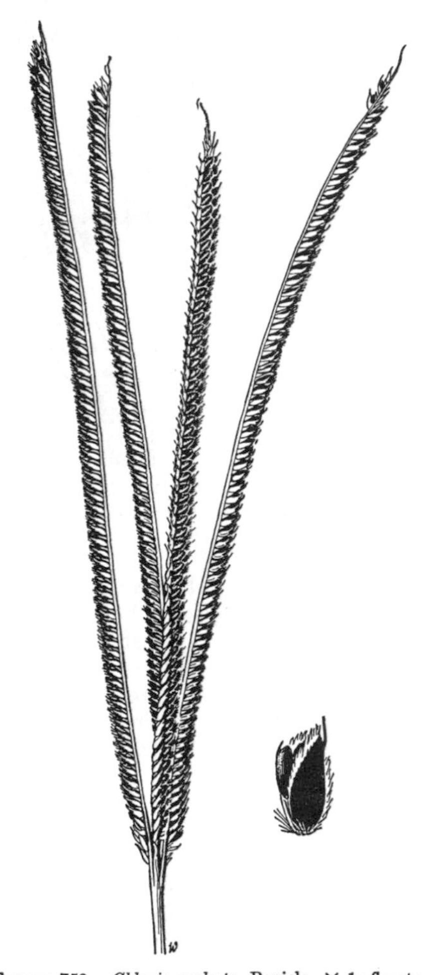

FIGURE 759.—*Chloris neglecta.* Panicle, × 1; florets, × 5. (Curtiss 3445, Fla.)

arcuate-spreading; spikelets about 2.5 mm. long, the glumes short-awned, the brown lemmas white-ciliate on the keel and margin, awnless. ♃ —Introduced from South Africa. Levy County, Fla. Probably escaped from cultivation.

FIGURE 760.—*Chloris gayana*. Plant, × ½; florets, × 5. (Hitchcock 13667, Ariz.)

SECTION 2. EUCHLÓRIS Endl.

Lemmas tawny to grayish or fuscous, awned; glumes acute to acuminate. Mostly perennial.

**5. Chloris gayána** Kunth. RHODES GRASS. (Fig. 760.) Culms 1 to 1.5 m. tall with long, stout, leafy stolons, the internodes compressed, tough and wiry; blades 3 to 5 mm. wide, tapering to a fine point; spikes several to numerous, erect or ascending, 5 to 10 cm. long; spikelets crowded, pale-tawny; lemma 3 mm. long, hispid on the margin near the summit, more or less hispidulous below, the awn 1 to 5 mm. long; rudiment commonly of 2 florets, the lower occasionally fertile, rather narrow, the awn usually somewhat shorter than that of the fertile lemma, the upper minute, broad, truncate. ♃ —Cultivated for forage in warmer regions, escaped into fields and waste places, North Carolina and from Florida to southern California and in tropical America. Introduced from Africa. A promising meadow grass in irrigated regions.

**6. Chloris chlorídea** (Presl) Hitchc. (Fig. 761.) Culms slender, 60 to 100 cm. tall; blades flat, 3 to 7 mm. wide, long-acuminate; spikes slender, few to several, mostly 8 to 15 cm. long, approximate on an axis 2 to 10 cm. long; spikelets appressed, not crowded; lemma narrow, glabrous, somewhat scaberulous toward the tip, about 6 mm. long, the awn 10 to 12 mm. long; rudiment very narrow, awned. ♃ (*C. clandestina* Scribn. and Merr.)—Open ground, Texas (Brownsville), Arizona, Mexico, and Honduras. Large cleistogamous spikelets are borne on slender underground branches, rather rare in herbarium specimens, either infrequent or readily broken off.

**7. Chloris andropogonoídes** Fourn. (Fig. 762.) Culms densely tufted, 20 to 40 cm. tall, the leaves mostly basal; blades about 1 mm. wide as folded; spikes slender, few to several, 5 to 10 cm. long, whorled, divergent, floriferous from base; spikelets scarcely overlapping; lemma minutely pubescent on midnerve and margin or

FIGURE 761.—*Chloris chloridea.* Terminal and subterranean inflorescences, × 1; florets, × 5. (Silveus 379, Tex.)

FIGURE 762.—*Chloris andropogonoides*. Panicles, × 1; florets, × 5. (Chase 6067, Tex.)

glabrous, 2 to 3 mm., usually about 2.5 mm. long, awned below the tip, the awn about 5 mm. long; rudiment narrow, the awn usually shorter than that of the lemma. ♃ (*C. tenuispica* Nash.)—Plains, Texas and northern Mexico.

**8. Chloris texénsis** Nash. (Fig. 763.) Culms taller and stouter than in *C. andropogonoides;* blades 2 to 3 mm. wide as folded; spikes slender, mostly about 15 to 18 cm. long, naked for 1 to 4 cm. at the base; spikelets appressed, not crowded; lemma about 4 mm. long, naked on the midnerve, minutely pilose on margin toward summit; awn about 1 cm. long. ♃ (*C. nealleyi* Nash.) —Plains, Texas, rare.

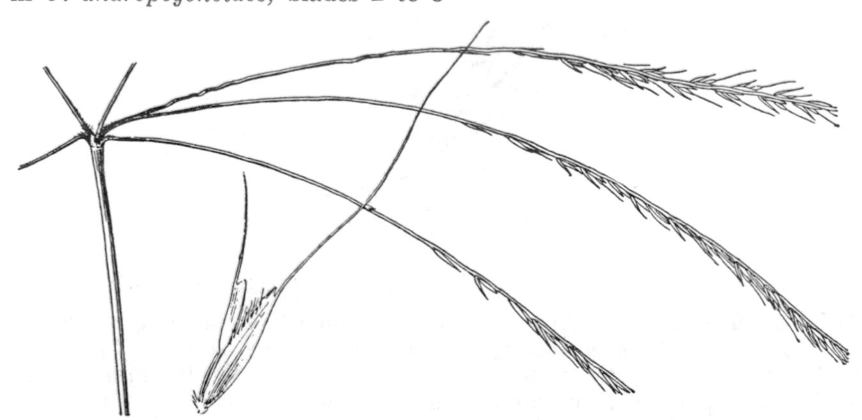

FIGURE 763.—*Chloris texensis*. Panicle, × 1; florets, × 5. (Thurow 8, Tex.)

FIGURE 764.—*Chloris virgata*. Plant, × ½; glumes and florets, × 5. (Tracy 8173, Tex.)

**Chloris prieúrii** Kunth. Annual; culms 30 to 60 cm. tall, often rooting at the lower nodes; blades 2 to 6 mm. wide, the upper sheath inflated; spikes 2 to 8, erect, 5 to 8 cm. long; fertile lemma 2.5 mm. long, narrow, ciliate near the summit, with a delicate awn 7 to 10 mm. long; rudiment narrow, of 3 or 4 reduced sterile lemmas each with a long delicate erect awn. ⊙ —Ballast, Wilmington, N. C., and Mobile, Ala.; West Africa.

**9. Chloris virgáta** Swartz. FEATHER FINGERGRASS. (Fig. 764.) Annual; culms ascending to spreading, 40 to 60 or even 100 cm. tall; upper sheaths often inflated; blades flat, 2 to 6 mm. wide; spikes several, 2 to 8 cm. long, erect, whitish or tawny, feathery or silky; spikelets crowded; lemma 3 mm. long, somewhat humpbacked on the keel, long-ciliate on the margins near the apex, the slender awn 5 to 10 mm. long; rudiment narrowly cuneate, truncate, the awn as long as that of the lemma. ⊙ (*C. elegans* H. B. K.)—Open ground, a common weed in fields and waste places; Nebraska to Texas and southern California; Maine and Massachusetts, on wool waste; introduced in a few localities in the Eastern States, Ohio, Indiana, and North Carolina to Florida; Louisiana and Missouri; tropical America.

**10. Chloris polydáctyla** (L.) Swartz. (Fig. 765.) Culms erect, wiry, 50 to 100 cm. tall; blades as much as 1 cm. wide; spikes several to many, mostly 10 to 15 cm. long, flexuous, nodding, tawny, feathery; spikelets crowded; lemma ciliate with long silky hairs; rudiment oblong, obliquely truncate, awns of lemma and rudiment about 3 mm. long. ♃ —Open sandy soil, southern Florida; West Indies to Paraguay.

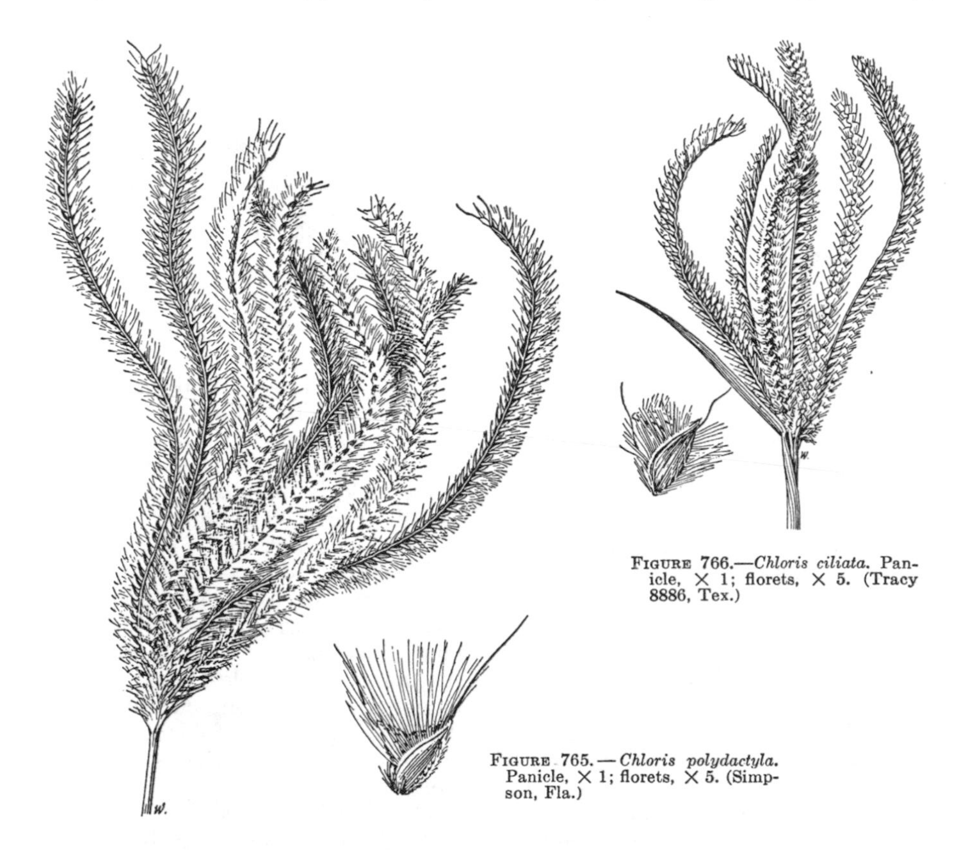

FIGURE 766.—*Chloris ciliata*. Panicle, × 1; florets, × 5. (Tracy 8886, Tex.)

FIGURE 765.—*Chloris polydactyla*. Panicle, × 1; florets, × 5. (Simpson, Fla.)

**11. Chloris ciliáta** Swartz. (Fig. 766.) Perennial; culms erect or ascending, 50 to 100 cm. tall; leaves not aggregate toward the base, sheaths not much compressed; blades 3 to 5 mm. wide, sharply acuminate;

FIGURE 767.—*Chloris verticillata*. Panicle, × 1; florets, × 5. (Ball 1112, Tex.)

spikes mostly 3 to 6, usually 5 to 7 cm. long, digitate or nearly so, erect to spreading, somewhat flexuous; spikelets crowded, about 3 mm. long; lemma densely long-villous on the keel and the middle of the margin, the awn shorter than the body; rudiment triangular-cuneate, about 2 mm. wide. ♃ (*C. nashii* Heller.) —Open grassland, southern Texas and Mexico.

**12. Chloris verticilláta** Nutt. WINDMILL GRASS. (Fig. 767.) Culms tufted, 10 to 40 cm. tall, erect or decumbent at base, sometimes rooting at the lower nodes; leaves crowded at base, 2 to 4, sometimes aggregate at lower nodes; sheaths compressed, blades 1 to 3 mm. wide, obtuse; spikes slender, 7 to 10 or even 15 cm. long, in 1 to 3 whorls, finally widely spreading; spikelets about 3 mm. long; fertile lemma pubescent on the nerves, the awn mostly 5 to 8 mm. long; rudiment (rarely fertile), cuneate-oblong, rather turgid, about 0.7 mm. wide as folded, truncate, the awn about 5 mm. long. ♃ —Plains, Missouri to Colorado, south to Louisiana and Arizona; introduced in Maryland, Indiana, Illinois, and California (Berkeley). The inflorescence at maturity breaks away and rolls before the wind as a tumbleweed.

**13. Chloris subdolichostáchya** C. Muell. (Fig. 768.) Similar to *C. verticillata*, but not more than 20 cm. tall, spikes mostly less than 6 cm. long, these more condensed and usually in one whorl or irregularly approximate; lemma 2 to 2.5 mm. long, the awns mostly less than 3 mm. long; rudiment oblong-cuneate, about 0.6 mm. wide as folded. ♃ (*C. brevispica* Nash.)—Plains, Kansas, Texas.

**14. Chloris latisquámea** Nash. (Fig. 769.) Culms densely tufted, 20 to 60 cm. tall, very leafy at base, sometimes rooting at the lower nodes; sheaths compressed, 2 to 4 often aggregate at the lower node; blades 2 to 4 mm. wide; spikes mostly 8 to 12, relatively broad, 4 to 10 cm. long, in 1 or 2 whorls, spreading; spikelets rather crowded, pale, turning fuscous at maturity; lemma about 2.5 mm. long, pubescent on the nerves, the awn 2 to 2.5 mm. long; rudiment (rarely fertile) triangular cuneate,

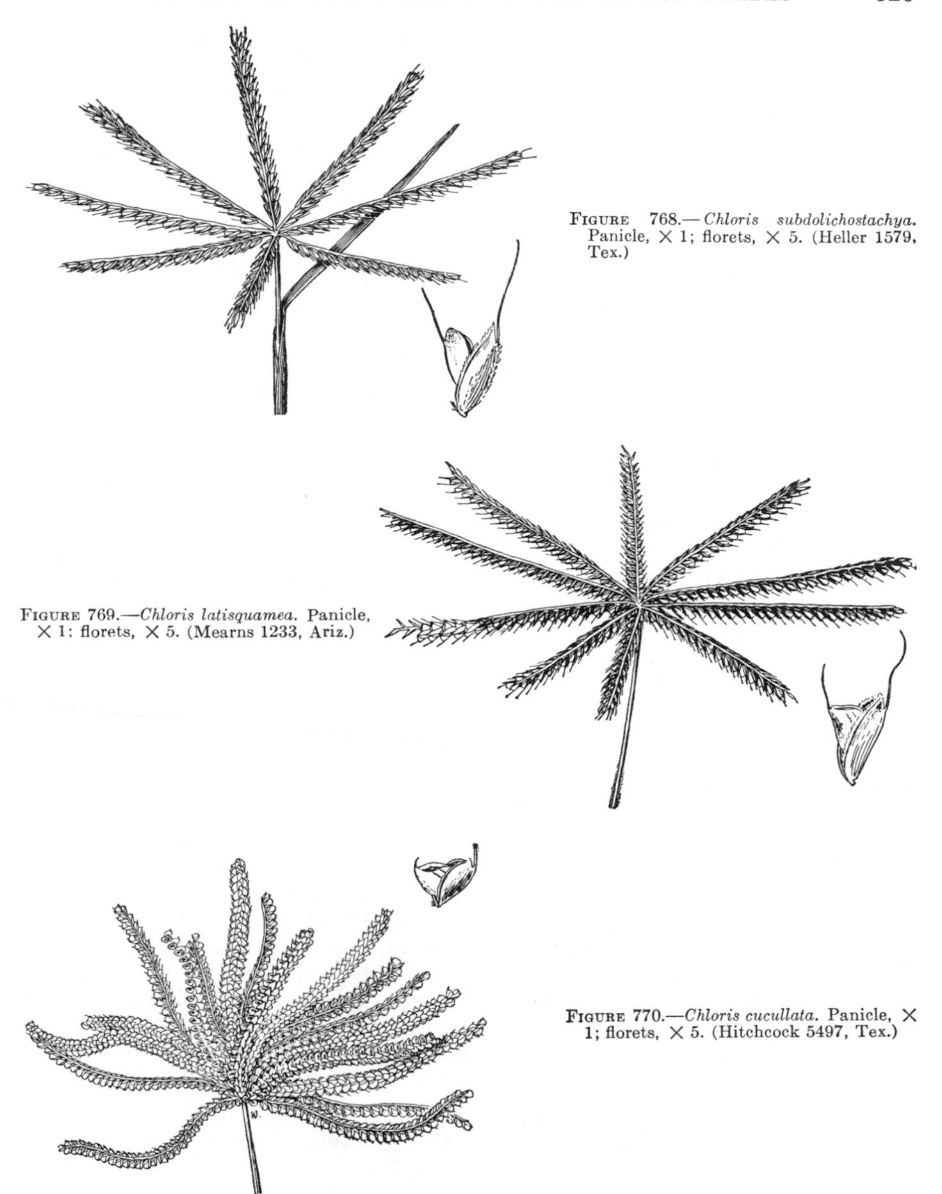

FIGURE 768.—*Chloris subdolichostachya.* Panicle, × 1; florets, × 5. (Heller 1579, Tex.)

FIGURE 769.—*Chloris latisquamea.* Panicle, × 1; florets, × 5. (Mearns 1233, Ariz.)

FIGURE 770.—*Chloris cucullata.* Panicle, × 1; florets, × 5. (Hitchcock 5497, Tex.)

about 1 mm. wide at summit as folded. ♃ —Plains, Texas, Arizona. Resembling *C. cucullata*, but commonly taller with longer spikes, the rudiment longer than broad, less inflated, the awns 2 to 2.5 mm. long.

**15. Chloris cuculláta** Bisch. (Fig. 770.) Culms tufted, erect or somewhat spreading at base, 20 to 50 cm. tall; sheaths compressed; blades 1 to 2 mm. wide as folded, the uppermost often much reduced; spikes numerous, 2 to 5 cm. long, digitate, radiating, flexuous or curled; spikelets crowded, stramineous, turning fuscous at maturity, triangular, about 2 mm. long and about as broad; rudiment prominent, compressed-cup-shaped, about 1.5 mm. wide, the awns of lemma and rudiment about 1 mm. long. ♃ —Plains and sandy barrens, Texas, Oklahoma, and New Mexico.

**Chloris radiáta** (L.) Swartz. Weedy branching annual; culms 30 to 40 cm. long, decumbent; blades thin, 2 to 3 mm. wide; spikes slender, several to many, 3 to 8 cm. long; lemma narrow, 2.5 mm. long, the narrow rudiment mostly included in its margins; awns of lemma and rudiment very slender, 5 to 10 mm. long. ⊙ —Ballast, near Portland, Oreg.; tropical America.

**Chloris submútica** H. B. K. Sparingly stoloniferous, culms 30 to 65 cm. tall; sheaths compressed-keeled; spikes 5 to 14, 3 to 8 cm. long, somewhat whorled on a short axis, spreading; spikelets 3 to 3.5 mm. long; fertile floret 3 to 3.5 mm. long, the callus bearded, the lemma obtuse, pilose toward the summit, awnless or mucronate; rudiment truncate, awnless ♃ —Dona Ana County, N. Mex., probably escaped from cultivation. Mexico.

CHLORIS BÉRROI Arech. Densely tufted, culms 40 to 65 cm. tall, leafy; spikes and spikelets much like those of *C. ciliata*, but the 2 to 5 spikes closely and permanently appressed, the rachises adhering, forming a subcylindrical silky inflorescence. ♃ —Occasionally cultivated, Oklahoma and Texas, introduced from Uruguay.

CHLORIS VENTRICÓSA R. Br. Culms straggling and rooting at the nodes, 40 to 90 cm. long; spikes 3 to 5, 7 to 10 cm. long, flexuous, spreading or drooping; spikelets about 5 mm. long; fertile lemma subindurate, brown, truncate, glabrous except for the pubescent callus, awn 4 to 5 mm. long, that of the truncate rudiment 1 to 2 mm. long. ♃ —Occasionally cultivated, Virginia and Oklahoma; introduced from Australia.

CHLORIS CANTÉRAI Arech. Perennial, resembling *C. polydactyla*, but blades only 2 to 5 mm. wide; spikes 2 to 4; spikelets slightly larger. ♃ —Spontaneous along roadsides and in uncultivated ground, Bexar County, Texas, introduced from Paraguay.

CHLORIS TRUNCÁTA R. Br. Stoloniferous perennial; culms erect, 10 to 30 cm. tall; spikes 6 to 10, 7 to 15 cm. long, horizontal or reflexed; spikelets 3 mm. long, the awns 6 to 12 mm. long. ♃ —Occasionally cultivated for ornament under the name star-grass. Australia.

## 111. TRICHLÓRIS Fourn.

Spikelets 2- to 5-flowered, nearly sessile, in 2 rows along one side of a continuous slender rachis, the rachilla disarticulating above the glumes and prolonged behind the uppermost perfect floret, bearing a reduced, usually awned floret; glumes unequal, acuminate, or short-awned, the body shorter than the lower lemma; lemmas narrow, 3-nerved, the midnerve and usually the lateral nerves extending into slender awns. Erect, slender, tufted perennials, with flat scabrous blades and numerous erect or ascending spikes, aggregate but scarcely digitate at the summit of the culms. Type species, *Trichloris pluriflora*. Name from Latin *tri*, three, and *Chloris*, a genus of grasses, the lemmas being 3-awned.

Spikelets 2-flowered, both lemmas with 3 long awns........ 1. T. CRINITA.
Spikelets 3- to 5-flowered, the lateral awns of the lemmas more or less reduced, sometimes obsolete........ 2. T. PLURIFLORA.

**1. Trichloris crinita** (Lag.) Parodi. (Fig. 771, *A*.) Culms 40 to 100 cm. tall; blades 2 to 4 mm. wide; inflorescence dense, feathery, the spikes 5 to 10 cm. long; spikelets crowded; fertile lemma about 3 mm. long, the second lemma much reduced, both with delicate awns about 1 cm. long. ♃ (*T. mendocina* (Phil.) Kurtz.)—Plains, canyons, and rocky hills, western Texas to Arizona and northern Mexico; southern South America. Rarely cultivated for ornament (as *T. blanchardiana* Fourn.).

**2. Trichloris plurifióra** Fourn. (Fig. 771, *B*.) Culms 50 to 100 cm. tall; blades 5 to 10 mm. wide; inflorescence looser and less feathery than in *T. crinita;* spikes 7 to 15 cm. long; fertile

FIGURE 771.—*A*, *Trichloris crinita*. Plant, × ½; glumes and florets, × 5. (Nealley, Tex.) *B*, *T. pluriflora*. Glumes and florets, × 5. (Griffiths 6484, Tex.)

lemma about 4 mm. long, the others successively shorter, the middle awns of all 5 to 15 mm. long, somewhat spreading, the lateral awns short or obsolete. ♃ —Plains and dry woods, southern Texas and Mexico; southern South America.

## 112. BOUTELÓUA Lag. Grama

Spikelets 1-flowered, with the rudiments of 1 or more florets above, sessile, in 2 rows along one side of the rachis; glumes 1-nerved, acuminate or awn-tipped, the first shorter and narrower; lemma as long as the second glume or a little longer, 3-nerved, the nerves extending into short awns or mucros, the internerves usually extending into lobes or teeth; palea sometimes 2-awned; rudiment various, usually 3-awned, the awns usually longer than those of the fertile lemma, a second rudimentary floret sometimes present. Perennial or sometimes annual, low or rather tall grasses, with 2 to several or many spikes racemose on a common axis, or sometimes solitary, the spikelets few to many in each spike, rarely solitary, pectinate or more loosely arranged and appressed, the rachis of the spike usually naked at the tip. The sterile florets forming the rudiment are variable in all the species and commonly in individual specimens. The general pattern of rudiment is fairly constant for each species, the variability being in the reduction or increase in number and size of the sterile florets, the reduction from 3 awns to 1, and in the amount of pubescence. Type species, *Bouteloua racemosa* Lag. (*B. curtipendula*). Named for the brothers, Boutelou, Claudio, and Esteban. The genus was originally published as *Botelua*.

The many species are among our most valuable forage grasses, forming an important part of the grazing on the western ranges. *B. gracilis*, blue grama, and *B. hirsuta*, hairy grama, are prominent in "short grass" regions of the Great Plains; *B. eriopoda*, black grama, and *B. rothrockii*, Rothrock grama, are prominent in Arizona. Two annuals, *B. barbata* and *B. parryi*, form a part of the sixweeks grasses of the Southwest; *B. curtipendula* is widely distributed and is much used for grazing and for hay; *B. trifida* is important from Texas to Arizona.

Spikelets not pectinately arranged (except in *B. chondrosioides*), the spikes falling entire at maturity .......... Section 1. Atheropogon.
Spikelets pectinately arranged, the spikes persistent, the florets falling from the persistent glumes .......... Section 2. Chondrosium.

*Section 1. Atheropogon*

Plants annual .......... 1. B. aristidoides.
Plants perennial.
  Spikes usually 20 to 50; awns short, inconspicuous.
    Spikes of 1 or 2 spikelets; culms very slender .......... 2. B. uniflora.
    Spikes of few to several spikelets; culms mostly stouter .......... 3. B. curtipendula.
  Spikes fewer; awns conspicuous.
    Glumes pubescent.
      Spikes rhomboid-oblong, as much as 2 cm. long, the spikelets somewhat pectinately arranged .......... 6. B. chondrosioides.
      Spikes cuneate-triangular, about 1 cm. long (including the awns), the spikelets appressed, not pectinately arranged.
        Culms 20 to 30 cm. tall; leaves crowded at base; spikes mostly 6 to 8. 4. B. rigidiseta.
        Culms mostly 30 to 50 cm. tall, leafy throughout; spikes mostly more than 10. 5. B. eludens.
    Glumes glabrous or scabrous, not pubescent.
      Base of plants hard, rhizomatous; culms mostly simple; spikes 2 to 3 cm. long. 7. B. radicosa.
      Base of plants not rhizomatous; culms branching; spikes usually about 1.5 cm., sometimes 2 cm., long .......... 8. B. filiformis.

*Section 2. Chondrosium*

Plants annual (see also *B. rothrockii*); densely tufted, spreading.
  Spike 1 .......... 9. B. SIMPLEX.
  Spikes 2 or more.
    Rachis papillose-pilose .......... 11. B. PARRYI.
    Rachis not pilose .......... 10. B. BARBATA.
Plants perennial.
  Plants decumbent or stoloniferous; culms white-lanate .......... 17. B. ERIOPODA.
  Plants erect or nearly so; culms tufted, not lanate.
    Spikes normally 2, sometimes 1 or 3.
      Rachis prolonged beyond the spikelets as a naked point; glumes tuberculate.
        Culms retrorsely hirsute below the nodes .......... 13. B. GLANDULOSA.
        Culms glabrous .......... 14. B. HIRSUTA.
      Rachis not prolonged; glumes not tuberculate (slightly so in *B. gracilis*).
        Culms herbaceous, the base not woody .......... 15. B. GRACILIS.
        Culms woody and perennial at base .......... 16. B. BREVISETA.
    Spikes normally 4 or more (see also *B. gracilis* var. *stricta*).
      Culms 25 to 50 cm. tall; awn 1 to 2 mm. long; glumes scabrous; spikes spreading. 12. B. ROTHROCKII.
      Culms 10 to 20 cm. tall; awn about 5 mm. long; glumes glabrous; spikes usually appressed .......... 18. B. TRIFIDA.

SECTION 1. ATHEROPÓGON (Muhl.) Endl.

Spikes deciduous from the main axis; spikelets not pectinately arranged (somewhat so in *B. chondrosioides*). (*Atheropogon* Muhl. based on *A. apludoides* Muhl. (*Bouteloua curtipendula*).)

**1. Bouteloua aristidoídes** (H. B. K.) Griseb. NEEDLE GRAMA. (Fig. 772.) Annual, erect or spreading, branching; culms slender, 10 to 30 cm. tall; blades small and few, in vigorous plants as much as 15 cm. long; spikes mostly 8 to 14 on a slender axis, reflexed, readily falling, the base of the rachis forming a sharp, bearded point; spikelets 2 to 4, narrow, appressed; rudiment of 3 scabrous awns about 5 mm. long, exceeding the fertile floret. ☉ (*Triathera aristidoides* Nash.)—Mesas, deserts, and foothills in open ground, Texas to Nevada, southern California, and northern Mexico; Argentina.

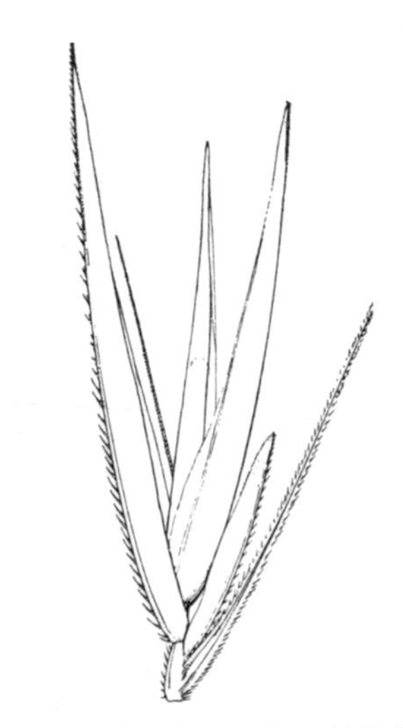

FIGURE 773.—*Bouteloua uniflora*, × 7. (Type.)

FIGURE 772.—*Bouteloua aristidoides*. Panicle, × 1; spikelet, × 5. (Griffiths 7308, Ariz.)

FIGURE 774.—*Bouteloua curtipendula.* Plant, × ½; spikelet and florets, × 5. (Chase 5408, Colo.)

BOUTELOUA ARISTIDOIDES var. ARIZÓNICA Jones. Spikes arcuate, to 2.5 cm. long, with 5 to 10 spikelets. ⊙ —Mesas and deserts, southern Arizona and northern Mexico.

**2. Bouteloua uniflóra** Vasey. (Fig.

773.) Resembles slender forms of *B. curtipendula*, culms slender, wiry, sometimes with slender stolons, the slender blades subinvolute, the spikes 8 to 9 mm. long, with 1 or 2 spikelets, the scabrous rachis mostly longer than the first glume; lemma awnless; rudiment reduced to a single awn appressed to the back of the palea. ♃ —Rocky hills and valleys, central and western Texas.

**3. Bouteloua curtipéndula** (Michx.) Torr. SIDE-OATS GRAMA. (Fig. 774.) Perennial, with scaly rhizomes; culms erect, tufted, 50 to 80 cm. tall; blades flat or subinvolute, 3 to 4 mm. wide, scabrous; spikes 35 to 50, 1 to 2 cm. long, purplish, spreading or pendulous and mostly twisted to one side of the slender axis, this 15 to 25 cm. long; spikelets 5 to 8, appressed or ascending, 6 to 10 mm. long; fertile lemma acute, mucronate; rudiment with 3 awns and subacute intermediate lobes, often reduced and inconspicuous. ♃ (*Atheropogon curtipendulus* Fourn.)—Plains, prairies, and rocky hills, Maine and Ontario to Montana, south to Virginia, Alabama, Texas, Arizona, and southern California; South Carolina (introduced); Mexico to Argentina.

**4. Bouteloua rigidiséta** (Steud.) Hitchc. (Fig. 775.) Perennial, tufted, leafy at base; culms erect, 20 to 30 cm. tall; blades narrow, flat or somewhat involute, 1 to 1.5 mm. wide, sparingly papillose-pilose; spikes 6 to 8, triangular-cuneate, spreading, about 1 to 1.2 cm. long including the awns; spikelets mostly 2 to 4, crowded, ascending; glumes pubescent; fertile lemma with 3 spreading awns, the intermediate lobes acute; rudiment with stout spreading awns, much exceeding those of the fertile lemma, the intermediate lobes firm, pointed, a second similar but smaller rudiment commonly developed. ♃ (*B. texana* S. Wats.; *Polyodon texanus* Nash.)—Plains and rocky hills, Oklahoma, Texas, and northern Mexico.

**5. Bouteloua elúdens** Griffiths. (Fig. 776.) Perennial, densely tufted, leafy at base; culms erect, 25 to 60 cm. tall; blades mostly 1 to 1.5 mm. wide; axis slender, flexuous, 6 to 8 cm. long; spikes 10 to 20, triangular, spreading, about 1 cm. long including the awns; spikelets about 5; rachis and glumes densely pubescent; fertile lemma pubescent toward the summit, the apex 3-cleft, the divisions awn-tipped; rudiment with stout pubescent awns about 5 mm. long, the long narrow intermediate lobes glabrous; a second similar but smaller rudiment usually developed. ♃ —Rocky hills, southern Arizona and Sonora, Mexico.

FIGURE 775.—*Bouteloua rigidiseta*. Panicle, × 1; spikelet, × 7; lemma and florets, × 5. (Griffiths 6370, Tex.)

FIGURE 776.—*Bouteloua eludens.* Panicle, × 1; spike and spikelet, × 5. (Type.)

**6. Bouteloua chondrosioídes** (H. B. K.) Benth. ex S. Wats. (Fig. 777.) Perennial, tufted, leafy at base; culms erect, 20 to 50 cm. tall; blades 2 to 3 mm. wide; axis 4 to 6 cm. long; spikes 4 to 6, rhomboid-oblong, ascending, 1 to 2 cm. long, the rachis densely pubescent, the tip 3-cleft; spikelets several, subpectinate; rachis broad, densely pubescent on the margin; glumes and fertile lemma densely pubescent, the lemma 3-cleft, the divisions awn-tipped; rudiment cleft nearly to the base, the middle awn broadly winged, the lateral ones slender, all spreading. ♃ —Mesas and rocky hills, western Texas to southern Arizona; Mexico and Guatemala.

**7. Bouteloua radicósa** (Fourn.) Griffiths. PURPLE GRAMA. (Fig. 778.) Perennial, tufted, from a stout rhizomatous base; culms erect, 60 to 80 cm. tall; blades 2 to 3 mm. wide, sparsely papillose-ciliate on the margin, mostly aggregate toward the lower part of the culm, the upper part naked; axis 10 to 15 cm. long; spikes mostly 7 to 12, oblong, 2 to 3 cm. long; spikelets mostly 8 to 11; glumes broader than in other species; fertile lemma indurate down the center, with 3 awns, the middle longest, and no intermediate lobes; rudiment with 3 awns 5 to 8 mm. long and no intermediate lobes, usually containing a palea and staminate flower sometimes a perfect flower, the lower floret being staminate. ♃ —Rocky hills, southern New Mexico to southern California and Mexico.

**8. Bouteloua filifórmis** (Fourn.) Griffiths. SLENDER GRAMA. (Fig. 779.) Resembling *B. radicosa;* culms erect or geniculate-spreading, sparingly branching, the base not rhizomatous; spikes ascending to spreading, mostly about 1.5 cm. long, sometimes as much as 2 cm.; spikelets mostly 6 to 10, very like those of *B. radicosa.* ♃ —Rocky hills, Texas to Arizona and Mexico; Panama.

FIGURE 778.—*Bouteloua radicosa.* Panicle, × 1; spikelet, × 5. (Griffiths 7181, Ariz.)

SECTION 2. CHONDRÓSIUM (Desv.) Benth.

Spikes persistent; spikelets crowded (looser in *B. eriopoda*), pectinate; florets falling from the glumes. (*Chondrosium* Desv. based on *C. procumbens* Durand (*B. simplex*).

FIGURE 777.—*Bouteloua chondrosioides.* Panicle, × 1; spikelet, × 5. (Type.)

FIGURE 779.—*Bouteloua filiformis.* Panicle, × 1; spikelet, × 5. (Griffiths 7199, Ariz.)

**9. Bouteloua símplex** Lag. MAT GRAMA. (Fig. 780.) Annual, tufted, prostrate or ascending; foliage scant; blades 2 to 3 cm. long, about 1.5 mm. wide; spike solitary, 1.5 to 2.5 cm.

FIGURE 780.—*Bouteloua simplex*. Plant, × 1; spikelet, × 5. (Griffiths 7362, Ariz.)

long, strongly arcuate at maturity; spikelets mostly 20 to 30, about 5 mm. long; fertile lemma pilose at base with stout awns and subacute intermediate lobes; rudiment bearded at summit of rachilla-joint, cleft to the base or nearly so, the awns equal, a second rudiment, broad and awnless, sometimes developed. ⊙ (*B. procumbens* Griffiths.)—Open ground, Texas to Colorado, Utah, Arizona, and Mexico; wool waste, Maine; Ecuador to Argentina.

**10. Bouteloua barbáta** Lag. SIX-WEEKS GRAMA. (Fig. 781.) Annual, tufted, branching, erect to prostrate, often forming mats with ascending ends, the culms as much as 30 cm. long; foliage scant; blades 1 to 4 cm. long, 1 to 1.5 mm. wide; spikes 4 to 7, 1 to 2 cm. long; spikelets 25 to 40, 2.5 to 4 mm. long, nearly as broad; fertile lemma densely pilose at least along the sides, usually throughout, the awns from minute to as long as the body, the intermediate lobes subacute to obtuse; rudiment from obscurely to conspicuously bearded at summit of rachilla joint, cleft nearly to the base, the intermediate lobes broad, subcucullate, the awns of rudiment and fertile lemma reaching about the same height, a second rudiment, broad and awnless, often developed. ⊙ (*B. microstachya* L. H. Dewey.)—Open ground, mesas, and rocky hills, Texas and Colorado to Nevada and southeastern California; Mexico. The awns vary in length. The form with shorter awns is that described as *B. pumila* Buckl.; the longer awned form is that described as *B. arenosa* Vasey.

**11. Bouteloua párryi** (Fourn.) Griffiths. PARRY GRAMA. (Fig. 782.)

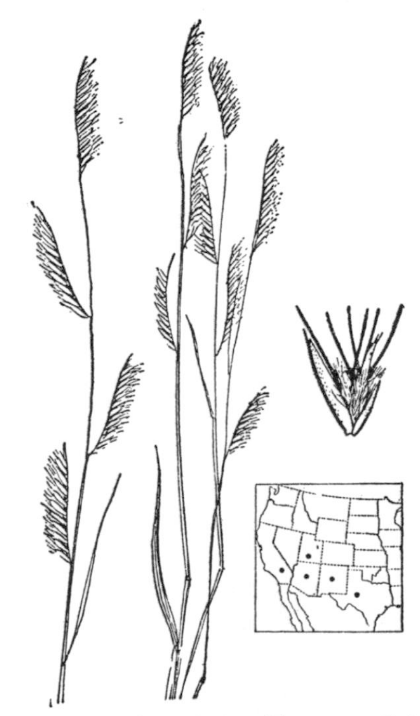

FIGURE 781.—*Bouteloua barbata*. Plant, × 1; spikelet, × 5. (Griffiths 6095, Ariz.)

Annual, resembling *B. rothrockii;* culms erect or geniculate-spreading, sometimes branching; blades papillose-pilose; spikes 4 to 8, often flexuous, commonly grayish purple, 2 to 3.5 cm. long; rachis papillose-pilose; spikelets 40 to 65, about 6 mm. long; second glume awned from a bifid tip, the keel papillose-pilose with spreading hairs; fertile lemma densely pilose, deeply cleft, the awns spreading, the oblong intermediate lobes fimbriate; rudiment densely bearded at summit of rachilla, cleft nearly to the base, the lobes obovate, fimbriate, the awns exceeding those of the fertile lemma; a second rudiment, broad, awnless or with a single awn, usually developed. ⊙ —Mesas and rocky hills, New Mexico, Arizona, and northern Mexico.

**12. Bouteloua rothróckii** Vasey. ROTHROCK GRAMA. (Fig. 783.) Perennial, sometimes appearing to be annual; culms tufted, erect, 25 to 50 cm. tall; blades 2 to 3 mm. wide; axis 10 to 25 cm. long; spikes 4 to 12, 2.5 to 3 cm. long, straight to subarcuate;

FIGURE 783.—*Bouteloua rothrockii.* Panicle, × 1; spikelet, × 4. (Griffiths 7185, Ariz.)

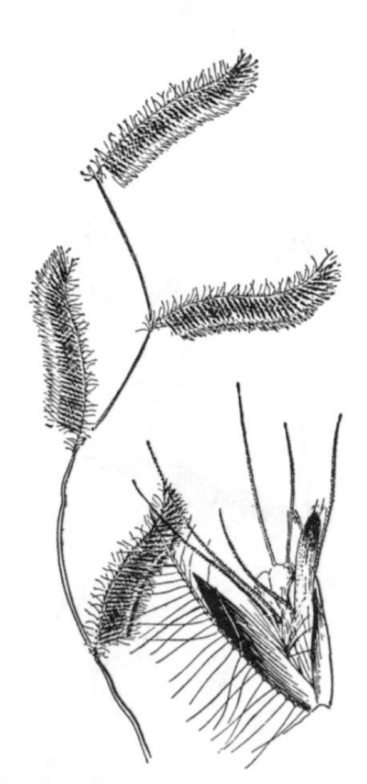

FIGURE 782.—*Bouteloua parryi.* Panicle, × 1; spikelet, × 5. (Griffiths 7277, Ariz.)

spikelets 40 to 50, about 5 mm. long; fertile lemma pilose at base, deeply cleft, the awns (1 to 2 mm. long) spreading, the intermediate and lateral lobes fimbriate; rudiment densely bearded at summit of rachilla joint, cleft nearly to the base, the lobes broad and rounded, the awns mostly exceeding those of the fertile lemma; a second rudiment, broad and awnless, usually developed. ♃ —Mesas, canyons, and rocky hills, in open ground, or among brush, Arizona and southern California (Jamacha), to northern Mexico.

**13. Bouteloua glandulósa** (Cervant.) Swallen. (Fig. 784.) Similar to *B. hirsuta;* lower part of the culms and the lower sheaths conspicuously papillose-hirsute with ascending or spreading hairs; blades flat, attenuate, 2 to 3 mm. wide, more or less ciliate or hairy toward the base; spikes 1 to 3, ascending to reflexed, the rachis prolonged beyond the spikelets as a prominent bristle, commonly 1 to 1.5 cm. long; spikelets similar to those of *B. hirsuta*, but the awns of the rudiment somewhat longer, the spikes more bristly. ♃ *B. hirticulmis* Scribn.—Rocky hills, prairies, and open ground, Arizona (Santa Cruz County); Mexico.

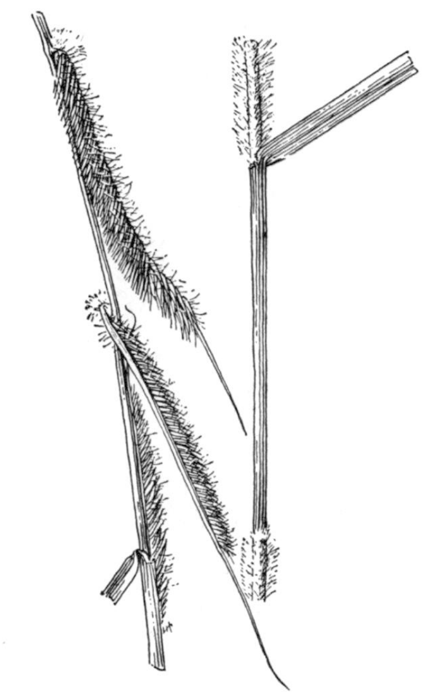

FIGURE 784.—*Bouteloua glandulosa*. Panicle, × 1. (Type of *B. hirticulmis*.)

**14. Bouteloua hirsúta** Lag. HAIRY GRAMA. (Fig. 785.) Perennial, densely tufted; culms erect, 20 to 60 cm. tall, leafy at base; blades flat or subinvolute, about 2 mm. wide, flexuous; spikes 1 to 4, usually 2, 2.5 to 3.5 cm. long, the rachis extending beyond the spikelets as a slender point 5 to 8 mm. long; spikelets 35 to 45, about 5 mm. long, second glume tuberculate-hirsute with spreading hairs, the tubercles black; fertile lemma 3-cleft, the divisions and margins of lemma pubescent, awn-tipped; rudiment from puberulent to bearded at summit of rachilla, cleft nearly to the base, the lobes firm, broad, spreading, the awns black. ♃ —Plains and rocky hills, Wisconsin and North Dakota to Texas, Colorado, Arizona, and California (Jamacha), south through Mexico; also peninsular Florida. *Bouteloua pectinata* Featherly was differentiated from *B. hirsuta* by taller more robust culms and by a rudimentary spikelet at the end of the rachis. Such a spikelet is rarely developed in *B. hirsuta*, but it is not correlated with robust plants.

**15. Bouteloua grácilis** (H. B. K.) Lag. ex Steud. BLUE GRAMA. (Fig. 786.) Perennial; densely tufted; culms erect, 20 to 50 cm. tall, leafy at base; blades flat or loosely involute, 1 to 2 mm. wide; spikes usually 2, sometimes 1 or 3, rarely more, 2.5 to 5 cm. long, falcate-spreading at maturity, the rachis not projecting beyond the spikelets; spikelets numerous, as many as 80, about 5 mm. long; fertile lemma pilose, the awns slender, the intermediate lobes acute; rudiment densely bearded at summit of

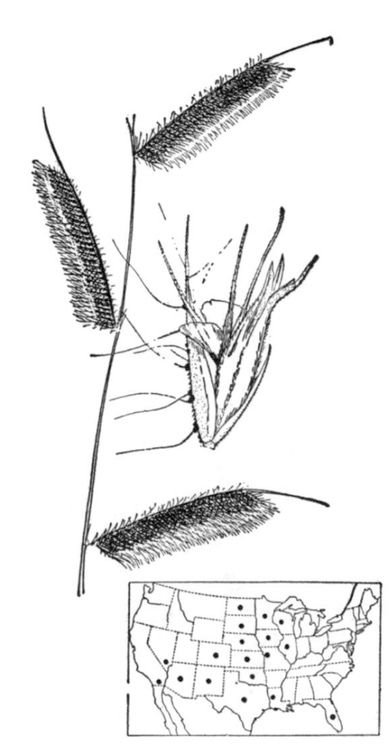

FIGURE 785.—*Bouteloua hirsuta*. Panicle, × 1; spikelet, × 5. (Griffiths 3371, Ariz.)

rachilla, cleft to the base, the lobes rounded, the awns slender, about equaling the tip of fertile lemma; one or two additional rudiments, broad and awnless, sometimes developed. ♃ (*B. oligostachya* Torr.) —Plains, Wisconsin to Manitoba and Alberta, south to Arkansas, Texas, and southern California; Mexico; introduced in a few places in the Eastern States.

Bouteloua gracilis var. strícta (Vasey) Hitchc. Spikes 4 to 6, usually ascending or appressed. ♃ —Rare, Texas and Arizona.

**16. Bouteloua brevisēta** Vasey. (Fig. 787.) Perennial, wiry, the base perennial, woody, loosely tufted; culms branching, 25 to 40 cm. tall; blades 3 to 6 cm. long, 1 to 1.5 mm. wide, flat or becoming involute, sharp-pointed; spikes mostly 2, sometimes 1, rarely 3, 2 to 3 cm. long; spikelets 30 to 45, about 4 mm. long; fertile lemma pubescent, with 3 awns and acuminate intermediate

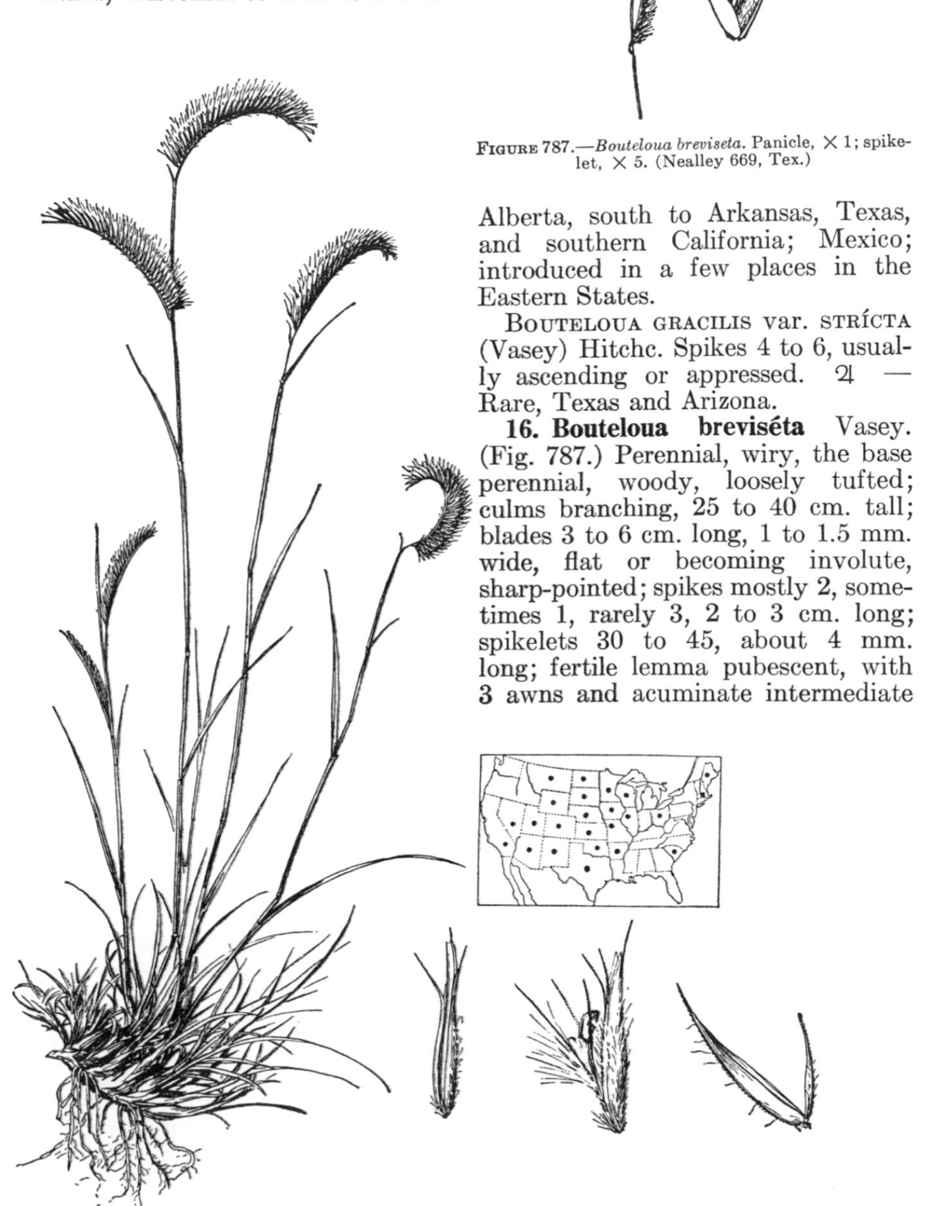

Figure 787.—*Bouteloua breviseta.* Panicle, × 1; spikelet, × 5. (Nealley 669, Tex.)

Figure 786.—*Bouteloua gracilis.* Plant, × ½; glumes and florets, × 5. (Amer. Gr. Natl. Herb. 384, Nev.)

FIGURE 788.—*Bouteloua eriopoda*. Plant, × 1; spikelet, × 5. (Hitchcock 13357, Tex.)

lobes; rudiment densely bearded at summit of rachilla joint, cleft nearly to the base, the rounded lobes obscured in the dense hairs. ♃ (*B. ramosa* Scribn.)—Gypsum sands and calcareous rocks, western Texas, New Mexico, and northern Mexico. Resembling *B. gracilis* but with loose, woody base and wiry culms; rachis prolonged and bearing a rudimentary spikelet at the tip.

**17. Bouteloua eriopóda** (Torr.) Torr. BLACK GRAMA. (Fig. 788.) Perennial; culms tufted, with swollen bases, slender, wiry, widely spreading with arched internodes or stoloniferous, white-lanate, 40 to 60 cm. long; blades 1 to 1.5 mm. wide, flexuous; spikes 3 to 8, commonly 4 or 5, loosely ascending, 2 to 3 cm. long; spikelets 12 to 20, not crowded and pectinate, 7 to 10 mm. long, narrow; fertile lemma acuminate, with a terminal awn, the lateral minute or obsolete; rudiment slender, cleft nearly to the base, the awns equaling the awn of the fertile lemma, the lobes minute, narrow. ♃ —Mesas, hills, and dry open ground, Oklahoma and Texas to Colorado, Utah, southern California, and northern Mexico.

**18. Bouteloua trífida** Thurb. (Fig. 789.) Perennial, tufted, leafy at base, rather delicate; culms erect, 10 to 20 cm. tall; blades usually only 1 to 2 cm. long; spikes 3 to 7, 1 to 2 cm. long, ascending or appressed; spikelets about 12, purplish, 7 to 10 mm. long; fertile lemma pubescent toward base, cleft more than half its length, with awns (5 mm. long) winged toward base and no intermediate lobes; rudiment cleft to the base, the awns similar to those of the fertile lemma, about as long. ♃ (*B. trinii* Griffiths; *B. burkii* Scribn.)—Mesas, ravines, and rocky hills, Texas to Nevada and Arizona; California (Death Valley); northern Mexico. Variable in length of the awns, the type of *B. trifida* being the longer awned form.

FIGURE 789.—*Bouteloua trifida*. Panicle, × 1; spikelet, × 5. (Amer. Gr. Natl. Herb. 669, Tex.)

FIGURE 790.—*Cathestecum erectum.* Plant, × ½; group of spikelets, central spikelet, and fertile floret, × 5. (Palmer 161, Mex.)

## 113. CATHÉSTECUM Presl

Spikes consisting of 3 spikelets, the upper or central perfect, the 2 lateral staminate or rudimentary, the spike falling entire; central spikelet with 1 perfect floret below and 1 or more reduced florets above; glumes unequal, the first a short, thin, nerveless scale in the central spikelet, narrow and acuminate in the lateral spikelets, the second about as long as the lemma, acuminate, all usually villous; lemma 3-nerved, the nerves extending into awns and the internerves into teeth; nerves of the palea extending into short awns; second and third floret with a fairly well developed lemma and palea, the fourth floret, if present, usually reduced. Low tufted or stoloniferous annuals or perennials, with short blades, and several to many short deciduous spikes approximate on a slender flexuous axis. Type species, *Cathestecum prostratum* Presl. Name from Greek *kathestekos*, set fast, stationary, the application not obvious.

**1. Cathestecum eréctum** Vasey and Hack. (Fig. 790.) Perennial with wiry stolons having arched internodes and hairy nodes; culms slender, 10 to 30 cm. tall; blades flat, about 1 mm. wide, mostly basal; spikes 4 to 8, ovoid, about 5 mm. long; lateral spikelets about two-thirds as long as the central spikelet; lemmas of all spikelets similar, the sterile ones more deeply lobed; awns from about as long as the lobes to twice as long, hairy at base. ♃ —Dry hills, western Texas, southern Arizona, and northern Mexico.

## 114. MUNRÓA Torr.

Spikelets in pairs or threes on a short rachis, the lower 1 or 2 larger, 3- or 4-flowered, the upper 2- or 3-flowered, the group (reduced spikes) enclosed in the broad sheaths of short leaves, usually about 3 in a fascicle, forming a cluster or head at the ends of the branches; rachilla disarticulating above the glumes and between the florets; glumes of the lower 1 or 2 spikelets equal, 1-nerved, narrow, acute, a little shorter than the lemmas, those of the upper spikelet unequal, the first much shorter or obsolete; lemmas 3-nerved, those of the lower spikelet coriaceous, acuminate, the points spreading, the midnerve extended into a mucro, those of the upper spikelet mem-

FIGURE 791.—*Munroa squarrosa*. Plant, × ½; group of spikelets, spikelet, and floret, × 5. (Zuck 43, Ariz.)

branaceous; palea narrow, enclosing the oval, dorsally compressed caryopsis. Low-spreading, much-branched annual, the short, flat, pungent leaves in fascicles. Type species, *Munroa squarrosa*. Named for William Munro.

**1. Munroa squarrósa** (Nutt.) Torr. FALSE BUFFALO GRASS. (Fig. 791.) Forming mats as much as 50 cm. in diameter, the internodes of the prostrate culms scabrous, as much as 10 cm. long, the fascicles at the nodes consisting of several short leafy branches, with 1 or 2 longer branches with slender internodes; blades stiff, mostly less than 3 cm. long, 1 to 3 mm. wide; fascicles of spikelets about 7 mm. long; lemmas with a tuft of hairs on the margin about the middle. ⊙ —Open ground, plains, and hills, at medium altitudes, common in old fields and recently disturbed soil, Alberta and North Dakota to Montana, south to Texas, Arizona, and Nevada. Occasional plants are found with a white floccose covering, the remains of egg cases of a species of woolly aphid. The variety *floccuosa* Vasey was described from such a specimen.

### 115. BÚCHLOË Engelm.

(*Bulbilis* Raf.)

Plants dioecious or monoecious. Staminate spikelets 2-flowered, sessile and closely imbricate, in 2 rows on one side of a slender rachis, forming a short spike; glumes somewhat unequal, rather broad, 1-nerved, acutish; lemmas longer than the glumes, 3-nerved, rather obtuse, whitish; palea as long as its lemma. Pistillate spikelets mostly 4 or 5 in a short spike or head, this falling entire, usually 2 heads to the inflorescence, the common peduncle short and included in the somewhat inflated sheaths of the upper leaves, the thickened indurate rachis and broad outer (second) glumes forming a rigid white obliquely globular structure crowned by the green-toothed summits of the glumes; first glume (inside) narrow, thin, mucronate, well developed to obsolete in a single head; second glume firm, thick and rigid, rounded on the back, obscurely nerved, expanded in the middle, with inflexed margins, enveloping the floret, abruptly contracted above, the summit with 3 green rigid acuminate lobes; lemma firm-membranaceous, 3-nerved, dorsally compressed, broad below, narrowed into a 3-lobed green summit, the middle lobe much the larger; palea broad, obtuse, about as long as the body of the lemma, enveloping the caryopsis. A low stoloniferous perennial with short curly blades, the staminate flowers in 2 or 3 short spikes on slender, erect culms, the pistillate in sessile heads partly hidden among the leaves. Type species, *Buchloë dactyloides*. Name contracted from Greek *boubalos*, buffalo, and *chloë*, grass, a Greek rendering of the common name, "buffalo grass."

**1. Buchloë dactyloídes** (Nutt.) Engelm. BUFFALO GRASS. (Fig. 792.) Gray green, forming a dense sod, the curly blades forming a covering 5 to 10 cm. thick; blades rather sparsely pilose, 1 to 2 mm. wide; staminate culms slender, 5 to 20 cm. tall, the spikes 5 to 15 mm. long; pistillate heads 3 to 4 mm. thick. ♃ —Dry plains, western Minnesota to central Montana, south to northwestern Iowa, Texas, western Louisiana, Arizona, and northern Mexico. Buffalo grass forms, when unmixed with other species, a close soft grayish-green turf. It is dominant over large areas on the uplands of the Great Plains, colloquially known as the "short-grass country," and is one of the most important grazing grasses of this region. The foliage cures on the ground and furnishes nutritious feed during the winter. The sod houses of the early settlers were made mostly from the sod of this grass. In 1941 it was planted at Boyce Thompson Institute, Yonkers, N. Y., and is proving to be an excellent cover for exposed dry banks.

FIGURE 792.—*Buchloë dactyloides.* Pistillate and staminate plants, × ½; pistillate spike and floret, × 5; staminate spikelet, × 5. (Ruth 156, Tex.)

## TRIBE 8. PHALARIDEAE

### 116. HIERÓCHLOË R. Br.

(*Savastana* Schrank; *Torresia* Ruiz and Pav.)

Spikelets with 1 terminal perfect floret and 2 staminate florets, disarticulating above the glumes, the staminate florets falling attached to the fertile one; glumes equal, 3-nerved, broad, thin and papery, smooth, acute; staminate lemmas about as long as the glumes, boat-shaped, hispidulous, hairy along the margin; fertile lemma somewhat indurate, about as long as the others, smooth or nearly so, awnless; palea 3-nerved, rounded on the back. Perennial, erect, slender, sweet-smelling grasses, with small panicles of broad, bronze-colored spikelets. Type species, *Hierochloë antarctica* (Labill.) R. Br. Name from Greek *hieros*, sacred, and *chloë*, grass, holy grass; *H. odorata* was used in parts of Europe for "strewing before the doors of churches on festival days."

Flowering culms with short blades only (rarely to 10 cm. long) with few to many long-leaved sterile shoots at base.
  Staminate lemmas bearing exserted awns.......... 1. H. ALPINA.
  Staminate lemmas awnless or nearly so.......... 2. H. ODORATA.
Flowering culms with blades 25 to 50 cm. long.......... 3. H. OCCIDENTALIS.

**1. Hierochloë alpína** (Swartz) Roem. and Schult. (Fig. 793.) Culms 10 to 40 cm. tall, tufted, with leafy shoots at base and short rhizomes; blades 1 to 2 mm. wide, the basal ones elongate, those of the culm shorter and wider; panicle contracted, 3 to 4 cm. long; spikelets short-pediceled, 6 to 8 mm. long; staminate lemmas ciliate on the margin, awned below the tip, the awn of the second lemma 5 to 8 mm. long, bent, twisted below, that of the first a little shorter, straight; fertile lemma acute, appressed-pubescent toward apex. ♃ —Arctic regions, Greenland to Alaska, south to Newfoundland and Quebec; alpine meadows and rocky slopes, high mountains, Maine, New Hampshire, Vermont, New York, and Montana; Europe.

FIGURE 793.—*Hierochloë alpina*. Plant, × 1; spikelet and floret, × 5. (Hitchcock 16058, N. H.)

**2. Hierochloë odoráta** (L.) Beauv. SWEETGRASS. (Fig. 794.) Culms 30 to 60 cm. tall, with few to several leafy shoots and slender, creeping rhizomes; blades 2 to 5 mm. wide, sometimes wider, those of the sterile shoots elongate, those of the culm mostly less than 5 cm. long, rarely to 10 cm. long; panicle pyramidal, 4 to 12 cm. long, from somewhat compact to loose with slender drooping branches; spikelets mostly short-pediceled, 5 mm. long; staminate lemmas awnless or nearly so, fertile lemma pubescent toward the apex. ♃ —Meadows, bogs, and moist places, Labrador to Alaska, south to New Jersey, Indiana, Iowa, Oregon, and in the mountains to New Mexico and Arizona; Eurasia. The Indians use the grass, known as Seneca grass, to make fragrant baskets. Also called holy grass and vanilla grass. A tall form with culm blades 12 to 17 cm. long, and a very loose lax panicle,

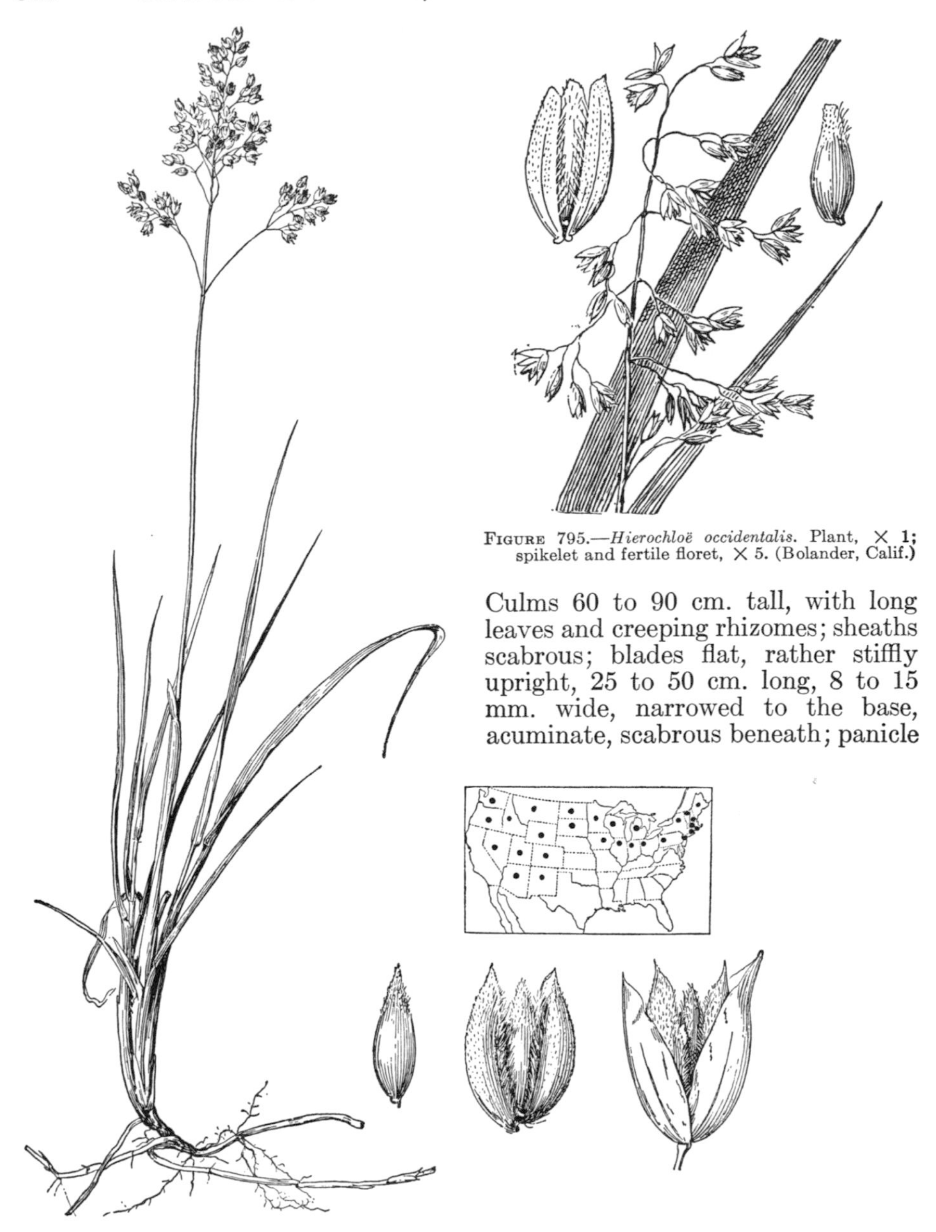

FIGURE 795.—*Hierochloë occidentalis*. Plant, × 1; spikelet and fertile floret, × 5. (Bolander, Calif.)

FIGURE 794.—*Hierochloë odorata*. Plant, × ½; spikelet, florets, and fertile floret, × 5. (Shear 437, Mont.)

found in Van Cortlandt Park, New York City, has been described as *Hierochloë nashii* Kaczmarek (*Savastana nashii* Bickn.).

**3. Hierochloë occidentális** Buckl. CALIFORNIA SWEETGRASS. (Fig. 795.)

Culms 60 to 90 cm. tall, with long leaves and creeping rhizomes; sheaths scabrous; blades flat, rather stiffly upright, 25 to 50 cm. long, 8 to 15 mm. wide, narrowed to the base, acuminate, scabrous beneath; panicle mostly open, 7 to 15 cm. long, the subcapillary branches drooping, loosely flowered or the spikelets aggregate toward the ends, the lower branches 2.5 to 7 cm. long; spikelets 4 to 5 mm. long, the glumes with a pale shining margin; staminate lemmas awnless or nearly so; fertile lemma appressed-pubescent toward apex. ♃ (*H. macrophylla* Thurb.)—Forests in the redwood belt, Oregon to Monterey, Calif.; Bingen, Wash.

## 117. ANTHOXÁNTHUM L. Vernalgrass

Spikelets with 1 terminal perfect floret and 2 sterile lemmas, the rachilla disarticulating above the glumes, the sterile lemmas falling attached to the fertile floret; glumes unequal, acute or mucronate; sterile lemmas shorter than the glumes, empty, awned from the back; fertile lemma shorter than the sterile ones, awnless; palea 1-nerved, rounded on the back, enclosed in the lemma. Sweet-smelling annuals or perennials, with flat blades and spikelike panicles. Type species, *Anthoxanthum odoratum*. Name from Greek *anthos*, flower, and *xanthos*, yellow, alluding to the yellow inflorescence.

Plants perennial.......... 1. A. odoratum.
Plants annual.......... 2. A. aristatum.

**1. Anthoxanthum odorátum** L. Sweet vernalgrass. (Fig. 796, *A*.) Culms tufted, erect, slender, 30 to 60 cm. tall, rarely to 1 m. tall; blades 2 to 5 mm. wide; panicle long-exserted, brownish yellow, acute, 2 to 6 cm. long; spikelets 8 to 10 mm. long; glumes scabrous, the first about half as long as the second; sterile lemmas subequal, appressed-pilose with golden hairs, the first short-awned below the apex, the second awned from near the base, the awn twisted below, geniculate, slightly exceeding the second glume; fertile lemma about 2 mm. long, brown, smooth and shining. ♃ —Meadows, pastures, and waste places, Greenland and Newfoundland to Louisiana and Michigan, and on the Pacific coast from British Columbia to California; introduced from Eurasia. Sometimes included in meadow mixtures to give fragrance to the hay, but the grass has little forage value.

**2. Anthoxanthum aristátum** Boiss. (Fig. 796, *B*.) Differing from *A. odoratum* in being annual, the culms lower, often geniculate and bushy branching; panicles looser; spikelets a little smaller. ⊙ —Waste places in several localities from Maine to Iowa; West Virginia; North Carolina; Florida; Mississippi and Arkansas; Vancouver Island to California; introduced from Europe.

---

Anthoxanthum grácile Bivon. Tufted annual; culms 20 cm. tall; blades pubescent; panicle silvery; spikelets about 12 mm. long, conspicuously awned. ⊙ —Occasionally cultivated for dry bouquets. Italy.

## EHRHÁRTA Thunb.

Spikelets laterally compressed with 1 fertile floret and 2 large sterile lemmas below enclosing the fertile floret; rachilla disarticulating above the glumes, the fertile floret and sterile lemmas falling together; glumes ovate, rather obscurely keeled; sterile lemmas indurate, compressed, 3- to 5-nerved; fertile lemma indurate, ovate, 5-nerved, obtuse. Erect or decumbent spreading annuals or perennials with flat blades and narrow panicles. Type species, *Ehrharta capensis* Thunb. Named for Friedrich Ehrhart.

**Ehrharta erécta** Lam. Culms erect or ascending from a decumbent base, branching, mostly 30 to 50 cm. tall; blades 5 to 12 cm. long, 4 to 9 mm. wide; panicles 6 to 15 cm. long, the branches narrowly ascending or sometimes spreading; spikelets 3 to 3.5 mm. long; sterile lemmas awnless, the first smooth, the second cross-wrinkled. ♃ —Escaped, Berkeley, Calif. Introduced from South Africa. Shows considerable competitive ability and may become of value in re-

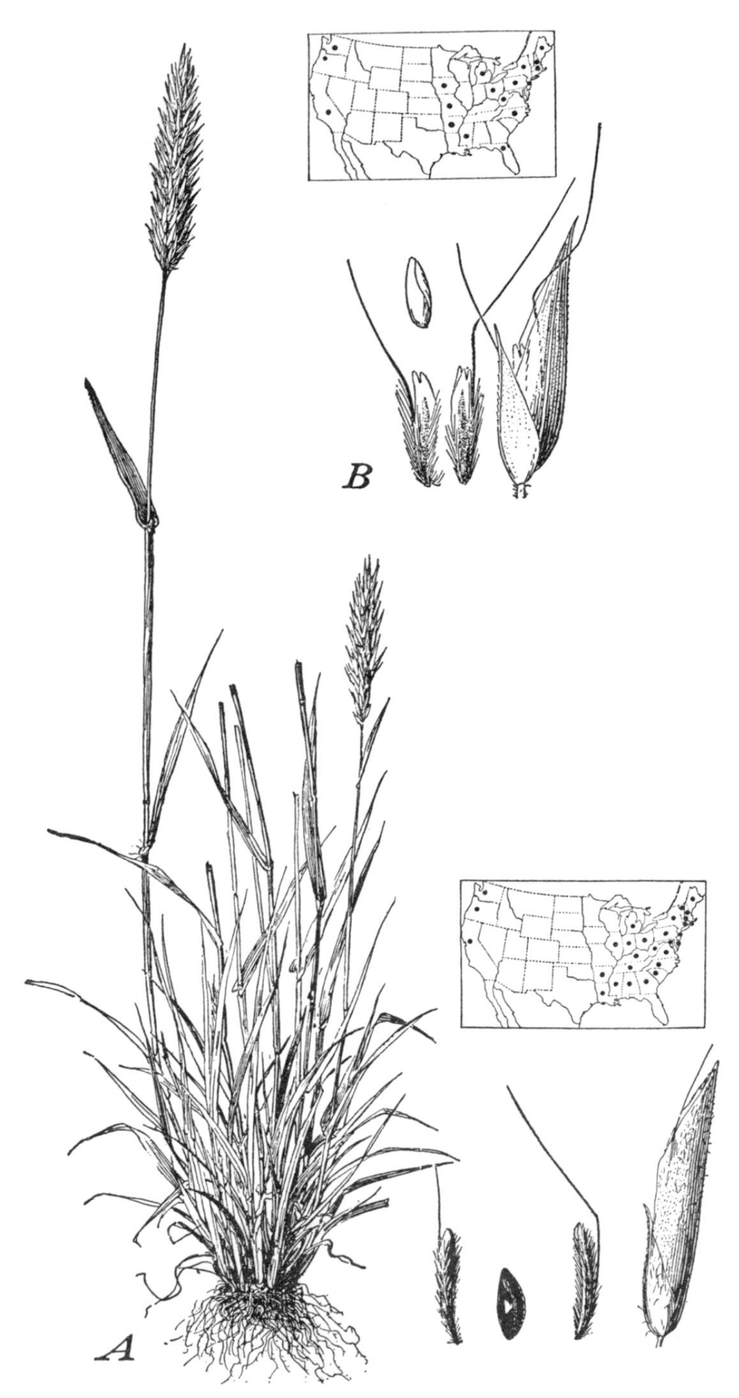


FIGURE 796.—*A*, *Anthoxanthum odoratum*. Plant, × ½; spikelet, sterile lemmas, and fertile floret, × 5. (Biltmore Herb. 74b, N. C.) *B*, *A. aristatum*. Spikelet, sterile florets, and fertile floret, × 5, (White 1591, N. Y.)

placing some of the troublesome weeds.

Ehrharta calycina J. E. Smith. Erect leafy perennial to 75 cm. tall; panicle 10 to 15 cm. long, branchlets and pedicels subcapillary; spikelets 7 to 8 mm. long, purplish; glumes nearly as long as the lemmas; sterile lemmas thinly silky-villous; fertile lemma silky on the nerves. ♃ —Grown at Davis, Calif., as a promising drought-resistant grass for nonirrigated range lands. Introduced; South Africa.

## 118. PHĂLARIS L. Canary grass

Spikelets laterally compressed, with 1 terminal perfect floret and 2 sterile lemmas below (obsolete in *Phalaris paradoxa*), the rachilla disarticulating above the glumes, the usually inconspicuous sterile lemmas falling closely appressed to the fertile floret; glumes equal, boat-shaped, often winged on the keel; sterile lemmas reduced to 2 small, usually minute, scales (rarely only 1); fertile lemma coriaceous, shorter than the glumes, enclosing the faintly 2-nerved palea. Annuals or perennials, with numerous flat blades, and narrow or spikelike panicles. Type species, *Phalaris canariensis*. *Phalaris*, an old Greek name for a grass.

Spikelets in groups of 7, 1 fertile surrounded by 6 sterile, the group falling entire. 1. P. paradoxa.
Spikelets all alike, not in groups falling entire.
  Plants perennial.
    Rhizomes wanting; panicle dense, ovate or oblong .......... 8. P. californica.
    Rhizomes present; panicle narrow, spreading during anthesis .... 9. P. arundinacea.
  Plants annual.
    Glumes broadly winged; panicle ovate or short-oblong.
      Sterile lemma solitary; fertile lemma 3 mm. long .......... 4. P. minor.
      Sterile lemmas 2, fertile lemma 4 to 6 mm. long.
        Sterile lemmas 0.6 mm. long or less .......... 3. P. brachystachys.
        Sterile lemmas half as long as fertile .......... 2. P. canariensis.
    Glumes wingless or nearly so; panicles oblong or linear, dense.
      Glumes wingless, acuminate; fertile lemma turgid, the acuminate apex smooth. 7. P. lemmoni.
      Glumes narrowly winged toward summit, acute or abruptly pointed; fertile lemma less turgid, villous to the acute apex.
        Panicle tapering to each end, mostly 2 to 6 cm. long (occasionally longer). 5. P. caroliniana.
        Panicle subcylindric, mostly 6 to 15 cm. long (occasionally smaller). 6. P. angusta.

**1. Phalaris paradóxa** L. (Fig. 797.) Annual, tufted, more or less spreading at base; culms 30 to 60 cm. tall; panicle dense, oblong, narrowed at base, 2 to 6 cm. long, often enclosed at base in the uppermost enlarged sheath; spikelets finally falling from the axis in groups of 6 or 7, those of the upper part of the panicle slender-pediceled, the central spikelet fertile, the subulate-acuminate glumes with a prominent toothlike wing near the middle of the keel, the others sterile, with smaller pointed glumes with toothed-winged keels; fertile lemma 3 mm. long, with only a few hairs toward the summit, the sterile lemmas obsolete; spikelets of lower part of panicle short-pediceled, the glumes of the outer 4 spikelets deformed, cuneate-clavate. ☉ —Occasional in grainfields and waste places, California and Arizona; ballast, Philadelphia, New Orleans; introduced from Mediterranean region.

Phalaris paradoxa var. praemórsa (Lam.) Coss. and Dur. Panicle mostly smaller, all the spikelets short-pediceled and with outer sterile spikelets having deformed clavate glumes, as in the lower part of panicle of the species; glumes of all spikelets subindurate. ☉ —Fields and waste

FIGURE 797.—*Phalaris paradoxa*. Plant, × 1; sterile (*A*) and fertile (*B*) spikelets, × 5. (Heller 11391, Calif.)

places, Washington to California and Arizona; ballast, Philadelphia; introduced from Mediterranean region.

**2. Phalaris canariénsis** L. CANARY GRASS. (Fig. 798.) Annual; culms erect, 30 to 60 cm. tall; panicle ovate to oblong-ovate, dense, 1.5 to 4 cm. long; spikelets broad, imbricate, pale with green stripes; glumes 7 to 8 mm. long, abruptly pointed, the green keel with a prominent pale wing, broadened upward; fertile lemma 5 to 6 mm. long, acute, densely appressed-pubescent; sterile lemmas at least half as long as fertile. ⊙ —Waste places, infrequent, Nova Scotia to Alaska, south to Virginia, Kansas, Wyoming, Arizona, and California, and occasionally southward; introduced from the western Mediterranean region. This species furnishes the canary seed of commerce.

**3. Phalaris brachýstachys** Link. (Fig. 799.) Differing from *P. canariensis* in having smaller spikelets, the glumes about 6 mm. long, the fertile lemma 4 to 5 mm. long, and especially in the short sterile lemmas not more than 0.6 mm. long. ⊙ —Texas (Asherton); California (Butte County); Oregon (ballast, near Portland); introduced from the Mediterranean region.

**4. Phalaris mínor** Retz. (Fig. 800.) Resembling *P. canariensis*; panicle ovate-oblong, 2 to 5 cm. long; spikelets narrower, not so conspicuously striped; glumes 4 to 6 mm. long, the wing of the keel narrower; fertile lemma lance-ovate, about 3 mm. long, acute; sterile lemma solitary, about 1 mm. long. ⊙ —Fields and waste places, New Brunswick to New Jersey, rare; Louisiana and Texas; Colorado; ballast, near Portland, Oreg.; Arizona; frequent in California; Mexico; introduced from the Mediterranean region.

**5. Phalaris caroliniána** Walt. (Fig. 801.) Annual; culms erect, 30 to 60

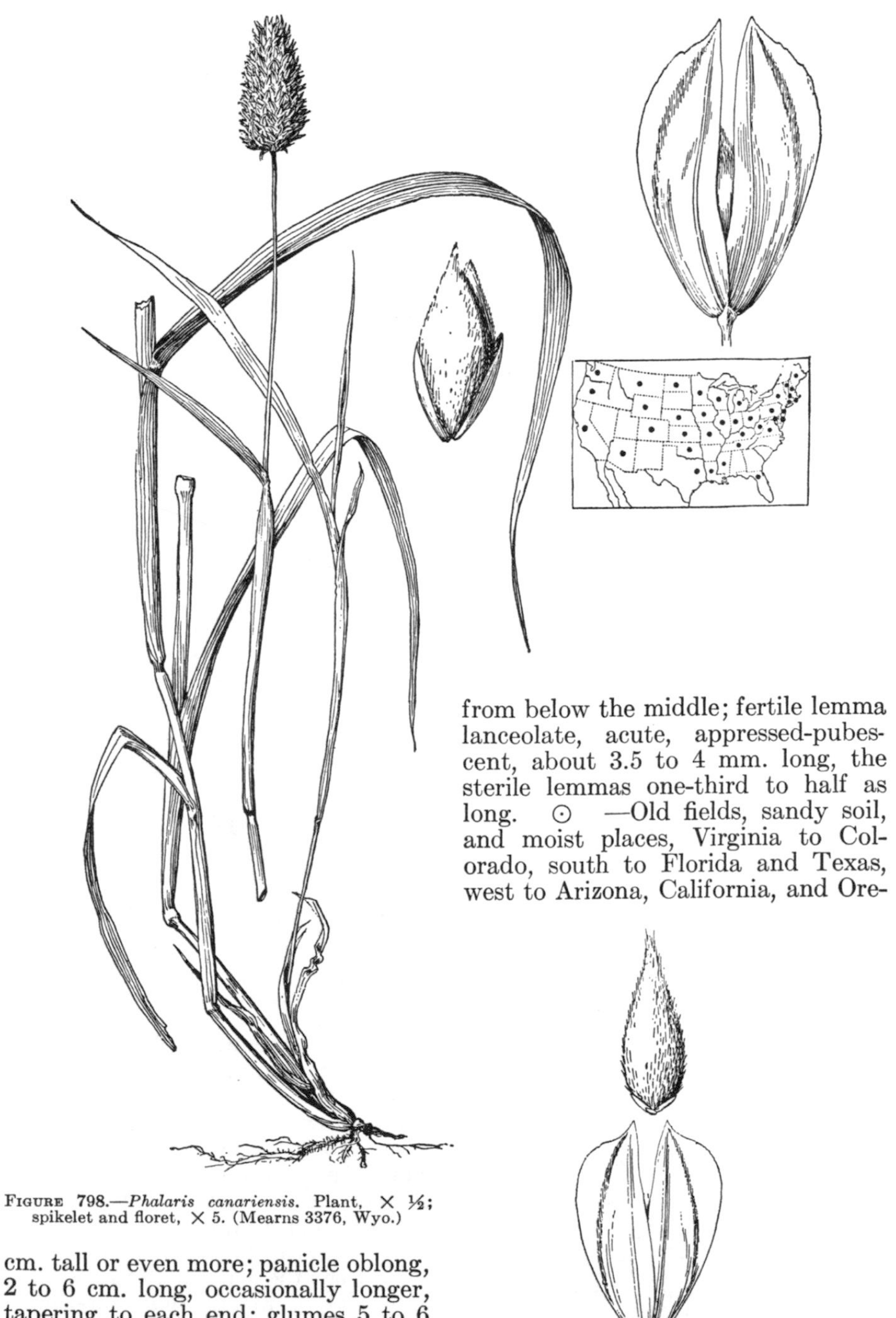

FIGURE 798.—*Phalaris canariensis*. Plant, × ½; spikelet and floret, × 5. (Mearns 3376, Wyo.)

cm. tall or even more; panicle oblong, 2 to 6 cm. long, occasionally longer, tapering to each end; glumes 5 to 6 mm. long, oblong, rather abruptly narrowed to an acute apex, the keel scabrous and narrowly winged above from below the middle; fertile lemma lanceolate, acute, appressed-pubescent, about 3.5 to 4 mm. long, the sterile lemmas one-third to half as long. ⊙ —Old fields, sandy soil, and moist places, Virginia to Colorado, south to Florida and Texas, west to Arizona, California, and Ore-

FIGURE 799.—*Phalaris brachystachys*. Spikelet and floret, × 5. (Suksdorf 1904, Oreg.)

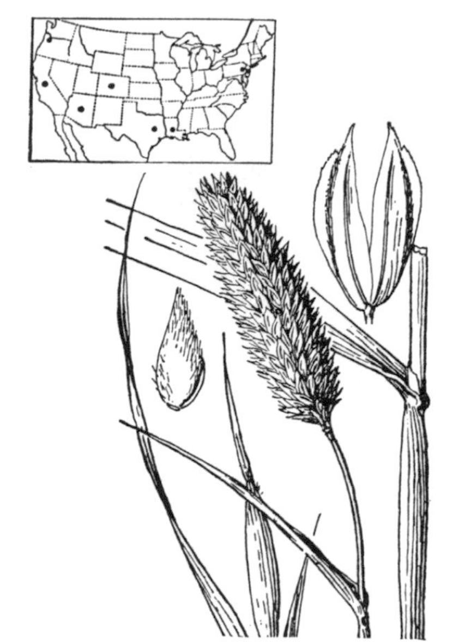

FIGURE 800.—*Phalaris minor*. Plant, × 1; glumes and floret, × 5. (Ball 1932, Calif.)

gon. A few specimens from the Pacific coast are relatively robust, up to 80 cm. tall, with panicles 3 to 8 cm. long,

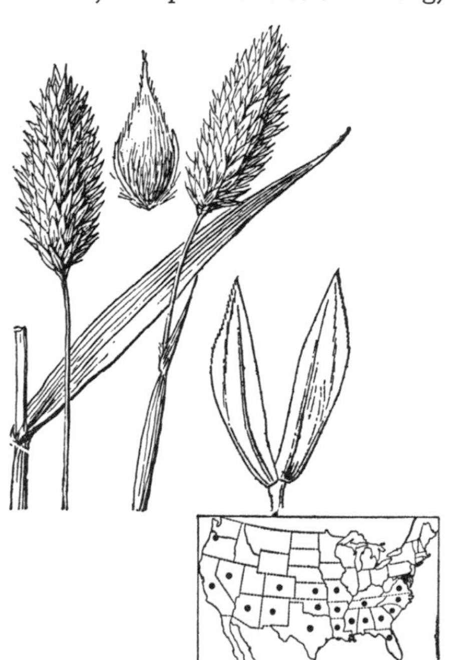

FIGURE 801.—*Phalaris caroliniana*. Plant, × 1; glumes and floret, × 5. (Hitchcock 1074, Miss.)

some of them slightly lobed and not tapering to the base, the spikelets 6 to 6.5 mm. long.

**6. Phalaris angústa** Nees ex Trin. (Fig. 802.) Annual; culms 1 to 1.5 m. tall; panicle subcylindric, mostly 6 to 15 cm. long, about 8 mm. thick; glumes 3.5 to 4 mm. long, narrow, abruptly pointed, the keel scabrous and narrowly winged toward the summit; fertile lemma ovate-lanceolate, acute, appressed-pubescent, 3 mm. long; sterile lemmas about one-third as long. ⊙ —Open ground at low altitudes, Mississippi, Louisiana, and Texas; Arizona and California; southern South America.

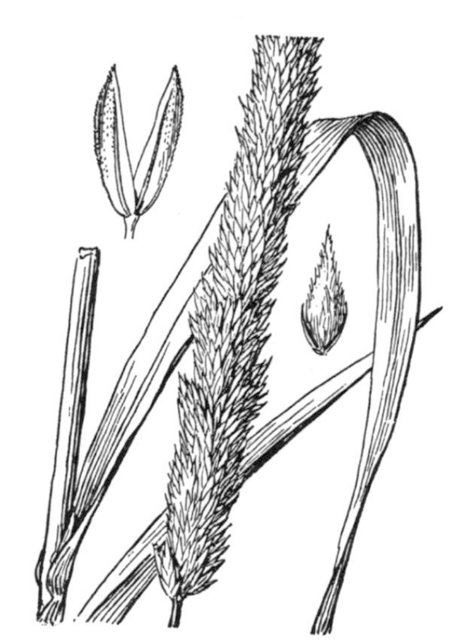

FIGURE 802.—*Phalaris angusta*. Plant, × 1; glumes and floret, × 5. (Suksdorf 32, Calif.)

**7. Phalaris lemmóni** Vasey. (Fig. 803.) Annual; culms 30 to 90 cm. tall; panicle 5 to 15 cm. long, subcylindric or lobed toward base, often purplish; glumes about 5 mm. long, narrow, acuminate, scabrous, not winged on the keel; fertile lemma ovate-lanceolate, acuminate, 3.5 to 4 mm. long, brown at maturity, appressed-pubescent, except the acuminate tip, sterile lemmas (1 or 2) less than one-third as long. ⊙ —Moist places, at low altitudes, in the coastal valleys, central and southern California.

**8. Phalaris califórnica** Hook. and Arn. (Fig. 804.) Perennial, often in dense tussocks; culms erect, 75 to 160 cm. tall; blades rather lax, 8 to 15 mm. wide; panicle ovoid or oblong, 2 to 5 cm. long, 2 to 2.5 cm. thick, often purplish-tinged; glumes 6 to 8 mm. long, narrow, tapering from below the middle to an acute apex, the keel smooth or nearly so, sharp but not winged; fertile lemma ovate-lanceolate, about 4 mm. long, rather sparsely appressed-pubescent, the palea often exposed, the sterile lemmas half to two-thirds as long. ♃ —Ravines and open moist ground in the Coast Range, southwestern Oregon to San Luis Obispo County, Calif.

FIGURE 803.—*Phalaris lemmoni.* Glumes and floret, × 5. (Type.)

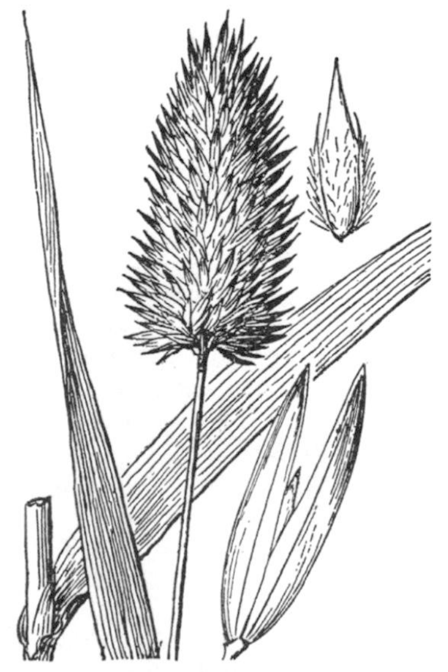

FIGURE 804.—*Phalaris californica.* Plant, × 1; spikelet and floret, × 5. (Heller 6677, Calif.)

FIGURE 805.—*Phalaris arundinacea.* Plant, × 1; glumes and floret, × 4. (Chase 7583, Md.)

FIGURE 806.—*Phalaris tuberosa* var. *stenoptera,* × 5. (McCrary, N. C.)

**9. Phalaris arundinácea** L. REED CANARY GRASS. (Fig. 805.) Perennial, with creeping rhizomes, glaucous; culms erect, 60 to 150 cm. tall; panicle 7 to 18 cm. long, narrow, the branches spreading during anthesis, the lower as much as 5 cm. long; glumes about 5 mm. long, narrow, acute, the keel scabrous, very narrowly winged; fertile lemma lanceolate, 4 mm. long, with a few appressed hairs; sterile lemmas villous, 1 mm. long. ♃ —Marshes, river banks, and moist places, New Brunswick to southeastern Alaska (also at Tanana Hot Springs, Alaska), south to North Carolina, Kentucky, Oklahoma, New Mexico, Arizona, and northeastern California; Eurasia. An important constituent of lowland hay from Montana to Wisconsin. PHALARIS ARUNDINACEA var. PÍCTA L. RIBBON GRASS. Blades striped with white. ♃ — Grown for ornament in gardens; also called gardener's garters.

PHALARIS TUBERÓSA var. STENÓPTERA (Hack.) Hitchc. (Fig. 806.) Perennial, with a loose branching, rhizomatous base; culms stout, as much as 1.5 m. tall; panicle 5 to 15 cm. long, 1.5 cm. wide, slightly lobed; glumes 5 to 6 mm. long, the keel scabrous, rather narrowly winged on the upper two-thirds; fertile lemma 4 mm. long, ovate-lanceolate, acute, appressed-pubescent; sterile lemma usually solitary, about one-third as long as fertile lemma. ♃ —About 1902 there appeared in Queensland, Australia, the source unknown, a species of *Phalaris* which gave promise of being a valuable forage grass. About 1907 it was distributed from the Toowoomba Botanic Gardens, Queensland. Stapf, of Kew Gardens, identified this grass as *P. bulbosa* L. Hackel described it as a distinct species, *P. stenoptera*. It has been grown at the California Experiment Station from seed from South Africa. It has also been cultivated in Oregon, in Washington, D. C., and in North Carolina, and is spontaneous in Humboldt County, Calif. This differs from typical *P. tuberosa* of the Mediterranean region in having short vertical or ascending, sometimes branching, rhizomes, the base of the culms little or not at all swollen. It has been called Harding grass. Burbank distributed it as *P. stenophylla* (error for *stenoptera*), calling it Peruvian wintergrass. The name *P. bulbosa* has been misapplied to *P. tuberosa* L., but true *P. bulbosa* L. is a species of *Phleum* (*P. tenue* Schrad.; *P. bulbosum* (L.) Richt.).

## TRIBE 9. ORYZEAE

### 119. ORÝZA L. RICE

Spikelets 1-flowered, laterally compressed, disarticulating below the glumes; glumes 2, much shorter than the lemma, narrow; lemma rigid, keeled, 5-nerved, the outer nerves near the margin, the apex sometimes awned; palea similar to the lemma, narrower, keeled, with a median bundle but with no strong midnerve on the back, 2-nerved close to the margins. Annual or sometimes perennial swamp grasses, often tall, with flat blades and spikelets in open panicles. Type species, *Oryza sativa*. Name from *oruza*, old Greek name for rice. The spikelet in *Oryza* and *Leersia* is interpreted by Stapf, Arber, and some others as consisting of 2 greatly reduced glumes and 2 subulate sterile lemmas below the single fertile floret. The true glumes, according to this interpretation, are represented by the minute cuplike expansion, sometimes distinctly 2-lobed, at the summit of the pedicel, persistent and showing no line of demarcation from the pedicel, the articulation of the spikelet being below the sterile lemmas, the latter wanting in *Leersia*. The problem deserves further study.

**1. Oryza satíva** L. RICE. (Fig. 807.) Annual, or in tropical regions sometimes perennial; culms erect, 1 to 2 m. tall; blades elongate; panicle rather dense, drooping, 15 to 40 cm. long; spikelets 7 to 10 mm. long, 3 to 4 mm. wide; lemma and palea papillose-roughened and with scattered appressed hairs, the lemma from mucronate to long-awned. ⊙ —Cultivated in all warm countries at low altitudes where there is sufficient moisture; one of the world's most important food plants; sometimes adventive near the coast from Virginia to Florida and Texas.

FIGURE 807.—*Oryza sativa*. Plant, × ½; spikelet, × 5. (Cult.)

## 120. LEÉRSIA Swartz

(*Homalocenchrus* Mieg.)

Spikelets 1-flowered, strongly compressed laterally, disarticulating from the pedicel; glumes wanting; lemma chartaceous, broad, oblong to oval, boat-shaped, usually 5-nerved, the lateral pair of nerves close to the margins, these and the keel often hispid-ciliate, the intermediate nerves sometimes faint; palea as long as the lemma, much narrower, usually 3-nerved, the keel usually hispid-ciliate, the lateral nerves close to the margins, the margins firmly held by the margins of the lemma; stamens 6 or fewer. Perennials, usually with creeping rhizomes, flat, scabrous blades, and mostly open panicles. Type species, *Leersia oryzoides*. Named for J. D. Leers.

Spikelets broadly oval, 3 to 4 mm. wide .......... 1. L. LENTICULARIS.
Spikelets elliptic, not more than 2 mm. wide.
  Panicle narrow, the branches ascending or appressed .......... 4. L. HEXANDRA.
  Panicle open, the capillary branches finally spreading.
    Spikelets glabrous, about 2 mm. long; culms tufted, erect; rhizomes wanting. 5. L. MONANDRA.
    Spikelets hispidulous; culms decumbent at base; rhizomes present.
      Lower panicle branches solitary; spikelets 3 mm. long, 1 mm. wide. 3. L. VIRGINICA.
      Lower panicle branches fascicled; spikelets 5 mm. long, 1.5 to 2 mm. wide. 2. L. ORYZOIDES.

**1. Leersia lenticuláris** Michx. CATCHFLY GRASS. (Fig. 808.) Culms straggling, 1 to 1.5 m. tall, with creeping scaly rhizomes; sheaths scabrous at least toward the summit; blades lax, 1 to 2 cm. wide; panicle open, drooping, 10 to 20 cm. long, the branches ascending or spreading, naked below, branched above, branchlets bearing closely imbricate spikelets along one side; spikelets pale, broadly oval, very flat, 4 to 5 mm. long, sparsely hispidulous, the keels bristly ciliate. ♃ —Ditches and swamps, Maryland to Minnesota, south to Florida and Texas.

**2. Leersia oryzoídes** (L.) Swartz. RICE CUTGRASS. (Fig. 809.) Culms slender, weak, often decumbent at base, 1 to 1.5 m. tall, with slender creeping rhizomes; sheaths and blades strongly retrorsely scabrous, the blades mostly 8 to 10 mm. wide; panicles terminal and axillary, 10 to 20 cm. long, the flexuous branches finally spreading, the spikelets more loosely imbricate than in *L. lenticularis;* spikelets elliptic, 5 mm. long, 1.5 to 2 mm. wide, sparsely hispidulous, the keels bristly ciliate; axillary panicles reduced, partly included in the sheaths, the spikelets cleistogamous. ♃ —Marshes, river banks, and wet places, often forming a zone around ponds and lakes, Quebec and Maine to British Columbia and eastern Washington south to northern Florida, Texas, Colorado, Arizona, and southeastern California; Europe. The late cleistogamous phase has been described as *L. oryzoides* forma *inclusa* (Wiesb.) Dörfl.

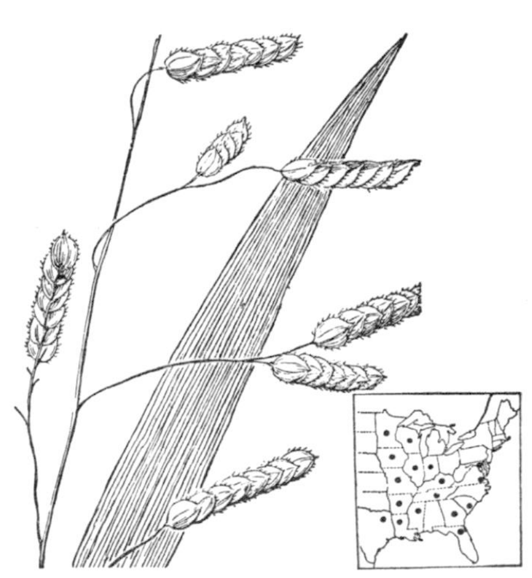

FIGURE 808.—*Leersia lenticularis*, × 1. (McDonald 68, Ill.)

**3. Leersia virgínica** Willd. WHITEGRASS. (Fig. 810.) Culms slender,

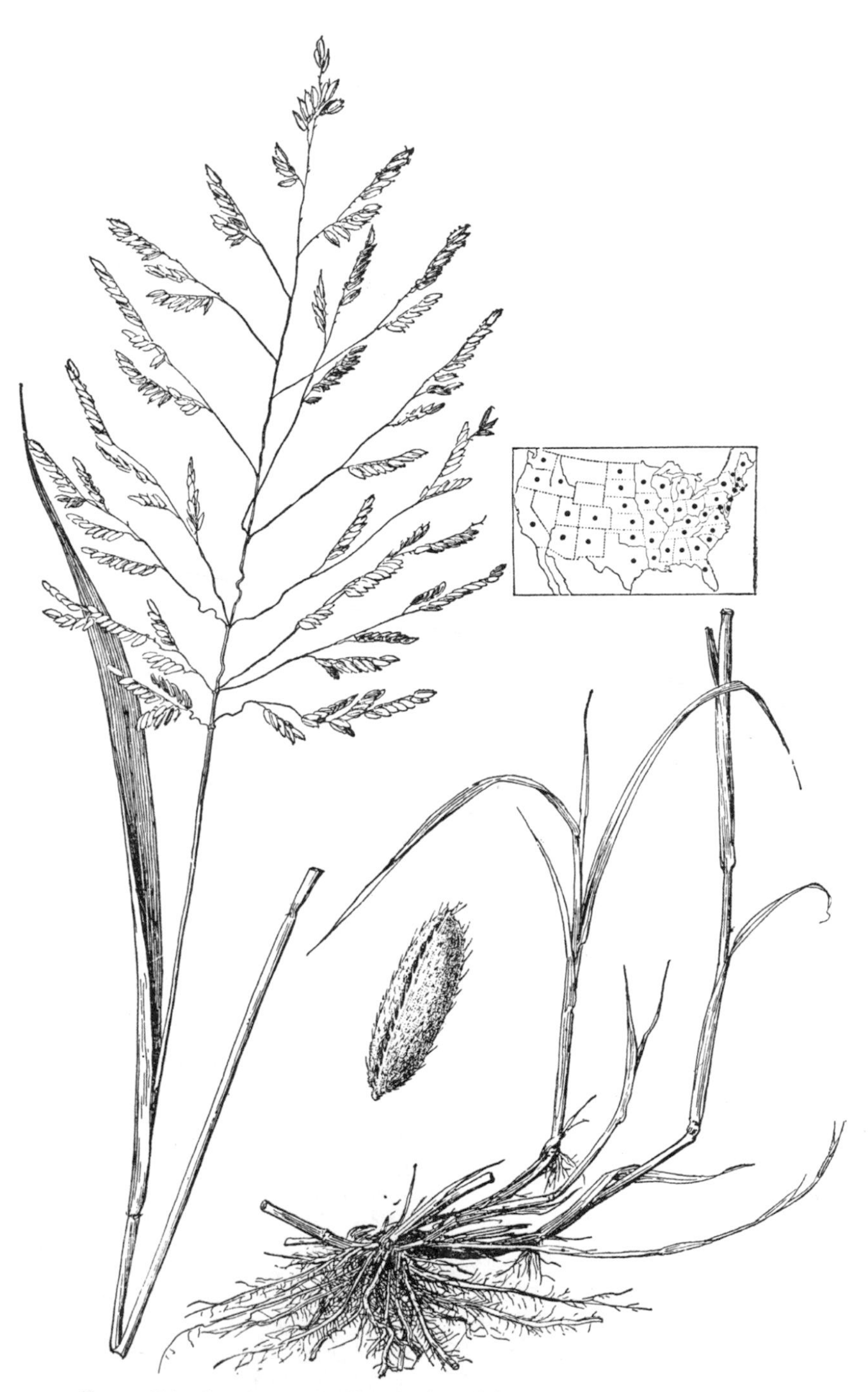

FIGURE 809.—*Leersia oryzoides*. Plant, × ½; spikelet, × 5. (Hitchcock 5317, Tex.)

FIGURE 810.—*Leersia virginica*, × 1. (French, Iowa.)

weak, branching, 50 to 120 cm. tall, with clusters of very scaly rhizomes much stouter than the culm base; blades relatively short, 6 to 12 mm. wide; panicle open, 10 to 20 cm. long, the capillary branches rather distant, stiffly spreading, naked below, those of the branches smaller, sometimes included in the sheath; spikelets oblong, closely appressed to the branchlets, about 3 mm. long and 1 mm. wide, sparsely hispidulous, the keels short-hispid. ♃ —Low woods and moist places, Quebec to South Dakota, south to Florida and Texas.

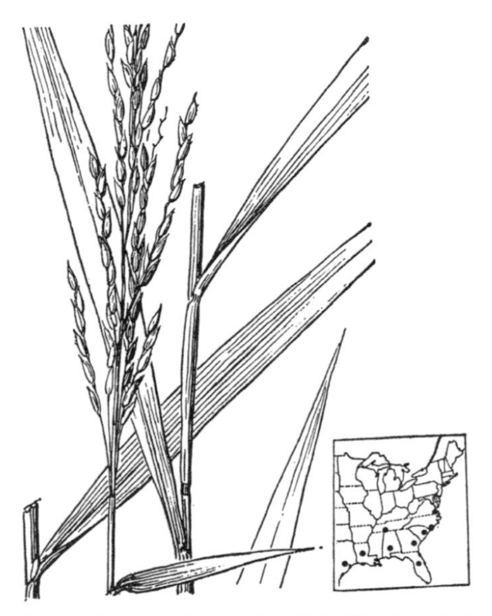

FIGURE 811.—*Leersia hexandra*, × 1. (Wurzlow, La.)

**4. Leersia hexándra** Swartz. (Fig. 811.) Culms slender, weak, usually long-decumbent from a creeping and rooting base, with slender rhizomes and extensively creeping leafy stolons; the flowering culms upright; blades rather stiff, 2 to 5 mm. wide; panicle narrow, 5 to 10 cm. long, the branches ascending or appressed, floriferous nearly to the base; spikelets oblong, about 4 to 5 mm. long, a little more than 1 mm. wide, often purplish, sparsely hispidulous, the keels bristly ciliate. ♃ —Shallow water, ditches, and wet places near the coast, Virginia to Florida and Texas; widely distributed in the tropics of both hemispheres.

FIGURE 812.—*Leersia monandra*, × 1. (Nealley, Tex.)

**5. Leersia monándra** Swartz. (Fig. 812.) Culms tufted, erect, wiry, 50 to 100 cm. tall, without rhizomes; sheaths smooth or nearly so; blades elongate, 1 to 5 mm. wide; panicle open, the capillary solitary branches spreading, naked below, the small spikelets near the ends; spikelets pale, broadly ovate, glabrous, about

2 mm. long. ♃ —Rocky woods and prairies, Florida Keys, southern Florida, and southern Texas; West Indies.

## TRIBE 10. ZIZANIEAE

### 121. ZIZÁNIA L. Wildrice

Spikelets unisexual, 1-flowered, disarticulating from the pedicel; glumes obsolete, represented by a small collarlike ridge; pistillate spikelet terete, angled at maturity; lemma chartaceous, 3-nerved, tapering into a long slender awn; palea 2-nerved, closely clasped by the lemma; grain cylindric, 1 to 2 cm. long; staminate spikelet soft; lemma 5-nerved, membranaceous, linear, acuminate or awn-pointed; palea about as long as the lemma, 3-nerved; stamens 6. Tall aquatic annuals or perennials, with flat blades and large terminal panicles, the lower branches ascending or spreading, bearing the pendulous staminate spikelets, the upper branches ascending, at maturity erect, bearing appressed pistillate spikelets, the staminate spikelets early deciduous, the pistillate spikelets tardily deciduous. Type species, *Zizania aquatica*. Name from *Zizanion*, an old Greek name for a weed growing in grain, the tares of the Scripture parable.

The seeds of wildrice were used by the aborigines for food and are still used to some extent by some of the northern tribes of Indians. Wildrice is important as a food and as shelter for waterfowl and is sometimes planted for this purpose in marshes on game preserves. The Chinese cultivate the Asiatic species, *Z. latifolia* (Griseb.) Turcz., as the source of a vegetable which they call *kau sun*. This consists of a thickened portion of the base of the culm, the point of incipient fruiting of a smut fungus, *Ustilago edulis*.

Plants annual, erect ........ 1. Z. aquatica.
Plants perennial, long-decumbent at base ........ 2. Z. texana.

**1. Zizania aquática** L. Annual wildrice. (Fig. 813, *B*.) Annual; culms robust, usually 2 to 3 m. tall; blades elongate, 1 to 4 cm. wide, scaberulous; ligule 10 to 15 mm. long; panicles mostly 30 to 50 cm. long, the branches mostly 15 to 20 cm. long; lemma and palea of pistillate spikelet about 2 cm. long, thin, hispid throughout. ⊙ —Marshes and borders of streams and ponds, usually in shallow water, Maine to Michigan and Illinois, south to Florida and Louisiana; Idaho.

Zizania aquatica var. angustifólia Hitchc. Culms usually not more than 1.5 m. tall; ligule 3 to 8 mm. long; blades usually not more than 1 cm. wide; lemma and palea of pistillate spikelet mostly larger, firm, shining, hispid only on the margin and nerves. ⊙ —Shallow water, Quebec and New Brunswick to North Dakota, south to New York and Nebraska.

Zizania aquatica var. intérior Fassett. (Fig. 813, *A*.) Closely resembling the species, or the blades narrower; pistillate spikelet as in var. *angustifolia;* intergrades in the Middle West. ⊙ —Michigan and Indiana to North and South Dakota; Idaho.

**2. Zizania texána** Hitchc. Texas wildrice. (Fig. 814.) Perennial; culm long-decumbent and rooting at base, 1 to 3 m. long; blades elongate, 3 to 15 or even 20 mm. wide; panicle 20 to 30 cm. long, narrow, the lower (staminate) branches ascending, 5 to 10 cm. long; staminate spikelets 7 to 9 mm. long, 1.5 mm. wide; pistillate spikelets about 1 cm. long, tapering into an awn 1 to 2 cm. long. ♃

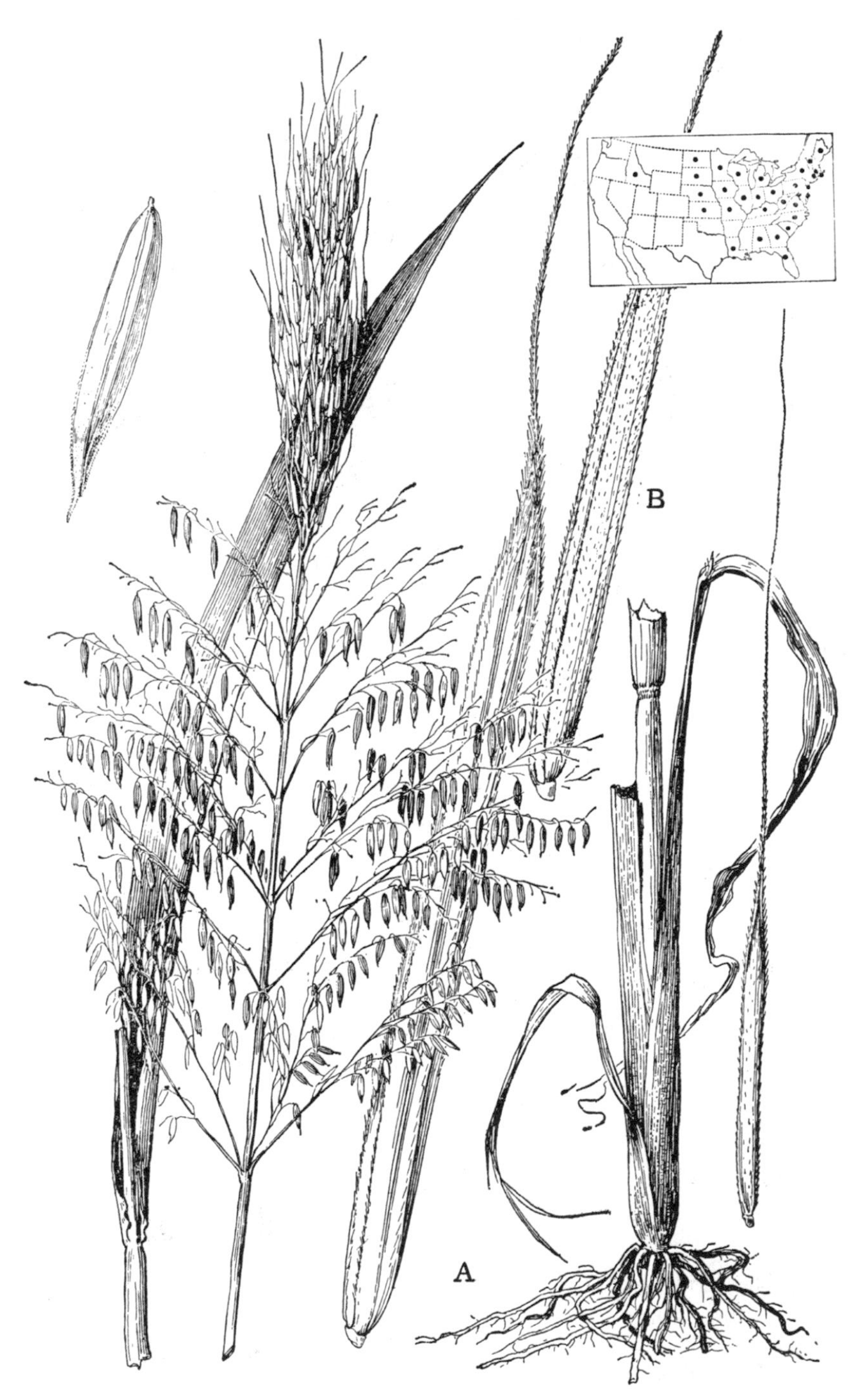


FIGURE 813.—*A*, *Zizania aquatica* var. *interior*. Plant, × ½; pistillate spikelet, × 2; second view, × 5. (Fink, Iowa.) *B*, *Z. aquatica*. Pistillate spikelet, × 5. (Hitchcock, Va.)

—Growing in rapidly flowing water, San Marcos, Tex. The grass grows in water 30 to 120 cm. deep, the lower part of the plant prostrate or floating on the water, the upper part erect. Flowers from April to November and at warm periods during winter. Said to be troublesome in irrigation ditches.

## 122. ZIZANIÓPSIS Doell and Aschers.

Spikelets unisexual, 1-flowered, disarticulating from the pedicel, mixed on the same branches of the panicle, the staminate below; glumes wanting; lemma 7-nerved, short-awned in the pistillate spikelets; palea 3-nerved; staminate spikelets with 6 stamens; styles rather long, united; fruit obovate, free from the lemma and palea, coriaceous, smooth and shining, beaked with the persistent style; seed free from the pericarp. Robust perennial marsh grasses, with stout creeping rhizomes, broad flat blades, and large open panicles. Type species, *Zizaniopsis microstachya* (Nees) Doell and Aschers. Name from *Zizania*, a generic name, and Greek *opsis*, appearance, alluding to the similarity to *Zizania*.

**1. Zizaniopsis miliácea** (Michx.) Doell and Aschers. (Fig. 815.) SOUTHERN WILDRICE. Culms 1 to 3 m. tall or even taller; blades glabrous except the very scabrous margins, 1 to 2 cm. wide, the midrib stout; panicle rather narrow, nodding, 30 to 50 cm. long, the numerous branches fascicled, as much as 15 to 20 cm. long, naked at base; spikelets 6 to 8 mm. long, short-awned, the staminate slender, the pistillate turgid at maturity. ♃ —Marshes, creeks, and river banks, Maryland to Kentucky and Oklahoma, south to Florida and Texas.

FIGURE 814.—*Zizania texana*. Plant, × ½; pistillate and staminate spikelets, × 5. (Type.)

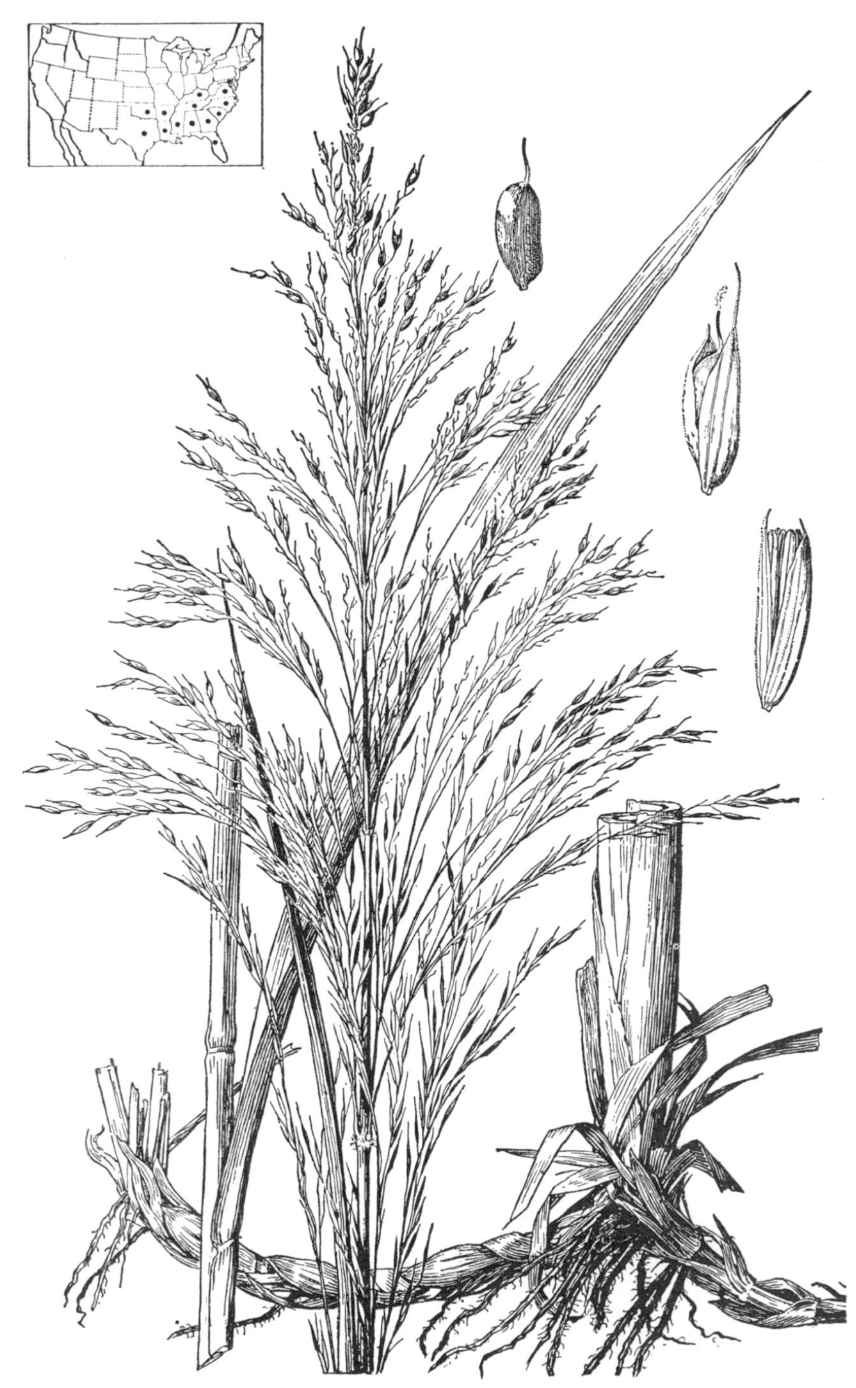

FIGURE 815.—*Zizaniopsis miliacea*. Plant, × ½; staminate spikelet, pistillate spikelet, and ripe caryopsis, × 5. (Chase 7121, S. C.)

## 123. LUZÍOLA Juss.

Spikelets unisexual, 1-flowered, disarticulating from the pedicel, the staminate and pistillate spikelets in separate panicles on the same plant; glumes

FIGURE 816.—*Luziola peruviana*. Plant, × ½; pistillate and staminate spikelets, × 5. (Curtiss 6871, Fla.)

wanting; lemma and palea about equal, thin, several to many-nerved, lanceolate or oblong; stamens 6 or more; stigmas long, plumose; grain free, globose, finely striate. Creeping, low or delicate perennials, with narrow flat blades and terminal and axillary panicles. Type species, *Luziola peruviana*. Name modified from *Luzula*, a genus of Juncaceae.

Pistillate spikelets ovoid, about 2 mm. long; staminate and pistillate panicles on the same shoot........ 1. L. PERUVIANA.
Pistillate spikelets oblong-lanceolate, 4 to 5 mm. long; staminate and pistillate panicles on different shoots........ 2. L. BAHIENSIS.

**1. Luziola peruviána** Gmel. (Fig. 816.) Culms slender, branching, the flowering shoots ascending, 10 to 40 cm. tall; blades 1 to 4 mm. wide, exceeding the panicles; staminate panicles terminal, narrow, the spikelets about 7 mm. long; pistillate panicles terminal and axillary, 3 to 6 cm. long, about as wide, the spikelets about 2 mm. long, ovoid at maturity, abruptly pointed. ♃ —Muddy ground and wet meadows, Florida (Pensacola) and Louisiana (vicinity of New Orleans); Mexico and Cuba, south to Argentina.

**2. Luziola bahiénsis** (Steud.) Hitchc. (Fig. 817.) Extensively stoloniferous, the flowering shoots not more than 15 cm. tall, mostly less; blades 2 to 4 mm. wide, much exceeding the panicles; panicles mostly terminal, the staminate few-flowered, the spikelets about 5 mm. long; pistillate panicles 4 to 6 cm. long, the few stiff branches finally spreading, with a few appressed oblong-lanceolate spikelets 4 to 5 mm. long, the lemma and palea much exceeding the caryopsis. ♃ —Lagoons and banks of streams, southern Alabama; Cuba, Venezuela, Brazil.

FIGURE 817.—*Luziola bahiensis*, × 1. (Mohr, Ala.)

## 124. HYDRÓCHLOA Beauv.

Spikelets unisexual, 1-flowered, disarticulating from the pedicel, the staminate and pistillate spikelets in separate panicles on the same plant; glumes wanting; staminate spikelets with a thin 7-nerved lemma, a 2-nerved palea, and 6 stamens; pistillate spikelets with a thin 7-nerved lemma and 5-nerved palea, the stigmas long and slender. A slender, branching, aquatic grass, probably perennial, the leaves floating; staminate spikelets in small few-flowered terminal racemes; pistillate spikelets in few-flowered racemes in the axils of the leaves. Type species, *Hydrochloa caroliniensis*. Name from Greek *hudor*, water, and *chloa*, grass, alluding to the habitat.

**1. Hydrochloa caroliniénsis** Beauv. (Fig. 818.) Culms up to 1 m. or more long, freely branching, leafy; blades flat, 1 to 3 cm. long, 1 to 2 mm. wide, in vigorous shoots as much as 6 cm. long and 5 mm. wide; spikelets inconspicuous and infrequent, the staminate about 4 mm. long, the pistillate about 2 mm. ♃ —Ponds and slow-flowing streams, sometimes in sufficient abundance to become troublesome. North Carolina to Florida and Louisiana. Eaten by livestock. Lemma 5- or 7-nerved; palea 4- to 7-nerved. (Weatherwax.)

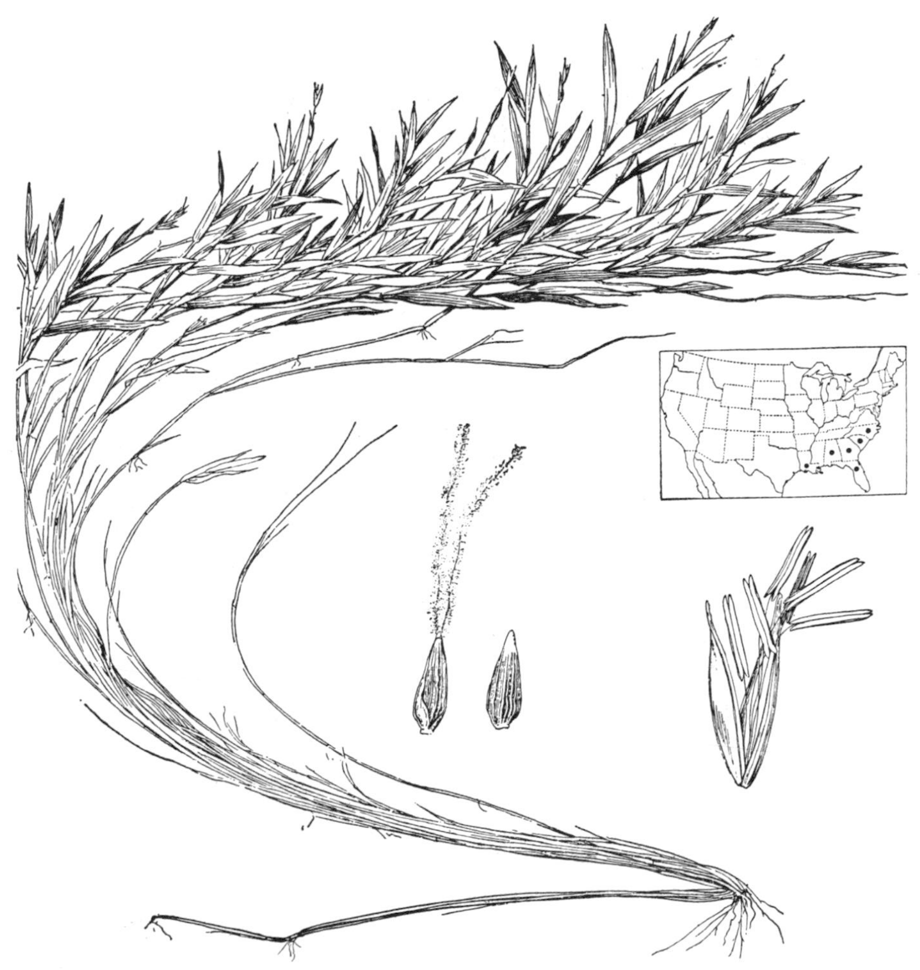

FIGURE 8 18.—*Hydrochloa caroliniensis.* Plant, × ½; two views of pistillate spikelet and staminate spikelet, × 5. (Nash 1152, Fla.)

## 125. PHÁRUS L.

Spikelets in pairs, appressed along the slender spreading, nearly simple panicle branches, one pistillate, subsessile, the other staminate, pediceled, much smaller than the pistillate spikelet; fertile lemma subindurate, terete, clothed, at least toward the beaked apex, with thick uncinate hairs; blades petioled (the petiole with a single twist reversing the upper and under surfaces of the blade), the nerves running from midnerve to margin, with fine transverse veins between the nerves. Perennials with broad flat elliptic or oblanceolate blades and terminal panicles with rather few stiffly spreading branches breaking readily at maturity, the terete pistillate spikelets appressed, the uncinate fruits acting like burs. Type species, *Pharus latifolius* L. Name from Greek *pharos,* cloth or mantle, possibly alluding to the broad blades.

**1. Pharus parvifólius** Nash. (Fig. 819.) Culms long-decumbent and rooting at base, the flowering shoot 30 to 50 cm. tall; blades elliptic,

FIGURE 819.—*Pharus parvifolius*, × ½. (Miller 1231, Dominican Republic.)

abruptly acuminate, 10 to 20 cm. long, 2 to 4 cm. wide; panicles mostly 10 to 20 cm. long, about as wide; pistillate spikelets about 1 cm. long, the glumes thin, brown, less than half as long as the lemma; staminate spikelets about 3 mm. long, the slender pedicels appressed to the pistillate spikelets. ♃ —Rocky woods, Florida, rare (Pineola; Orange Lake); West Indies to Brazil.

### TRIBE 11. MELINIDEAE

## 126. MELÍNIS Beauv.

Spikelets small, dorsally compressed, 1-flowered with a sterile lemma below the fertile floret, the rachilla disarticulating below the glumes; first glume minute; second glume and sterile lemma similar, membranaceous, strongly nerved, slightly exceeding the fertile floret; fertile lemma and palea subhyaline toward summit. Perennials with slender, branching, decumbent culms and narrow many-flowered panicles, with capillary branchlets and pedicels. Type species, *Melinis minutiflora*. Name from Greek *meline*, millet.

**1. Melinis minutiflóra** Beauv. Molasses grass. (Fig. 820.) Culms ascending from a tangled much-branched base, as much as 1 m. tall; the foliage viscid-pubescent; blades flat, 5 to 15 cm. long, 5 to 10 mm. wide; panicle 10 to 20 cm. long, purplish; spikelets about 2 mm. long; sterile lemma 2-lobed, with a delicate awn 1 to 10 mm. long from between the lobes. ♃ —Introduced from Brazil, though native of Africa. Cultivated for forage and spreading in open ground through Central and South America and the West Indies. It has been tried successfully in southern Florida. The grass has a heavy sweetish odor when fresh. Called in Brazil capím gordura.

Figure 820.—*Melinis minutiflora*. Plant, × 1; spikelet, × 10. (Moldenke 453, Fla.)

---

Thysanolaéna máxima (Roxb.) Kuntze. Robust perennial, 1 to 3 m. tall; blades 3 to 7 cm. wide; panicle commonly 1 m. long, the slender flat densely flowered branches drooping; spikelets about 2 mm. long, pointed; fertile lemma long-ciliate. ♃ —Introduced in southern Florida and southern California as an ornamental.